Coastal Ocean Space Utilization III

Edited by

Norberto Della Croce

Institute of Marine Environmental Sciences
University of Genoa, Italy

Shirley Connell

Director of Policyholder Services
Medical Liability Mutual Insurance Company
USA

and

Robert Abel

Research Professor
Texas A & M University
and
Senior Scientist
Stevens Institute of Technology
New Jersey
USA

CRC Press
Taylor & Francis Group
Boca Raton London New York

CRC Press is an imprint of the
Taylor & Francis Group, an **informa** business
A TAYLOR & FRANCIS BOOK

CRC Press
Taylor & Francis Group
6000 Broken Sound Parkway NW, Suite 300
Boca Raton, FL 33487-2742

First issued in paperback 2019

ISBN-13: 978-0-419-20900-3 (hbk)
ISBN-13: 978-0-367-86552-8 (pbk)

A catalogue record for this book is available from the British Library

Visit the Taylor & Francis Web site at
http://www.taylorandfrancis.com

and the CRC Press Web site at
http://www.crcpress.com

Table of Contents

Regional Seas & Embayments

New Concepts in the Governance of Ocean Space

The Coastal Environment: Assessment Standards & Issues

Ocean Resources & Sustainable Development

Ocean Space Development & Related Technologies

Acknowledgements

The editors gratefully acknowledge the sponsorship of the following organisations:

The Max and Victoria Dreyfus Foundation, Inc.
The National Oceanic & Atmospheric Administration
US Department of Commerce
The US National Science Foundation
The University of Genoa, Italy and several agencies and affiliates of the Government of Italy

Editor's Preface

Editing the scientific papers that make up this volume was both *challenging* an *enlightening*. It was challenging because of the disparate nature of each paper, e.g fonts, formats, figures, and use of language. It was enlightening because of the ve creative and dynamic nature of the ideas presented.

We would like to thank the contributing authors for their time and assistance in t convoluted process. Where it was necessary, graphics were redrafted to meet produc standards of clarity; however in a few instances, certain pieces of tone art had tc eliminated.

A great deal of gratitude is owed to Ms. Barbara Brengel and the staff of Other Office for their professionalism and honest concern. Finally, this Editor woul to thank Dr. Robert Abel for his abiding sense of Humor.

Shirley H. (

Introduction

It is perhaps indicative of the perverseness of mankind that only as we attained the ability to leave our planet did we exhibit any interest in what lay within it—particularly its embracing seas.

Now, at last, scientists and engineers the world over discover—almost on a daily basis—new illustrations of the ocean's resource potential. Of even greater relevance, they continue to find new uses of this power, for food, health, and even fun.

The power of aquatic alchemy, i.e., to transmute seawater into metals, feeds, and fertilizers, is by no means, however, limited to a few large industrial nations. It is found in small, even landlocked, countries where the only raw material necessary to discovery is an unfettered mind.

It was this remarkable spread of accomplishment which lead a few dedicated entrepreneurs to design and nurture the concept of an International Symposium on Coastal Ocean Space Utilization, or "COSU." COSU allows anyone with an idea to test it on an assemblage gathered biennially from more than a dozen ocean-active countries. The first two Conferences were held in 1989 and 1991 in New York City and Long Beach, California, respectively.

COSU III departed American shores for Genoa, Italy, partly owing to the historical significance of the time and place and partly in recognition of that nation's intellectual leadership in our field.

In four days of discussions nearly fifty papers, ranging broadly over Coastal Zone Management, man-made islands and large platforms, new sources of energy, and inter-country cooperation, were presented by academicians, industrialists, and government officials from fifteen countries.

The relationship of papers included herein to those actually given is not precisely one-to-one. A few of the participants were apparently prevented by one thing or another from submitting completed papers. In two other cases, papers submitted by nonparticipant were deemed sufficiently pertinent for inclusion.

Viewed in a certain perspective, possibly the most valuable aspect of the COSU series is that each conference in the series acts as a sort of training ground for its successor, as technology advances and the ocean's service to man continues to grow in both real and perceived importance. Looking back to 1989, several such advances in thoughts and deeds are discernible, and hopefully, COSU III will breed inspiring contributions to COSU's IV, V, and VI, already in planning stages.

Finally, by its nature, COSU probably needs larger and more varied input than any other meeting of its type, and the officials responsible for COSU III were fortunate in finding and receiving much help from a number of sources. In this connection it is important to recognize the sponsors of the conference: The Consiglio Nazionale dell Recherche (Italy); Consortium Telerobot (Italy); Ente Nazionale per L'Energia Elettrica, S.p.A. (Italy); the U.S. National Science Foundation; U.S. National Oceanic and Atmospheric Administration; Regione Liguria (Italy); and Universita Di Genova (Italy), who furnished the resources to plan, coordinate, and execute the conference; and the Max and Victroia Dreyfus Foundation, which made possible the printing and publication of this book, and Rachel Jones, E&FN SPON, Chapman & Hall, Ltd. who arranged for its publication.

The staff of the Institute of Marine Environmental Sciences of the University of Genoa, directed by Dr. Mario Petrillo, Conference Coordinator, administered the conference superbly and the Davidson Laboratory of Stevens Institute of Technology translated the myriad software inputs into a common standard for publication. A special note of gratitude is due Mr. Joseph Vadus of NOAA—clearly the sparkplug of the COSU series. His hands are in evidence during all phases of the meeting's preparation and conduct.

Finally, the Co-Chairmen wish to acknowledge with deepest appreciation the intensive and dedicated effort put forth by our colleague, Editor Shirley Connell, without which, this book, once born, could never have left the delivery room.

Norberto Della Croce
Robert A. Abel
February 1995

PROGRAM ANNOUNCEMENT

THE THIRD INTERNATIONAL SYMPOSIUM

ON

COASTAL OCEAN SPACE UTILIZATION

MARGINAL SEAS: PROBLEMS & OPPORTUNITIES

HOSTED BY:

UNIVERSITÁ DI GENOVA

INSTITUTO DI SCIENZE AMBIENTALI MARINE

THE COSU II SPONSORS & COMMITTEES

SYMPOSIUM SPONSORS:

Consiglio Nazionale delle Ricerche (Italy)
Consortium Telerobot (Italy)
Ente Nazionale Per L'Energia Elettrica, S.p.A. (Italy)
Max & Victoria Dreyfus Foundation (U.S.A)
Municipality of Santa Margherita Ligure
National Oceanic and Atmospheric Administration (U.S.A.)
National Science Foundation (U.S.A.)
New Jersey Marine Science Consortium (U.S.A.)
Regione Liguria (Italy)
Universita' Di Genova (Italy)

SYMPOSIUM CO-CHAIRMEN:

Professor Norberto Della Croce, Director
Institute of Marine Environmental Sciences
University of Genoa
S. Margherita Ligure, Italy

Dr. Robert B. Abel, President
New Jersey Marine Science Consortium
Fort Hancock, New Jersey, U.S.A.

COSU III ADVISORY COMMITTEE:

Chairman:
Professor Sandro Pontremoli
University of Genoa, Italy

Dr. WillardBascom
Ocean Engineer & Scientist, USA

Mr. Charles Bookman
Marine Board, National Research Council, USA

Mr. Charles Ehler
Ocean Resources Conservation & Assessment
NOAA, USA

Senator S. T. Gagliano
President, Jersey Shore Partnership, Inc.
Giordano, Halleran & Ciesla, USA

Dr. Hyung Tack Huh
Korea Ocean Resch. & Develp. Institute, Korea

Mr. Hajime Inoue
Ship Resch. Inst., Ministry of Transport, Japan

Dr. Chia Chuen Kao
National Science Council, ROC

Dr. C.Y. Li
Executive Yuan, ROC

Dr. Fabio Morchio
Regional Counsellor for the Environment, Italy

Professor Leonardo Santi
Faculty of Medicine, University of Genoa, Italy

Dr. Maurizio Scajola
Union of the Ligurian Chambers of Commerce, Italy

Ing. Armando Sodaro
Selenia Elsag - Naval Systems, Italy

Dr. Isamu Tamura
Ports & Harbours Bureau, MOT, Japan

Dr. Hajime Tsuchida
Coastal Development Inst. of Tech., Japan

Professor Ezio Volta
Faculty of Engineering, Univ. of Genoa, Italy

COSU III PROGRAM COMMITTEE:

Chairman:
Professor Norberto Della Croce
Institute of Marine Environmental Science
University of Genoa, Italy

Dr. Robert B. Abel
New Jersey Marine Sciences Consortium, USA

Dr. Romano Ambrogi
Environ. Biology, Nuclear & Thermal Resch. Ctr.
ENEL S.p.A., Italy

Dr. Marco Berta
Section Research & Development
Technomare S.p.A., Italy

Mr. Norman Caplan
Environmental & Ocean Systems, NSF, USA

Ing. Giuseppe Mosci
Consortium Telerobot, Italy

Arch. Lino Tirelli
Department of the Environment, Regione Liguria
Italy

Mr. Joseph R. Vadus
National Ocean Service, NOAA, USA

Dr. Don Walsh
International Maritime Inc., U.S.A.

COSU III CONFERENCE COORDINATORS:

Dr. Mario Petrillo
Inst. Marine Environmental Sciences
University of Genoa, Italy

Mrs. Joan Sheridan
New Jersey Marine Science Consortium, USA

Dr. Riccardo Cattaneo-Vietti
Inst. Marine Environmental Sciences
University of Genoa, Italy

COSU III

Coastal Ocean Space Utilization (COSU III) is the first in this symposium series to be held outside the United States. The theme of "Marginal Seas: Problems and Opportunities" encompasses and emphasizes harmonious, multiple uses of the world's many enclosed ocean regions. This symposium will present significant international developments which apply to coastal nations and regions facing similar environmental, political, social and economic challenges.

TUESDAY 30 MARCH, 1993

SYMPOSIUM REGISTRATION

Registration opens at 7:30 AM in the lobby area adjacent to the meeting room. Coffee and tea service will be available in the registration area throughout the symposium.

CONVENE THE SYMPOSIUM & INTRODUCTIONS

09:00-10:00

Remarks:

Joseph R. Vadus (*session moderator*)
Senior Advisor
National Ocean Service
National Oceanic and Atmospheric Administration (NOAA)
Washington, D.C.

OPENING ADDRESSES

Professor Norberto Della Croce

COSU III Co-Chairman
Director
Institute of Marine Environmental Sciences
University of Genoa

Mr. Norman Caplan

Head
Environmental & Ocean Systems
National Science Foundation
Washington, D.C.

Professor Sandro Pontremoli

Rector
University of Genoa

KEYNOTE ADDRESS

10:15-11:30

Inner and Outer Space: Technology for Ocean Observation and Assessment

Dr. Sylvia A. Earle
Hydronaut and former Chief Scientist
National Oceanic and Atmospheric Administration
Washington, D.C.

Protection and Management of the Coastal Area Among Bordering Countries (Ramoge Agreement)

Ing. Giovanni Gallino
Assessorship for the Environment
Regione Liguria
Genoa, Italy
&
Prof. Norberto Della Croce
University of Genoa

12:00 - 14:00

LUNCH

SESSION I. REGIONAL SEAS & EMBAYMENTS

14:00 - 18:00

Moderators: Professors Della Croce & Abel

First Approach for an Integrated Environmental Planning at Regional Level of the Coastal Marine System (Regione Liguria)

Arch. Lino Tirelli
Assessorship for the Environment
Regione Liguria
Genoa

The Baltic Sea

Dr. Gotthilf Hempel
Director
Baltic Sea Research Institute
Warnemünde, Germany

Egypt-Israel Cooperative Program

Dr. Robert B. Abel
Program Manager
New Jersey Marine Sciences Consortium

Shared Solutions for Land-Based Marine Pollution: Case of the Black Sea

Dr. Raphael Vartanov
Director
Institute of World Economy & International Relations
Russian Academy of Sciences
Moscow, Russia

Shared Solutions for Land-Based Pollution: Case of the Gulf of Mexico

Dr. James Broadus
Director
Marine Policy Center
Woods Hole Oceanographic Institution
Woods Hole, Massachusetts

Rio de la Plata Regional Maritime System: Potential for an Integrated Multifunction Offshore Complex

Ascensio C. Lara & Esteban L. Biondi
Catholic University of Argentina
&
Albina L. Lara
University del Salvador & Consejo Nacional de Investigaciones
Cientificas y Tecnicas, Argentina
&
Joseph R. Vadus
National Oeanic and Atmospheric Administration
Washington, D.C., U.S.A.

The Special Case of the Gulf of Aqaba

Dr. Mohammed Wahbeh
Director
Marine Science Center
Aqaba, Jordan

Latin America and the Caribbean: Utilization and Development of the Coastal Zone

Dr. Alberto G. Lonardi
Coordinator, Multinational Project on the Environment & Natural Resources
Department of Scientific & Technical Affairs
Organization of American States
Washington, D.C.

ADJOURN

18:00

INTERNATIONAL RECEPTION

18:00-20:00

Hors D'Oeuvres and Complimentary Wine Sampling at the Park Hotel Suisse

Sponsored by the hotel and the COSU III Committees

WEDNESDAY 31 MARCH

KEYNOTE ADDRESS

08:30 - 09:15

Russian Technology for Coastal Ocean Observations, Measurements and Assessments

Academician Vladimir E. Zuev
Chairman, Department of Oceans
Atmospheric Physics, and Geography
Russian Academy of Sciences
Moscow, Russia

SESSION II. NEW CONCEPTS IN THE GOVERNANCE OF OCEAN SPACE

09:15-11:30

Moderator: Prof. Nicola Greco, Edistudio, Rome

Management Issues in Coastal Lagoon Ecosystem: the Case of Venice Lagoon

Arch. Francesco Bandarin
Consortium "Venezia Nuova"
Palazzo Morosin
Venice

The Jersey Shore Partnership Inc.: The Role of the Public/Private Partnership in Governance of the Coastline

Senator S. T. Gagliano
President, Jersey Shore Partnership, Inc.
Giordano, Halleran & Ciesla
Middletown, New Jersey

Legal Rules, Administrative Planning and Negotiation to Solve Clashing Interests in Coastal Zones: Italian and Mediterranean Perspectives

Prof. Nicola Greco
Edistudio
Rome

Creative Financing for Coastal Ocean Ventures

Dennis de Vito and Dan Skowski
Chemical Bank of New York
&
Giuseppe Spotoforro
Chemical Bank of Italy

The Coastal Use Framework as a Methodological Tool for Coastal Area Management

Prof. Adalberto Vallega
Institute of Geographical Sciences
University of Genoa

Improvement in Mathematical Modelling for Shore Line Changes Evaluation: An Application to the Veneto Coast Case

Ing. Piero Ruol & Ing. Massimo Tondello
Institute of Maritime Constructions & Geotechnics
University of Padua
Padua, Italy

ECO^2: A New Concept in Coastal Management

Ing. Alfredo Fanara
Alenia Elsag Naval Systems
Genoa, Italy

12:00 - 14:00

LUNCH

SESSION III. THE COASTAL ENVIRONMENT: ASSESSMENT STANDARDS & ISSUES

14:00 - 18:00

Moderators: Mr. Charles N. Ehler, NOAA & Ing. Mario Tomasino, Head Hydrologic Service, ENEL, S.p.A., Mestre, Venice

Coastal Ocean Space Management: Challenges for the Next Decade

Mr. Charles N. Ehler, Director
Office of Ocean Resources Conservation & Assessment
NOAA
Rockville, Maryland

Reasonable Approach for Special Regulations and Standards to Enable Environmentally Sustainable Development in the Coastal Zone

Francisco Jose Montoya Font
Head, Coastal Service
Ministry of Public Works and Transports
Tarragona, Spain

A Sea Change for Oil Tanker Safety

Charles A. Bookman
Director, Marine Board
National Research Council
Washington, D.C.

Scientific Approach for Evaluating the Sites of Coastal Thermoelectric Power Stations

Ing. Mario Tomasino
Hydraulics & Structure Research Center
ENEL, S.p.A.
Mestre, Venice
&
Dr. Romano Ambrogi
Nuclear and Thermal Research Center
ENEL S.p.A.
Milan
&
Dr. Edmondo Ioannilli
Central Laboratory Unit
ENEL, S.p.A., Piacenza

Estuarine Dynamics and Global Change

Dr. Norbert Psuty, Associate Director
Institute of Marine and Coastal Sciences
Rutgers University
New Brunswick, New Jersey

Coastal Management: A New Approach for the Evaluation of the Beach Trend

Prof. Antonio Brambati
Institute of Geology and Paleontology
University of Trieste

Oceanography and Coastal Environmental Assessment: Two Case Studies of Different Areas in the Mediterranean and Adriatic Seas

Dr. Romano Ambrogi
ENEL, S.p.A., Milan
&
Dr. Giulio Queirazza
Nuclear and Thermal Research Center
ENEL, S.p.A., Milan
&
Prof. Tecla Zunini Sertorio
Institute for Marine Environmental Sciences
University of Genoa

Coastal-Dependent Marine Leisure Activities: Considerations of Economic Development and Environmental Conservation

Dr. Don Walsh
President
International Maritime Incorporated
San Pedro, California

Waste Management in the Coastal Ocean

Dr. Jerry Schubel, Dean
Marine Sciences Research Center
State University of New York
Stony Brook, N.Y.

IFREMER's Activities in Environmental Monitoring

Mr. Jean Jarry
Director
IFREMER Mediterranean Center
Toulon, France

Monitoring and Testing Programs for Identifying Ciguatera Fish Poisoning Risks

Douglas Park
&
Catherine Goldsmith
Hawaii Chemtect International
Pasadena, California

Large Marine Ecosystems: A New Concept in Ocean Management

Thomas Laughlin & Kenneth Sherman
National Oceanic and Atmospheric Administration
(To be read by Charles Ehler, time permitting.)

ADJOURN

18:00

THIS EVENING THERE WILL BE AN OPTIONAL SPECIAL DINNER FOR SPEAKERS AND PARTICIPANTS FOR AN ADDITIONAL CHARGE OF $25.

THURSDAY 1 APRIL

KEYNOTE ADDRESS

08:30 - 09:15

An Innovative Power Generation System From Sea Currents in the Messina Straits

Ing. Dario Berti
Tecnomare S.p.A.
Venice

SESSION IV. OCEAN RESOURCES & SUSTAINABLE DEVELOPMENT

09:15-11:30

Moderators: Mr. Terry Veness, DTI & Mr. Joseph R. Vadus, NOAA

A Multi-Purpose Approach to Coastal Zone Development: A Case Study

Dr. C. Y. Li
&
Dr. Karl Pan
Science & Technology Advisory Group
The Executive Yuan
Republic of China
Taipei, Taiwan

U.S. Ocean Resources & Technology Development

Dr. Patrick Takahashi
Director, Hawaii National Energy Institute
University of Hawaii

Wealth from the Oceans: A U.K. Program

Mr. Terry Veness
Department of Trade & Industry
United Kingdom
London

Coastal Ocean Space Development in Korea

Dr. Hyung Tack Huh
Korea Ocean Research & Development Institute
Seoul, Korea

Potential Impact of Algae Biotechnology on the Utilization of Coastal Areas

Dr. Riccardo Materassi
Autotrophic Microorganisms Study Center
National Research Council

12:00 - 14:00

LUNCH

SESSION IV. OCEAN RESOURCES & SUSTAINABLE DEVELOPMENT (CONTINUED)

14:00 - 18:00

Moderators: Mr. Terry Veness, DTI & Mr. Joseph Vadus, NOAA

Lobster Ranching in Coastal Waters

Prof. Jens G. Balchen
Division of Engineering Cybernetics
Norwegian Institute of Technology
Trondheim, Norway

Derivation of Beta-Carotene from Marine Organisms

Prof. Ami Ben Amotz
Institute of Oceanographic & Limnological Research
Haifa, Israel

Marine Biotechnology Applications in the Coastal Oceans

Dr. Rita Colwell
President
Maryland Biotechnology Institute
University of Maryland System

Potential of Wave Energy for Coastal Applications

Dr. Even Mehlum
President, Norwave A.S.
Lysaker, Norway

Singling Out of New Breeding Species for a Better Strategy in Aquaculture and Mariculture

Prof. Francesco Faranda
Institute of Marine Environmental Sciences
University of Genoa

Artificial Habitats for Rearing Slow-Growing Marine Invertebrates

G.C. Albertelli, G. Bavestrello, R. Catteneo-Vietti, E. Olivari, & M. Petrillo
Applied Marine Studies Center
University of Genoa

Sustainable Coastal Development Through Integrated Coastal Policy, via Building With Nature

Dr. Ronald Waterman
The Hague, Netherlands

The Environmentally Sound Disposal of Waste in an Offshore Island Developed for Multipurpose Uses

Walter Tengelsen
Chairman, MACROSYSTEMS Inst.
Fort Lee, New Jersey

ADJOURN

(EVENING FREE)

FRIDAY 2 APRIL

KEYNOTE ADDRESS

08:30 - 09:15

Coastal Development in Harmony with the Environment

Dr. Willard Bascom
Ocean Engineer & Scientist
Long Beach, California

SESSION V. OCEAN SPACE DEVELOPMENT & RELATED TECHNOLOGIES

09:15 - 11:30

Moderators: Mr. Norman Caplan, NSF & Professor Ezio Volta, D.I.S.T. Faculty of Engineering, University of Genoa

Very Large Platforms for Floating Offshore Facilities

Dr. Hajime Inoue
Director-General
Ship Research Institute
Ministry of Transport
Government of Japan
Tokyo

Technology for Coastal Development Activities in Japan

Dr. Hajime Tsuchida
Director
Coastal Development Institute of Technology
Tokyo, Japan

Advances in Coastal Ocean Space Utilization: Artificial Islands and Floating Cities

Prof. Takeo Kondo
Nihon University
Tokyo

Large Scale, Multi-Purpose Artificial Island Development Near the Gaza Strip

Prof. Ernst Frankel
Department of Ocean Engineering
Massachusetts Institute of Technology
Cambridge, Massachusetts

Changing Nature of Ports in Coastal Management

Mr. Gerhardt Muller
Port Authority of New York & New Jersey
New York, N.Y.

Technological Innovations for Marine Geology: General Views and Apllications of Advanced Robotics

Ing. Ettore Gallo
Consortium Telerobot
Genoa, Italy
&
Com. Gian Luigi Figari
Environmental Division
Ansaldo Industries
Genoa, Italy

Development of Artificial Reclaimed Lands and their Integrated Planning in Taiwan, ROC

Dr. Ho-Shong Hou
Institute of Transportation
Ministry of Transportation & Communications
Republic of China
Taipei, Taiwan

Underwater Road Link in the Gulf of Naples: A Case Study for EIA in Coastal Zone

Dr. Marco Berta
Tecnomare S.p.A.
Venice

12:00 - 14:00

LUNCH

SESSION VI. SUMMING UP COSU III

14:00 - 15:30

Moderators: Professor Della Croce & Dr. Abel

Agenda:

- Reports by co-chairs of each session
- Open discussion of the program, questions and answers
- COSU III critique for future meeting planning
- Closing remarks by Professor Della Croce

COSU III ADJOURNS

FAREWELL WINE AND CHEESE RECEPTION TO BE HELD IN THE CONFERENCE AREA, HOTEL PARK SUISSE

HOSTED BY THE HOTEL AND COSUIII ORGANIZERS. 17:00 - 18:30

Opening Address

The Coastal Ocean - Challenges and Opportunities

Norman Caplan, Head
Environmental and Ocean Systems
National Science Foundation, Washington, D.C.

On behalf of the United States and the National Science Foundation, I would like to welcome you to this extremely important meeting.

The COSU International Symposium series was originated and continues because of the deep concern that many of us have for the coastal ocean. A concern that goes beyond a longing for the pristine environment that existed in bygone days, and extends to the opportunities that are available to use the ocean for the benefit of mankind. This is where we discover the first problem: 'for the benefit of mankind' has many meanings and different interpretations. One hundred years ago, at the height of the industrial revolution, the coastline was used to play host to piers, harbors, warehouses that soon became rat-infested, and other facilities involved with shipping and the transportation of goods. Soon, coal burning power plants, mills, and other industrial facilities filled the waterfront, since these areas were undesirable, inexpensive, and provided quick and cheap access. The harbors and coastal ocean deteriorated rapidly for the 'benefit of mankind'.

Today we have an opportunity and a challenge to reverse that condition and move forward in a responsible manner to use and enjoy our coastal ocean. In the rush to use the coastline irresponsibly, the real wealth that could have been derived from this natural resource was neglected. Everyone has their own list of natural, mineral, and living resources that represent wealth and jobs, as well as beauty and recreation. This challenge and opportunity is being referred to as *sustainable development.*

In the United Nations report entitled, "Our Common Future," published by the Oxford University Press in 1987, a definition of sustainable (here somewhat shortened and paraphrased) was put forth as follows:

> Sustainable Development is a process of change in which the direction of investment, the orientation of technology, the allocation of resources, and the development and functioning of institutions meet present needs and aspirations without endangering the capacity of natural systems to absorb the effects of human activities, and without compromising the ability of future generations to meet their own needs and aspirations.

I believe that this is the challenge and opportunity that we now have as leaders in the planning and use of the coastal ocean. This is the theme that I want to establish, and this is the reason to support the COSU series of symposia, as another small step in a very difficult process of change. A great deal of the problem is the process to change; changes in the attitude of governments and their allocation of financial resources, change in the philosophy of industry in a market driven economy, change in the understanding of the ordinary people to give them a better appreciation of the problem, and changes in the international perception of developing countries versus industrialized countries. Everyone knows the list of problems that will be present. The process is capital intensive; investment must be attracted (and that means return on investment); human resources are required; government/industry/university partnerships must be forged; and dedicated people are needed to influence the policymakers. People, communities, and policymakers are rigid and reject change.

Table 1

Role of Engineering in Sustainable Development

* Remediation technology for existing problems
* Life cycle product design
* Sustainable use of the water column
* Sustainable resource recovery
* Monitoring instrumentation and vehicles
* Environmental compatible structures
* System modeling and risk management
* Transportation and recreation with no environmental insult
* Economic analysis

What can we do to contribute to this very important crusade? Most of us are not politicians or policymakers, but we are scientists and engineers. Therefore we have a better than average appreciation and understanding of the problems involved with developing and utilizing the coastal ocean. At this point in time, we have to develop the information and the technology and, most important, we have to articulate the case for coastal ocean space utilization. Table 1 shows a few of the activities that are of direct interest to the National Science Foundation (NSF) and the programs in our Engineering Directorate. Our long-range planning committees have cited environmental technology and sustainable development as two areas that are high-priority in the next five years. Several other organizations within

NSF, including Ocean Sciences, Marine Biotechnology, and our economics groups, have expressed an interest.

That brings us back to this meeting and the challenge that we face. I believe that most of the people in this room agree with the need for a new approach to our coastal oceans. In addition, we have gathered here the scientists and technologists who understand the problems and are capable of generating realistic plans that can be presented to policymakers at all levels. In my view the key to this task is the realistic and intelligent use of modern, high technology to provide the tools and the systems for sustainable development of the coastal ocean. At the National Science Foundation, we are chartered with the task of developing new knowledge in science and engineering. In my Directorate (Engineering), we are in the process of generating strategic plans and highlighting strategic societal needs. One of these needs is sustainable development and another is civil infrastructure. Each of our programs: electrical, mechanical, materials processing, chemical engineering, manufacturing, and engineering education, will contribute by supporting research that develops the new knowledge to promote the technology needed to develop the coastal ocean. This is a long and difficult task, but we are determined to make a contribution and help get our message across by supporting meetings such as this, along with our Italian colleagues.

Which brings me, in closing, to say that the experience of working with Professor Della Croce has been outstanding. He has been patient and cooperative over the difficult months that preceded this day. I wish to thank him and also Dr. Able for their cooperation and hard work in developing what should be a stimulating and productive meeting. I am looking forward to the next few days, interacting with all of you and learning a great deal.

Opening Address

Greetings from the University of Genoa

Sandro Pontremoli, Rector
University of Genoa, Genoa, Italy

I am delighted to have the honor of opening and hosting this Third International Conference on Coastal Ocean Space Utilization, especially considering that this is the first of the series to be convened outside of the United States.

In view of the myriad of shared problems and opportunities confronting our two great nations, it is most appropriate that Genoa be selected as the site for this Conference. First, the University of Genoa under Dr. Della Croce's inspired guidance has taken a commanding lead in approaching Italy's coastal and offshore issues. Secondly, these issues -- living and mineral resources, environmental protection, aquatic recreation and tourism, to name a few -- penetrate all sectors of Italy's economic growth, cultural enhancement, and general welfare. Third, as it must be quite apparent to all of us here, Genoa and the year 1992 enjoy particular significance in American history.

Thus for those and many more reasons, I am pleased beyond words to welcome you all and hope that you will avail yourselves to the fullest of our famous Italian hospitality. Again, thank you for inviting and honoring me on this most significant occasion.

Keynote Address

Protection and Management of the Coastal Area Among Bordering Countries (Ramoge Agreement)

Dr. Ing. Giovanni Gallino
Assessorship for the Environment, Regione Liguria.

Prof. Norberto Della Croce
Director - Institute of Marine Environmental Sciences, University of Genova.

INTRODUCTION

During the opening session of the 32nd Convention, Plenary Assembly of the Commission Internationale pour l'Exploration Scientifique de la Mer Méditeranée (CIESM), held in Rome in December 1970, Prince Ranieri III of Monaco, Commission Chairman, suggested that a test area for the elimination of pollution be created within the framework of the Etudes en Commun de la Méditerranée (ECM).

Also in 1970, the Italian Commission for Oceanography stated that the correction of pollution discovered in the Upper Tyrrhenian Sea was one of its main goals for 1971 and 1972: this resulted in the proposal of the RIMAT project.

The Monegasque initiative was favorably accepted by the French and Italian governments and the project was called RAMOGE, taking the first syllables of St. Raphael, Monaco, and Genoa; cities located in the test area.

In 1972, the proposal by Prince Ranieri III found response in a meeting held by Italian-French-Monegasque experts, who drew up a report on coastal conditions in the RAMOGE test area.

On May 10th, 1976, the three governments signed the RAMOGE agreement, concerned with preserving the quality of Mediterranean coastal waters, preventing their pollution as much as possible, and improving their current conditions.

The initiative involves an area of considerable tourist and economic importance and it points out the top priority of environmental protection which requires the combined and coordinated efforts of the three, Riviera towns.

First and foremost, the RAMOGE agreement, which was ratified and became effective in 1981, defined its organization, the goals to pursue, and the development of its actions, depending on the methods used.

The time lapse between the agreement's ideation and enactment points out how, in an international situation, operative mechanisms are considerably complex; in this case, this complexity is increased by the fact that the issue is relatively new.

THE AGREEMENT'S STRUCTURE

The institutional body governing the Agreement is the trilateral, international Commission with the help of a Technical Committee and several Work Groups.

Briefly, the Commission consists of three Delegations. Each Delegation is comprised of a Head Delegate and six Delegates. Each Country adds to these Delegates, representatives of the territorial bodies concerned (Regions). The Commission has a Chairman who remains in office for two years; the chairmanship is assigned, in turns, to the Head Delegate of each country. Originally, the Technical Committee consisted of fifteen experts, five from each country. From March 30, 1990, however, the organization of this Committee foresees the presence of four experts from each country who are to be appointed, preferably from among the representatives of the technical services of the local Bodies. After the Commission's consensus, the Technical Committee may benefit from the cooperation of experts who make up part of the work Groups, depending on the initiatives undertaken within the range of its activity.

Since the Agreement's first phases, Secretariat services have been taken care of by the Principality of Monaco.

The Delegations and the Technical Committee members are appointed by the Government; however, the Government does not appoint the experts who are going to be part of the Delegations.

The Commission's regulations mandate that each Delegation has one vote to be expressed by the Head Delegate. All the Commission's decisions must be voted unanimously.

Common interest expenses are divided as follows: 45% France, 45% Italy, and 10% Principality of Monaco, whereas the expenses corresponding to extraordinary research are subdivided depending on the case.

During 1992, the Commission met seventeen times.

GOALS PURSUED BY THE AGREEMENT

At the time of the Commission's creation, its scope was defined as establishing closer cooperation between the existing services of the three Governments in order to fight pollution in territorial and continental shelf waters. This goal must be achieved by means of the following activities:

1. Analysis of all common problems as far as water pollution is concerned
2. Definition of the competent, administrative services of the three countries in order to
    a) point out polluted areas
    b) obtain mutual information on territorial re-organization projects that entail serious pollution risk
    c) analyze, in economic terms, the necessary infrastructures and equipment to fight pollution
3. Promotion of different forms of goal-oriented, scientific cooperation, bearing in mind already existing work and material means
4. Propose to the three Governments, all the suitable measures for water protection by means of special agreements

THE COMMISSION'S ACTIVITY

Since this is an International Commission dealing with problems inherent to the pollution of the waters between bordering countries and therefore, the aspects of the topics being discussed are rather complicated and susceptible to change, it is easy to understand the confusion that has characterized the Agreement's evolution.

At the initial, informal meetings, the Delegations presented and discussed various topics and tasks, among which the following had priority:

1. Coast reclamation
2. Reclaimed area surveillance
3. Hygienic conditions and cleanliness of beaches
4. Information

Five Work Groups were also planned.

1. Marine environment surveillance
2. Reclamation program
3. Legal matters
4. Fight against accidental pollution
5. Information and enhancement of public awareness

It will be seen that the Work Groups that were finally set up were quite different from those just mentioned.

The Delegations discussed whether or not the Work Groups should be set up from the beginning of the Commission's work, and, above all, highlighted tasks and goals that did not find easy or unanimous approval.

At this point, is seems correct to mention the goals stated and pursued, following the first meeting held by the Commission in February 1982, ten years after the first meeting held by the Italian-French-Monegasque experts.

Having defined the range of its activity, as well as that of the Technical Committee and Work Groups, the Commission established the following priority goals:

1. Mutual information regarding the prevention and fight against pollution on administrative, technical, and scientific levels
2. Inventory of marine environment polluting sources coming directly or indirectly from either the continent or the sea
3. Assessment of marine environmental quality on the basis of the scientific work carried out in various applicable disciplines
4. State of the programs for action against pollution, pointing out already existing and future structures, as well as action programs in case of accidental pollution
5. Comparison between all the regulations concerning marine environmental protection

Furthermore, while the Commission believes that particular attention should be devoted to spreading information in order to enhance public awareness of Mediterranean coastal protection, it recommended that the utmost care be taken to prevent uncontrolled use of collected data exchanged between the countries. The Commission entrusted the Technical Committee with the task of initially overseeing these priority activities.

GOAL 1: MUTUAL INFORMATION

In 1972, the data available on the RAMOGE area were summarized in a general document, presented according to the following outline:

1. Administrative and Legal Aspects

2. Scientific and Technical Aspects
   a) Oceanographic information
   b) Information on pollutants and their origin
   c) Influence of pollution on fauna and flora
   d) Consequences of marine pollution on health
   e) Analysis of maps and documents drawn up by the administrative authorities of the test area
   f) Sewage plants
   g) Comments on coastal reclamation projects

The document drawn up in 1972 by groups of Italian-French-Monegasque experts for the administrative and scientific parts and which represents the basis of the Agreement, was analyzed during the first meeting held by the Commission in March 1982. This document was adequately integrated with notes supplied by the corresponding Delegations.

In order to pursue the defined goals, the Commission entrusted the Technical Committee, helped by several Work Groups, with the task of working in the various fields of research on both technical and scientific levels.

Considering that the activities necessary to achieve the goals stated in Goal 1 may be considered as a compendium of those activities geared towards pursuing the objectives of the following points, the undertaken actions are described below.

GOAL 2: INVENTORY OF MARINE ENVIRONMENT POLLUTION SOURCES

The first activity consists in measuring sewage dumped from public sewers into the sea, regardless of whether it was directly conveyed or arrived through surface waterways.

The inventory of the coast's industrial waste network was postponed.

In the case of the coast's urban sewage being directly dumped into the sea, the following data was collected:

1. Pipeline
   a) Length (in meters)
   b) Cross-section area (in square cm)
   c) Discharge depth (in meters)
2. Polluting Load
   a) Mean flow rate (l/s)
   b) Polluting content in BOD5 (mg/l)
3. Treatment prior to dumping
   a) Primary
   b) Secondary
   c) Tertiary

Concerning the waste directly dumped into the sea through surface waterways, it was deemed adequate to characterize each waterway with a hydrographic basin of a certain magnitude by gathering the data listed in the following diagram. These data give a general idea on the relevance of the polluting contents conveyed to the sea by each waterway.

However, it must be noted that, apart from the date of the last population census (1987 in Italy and 1982 in France), BOD5 calculation does not lend itself to particular comparisons if one considers that for the Regione Liguria the entire population of the emptying basin was considered with the sole exception of cities on the coast, whereas for France and the Principality of Monaco, the calculation failed to include the coast cities and the populations which are already linked to the sewage network that dumps directly into the sea. The lack of uniformity of the data in this instance makes it difficult to estimate the inputs into the marine system and, therefore, it is hard to assess the environmental impact, should one want to examine other aspects of the emptying basins.

Because of this, the data sheet that groups together emptying basins data shows differences in the Italian and French versions, a fact that underlines how different approaches to the same problem may lead to difficult data interpretation and evaluation. The differences are even more marked when considering the relevance of seasonal population fluctuations in the RAMOGE area, since the Italian and French Rivieras are popular tourist and sea-bathing resorts.

Table 1

Emptying Basin

Physical Aspects
1. Emptying basin maps in 1:100,000 or 1:200,000 scale
2. Resident population estimate
3. Fluctuating population estimate
4. Classified flow-rate curve
5. Flow-rate measuring stations yearly chart
6. Monthly flow rates mean curves
7. Purification plants inventory
8. Localization of direct and indirect sewage sources (optional)

Theoretical Polluting Loads Estimate Expressed in BOD5
1. Produced:
   a) By inhabitants (54 grams/inhabitant/day)
   b) By industrial equivalents

Measurements at River Mouths
1. Flow rates
2. Temperatures
3. Suspended matter
4. Nutrient salts:
   a) Nitrates
   b) Nitrites
   c) Ammonia
   d) Phosphates
   e) Silicates
5. Heavy metals
   a) Cadmium
   b) Copper
   c) Lead
   d) Mercury
6. Anionic detergents

Table 2

Italian and French Questionnaire

General Data
1. Name of the Municipality
2. Resident population
3. Seasonal population
4. Duration of the population fluctuation period

Garbage Collection
5. Number of inhabitants served by garbage collection service
   a) During a normal period
   b) During a population fluctuation period
6. Overall yearly quantity of collected garbage
7. Number of inhabitants served by a bulky garbage collection service
   a) During a normal period
8. Is there a special collection service for other types of waste?

Waste Treatment
9. Household waste destination
   a) Incinerating unit
   b) Composting
   c) Authorized dumping sites
      (1) Public
      (2) Private
   d) Other
10. Purification plant sludge destination
11. Industrial waste destination
12. Bulky garbage destination
13. Existence of illegal dumping sites
    a) Number
14. Existence of programs to eliminate illegal deposit areas
    a) Number
15. Existence of projects for waste-treatment units creation
    a) Type
    b) Deadline
    c) Description

Also within the range of activities concerning pollution sources, the potential quality of solid urban waste dumped into the sea was assessed by calculating the quantity of waste produced by the single user basin involving the RAMOGE area (identifying them with main waterways basins). Guidelines were then drawn up, listing the following parameters for each basin:

1. Emptying basin
2. Surface area in sq. km.
3. Number of Municipalities grouped in the emptying basin
4. Kilometers of coastline of the emptying basin
5. Resident population

6. Maximum population
7. Duration, in days, of population fluctuations
8. Percentage of inhabitants served by the garbage collection service during normal periods
9. Percentage of inhabitants served by the garbage collection service during periods of fluctuations
10. Percentage of inhabitants served by bulky garbage collection service
11. Yearly quantity of collected waste (in tons)
12. Emptying basin specific waste production (in kg/day/inhabitant)
13. Yearly theoretical quantity of produced waste (in tons)
14. Number of incinerating plants
15. Number of composting plants
16. Number of authorized dumping sites
17. Number of other plants
18. Number of projects for the creation of waste treatment units

Part of the data came from a study carried out using the following questionnaire on waste which was especially drawn up for Municipalities, which when completed highlighted the fact that the Italian and French experts adopted slightly different items for the qualification of certain parameters.

In order to gather additional information on the quantity of solid waste that may reach the sea in the RAMOGE area and to study its possibility of shifting towards the above-mentioned area, data were collected on:

1. Beached macro-waste
2. Waste floating in coastal waters

Concerning item 1., observations were made during a campaign where waste found on the beaches of six Italian towns, six French towns, and Montecarlo was collected and analyzed on pre-set days, determining its quantity, origin, source (where possible), and type, as a function of the surface area of the beaches used for the test (Italy was not able to provide an analysis of the origin of the waste collected on its beaches).

As for point 2., an identical type of analysis was carried out on the quantity and quality of floating waste collected at sea by special boats during the summer periods.

Both campaigns took place during the 1986/1988 period.

Data was collected using the following outline that foresees the collection of the listed information and for the campaign on floating waste, the item algae was added.

Table 3

Data on Beached Solid Waste

1. Municipality
2. Date
3. Coast
   a) Length in meters
   b) Surface in square meters
4. Total solid waste (in 100 kg units)
   a) Plastic (in 100 kg units)
   b) Wood (in 100 kg units) and in %
   c) Other (in 100 kg units) and in %

These analyses achieved results that may be considered revealing concerning the type of waste, but were not very accurate as far as quantity and distribution are concerned.

The main conclusions may be summarized as follows:

1. Wastes of manmade origin represent approximately 38% of the total and, generally speaking it consists of plastic materials (approximately 20% of the total).
2. In Liguria, waste of natural origin (wood) adds up to approximately 26% of the total against 47% in the District of the Maritime Alps.
3. Other waste is present in both areas in similar quantities (35 -36%).

4. Beached waste of Italian origin rapidly decreases as one moves along the Monegasque and French territories.

For the time being, maritime traffic has not yet been analyzed, whereas there is a census of the ports where each port's activity and size is specified.

GOAL 3: MARINE ENVIRONMENTAL QUALITY ASSESSMENT

In this case, a joint campaign for data collection on water quality in the sample area of sea between the mouths of the Var and Roja rivers was carried out. Data was collected according to what is described in the following chart and it was interpreted in the report supplied to the Governments by the Member Countries.

During this campaign, marine environment quality was assessed as follows:

Table 4

Marine Environment Quality Assessment

1. Hydrology
2. Marine organisms contamination by heavy metals or enteric bacteria
3. Nutrients distribution
4. Discharge of suspended matter and corresponding heavy metal contamination resent in the suspended matter
   a) suspended matter discharge
   b) metal contamination of suspended matter
   c) comparison between the concentrations of heavy metals dissolved in seawater and heavy metals associated with seawater particles in coastal areas
   d) plankton characteristics and distribution in the RAMOGE seawater

GOAL 4: STATE OF THE ACTION PROGRAMS AGAINST POLLUTION

The assessment of action programs has provided, through the collection of the Region's planning documents, a sense of the existing and planned purification structures for this area.

In particular, information on the existing and planned structures which discharge directly into the sea or into major waterways flowing into the RAMOGE area was collected and the data presented in map form.

Unfortunately, it should be noted that the RAMOGE activity did not sufficiently stimulate Governments to construct purification infrastructures in the area in question, even though during the period of RAMOGE'S activity, relevant works were carried out, such as water-purification plants in Savona, Sanremo, Ventimiglia, Bordighera, MonteCarlo, and Nice.

Among the action programs to be used in case of accidental pollution in the RAMOGE area was a Commission draft of a joint Emergency Plan. This plan, which was about to be implemented during the Haven oil tanker accident, was proposed to the Governments concerned for its expeditious adoption. The three Countries are working to complete all the respective administrative requirements in order to make this plan operative as quickly as possible.

It may be said that this last activity will certainly prove to be the most impactful one from a practical standpoint.

GOAL 5: COMPARISON BETWEEN ALL THE REGULATIONS

One of the Work Groups drafted the following Summary Note which proved to be extremely helpful in comparing the various regulations in force concerning marine environmental protection.

Unfortunately this exercise stressed the differences of the laws and, as a consequence, the adoption of homogeneous regulations in the three countries was not achieved. It is true nevertheless, that there will have to be a form of homogenation, at least as far as Italy and France are concerned, in accepting European Economic Community (EEC) directives.

Table 5

Summarizing Note: Comparison between Regulations

TITLE 1 SAMPLING
TITLE 2 GENERAL DUMPING RESTRICTIONS
TITLE 3 SEWAGE DUMPING REGULATION
- CHAPTER 1 Coming from the coast
  - Sec. 1 General provisions
    Preliminary authorization
  - Sec. 2 Specific provisions
    Industrial waste
    Other waste
- CHAPTER 2 Coming from ships
  - Sec. 1 Dumping of waste and dregging products into the sea, including accidental dumping
  - Sec. 2 Dumping of hydrocarbons and toxic substances
  - Sec. 3 Incineration
  - Sec. 4 Pleasure boats
- CHAPTER 3 Coming from the continental shelf
- CHAPTER 4 Beaches and bathing water
  - Sec. 1 Beach cleanliness
  - Sec. 2 Bathing water quality
- CHAPTER 5 Marine reserves

COASTAL OCEAN SPACE UTILIZATION

To supply information and promote environmental protection, the Commission produced an audiovisual document for schoolchildren that was not very successful, perhaps because of the different ways in which schools are organized in the three countries and the different equipment available. Likewise, a text for lower high school teachers was produced, hoping it may be used for short lessons to foster awareness on environmental problems typical of this area.

Additionally the Commission promoted the A. Vatrican Award in memory of RAMOGE'S first Secretary General, for the implementation of a study protocol on marine pollution in the RAMOGE area. This prize was awarded for the first time in 1991.

Currently, the possibility of organizing a contest for compulsory-school students (primary and lower high schools) on marine environment protection is being studied.

GENERAL OBSERVATIONS ON THE ACTIVITIES CARRIED OUT THUS FAR

The essential and decisive activity contemplated by the RAMOGE agreement is the measurement of pollutants in coastal water, originating from household waste dumping which is insufficiently treated or inadequately disposed of into waterways or conveyed to the sea by means of pipelines.

A goal of considerable importance, attained in full, deals with pipeline inventory and localization, thus making it possible to calculate the polluting content. Waterways situations are more complex, such as in the Var and Roja Rivers, whose contents, when dumped into the sea, are the result of the combination of various factors, and are characterized by the great variability of their water volumes and, therefore, the great variability of polluting element concentrations.

Apart from the results obtained by the Work Groups, discussion is still open on the assessment method of the loads carried by basins and waterways into the sea of suspended matter, heavy metals, and nutrients concentrations and on possible quali/quantitative waterways standardization. The organic load, expressed in BOD5, has been theoretically estimated on the basis of emptying basin populations (in the measure of 54 grams/day/inhabitant), bearing in mind the industrial equivalents and rivers' flow rates. In any case, it was found that, notwithstanding the great effort to collect the existing data on basins emptying into the RAMOGE area, the information gathered was deemed insufficient to give a precise idea on their discharged into the sea. In this instance, too, it is being debated whether the collected data ought to be completed by means of systematic reading of certain qualitative parameters of a well-defined series.

Generally speaking, it is believed that information integration would allow for the possible orientation toward actions already marked by a commitment in fighting pollution.

RAMOGE's activity in determining marine environmental quality was carried out through a series of campaigns, initially defined as joint area surveillance campaigns and then as joint campaigns. During these campaigns, a series of analyses was carried out on mussels and sea urchins gathered on the coasts of the three countries, as well as a series of observations in correspondence to the mouths of the RAMOGE area's main rivers (Roja and Var) and along the coast between these two mouths. The specific site of these observations, like the topic dealt with in this report, exclude the possibility of examining the ad hoc technical report which analyzes the results.

The program foresaw the continuation of the study at the mouths of the RAMOGE area's main rivers.

MAIN ACHIEVEMENTS

It was impossible to carry out a global analysis of the state of the three countries' action programs to fight marine pollution as was originally envisaged. Without returning to the pipelines and purification plants issues, it may be more interesting to mention the main achievements, making reference to what is outlined in the notes presented by the individual, participating countries.

The front of the RAMOGE area, as is well-known, is essentially dedicated to activities such as tourism, sea bathing, and pleasure boating.

The French aimed mainly at improving the bacteriological quality of coastal waters and in solid waste management. Concerning water quality control of the waters subject to intense bathing activity, in the District of the Maritime Alps which includes sixteen towns along a 120 kilometer-long coastline (forty of which are beaches), 164 check points have been set up, making it possible to classify hygienic conditions.

The reclamation of the RAMOGE coastal area, i.e., the reduction from 40% to 8% of the beaches classified as no bathing, is linked to the restructuring of municipal waterworks, the creation of purification plants, the recording of dry season flow rates, the elimination of 46% of illegal sewage dumping sites, and the mechanical removal of floating waste from water sheets. Furthermore, during the summer period, coastal surveillance is carried out from the air, thus making it possible to pinpoint, for example, illicit dumping along the coast from ships, to check the working conditions of pipelines, and adequately direct scavenger boats. As for household waste, it is reported that only 3% of the municipalities do not have this type of garbage collection service and that more that 200 irregular dumping sites and deposits have been closed or cleaned up; whereas thirty intermediate deposit units for bulky garbage and for the collection of recyclable waste have been set up.

On the Italian side, the antipollution plan in the RAMOGE area mainly deals with the purification of urban effluents, giving priority to the building of purification plants in the cities close to the French border (Ventimiglia, Bordighera, and Sanremo) and for large urban and industrial centers like Savona and Vado Ligure. Concerning sea bathing in the RAMOGE area, 197 check points to monitor water quality have been set up along an approximately 170 kilometer-long coastal front which includes thirty-eight towns. Coastal strip reclamation has reduced to 14% the number of points where sea bathing is forbidden, according to the standards set by the Italian laws. Scavenger boats for floating waste collection were not used on a large scale.

In the Principality of Monaco, initiatives mainly concentrated on urban effluents and those of peripheral municipalities which were planning pre-treatment and purification plants as well as pipelines directly dumping into the sea. Boats for floating waste removal were also used in the waters of the Principality.

During the 1982-1990 period, apart from press conferences and within the limits of the means assigned for this purpose, the Commission promoted a series of joint initiatives to enhance public awareness on coastal water protection. Along with propaganda events, in order to enact the guidelines outlined during the eighth meeting (1987), the Commission decided to promote the Alain Vatrican Award, in memory of RAMOGE's first Secretary General, for a student or young researcher who is involved, either directly or indirectly, with marine pollution and its consequences in the RAMOGE area. The award, 30,000 French Francs (1990), was given for the first time during the 1990-1991 academic year. The Commission's Report states that the analysis of predicted and already taken actions reveals the inadequacy of the methods and means available, as compared to the importance of enhancing public awareness.

An instructional instrument for lower high school teachers was produced. It will be distributed during the course of the next school year.

COASTAL OCEAN SPACE UTILIZATION

Among the initiatives undertaken, the most recent, and certainly most important, consists in the drafting of a joint action Plan in case of accidental seawater pollution and the proposal to make the Governments in question adopt it.

The recently approved extension of RAMOGE's boundaries from La Spezia to Marseille implies, as a consequence, that measurements and data collection must be extended in the near future to the areas that have not yet been investigated.

RECOMMENDATIONS FOR THE GOVERNMENTS

It has been suggested by the Commission that the countries take most seriously the importance of preparing the complete inventory of industrial waste dumping, both direct (coastal) and indirect (waterways), and the inventory of all direct and indirect dumping into waterways on the RAMOGE area's basins.

The three countries are encouraged to disseminate information to the general public to reduce the dumping of objects (particularly plastic ones) into the sea and onto beaches by either boaters or sea bathers. Additionally, they are encouraged to reclaim valleys and to clean up illegal dumping sites.

On the basis of the results obtained from the investigations carried out on emptying basins, it is suggested that the countries take necessary measures to proceed, in a coordinated manner, with measurements at the mouths of the waterways to better understand the importance of pollution draining into the sea from RAMOGE's emptying basins.

As for determining marine environmental quality, the Commission has no recommendations to address to the governments of the three Countries; merely guidelines and encouragements.

The Commission feels that the achievements of the three participants in the RAMOGE area over the 1972-1990 period are important and, therefore, recommends that the countries pursue this effort in terms of purifying both coastal dumping sites and the sewage that flows into the waterways of the RAMOGE area's emptying basins.

Considering the analysis and regulations comparison, the Commission recommends that the following be encouraged:

1. In-depth study of shared information on how to organize prevention and fight against pollution;
2. To look for points in common for regulation compatibility;
3. Formulation of concrete proposals for achieving shared goals as quickly as possible (especially for Italy and France) within the time limits indicated in the corresponding EEC directives.

The Commission, acknowledging the importance of the progress made so far, agrees that this task is far from being completed. And while acknowledging that the inventories, studies, and conclusions produced are important and full of information, the Commission admits that it took too long to achieve them and this limits their relevance and possible use. Likewise, the Commission acknowledges that the outcome of the work carried out by the experts that may be directly exploited by the countries at the end of the studies, is limited.

The Commission points out that the slow progress of the works, mainly caused by lack of ad hoc personnel, may also be partly explained by a lack of clarity in the definition of the goals to be attained, tasks to be carried out, and methods to be applied for their achievement, as well as by insufficient rigor in carrying out the actions and in deadline observance.

CONCLUSIONS

This experience should reveal, more indirectly than directly, the difficulties that arise between bordering countries even when it is clear that they are willing to cooperate when facing shared coastal strip pollution problems.

In its Report to the French, Italian, and Monegasque governments, the Commission points out the need to redirect the efforts of the partners of the RAMOGE project.

Regional Seas and Embayments

1

First Approach for an Integrated Environmental Planning at Regional Level of the Coastal Marine System (Regione Liguria)

Arch. Lino Tirelli
Head, Department of Assessorship for the Environment
Regione Liguria, Genova, Italy

INTRODUCTION

This report is aimed at providing information about the first steps that the Ligurian Region is taking concerning the planning of its activities for the defense of the marine environment; both to preserve it from pollution caused by economic operators and by the Ligurian community which dwells near its banks and to promote improvement of environmental conditions that impact the marine resources of the Ligurian region.

As commonly happens in coastal regions, it is evident that the relationship between the sea and land is not only extensive, but structurally strong due to the morphology of the Ligurian territory which is distinguished by a stretch of harsh mountains reaching down towards the sea along a fine borderline. Additionally, this relationship is fragile, due to the precarious environmental equilibria involved, as it is full of settlements and connecting infrastructures. The most relevant cities and their confused urban sprawl are developed in the sections of the coastline which assume, for a minimum width (several hundred metres), the characteristics of an alluvial plain or a hill with a minor acclivity. This same stretch is also the home of a system of channels of a local, national, and international character. Due to the tourist industry, which in turn is directly related to the environmental quality of the territory, infrastructures and settlements are seasonally burdened by tourists, especially along the narrow stretch of coastline. Moreover, the presence of important harbor outlets sets the stage for further exchange of goods (including very hazardous energy products) which present potential dangers to the marine spaces of access, as well as to the valleys running perpendicular to the coast which flow into the same outlets. The scarce environmental resources of the narrow coastline are utilized apart from any reasonable standard of utilization, by extra-regional energy plants and primary industrial iron and steel factories, clearly incompatible with the environmental equilibriums of the sites and with the presence of inhabited areas.

This outlined view of environmental stress does not, however, hinder one from recognizing the significant quality characteristics of the favorable relationship that exists between the resources of the territorial environment and the resources of the marine environment. It is really the unique qualities of both which limit the environmental degradation in progress and justify the efforts for a two-fold action of defense and requalification. From a physical point of view, the marine resource has been naturally defended by the depth of the water itself, which is quite considerable right up to the shore; while a large area of the land resource has been protected by the morphological inaccessibility of the mountainous peaks and valley inclinations. All this is true, despite the fact that some sections have been put to harsh tests because of landscape constraints and, for some time, from military constraints.

From these introductory notes, it is possible to perceive the importance which for historical, cultural, and economic reasons, the Ligurian Region intends to bestow upon the marine resource. This is done in the hopes of improving the direct management of its own administrative activities, as well as the understanding and comprehension of the actual phenomena and their evolution, in order to ask other national, EEC, and international bodies to honor their own commitments towards cooperative efforts regarding the coastal zone's fragile environment. This would also include any requests of the scientific community for financial contributions.

"ENVIRONMENT PROJECT" - GUIDELINES FOR STRATEGIC PLANNING

The guidelines for such an approach have been identified as the "Sea-coast sub-project" of the so-called "Environment Project": a coordinated series of relevant, integrated, environmental impact actions which the Ligurian Region intends to carry out in conformity with the regional law 26/91. For the strategic planning, those actions which in the meantime have been initiated and those which were previously envisaged have been pivotal. For definition of this context, requests have been made for support from national and EEC authorities, as well as requests for collaboration with institutions, associations, and public and private enterprises. In the same way, efforts are being made to understand and follow the initiatives regarding other subjects and to attempt to amalgamate wherever possible, their results and investments.

Such guidelines will probably move according to integrated paths among the coast urban-territorial plan, programming of financial activities from a tourist, pleasure craft, fishing and coasting trade profile, and design of defense interventions, by attempting to create a complex preventative evaluation system for environmental impact and continuous monitoring. These efforts are directed towards optimal preservation of the marine resource by keeping alterations to a minimum, especially irreversible ones.

Regarding the major part of the population and economical activity of the sea (or coastline), it is not possible to think in terms of a purely conservational policy; one must envision developing a dynamic control over set transformations, which will afford a rigorous control system to handle the environmental impacts.

Under this aspect, the "environment project" will identify three types of coast:

1. Pure conservation zones, based on their naturalistic and historical-cultural character
2. Zones susceptible to transformation with set utilitarian additions
3. Degraded zones which need to be reclaimed and reused

Among the first type, preference will be given to protected coastal areas and those with marine areas which are to be biologically and physically protected from a naturalistic view point, e.g. Portofino Parks, "Cinqueterre", and areas surrounding Gallinara Island.

Among the second type there are the inhabited coastal areas, tourist and pleasure craft areas, and those used for other economic activities. There are many utilization conflicts and obvious incompatibilities present in these areas, as well as biological and chemical-physical pollution effects which are well above the levels allowed for bathing areas and hazardous, to a certain degree, to the marine life. Particular importance is also given to the problems concerning navigation and the commercial and tourist ports.

In the last type there are the sections of coast involved with industrial buildings or ports which are partially used or underused, due to the evolution and concentration of specialized traffic. The Genoa Harbor area and the Stoppani area between Arenzano and La Spezia roads are interesting cases to be studied and carefully planned. Involved here are considerable recovery and reuse potentials as well as archaic, unresolved problems of environmental requalification (for the coast as well as the sea floor) which are interesting from a technological, scientific, and economic viewpoint.

It must be pointed out that the natural equilibria of the beaches have been changed due to the altering of the water system of the hydrographic basins, as well as their banks and inclinations, for urbanization of the outlets. Additionally, the sea floor at the coastline, because of the dispersal of all types of waste in the river floors or the wash-out of polluting soils, is subject to aggression of such magnitude that its characteristics are being drastically changed. Accretion of aggregates and harbor or coastal defense works create alterations whose environmental impacts must be understood and channelled to eliminate the inconveniences connected with any profound alteration of the natural assets.

Therefore, a qualitative control of the marine body of water assumes a strategic importance in the regional environmental politics as a means of measuring the:

1. Condition of the internal water which drains into the sea
2. Efficient control of civil and industrial discharges
3. Condition of atmospheric purification
4. Release of pollutants from intensive agricultural areas
5. Impact of maritime traffic
6. Release of hydrocarbons from ordinary discharge and loading cycle, stocking, transport, and distribution
7. Different incidents for events connected with industrial, civil, and natural risks in an urban environment

Amongst the actions and interventions which the "environment project" guidelines have identified, the following mainly involve the marine environment and coastal profile:

1. Conservation of naturalistic interest areas will be pursued by means of actions which favor
   a) Institution of marine and coastal parks
   b) Repopulation and reforestation of the sea floor
   c) Initiative for accretion and conservation of beaches
   d) Diffusion of knowledge for characterization of sea floor
2. For requalification of degraded environments and reconversion interventions such as:
   a) Reutilization of underused harbor areas
   b) Reutilization of unused coastal settlements
   c) Recovery of degraded, industrial coastal areas
   d) Recovery of polluted sea floors
3. To reconcile economic development with the protection of marine resources, specific projects will be formulated (in agreement with competent authorities) for
   a) Support in the form of compatible fishing methods
   b) Reorganization of access channels and safety of traffic
   c) Compatibility of nautical/pleasure-craft activities
   d) Starting up of an ecological vessel for marine monitoring
4. To improve marine-coastal utilization, projects will be developed for
   a) Quantity-quality adaptation for efficiency of purifiers
   b) Maintaining efficiency of purifiers
   c) Development of sea transport as an alternative and integration of coast system

d) Improvement of tourist and leisure activities through
  (1) Creation of nautical berths other than marinas
  (2) Creation of pedestrian and bicycle ways
  (3) Temporary receptivity
  (4) Culture and practice of marine activities

VARIOUS REGIONAL ACTIVITIES IN PROGRESS AND DECISIONS AWAITING THE "ENVIRONMENT PROJECT"

The proposal of the so-called "environment project" will not die from benign neglect because it is being implemented in an area which is already involved with various activities to be developed and reformed. It is therefore of no surprise that during preparation of the guidelines from which the "environment project" is to be produced, certain regional activities in progress and others to be carried out shortly, have not been interrupted. Initiatives may be considered in the manner of anticipation and preliminary control of the "project" itself. These are aimed at defending and safeguarding the sea from pollution and, therefore make up part of the competent, regional office activities for planning and improvement of the water, particularly the bodies of water which are used for domestic, civil, and production purposes as well as for discharge activities.

Census of Bodies of Water in Liguria

In conformity with article 7 of law 319/76 concerning measurement of quantity-quality data for bodies of water, the Ligurian Region has arranged for an update of a census of water masses in Liguria by means of an agreement stipulated with the Genoa Gas and Water Board (A.M.G.A.).

The assignment, which is going through its final stages, plans for several activities, among which is the monitoring of coastal waters by means of sampling seawater at eighty-four stations situated 100 meters from the shoreline and arranged along the entire Ligurian coast.

More than 670 samples are scheduled to be taken in eight different lots within the period of twelve months, including the following assessments:

1. Transparency
2. Nutrient substances (ammonia nitrogen, nitrous nitrogen, nitric hydrogen, orthophosphates, total phosphide)
3. Mineral oils
4. Bacteriological indexes (fecal and total coliform bacteria)

Environmental Engineering Study of Sea Discharge Pipelines Along the Ligurian Coast

The discharge waters of an underground pipeline, due to the amount of motion and lack of density compared to that of the receiving body of water, may cause, notwithstanding the initial mixing, a large concentration of pollutants near to the inlet point.

After the initial mixing, which is limited to the inner parts of the inlet point, the discharge waters are dispersed with the coastal current and the pollutants are further diluted. The study of dispersion of such pollutants is therefore proposed to provide data on their concentration, both near to and at a certain distance away from, the discharge outlet point.

In order to safeguard the marine coastal environment, the Regional Administration has stated the need to arrange for an adequate plant in order to evaluate the impact of numerous manmade discharges into the sea by means of a convention with ENEA at S. Teresa near La Spezia.

The project is, at present, under way and is divided into three phases which are partially overlapping:

1. Preliminary investigations
2. Environmental studies and monitoring
3. Creation of an information system

Preliminary Investigations

Any type of investigation that involves coastal waters requires a complete understanding of the situation enabling researchers to predict the destination of the pollutants emitted into the sea and to properly evaluate their effects.

An investigation of already existing data also allows a more efficient and accurate planning of the envisioned activities and makes it possible to identify known shortages and mark out new environmental studies for further projects.

Identification of the Environmental Situation

An investigation of the coastal waters in order to be accepted as valid and efficient, must take into account the influence of all factors, both singularly and as a multifaceted variable. Therefore, it will be necessary to analyze the large amount of existing information in reference to the chemical and biological characteristics of the marine waters to localize the areas at risk where further pollutant emissions may cause eutrophication, and study the situation of the sea floors in relation to their morphological, sedimentological, chemical, and biological characteristics.

The above existing data (available at Regional, Provincial offices, Institutions, Consortiums, Universities, etc.) will be gathered and reviewed under the following rubrics:

1. Type of sea floors for the marine platform (morphology, granulometry, biology, chemistry, microbiology)
2. Identification of coastal inflows
3. Quality of marine coastal waters (chemical-physical and biological parameters)
4. Benthic population.

Census of Sea Pipelines

During the course of this activity, data and information relative to sea discharge pipelines will be collected. A census will be taken of the operating, underwater pipelines along the Ligurian coast and the data relative to technical and functional characteristics of the pipelines; shore equipment; and the types, quantity, and quality of disposed waste.

Collection and Validation of Existing Oceanographic Information

Data collection will be extended for this activity to meteorological-oceanographic information relative to the Ligurian Sea, which is available from the ENEA databank and other sources to be identified.

In particular, the following data will be used:

1. Hydrographic data
2. Hydrologic data (temperature, salinity, and density) for assessment of the bodies of water and their seasonal variability)
3. Climatic and current data, and above all, summer surface currents when the pollution phenomena are more intense due to stratification of the water, winds, and weaker currents

Environmental Studies and Monitoring Activities

The forecast of initial and successive dispersion of the pollutants from a marine pipeline may be made by means of hydrographic data relative to a time series aimed at representing the seasonal variations.

Quite frequently it happens that the hydrographic data is not available or it refers to the open sea or an area which is different from that in question. This is true above all for the Ligurian territory where the structure of the coast makes it difficult to describe the circulation of the bodies of water.

From an examination of existing and valid measurements to satellite and aerial photos, and on-site readings tailored for the aims of this project, an adequate knowledge will be acquired during this phase regarding the two selected pilot areas so that operations will be efficient and allow an acceptable level of intervention within the limits of this study.

Identification and Processing of Satellite Images

During this activity, satellite images will be acquired and processed in order to obtain geo-referential maps for surface temperature and turbidity. This will show the time and space evolution of the previously cited parameters which will be useful in identifying local circulation patterns and dispersion of the pipeline discharges. The areas of interest for this study will be those identified in detail by the preliminary investigations.

After analyses and evaluation of the characteristics of the on-board sensors already in operation, it was decided to make use of 4 AVHRR images of the NOAA satellite which should represent a time evolution during the same month of the phenomena on a one month scale.

Moreover, 4 TM images will be used from the LANDSAT 5 satellite since these are more suited in describing small-scale phenomena and therefore capable of identifying the marine areas where certain discharges are present.

Comparisons between the thematic maps obtained by AVHRR and those obtained by TM will depict relations between the conditions at sea and those chosen coastal installations.

For this purpose within the due limits of the cycle times for the satellites and requirements for the on-site measurements, the processed images must refer to the same period. Processing of the images will be carried out by means of algorithms for geometric and radiometric correction of all spectral bands and for determination of temperature and turbidity parameters. During the course of the activity it will also be determined whether it is possible to process the maps for chlorophyll.

Validity and accuracy of the results depends on the algorithms which are used for processing. Therefore, two sets of measurements will be carried out at the pilot sites ("sea verity" lots) at the same time the satellite passes over.

The "sea verity" sets of measurements will make it possible to select algorithms most suited to the Ligurian coastal environment.

Collection and Processing of Images from Aerial Platforms

Since the project also has the specific aim of controlling marine pipelines of a limited capacity, it will be necessary for a more detailed image of the ground than that obtained by satellite. Therefore, an airborne system will be used, supported by a multi-spectral scanner Daedalus, providing for a 4x4 meter pixel.

The system acquires the rays emitted from the body of water over twelve channels, from invisible to thermal infrared, allowing efficient operation for identification of the discharges and areas of influence for each discharge.

By means of an air flight during the summer, images will be acquired for the coastal areas, including the pilot areas which extend for about 15 km along the coast and for about 6 km offshore. During the flight, measurements will be taken from the air as illustrated, in order to validate the algorithms used for processing the images.

Sets of Measurements for "Sea Verity"

In order to validate the results obtained by applying the processing to the distance-measured data and to calibrate the thematic maps, four sets of measurements will be carried out in the marine areas selected as pilot sites.

These sets of measurements will make it possible to acquire from moving naval craft, surface temperature measurements from radiometry and turbidity, and chlorophyll measurements from continuous sampling of water. Moreover, temperature salinity, oxygen, turbidity, and chlorophyll measurements will be carried out for the column of water over a sample grid by means of an appropriate, multiparameter probe correlated with current and meteorological measurements.

Continuous Monitoring of the Sea Level

The level of the sea is an important factor of marine circulation; its trend during time is the result of complex interactions between numerous factors, among which, for marine areas with a low tide ebb effect, meteorological events over a one-month scale and local scale may be the primary factors. The level data collected and validated will be used as input data for mathematical models to simulate the circulation of the marine areas in question.

Regarding the Ligurian Sea, the general trend of the level of the open sea may be described as the overlapping of harmonic functions having considerable amplitude and frequency. Furthermore, for coastal areas, since their level is affected by numerous local factors, i.e., bathymetry of sea floor, wave motion, and the shape of the coast, it is possible to observe oscillations with amplitude and frequency connected with the geometrical characteristics of a particular area. These determine a field of local velocity which influences, in certain conditions in a positive sense and in others in a negative sense, the dispersion of a marine discharge, and it is thus necessary to measure their amplitude and frequency. For this reason, and to integrate and complete an already existing network in Liguria, three marigraph stations will be made operative for the areas in question.

The measurements will be controlled by means of calibration parameters, operational tests, and laboratory calibration operations. The data will be processed in order to identify the amplitudes of the main periodic factors and the amplitudes and frequency of eventual local oscillations connected with overlapping of the fundamental factors. By means of analyzing the level data together with other meteorological-marine parameters, it will be possible to investigate the causes and origins of the oscillations.

Treatment, Analysis and Validation of Measured Physical Data and Utilization in the Database Structure

Interpretation and integration of the measured physical data will be used to do a database structure.

Selection of Area to Be Studied

On the basis of the data and information obtained from previous phases, the zones for investigation and the pilot sites will be identified. Here, sets of measurements will be carried out and at one of these sites, a continuous monitoring system will be positioned.

The criterion of choice must take into consideration the fact that for a study of the impact of a marine pipeline, two different realities must be appraised. First, the study is directed towards a marine area influenced by an elevated polluting load (a situation which is typically caused by the discharge from a large urban center); and second, towards a marine area influenced by a limited polluting load but, due to an elevated touristic-environmental interest, the quality of the water in particular areas, i.e., certain bathing areas, must be safeguarded to the maximum.

Sets of Measurements

The on-site measurement activity is aimed at collecting all information necessary to identify the zones where a pipeline has its most pronounced effects. The investigations shall deal with the main elements involved in the health of the marine environment, and in particular for the two pilot sites, the chemical and bacteriological quality of the water, sedimentology, and the observation of the biocenolosis of the sea floor.

Continuous Monitoring

A study of the dispersion of urban discharges which have been treated and emitted into the sea by means of suitable channels would be closely linked with the realization of a map showing the field of velocity of the site under examination by means of sea current data. To accomplish these things, continuous automatic measurements will be carried out at these sites for significant time periods. These measurements will be carried out at different depths, in the direction and speed of the currents.

At one of these sites, a mooring buoy fitted with a multiparameter probe, will be positioned. This will be remotely operated from the surface at a given frequency for the accumulation of data relative to temperature, conductivity, dissolved oxygen, turbidity, and pH along the column of water.

Information System

The data collected within the scope of this study will be organized in a manner suitable for computer-based management, compatible with the Ligurian Region environmental information system.

During this phase, a preliminary project, involving the establishment of the marine data bank will be made, taking into account the following:

1. Future requirements for data processing
2. The use of the data for calibration of the information systems used for decisionmaking (management module for the discharges)
3. Requirements deriving from the methods of presentation of information to the user (graphic synthesis, charts, etc.)
4. Possibility of future updates and territorial extensions of the monitoring

During this phase, the management software for the data banks will be developed with reference to the census carried out on the pipelines, as well as the environmental data derived from the bibliographic research and field investigations.

A further objective of this activity is the setting up of the "discharge management module". This module is made up from the preliminary version of a diffusion model for the outflow from a discharge pipeline.

Research Into the Control of the Level of Nutrition in the Gulf of La Spezia

An agreement with Ansaldo Industria S.p.A. has allowed a monitoring program to be carried out, both in manual and automatic, for the Gulf of La Spezia to acquire data for a report describing the condition of pollution in the water, with particular reference to the eutrophication of the bodies of water.

The one-year research period has been divided into eight sectors in order to accurately evaluate the variation during time of the nutritional load and phytoplankton activity at the fifteen stations where the main chemical-biological parameters are measured in order to define the nutritional level of the water. Moreover, at these same stations, sediment samples are taken for their stability characterization and to check for seasonal trends.

The main objectives of the intervention may be summed up as follows:

1. To interpret the ecological dynamics present at La Spezia roads
2. To point out and pinpoint the environmental dangers
3. To suggest rational proposals for management and improvement

Study Concerning Identification of Polluted Areas Along the Ligurian Coast through the Use of Bio-Indicators (Mollusk)

This project, developed within the limits of an agreement between the Region and The National Institute for Cancer Research (IST), is split up into four complementary components designed to evaluate the concentration of organic and inorganic pollutants in the water, particulated material, and mollusk fibers and to study the effects induced by the pollutants on the mollusk.

1. Biological effects of the pollutants on mollusk (Institute of General Physiology - National Institute for Cancer Research)
2. Evaluation of the accumulation of hydrocarbons on the mollusk tissues (National Institute for Cancer Research)
3. Evaluation of the accumulation of inorganic pollutants in the mollusk tissues (University of Analytical Chemistry)
4. Evaluation of the quantity of particulate present in seawater and its content in pollutants (Institute of Marine Environmental Sciences)

At the conclusion of the first year's research by IST, the Region will evaluate the logical follow-up to this research to be performed during the second and third year of the program. These activities will be designed to obtain a low cost tool, (via bioindicators) for evaluation of marine pollution along the coast.

Informative Register Concerning Discharges

The informative register of discharges (competence assigned by law 319/76 to the District Council) allows a detailed understanding of the condition of surface, internal, and marine waters. An update of such a register has been recently started by the District Council with the aid of state financing (Decree by Ministry of Environment 2.10.1990).

Due to the exiguity of state financing, the Region has given the following indications to the District Council regarding priorities to be considered:

1. Productive settlements
2. Public sewages
3. Civil settlements

Within the activities connected with the register, the Region has provided a card which the District surveyors will use for data acquisition and also for the establishment up of the management automation of the register itself.

The dedicated system which is being carried out, will make it possible to acquire on a magnetic support, in accordance with the data base organization, the relevant data on paper cards, thus making sure that results are not lost. The system will also allow a quick management and selection, representation, and print-out, in accordance with defined research methods.

Regional Participation in the "Integrated Management of the Gulf of Paradise Ecosystem"

Within the limits of responsibility at a regional level concerning environmental defense, the Ligurian Region has seen the need to arrange for a detailed study of the marine environment which will allow an integrated understanding of the phenomena at both a local and regional level.

In conformity with a deliberation in November 1992, the regional Council has authorized the financing required for the realization of a part of the research activities envisioned in an ample and articulated project aimed at integrated management of the Gulf of Paradise ecosystem. This project, of approximately two-years duration, has been allowed within the limits of an EEC program called NEDSPA-AO 91-1, at a community financing equal to 40% of its total value, estimated to be about

1635 million ECU. The costs not covered by the EEC financial aid and by the Ligurian Region - as an investment for research - shall be met by a temporary association of Companies which have arranged the research in question. This consortium is made up of Castalia-Società Italiana per L'Ambiente, as the leading group, CLOE, and Klaga and Agroplantec. The Gulf of Paradise Town councils interested in the study will contribute in a nonfinancial manner through dedicated work plans.

Due to the particular morphology of the regional territory and urban development which have engendered a heavy concentration of inhabitants along the coast and, in particular, in the main town, the Ligurian coastal marine environment is subject to a considerable anthropic pressure. This pressure, in the form of both civil and industrial discharges, has over time caused a reduction of the biological species, pelagic and benthic, and is closely linked with the quality of the water, a rarefaction of algae population and above all, a regression of the marine phanerogam grasslands made up from *oceanic Posidonia and Cymodocea nodosa*.

The area of the coast that stretches from the city of Genoa up to Punta Chiappa in the Gulf of Paradise, while having undergone environmental stress, still might be saved. It is therefore necessary to intervene with a specific program of integrated management in order to restrain the expansion of degradation factors as well as to conserve and, as far as possible, to recover its original character.

The marine waters subject to the study are those contained between the foreshore and isobath of about 50 meters for the section of coast close to Bogliasco, Sori, Pieve Ligure, Recco, and Camogli, as well as the area of those towns whose hydrographic basin directly or indirectly lies in the coastal waters mentioned above.

The project in question is divided into three distinct phases, each based upon the results of the previous one: (1) a formation phase for the required basic understanding; (2) an identification phase for the shore improvement interventions, with respective priority and technical-economic feasibility; and (3) a final phase represented by direct actions on the coastal ecosystem for restoration of the conditions required for protection and development of the *Posidonia* grasslands and *Cymodocea n*. These two grass species create the optimal condition characteristic of the Mediterranean for development and growth of numerous marine organisms.

The project's thrust is to provide an overall picture of the territorial environment and a basic knowledge of the existing environmental situation and its evolution over time, other than defining the condition of the sewage and purifying plants.

The following are also planned for this activity: (1) analyses of quality, nutritional condition, and self-purifying power of the coastal waters by means of the marine ecosystem; and (2) research into dynamics and diffusive property of the water, the condition of conservation of the benthic community, quality of water, presence of pollutants in the sediments, and the extension and condition of conservation of the phanerogam grasslands.

The data generated will be used in suitable simulation models for the processes of diffusion and dispersion of effluents emitted into the sea. The results will be used in the planning process to define correct engineering solutions for the verification of the release and treatment hypothesis of liquids which should decrease the risk of euthrophication, and general environmental degradation. This method is also the most economical in terms of management, maintenance and completion costs.

Evaluation of existing and planned purification plants designed to depollute discharges and evaluation of self-purifying characteristics of marine water will allow, during the successive phase proposed by the plan, for the formulation of a program of interventions concerning engineering plant proposals to safeguard the water and minimize the risks of marine environment degradation.

Finally, the program in question foresees biotechnological interventions aimed at restoring the marine environment mainly through the placement of suitable structure capable of allowing the settlement of marine grasslands made up of *Posidonia a* and the *Cymodocea n*. This phase will be carried out in the following manner: (1) multiplication (on land) of the material to be propagated; (2) selection of the sites for thickening-out operations and the new plants; (3) on-site laying of the pre-planted laths; and (4) verification of the results.

The material to be transplanted, contrary to that which normally takes place, will not be collected in natural grasslands, but propagated in an agamic way in suitable laboratories. The advantages of this procedure are : (1) no existing grasslands will be altered following collection of the plants; (2) unlimited availability of plants; (3) low production cost; and (4) the possibility of intervention through selection programs and genetic improvement. The laths arranged on land shall be laid in a different way for the thickening-out interventions or a new plant.

Moreover, in order to verify the success of the interventions during time and their effects on the surrounding environment, a series of measurements is planned, up to 600 days from the plant, for the main vegetation parameters such as survival of the plants, speed of growth, and that concerning covering of the sea floor.

Development of a Means of Analysis of the Ligurian Coast Based on Bio-Indicators

The study which is projected by a contract between the Region and the Institute of Marine Environmental Sciences at the University of Genoa, intends to develop a means of analyses aimed at characterizing – in an environmental sense – the Ligurian coastal regions using an "experimental module".

This module concerns the benthic ecosystem, starting from the shoreline to a depth of 200 m, the limit of the continental shelf. The benthic ecosystem was chosen because of its capability of "recording' the environmental events which have been verified, not only on the sea floors, but also in the column of water above them.

The experimental module will be structured into four distinct phases.

1. "sampling structure": in which both biological and abiological samples will be taken in the underwater and subair zones, with a greater number taken within 50 m from the isobath, since this is considered to be an area of major environmental variability and therefore more prone to impact from anthropic activities
2. "analyses of samples": in order to acquire the quality-quantity picture of the macrobenthic populations (gathered by cage, basket, and bucket), of benthonic populations (gathered by drag nets), melofauna and microbic communities, as well as the characteristics of sediments as they pertain to situations connected with allochthonous transfer
3. "data processing": relative to benthic and benthonic populations, to draw up biocenelogical maps indicating the presence of rare species to be safeguarded, and to calculate, for the various systematics components and for the community overall, the biomass values, in order to define the main components of melobenthos, microphytobenthonic, and microbic communities
4. "environmental evaluation": through the summation of the results and their processing, the natural resource will be verified, together with the fishing resources and the environmental condition of the coastal area.

Monitoring of the Water to Understand the Marine Ecosystems

In order to understand the marine ecosystems, coastal water monitoring activities will be entrusted to Arcatom S.r.L. by agreement with the Ligurian Region. This arrangement will be done in accordance with the methods established by Sea defense Ministry. The area to be investigated has been identified as the entire Ligurian coastline with monitoring stations situated at intervals of 10 km.

The stations will be located in those areas which are considered environmentally sensitive, due to their proximity to anthropic, industrial, and port settlements, or due to their use for such activities as bathing, natural reserves, and mollusk-culture.

For each station there are three, planned sampling posts at 150, 500, and 1.500 m from the coast. At each post, samples and measurements will be taken at the surface and close to the sea floor.

The seasonal characterization of the parameters will be guaranteed by quarterly sampling.

The following parameters will be measured:

1. Temperature
2. Salinity
3. Dissolved oxygen
4. pH
5. Transparency
6. Presence of tar residuals
7. Presence of film of oil
8. Coloration
9. Chlorophyll
10. Ammonia nitrogen, nitrous nitrogen, and nitric nitrogen
11. Orthophosphates, total phosphides
12. Phenols
13. Hydrocarbons
14. Bacteriological indexes (fecal and total coliform bacteria, fecal streptcocci, and salmonella).

At the same time, oceanographic, and meteorological observations will be reported. Moreover, upon completion of the above activity, it is envisaged to integrate this with the more extensive program planned by the agreement with the Merchant Navy Ministry described in following subsection.

Agreement with the Ministry of Merchant Navy for Marine Ecosystem and Eutrophication Monitoring

This agreement, whose administrative steps are now being completed, was signed by the Ministry of Merchant Navy and the Regional Council. It provides for a three-year survey to monitor the marine ecosystems involved and the area's eutrophication.

This agreement stems from Art. 3 of Law Act n. 979 of 31.12.1982, which sets forth that "oceanographic, chemical, biological, microbiological and commodity-related data and any other information required to fight against all forms of pollution, to manage coastal areas and to protect marine resources also from an ecological point of view" should be collected. In addition, the implementation of such a monitoring system falls within Italy's duties within UNEP (Med-Pol Phase II), in which the member countries undertake to set up a coastal water monitoring network and to forward the results to UNEP.

As provided for this agreement, the survey will:

1. Monitor waters to acknowledge the conditions of marine ecosystems by sampling water at 100, 500, and 1.500m from the shoreline. Sampling will be performed from stations located along transects perpendicular to the coastline and erected at a distance of approximately 10 km from one another along the Ligurian coastlines.
2. Monitor waters to control eutrophication through samplings at 500 and 3,000 m from the shore. Sampling will be performed from transects located in the proximity of polluting sources such as harbors, canals, rivers, or coastal settlements identified in the territory.
3. Monitor bivalves by sampling mollusks as indicators of coastal water pollution levels, with particular reference to the following parameters: total and fecal coliform count, fecal streptococci, salmonella, mercury, cadmium, and high-molecular-weight chlorinated hydrocarbons.

Participation of the Regional Council to a Probe-Equipped Buoy Trial

For 1993, the Regional Council has planned to participate in the trial, automatic, marine water testing system, consisting of a probe-equipped Meda and an already existing monitoring and data collecting system owned by and located within Alenia Elsag Sistemi Navali premises in Genoa.

The sampling and measuring station shall be located in the sea area in front of Cogoleto (1 mile westward of Capo Arenzano, at a distance of approximately 100 m and in approximately 50 m depths).

The station's position was selected based on the following principles:

1. Tourist area with the presence of high environmental risk factories in the surrounding areas.
2. Borderline between an area characterized by a high concentration of anthropical/industrial activities and the western tourism-oriented Riviera.
3. An area which, considering the constant flow due to the Ligurian/Provençal stream, is a transitional check point downstream from the main polluting sources.
4. Proximity to Haven's wreck (undercurrent), as a monitoring station for any residual hydrocarbon leaks.

This system shall consist of a pilot center for integrated, water pollution monitoring made up of:

1. Fixed sampling station including:
   a) A platform equipped with electro-hydraulic services
   b) A probe package
   c) A remote control, special purpose computer
   d) Communication interface devices
2. Communications system
3. Data collecting, monitoring, and control system including:
   a) A display large screen
   b) A workstation
   c) A general-purpose computer
   d) Computing peripherals
   e) Communications interface devices

At least initially, the Meda station will be equipped with the following probes:

1. Hydrocarbon content probe, with particular reference to (soluble) aromatic compounds, sampling in the surface stratum or at 5 m depth
2. Anemometer to determine wind intensity and direction at 10 m above sea level
3. Current meter to determine current intensity and direction at approximately 5 m depth
4. Probe to determine water temperature in the surface stratum

Other Initiatives

Talks are presently underway with International Center for Coastal and Ocean Policy Studies (ICCOPS) to develop through an ad hoc agreement, a qualified involvement of the Ligurian Regional Council on a national and international level, to strengthen the "Sea Technological Pole" in Genoa in order to set up a European Mediterranean policy reference center. As a matter of fact, in the framework of the European FESR Target Two program, the Production Activity Regional Office has launched the setting up of the "Sea Technological Pole" with the participation of various business professionals. In planning the second phase, the construction of a laboratory was designed; this laboratory will be the basis of the regional project for an ecological boat.

The Regional Town Planning Office has approved the coastal management coordination master plan. The "Coast Plan" refers to the "Environment Project" designed to carry out systematic environmental impact assessments relating to the interventions planned. It will also provide important and valuable references for E.I.A. layout.

The active involvement of the Ligurian Regional Council in setting up the "Cetacean Sanctuary" in the Ligurian-Corsica-Provençal sea by Sea and Environment Ministries of Italy, France, and Monaco will be another important drive for heightening the struggle against coastal pollution. At the same time, it will also provide the opportunity to strengthen the role of the Ligurian coastal centers as far as sea access and the localization of any services linked to sea protection and exploitation operations are concerned.

Owing to its historical traditions, to its cultural background, and to the presence of top-grade technological and scientific skills in this town, the wish that Genoa may be the venue of the Mediterranean Marine Environment Agency seems to us to be worth a concrete, political-institutional proposal.

COASTAL REQUALIFICATION INTEGRATED PROJECTS: A TEST-BENCH INTERVENTION INSTRUMENT

Among the firm steps to be taken and submitted to testing on a technical-procedural and financial level, the regional law act n. 28/92 provides a first operating support to be tested and perfected based on previous experience. As a matter of fact, this law act envisages the funding of "integrated projects of water resource requalification". The application field of choice for that instrument can be the closeness of land-based and marine parks with the adjacent surrounding and proximal areas.

For instance, it is clear that though environmental conditions are critical in some areas, (the depths around Portofino's Promontory, those in front of Cinqueterre, or those surrounding the Gallinara Island), by observing the meteo-marine phenomena, one can see that no sea segments can be isolated from the whole, due to the pollution scattering effect produced by streams, winds, and rain along the whole coastline. As is well known, the difference in temperature, hence in density, of superficial waters and the difference between these and deep waters create horizontal, vertical, and combined movements of water bodies which highly affect the preservation of the natural characteristics of the sea, of depths, and of underwater rock walls, the latter being the dwelling of the marine biological world. The preservation of marine environment's vital conditions is strictly connected to a balance within which the various components of its natural world can survive.

The release of organic and inorganic substances into this environment and/or any changes in physical conditions (e.g. temperatures and noise) upsets the natural balance. Thus, these uncontrolled inputs, which fortunately are not generalized, have detrimental effects upon the marine environment by altering its chemical-physical characteristics (including the organoleptic characteristics). It also causes precious fish to move away or to disappear, causes water to become cloudy, and silt, in general, to deposit in the depths.

Besides the sewage waters discharged directly or through pipelines at the distances and depths provided for by the law, it is clear that any substances that cannot be retained by solid, fluid, or gas waste treatment plants end up in the sea. In case of violent rains, the larger-sized garbage left unlawfully near watercourses or in the valleys also ends up in the sea.

Therefore, knowing the quantitative and qualitative aspects of the substances that end up in the sea, the dynamics of moving water bodies, and the changes that gradually affect the sea's biological characteristics become a must for laying down a protection policy against seawater pollution. This should be coupled with a natural environmental protection policy targeted at some interesting areas identified as future "underwater parks".

As a matter of fact, the coastline is a fragile border along and through which very different ecosystems coexist and interface, becoming a single-relation system which has characterized the Ligurian landscape through its history, culture, and development.

The borderline between fresh- and salt water is almost impalpable to nonprofessionals; however, there are clear-cut differences between these two worlds. Any shift in the separation line affects land-based life and the use of resources. However, as already stressed, all land-based activities add a pollutant load which highly affects the overall balance to the fresh water flowing in the area. Knowing the exchanges and understanding the mutual effects that take place, become important planning and control instruments for all implementation and management activities.

To give another example, the setting up of an underwater park in Portofino would require that the following activities were coordinated in the Golfo Paradiso area which links Genoa's harbor and city area with Portofino:

1. Protection of sewage water
2. Containment of stream pollutant supplies
3. Renaturalizing of water courses and of their catchment areas
4. Reclaiming of illicit dumps
5. Planning of small (diffuse and accessible) aggregate dumps
6. Diversifying access roads leading from the city to the Riviera
7. Alternative coastal, mountain, and seaside ways

Similar examples could be given for Cinqueterre and for the area surrounding the Gallinara Island.

The importance of struggling against coastal pollution through integrated, environmental resource planning instruments is stressed by a stringent requirement which is at the basis of the previous-mentioned "Cetacean Sanctuary". Once again, the large-scale, environmental importance of the area considered stands out - the area where Liguria is and lives both functionally and culturally. If most of the actions and interventions outlined so far were developed with the appropriate national and EC coordinating bodies, they would significantly contribute to safeguarding and protecting the environment.

THE NEED FOR AN INFORMATION SYSTEM: THE REGIONAL SEA DATA BANK

The knowledge and decisionmaking models concerning the marine and coastal environment require a specific vehicle within the regional, environmental information system that must be designed. It will be treated as a regional branch of SINA, the national environmental information system.

Please consider that the Region is operating on two levels to design its own environmental information system.

1. On a local/regional level, jointly with SIR, the Regional Computing Service, and Datasiel, a region-funded company, a sea data bank is being developed that should contain all information collected on waste disposal and on the water body in order to form new registers and carry out new censuses. It will also include data concerning present and future monitoring activities, as well as data on bathing and sea depth characteristics. These data will be geo-referenced. It will be also possible to retrieve and to use such data within knowledge- and decision-making models. The design and implementation of this information system on a local level makes it possible for it to be shared by the Provinces performing activities in this sector.
2. Let us consider now the regional branch of SINA, the national environmental information system, to be designed by Datasiel following the guidelines and under the supervision of the Ministry of the Environment as far as national standards are concerned. In this case, the Region and Provinces will be networked and the latter will increasingly take over the Regions' and Municipalities' role in controlling and governing environmental phenomena and their monitoring on a local level, with particular reference to water resources. Sea data are included in this system because they concern the water body receiving household and industrial waste and, since this will be an open-type system, it will also be possible to access - in principle at least - the whole sea data bank.

The regional meteorological service - presently under study, except for a few special operations that are already underway - may have a positive impact on future developments.

The sea information system will become significant when individual and aggregate data are made available (and also printed out graphically) according to several models that might be accessible to different users; e.g. in the framework of environmental information and education programs (INFEA), or on the tourist telematic network, or to the public in general through Televideo service or similar initiatives.

The INFEA program also includes a first, regional support step targeted at the initiative undertaken by Portovenere Municipality to set up a permanent, environmental education center whose headquarters will be located in refurbished buildings that are presently abandoned on the Palmaria Island. This center could also become a precise reference point to access the Cinqueterre Park and, in general, to make the Ligurian marine environment more widely known.

CONCLUSION

To conclude this brief outline on current and planned activities in the regional environmental field that are targeted at coastal sea resource protection, it is possible to maintain that the present goals considerably support the naturalistic protection of the major scenic and culturally rich areas bordering the sea and of the sea itself. However, one should not forget that these goals are being pursued despite the considerable operating difficulties of Italy's Public Administration and despite the unavoidable defects of the approach, which are still too much sector-oriented and poorly tested.

The development of these models is particularly meaningful, both today and in the future, for Liguria, for by reversing a century-old trend, the Region's health is beginning to be based more on the primacy of life and environmental quality rather than the quantitative growth of territorial modifications.

2

The Baltic Sea

Gotthilf Hempel
Head, Institute for Baltic Sea Research, Warnemünde, Germany

INTRODUCTION

The Baltic Sea is an area of multiple-use conflicts of more than 20 million people living permanently at its coasts and further millions spending their money there for swimming and sun-bathing. There is an obvious conflict in using the Baltic Sea as a source of fish and recreation and as a sink for nutrients and pollutants.

All this is embedded in a multi-national political arena which has witnessed many changes. For about 800 years, the Baltic has been a trade way between western and eastern Europe, but for the last forty years, it has been divided by the "iron curtain", its eastern and southern coasts belonging to the socialist bloc, the northern and western coasts to a more capitalistic camp. Since 1990 three Baltic states were reestablished while the German Democratic Republic (GDR) disappeared. Now nine independent states are bordering the Baltic, more than at any other sea. Each of the states has its specific interests in the Baltic.

International cooperation in Baltic marine research has a long tradition, and rather powerful international conventions are in force both for pollution control and fisheries management. But with the rapid and still ongoing fragmentation of the "Eastern Bloc", the political and economic framework for the international cooperation in environmental research and monitoring is changing. Along with the number of countries the range of national policies has increased and the economic situation has deteriorated in parts of the region.

NATURAL HISTORY OF THE BALTIC SEA

The nature of the Baltic Sea as the largest brackish water area in the world, puts it in a unique position amongst the large marine ecosystems. It is Europe's only intra-continental Mediterranean Sea. With 0.4 Mio km^2, it has a third of the surface of Hudson Bay and is about the size of the Persian Gulf or the Red Sea. In terms of surface, the Baltic amounts to 1 per million of the Worlds Ocean, and to one per million in terms of volume.

The Baltic Sea is a series of basins separated by shallow sills. The deepest parts are up to 450 m deep, but the average depth is 55 m. The shallowest sill in the entrance to the Baltic proper (Darßer Schwelle) is only 18 m.

The entrances of the Baltic are very narrow. The Baltic is, oceanographically speaking, a large fjord in a humid climate with a very strong haline stratification and no tides.

The Baltic Sea is a product of the last glaciation of northern Europe. After the retreat of the ice cap, a huge lake was formed about 14,000 years ago. Then it became a sea canal (Yoldia Sea) for a short time. But soon it turned again into a freshwater lake (Ancylus Lake). Only since about 5100 B.C. has the Baltic become linked with the North Sea through the Danish Straits. All those changes in the recent history of the Baltic are related to the retreat of the ice cap, the continuous uplifting of Scandinavia, and the rising of the sea level due to the melting of large amounts of continental ice.

HYDROGRAPHY

The hydrographic budget of the Baltic Sea is determined by the river inputs (430 km^3/a) and the surplus of precipitation over evaporation (45 km^3/a). The inflow of water from the North Sea is of the same order as the freshwater input, but highly fluctuating from year to year. Therefore, the total outflow amounts to about 1.000 km^3/a, i.e., less than 5 % of the volume of 20.900 km^3. The exchange is mostly confined to the upper layer. From the Belt Sea in the west to the Finnish Bight and Bothnian Bay in the northeast the surface salinity decreases from ca $15^0/00$ to less than $3^0/00$. Year round there is a steep, permanent halocline at 40-70 m depth separating the low salinity surface water of about $8^0/00$ S in the central Baltic from the deep water of 10 °/oo or more in the basins. The upper, less saline water is subject to seasonal changes with a thermocline at 10-20 m depth in summer and with a widespread ice cover in winter and spring in the north eastern parts.

The deep water is relatively independent of seasons, and is renewed occasionally after storm events in the North Sea. Only the strongest inflows finally reach the farthest and deepest basins in the Gotland and Åland Seas. For 16 years, from 1976 to 1992, no major inflow took place, while in earlier decades intrusions of up to 200 km^3 per event occured at least every few years (Matthäus, 1993; Nehring et. al. 1993). As a consequence, salinity and temperature of the bottom water decreased.

In early February 1993, a heavy gale in the Southern North Sea resulted in a large inflow of highly saline water into the Kattegat (Matthäus and Lass 1993). It reached the Darß sill at the entrance of the Baltic proper in mid-February and cascaded down into the deep basins, replacing part of the stagnant, less dense bottom water of the central Baltic.

EUTROPHICATION

The deep water stagnation resp. circulation determines the extent of oxygen deficient zones and layers. Their primary cause lies in the recent eutrophication of the Baltic. In the 1970s the concentration of phosphate and nitrate in the surface waters increased sharply and stabilized in the 1980s at a level about three times as that experienced in 1969.

The Baltic drainage area has four times the surface area of the Baltic itself. It is populated by more than 70 million people. Twelve million Russians, Finns, and Estonians discharge into the Gulf of Finland while 2.5 million Swedes and Finns do likewise into the Bothnian Bight. The Riga Bight is heavily polluted by 4 million Latvians, Estonians, and Bjelorussians. The Central Baltic receives the sewage of 44 million people, mainly from Poland. The narrow entrances of the Belt Sea and the Kattegat are affected by 10 millions Danes, Swedes, and Germans.

The annual input of nutrients is by far the heaviest in the shallow Belt Sea, where about 0,5 t phosphorus and 4 t nitrogen per km^2 (about 7 times more per km^3) are poured into the sea. Nutrient input is also high in the Finnish and Riga Bights where exchange is very poor.

In 1980, all together about 50,000 t P and 500,000 t N reached the Baltic via the rivers. Further 400,000 t N are air borne (HELCOM 1987, 1991). A major part of the nutrients stem from municipal sources and industry. Most of the cities in eastern Europe and Poland have no proper sewage treatment and particularly no third purification step. Paper mills and food industry produce large amounts of organic waste, much of it reaches the sea untreated. Further nutrient inputs, mainly nitrogen, consist of artificial and natural fertilizer. Overfertilization was one of the main characteristics in socialist agriculture but was also common with western farmers.

Only about 10% of the anthropogenic nutrients leave the Baltic through the Danish Straits (Wulff and Stigebrandt 1989). The rest is accumulated in the Baltic ecosystem which is a far more closed system than the arid Mediterranean with its outflow of nutrient rich deep water through the Strait of Gibraltar. Denitrification at low oxygen concentrations is also important in the Baltic Sea.

Originally the Baltic was rather oligotrophic compared with the North Sea. Now the phosphate concentrations in winter are about the same in the Central parts of the North Sea and of the Baltic Sea. Primary production has increased substantially over the past decades, the annual period of algal blooms has been prolonged, and the composition of phytoplankton has changed; cyanophytes becoming more abundant.

Part of the surplus primary production found the "right" path ways in the food web, resulting in increased fish production. Fish catches doubled from 1966 to 1980 (Nehring et al. 1984), but are now stagnant or even decreasing due to deterioration of the oxygen conditions and to overfishing.

Most of the organic production, however, is not harvested by fisheries but finally sedimented - often after one or more rounds of recycling in the food web. Decomposition in the almost stagnant, aphotic deep water and at the sea bed results in oxygen depletion to an extent that major, deep parts are affected by H_2S year round; while in some shallow areas, oxygen deficiencies occur occasionally after heavy summer blooms of phytoplankton, stagnant conditions in the deep water, and after wind-driven upwelling events carrying oxygen deficient water into shallow bays.

Increased abundance of phytoplankton means decrease in photic depth and hence reduction in phytobenthos. Therefore the areas covered by macroalgae along the coasts are shrinking.

The decrease of salinity and the oxygen depletion in the deep water had a serious effect on the cod stocks (BERNER et al 1989). Their eggs are buoyant at salinities of at least 10^0/00S. In the central Baltic therefore, cod eggs cannot float near the surface, but only in the deep water or may even sink to the bottom where they might die of H_2S. Sprat eggs float at 6^0/00 S and are therefore less affected. Herring spawn demersally at very shallow depth.

There is another threat to fish due to mass occurence of medusae, mainly of *Aurelia aurita* (Nehring 1992), which consume much of the fish larvae and their copepod food. Possibly the polyp stages of *Aurelia* benefit from the increased productivity of the near shore waters.

TOXIC SUBSTANCES

The Baltic Sea has been and is still a sink for pesticides, chlorinated hydrocarbons, and heavy metals. National and international regulations have drastically reduced those forms of pollution, e.g., the use of DDT and PCB's is forbidden in all riparian countries. The industrial discharges, particularly from the former USSR and from Scandinavian pulp and papermills, are heavily loaded with toxic substances.

The total amount of money spent on pollution control is enormous and will increase sharply if the former Eastern Bloc countries will develop sewage systems similar to the ones existing in western countries. Presently it would be most beneficial to the health of the Baltic if most of the money allocated for new pollution treatment plants would be channelled into the "eastern" countries rather than being used for further improvements in the West. Recent decreases in the pollution load of the eastern rivers are not due to better pollution control but to the closing down of some uneconomic industries.

INTERNATIONAL POLLUTION CONTROL

International combating of pollution started relatively early in the Baltic Sea, in spite of severe restrictions in east-west information transfer. The Helsinki Convention for the protection of the marine envrionment of the Baltic Sea was signed in 1974 and ratified by 1980. The Helsinki Commission (HELCOM) is a powerful organization which advises member states on all questions related to international monitoring and management of pollution in the Baltic. Regular baseline studies and periodic reports on the health of the Baltic are produced. A major part of the marine research of the coastal states are directly or indirectly related to HELCOM and its Baltic Monitoring Programme (BMP). The Programme Implementation Task Force of HELCOM has identified more than 100 'hot spots' of pollution in all regions of the Baltic Sea but mainly in Russia, the new Baltic States, and Poland (Svenson 1993).

The Task Force recommended ways and means for eliminating those hot spots.

LIVING AND NON-LIVING RESOURCES

Nine countries are eagerly fishing in the Baltic Sea which is almost completely split into national EEZs. According to the official International Council for Exploration of the Sea (ICES) statistics, overall fish production per unit area is less than half of the North Sea values. Only a few species of fish were able to adapt to the fluctuating, brackish water conditions. The total catch of about 1 Mio t consists mainly of herring, sprat, and cod. Each of those species reacts in a different way to eutrophication and to shifts in the hydrographic conditions and to fishing pressure from the different national fleets. So far, no well-balanced multispecies management of the Baltic fish population as a whole has been achieved under the International Warsaw Convention which was founded in 1974. This is more a political than a scientific problem.

The nonliving resources of the Baltic Sea are very limited. In some areas gravel and small oil deposits are exploited; formerly extraction amber was productive in certain places.

Sea traffic is an important economic asset of the Baltic Sea. Oil supplied by pipelines from Russia might replace part of the risky sea transport of 100 Mio tons of oil per year. But the Baltic will remain one of the busiest seaways in the world since more than 10% of the world's industrial production takes place in Baltic coastal regions. There are strict international regulations by the International Maritime Organization (IMO) and HELCOM regarding safety and pollution prevention in sea transport, ferry traffic, and yachting.

TOURISM

Far more important than fish and amber are the millions of tourists, mostly concentrating in Germany and Poland at the Baltic south shore and at places outside the cities of the other Baltic countries. Sun-bathers, surfers, and hobby-sailers are a major economic factor and become a powerful pressure group. There is also an increasing conflict between the tourist industry and nature conservation about coastal areas put off-limits to campers or to building marines and summer houses. In the former GDR, the conversion of a primitive mass-tourism, with very little infrastructure, into a luxury, space demanding, highly motorized tourism, accentuates those conflicts with its complex socioeconomic and environmental implications. Hopefully some lessons now learned in eastern Germany can be used in Poland and in the Baltic republics in future.

MARINE RESEARCH

The Baltic Sea has been the subject of some of the early studies in marine science, initiated for both basic scientific curiosity and for the understanding of the fluctuations in fish stocks.

International cooperation in the Baltic Sea research is almost 100 years old. It commenced with the founding of the International Council for the Exploration of the Sea (ICES) in 1902. ICES initiated regular, seasonal oceanographic surveys in the Baltic and in the North Sea, including assessments of fish stocks. ICES has always combined studies of basic science with surveillance of fish stocks. Rather early, it iniated studies of pollution and eutrophication, paving the way for the Warsaw and Helsinki conventions. Both of them still benefit from the scientific advice by ICES.

With several hundred scientists and more than a dozen research vessels, the research potential in the Baltic Sea is presumably the highest in the world-relative to its area. In most countries different institutes deal with fisheries research, with basic marine science, and pollution studies respectively. Until 1990 Baltic research of the Eastern Bloc was fairly well coordinated. Within the USSR, the central authorities in Moscow (Academy, Fisheries, Hydrography) allocated specific tasks to each of the Baltic republics and to the institutes in Kaliningrad and Leningrad. Now each place is developing its own program and each republic wishes to cover most of the fields of Baltic Sea research.

In August 1992 the research vessel of the Institute for Baltic Sea Research in Warnemünde toured all major places of marine research along the coasts of the former Eastern Bloc and found everywhere strong communities of marine scientists. Some, but by no means, all of the institutes were already fairly well equipped with modern instrumentation, but all were in very difficult economic conditions. The libraries in the Baltic states are still lacking most of the modern western literature and the laboratories are very short of glassware and chemicals. Only the Polish and eastern German institutes fare better. Support is mainly coming from Scandinavia and Germany. Some of the research vessels are in good shape and well equipped, but money is short for covering the running costs. In most formerly socialist countries, research institutions are subject to reorganization and reduction in manpower, and all of them look for new international ties and cooperation.

Joint projects are initiated, e.g., in the Gulf of Riga and the Oder-Bight. The European Science Foundation (ESF) and the Commission of the European Communities (CEC) are jointly sponsoring the planning for a major experiment in the Central Baltic in 1994 and the following years.

The experiment is related to the Joint Global Ocean Flux Study (JGOFS) by following the transport from the coastal waters into the open Baltic and there, from the surface to the sea floor. The fixation and release of carbon, nutrients, and pollutants in the sediments will be studied by long-term observations.

Furthermore, in preparation of Global Energy and Water Cycle Experiment (GEWEX), the Baltic Experiment (BALTEX) shall produce reliable figures on the energy and water budget of the Baltic.

All Baltic states have established a joint monitoring and information network for the early identification of saltwater intrusions. The network shall provide a data set sufficient to model and predict the changes in temperature, salinity, oxygen, and nutrients in the different compartments of the Baltic.

OUTLOOK

After a period of political separation and restriction in the Baltic region, there is a strong movement towards close cooperation in marine research between all countries bordering the Baltic. A council of foreign ministers was established in March 1992 which, inter alia, focussed on environmental protection in the region. Changes for a meaningful and productive management of the Baltic marine ecosystem as a whole are rather good - in spite of the multi-use conflicts - provided sufficient international support will be given to those countries which are in need of building up their democratic systems, economy, and science.

LITERATURE

Berner, M., H. Müller, and D. Nehring. 1989. The influence of environmental and stock parameters on the recruitment of cod stocks to the east and west of Bornholm described by regression equations. *Rapp. P.-V. Réun. Cons. int. Explor. Mer* 190:142-146.

HELCOM. 1987. First Baltic Sea pollution load compilation. *Baltic Sea Environment Proc.* 20:1-53.

HELCOM. 1991. Airborne pollution load to the Baltic Sea 1986-1990. *Baltic Sea Environment Proc.* 39:1-1959.

Matthäus, W. 1993. Salzwassereinbrüche und ihre Bedeutung für die Meeresumwelt der Ostsee. *Wasser und Boden* 12:922-928.

Matthäus, W., and H.-U. Lass. 1993. The major Baltic inflow in January 1993. *ICES* C.M./C:5: 1-9.

Nehring, D. 1992. Hydrographisch-chemische Langzeitveränderungen und Eutrophierung in der Ostsee. *Wasser und Boden* 10:632-638.

Nehring, D., S. Schulz, and W. Kaiser. 1984. Long-term phosphate and nitrate trends in the Baltic Proper and some biological consequences: A contribution to the discussion concerning the eutrophication of these waters. *Rapp. P.-V. Réun. Con. int. Explor. Mer* 183:196-203.

Nehring, D., W. Matthäus, and H.-U. Lass. 1993. Die hydrographisch-chemischen Bedingungen in der westlichen und zentralen Ostsee im Jahre 1992. *Dtsch. Hydrogr. Z.* 45:281-304.

Svensson, G. 1993. Ecological balance to be restored by 2010. *Water Front* (Sweden) 2:8-9.

Wulff, F., and A. Stigebrandt. 1989. A time dependent-budget model for nutrients. *Global Biochemical Cycles* 3:63-78.

3

Cooperative Ocean Science For Advancing World Peace: An Eastern Mediterranean Example

Robert B. Abel
Director, International Programs, Davidson Laboratory
New Jersey Marine Sciences Consortium, Hoboken, New Jersey

ABSTRACT

Egypt, Israel, and the United States are conducting a cooperative program of marine technologies under the auspices of the U.S. Agency for International Development's (USAID) Middle East Regional Cooperation (MERC) Program. The Program, which began in 1980, has encompassed a dozen projects in over twenty laboratories in the three countries. They include fisheries and mariculture, shore processes and shoreline protection, lakes management, climate prediction, seafood toxins, wastewater recycling, and primary productivity of the Eastern Mediterranean Sea. The Program is designed and coordinated by a steering committee which includes representatives from the Egyptian Academy of Scientific Research and Technology, the Israeli Institute for Oceanographic and Limnological Research, Texas A&M University, and the New Jersey Marine Sciences Consortium.

BACKGROUND

Throughout history, man has viewed the ocean - when he's though about it at all - as an infinitely broad highway on which to transport people and things; as a source of food; and depending on the viewer's perspective, as a protective shield or convenient battle zone.

For the past three decades, the world's population has been increasingly sensitized to our surrounding seas through two, additional, major issues

1. Whether recovery of the ocean's mineral resources is economically feasible
2. Whether the ocean's capacity as a garbage repository is really unlimited

Very recently, a new concept has been introduced to peoples in certain parts of the world, i.e., use of the ocean as a persuasive instrumentality for peaceful cooperation. Typically, it was President Harry S. Truman who observed that nations working together were less likely to be attacking one another. At the time, he was referring to the possibility of persuading Israel and her Arab neighbors to cooperate on some major engineering projects of mutual gain.

It took three decades to translate those thoughts into deeds. Then, in September, 1978, the U.S. Congress passed, and the President signed, the International Security Assistance Act of 1978, PL94-224, amending the Foreign Assistance Act of 1961. This legislation included for the first time (Section 48 [C] [5]), establishment of a program and fund for Regional Cooperation in the Middle East. It became known as the "Regional Fund", later, "Middle East Regional Cooperation" (MERC) and was directed to cooperative projects between Israel and her neighbors. Responsibility for the Program's implementation was assigned to the United State Agency for International Development (USAID).

During the period October 1978-August 1980, a small group of American, Egyptian, and Israeli oceanographers, working together very informally, developed a series of projects in science and technology which became known as "The Cooperative Marine Technology Program for the Middle East." The Program was officially accepted by USAID on August 23, 1980, as the first endorsed enterprise under the Regional Fund's auspices. Actually, on that date, a group composed of a dozen scientists each from Israel, Egypt, and the United States was quietly conducting an historic meeting in San Diego.

PROGRAM DESCRIPTION

The group deliberately based their Program on fundamental needs: food, water, and land protection; specifically including, ocean productivity, seafood toxins, mariculture (i.e., seawater aquaculture), wastewater usage, shoreline protections, climate prediction, and lakes management. Teams of scientists and engineers from two dozen institutions in the three countries have arrived at an interrelationship where all projects are conducted cooperatively, either trilaterally with the United States or bilaterally between the Middle Eastern countries. Coordination points are the Egyptian Academy of Scientific Research, the Israeli Institute for Oceanographic and Limnological Research, Ltd., the New Jersey Marine Sciences Consortium, and Texas A&M University.

Tilapia

Tilapia is one of the Third World's two or three most important food fish. It spawns easily, grows fairly quickly, and lends itself to cross-breeding for purposes of regional adaption.

The Israelis and Egyptians began their work on Tilapia at the Program's commencement a dozen years ago. The Egyptians now appear able to grow a Tilapia to market size (125 grams) in four to six months. The implications for the nation's food supply are obvious.

Induced Spawning of Grey Mullet

Mullet is also one of the most important, if not *the* most important food fish in the third world because of its ability to survive on a low protein diet, feeding on phytoplankton. Unfortunately, mullet is difficult to farm, and success has eluded aquaculturists, mainly relating to spawning. The Israelis and Egyptians are conducting intensive sets of experiments to induce mullet to spawn. The Israelis approach the problem through examination of the role of hormones, especially gonadotropin. The Egyptians are studying all characteristics of the eggs as indicators of breeding success.

The groups have now published half a dozen papers on mullet reproduction, illustrating a number of factors which appear to influence the mullets' reproductive processes, the most surprising of which is longitude! The scientists have already made clearly defined gains in influencing the reproductive cycle, and in a closely related project, Brackish Water Fish Studies, have accelerated mullets' growth rates appreciably. For instance, they have discovered that mixtures of egg yolk and rice bran as protein/starch combinations, are more effective as feed than protein products themselves.

The Israelis, by determining the rate at which the mullets can synthesize various acids, and the Egyptians, by the above-mentioned methods applied to various types of enclosures, are close to optimum techniques to turn these enclosures into managed fish farms, with particular reference to Tilapia, mullet, and carp.

Waste Water Reuse

This study addressed two issues of major concern in the Middle East, i.e., the management and possible reuse of waste water, especially as a protein-enhanced medium in which to farm fish. Relating to the Lakes Management project, the project's surveys yielded the first-ever, quantified assessment of the kind and distribution of pollution in the important Egyptian lake, Manzella. This information has been conveyed to the authorities, who are currently in the process of designing and imposing regulatory measures, to bring about the Lake's ecological recovery. This action has achieved national priority status in Egypt. Further, it has attracted the attention of several other governorates in that country.

Construction of the first wastewater recycling plant was completed in 1991, outside of Suez. The pond system in enabling the Egyptians to determine the relative values of the French and American systems. It is estimated that this pilot plant is potentially capable of handling 30% of Suez' wastewater outflow.

Lakes Management

The first series of hydroacoustic assessments has just been completed of the fish stocks at Lake Kinneret. This method quantifies fish stocks more reliably than the classic, catch-and-effort statistics. The technology is currently being transferred to the Egyptian Academy of Scientific Research & Technology for use in Lakes Burullus and Manzellah.

In this regard, the investigations done under the auspices of this Program of these two lakes (which yield more fish than the aggregate of all of the others), have produced more pertinent data than in all previous years combined. For instance, the eight reports on Lake Burullus reflect, for the first time, a truly multidisciplinary ecosystem approach.

The Israelis had achieved meaningful results early in the Program when the Kinneret food chains were delineated with complete accuracy. Further, the prediction techniques achieved during the project, caused the Israeli government to change regulations pertaining to fishing, to ensure the viability of the desirable stocks.

Predictive Model for Shoreline Changes Along the Nile Littoral Cell

Long ago in geologic time, Egypt's Mediterranean coast was determined as the equilibrium between sediment furnished by the Nile River and sediment removed by the sea. Since the Aswan Dam complex reduced the flow of sediment by 90%, Egypt now possesses the fastest eroding coastline in the world. The Israeli coast is also affected, but more so in prospect.

Israeli and Egyptian engineers are concentrating on techniques to predict this erosion, and, in particular, the probable efficacy of structures either planned for installation at or near the shore or structures actually intended to reduce or halt the erosion itself.

In cooperation with the University of California, San Diego, the team has:

1. Determined the wave climate of the Southeast Mediterranean
2. Determined the sediment transport budget from Alexandria to Haifa
3. Constructed the first predictive model for the purpose of the Program

Eastern Mediterranean Circulation

This project was formed for the purpose of constructing a model of air-sea energy interchange and oceanic circulation for assistance in predicting storm surges and current patterns, and ultimately, the climatology (e.g., rainfall) for the surrounding land masses.

This project was delayed for approximately half its contracted lifetime, owing to equipment purchase problems under existing regulations. However, the models are now beginning to take shape and will hopefully be ready for testing in the near future.

SOCIOLOGICAL AND ECONOMIC ACHIEVEMENTS

This Program differs significantly from conventional USAID programs in that social progress (i.e., cooperation between Egyptian and Israeli scientists and institutions) is considered to be at least as important as the economic and intellectual accomplishments. A few highlights of cooperation may serve to demonstrate the Program's objectives.

The Egyptian an Israelis have conducted to date, twenty-four, joint and reporting conferences, mainly in Cairo and Alexandria to begin with, but increasingly, in Haifa. A full-scale workshop is held each year in which all of the project Principal Investigators participate. With a few exceptions, American participation has been limited to the Program coordinators. Planning and operations procedures have been developed increasingly by representatives of the Israeli and Egyptian Institutions vis-a-vis their American colleagues.

At this point in time, fifty Israeli person-trips have been made to Egyptian laboratories and at least seventy-five Egyptian person-trips to Israeli laboratories, where the scientist have cooperated in the research and have assisted with students.

The Israeli aquaculturists have entered an agreement to transfer technology - as it's developed -to a coalition of four kibbutzim (collectives) who have already entered marketing arrangements with French and Italian consumers. An Egyptian entrepreneur is currently exploring the possibilities of a commercial aquaculture venture using the Nile drainage lakes.

More than forty doctors and masters degrees have been obtained under the Program's auspices, and the projects have resulted in over seventy papers. The Primary Productivity, Waste Water Utilization, Management, and Shore Processes Projects produced the first coauthored publications.

In accordance with general agreements between the two countries and under USAID guidelines, the program seeks to balance oceanography with more traditional technologies in food resources and health control.

Perhaps the highlight of the Program, to date, occurred during September, 1983, when Dr. A.R. Bayoumi and Admiral Yohay Ben Nun (original Egyptian and Israeli coordinators respectively) were honored for their contributions to the Program by being designated as the firsts Co-recipients of the International Compass Award given by the Marine Technology Society for distinguished service in international marine affairs. In August, 1985, Dr. El-Sayed received the Distinguished Service Award from the American Institute of Biological Sciences for his role in developing the Program.

November 1988, marked another milestone in the Program's career, when a U.S. Congressional Delegation, led by Representative James Scheuer, met with the Israelis and a large Egyptian delegation in Elat, Israel, to reflect upon the Program's achievements and consider its future. Owing to aquaculture's central role in Middle East technical development, the entire group then inspected Israel's national mariculture center.

Symbolic of the burgeoning Egyptian interest in the Program, the nine-person Egyptian delegation (the largest yet to attend such a meeting) included Dr. Ibrahim Gohar, the pioneering leader of Egypt's marine program. His prominent role dates back to King Farouk's regime. Upon learning of his impending, eighty-second birthday, the Israelis held a party featuring a gigantic cake, to celebrate. The emotional impact was enormous.

In summary, social gains seem to be self-catalyzing and progress to be exponential. Closer working relationship lead to better results. Better results awaken interest by scientists outside the Program. The consequently improved recruiting opportunity offers more selectivity and more competent participation to the coordinators. Increasing competence leads to closer working relationships, better results, etc. In effect, the social machinery appears to be fueled by its own achievements.

PROGRAM MANAGEMENT

Very early in the Program's development, in fact, nearly from its inception, it became evident that regardless of the manner in which the sponsoring agency, USAID, viewed the Program's management, traditional doctrines of management simply had to be discarded. For instance, while total authority over, and responsibility for, the Program was vested in the Prime Contractor, i.e., the New Jersey Marine Sciences Consortium (Dr. Abel - Corporate President, and the Program's principal investigator for the first decade), the traditional doctrine of authoritarianism would simply not work (this assumption has been amply proved in the intervening years in other projects).

Accordingly, Abel adopted a laissez-faire approach in which his authority was delegated to the two country coordinators to the maximum degree and then, with them and Dr. El Sayed, he formed a Steering Committee. Thus Abel

1. Submitted all planning documents, even including meeting agendas, in draft to Cohen and Eisawy
2. Encouraged maximum review and discussion of the budgets that he introduces for the respective projects
3. Attempted wherever possible to maintain the Consortium's role as sort of service agency

The Program's leaders did not bind themselves to a set technique for developing the Program and its proposal. Israel adopted a methodology early in the tenure which appears to be more or less emblematic of the biennial forging of a Program package.

1. The country issued a call-for-proposals to the major institutions in Israel.
2. Applicants sent idea papers to the coordinator.
3. The coordinator convened a meeting at the Hebrew University in Jerusalem, hosted by a prominent member of that faculty. Dr. Abel attended.
4. Intensive discussion guided adroitly but dispassionately by the coordinator led to consensus, selecting a dozen of the most promising projects. The participants also deputized two additional scientists to accompany the coordinator to the Steering Committee meeting in Cairo, the following week.
5. The Egyptians having completed a similar process, the Steering Committee met to refine the package further, leaving in only the projects in which both Israel and Egypt possessed at least minimal technical capability.

STEERING COMMITTEE

More than any other aspect of the Program, the Steering Committee reflects the spirit of cooperation so central to the Program's success. The Committee's functions include, inter alia:

1. Stimulating thought towards project initiation in the three participating countries.
2. Assisting prospective principal investigators in preparing their projects, including identifying partners in the other countries.

3. Facilitating communications among the prospective partners.
4. Screening the projects at first and second levels (this normally involves reducing twenty to thirty proposals to a package of between five and eight.)
5. Preparing the final proposal package, including management. Proposals are generally prepared at approximately two-year intervals.
6. Negotiating with the Agency for International Development.
7. Meeting periodically with the principal investigators to assess progress and assist in the administrative phases.
8. Meeting with senior officials in the three participating countries to brief them on the nature and activities of the Program.
9. Preparing the final technical reports to be submitted each year.

One of the most heartening aspects of the Program's conduct has been the demonstrated high-level support. In Egypt, the Deputy Prime Minister, Yousef Walli, has been particularly outspoken in his support for intensified cooperation between the two nations, generally, with particular reference to the Program. The Egyptian Minister of Science, Adel Ez, and the President of the Egyptian Academy of Scientific Research and Technology, have expressed the strongest possible support for the Program. In Israel, both previous Prime Ministers Yitzak Shamir and Shimon Peres, have expressed their support for the Program, with Mr. Peres displaying special enthusiasm for the principle.

In the United States, support has been generated within the Congress, particularly within the Senate Foreign Relations Committee and the House Committee on Foreign Affairs.

MANAGEMENT ISSUES

The unusual and pioneering nature of this Program has naturally surfaced a number of managerial issues, many common to all scientific programs, some unusual, or even unique to this Program.

The *first* and most obvious issue relates to support for the Program. It must be emphasized, in this regard, that the Regional Cooperation Program was spawned by the legislative process. It is conceived, inspired, and fostered by the Congress. The U.S. Agency for International Development (USAID) is charged with responsibility for the Program's implementation. A group of farsighted Senators and Representatives asked the question: "In an era when the United States gives $5 billion/year to Israel and Egypt, a large portion of which goes for munitions, what's wrong with allocating 0.1% of that sum to encourage those countries to cooperate in their quest for food and water resources, health benefits, and general economic improvement."

At the moment, Regional Cooperation includes Programs in Arid Lands, Disease, and Agriculture, in addition to the Marine Program which was the trail blazer.

Viewed as objectively as possible, Congressional oversight appears to be supportive, far sighted, and normally non-interfering.

The *second* management question, relates to just how fast and far a Program like this can be pushed. It is handled jointly by the implementing agency, USAID, and by the Steering Committee, which, as previously mentioned, is comprised of those who have been immersed in the Program long enough to know what is good science and what is of lesser quality, and what is good politics and what is less efficacious. The present consensus is that all of the money that is being spent in the Program at the present time is in the Program's best interest. Further expansion, however, depends on the Program's attractiveness to excellent scientists in Egypt and Israel. The question may be asked: "Are there enough competent scientists and engineers in the two cooperating Middle East countries who are really interested in joining this Program and who would subscribe wholeheartedly to its fundamental tenets of useful cooperative endeavor?" At this point in time, the candid answer is probably "few, but growing."

A crucial *third* issue, however, relates to the younger folks. One can't help but wonder whether all of us are making the strongest possible effort under the circumstances, to search for, identify, and recruit willing and competent graduate students into this Program. The success of our Program must lie with this next generation. While the cadre of eminent scientists who have chosen to devote their careers to peace in the Middle East is, of course, the sine qua non of our Program, without whom we could never have gotten started, expansion to a recognizable regional effort will rest with the next generation, i.e., the younger scientists.

The *fourth* issue is often debated by the Program's participants, some of whom feel that for the Program to reach its objectives, it is necessary to search constantly and recruit into the Program new institutions, new people, and new topics. Others feel that best cooperation will be achieved with the old hands who have learned to work with one other. Current consensus favors an optimal combination of the two, i.e., a continual blending of new subject areas and new organization, with a constant leavening of people experienced in the Program's unusual philosophy and who have learned to work well with one another.

The *fifth* issue concern the relative effectiveness (towards the Program's fundamental goals) of the individual project vs. the packaged Program. The "Cooperative Marine Technology Program for the Middle East" has to date encompassed twelve separate and distinct projects, seventeen, if the aquaculture projects are categorized individually. In theory, any one of them could have been funded, sponsored, and managed directly from USAID. In one sense, therefore, the Manager could be considered dispensable. This is not really a clear cut issue. Were the Program's goals purely technical, i.e., breeding better fish, establishing better erosion control, etc., there's really not much reason why the USAID staff could not administer each project separately. In turn, this would depend to a degree upon the wishes and philosophy of the Egyptians and Israelis.

Because the Program's fundamental goals are at least as much social as they are technical, The Steering committee (composed of the Program's managers, who conduct the communications, coordination, advanced planning, and composite reporting), believes itself to play a necessary and beneficial role in the Program's development. As the Program progresses, however, and its personnel grow to know each other and to attain comfortable working relations, management's role may diminish somewhat. Of course, management's function in seeking out new topics, new players, and hopefully, new countries to involve in the Program, must continue. Put another way, although the American "technician" roles ought to diminish in time; our "ambassadorial" roles ought probably never to be relaxed.

A *sixth* closely related issue concerns the relative importance of the Americans vis-a-vis the Middle Easterners in the Program. The Program began as a parallel bilateral operation with the American role dominant, as it was necessary to relate to the Egyptians and Israelis partly independently. As working relations rapidly improved, however, the trilateral aspects became more important, and more recently, as we have achieved bilateral - between Israel and Egypt - relations, the American role has diminished to that of consultants. In any case, regardless of how a project is organized at the start, the American role is progressively reduced as the project develops.

The *seventh* issue of local management probably ought to depend on the respective Programs' individual styles. In the present case, the authors constituted ourselves the Program's steering committee with the President of the Egyptian Academy, tacitly accepted as our guide. We have found this arrangement to be convenient and effective in screening original projects to a workable package to be submitted to USAID. Perhaps looking at it from USAID's viewpoint, its administrative efficiency may be somewhat offset by the chance that a project in which USAID could conceivably be interested might be screened out at our local level. We try to overcome this by informal contact with USAID prior to tying together the final package.

The *eighth* issue relates to recruitment of other countries. Clearly, the "moderate" nations ought to be courted. Two recommendations are offered: First, timing is important; attempting to rush these groups into cooperation with Israel will become a self-defeating movement. Secondly, collaboration with other Regional Cooperative Programs, e.g., Infectious Diseases, Arid Lands, and Agriculture, is imperative. The principals of the respective organizations need to work out long-range strategies together.

The *ninth* issue, relating to the previous, concerns coordination, cooperation, and communications among the major Regional Cooperation Programs. There is none. This is ironic, considering the pioneering nature of the overall Program. It would seem logical that management innovations would hardly be limited to any one or two of these four programs and that the overall program would benefit from opportunities to compare notes, etc.

In 1983, at Abel's suggestion, USAID convened a meeting of Program coordinators. Although it was clearly beneficial, it was not repeated. In 1992, USAID awarded a contract to an external corporation to assist in managing and evaluating the Program.

The *tenth* issue concerns the Program's public image. For obvious reasons, our early meetings and negotiations were cloaked in secrecy and the Program's participants were extremely cautious about describing their work, especially in public. This situation has relaxed, respecting Egypt and Israel, but has become even more sensitive as other Arab nations have started to explore the possibilities of collaboration.

In the United States, we encounter a strange sort of acceptance, probably owing to the American Man-in-the-Street's concept of his country in relation to the rest of the world. In the first place, an appalling number of Americans have no idea where the Middle East is. In the second place, those who do, appear to harbor an attitude of superiority explainable only in terms of ignorance.

The *eleventh* issue relates to a reordering of USAID's priorities, switching emphasis from the social to the scientific. Emphasis appears to be more and more on highly meritorious science vis-a-vis projects which brought together persons of diverse backgrounds. Probably under this rationale, USAID has encouraged American universities to resume more prominent roles in the Program.

PROSPECTS FOR INTERNATIONALIZATION

Three years ago, the Egyptian coordinator initiated a dramatic document of possible historical significance. Known as the "Aqaba Plan," it deals with the Gulf of Aqaba, a small, economically crucial, ecologically stressed, semi-enclosed body of water common to Jordan, Saudi Arabia, Egypt, and Israel. The Egyptian Academy has proposed a collaborative program with the other Arab nations, without giving up, but keeping separate, its cooperative projects with Israel. This "parallel bilateral" approach would effectively bring about broad de facto cooperation among Israel and her Arab neighbors, and would seemingly advance world interests significantly in the Middle East.

While apparently imaginative, farsighted, and courageous, however, the plan may not be colinear with current U.S. policy. The Program's practitioners and Congressional supporters would very much like to expand the Program to encompass other Arab nations. Several factors, however, inhibit such progress.

First, most of these countries are still technically at war with Israel. This limits communications severely, if in fact not completely precluding them. Second, USAID adopting a strict interpretation of the law, prohibits the use of Regional Cooperation funds for any uses external to Egypt and/or Israel. This curtails the practical movement of the "missionaries" to other countries. On the other hand, USAID's scarce funds might be overly stretched, were the Program to be extended to other countries. Third, additional funding, which would be mandatory, appears unlikely at this time. Some supporters, strongly advocate external, i.e., matching, funding for this purpose.

About two dozen private foundations have thus been approached. The universal response is that the Program, while "fantastic" in appearance, is too far removed from anything they have been accustomed to support. They are thus unwilling to stretch their characters, particularly in an era when they are under such stress. Ultimately, however, such foundation support will have to be located, if the Program's viability is to be maintained.

THE FUTURE

Future aims include

1. Adding new technologies, institutions, and people to the Program, to spread its beneficial influence throughout as many communities as possible in the two countries.
2. Encouraging as many Egyptians as possible to visit Israel: Dr. Eisawy is attempting to "sell" a team, i.e., to the Egyptian Foreign Office, of up to twenty personnel.
3. Conveying the Program's benefits to other Middle Eastern and African countries in an effort to persuade them to join the Program.
4. Translating the scientific achievements into economic and cultural gains.
5. Persuading U.S. Government official that this Program's charter, motivation, and progress *merit at least one fifth of one percent of what this country spends on aid to the Middle East*.

As we look to the future, the Program's leaders don't envision a smoothly rising curve of acceptance and participation. We would prognosticate, rather, a step-wise motion as one after another, the social and financial barriers give way to good fellowship, beneficial technology, and, above all common sense. The personal aspirations which we share include

1. Acceptance of the Program in USAID and an attitude of - if not unbridled enthusiasm - at least straightforward encouragement.
2. Acknowledgment of the Program's achievement and recognition of its extraordinary potential at the top management levels of both the Executive and Legislative branches of Government.
3. Continuing willingness of the Middle East partners to appreciate each other's willingness to cooperate and recognize each other's capability.

In summary, we, the Program's practitioners, believe we are in process of demonstrating what history may term the ocean's greatest gift to mankind: Peace!

4

Environmental Security and Shared Solutions: Land-Based Marine Pollution in the Gulf of Mexico and the Black Sea

James M. Broadus
Director, Marine Policy Center
Woods Hole Oceanographic Institution

Raphael V. Vartanov
Head of Section, Department of Oceans and Environment
Institute for World Economy and International Relations, Russian Academy of Sciences, Moscow
and
Senior Fellow, Marine Policy Center
Woods Hole Oceanographic Institution

ABSTRACT

A collaborative research project by the Institute for World Economy and International Relations (IMEMO) of the Russian Academy of Sciences and the Marine Policy Center of the Woods Hole Oceanographic Institution (WHOI) investigated the concept of environmental security as applied to the world's oceans. This presentation addresses land-based marine pollution in the context of two regional seas: the Gulf of Mexico and the Black Sea. Environmental security is defined as: "the reasonable assurance of protection against threats to national well-being or the common interests of the international community associated with environmental damage." While most land-based marine pollution remains within domestic jurisdictions, it may raise issues of international environmental security in several ways: transboundary pollution, shared resource stocks, damage to export goods, pollution and tourism, pollution and non-use values, and emissions "export" through capital mobility. Drawing on the work of the IMEMO-WHOI project participants, the Gulf of Mexico and Black Sea cases are reviewed in terms of: coastal development, nutrient enrichment, public health, industrial and municipal sources, national management approaches, and regional cooperation. The U.S. approach reflects its federal form of government, while coastal management in the former Soviet Union is currently in disarray.

INTRODUCTION

Environmental security is an aspiration for all countries of the world, and the world's oceans figure prominently in its attainment. This presentation introduces the concept of environmental security, explains how it applies to the problem of land-based marine pollution, and reviews the cases of two regional seas of special interest to the United States and Russia.

ENVIRONMENTAL SECURITY AND THE OCEANS

The concept of environmental security is a way of thinking about international environmental management (Broadus and Vartanov [in press]). It draws on the widely understood notion of international, strategic interdependence (in facing threats of nuclear war or economic collapse) to focus attention on the similarly shared exposure to threats from global environmental degradation. The concept also links directly to conventional international security, in the potential for conflict over both natural resources and environmental practices.

For over two years, the Institute for World Economy and International Relations (IMEMO) and the Woods Hole Oceanographic Institution (WHOI) have combined forces to compare thinking on the concept of environmental security as it applies to the world's oceans, to define the concept more precisely, and to identify opportunities for cooperative US–Russian actions (Broadus and Vartanov 1991). Within the Russian Academy of Sciences, IMEMO is a major center of scholarship on world trends in economics, politics, organization, and strategic relations. IMEMO's Department of Oceans and Environment is a counterpart to WHOI's Marine Policy Center in terms of disciplinary orientation (law and economics), research emphasis (oceans, environment, and international relations), and location within a larger research organization.

The collaborative project was initially suggested by IMEMO researchers following a 1987 visit by one of the authors (Broadus) sponsored by the USSR Academy of Sciences. The project, sponsored by the John D. and Catherine T. MacArthur Foundation and the Peace Research Institute of the Academy of Sciences, began in late 1989. More than thirty Soviet and American scholars took part in four joint workshops (two in Moscow and two in Woods Hole), numerous exchanges, and collaborative analyses of selected cases.

The project is unusual since it was truly a joint effort. All elements of the planning, research, and writing were shared. Our interdisciplinary research combined the efforts of specialists in economics, law, international relations, ecology, and ocean science. Among cooperative US-Russian bilateral ocean studies, this one is distinctive in its emphasis on social science rather than natural science. Through our work we reached the conclusion that cooperation in both of these areas of science is vital.

Early in our collaboration we formulated a working definition:

> *Environmental security is the reasonable assurance of protection against threats to national well-being or the common interests of the international community associated with environmental damage.*

Critical problems of international environmental security were determined to be those that are *likely to destabilize normal relations between nations* and *provoke international countermeasures*.

Using this definition and these criteria for guidance, our joint research team identified seven problems of ocean environmental security that are of great mutual interest to our two countries. The reader must look elsewhere for six of these cases: the Law of the Sea, hazardous materials transport, nuclear contamination, North Pacific fisheries depletion, Arctic Ocean sensitivities, and the Southern Ocean (Broadus and Vartanov [in press]). The seventh, land-based marine pollution, is reviewed briefly here.

LAND-BASED MARINE POLLUTION AND ENVIRONMENTAL SECURITY

Land-based marine pollution is the world's worst marine pollution problem. The United Nations Joint Group of Experts on the Scientific Aspects of Marine Pollution (GESAMP) has estimated that land-based sources contribute more than 75% of the pollutants entering the sea (although the unit of measure is left undefined) (1990). Land-based marine pollution–from agricultural runoff, sewage discharges, industrial emissions, and atmospheric deposition–presents a much more complex problem than do the relatively minor sources of pollution from vessels. The fundamental obstacle is that land-based marine pollution arises throughout the very fabric of daily life and from virtually all economic activity. Measures to address land-based marine pollution must reach practically all the polluting aspects of society's activities. The problem's complexity thus arises from its broad range of sources, large variety of pollutants, constant and daily discharge modes, cumulative effects, and far-reaching impacts on national economic and social developments (Meng 1987).

Unlike widely dispersing global phenomena such as global warming or hazardous materials transport, land-based marine pollution may seem largely a domestic problem and not an issue of international environmental security. Indeed, regarding land-based pollution, GESAMP stated, "Only a small part of those contaminants has spread beyond the limits of the continental shelf. The bulk remains in coastal waters and, in places, particularly in poorly flushed areas, has built up to significant levels" (1990). Despite widespread problems of land-based marine pollution, instances of significant transboundary damages do appear to be rare exceptions.

Yet there are several ways, some indirect or subtle, in which land-based marine pollution does express itself as an issue of international environmental security:

1. *Transboundary pollution*. This is the most obvious mechanism. While uncommon, it is not unheard of; and international institutions to address this problem are just emerging.
2. *Shared resource stocks*. While pollution may remain in one country's jurisdiction, shared resource stocks may become contaminated or damaged there and thus hurt a neighboring country's well-being.
3. *Damage to export goods*. Similarly, export goods, such as seafood products, damaged by domestic pollution may be consumed in another country whose well-being would be reduced.
4. *Pollution and tourism*. Or, consumers from another country might themselves travel to the polluting country and suffer directly from the marine pollution there.
5. *Pollution and non-use values*. More subtly, citizens in a foreign country may feel a real loss because of their concern about damages to wildlife or other natural resources (coral reefs, endemic species, marine mammals, etc.) in the polluting country. This is a genuine loss in well-being even though they may have no intention ever to visit the polluting country or to use its natural resources first-hand.
6. *Emissions "export" through capital mobility*. Perhaps the most subtle international connection occurs when investors from one country locate their polluting facilities in another country. By this means they indirectly export the polluting emissions and threaten the environmental well-being (though not necessarily the net economic well-being) of the country whose waters they pollute.

The 1992 UN Conference on Environment and Development (UNCED) recognized and called special attention to the importance of land-based marine pollution. Agenda 21 of UNCED, for example, calls on states "to assess the effectiveness of existing regional agreements and action plans, where appropriate, with a view to identifying means of strengthening action, where necessary, to prevent, reduce and control marine degradation caused by land-based activities."

REGIONAL COOPERATION

In contrast to vessel source pollution and dumping, there is no comprehensive, global legal regime for land-based marine pollution. Within the global legal structure, provisions for land-based pollution are made only through very general obligations or guidelines (U.N. Convention on the Law of the Sea, Montreal Guidelines on Land-Based Pollution, and ILC draft Articles on International Watercourses). Regional regimes for the control of land-based marine pollution, on the other hand, typically are more detailed and inclusive by design (i.e., 1992 Convention on the Protection of the Marine Environment of the Baltic Sea Area, 1992 Convention for the Protection of the Marine Environment of the North East Atlantic, 1976 Convention for the Protection of the Mediterranean Sea Against Pollution).

Control may work best within a regional framework because of the need to resolve potentially conflicting national and international interests. Moreover, the problems of land-based marine pollution are mostly felt regionally. States sharing a regional sea discharge pollutants into it as a common pool, and they share in the resulting damages (or benefits of remediation) on a common basis, if in varying degree. Moreover, states in the same region often share similarities in culture, economic geography, public preferences, and trading relationships.

Still, there appear to be shortcomings in attempts to address marine pollution at the regional level. In several regions, organizational efforts "have not yet led to concrete results" (United Nations Secretary General 1989). Also, support at the domestic level for regional programs is sometimes quite limited. If a state perceives that its costs of participating in collective action exceed its benefits, it has a strong incentive to avoid regional cooperation, even if region-wide benefits are greater than costs.

In view of widespread perception and complaint that regional agreements on land-based marine pollution have been notably ineffective, the WHOI Marine Policy Center intensively examined and compared three pioneering regional programs to control land-based marine pollution. Sponsored by the US Environmental Protection Agency, this research identified the efforts made in the Baltic, North Sea, and Mediterranean regions, and sought to show what has worked best, what has not worked, and why (Broadus et al. 1992). A number of relevant lessons learned from the experience in those three regions were reported (Table 1).

Several noteworthy observations were made about all three European programs. In many instances, it was hard to show that active investment or strict compliance by all the parties is in their own self-interest. None of the programs included explicit mechanisms for trade (of emission quotas, *quid pro quos*, know-how, etc.) among parties, with the exception of the Med Plan's foreign assistance and technology transfer activities. All of the programs have been hindered by inadequate compliance reporting and lack of transparency, though this seems to be improving markedly for the Baltic and the North Sea. All three programs have been subject to, and have apparently benefited from, the superimposition of high-level, political oversight structures.

COMPARATIVE CASES: GULF OF MEXICO AND BLACK SEA

Two of the regional seas of greatest interest to the United States and Russia have only recently been included within regional, international cooperative programs. Land-based marine pollution in both those seas remains almost entirely a domestic concern, although clearly exhibiting many of the aspects of international environmental security cited above.

To gain insight into the similarities and differences at work on land-based marine pollution in our two very different systems, our joint IMEMO-WHOI team developed comparative descriptions of the Gulf of Mexico and the Black Sea. Both are large, semi-enclosed seas, respectively important to our countries economically and shared with neighboring states. In each instance, one of our countries is, in turn, a major economic and political presence as well as a significant contributor to coastal pollution.

The cases were developed and reported at length together with our colleagues Anna K. Bystrova, Suzanne M. Demisch, Mark E. Eiswerth, Arthur G. Gaines, Kristina M. Gjerde, Yoshiaki Kaoru, Anna Korolenko, Elena N. Nikitina, Mary Schumacher, and Tom Tietenberg (Broadus and Vartanov [in press]). In the descriptions below, we draw from this work and merely summarize some of the findings.

THE GULF OF MEXICO

The Gulf of Mexico (Figure 1) is an important industrial and recreational center for both its bordering countries, the United States and Mexico. In the United States, the natural resources and commercial and recreational activities supported by the Gulf of Mexico are of vital economic significance to the nation, and especially to its five bordering states: Florida, Alabama, Mississippi, Louisiana, and Texas. The Gulf region, with its warm waters and extensive beaches, supports an important tourist industry for each of the U.S. Gulf coastal states. The Gulf's coastal wetlands comprise about half of the national total and, together with its barrier islands, provide critical habitat for 75% of the migratory waterfowl traversing the continent (AMS and Webster 1991). The Gulf of Mexico accounts for 2 billion pounds of commercial fish landings annually (or about 21% of the U.S. total) and for 40% of all U.S. recreational fishing (AMS and Webster 1991).

Two-thirds of the contiguous United States lies in the drainage basin of the Gulf of Mexico. The Mississippi River is the largest carrier of water and associated materials into the Gulf of Mexico (Figure 2). Water exchange between the Gulf of Mexico and the Atlantic is limited, and pollutants tend to accumulate and concentrate in the Gulf.

Impacts

Impacts of Coastal Development

Coastal development and the subsequent rise in population density is commonly associated with habitat loss and increased sewage and toxic runoff entering coastal waters. In 1988, 14 million people lived in the ninety-nine U.S. coastal counties of the Gulf of Mexico – 13% of the entire U.S. coastal population. Between 1970 and 1980, the Gulf-area population increased by 33%.

Impacts of Nutrient Enrichment

Human sources of aquatic nutrients, such as nitrogen and phosphorous, include both: "point sources" (10–30%) such as on-site, domestic septic systems, discharge pipes from municipal sewage treatment pipes, and industrial sources; and "non-point sources" (70–90%) such as agricultural runoff (fertilizer and animal wastes), automotive exhaust, and stack emissions. The combined drainage of the Mississippi–Atchafalaya River system contributes 70–75% of the total nitrogen and phosphorous concentrations entering the Gulf from all US sources. Oxygen depletion, or hypoxia, puts a stress on commercial fish and shellfish species, and it is suspected that excessive organic productivity from over-enrichment by nutrients is responsible. Hypoxic conditions on the Louisiana continental shelf were first monitored in 1986, but it is not known when they first developed or whether they pre-date human intervention in the nutrient cycle.

Table 1
Lessons Learned from the Experience in the Baltic, North Sea, and Mediterranean Regional Programs

1. **Clarity.** Costly misunderstandings, false stars and wasted effort can be avoided by seeking maximum clarity in defining program goals, the relations among parties, measures and recommendations, and the role of the secretarial. Particular effort should be made to excise hidden agendas (e.g., if income transfer or lobbying of governments is to be a program objective, this should be set out clearly and incorporated explicitly into program design).
2. **Capacity to evolve.** The ability to adjust the program to changed circumstances and improved knowledge is vital to its effectiveness over time. The method of keeping the framework convention quite spare and leaving program elaboration to subsequent protocols, directives, recommendations, or flexible action plans is useful in this regard. So, too, is the periodic oversight and intervention of high-level political authority from among the parties.
3. **Political commitment.** Program effectiveness will depend inevitably on the political commitment of the parties. Examples of movement toward enhanced effectiveness were observed in all three regions as a result of collective political intervention by parties at a ministerial level. The device of an overarching, high-level political forum above the regional program, as in the North Sea Conferences, appears to have been useful in this regard. Such a forum also provides a highly-visible focal point for public pressure, and it contributes to program transparency, which enhances the relevance and effectiveness of public pressure.
4. **Specificity of objectives.** Demonstrating effectiveness and clarity in its measures is important both in sustaining program support and in operational program implementation. For this it is necessary to specify objectives whose accomplishment can be measured and demonstrated. Targets and timetables are useful, especially if they are sensitive to differences in the stakes and economic capabilities among the parties. Black lists and grey lists are useful in specifying the scope of concern, but experience has pointed to the need to narrow program focus in practice to priority targets (sometimes called "red lists").
5. **Scientific involvement.** A mechanism for expert scientific input and advice is essential, to clarify the nature and magnitude of problems and to monitor changing conditions, but is important to assure that scientific research interests do not altogether run away with program resources. The persuasive establishment of baseline conditions, both in terms of ambient measures and emissions, must be a priority, as demonstrated by the hindrances created by shortcomings in doing so in all the cases examined. It should be recognized that the baseline profiles can be useful even if somewhat crude, and that assembling credible baselines will be expensive and time consuming.
6. **Self-interest test.** States can be expected to serve their own self-interests, but program effectiveness will be improved to the extent that program design takes account of this and explicitly seeks to accommodate the genuine (rather than presumed) self-interest of its parties. Program structures that encourage or accommodate agreements and actions by sub-groups of parties, that allow for flexible financial or technical assistance, or that provide explicitly for variable schedules or an "op out" option on some measures, may also be helpful in this regard.
7. **Mechanisms for trade.** Because the self-interests of states will differ, and because one of the principal rationales for collective action is to facilitate sharing and exchange, program effectiveness will be enhanced by provision of mechanisms for explicit "trade" (quid pro quos, specialization) among participants. Similarly, efficiency is served by implementation measures that take into account differences among states and that facilitate trading to accomplish mutual objectives (such as tradable quota schemes). Implicit trades or explicit quid pro quos may be achieved through program funding schemes.
8. **Reporting and compliance.** Assessing the degree of state compliance with program recommendations has been obviated in all three cases by the poor performance of state reporting. Special emphasis in program design should be placed on specifying expectations for compliance, means for monitoring compliance, and particularly, procedures that encourage accurate and timely state reporting. In this regard, keeping the reporting burden simple and at a minimum is important. It may also be useful to assure that the results of reporting are useful to all parties and to suggest a reporting system that meshes routinely and automatically into the states' own practices. Some provision for non-intrusive, third party inspection may also be useful, though this is only implicit in the cases examined through their lack of success with more conventional, passive means of collecting state reports.
9. **Transparency.** Most of the lessons proposed above speak to the value of transparency in program design and execution, but it warrants explicit inclusion. Program effectiveness is enhanced (both in terms of party state support over time and of verifiability by other interests) by the maximum transparency comparable with the protection of proprietary interests and the rights of national sovereignty.
10. **Limit expectations.** Regional programs in the real world can at best be catalytic. Measures to control land-based marine pollution will necessarily occur almost exclusively through domestic actions. The programs may succeed as a medium for exchange and consensus among members and as a promoter of external funding support, which can be used to reinforce internationally agreed program objectives and reward compliance. The effectiveness of the regional trust fund device is not clear from the cases examined and must be carefully questioned. Again, clarity of purpose and transparency cannot be overemphasized.

Source: J.M. Broadus et al., Comparative Assessment of Regional International Programs to Control Land Based Marine Pollution: The Baltic, North Sea, and the Mediterranean. Report prepared for the U.S. Environmental Protection Agency.

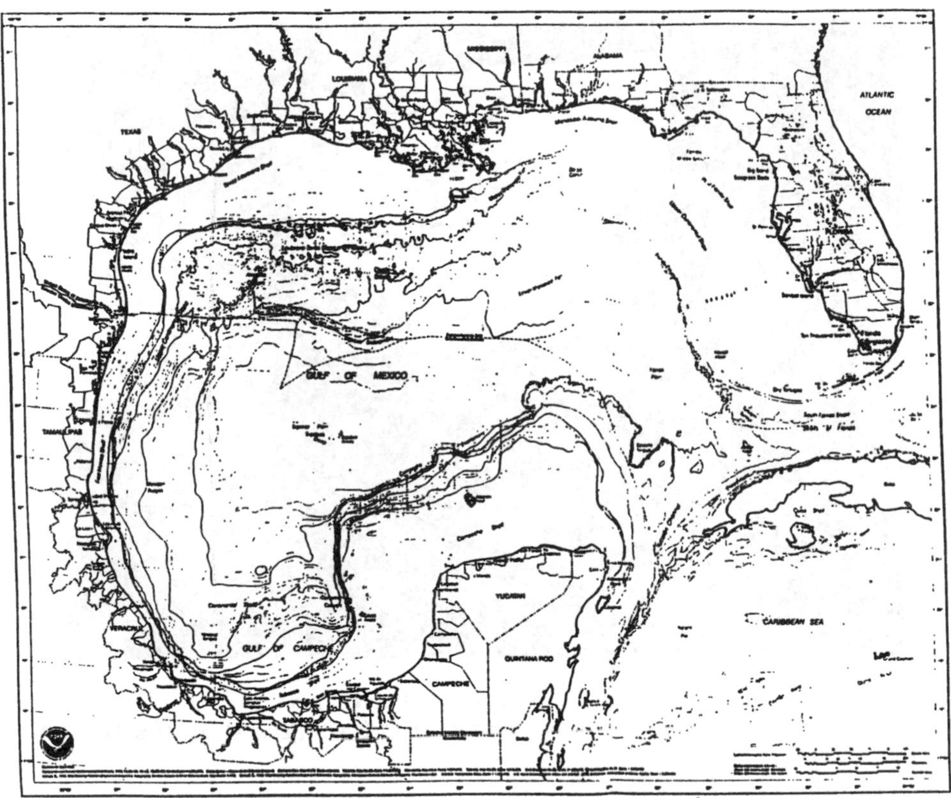

Figure 1. Gulf of Mexico map compiled by Strategic Assessment Branch, Ocean Assessments Division, Office of Oceanography and Marine Assessment, National Ocean Service, NOAA; and Southeast Fisheries Center, National Marine Fisheries Service, NOAA.

Source: National Oceanic and Atmospheric Administration, Gulf of Mexico Coastal and Ocean Zones Strategic Assessment: Data Atlas.

Public Health Impacts

Red tides are often cited as one impact of nutrient over-enrichment, but the relationship has not been sufficiently validated. It is likewise unclear whether red tides have become more frequent or intensive than in the past. Red tides are known to have occurred in the Gulf for centuries, but the danger to human health appears to be minimal. Closures of inshore waters due to the presence of red tides, however, can have significant economic impacts on tourism, recreation, and commercial fishing.

Human exposure to toxic chemicals released into the environment is a matter of widespread public concern, yet data and risk assessments bearing on this problem are scanty. No clear problems appear to have arisen along the U.S. Gulf Coast despite some heavy concentrations of petrochemical activity.

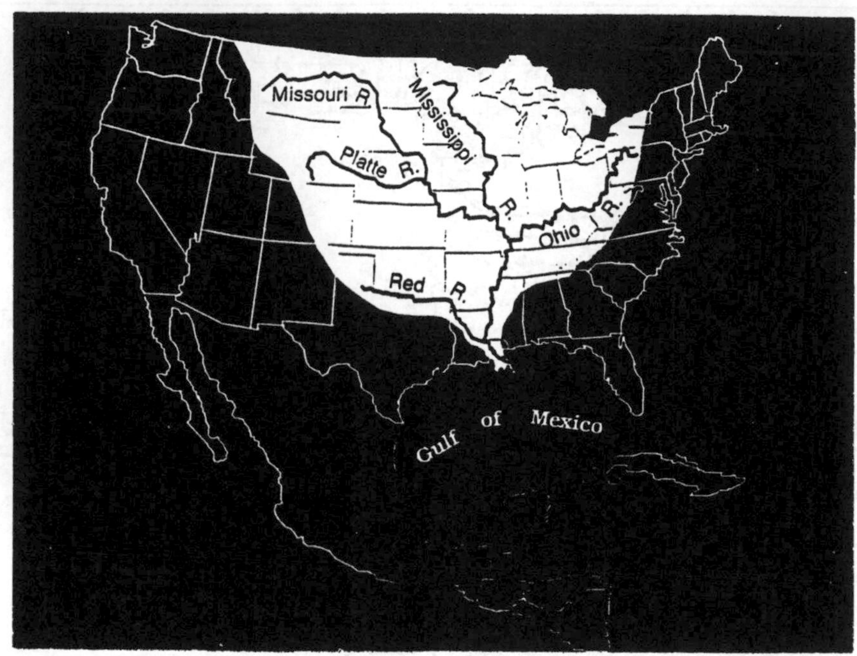

Figure 2. Mississippi River Drainage Area.
Source: The Louisiana Land and Exploration Company, Louisiana's National Treasure, LL&E, New Orleans, undated.

The risks of pathogen-related diseases have diminished greatly since the advent of enforcement of sanitary standards for on-site domestic septic systems; installation of municipal sewage treatment; laws prohibiting discharge of raw sewage into U.S. waters; and use of on-board disinfectant equipment.

Industrial and Municipal Sources

Three hundred forty-seven U.S. industrial facilities discharge through pipelines directly into the Gulf of Mexico or its estuaries. Most are petroleum, forest, or fish processing plants. Still others discharge into municipal treatment facilities.

One hundred thirteen cities and towns discharge wastes directly into Gulf estuaries and coastal waters. Urban runoff puts oil and grease, lead, chromium, and many other contaminants into marine waters. Municipal septic systems are also significant contributors of nutrients and human pathogens into the Gulf marine environment.

U.S. Management Approach

U.S. mitigation efforts in response to land-based marine pollution in the Gulf of Mexico are embodied in local, state, and national efforts that reflect the nation's federalist form of government. These efforts include both information gathering on pollution and the legislation and implementation of regulations.

Information Gathering

In addition to some longstanding data systems such as fishery and water resources statistics, new environmental information sources are being developed at the federal level for use as resource management tools.

The National Oceanic and Atmospheric Administration's (NOAA) National Status and Trends (NS&T) program monitors the concentrations of toxic chemicals and trace elements in bottom-feeding fish, shellfish, and sediments at over 200 coastal and estuarine locations, fully a third of which are on the Gulf coast region (1992). Among its many data bases, the National Coastal Pollutant Discharge Inventory Program (NCPDIP) documents pesticide transfer from streams and rivers into marine systems and measures toxicity, contamination levels, and bioconcentration factors.

The federal interagency National Ocean Pollution Research and Development and Monitoring Planning Program was established to encourage coordination among federal, state, and local agencies; to determine national activities and problem areas for action; and to make recommendations for improving efficiency and effectiveness. Its record of accomplishment, however, has been disappointing.

The U.S. Environmental Protection Agency (EPA) introduced a Risk Reduction and Assessment Program to integrate governmental and popular environmental concerns and then target protective efforts at the highest-risk problems. EPA has also launched an Environmental Monitoring Assessment Program (EMAP) to provide statistical data summaries and periodic interpretive reports on ecological status and trends.

Legislation and Regulation

Under the Clean Water Act (CWA), originally enacted by Congress in 1972 as the Federal Water Pollution Control Act, EPA or its authorized agencies in certain states issue National Pollutant Discharge Elimination System (NPDES) permits to discharge legally accepted material into navigable U.S. waters. Standards are based on an assessment of the achievability of control technologies by individual categories of discharge.

The 1987 amendments to the CWA direct the states to identify water bodies where toxic pollutants prevent the attainment of water quality standards or of designated beneficial uses. For each identified water body, "Individual Toxic Control Strategies" to reduce point-source discharges must be developed for implementation within the NPDES permitting process. Another program under the 1987 CWA Amendments requires each state to develop a Management Program to control non-point pollution sources, identifying actions needed to mitigate the problems caused by each source category.

The National Estuary Program (NEP), established in 1987 under CWA authority, addressed the need for estuarine conservation. Four, Gulf bay areas are sponsored under this program, which relies on a complex network of support from state and federal agencies, local communities, and interest groups. NEP status is strictly voluntary; states nominate their estuaries to be considered for participation in the Program. Mitigation efforts often differ from region to region and even from state to state because of the specificity of problems in a given area (Environmental Protection Agency 1990).

The 1972 Coastal Zone Management Act (CZMA) established a cooperative state-federal program for comprehensive coastal zone management. Administered by NOAA, the program seeks to ensure wise management of future development through the establishment of federally supported, state coastal management plans. In implementing its program, a state must attend to several issues of national interest: (1) protecting natural resources; (2) managing coastal development; (3) giving development priority to uses that are coastal-dependent; and (4) concentrating new development in existing developed areas. As long as a state's coastal zone management plans conform to national guidelines, all federal activities in the state are subject to consistency provisions and, if they affect natural resources, land and water uses in the coastal zone must conform to state plans.

Still, state and federal environmental legislation are frequently inconsistent, and gaps in authority may imply little action. In the Gulf of Mexico, coastal states sometimes respond to this problem by pushing for measures that allow for more comprehensive state jurisdiction over sensitive coastal areas. Louisiana, for example, which contains nearly half of all U.S. wetlands, passed the Louisiana Wetlands Conservation and Restoration Act of 1990 to enhance protection of the fragile and endangered wetlands resources that were ignored by the Clean Water Act. Similarly, gaps in legislative/regulatory programs and sluggish administrative mechanisms have led to the recent growth of Citizen Advisory Committees in each of the Gulf states, local volunteer clean-up programs, and an increase in voluntary waste reductions by industrial companies.

The Gulf of Mexico Program, developed by EPA and NOAA's Sea Grant Programs in the five Gulf coastal states, is an example of an intraregional initiative launched in recognition of the need to address comprehensively the problems of the marine environment. The Gulf of Mexico Program is administered by federal, state, and local agencies. A Citizens Advisory Committee, representing environmental, agricultural, business/industry, development/tourism, and fisheries interests in each state, provides public input and assistance in disseminating information relevant to the goals of the program. The purpose of the Gulf of Mexico Program is twofold: (1) to develop an infrastructure capable of addressing interjurisdictional and multijurisdictional environmental problems and (2) to establish a strategy and framework for action. The Program has succeeded in defining the problems facing the Gulf marine environment. The next steps are to develop options for action and to implement them. That is a much greater challenge.

International Regional Cooperation: The Wider Caribbean

Coordinated through the United Nations Environment Programme's (UNEP) Regional Seas Program, the Action Plan for the Wider Caribbean was adopted by representatives of twenty-three states in April 1981 at Montego Bay, Jamaica, in response to the growing concern for conservation, protection, and development of the marine and coastal resources of the region. "The Convention for the Protection and Development of the Marine Environment of the Wider Caribbean Region" (Cartagena Convention), called for by the Action Plan, was signed by representatives of sixteen states on 24 March 1983 at a conference in Cartagena, Colombia. Implementation of the Caribbean Environmental Programme's strategy includes the development of several, regionally coordinated comprehensive programs between 1990 and 1995. As part of the program for Assessment and Control of Marine Pollution (CEPPOL), a regional Protocol for land-based sources of marine pollution is currently being negotiated (United Nations Environment Programme 1990). The United States is a leading participant in these negotiations. As the region's dominant economy and one of its most progressive environmental managers, the United States has a strong interest in assuring that the negotiations result in a protocol that is both realistic (in matching the priorities and capabilities of the region's states) and effective (in achieving its marine environmental objectives).

THE BLACK SEA

The Black Sea (Figure 3) is an important industrial and recreational center for the former Soviet Union and the other bordering nations of Bulgaria, Romania, and Turkey. Of the former Soviet republics, Russia, Ukraine, and Georgia border the Black Sea. Rich in marine resources, the Black Sea's total annual commercial landings amounted by the end of the 1970s to about 880 million pounds, of which the former USSR's share was about 550 million. The Black Sea is the most popular coastline with vacationers and tourists in the former USSR. A leading recreational resource for all the other bordering states as well, the Black Sea coastal region is nonetheless seriously threatened by land-based marine pollution.

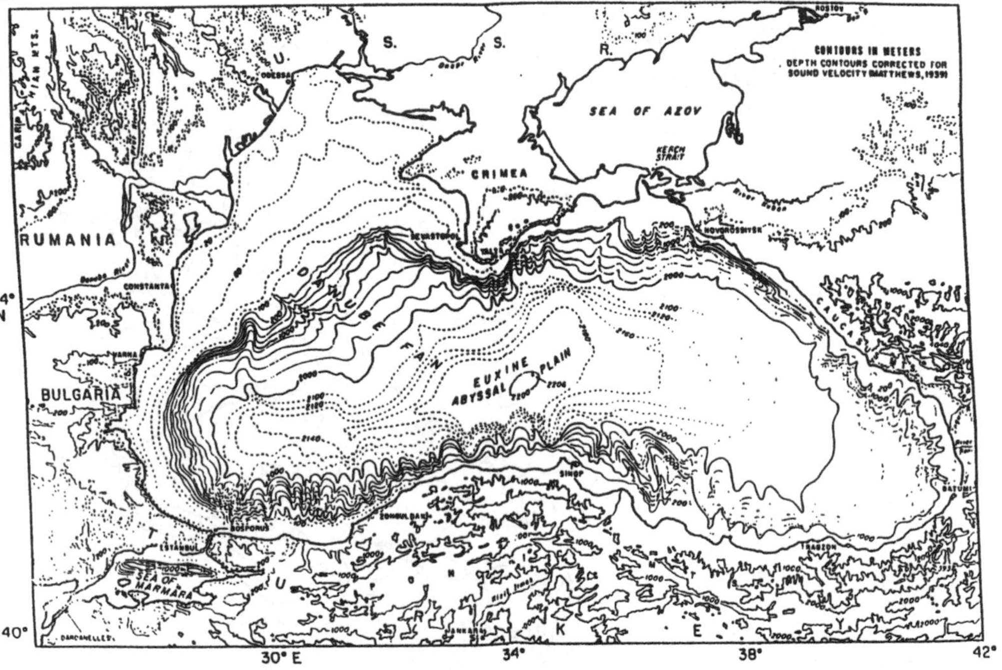

Figure 3. The Black Sea.

Source: D.A. Ross, E. Uchupi, K.E. Prada, and J.C. MacIlvaine, "Bathymetry and Microtopography of Black Sea," In: The Black Sea – Geology, Chemistry, and Biology, The American Association of Petroleum Geologists, Tulsa, OK, 1974, pp. 4–5.

The Black Sea's watershed is 5.6 times as large as its water surface, encompassing fifteen European states and Turkey, or more than 1.1 million square miles (Figure 4). Narrow straits virtually shut off the Black Sea from the neighboring Mediterranean, allowing a complete water exchange only once in 2,250 years. In addition, because of persistent stratification of water layers, complete vertical mixing requires several decades.

Impacts of Coastal Development

The population of the areas adjacent to the Black and the Azov Seas is roughly 160 million (as compared to 90 million for the Baltic Sea coast), including 68 million inhabitants in the Russian and Ukrainian portions of the Black Sea Basin. More than 8% of world industrial output is produced here, and Black Sea ports serve as connections to international trade and interstate transport facilities for all the republics of the former USSR. Growing populations and coastal development have had major adverse impacts on the coastal marine environment.

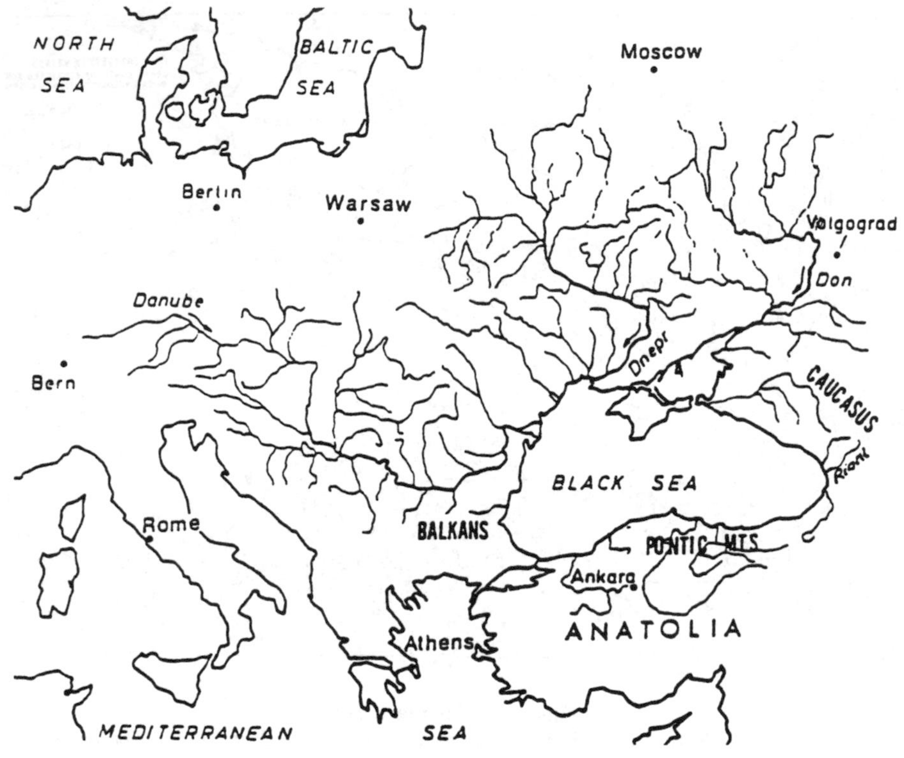

Figure 4. Drainage Area of the Black Sea.
Source: G. Müller and P. Stoffers, "Mineralogy and Petrology of Black Sea Basin Sediments," In: Degens and Ross, op. cit., p. 202.

Impacts of Nutrient Enrichment

One-quarter of the Black Sea suffers from hypereutrophication because of the uncontrolled input of nitrogenic and phosphoric compounds. In the agricultural sector, widespread and uncontrolled use of chemical fertilizers steadily increases. The use of mineral fertilizers in Georgia, for example, increased fourfold in the last thirty years. Animal wastes continue to go unmanaged. Purification installations are available at all fifty-five stock-breeding plants in Georgia, but at least twenty-one of them do not work. (These conditions prevailed even before the outbreak of civil war in Georgia, where the Black Sea coast has been a major area of confrontation.) Red tides in the northwestern part of the Black Sea shelf are common, but a direct relationship with nutrient loading has not been documented. In general, there has been mounting anxiety as the hypoxic, lifeless water layers have accelerated their rise into the upper layers. Recent research however, has somewhat calmed these anxieties.

Public Health Impacts

Communal sewage from Black Sea cities typically is only slightly purified or not purified at all before discharge. In resort areas, the number of peak-season vacationers surpasses the capacity of purification facilities by a factor of five to seven. Increased rates of illness are associated with elevated pollution in recreational coastal areas and with unsafe drinking water.

These public health impacts entail other economic and social consequences. Many Black Sea sanatoria resorts, for example, have lost business and revenue. In 1987, some 300,000 reservations for accommodations in the seaport resort of Batumi were cancelled because of threats to public health, reducing tourism income by 30%. In 1990, outbreaks of intestinal bacillus on Black Sea beaches resulted in numerous temporary beach closures.

Industrial and Municipal Sources

One-quarter of all household and municipal sewage discharged into the Black Sea is unpurified. So are livestock and railway sewage, the latter containing high concentrations of strong disinfectants. The waters around the naval fuel-storage facilities in Sevastopol have, for many years, contained levels of petroleum-related substances that far exceed, sometimes a hundred-fold, the maximum permissible concentration. Airborne pollutants reach the Black Sea via automotive and other exhausts generated in connection with industrial activities. Chemical and steel plants are heavily concentrated in the coastal regions. Along the banks of the Prypjat River, not far from the Black Sea itself, stands the Chernobyl nuclear power plant, which spread radioactive pollution to the Black Sea from its infamous reactor disaster in April 1986.

Former Soviet Management Approach

The management approach of the former Soviet Union reflects its centrally planned form of government, and land-based marine pollution is among the notorious environmental abuses engendered by that system (Vartanov 1991). It is unfair, however, to say that the Soviet Union did nothing to protect the Black Sea from land-based pollutants. An extensive monitoring system has long been in place, and much legislation concerning industrial practices, recreational uses, and other land-based sources in the Black Sea has been introduced and enacted over the past twenty years. Sadly, implementation and enforcement of these measures left much to be desired.

Information Gathering

The State Oceanographic Institute (SOI) monitored Black Sea pollution levels for many years, as part of the "Seas of the USSR" project, in cooperation with the All-Union Service of Monitoring and Environmental Pollution Control under the auspices of the USSR State Committee on Hydrometeorology (*Goskomgidromet*). The SOI succeeded in obtaining more than 3,000 water quality analyses annually from 242 monitoring stations. Criteria included levels of oil and chlorinated hydrocarbons, detergents, mercury, and phenols. Sanitary and Epidemiological Stations (SES) were responsible for water quality in recreational areas. The sampling frequency for each index was determined by the local SES, but was not to occur less than twice a month during the swimming season for any one index.

The system suffered from organizational and financial fragmentation, however, which resulted in information gaps between national and local services. Moreover, *Goskomgidromet* did not always fulfill its enforcement responsibilities with respect to the marine environment. As a result, its regional branches often did not receive data from the enterprises discharging wastes into the Black Sea, and thus they were unable to pass that information to the services responsible for water-quality mitigation efforts.

Monitoring activities have become even more fragmented since the dissolution of the USSR, and it will take much time and effort to establish cooperative and routine interactions among the cognizant authorities in the newly independent states.

Legislation and Regulation

The December 1972 Decision of the Central Committee of the Soviet Communist Party and the USSR Council of Ministers restricted the use of valuable recreational areas for agricultural and industrial purposes, but with little effect on Black Sea coastal areas (or the coastal areas of any other major seas in the former Soviet Union). The 1976 Decision of the Central Committee and the USSR Council of Ministers, "On Measures of Pollution Prevention in the Black and Asov Seas' Basin," charged the Councils of the Ukraine, Byelorussia, Georgia, and Moldavia, along with the interested Soviet ministries and agencies, with developing measures to reduce, by 1985, input into the Black Sea watershed of untreated wastewaters from the USSR. A number of decisions concerning the Black Sea were adopted by the republics as well. Unfortunately, all these measures were ineffective. For example, 80% of the enterprises affected by the 1976 decision were unable to make the planned investments, and those that were financially able could not find contractors to build the necessary installations.

Shortly before the Soviet collapse, the USSR and the republics attempted to create an infrastructure of environmental quality-control and management branches in the Black Sea region, in effect planting the first seeds of transfer to independent republics. This network included the State Committee on Hydrometerology (*Goskomgidromet*); republican representation from the Ukraine, Moldavia, and Georgia; the USSR Ministry of Nature (formerly *Goskompriroda*); republican environmental protection committees; and numerous corresponding local branches.

Agency fragmentation, however, was and continues to be a characteristic problem of management structures in the Black Sea region, with the separation between regional and central management offices causing severe disruption in the flow of information. This became especially clear during the Chernobyl disaster, when *Goskomgidromet* did not pass vital operational information to *Goskompriroda* in time for action.

The lack of effective implementation is the pervasive problem. Legally fixed, permissible concentrations of pollutants are routinely exceeded by tens to hundreds of times in the Black Sea. No integrated coastal zone management program has been established, and local authorities have lacked the clout and the technical skills to regulate or otherwise influence the activities of huge, state-sponsored industrial and agricultural enterprises.

Citizen organizations in the Black Sea region have had little influence over state environmental control structures, much less over polluters themselves. To a large extent, state bureaucracies continue to be driven by the old governing structure and by the entrenched industrial complex.

International Regional Cooperation: The Black Sea

Preparatory work began in 1969 on the International Convention on Protection and Preservation of the Marine Environment of the Black Sea with the support of the Soviet Union, Bulgaria, Romania, and Turkey. Following the withdrawal of Turkey, however, cooperation was suspended until 1988. The break-up of the Soviet Union further postponed development of the Convention until April 1992, when it was adopted in Bucharest by Bulgaria, Georgia, Romania, Russia, Turkey, and the Ukraine. It calls for the establishment of a Black Sea Commission in Istanbul. Of its three legal protocols, that dealing with land-based marine pollution receives the most attention.

After the collapse of the Soviet Union, Russia, Georgia, and the Ukraine made a number of very general declarations about the need to establish strong collaborative measures to preserve their common ecology and traditions in the Black Sea region. Although cooperation among their governments has deteriorated since then, the scientists and technicians who have long worked together at the local level continue to enjoy good communication and cooperation. This working relationship can serve as a foundation for a newly conceived program of regional cooperation in the management of Black Sea resources and the protection of the Black Sea ecosystem. In addition to providing sorely needed funds, international support for such a program can help the newly independent states fill important gaps in technical skills and establish the kinds of private companies and nonprofit organizations that have played such a vital role in environmental protection efforts elsewhere around the world.

CONCLUSIONS

Although the widespread problem of land-based marine pollution is largely confined to coastal waters within national jurisdictions, it does raise issues of international environmental security. This occurs not only where there is direct transboundary pollution, but also more indirectly where shared resource stocks are affected and through trade, tourism, and capital mobility.

The United States and Russia each borders a major, semi-enclosed regional sea that it shares with neighboring states. The Gulf of Mexico is economically and culturally important to the United States, which is a dominant economic and political presence in the region. Russia, as a successor state to the former Soviet Union, has a similar relationship to the Black Sea, with a few important distinctions. Since the collapse of the Soviet Union, Russia has lost its claim to vast areas of the Black Sea, including the traditionally vital ports of Sevastopol and Nikolayev, and it now ranks as only the third largest of the Black Sea coastal states (after Turkey and the Ukraine). Russia is therefore acutely concerned with maintaining social and political stability in the region, since the conduct of its marine and coastal activities depends upon international cooperation.

Both the U.S. coast of the Gulf of Mexico and the Black Sea coast of the former Soviet Union are subject to serious land-based marine pollution, though perhaps more is known about the U.S. case. Management approaches in the United States reflect the federal form of government and involve a complex of interactions among local, state, and national programs. Coordination is at a premium, and efforts to enhance this are still largely nascent. Citizen action is influential. In the Black Sea region, on the other hand, coastal zone management approaches and skills remain highly fragmented.

In both cases, new efforts are under way to coordinate regional control of land-based marine pollution at an international level. Constructive experience is available from efforts in other regions, but it remains to be seen how effectively those lessons will be transferred to the Wider Caribbean and Black Sea regions.

Perhaps the most striking conclusion of the IMEMO-WHOI case study is not surprising. It merely notes one symptom of the disintegration of the previous centralized state and economy. As Bystrova *et al.* have written in the full case study (Broadus and Vartanov [in press]):

> *The severe and very complex political and socioeconomic problems that now confront the post-Soviet republics mean that no significant environmental activity is likely to be undertaken in the region in the near future. Most republics simply cannot afford to participate, and so far no inter-republic connections have been forged to replace the hierarchical Soviet structure in which regional cooperation was managed, however inadequately.*

ACKNOWLEDGMENTS

We gratefully acknowledge the development of these case studies by our colleagues Anna Bystrova, Suzanne M. Demisch, Mark E. Eiswerth, Arthur G. Gaines, Kristina M. Gjerde, Yoshiaki Kaoru, Anna Korolenko, Elena N. Nikitina, Mary Schumacher, and Tom Tietenberg. We have drawn on their text in summarizing the findings. We also appreciate the ideas and suggestions of all the other participants in the joint IMEMO-WHOI project, "Environmental Security and the World Ocean:

Analytical Approaches and Shared Solutions," generously supported by the John D. and Catherine T. MacArthur Foundation. WHOI Research Assistants Mary Schumacher and Suzanne Demisch provided vital assistance. Thanks to the organizers of COSU III and our Italian hosts for the opportunity to share these thoughts with our international colleagues. WHOI Contribution No. 8321.

REFERENCES

AMS (American Management Systems, Inc.) and G.L. Webster. 1991. *Marine debris action plan for the Gulf of Mexico*. Report prepared for the Gulf of Mexico Program, Stennis Space Center, MS.

Broadus, J.M., S. Demisch, K. Gjerde, P. Haas, Y. Kaoru, G. Peet, S. Repetto, and A. Roginko. 1992. *The comparative assessment of regional international programs to control land based marine pollution*. Report prepared for the Office of International Activities, U.S. Environmental Protectio Agency.

Broadus, J.M., and R.V. Vartanov. 1991. The oceans and environmental security. *Oceanus* 34(2): 14–19.

Broadus, J.M., and R.V. Vartanov. eds. In press. *The oceans and environmental security: Shared U.S. and Russian perspectives*. Washington, DC: Island Press.

Environmental Protection Agency. 1990. *Progress in the National Estuary Program: Report to Congress*. EPA 503/9-90-005. Washington, DC: U.S. EPA.

Joint Group of Experts on the Scientific Aspects of Marine Pollution (GESAMP). 1990. *The state of the marine environment*. UNEP Regional Seas Reports and Studies no. 115. United Nations Environment Programme.

Meng, Q-n. 1987. *Land-based marine pollution: International law development*. London: Graham & Trotman, Kluwer Academic Publishers.

National Oceanic and Atmospheric Administration (NOAA), National Ocean Service Office of Ocean Research, Conservation, and Assessment. 1992. *National Status and Trends Program:Marine environmental quality*. Rockville, MD: NOAA.

United Nations Conference on Environment and Development (UNCED). 1992. Agenda 21:17.25b.

United Nations Environment Programme. 1990. *The strategy for the development of the Caribbe an Environment Programme*. CEP Technical Report No. 5. Kingston, Jamaica: UNEP Caribbean Environment Programme.

United Nations Secretary-General. 1989. *Law of the Sea: Protection and preservation of the marine environment. Report of the Secretary-General*. United Nations.

Vartanov, R. 1991. Greening of the USSR. *The Christian Science Monitor*, August 12.

5

Rio de la Plata Regional Maritime System: Potential for an Integrated Multifunction Offshore Complex

Ascensio C. Lara and Esteban L. Biondi
Catholic University of Argentina

Albina L. Lara
University del Salvador & Consejo Nacional de Investigaciones Cientificas y Tecnicas, Argentina

Joseph R. Vadus
National Oceanic and Atmospheric Administration, Washington, D.C.

ABSTRACT

A United States-Argentine Workshop held in December 1991 in Buenos Aires and in January 1993 in Miami, focused on present and emerging problems in coastal ocean space utilization in Argentina and the application of knowledge and experience gained in the United States, to provide a basis for collaborative studies. This paper focuses on the Rio de la Plata regional fluvial maritime system and the potential for economic growth and societal benefit by improvement and expansion of the marine infrastructure through an integrated, multifunction offshore complex. This paper addresses social and commercial aspects; regional ports and waterways; the bridge across the Rio de la Plata; an offshore airport; water pollution; hydraulics of the total system; and the need for an operations research (OR) Analysis-type of study to address the multinational interests and multi-disciplinary factors involved. All of these factors must be considered in order to make tradeoffs among technical, economical, and environmental factors to optimize an approach for an integrated complex that can provide the basis for strategic planning and ultimate implementation. The objective is to enhance opportunities for sustainable development and economic growth for the del Plata Basin.

INTRODUCTION

Worldwide population and industrial activity is rapidly expanding, especially in the coastal regions that typically have 50% of the population and supporting coastal infrastructure that is within 100 km. of the coast. This is especially true of major coastal cities that have grown in a relatively unplanned manner and are faced with major growth problems and stymied by the inefficiencies of the present infrastructure for sea, air, and land transportation, port and harbor facilities, and related marine infrastructure.

During the first United States-Argentine Workshop held in Buenos Aires in December 1991, some of the most significant social and economic needs and technical and environmental problems focused on the Rio de la Plata and its surrounding coastal environs (Figure 1).

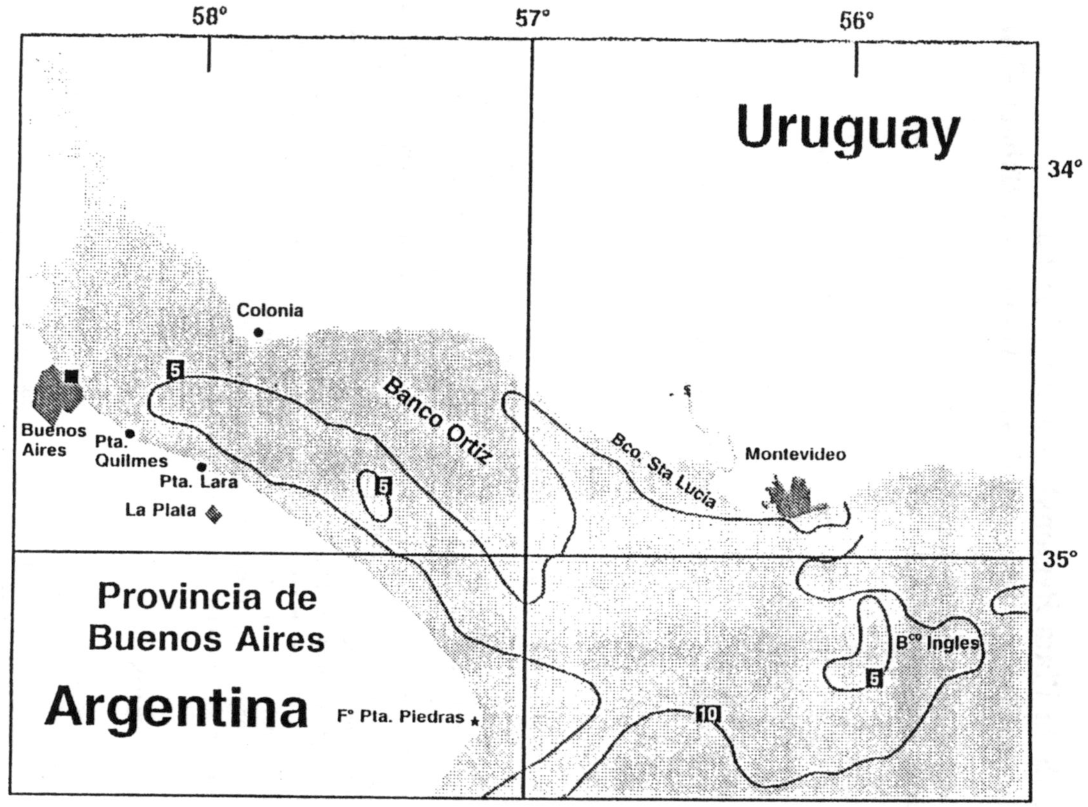

Figure 1. Rio de la Plata Region

The Rio de la Plata is an enormous body of water that is fed by the Uruguay and Paraguay-Parana rivers and provides the only outlet to the Atlantic Ocean. This body of water is larger than the Chesapeake Bay or Long Island Sound in the United States and Osaka Bay in Japan, and has many common problems and developing needs. The Paraguay-Parana and Uruguay rivers are part of the Rio del Plata fluvial maritime system. These rivers are comparable in size to the Mississippi River in the United States and provide navigable marine transportation outlets for handling cargo, fuel supplies etc., for the Rio de la Plata region and especially for international trade by marine shipping. The Rio de la Plata provides ocean access and marine-related infrastructure to: the seventh largest metropolitan city in the world, Buenos Aires, and to major coastal cities of Uruguay, mainly the capital city of Montevideo and the cities of Colonia and the region of Punta del Este. The population of the Rio de la Plata region is about 25 million.

The Rio de la Plata and the del Plata fluvial maritime system is part of the vast del Plata Basin. This region encompasses parts of Argentina, Uruguay, Brazil, Paraguay, and Bolivia in an area that is approximately 2,500 km. long and 2,200 km. wide. The population of this region is about 100 million. The Gross National Product (GNP) of the del Plata Basin is more than 50 % of the total GNP of the countries represented. The MERCOSUR, a common economic market of South America, recognizes the importance of this region. Its potential for future economic growth are obvious. This paper examines the Rio de la Plata Regional Maritime System of the del Plata Basin and its potential for economic growth and societal benefit by improvement and expansion of the marine infrastructure through the development of an integrated, multifunction, offshore complex.

PHYSICAL DESCRIPTION

In his book "Geografia y Unidad Argentina" ("Argentine Geography and Unity"), Federico A. Daus (1978) tells us that in the extensive South American littoral, the Rio de la Plata is the most important geographical feature from the San Roque Cape towards the south.

The Rio de la Plata is the most important outlet to the sea of the vast and rich Plata Basin that includes a significant area of five countries with the important, Paraguay-Parana Waterway (longer than the Mississippi River in the United States) and the Uruguay river, which are significant waterways. A projected Hidrovia Paraguay-Parana (HPP), 3,440 km. long, is strongly promoted.

The Rio de la Plata is a coastal body of water where current technology may be applied for infrastructure improvement and expansion (for transport purposes, exploitation of resources, water supply, purification of effluents, urban expansion, recreational resorts, etc.).

The Rio de la Plata is a river, but may be considered as a coastal ocean space, where technologies developed for ocean applications may be applied advantageously, given that its waters are not deep, are less agitated, less aggressive from a chemical point of view, and have coastal support nearby.

The Rio de la Plata, which is trumpet shaped, is 317 km. long; its mouth is 220 km. wide and has an area of more than 35,000 square kilometers, approximately the same area as the Netherlands. Its northern coast is 416 km. long and the southern coast is 393 km. long. It's main tributaries are the Parana and Uruguay international rivers. From these two rivers it gets about 16,000 and 23,000 cubic meters of water per second and about 60 million cubic meters of solid sediments yearly (Iglesias de Cuello 1985). The Parana Delta is formed where the Parana River meets the Rio de la Plata, about 300 km. before it reaches the Atlantic. It is basically formed by the amount of sediments brought down by the Parana River as well as by the tides of the Rio de la Plata. The sediments increase the size of the Delta by an advance estimated between 60 to 70 meters per year (Bonfils, 1962; Siragusa 1974). The Parana-Paraguay and Uruguay Rivers, the Parana Delta and the Rio de la Plata are strongly interrelated and form a particularly important fluvial system. The Rio de la Plata that flows into the ocean, is the ultimate outlet for the maritime transport of cargo carried along this fluvial system.

SOCIAL AND COMMERCIAL ASPECTS

The Rio de la Plata serves major cities, including Buenos Aires, La Plata, Montevideo, and Colonia and is an essential part of the del Plata fluvial maritime system. It serves a great part of Argentina and all the del Plata Basin. This Basin (Fraga 1893) has an area of 3,200,000 square km., is 2,500 km. long and 2,100 km wide. It comprises 18% of South America and is the fourth largest basin in the world after that of the Amazonas, the Congo, and the Mississippi.

The area percentage of the five countries that form the basin is as follows:

	% of Basin	% of its territory
Argentina	29.9	34
Bolivia	6.4	19
Brazil	45.9	17
Paraguay	13.0	100
Uruguay	4.7	80

The population (in millions) at the city, country, and regional levels are as follows:

Buenos Aires city and outskirts	11 M.
Argentine basin area approximately	25 M.
Five nations regional basin (del Plata basin)	100 M.

The GNP of the del Plata Basin is more than 50% of the GNP of the five countries that make up the Basin. In Argentina, in 1990, the cargo exported by maritime transport was 37 M tons (94% of total exports), consisting mainly of grains and by-products, oils, meats, leather, and fish. In 1990, the imports by sea were 6 M tons, which accounted for 73% of the total amount. In regard to its foreign trade, Argentina transports two-thirds of its grain by the Parana-de la Plata river systems. The industrial fluvial front, which is the most dynamic center of Argentina, is located on the right bank of the lower Parana and the Rio de la Plata. This front, situated between the cities of Rosario and La Plata, is becoming a "megalopolis".

The riverside of the Rio de la Plata and the quality of the natural system is the most important open space heritage of Buenos Aires city. This space includes recreation facilities and transportation facilities including: harbors, airports, bus and train terminals, and north-south access routes. The Rio de la Plata provides drinking water for the city. The metropolitan area of Buenos Aires is the most important region of Argentina especially in regard to population, economy, finances, and culture. "As is the case in all national big cities, its functions are highly complex and interdependent, some of them derived from the fact of being a capital city, seat of the government and, consequently, a control and decision-making center" (Gomez Insausti 1992).

The development of the Parana-Paraguay Hidrovia(HPP) will strongly increase the contribution of the influential areas of Uruguay, Brazil, Paraguay, and Bolivia, which will join Argentina's influential area, especially in regard to maritime transport via the Rio de la Plata.

The del Plata Basin is an extraordinary basin due to its geographical expanse, many tributaries, its large population, its agriculture and cattle value, and its economic development. Its area is slightly bigger than the combined areas of Italy, Austria, France, Spain and Portugal. Relative to Argentina, the Rio de la Plata is a strategic resource in itself because of its expansive fluvial system and rich soil area and also because it contains the 70% of Argentina's population in 34% of its area. A very high percentage of Argentine exports go out from the ports linked to this basin.

The most important waterways of the country belong to the del Plata basin, which comprises three main routes:

1. The Rio de la Plata route itself
2. The Parana-Paraguay route
3. The Uruguay route

More than the 80% of the Argentine ports are located along these routes. The great potential of these rivers is not fully used at present and is a major factor in strategic planning for the future.

REGIONAL PORTS AND WATERWAYS

The Rio de la Plata is the principal means of transport for most imports and exports coming to and from Argentina as well as several countries of the Plata Basin. A considerable part of the cargo of this fluvio-maritime system originates in the lower Parana River ports. The Rio de la Plata provides the major linkage and distribution function for the regional maritime transport system but, because of its shallow depth and great load of sediments, it lacks the optimal features needed for a major waterway. The main need is to optimize fluvio-maritime traffic dealing with the limited natural depths and harmoniously integrating the different modes of transport. To enter the Rio de la Plata, oceanic ships must use the dredged channels, the first one of which begins at "Ponton Rocalada." This channel, which is successively named "Punta Indio" and "Canal Intermedio", and the access channel to the port of Buenos Aires (North and South) are all dredged at 9.9 m. (32.5').

To enter the Parana River, two channels come out of the Canal Intermedio: the "Martin Garcia" Channel, which is connected to the Parana Guazu (9.7m.) and the "Emilio Mitre" Channel, linked to the Parana de las Palmas (9.7) m.. Up to the region of Rosario, the dredged depth of the Parana is kept at 8.5 m.. (Guia Prtuaria Argentina 1990) These channels presently lack adequate maintenance and navigation signals and, in general, they do not safely maintain the depths declared. The Paraguay-Parana Hidrovia project is now underway and is expected to make up a permanently operating, 3,440 km-long route with a minimum draft of 3 m. between Caceres Port (Brazil) and the outlet to the Rio de la Plata.

The total movement of goods of the port of Montevideo between 1984 and 1988 reached an average of 1.3 million tons annually (Guia Prtuaria Argentina 1990). The port of Buenos Aires handles an annual volume of 15 million tons. The port of Rosario is one of the most important cereal-exporting ports in Argentina. The facilities (with twenty-four berths) extend along 10 km on the right shore of the Parana River. The area of San Martin-San Lorenzo covers several terminals located about 30 km to the north of Rosario on the right shore of the River. There are seven modern and efficient terminals with good land access for loading of cereal, by-products, and vegetable oil as well as seven terminals for chemical products, oil, and other derived products (distillery of San Lorenzo - YPE). Up to this region, the determinant depth of the river is kept at about 8.5 m although, in several terminals, depth by the piers is greater. This group of terminals, added to the Port of Rosario, concentrates two thirds of Argentine cereal and by-products exports. All along the Parana from the Delta up to San Lorenzo, there are other port terminals of siderurgical and chemical industries and of industries related to the distillation of oil: Campana, San Pedro, San Nicolas, and Villa Constitucion. The port of Nueva Palmira (Uruguay) is located on the outlet of the Uruguay River. It is developing facilities to transfer agriculture bulk originated on the waterway and headed for a transoceanic destination, although the determinant depth there is that of the Martin Garcia Channel (6.7 m.).

BRIDGE ACROSS THE RIO DE LA PLATA

The 45 km span between Buenos Aires and Colonia is presently traversed by water transport (little ships, ferry, hydrofoils, hover craft), mainly for traffic of passengers. The proposed bridge would be for cargo and passengers between Argentina and Uruguay. Traffic at the different frontier passes shows substantial growth in the past twenty years. In 1971 there were 1,166,602 passages; in 1991, 3,281,970.

The idea of the bridge is very popular but needs market justification and attraction of private investment. In making marketing studies, it is important to include estimates for increased usage once the public becomes exposed to the convenience. There are many examples of bridges and roads, especially in the United States, that have exceeded their predicted capacities for traffic flow within a very short time.

The Bi-national Commission for the Buenos Aires-Colonia Bridge calculated that the levels of average daily traffic of 6,400, 8,900 and 11,700 could yield a rate of income return of 14%, 18% and 22% respectively. These figures are then compared with the number of vehicles which would go by land routes.

A prospective bridge should arch over the present channels, to provide more span and clearance above vessel traffic.

In order to take more advantage of the multiple potentialities of the Rio de la Plata, the design of a bridge must consider fluvio-maritime navigation, a potential offshore or hybrid airport, fishing, recreation, and more.

The study of an Integrated Multifunction Offshore Complex may stimulate the identification of complementary projects and , of course, avoid any possible chaotic development of this important resource; the territory covered by the Rio de la Plata. The bridge project appears to be suited for developing a government - industry partnership promoted by the governments of Argentina and Uruguay. Preliminary estimates for a bridge range from $800 million to $ 2 billion.

OFFSHORE AIRPORT

One of the early proposals for an offshore airport was offered in 1945 by Amancio Williams, a noted and respected Argentine Architect, who had great vision for the future advancement of his country. He developed a convincing rationale for such an endeavor. The airport should be located outside of the city but near enough for easy access. A land airport at the city's outskirts would be rapidly engulfed by the spreading population, resulting in complaints of noise, air pollution, and safety. A coastal city such as Buenos Aires has a shoreline limit and thus provides a natural buffer of space to prevent encroachment of an offshore airport. There are many economic and technical factors to be considered including: airport projected needs for capacity and support facilities; nearby ports and transportation facilities and routes; airport access by bridge, tunnel and rapid transit; geological considerations for seabed foundations and stability; water circulation and sediment transport; and meteorological conditions and environmental considerations. A. Williams proposed an elevated structure supporting air strips above the River which could be easily drained and radiant heated. His approach would not require cost for land or expropriation.

In the Argentina-United States Workshop held in Buenos Aires in December 1991, O.C. Grimaux prepared an internal paper supporting a proposal for an offshore airport and recommended design studies for a future AEROISLA and freeway access to the city. Studies would include noise predictions, seafloor structure analysis, and meteorological factors including fog and wind shear considerations. An internal paper presented by J. Vadus described considerable relevant work pertaining to construction of artificial islands, and also offshore airports such as Kansai and Haneda Airports in Japan and Honolulu Airport in the United States. Other factors that should be considered include the need to account for future supersonic aircraft in the next two decades that have the potential of reducing air travel times by factors of 2 to 3. Instead of twelve hours from New York, four to six hours would be possible. Such improvements could enhance future economic growth. Also the incorporation of an airport mini-city is desirable and co-location of an International Communications Trade Center with conference facilities, hotels, etc. An offshore airport could serve both Argentina and Uruguay. Hence, the idea of integrating the proposed bridge across the Rio de la Plata may be worth exploring.

WATER POLLUTION

The southern shore of the Rio de la Plata, between the Parana Delta and the city of La Plata, is a densely populated area with important industrial concentrations. This area, which includes the city of Buenos Aires, its suburbs, as well as La Plata, covers 4,100 sq. km. and has a population of more than 12 million.

The industrial waste of metallurgical plants, tanneries, cold-storage plants, dairy industries, paper industries, shipyards, chemical and pharmaceutical plants, oil refineries, etc., does not usually receive any treatment and is poured into the River or into its tributaries in the whole region. Less than 10% of the industries have plants of effluent treatment (Pascuma and Guaresti 1991). To these sources of pollution, add sewage discharge (sometimes combined with industrial waste) and pollution due to oil spills and leaks. Some highly polluted rivers and streams which are affluents to the Rio de la Plata in this area are: Reconquista, Lujan, Mantanza-Riachuelo, Sarandi, Rio Santiago, etc.

The effluents of 7,300 factories in the area are estimated to be about 300,000 tons a year of hazardous solid waste; 250,000 tons of toxic sludges; 500,000 tons of dilute solvents; and 500,000 tons of effluents with heavy metals (Simposio Latin Americano de Ambiente y Desarrolo 1990). These discharges go into the Rio de la Plata either directly or through its affluents.

The present state of environmental conditions is serious on account of the degree of pollution in shallow watercourses and in the phreatic layer. Because of its huge volume and surface, the Rio de la Plata acts as a great purifier and keeps good general conditions in its main bed (CARP 1989). However, there are punctual sources of pollution in the outlets of watercourses, in sewage discharges and in ports, and this entire riverside presents signs of pollution due to the River's flow of currents.

Opposite near shore the city of Buenos Aires, the water shows high figures of microbiological pollution, of biochemical oxygen demand (BOD of 50mg/1), and of concentrations of metals such as lead, chrome, and cadmium, etc. In rivers such as the Reconquista and the Riachuelo, these figures become much more critical (BOD of 200 and 130 respectively) (Bauer and Ballester 1991).

There are mathematical models of currents in the Rio de la Plata which can serve as a basis to specific models of pollution control. Studies were also carried out on the dispersion of pollutants in the sewage discharge of Berazategui (and on their influences upon the water intake in Bernal, situated upstream), by a request of "Obras Sanitaries de la Nacion" (the National Water Service Company) to INCYTH.

Three campaigns were developed during 1983-84, 84-85 and 89-90, taking water samples at 500 m, 1000 m and 1500 m from the shoreline. As far as the problem of pollution is concerned, needs are mainly related to the control of the discharge of pollutants in watercourses by the authorities and to the studies of monitoring and following levels of pollution through a program which should include samples and models of simulation. Of course, every new project – like the ones mentioned – implies studies of its environmental impact.

HYDRAULICS OF THE TOTAL SYSTEM

The analysis of the different prospective projects included in an Integrated Large Scale Multifunction Offshore Complex require hydrodynamic studies of the site where structures and operations would be developed.

Basic information exists on the hydrodynamic behavior of the Rio de la Plata as a whole, including outfalls of its tributaries, boundary conditions in the region of the Parana Delta, and in the mouth with its interaction with the ocean. The hydraulic regimen is complex and includes tidal currents, wind driven currents, very shallow waters, and pile-up during the "sudestada" (strong and sustained winds from the south east). Data and information is available on tidal effects, water levels, currents, and sediment transport, etc. Mathematical models and in-situ measurements were used for studies of variables such as circulation and wave climate. These hydraulic studies were conducted mainly by the following institutions: Servicio de Hidrografia Naval (SHN-Naval Hydrographic Service), Applied Hydraulics Laboratory of INCYTH (National Institute of Water Science and Technology) and GC Cespedes Laboratory of the Hydraulics Dept. of University of La Plata, from Argentina; and Servicio de Oceanografia, Hidrografia y Meteorologia de la Armada (Oceanographic, Hydrographic and Meteorological Service of the Navy - SOHMA) from Uruguay.

Data and information related to dredging of navigation channels, diffusion of sewage outfalls, etc., were also available for specific locations.

The study of an integrated complex includes the analysis of deepening of channels, transhipment facilities, artificial island ports, an offshore airport, and a bridge across the river. It will demand separate studies of structures, sediment transport and stability, and contamination, etc. All of them need physical and mathematical modeling, environmental measurements, verification tests, and periodic monitoring. Moreover, the analysis of sedimentation processes may result in the creation of man-induced, nature-built islands.

OPERATIONS RESEARCH ANALYSES

Preceding sections covered the overall physical aspects of the Rio de la Plata fluvio maritime system; problems of population growth and environmental degradation; social and economic needs; and potential for future development. Also covered were various proposals for infrastructure improvement of ports and waterways and for new marine facilities such as: a bridge, offshore airport, lightering facilities, a deep water port, and facilities for marine recreation. A major driving force is the importance of this region for sustainable development and economic growth of the MERCOSUR, a common market of South America. Another major, long range consideration with great potential for future growth is the proposed linking of the Pacific Ocean to the Parana River. This includes an improved access way through the Andes. Based on the above, multinational interests, multidisciplinary factors, and multiproposals, it appears obvious that an operation research (OR) - Type analysis which uses an integrated system approach to weigh all the dominant items, is needed to develop a balanced perspective that could provide the basis for strategic planning by the respective nations involved. It is also obvious that an OR analysis will make the necessary tradeoffs, assess risks, identify options, establish priorities, and provide a time-line for future implementation. This would be invaluable for decision makers to add political and budget availability factors before proceeding along a preferred path for implementation. Because of the complexity of the problem it may be desirable to conduct an OR-Type analysis in phases, whereby the first phase provides an overview defining the major problems and needs, priorities, options and recommendations. The second phase would benefit greatly from the knowledge gained in phase 1, enabling researchers to more sharply and define separate ancillary or supportive studies that are needed, e.g., in hydrodynamic and environmental modeling, port and waterway improvement, as well as focused design studies for new marine facilities previously mentioned. The second phase OR analysis would proceed in more detail by analyzing major technical, environmental, and economic factors, taking advantage of any separate studies conducted in parallel.

An overall OR-Type analysis would help business leaders and financial institutions to invest in present and planned projects and prepare for future projects.

The major focus or heart of the OR-Type analysis pertains to examining an approach for an integrated, multifunction, offshore complex that is described in the next section.

INTEGRATED OFFSHORE MULTIFUNCTION OFFSHORE COMPLEX (15)

Each of the marine facilities discussed herein represents a major research and design effort, including specific alternatives. Of course, each of these facilities would have varying degrees of interaction or conflict with the other. An optimum solution for one may be detrimental to another. Therefore, it is difficult to predict interrelated impacts unless they are based on an integrated OR-Type of analysis. Before suggesting examples of integrated approaches that would be considered in OR-Type analyses, a partial listing of needs, problems, and concepts for facilities in various stages of investigation are listed below.

1. Need for improving navigation access and control in channels, ports, and harbors; and addressing silting and dredging problems
2. Lightering facilities, in the form of an artificial island, for transferring river borne cargo to deeper draft vessels for ocean transport and export; and the reverse process for import
3. Deep water, offshore terminals to enable large vessel access into the Rio de la Plata
4. Improved port and harbor facilities and traffic control in the Rio de la Plata
5. A bridge spanning the Rio de la Plata to link e.g., Colonia, Uruguay with Buenos Aires
6. An offshore airport near Buenos Aires which would handle increased traffic, provide better safety, reduce noise levels for urban residents, and accommodate future supersonic aircraft
7. Need to address the present marine pollution problems and controls, and the environmental impact of introducing new facilities in the marine environment

One integrated approach would be to create an artificial island about 10 km offshore from Buenos Aires, just west of the junction where the deep water channel splits into two waterways; one heading toward the Parana Delta and the other feeding the Buenos Aires side . This island could serve several functions and could be the hub of a multifunction, transportation facility. One part of the island could accept deep water vessels and provide for lightering of cargo to barges. Docking space for cruise ships could also be provided. Trade off studies may reveal that a lightering facility for the Hydrovia waterway may be better located at the Parana Delta for greater efficiency and in order not to overburden the main artificial island because of its greater potential for other functions. The Central part of the island could serve as an airport and a portion of the runways could be projected over water on elevated pilings. The air traffic approach would be lengthwise along the flow of the Rio de la Plata. The artificial island could also serve as a way station for a bridge linking Buenos Aires with Colonia. Because of the airport, this portion of the bridge link could be a tunnel on the seabed under the airport. This would avoid any major obstruction in the landing approaches. The bridge could have three major segments: a 10 km bridge form Buenos Aires to the artificial island with sufficient span to allow for waterway access; a 2 km bridge tunnel would cross through the artificial island but under the airport; and a 30 km span from the artificial island to Colonia. A small portion of the 30 km span would be elevated to allow for vessel passage along the canal waterway. Since the Rio de la Plata is very shallow, the foundation pilings and bridge supporting structure would not be very massive and not require a suspension-type design, except perhaps over a channel waterway. The bridge should be dual layered to accommodate high-speed rail transit such as MAGLEV on the lower level and auto traffic on the upper. This approach has resulted in a central transportation hub for air, rail, ships, and auto traffic. Another possibility for the island is to co-locate a small airport city in addition to a main air terminal complex. The airport city would provide hotels, restaurants, and condominiums and could include a centralized communications and trade center to handle international banking and finances as well as a center for international conferences.

Advantages of an Integrated Complex

An integrated approach as just described, has several major advantages:

1. A large artificial island common to several functions.
2. A multifunction transportation hub for greater intercarrier transport of people and cargo including a high speed access to the airport from both sides (Vadus, Kondo, and Okamua 1989).
3. The artificial island and airport is the only stop on the bridge, shortening the span on both sides.
4. An integrated International Trade and Communications Center with a supporting airport city with multimode transport capabilities. (Vadus, Kondo, and Okamua 1989).
5. International financial support could be raised by inviting potential tenants to provide for start-up capital and long-term commitments which would greatly appeal to further private investment.
6. An integrated complex would draw considerable bridge traffic which would produce payback capital through tolls and other property lease arrangements.
7. The integrated complex would be of social and economic interest not only to Argentina and Uruguay but to Brazil, Paraguay, and Bolivia and served by the vast linking river waterways. Hence, the MERCOSUR, a common market of South America, would be a great beneficiary.

There are many more variations for this integrated offshore complex (Vadus and Kondo 1990; Kondo and Vadus 1991). Another variation could address two separate complexes. Trade-offs and comparisons can be made via mathematical parametric analyses and application of statistical methods and game theory, all of which are tools used in an overall OR-Type analysis.

CONCLUSIONS

The Rio de la Plata Basin has great potential for implementing an integrated, multifunction, offshore complex or complexes based on the findings of operations research type analyses and supporting studies. The importance of this Basin for the social and economic benefit of this vital region are clearly apparent. The GNP of the del Plata basin is more than 50% of the GNP of the five countries that make up the basin. The proposed, integrated complex and related improvements and coastal developments would have many other spin-offs that would greatly enhance the present and future social and economic outlook of the region. The individual countries and their alliance in the MERCOSUR would greatly benefit through: greater international visibility; greater opportunities for international investment; facilities for increased transportation, communications, and international trade; and increased potential for social and economic growth. The MERCOSUR would have the opportunity to grow in stature as a world-wide economic block and begin to position itself as a major, sustainable growth area in the 21st century.

ACKNOWLEDGMENTS

The authors would like to acknowledge the valuable contributions received from: Gloria Vadus on societal and environmental aspects; Stephen Morrison for preparation of the manuscript; and Augustin Biondi for his inspiration about the future of Argentina.

Acknowledgments are also given to the participants of the United States-Argentina workshop for their valuable suggestions. Special recognition goes to the proponents and/or potential research team leaders that are interested in conducting Operations Research-type analyses of an integrated approach for the proposed offshore complex, including: A. Lara, R. Humar, A. Lonardi, A. Yung, R. Lopardo, J. Alverez and C. Bauer from the Argentine side; and R. Abel, N. Caplan, S. Kikuchi, M. Bruno, J. Flipse, H. Haar, and J. Trefry from the United States.

REFERENCES

Bauer, C., and R. Ballester. 1991. Workshop USA-Argentina Coastal Ocean Space Utilization: Internal Report.

Bonfils, Constante "Los Suelos del Delta del Rio Parana" - Factores Generadores Clasificacion y uso Revista de Investigaciones Agricolas No. 16, 1962

CARP, "Estudio para la Evaluacion de la Contaminacion del Rio de la Plata, Informe de Avance", Comision Administradora del Rio de la Plata, 1989.

Daus, F.A., "Geografia y Unidad Argentina", Ed. Instituto de Publicaciones 'Navales, Buenos Aires, 1978.

Echichurri, J.L. Giudice, y N. Prudkin, "La Ciudad y el Rio", "Medio Ambiente y Urbanizacion", IIED-AL, Ano 9, p. 37-54, 1991.

Fraga, Jorge A., "La Argentina y el Atantico Sur", Ed. Instituto de Publicaciones Navales, 1983.

Gomez Insausti, R., "La Region Metropolitana de Buenos Aires. Una Desproporcionada Concentracion", "La Argentina. Geografia Regional y los Marcos Regionales", A. Roccatagliata, Ed. Planeta, p. 453-476, 1992.

"Guia Portuaria Argentina", GPA Ed., 1990.

Iglesias de Cuello, Alicia, "El 'Mar Dulce' de Solis", "Atlas Total de la Republica Argentina", 1985

Kondo, T., and Vadus, J.R.. 1991. Ocean space utilization: Technology trends and future concepts. In *Proceedings of the international coastal ocean space utilization symposium II*, Long Beach, California.

Pescuma, A. y M.E. Guaresti, "Gran Buenos Aires: Contaminacion y Sancamiento", "Medio Ambiente y Urbanizacion", IIED-AL, Ano 9. 1991

Santamaria, Pedro, "Situation de las Vias Navegables", "El Deterioro del Ambiente en la Argentina", FECIC, 1988.

Simposio Latinoamericano de Ambiente y Desarrollo, Bariloche, Argentina, 1990.

Siragusa, Alfredo, personal interview, 1974.

Vadus, J.R. Integrated large scale, multi-function offshore complex: Concept analysis, modeling and technical and environmental assessment. Paper for USA-Argentina Workshop on Coastal Ocean Space Utilization, February 1992.

Vadus, J.R., T. Kondo, and K. Okamura. 1989. Triportopolis: A concept for an ocean-based multimode transportation and communication complex. In Coastal ocean space utilization, 315-21. Elsevier Science Publishing, Inc., New York, NY.

Vadus, J.R., T. Kondo, "Maritime Media Port Complex: Conceptual Design." In: *Proceedings of the Fourth Pacific Congress on Marine Science and Technology*; July 1990; Tokyo, Japan. pp. 358-361.

6

The Special Case of the Gulf of Aqaba

Mohammad I. Wahbeh
Director, Marine Science Center
Aqaba, Jordan

INTRODUCTION

It must be realized that the sea is not a sink of unlimited capacity for all human refuse (UNESCO 1982). The marine system and the organisms within it are constantly adjusting to stresses and variations. Such elasticity of the ecosystem allows for a certain capacity to absorb stress and tolerate management. The maintenance of such capacity requires the determination of targets most at risk as a result of the input of pollutants and the adoption of appropriate protection standards.

The marine environment is a complex system controlled by a variety of physical, chemical, and biological processes (UNESCO 1977). The understanding of these processes is a prerequisite of any consideration of man's past or future impact on the sea. For example, a pollutant introduced into the sea at one location will be transported to other areas where it may cause harmful effects to man or the environment. Interpretation of limited data on the pollutant's initial spread or prediction of its long-term transport, is only possible if the processes of its transport are reasonably well understood.

Of particular interest is the description of the processes and their spatial and temporal scales. Along the coast, where man's impact is greatest, inputs of pollutants influence many aspects of the environment, physical processes are relatively fast and, in general, biological activity is high. Some of the introduced material is transported to the deep sea where it is added to that which has crossed the air-sea interface. Within the sea, material is swept by currents to other regions while being mixed into nearby water masses. Much will reach the sea floor, sometimes relatively quickly by adhering to sinking particulate matter. Thus, in the sea's interior and at its boundaries with land, atmosphere, and deep sediment, material introduced into the marine environment is subjected to various physico-chemical and biological processes that determine its ultimate fate.

Though the protection of human life from the effects of environmental pollution is a major concern, there are equal concerns of the ecological impact of pollution and aesthetic considerations (UNESCO 1982). One of the most important effects of pollutants in the marine environment is the ecological imbalance created between organisms and their environment. This effect is often insidious and long-term and may lead to large changes in populations of commercially important marine species.

All pollutants exhibit toxicity to marine organisms in various degrees. Some may be acutely toxic in low concentrations, whereas others may have a slow effect, and it is these that are most important from a long-term, ecological point. Even materials not characterized as toxic may have pollutional effects on marine organisms. Suspended particulate matter may retard light penetration and thereby inhibit photosynthesis. Excess nutrients may cause dense algal blooming which adversely affect higher forms of life such as corals. The input of heat with cooling waters not only affects organisms, but alters conditions in which they live.

POLLUTION IN THE GULF OF AQABA

The situation in the Gulf of Aqaba has changed rapidly in recent years, as intense commercial and industrial development has taken place along the coast. This expanded development has triggered a population increase of more than tenfold in the last two decades. These developments are now at a stage when many signs of impact and degradation are becoming common. This calls for urgent and effective conservation and environmental management measures to protect and conserve the unique resources of the Gulf.

Perhaps the most serious aspect of the current expansion of urbanization is the strong tendency for this development to be in a strip-like fashion along the coast (Edwards and Head 1987). This inevitably leads to the slow degradation of the littoral and sublittoral environment. It would therefore by more desirable to direct development away from the coast, protecting the shore and marine habitats for recreation, tourism, and fisheries.

CHARACTERISTICS OF THE GULF

As a semi-isolated basin, the Gulf of Aqaba, like the Red Sea, is especially vulnerable to the effects of pollution. The Gulf of Aqaba is an evaporative mechanism and its circulation is dependent upon the exchange that occurs at the air-sea interface (Hulings 1989). If the steady state of exchange is interfered with, it is reasonable to expect changes in the patterns of circulation. Equally important will be the effect on local economics and living conditions through alteration of the climate which, in turn, is dependent on the evaporative mechanisms.

The fact that the flow of bottom water from The Gulf to the Red Sea is very reduced and the exchange thus limited, is another cause for concern in terms of pollution. This implies that the potential for dispersion of pollutants will be reduced and that pollutants reaching the deeper parts of The Gulf would accumulate with time and, in turn, have an eventual effect on the shallower and surface waters. One effect that would be expected is the reduction of oxygen content of the bottom waters, perhaps to the degree of becoming completely deoxygenated and resulting in reducing conditions and the production of hydrogen sulphide. Because the oxygenation of deeper water depends on surface and shallow water oxygenation processes, including air-sea exchange and photosynthesis, any alteration in these processes would lead to deoxygenation of deeper waters.

The Gulf of Aqaba is characterized by rich and diverse habitats which comprise extensive sand and rocky outcrops, shallow coastal lagoons, seagrass beds, and fringing coral reefs (Edwards and Head 1987). These habitats together form a rich and productive coastal ecosystem on which many fisheries depend; which are of significant scientific and educational value; in which coral reefs produce a coastal barrier against wave action and erosion; and which are of potential touristic attraction.

Of all marine habitats, coral reefs are the most complex and probably the most sensitive to pollution. This is due, in part and to a large extent, to their being self-supporting (Hulings 1989). Any factor causing a reduction of light penetration such as a continuous, thick film of oil and increased amounts of suspended matter (Table 1) or sediment, would interrupt the photosynthetic cycle of zooxanthellae, leading to secondary effects.

A constant decrease in the density of living coral in the northern Gulf of Aqaba has been noted (Wahbeh and Hülings 1987-1989), and this decrease has closely paralleled industrial development. Specifically, the decrease has been attributed to the northerly winds which concentrate pollutants in the south. Not only are corals affected, but so to the associated invertebrates and fish fauna. Many of the fish leave the reefs because of lack of food and shelter.

Table 1

Suspended matter (mg/1) in seawater from the areas of the Marine Science Station (MSS), the Phosphate Loading Berth (PBL) and the Jordan Fertilizers Industry (JFI)

Date		1991			1992	
	MSS	PLB	JFI	MSS	PLB	JFI
Jan	0.63	1.05	0.33	0.39	7.39	0.82
Feb	0.29	5.33	0.57	0.49	-	0.43
Mar	0.57	2.22	0.95	0.55	4.83	0.54
Apr	0.40	1.48	0.42	0.60	4.17	0.43
May	0.80	1.18	0.83	0.93	4.90	0.68
Jun	2.17	0.93	0.70	0.92	6.42	0.61
Jul	2.47	3.44	1.09	0.59	0.90	0.72
Aug	1.10	1.37	1.07	0.37	1.03	1.70
Sep	1.20	1.40	0.57	2.75	1.93	2.10
Oct	0.57	0.83	0.84	0.34	1.88	0.78
Nov	1.31	1.55	1.20	0.15	1.81	0.32
Dec	1.26	1.06	1.04	0.58	0.61	0.58
Mean±SE	1.06±0.20	1.32±0.38	0.80±0.08	0.72±0.20	3.26±0.72	0.81±0.16
ANOVA		$F2,33 = 4.48^{NS}$			$F2,32 = 11.65^{NS}$	

(Wahbeh, unpublished data)

SOURCES OF POLLUTION AND EFFECTS ON RESOURCE

Sources of pollution and human impact in the Gulf may be divided into the following general categories.

Urbanization, Tourism, and Shipping

A major input from conurbation to the sea is sewage. Up to June 1987, untreated sewage was discharged directly into the Gulf in the area of the port in Aqaba. Subsequently and at present, the sewage is treated in a plant located inland. There are areas in the north, however, where small amounts of untreated sewage and waste water are periodically discharged. The problems here are nutrient enrichment, possibly leading to eutrophication; increased suspended matter (Table 2) resulting in reduced light penetration; and human health problems associated with coliform bacteria on recreational beaches.

Table 2

Organic Content (%) in Suspended Matter Collected from the Marine Science Station (MSS), the Phosphate Loading Berth (PLB) and the Jordan Fertilizers Industry (JFI) areas

Date	1991			1992		
	MSS	PLB	JFI	MSS	PLB	JFI
Jan	38.1	15.2	51.5	33.3	25.0	41.7
Feb	41.2	3.2	25.2	51.0	-	95.6
Mar	14.8	32.9	26.1	36.4	10.8	37.2
Apr	11.2	20.7	29.4	40.0	55.3	45.0
May	43.7	24.7	33.7	45.2	14.9	38.9
Jun	31.3	32.9	25.7	40.5	59.2	46.9
Jul	24.7	72.2	24.2	37.1	68.8	36.8
Aug	16.1	18.2	18.6	24.4	25.2	22.8
Sep	33.8	28.7	47.3	43.3	32.1	35.4
Oct	42.1	27.7	39.5	76.0	30.6	46.9
Nov	41.9	40.6	44.6	53.3	48.6	40.1
Dec	57.9	62.2	61.4	24.1	74.2	43.6
Mean±SE	33.1±4.0	31.6±5.6	35.6±3.8	42.1±4.0	40.4±6.6	41.2±2.5
ANOVA		$F_{2,33} = 0.20^{NS}$			$F_{2,32} = 0.07^{NS}$	

(Wahbeh, unpublished data)

In addition to sewage, considerable amounts of garbage (Table 3) enter the sea from urban and recreational areas and from ship traffic. While these are a great nuisance and aesthetically unpleasant, they do little biological damage.

Coastal construction also affects shore zones and nearshore waters due to increased sediment loading (Table 4) and covering of sea areas. Coral reefs, and to a lesser extent seagrass beds, are not able to tolerate heavy sedimentation generated by coastal infilling and construction. Beside the direct destruction of corals and intertidal habitats, the construction of solid jetties causes the area down current to become stagnant. As a result of this, marine life succumbs to sedimentation and raised temperatures that occur in such enclosed shallow areas (Edwards and Head 1987)

Tourism is important to Jordan because of its need for foreign exchange earnings. And because the coral reefs have an international reputation that attracts divers from all over the world, the number of tourists coming to Aqaba for recreation is increasing rapidly. Dive tourism is, however, restricted in Jordan due to its very small coastline, much of which is occupied by essential industrial developments.

Table 3

Quantities of Garbage Collected from the 0.5km Beach of the Marine Science Station During the Period September 1990 and April 1991

Date	Plastic Kg	Metal Kg	Wood m^3	Miscel Kg
Sep. 1990	66.0	4.0	-	23.0
Oct.	21.0	-	0.5	6.5
Nov.	33.5	5.5	0.3	7.0
Dec.	57.0	2.0	0.3	6.0
Jan. 1991	44.0	5.0	0.2	5.0
Feb.	48.0	5.0	-	7.0
Mar.	60.0	4.0	0.5	60.0
Apr.	28.0	4.0	0.5	8.0
Mean	44.7	4.2	0.4	15.3

(Wahbeh, unpublished data)

Tourism may, however, produce both direct and indirect effects on coastal resources. The uncontrolled use of these resources can, in itself, result in serious impacts (Edwards and Head 1987). Extensive collection of corals, shells, and other marine animals; spearfishing; coral damage by swimmers and anchors; and proliferation of waste garbage cause the loss of diversity and degradation of reefs.

Table 4

Sedimentation Rate (mg/cm2/day), Organic Matter (%) in Sediment and Chlorophyll a (mg/kg) in Sediment Collected in Traps

Date		1991			1992	
	SR	OM	Chl.a	SR	OM	Chl.a
Jan-Feb	0.7	8.50	90.0	3.4	3.4	19.2
Feb-Mar	3.6	16.30	10.0	4.0	4.5	25.4
Mar-Apr	1.4	7.40	52.6	2.8	6.6	37.6
Apr-May	1.2	7.90	41.1	0.8	23.3	164.8
Jun-Jul	1.0	8.20	40.0	1.0	49.7	28.3
Jul-Aug	0.8	8.00	49.0	0.4	30.3	1.6
Aug-Sep	0.8	7.50	48.4	1.4	52.8	3.1
Sep-Oct	0.6	7.0	68.2	0.8	9.1	2.6
Oct-Nov	0.6	18.7	53.7	0.9	22.3	2.9
Nov-Dec	0.8	6.7	84.4	0.4	7.1	150.5
Mean±SE	1.2±0.3	9.6±1.3	53.7±7.4	1.6±0.4	20.9±4.84	3.6±19.5

(Wahbeh, unpublication data)

Oil Pollution

Oil pollution in Aqaba is still a minor problem (Ormond 1978). Most persistent and chronic is the oil from cargo vessels anchored in the ports. Though Jordanian law prohibits discharge of oil into the sea, the cargo vessels discharge bilge water which, in most cases, contains variable types and quantities of processed oil. While such discharge is minimal, because it is done on a continuous basis, the effects on the marine habitats, as well as on physical and chemical parameters may be cumulatively important.

Industrial Inputs to Coastal Waters

Phosphate is a major type of pollution in Aqaba. At present, more than eight million tons of phosphate per year is exported and this is expected to increase (Figure 1). During the loading, approximately 1% is lost as dust into the atmosphere. Much of this dust settles into the sea and the solubility of phosphorite in seawater has been found to be 20-56 ug/1, depending on the particle size. This would contribute to the water-borne phosphate concentration with two possibilities; eventual eutrophication (Table 5) and "phosphate poisoning". The latter is more serious with the result being the prevention of calcification of corals, molluscs, and other organisms. Death of corals resulting from phosphate dust has been related to stress caused by reduced light intensity and increased sediment load (Wahbeh and Hulings 1987-1989).

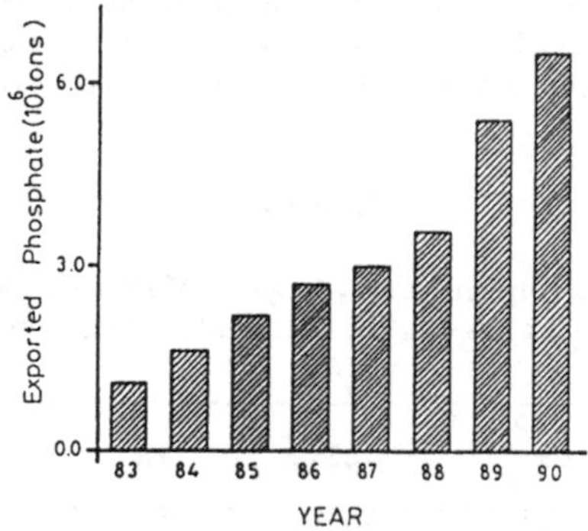

Figure 1. Amounts of exported phosphate from Aqaba during 1983-1990

Though not a serious problem at present, thermal pollution may become one in the future. Considering the fact that Aqaba marine organisms live within a few degrees of their upper thermal limits, an increase in the temperature of the natural environment of one to a few degrees could have drastic effects on these organisms. As significant would be the effect on the evaporative mechanisms discussed previously.

Sources of heated effluents to coastal marine water in Aqaba are directly related to power generation and fertilizers industries. Seventy thousand cubic meters of heated water is being discharged per hour in the form of jets 180 m offshore and at a depth of 20 m. The temperature of the discharged water at the point of the jet is 3°C higher than that of the surrounding water. The fate of these effluents depends upon the rate of thermal emission and the characteristics of the receiving waters, such as turbulence, stratification, and surface cooling. To be noted is the stratification of Aqaba waters during summer, with the difference in temperature between the upper and lower layers not exceeding 3°C.

Table 5

Chlorophyll a (mg/1) in Seawater from the Areas of the Marine Science Station (MSS), the Phosphate Loading Berth (PLF) and the Jordan Fertilizer Industry (JFI)

Date		1991			1992	
	MSS	PLB	JFI	MSS	PLB	JFI
Jan	0.29	0.23	0.28	0.16	0.22	0.22
Feb	0.40	0.35	0.29	0.19	0.28	0.14
Mar	0.41	0.56	0.61	0.30	0.18	0.18
Apr	0.08	0.19	0.36	0.31	1.01	0.44
May	0.20	0.18	0.05	0.77	1.03	1.09
Jun	0.22	0.09	0.15	0.25	0.49	0.30
Jul	0.07	0.05	0.07	0.31	0.36	0.42
Aug	0.22	0.35	0.14	0.13	0.15	0.16
Sep	0.14	0.07	nd	0.18	0.26	0.09
Oct	0.18	0.18	0.05	0.30	0.27	0.23
Nov	0.22	0.31	0.02	0.18	0.26	0.17
Dec	0.21	0.38	0.16	0.22	0.24	0.21
Mean±SE	0.22±0.03	0.25±0.04	0.20±0.05	0.28±0.05	0.40±0.09	0.03±0.08
ANOVA		$F_{2,33}$= 1.03^{NS}			$F_{2,32}$= 0.74^{NS}	

(Wahbeh, unpublished data)

The heated effluents contain suspended particles, residual oxidant products due to biocide treatment, and metal corrosion products. The chemical form used in Aqaba industry and power plant is chlorine. The reaction of this biocide with seawater and some of its constituents results in the production of toxic derivatives which are harmful to marine organisms such as algae and mussels. It is important to identify the effects of these contaminants on the marine environment.

Pollution by metals arises from various land-based operations such as export of klinker and fertilizers industry (Table 6) in Aqaba, as well as desalination of seawater in Elat. Some metals enter the sea via the atmosphere as dust particles carried by the northerly, prevailing winds from Wadi Araba and heavy traffic emissions. Almost all metals are bioaccumulated (Table 7) in one or more components of the marine food chain. They are the most persistent of all substances in the environment. It is important to note that certain metals can combine with organic substances, producing highly toxic complexes.

It is important to emphasize the individual as well as the syner-getic effect of pollutants. In addition, though the effect of an individual or a group of pollutants may be obvious biologically, the effects on the physical and chemical environment though not as obvious, may be of great importance. The point is that the totality of the marine environment - its physics, chemistry, and biology - has to be considered as one unit, also taking into account the interaction of the atmosphere and the terrestrial environment with that of the marine. Many pollutants reach the marine environment via the atmosphere and land.

Table 6

Trace Metals (ppm) in Sediment from the Areas of the Marine Science Station (MSS), the Port Area (PA) and the Jordan Fertilizers Industry (JFI)

Metal	MSS	PA	JFI	ANOVA
Cd	nd	nd	nd	
Co	2.47±0.10	3.25±0.71	2.19±0.04	$F_{2,8}=1.40^{ns}$
Cr	3.10±0.79	5.81±2.15	1.52±0.16	$F_{2,8}=2.19^{ns}$
Cu	3.03±0.54	5.13±1.12	2.00±0.14	$F_{2,8}=3.86^{ns}$
Ni	3.19±0.36	4.03±0.99	2.25±0.14	$F_{2,8}=1.63^{ns}$
Pb	8.44±3.47	13.97±3.74	2.71±0.11	$F_{2,8}=5.59^{ns}$

(Wahbeh, unpublished data)

Table 7

Trace Metals (ppm) in the Tissue of the Mussel, Modulus Auriculatus Collected from Three Different Areas

Metal	MSS	PA	JFI	ANOVA
Cd	0.08±	0.08±	0.25±	
Co	1.50±0.41	2.10±0.57	1.67±0.92	0.83^{ns}
Cr	5.95±2.50	5.15±2.83	5.67±0.67	0.02^{ns}
Cu	4.40±1.37	9.35±3.36	6.75±2.31	1.08^{ns}
Mn	55.33±11.20	43.30±9.87	8.75±3.75	3.45^{ns}
Ni	3.85±0.24	3.35±0.16	4.83±0.52	3.64^{ns}
Pb	2.05±0.46	14.00±3.84	1.83±0.72	7.62^{ns}

(Wahbeh, unpublished data)

RESEARCH RECOMMENDATIONS

There are numerous gaps in our knowledge of pollution effects in the Gulf of Aqaba. In order to quantify and evaluate the present status of these effects, there is a pressing need for a comprehensive investigation in the Gulf region to provide a sound, scientific basis for the assessment and regulation of pollution problems, including a sensibly planned monitoring program.

The objectives of the proposed investigation are

1. Execution of baseline studies in order to provide valuable data on inputs, distributions, and pathways of pollutants
2. Identification of the major sources of pollution, determining-rates of influx and efflux from various pathways, and constructing a mass balance
3. Studying transfer processes and establishing basic standards for exposure of the biota to pollutants
4. Establishing a data bank
5. Training of technicians and scientists

EXECUTION OF BASELINE STUDIES

As an early start in the monitoring program, priority should be given to the conducting of baseline studies, both at the national and regional levels (UNESCO 1977). The aim of these studies is to provide valuable data on inputs, distributions, and pathways of pollutants and help in studies of mass balance. The following sequence of operations provides the basis of baseline.

Evaluation Of Our Present Knowledge

The evaluation of our present knowledge requires drawing up a list of all scientific and technical institutions dealing with marine pollution in the region. Those interested in participation in the studies should be asked to provide all available data including those relating to sampling and analytical capacities. It is desirable then to retain a consultant to review the gathered information as part of the evaluation process.

The Marine Science Station in Aqaba, the only marine research center on the eastern coast of the Gulf of Aqaba, has already accomplished a comprehensive review of the marine research in the Gulf updated to 1987. The review has been published in 1989 in a bound, 267-page volume. Major gaps in our knowledge of marine science in the Gulf have been identified.

Identification Of The Major Pollutants

Major pollutants which require most urgent attention must be identified at the regional level taking into consideration:

1. The evaluation of the present knowledge
2. The existence, in the region, of reliable sampling and analytical methods
3. The availability of appropriate equipment and technical staff for applying the methods

Identification Of Inputs Of Pollutants

Locations of sources and magnitudes of pollutants should be specified, as a prerequisite for designing a sampling strategy. All domestic and industrial waste discharge outlets, as well as all dumping sites, should be mapped. For each site, an estimate of the gross annual quantity of each pollutant discharged as well as the rate of this charge and its nature, must be made (UNESCO 1977).

Selection Of Sampling Procedures

Pollutants are normally present in seawater, in sediment, in living organisms, and in the atmosphere at sea surface. All four media must be sampled. However, difficulties may arise in connection with the inadequacy of the sampling and analytical techniques or with the insufficiency of laboratory facilities and logistics.

The sampling program should take into account the method of introduction of pollutants into the sea, the character of their introduction, and the mechanism of their transport within the marine environment.

MASS BALANCE OF POLLUTANTS

The construction of a mass balance for individual pollutants in the marine environment requires the determination of inputs and outputs of all net polluting material (UNESCO 1977).

The primary modes of input of terrestrial and human mobilized material into the sea are through atmospheric transport (Table 8), deposition, flood discharge, direct discharge from land via pipelines, discharges from ships, and sea dumping activities. The relative importance of the various pathways of pollutants entering the sea varies with location and type of pollutant. In addition, pollutants may be transferred from one place to another in the sea by physical or biological transport. An understanding of the modes of entry of pollutants into the sea is essential and the transfer processes of these pollutants through different compartments of the marine environment are of considerable ecological interest and a requirement for mass balance construction. The ultimate fate of the pollutants, whether degraded or locked up in the sediments, must be known in order to complete the budget information needed.

Table 8

Amount of Emission of Gases (Tons/Year) from the Aqaba thermal power station (ATPS)

Gases	1986	1987	1988	1989
CO^2	390306	859968	794576	988300
CO	98	214	198	247
SO_x	6578	14493	13391	16657
NO_x	964	2125	1964	2443
N_2	2427354	5348233	4941549	6146240

The outputs of pollutants from the sea are also needed for constructing the budget. It is important to identify potential marine sinks for pollutants and to estimate the corresponding rates of removal. The major modes of pollutants removal from the sea are through sedimentation, volatilization, and aerosol production processes.

POLLUTANTS TRANSFER PROCESSES

The study of the physical, chemical, and biological processes in which pollutants take part is important. A pollutant originates from a source in the atmosphere or on land and passes through the sea reservoir until it is removed to a sink. During this time, the pollutant takes part in a number of processes.

Types of processes include photo-oxidation, bioreduction, sedimentation, flocculation, dissolution, bioturbation, and mixing. These processes can affect the nature of the pollutants before entering the sea, govern the rates of their influx and efflux, control their mass transport between various zones of the sea, and govern their interaction with organisms as well as their distribution within the marine biosphere.

ESTABLISHING BASIC STANDARDS FOR EXPOSURE

Sensible control over the introduction of pollutants into the marine environment depends on establishing basic exposure standards for man and marine organisms. In order to do this, dose/response relationships for various categories of pollutants and organisms must be considered.

There are great difficulties in obtaining dose/response information for man or for organisms in the natural environment. Exposure of organisms can be made under laboratory conditions, which usually uses higher rates of exposure than those in the natural environment. Thus, there is a need for developing methods for estimating the effects of chronic exposure of individual organisms, populations, and ecosystems, without which the definition of appropriate standards would be difficult. Studies conducted in large impoundments in which pollutant levels are similar to those presently observed in the marine water, are needed.

ENVIRONMENTAL DATA BANK

A basic requirement for the assessment of the extent of marine pollution is a reasonably well-organized data bank on the conditions in the environment. The data bank must consist of

1. Data on input of pollutants, including information on the distribution of sources and the quantities introduced
2. Data on the natural conditions in the region as regards physical, biological, chemical oceanography, and sedimentary conditions
3. Data on the interaction between the pollutant and the marine environment, including information on the geochemical and biological processes which may lead to its removal or its concentration in the marine environment
4. Data on the concentration levels in various compartments of the environment

The data bank must be designed to satisfy the following functions

1. Serve as a regional depository of information
2. Able to assimilate and disseminate information in a format convenient to the user
3. Assess the reliability and validity of the retrieved information
4. Capable of handling large data volumes

A quality control of acquired data is a must.

TRAINING REQUIREMENTS

Sampling and analysis of pollutants from the three compartments of the marine environment require skills which are rare in the Gulf area. It is even more rare when research on the effects of pollutants on the marine organisms and ecosystems are to be carried out. There will be, therefore, a great need for training of technicians as well as scientists.

REFERENCES

Edwards, A.J., and S.M. Head 1987. *Key Environments: Red Sea, Exeter*: Pergamon Press.

Hulings, M.C. 1989. A review of marine science research in the Gulf of Aqaba. Publications of the Marine Science Station, Aqaba. Amman: Jordan University Press.

Ormond, R.F.G. 1978. A marine park for Jordan. Report on the feasibility of establishing a marine park in Jordan. ALESCO.

Wahbeh, M.I., and N.C. Hulings 1987-1989. Collected reprints of station contributions to marine research in the Gulf of Aqaba, 5 vols. Publications of the Marine Science Station, Aqaba, Amman, Jordan: University Press.

UNESCO. *Global marine pollution: An overview*. Tech. Series 1OC(18). 1977.

UNESCO. *Review of the health of the oceans*. GESAMP Report No. 15. 1982.

7

Utilization and Management of the Coastal Zone in Latin America and the Caribbean

Alberto G. Lonardi, Coordinator
Multinational Project on the Environment and National Resources
Department of Scientific and Technical Affairs
Washington, D.C.

NATURAL CAPITAL OF THE REGION

The Latin American and Caribbean region has a privileged combination of valuable natural resources and scenic beauty. In terms of biological diversity, it can be considered one of the world's reservoirs of genetic riches. The quality and quantity of its marine resources and the present state of the coastal environment can be better described according to climate and physiography. Where the population has concentrated in coastal cities in cold and temperate zones, the high energy of coastal processes and other environmental conditions have limited, to a certain degree, the nature and frequency of occupations along the oceanic coastline. On the other hand, in countries located in tropical areas along the Central and North American Pacific mountain belt, the population has historically settled in higher altitudes to avoid the discomfort of coastal heat and unhealthy environments. The smaller Caribbean island states can each be considered as a coastal zone like to itself, because ranging in size from 305 (Saint Kitts) to 10,830 square kilometers (Jamaica), their entire populations are in essence living in the coastal region.

One measure of the natural capital of the hemisphere and the early concern which many countries within the region had in protecting coastal resources, is illustrated by the fact that almost all of them have at least one marine reserve or protected area. The first one was created in 1935 (Silva and Desilvestre 1990) and there are many being developed as of this writing. (Table 1).

Since the adoption of the 1982 Third UN Convention on the Law of the Sea (UNCLOS), while not yet in force, most of the Latin American and Caribbean countries have expanded their sovereignty and improved their resource base with claims over vast offshore areas. The Caribbean Island nation of Grenada, for example, with a land area of 344 sq.km., has increased its national jurisdiction nearly 80 times, to 27,000 sq.km. After several years, there is general understanding at all governmental levels in the countries of the region that UNCLOS has been instrumental in creating the national need for a real occupation of this space and for the effective utilization of resources contained within the

Table 1

Latin American and the Caribbean Marine Protected Areas (Ha)

	NUMBER	AREA
Ecuador	2	846,683
Brazil	9	799,945
Venezuela	19	496,587
Colombia	6	186,600
Suriname	3	105,000
Argentina	21	87,517
Costa Rica	9	67,077
Panama	4	60,675
Honduras	3	10,000
Guatemala	2	Proposed
Mexico	19	Proposed
Belize	2	Proposed
TOTAL:		2,660,084

Sources: Silva and Desilvestre, 1990, Ministry of Environment, Venezuela, 1988

boundaries of the Exclusive Economic Zone (EEZ), which amounts to 15% of the total area claimed by the world's countries. On the other hand, the concept, dimension, and claim of an EEZ was first introduced in the international arena during the 50s by four Latin American countries.

More recently, the preparatory activities related to the UN Conference on Environment and Development, held recently in one of the countries of the region (UNCED, Rio de Janeiro, June 1992), and its Agenda 21, were a very helpful national exercise for revision and reassessment, particularly with regard to the state of the coastal environment in the region.

ECONOMICAL RELEVANCE OF MARINE RESOURCES

Table 2 shows some of the indicators which are most frequently used to determine the economic potential of marine and coastal resources. For many years, the high productivity of the Southeast Pacific Ocean helped to develop important industrial fisheries in Peru, Chile, Ecuador, and other countries of the world. The maximum yield of 10 million tons was attained by the Peruvian fisheries more than ten years ago (mostly anchovy). At present, and after two strong El Niño events occurring in 1973 and 1983, catches of the most commercially valuable fish have declined, indicating that a maximum sustainable yield has been reached in this area. During the span of years just mentioned, the ecosystems were altered by new oceanographic conditions. As the Peruvian fisheries decreased, Chile was able to take advantage of an increase in the fisheries inside its own coastal waters. New species replaced part of the anchovies and shifted to the south, occupying in a few years the ecological niches left vacant by the massive mortality caused by El Niño. The Chilean fisheries increased during that period, from about 300,000 tons per year to more than 5 million tons per year in the late 80s.

Table 2
Coastal Resources Indicators - Latin America and the Caribbean

	CL	EEZ	FC	AQ	MA	LP	LDC	OP	OR	GP	GR
N.AMERICA											
Mexico	9330	2851	1181	59	6600	68630	15534	82979	5168	1115	1303
C.AMERICA											
Belize	386	--	--	--	--	63	211	0	0	0	0
C.Rica	1290	259	20.1	0.3	390	676	2401	0	0	0	0
El Salvador	307	92	15	1	450	679	1099	0	0	0	0
Guatemala	400	99	2.2	0.6	500	889	3906	0	0	0	0
Honduras	820	201	12.1	2.1	1450	636	1947	0	0	0	0
Nicaragua	910	160	3.8	--	600	634	1115	0	0	0	0
Panama	2490	307	193	3.5	4860	2069	1201	0	0	0	0
CARIBBEAN											
Antigua	153	--	--	--	--	61	82	0	0	0	0
Bahamas	3542	759	--	--	--	26301	3370	0	0	0	0
Barbados	97	167	3.9	0.2	--	287	573	0	0	0	0
Cuba	--	--	209	17.4	4000	9450		0	0	0	0
Dominica	148	20	--	--	--	4	78	0	0	0	0
Dominican R.	1288	269	17	0.5	90	2212	3464	0	0	0	0
Grenada	121	27	--	--	--	19	58	0	0	0	0
Haiti	1771	161	7.6	--	180	111	899	0	0	0	0
Jamaica	1022	298	9.2	1.5	70	2081	7959	0	0	0	0
S.Kitts	--	--	--	--	--	--	--	0	0	0	0
S.Lucia	158	121	--	--	--	--	--	0	0	0	0
S.Vincent	--	--	--	--	--	--	--	0	0	0	0
Trinidad	362	77	3	--	40	12012	5198	5837	79	3872	256
S.AMERICA											
Argentina	4989	1164	454	0.3	--	--	36583	--	34	--	5
Brazil	7491	3168	608	17.6	25000	37465	150779	18725	639	4693	116
Chile	6435	2288	5063	4.8	--	2037	14682	450	71	934	65
Colombia	2414	603	25	3.3	4400	4563	9563	--	10	2533	40
Ecuador	2237	1159	922	73	160110	698	2634	--	10	--	30
Guyana	459	130	41	--	1500	446	1586	0	0	0	0
Peru	2414	1027	4746	4.3	280	2452	11818	5156	30	--	3
Suriname	386	101	4.2	--	1150	700	6757	0	0	0	0
Uruguay	660	119	138	--	--	1271	1127	0	0	0	0
Venezuela	2800	364	273	0.6	--	73823	20944	49252	1599	6213	854

CL: Length of Coastline (km); **EEZ**: Exclusive Economic Zone (sq km); **FC**: Average Annual Marine Fisheries Catch, 1985-1987 (000 metric Tons); **AQ**: Average Annual Aquaculture Production, 1985-1987 (000 metric Tons); **MA**: Mangroves Area, 1980's (sq km); **LP**: Average Annual Volume of Crude Oil and Oil Products Loaded and Unloaded, 1983-1985 (000 metric Tons); **LDC**: Average Annual Volume of Dry Cargo Loaded and Unloaded, 1983-1985 (000 metric Tons); **OP**: Offshore Oil Production, 1988 (000 metric Tons); **OR**: Offshore Oil Proven Reserves, 1988 (million Tons); **GP**: Offshore Gas Production, 1998, (million cubic meters); **GR**: Offshore Gas Proven Reserves, 1988 (billion cubic meters). (--): Data not available

Sources: World Resources 1990-1991, WRI 1992.

OCCUPATION OF THE COASTAL ZONE

Population dynamics is a valuable tool for coastal zone management and a sensitive index of sustainable development. Latin America and the Caribbean follow similar trends observed in other parts of the developing world in the sense that the coastline population is increasing at a faster rate than in the hinterland. However, geography, climate, resources, and historical factors have created a diverse pattern for the region, as shown in Table 3. On the average, coastal populations of Latin America and the Caribbean amount to 27% of their totals, while the proportion for the world is 15% (index based in urban cities of 100,000 or more) (World Resources Institute 1992).

Table 3

Coastal Population in the Americas (1990)

	INLAND POPULATION X1000	COASTAL POPULATION X1000	TOTAL POPULATION X1000	PER CENT COASTAL
BAH	0	255	255	100
SL	0	147	147	100
SV	0	100	100	100
DOM	0	100	100	100
GRE	0	91	91	100
ANT	0	79	79	100
SK	0	44	44	100
BEL	190	150	40	79
CUB	2810	7800	10610	74
TRI	380	900	1280	70
DRE	2870	4300	7170	60
PAN	1020	1400	2420	58
JAM	1060	1400	2460	57
COS	1320	1700	3020	56
URU	1390	1700	3090	55
NIC	1870	2000	3870	52
PER	10850	10700	21550	50
BAR	137	123	260	47
ELS	2850	2400	5250	46
ARG	18862	14458	33320	43
SUR	240	180	420	43
GUY	481	319	800	40
VEN	12440	7300	19740	37
CHI	9070	4100	13170	31
HAI	4510	2000	6510	31
USA	181820	67400	249220	27
ECU	7890	2700	10590	25
HON	3840	1300	5140	25
BRA	112970	37400	150370	25
CAN	23020	3500	26520	13
COL	29480	3500	32980	11
GUA	8300	900	9200	10
MEX	80600	8000	88600	9
BOL	7310	7310	0	0
PAR	4280	0	4280	0
	531710	188446	720156	26

Source: World Resources 1990-1991, WRI 1992, Organization of American States, 1992

COUNTRY PROFILES

North America

Mexico

The country has a record of strengthening its marine and coastal capabilities since 1970, with initial emphasis in building marine research and technology. Towards that end, ample use was made of opportunities offered by international organizations like Organization of American States (OAS), IOC-UNESCO, and UNDP. The center of gravity for this development has been the Institute of Marine Science and Limnology of the Autonomous National University of Mexico and the National Institute of Fisheries of the Fisheries Secretariat. In more recent years, a general increase in the awareness of environmental matters triggered the creation of new institutions at the ministerial level, such as the Secretariat for Urban Development and Ecology. Under its leadership, the Law on General Ecological Equilibrium and Protection of the Environment was enacted in 1988. The law created a new Ordinance named Ecological Territorial Organization, aimed at decentralization of administrative responsibilities and coparticipation of public and private organizations. The Ordinance became the main legal instrument for the management of the coastal zone. One example of the possibilities offered by the new legal tool was demonstrated by the implementation of the Terminos Lagoon Coastal Zone Management Plan in the Bay of Campeche. The importance of the site is based on the fact that the Bay sustains 40% of the fisheries of Mexico, as well as 96% of the oil, and 80% of the total national production of gas. The Law introduced for the first time, the notion of the need for using ecological modeling at the decision making level of the country, when dealing with coastal zone management issues. The modeling includes cost/benefit evaluation and cost of environmental losses. Mexico is one of the first countries in establishing protected coastal areas, most of which are located at the Peninsula of Baja California, Yucatan Peninsula, Sonora, Veracruz, Guerrero, Nayarit, and Oaxaca.

In 1983, fisheries ($1,362 million), production of offshore oil and gas ($27,000 million), maritime transportation ($149 million) and tourism (approximately $731 million), were the main coastal resources of economical importance to Mexico. The fact that most of the oil has been recovered from the coastal zone has been a major factor in the creation of a national awareness in the coastal zone (Merino Ibarra 1990).

Central America

Costa Rica

An extensive coastline of more than 1,300 km along the Atlantic and Pacific Oceans and abundant marine resources characterize the Costa Rican coastal zone. Among the Latin American nations, their was one of the first efforts to regulate the use of the coastal zone with the enactment of Law 6.043 in 1977. The Law assigned the main responsibility for its implementation to the Costa Rican Institute of Tourism, but also stated that the Municipalities were charged with the supervision and control of coastal activities, including those related to tourism. Since 1979, the Institute started several studies related to coastal zoning, application of remote sensing to coastal management, modeling and environmental impact assessment.

From 1982 to 1990, there were an average of five to seven Regulating Plans for Management of the Coastal Zone completed every year. It is estimated that a total of 141 Regulatory Plans will be required to cover the most important coastal sectors of the country (Chaverri Pattison 1990).

The Costa Rican experience is very rich in examples of successful implementation of practical ideas which could be shared by other countries of the region; among them were: (1) assign the main regulatory responsibility to a single governmental agency; (2) reduce to the bare minimum, the number of public institutions taking part in the management of coastal resources; (3) start the Program with an adequate mapping and aerial photographic coverage of the coastal zone (ideally at a scale of 1:10:000, as satellite imagery does not provide a convenient scale for resource management); (4) freeze development in areas where not enough information is available; (5) define with precision and confine to specific sectors, those areas of greater potential for conflict, those which deserve special consideration for recreation and tourism development, and those which ought to be declared protected areas; (6) create a special entity in charge of supervising and controlling the coastal zone, with functions similar to a coastal police; (7) provide the regulatory body with independent authority to impose laws and regulations; (8) create a registry of property and concessions; and (9) substitute long judicial procedures in case of violation, by fast and energetic administrative decisions.

Honduras

Honduras possesses 683 km of coastline in the Atlantic and 153 km in the Fonseca Gulf in the Pacific, shared with El Salvador and Nicaragua. For the last twenty years, the Gulf has been under strong environmental pressures due to an increase in the exploitation of wood, salt (extraction), and fisheries. In 1973, the development and construction of shrimp farm ponds was instrumental in the destruction of mangroves. Mangroves and coral reefs are the most environmentally vulnerable areas, and therefore, of great concern to the authorities.

Some of the main tourist development projects of the country are centered in the Islas de la Bahia (Bahia Islands), known for their beautiful beaches surrounded by pristine waters and coral reefs. The islands are also the main center for Honduras' fishery activity.

Historically, the population has concentrated in the cooler highlands of the central massif, and the coastal region has been mainly a bridge of communication and trade with the outer world. In recent years, and particularly after the Rio Conference on Environment and Development of 1992, national authorities are starting to take systematic action directed toward the preservation and management of the coastal zone. For many years, there was a special branch of Government charged with the administration of natural resources (DIRENARE), which lacked the resources and political leverage necessary to create adequate legislation and regulations regarding the environment and the coastal zone, let alone to enforce them. The creation, in July of 1990, of the National Commission for the Environment (CONAMA), placed under the direct supervision of the Office of the President, may trigger new initiatives for the establishment of a National Coastal Zone Management Program or Plan. In fact, the Department of the Environment under the Secretariat for Planning, Coordination, and Budget, prepared a detailed plan of activities related to the coastal zone, dealing with the preparation of basic information for management: an inventory of natural resources and zoning of the coastal zone for future exploitation, development of tourism, and creation of protected areas. Regarding cooperation efforts with other countries of the region, Costa Rica is particularly interested in integrating with the tourist system of the Gulf of Mexico and the Caribbean, as well as working with Nicaragua and El Salvador towards the joint exploitation and preservation of the Gulf of Fonseca (Alduvin 1993).

Panama

The Panamanian coastline extends for 2,490 km. along the Atlantic and Pacific Oceans, where coral reefs alternate with mangroves, sandy beaches, and coastal scarpments, and is bordered by 1.600 islands in both oceans, many of which are important tourist attractions. The Panama Canal, which remains the major man-made modification of nature in the region, is a major economic asset of the country (more than 12,000 ships per year) as well as a potential source of pollution. In addition, both the giant Oleoduct running coast to coast in the west, and the large Panama Oil Refinery nearby are matters of great concern because of the threats they pose to the health of the entire coastline of Panama and its adjacent ocean waters. The authorities have prepared a detailed mapping of the ecological sensitivity of the coastline and a well designed oil spill contingency plan, both of which are powerful tools for coastal management.

The Pacific Ocean near Panama is rich in nutrients due to upwelling, because of which, and the fisheries have become the most important in Central America. The average annual marine catch amounted to 155,500 tons during 1987-1989, which, however, has shown a decrease of about 14% in the last ten years (World Resources Institute 1992) due to overfishing and to the El Niño Oceanographic Phenomenon. With its 4,860 sq.km. of mangroves along both coastlines, habitats for abundant fish and shellfish larvae, Panama has become the leader in aquaculture production in Central America, totalling 3,800 tons in 1989.

Unfortunately, the legal framework for coastal and ocean management is not fully established (Suman 1990). Despite the fact that Panama has claimed by Law Nbr. 31 of 1967 that its national sovereignty extends to the 200 nm territorial sea, and that it is a signer of 1982 UNCLOS III, Panama has still not officially declared and claimed its Exclusive Economic Zone. In addition, national environmental laws and regulations are sectorial and diverse, penalties are insufficient or lacking, and there is no comprehensive regulation of the coastal zone.

The rest of the Central American countries (Belize, El Salvador, Guatemala, and Nicaragua), are in the process of preparing coastal environment inventories.

Caribbean

Antigua and Barbuda

These are the two main islands of the Caribbean chain and are located 30 nm apart. While they are of volcanic origin, they are nonetheless, but low-lying and flat, with a total area of 440 sq. km., and a deeply indented coastline extending 153 km, providing many natural harbors ideal for yachting. Their total population is 79,000, and the main industry is tourism, with 188,000 visitors per year. There is no comprehensive management plan or an specific set of regulations for the coastal zone.

Bahamas

The Bahamas is composed of fourteen major islands and fifteen smaller ones, with a total area of 11,401 sq.km., and a coastline 3.542 km. It has the largest EEZ area of the Caribbean with 759,200 sq.km. (World Resources Institute 1992), ranking sixth among all Latin American and Caribbean countries. While it did ratify the 1982 UNCLOS III in 1983, it is only claiming a 200 nm Fisheries Zone, and has not yet declared the EEZ.

The economy is mainly based on tourism (3.5 million visitors per year), offshore banking, and transshipment of goods and other commodities. The per capita income exceeds $9,000, being one of the highest in the region. Although tourism is the major beneficiary of the coastal zone, there is good potential for long-term development of marine aquaculture and a strong fisheries industry based in its large EEZ.

The management of the coastal zone is not based at present in a comprehensive national plan. However, the authorities are considering the initiation of research activities in the coastal zone to prepare an inventory of marine resources and to evaluate the coastal impact of pollutants near the major human settlements and the long term effects of beach erosion.

Barbados

Barbados has a relatively regular coastline 97 km long, encircling a land area of only 430 sq.km., possessing beautiful beaches and a pleasant climate. On the other hand, the EEZ of this most easterly island country of the Caribbean covers 167.300 sq.km. Geologically, Barbados is the only non-volcanic island of the Eastern Antilles, having grown out of a folding of the ocean floor. Historically, its economy has shifted from cultivation of sugar cane to manufacturing and tourism (800,000 visitors in 1990), although the former still represents the highest revenue from exports. Fisheries contribute to less than 1 % of the GNP. In the past, mangroves were abundant in the swamps of the river mouths, but most of them became extinct due to coastal occupation. Although 45% of the coastline have sandy beaches, tourism has developed on the west side of the island, where wave energy is low. Residential construction is destroying the dynamic balance of the beaches of higher tourist value. In addition, destruction of coastal vegetation and runoff after summer storms, are depriving the beaches of their valuable sand.

Other major environmental problems of Barbados include the destruction of fringing coral reefs from pollution created by organic waste derived from the rum distilleries and oil refinery, as well as thermal wastes created by the power company.

There is no single approach to coastal zone management in the country. The main achievement of the Government took place in 1983 with the creation of the Coastal Conservation Project Unit. However, enforcement of existing regulations is practically non existent. Legislation needs to be revised to accommodate modern principles and regulations concerning the preservation and adequate management of the coastal zone and to devise adequate institutional arrangements like the possible creation of a Department of the Environment (Nurse 1993).

Jamaica

With a total area of 10,830 sq.km. and a coastline extending 1,022 km, Jamaica is the third largest island of the Caribbean (World Resources Institute 1992). About 40% of the perimeter of the island have well-developed sandy beaches (prevailing white sand in the north and black sand in the south). Coral reefs are also well developed, but unfortunately, many of those paralleling the northern and most frequented beaches (Montego Bay and Ocho Rios) are being killed by coastal pollution and abnormally high ocean temperatures (coral bleaching). Mangroves cover 30% of the coastline length and three species of sea grasses are found to be abundant. The irregularity of the coastline has created 15 natural harbors, with active ports moving a dry cargo of almost 2 millions Tm a year. Coastal erosion caused by weathering, but most significantly by large scale illegal removal of sand, is a major environmental problem requiring urgent attention and effective enforcement of relevant, current regulations (Young 1993).

In 1991, Jamaica claimed an EEZ extending 297,600 sq.km., one of the largest of the region, but did not declare a Fisheries territorial zone. About 20,000 people engage in fishing activities, employing 9,000 boats. It is estimated that 1992 production will reach 8,500 tons, at a price of about $110 million. The major increase in productivity has been experienced by inland fisheries, and Jamaica has become a leading producer of aquaculture products in the region. Inland fisheries yield amounted to about 3,000 tons in 1998, but it is estimated that 1992 production will reach 8,900 tons., nearly ten times as much as achieved in 1986 (Organization of American States 1992).

There are many resource management institutions, such as the Town and Planning Department, the Town and Planning Authority, the Environmental Control Division, and the Natural Resources Conservation Division of the Ministry of Agriculture. The latter is the chief resource management and conservation agency of Jamaica, responsible for the application of among others the three major environmental laws; such as the Wildlife Protection Act (1945), the Beach Control Act (1955), and the Watershed Protection Act (1963). It also coordinates the activities of the Beach Control Authority, Watersheds Commission, Wildlife Protection Committee, Natural Resource Planning Unit, Marine Authority Committee, and Kingston Harbor Quality Monitoring Committee.

Despite the existence of important regulatory and management agencies, one of the major environmental problems is the destruction of wildlife habitat, as well as the overexploitation of species for profit or sport. Other visible problems arise from the lack of environmental impact assessment for agricultural projects, lack of enforcement and regulations for land use, lack of coordination among the users groups, and lack of environmental education in the general public (Young 1993).

Saint Lucia

The mountainous island nation of San Lucia has an area of 620 sq.km and a coastline of 158 km (The Software Toolworks World Atlas 1990). In 1984, Saint Lucia claimed an EEZ extending 12.170 sq.km., and in 1985 ratified the 1982 UNCLOS III. Having a relatively small EEZ, pelagic fisheries landings amount to 70% of the total landings (less than 500 tons per year).

Its production is mainly agricultural (predominantly banana, but also coconut, sugar, cocoa, and spices). In recent years, tourism has become a growing source of revenue, as have some manufacturing industries (clothing, electronic components, etc) (The Software Toolworks World Atlas 1990).

Reefs, in association with mangroves and seagrass, provide excellent habitats for fishes and spiny lobsters and help to reduce the rate of erosion of the coastal zone. The coral formations to the west of the Island are attracting an increased number of divers and other tourists. Historically, human settlements concentrated in the narrow coastal plain. Together with Jamaica, Saint Lucia has exceptional natural harbors and one of the few ports in the West Indies capable of receiving ocean going vessels along wharves. On the other hand, the most pressing environmental problems include illegal sand mining of beaches which is causing coastal erosion and loss of tourist value, degradation of the coastal environment by the discharge of domestic sewerage, lack of adequate solid waste disposal, and environmental risk associated with the storage and transshipment of millions of barrels of oil through the Hess Oil Company facility in Saint Lucia.

At present, a large number of government agencies have jurisdiction and management responsibilities over coastal and marine areas: Ministries of Agriculture and Fisheries; Communications and Transport; Health, Housing and Labor; Trade, Industry and Tourism; Air and Sea Authority; and the National Development Corporation. There are also a profusion of Laws spanning from 1967 to 1986 and including the Water and Sewerage Act, the Pesticides Control Act, the Tourist Industry Development Act, the Park and the Beaches Commission Act, the Beach Protection Act, and the Fisheries Act. However, there is no formal mechanism in place for overall coordination and regulatory control (Walters 1993).

Trinidad and Tobago

Trinidad and Tobago is an Archipelagic State comprising twenty-three islands, with an EEZ claimed in 1983 extending 58,000 sq.km. The major island, Trinidad, is only 12 km away from the Orinoco River Delta of neighboring Venezuela, and comprises an area of 4,828 sq.km. The second major island, Tobago, lies 34 km away and has an area of only 300 sq.km., but is privileged by having beautiful beaches and a large coral formation (Bucoo Reef) which holds a great attraction for tourists.

Trinidad has evolved from being a plantation-type economy to an oil-based economy. After the oil crisis of 1986, Trinidad reverted to the tourist industry as the major source of revenue, keeping a strong agricultural base. Over 88% of the population lives in the coastal area of Trinidad, heavily concentrated in the western part, where 60% of the agricultural area is located. Oil is the major coastal resource, with 72% being produced in marine fields located to the south and west of the main island (total probable reserves are estimated at 91.7 million cubic meters of oil and 197,300 cubic millions meters of natural gas). Forests cover 50% of the land area of Trinidad and Tobago, of which about 5% is under intensive management. While fisheries and aquaculture are growing sources of employment and revenue, they are economically much less important than oil.

The process of development started in the 70s has resulted in rather severe environmental pollution and degradation (McShine 1993). The environmental problems of Trinidad do not, however, differ from those occurring in other island states. Due to space limitations, population growth, and economical transformation, the Islands suffer from pollution by wastes from oil wells and refineries, industrial toxic materials, agricultural chemicals, and domestic effluents; as well as degradation of their forests, mangroves, and beaches caused by the uncontrolled human intervention.

The Ministry of Planning has been the agency with major responsibility in preparing national development plans, which have been basically concerned with economic growth, but not necessarily with balancing development and environment. Among several agencies with environmental responsibilities, the Town and Country Planning Division is concerned with controlling land development, which do not exclude the marine environment. The National Physical Development Plan prepared by the Division, includes coastal zone management requirements and suggests the need for having an improved legal framework related to environmental conservation. Of fifteen extant Ministries, seven have clear environmental responsibilities and, in addition, three special units also have responsibilities: the Institute of Marine Affairs, the Solid Waste Management Company and the Caribbean Industrial Research Institute. The former was created with UN support several years ago and one of its main, initial projects was the preparation of a comprehensive Coastal Zone Management Project. At present, forty-two pieces of legislation relate directly or indirectly to the coastal environment, but enforcement is not fully effective due to inadequate regulations and staffing as well as extremely low fines (McShine 1993).

(At the time of this writing, information available was incomplete to describe the situation related to coastal zone management in Dominica, Dominican Republic, Grenada, Haiti, Saint Kitts and Nevis, and Saint Vincent and the Grenadines).

SOUTH AMERICA

Argentina

Argentina ranks 6th among the countries of the hemisphere in length of coastline (4,989 km), and 4th in terms of land surface (2,736,690 sq.km). A fisheries zone of 200 nm was claimed as early as 1967, later superseded by EEZ legislation, which extended its national jurisdiction offshore to 1,164,500 sq.km.

Traditionally, Argentina has been a nation with mixed or cyclic interest in coastal and marine resources development, subject to world and national changing economic conditions. After WWII, for example, marine transportation became a major oceangoing activity. Oceanographic research and hydrographic activity has played a significant role in creating a scientific basis for future development of ocean and coastal resources, particularly after Argentina's participation in international programs started at the time of the International Geophysical Year. The coastal zone extending from the Rio de la Plata mouth to the southern part of the Buenos Aires Province has been used since the beginning of the century as the major seaside recreational region of the country, as well as the major center of fishery activities and port development. The rest of the coastal zone, extending south to Tierra del Fuego Island, comprises 66% of the total length, but holds only 1.6% of the population of the country (Brandami 1990).

The capital city, Buenos Aires, is located in the coastal zone of the major river mouth of the hemisphere. The city and its surroundings concentrate almost 50% of the country population. A similar demographic trend is found in the neighboring country across the river, where the capital city of Uruguay, Montevideo, holds the majority of the country's inhabitants. The dependency of both countries on the river is great and there has been a growing concern regarding the health of this environment, as well as that of the gigantic Plata Basin. These two nations also stand the potential share in the gains which can derive from this environment's multiple utilization.

One main response for this common interest has been the mutual agreement reached in 1973, entitled the Argentina-Uruguay Treaty of the Maritime Front of the River Plate. This treaty was preceded in 1996 by the Treaty on the Boundaries of the Uruguay River. The creation of three, binational administrative Commissions, one for the Uruguay river, one for the Rio de la Plata and the other for the Maritime Front, constitute one of the first examples of coastal zone management arrangements between two countries of the region.

At the national level, however, there is no formal program directed to the management of the coastal zone, with the exception of one short-lived attempt started by the government of the Buenos Aires province in 1984.

As is the case in other countries of the region, environmental laws and regulations concerning the coastal zone and its resources are numerous but disperse. On the other hand, issues like conservation of natural resources are generally decided by the governments of the provinces concerned. However, there are many grounds for legal conflict between the central governments and the provinces. Marine resources found inside the national jurisdiction are property of the nation (Law 20136 of 1973), while the Law 18,502 of 1969 dictates that the provinces have jurisdiction over 3 nm offshore.

Brazil

The coastal zone of Brazil is characterized by its length (7,491 km) and its environmental richness, particularly with regard to the high number of white-sand beaches, its mangroves, coastal lagoons, and an attractive morphology formed by outcrops of Paleozoic basement in its central region. Brazil ranks second to the United States in the hemisphere, regarding the area of its EEZ. Claimed in 1998, it amounts to 3,168,400 sq.km. In 1971, Brazil claimed a Fisheries Zone of 200 nm.

The coastal region of Brazil remains the main area of human settlement. Sixty percent of its inhabitants (more than 90,000,000) live in its seventeen coastal States, a little less than half of them concentrating in only two States (Rio de Janeiro and Sao Paulo).

To cope with increasingly difficult environmental and social problems associated with coastal areas, several years ago the central Government started a program of orientation and participation with States and local authorities, directed to the preparation of a National Program for Coastal Zone Management. This initiative followed others of similar importance started in the 70s, such as the Hydrographic Basin's Committees. However, in that particular instance, some errors were made; like the creation of industrial conglomerates (e.g., chemical plants) in the vicinity of ecologically sensitive estuaries such as Aracaju, Cubatao, and Guanabara. The criteria for the selection of these sites was based more on classical economics than on environmental impact assessments (Brazil 1991).

The most important development prior to the preparation of the Coastal Zone Management Program has been the creation, in the 80s, of the National System for the Environment and the establishment of the Interagency Commission of Marine Resources (CIRC). In 1988, a Chapter on the Environment was introduced in the revised Constitution of that year, and in 1989 the Brazilian Institute for the Environment and Natural Resources was created (Hertz 1990).

The size and importance of the National Program encouraged Brazilian authorities to request financial assistance from the World Bank for its implementation, which was granted for the period 1990-1995.

Colombia

Colombia has an extensive coastline of more than 2,952 km distributed along the Atlantic and Pacific Oceans. The area claimed for the EEZ by Law 10 of August 4, 1978, amounts to 603,200 sq.km. (World Resources Institute 1992). In the other hand, Colombia has not declared or claimed a Fisheries Zone within the framework of 1982 UNCLOS III (Fenwick 1992).

Since colonial times, population has settled in the highlands of the Andes. Although of lesser importance, historically, coastal development has been much greater in the Caribbean sector. In recent years, the Colombian government has started several programs directed to the development of the Pacific coast. Despite the great expanse of the ocean under national jurisdiction, the Colombian fisheries only contribute with 0.4% of the GNP, having more a social, than an economical value.

The present institutional strength of the country in marine related areas, is due to the leading role played in the early 70s by the Navy and its Dirección Marítima y Portuaria, the National Committee of Oceanography, created in 1969 by Presidential Decree, and by the National Research Council

(COLCIENCIAS), working together during several years for the preparation of a National Plan for Development of the marine sector.

The environmental problems of the country are many and diverse. The tropical nature of the coastal zone makes it more sensitive to certain pressures, and resources such as mangroves, are decaying at an alarming rate. As is the case in other countries of the region, there are many laws, but little application and little management. There is a wide dispersa of responsibilities and duplications among several national agencies. One hope of reorganization was recently given by the sanction of the new Constitution, devoting one entire Chapter to the Environment, while another is the solid institutional basis which already exists at the national level to put into force a comprehensive system of coastal zone management in the country (Sanchez 1993).

Ecuador

A country with strong marine orientation, Ecuador was a pioneer in identifying the need for the establishment of laws, regulations and plans capable of saving the coastal zone from improper use and degradation. The country has an important coastal plain located between the Andes and the Pacific Ocean, with a coastline stretching along 2,237 km. The Fisheries Zone was claimed in 1951 (200 nm) but so far, there has been no claim of the EEZ.

The Galápagos Islands, located 1,000 km to the west of continental Ecuador, comprise nineteen islands which constitute one of the few natural ecological laboratories left on the face of the Earth (Broadus and Gaines 1990).

Fifty percent of the population lives in the coastal provinces along a strip of land approximately 60 km wide. The coastal zone shows high ecological diversity and is irrigated by seventy-six river basins, although a large percentage of the rivers run dry from December to May (Pozo Haro 1993). Biological productivity in several major estuaries along the coast is very high and they serve as a source of shrimp larvae for aquaculture and fish.

One of the main environmental problems of Ecuador is the destruction of mangroves by the shrimp industry and related activities (Table 4). Shrimp culture was started in 1975 and became the second source of revenue from exports, after oil. The path followed by Ecuador in its coastal development is rather similar to the one which occurred in Colombia during the 70s. Oceanographic institutions took the lead in assessing the major, marine environmental problems by doing coastal and ocean research and creating basic inventories of natural resources. During the period, the country took advantage of the opportunities and aid offered by international and regional organizations like OAS and IOC/UNESCO.

In 1984, coastal problems were increasing with the installation of hundreds of new shrimp cultivation ventures along the coast, and in 1986 the Government signed the first draft of an Agreement of Cooperation with USAID and the University of Rhode Island, for the creation and implementation of an integral national plan for the management of coastal resources. As a result of this effort, implementation of the Plan is well advanced, and progress has been made in determining Special Management Areas, organizing local communities along the coast, creating new regulations and devising new ways for enforcing the law (Olsen and Figueroa 1990).

Table 4

Ecuador - Evolution of the Area Covered by Mangroves, Salt Marshes and Shrimp Ponds Between 1969, 1984, and 1987 (Ha)

YEARS	MANGROVES	SALT MARSHES	SHRIMP PONDS
1969	203,626	51,496	0
1984	182,159	20,008	89,078
1987	175,350	12,420	116,796
DIFFER. 69/87	28,276	39,076	116,796

Source: (del Pozo Haro, 1993)

Venezuela

According to The World Resources Institute Venezuela has a coastline extending 2,800 km along the Caribbean Sea and the Atlantic Ocean. Other authors increase this value to 4,006 km (2,740 km in the Caribbean, 1,006 km in the Atlantic and 260 km being internationally claimed) (Perez Nietro 1990). Venezuela also has national jurisdiction over 311 islands of varying size, three of which became the State of Nueva Esparta. Venezuela is one of the few Latin American countries with a long record of environmental laws and institutions. Among the latter, those with broader responsibility in coastal management are the Ministry of the Environment and Natural Resources (created in 1977); the National Commission for Territorial Zoning and Management (Ordenación), (created in 1983); and the National Commission of Oceanography (1985). The main instrument related to coastal zone management planning is the National Plan for Territorial Zoning and Management, which is in the process of revision, and which includes provisions related to maritime areas. In addition, there are several laws and Executive Decrees which regulate coastal activities, having defined nineteen categories of coastal management. The most recent initiative in this field is the Draft Law for the Coast and Adjacent Regions, which will regulate the use of the coast in more detail than before. Venezuela has also produced very specific legislation with regard to protected marine areas, and the National System of Natural Protected Areas (ABRAES) has been created to protect specific areas of high ecological sensitivity. Recent reports on the state of the coastal environment indicate, however, that severe degradation is occurring in many coastal areas and that implementation of coastal regulations is still lacking (Venezuela, 1988).

OAS ACTIVITIES IN SUPPORT OF COASTAL ZONE MANAGEMENT IN LATIN AMERICA AND THE CARIBBEAN.

International assistance played a major role in the initiation of coastal zone management efforts in Latin America and the Caribbean, as well as in other developing countries, either by supporting information exchange such as meetings and reports or feasibility studies and pilot projects (Soresen and Brandani 1990).

The Multinational Project on the Environment and Natural Resources and the Multinational Project of Marine Science and Marine Resources of the Regional Program for Development of Science and Technology in Latin America and the Caribbean (a component of the international technical cooperation Program of the OAS) has been promoting and supporting several activities related to the coastal zone in the region since 1970. The first OAS meeting specifically devoted to the issue of coastal zone management was the I Course/Seminar on Coastal Zone Management and Management of the Exclusive Economic Zone, which was organized from April 8 to 18, 1984, in Mar del Plata, Argentina, in cooperation with the Secretariat of Science and Technology and the University of Mar del Plata, (Lonardi, 1990). The concept of coastal zone management has been rapidly adopted as convenient and desirable - although not necessarily implemented - by many developing countries, because it is seen as a final solution to the unsolved task of achieving inter-sectorial coordination in the midst of complex and sometimes conflicting environmental legacies from the past.

The II OAS Seminar on Coastal Zone Management and Management of the Exclusive Economic Zone and Aquaculture Applications in Latin America and the Caribbean, was held in Miami, Florida, December 12-16, 1988, in cooperation with the Rosenstiel School of Marine and Atmospheric Sciences of the University of Miami and the U.S. National Oceanic and Atmospheric Administration (NOAA).

The III OAS Seminar was held from March 26 to April 2, 1990, but this time dealt with the problem of beach erosion and coastal pollution and recovery. It was held in Buenos Aires, Argentina, with the decided support and encouragement of the Commission of the Environment and Natural Resources of the Argentine Senate. Hundreds of national and foreign participants confirmed the interest of different sectors of society in learning and sharing experiences related to the coastal zone from a new perspective, in being more constructive than in the past, and more individually responsible; and in being deeply concerned with the environmental legacy which might be left to future generations.

A fourth Interamerican Seminar on Management of the Coastal Zone: Science and Technology, Preservation of the Environment and Coastal Development, was organized by the OAS and held from November 25th to 29th, 1991, in Montevideo, Uruguay, under the sponsorship of the Ministry of the Environment and Housing and the Ministry of Tourism.

In 1992, as a continuing effort to strengthen local capabilities in dealing with coastal processes, and particularly with the preservation of beaches of high tourist value, the OAS Multinational Project on the Environment and Natural Resources organized the first Cooperative Course on Coastal Geology, Coastal Erosion, and Coastal Management. It was held in conjunction with the Department of Geology of Duke University, North Carolina and tailored for the needs of Caribbean Island Countries. The success attained by the first Course has encouraged both institutions to organize a second one and this time it will be held in two island States of the Caribbean to maximize attendance.

In addition, from 1990 to 1995, several Latin American and Caribbean countries are participating in the OAS Multinational Project on the Environment, in different aspects related to coastal zone management.

REGIONAL NEEDS AND PROPOSALS

The countries of the region are experiencing rapid changes under varying economical and political pressures. There is a need for permanent consultation and exchange. The OAS scientific and technological assistance program is providing a forum which produces a helpful flow of ideas, suggestions, and specialized information which can be useful to other regions of the world. The most recent encounters, attended by distinguished scientists and professionals provided the following recommendations and suggestions.

Some of the Facts

1. In the region, environmental regulations are not applied properly.
2. In general, agencies with primary responsibility in environmental issues are positioned too low in the government structure and are generally too disperse.
3. There is a lack of commitment at high political level, although member countries are fully aware of the high complexity of the coastal processes and of the methodologies and plans which are necessary to apply in order to guarantee proper and responsible management of the coastal zone.
4. Although human beings are part of the environment and ought to adapt to it, they generally impose arbitrarily their own conditions to the environment.
5. There is a lack of conceptual and operational integration between science, technology, education, culture, economy, society, legislation, and regulations regarding the use and management of the coastal zone.
6. Neither the marine resources nor the conflicts arising from its use, are evenly distributed throughout the region.

Policy Issues, Institutional Arrangements and Relevant Aspects

1. Promote more active participation of legislators in the discussions related to the formulation of environmental policies and legislation in regional and international level.
2. Promote the preparation of comprehensive National Management Coastal Plans.
3. Create new agencies at the ministerial level, empowered with sufficient authority and resources to coordinate, take high level environmental decisions, and apply adequate methodologies in the implementation of coastal zone management programs at the national level.
4. The number of governmental agencies with responsibility in the administration of coastal resources ought to be reduced to a minimum, preferably to a single one, but the income produced ought to be distributed among local governments.
5. Provincial governments ought to be those who preferably make decisions related to coastal zone management, securing wide participation of public and private sectors in the process.
6. Assign high priority to the completion of a national territorial zoning, defining from an early phase, all public and restricted zones, using as much scientific input as possible.
7. The coastal zone is considered to be a unique area, yet for management purposes, particularly in relation with river basins and estuaries, it is essential that the coastal zone be considered only as part of a larger environmental system.
8. Make sure that representatives of Municipalities take part in regional meetings and seminars related to coastal zone management.
9. Integrate legal aspects with science and technology and other sectors, to develop a systemic coastal perspective in the region.

Preservation of the Coastal Zone

1. Avoid the introduction of modifications of any kind in the coastal zone or reduce them to the minimum possible, subject to prior completion of scientific studies of coastal dynamics and its long term effects.
2. For coastal areas were zoning for management purposes is not an imminent need, it is advisable to classify them as protected or restricted areas, until more environmental information is available.

3. Assign to the central environmental agency, the responsibility of assuring that each coastal zone development project proposal or industrial activity proposal, has complied with the preparation of an adequate environmental impact assessment.
4. Every living resource intended to be exploited in the coastal zone ought to be previously scientifically evaluated within the framework of its marine ecosystem.
5. For smaller island countries in particular, it is essential to determine the carrying capacity of the environment related to tourism, as well as to other types of development.
6. Educate the public, particularly children.

Legal Aspects

1. Foster the creation of the Coastal Law, as a new branch of public law.
2. Legislation for the coastal zone has to maintain public trust and be flexible enough to accommodate changes which are common in any area of transition.
3. Make a clear distinction between technical and administrative responsibilities when preparing coastal zone management laws.
4. Consider the inclusion of a Chapter on sanctions and violations in the same body of the environmental law and its regulations.
5. Promote the creation of Environmental Courts of Justice in each member country.

REFERENCES

Alduvin, Rafael, 1993. Manejo y Planificacion de la Zona Costera de Honduras, in: *El Manejo de Ambientes y Recursos Costeros en América Latina y el Caribe, Volumen III,* Organización de los Estados Americanos, Washington, DC. (in press).

Brandani, Aldo, 1990. La Zona Costera de Argentina: Perfil Ambiental e Institucional, in: *El Manejo de Ambientes y Recursos Costeros en América Latina y el Caribe, Volumen I*, Organización de los Estados Americanos, Buenos Aires, Argentina, Diciembre 1990, pp. 37-53.

Brazil, 1991. Subsídios Técnicos para Elaboraçao do Relatório Nacional do Brazil para a CNUMAD *(Brazil National Report for the UN Conference on Environment and Development of Rio de Janeiro, June 1992)*.

Broadus, James y Gaines, Arthur, 1990. Caso de Estudio: Manejo del Area Costera y Marina para las Islas Galápagos, in: *El Manejo de Ambientes y Recursos Costeros en América Latina y el Caribe, Volumen I*, Organización de los Estados Americanos, Buenos Aires, Argentina, Diciembre 1990, pp. 257-269.

Chaverri Pattison, Robert, 1990. Iniciativas de Manejo Costero en Costa Rica, in: *El Manejo de Ambientes y Recursos Costeros en América Latina y el Caribe, Volumen I,* Organización de los Estados Americanos, Buenos Aires, Argentina, Diciembre 1990, pp. 105-119.

Fenwick, Judith 1992. *International Profiles on Marine Scientific Research, National Maritime Claims, MSR Jurisdiction, and U.S. Research Clearance Histories for the World's Coastal States, Woods Hole, MA.:* WHOI Sea Grant Program.

Hertz, Renato, 1990. Programa Brasilero para el Manejo Costero, in: *El Manejo de Ambientes y Recursos Costeros en América Latina y el Caribe, Volumen I,* Organización de los Estados Americanos, Buenos Aires, Argentina, Diciembre 1990, pp. 55-78.

Lonardi, Alberto G., 1990. La Cooperación Internacional y el Manejo con Base Científica de la Zona Costera de las Américas, in: *El Manejo de Ambientes y Recursos Costeros en América Latina y el Caribe, Volumen I*, Buenos Aires, Argentina, Diciembre 1990, pp.7-11.

Merino Ibarra, Martin, 1990. El Manejo de la Zona Costera Mexicana: Una Evaluación Preliminar, in: *El Manejo de Ambientes y Recursos Costeros en América Latina y el Caribe, Volumen* I, Buenos Aires, Argentina, Diciembre 1990, pp. 137-154.

McShine, Hazel, 1993. Trinidad and Tobago National Essay: Management and Planning of the Coastal Zone and the Exclusive Economic Zone and Aquaculture Applications, in: *El Manejo de Ambientes y Recursos Costeros en América Latina y el Caribe, Volumen II*, Organización de los Estados Americanos, Washington, D.C. (in press).

Nurse, Leonard, 1993. The Barbados Coast: A Planning and Management Evaluation, in : *El Manejo de Ambientes y Recursos Costeros en América Latina y el Caribe, Volumen II*, Organización de los Estados Americanos, Washington, D.C. (in press).

Olsen, Stephen and Figueroa, Eduardo, 1990. Edificando la Base de un Programa de Manejo de Recursos Costeros en Ecuador, in: *El Manejo de Ambientes y Recursos Costeros en América Latina y el Caribe, Volumen I*, Buenos Aires, Argentina, Diciembre 1990, pp.155-170.

Organization of American States 1992. *Country profiles for aquaculture development in the Caribbean, OAS Suregional Project for Aquaculture in the Caribbean*. Washington, D.C. 44 pp. Multinational Project on the Environment and Natural Resources.

Pérez Nieto, Hernán 1990. La Ordenación de las Areas Marinas y Costeras de Venezuela, *paper submitted to the XXI Convention of the Panamerican Union of Engineers, UPADI 90*, Washington, DC, 19-24 August, 1990.

Pozo Haro, Magdalena 1993. Importancia del Programa de Manejo de Recursos Costeros en el Ecuador, in: *El Manejo de Ambientes y Recursos Costeros en América Latina y el Caribe, Volumen II*, Organización de los Estados Americanos, Washington,DC. (in press).

Sánchez Moreno, Hernando, 1993. La Zona Costera Colombiana, Descripción, Manejo, Problemática y Marco Legal, in: *El Manejo de Ambientes y Recursos Costeros en América Latina y el Caribe, Volumen III,* Organización de los Estados Americanos, Washington, DC. (in press).

Silva, Maynard and Desilvestre, Ingrid, 1990. Análisis de Areas Marinas y Costeras Protegidas de Latinoamérica, in: *El Manejo de Ambientes y Recursos Costeros en América Latina y el Caribe, Volumen I*, Buenos Aires, Argentina, Diciembre 1990, pp. 225-255.

The Software Toolworks World Atlas, (c) 1990.

Sorensen, Jens and Aldo Brandani 1990. Esfuerzos de manejo Costero en Latinoamérica, in: *El Manejo de Ambientes y Recursos Costeros en América Latina y el Caribe, Volumen I*, Buenos Aires, Argentina, Diciembre 1990, pp. 155-170.

Suman, Daniel, 1990. El Manejo de la Zona Costera en Panama, in : *El Manejo de Ambientes y Recursos Costeros en América Latina y el Caribe, Volumen* I, Buenos Aires, Argentina, Diciembre 1990, pp.155-170.

Venezuela, Ministry of the Environment and Natural Resources 1988. National report on environmental problems of the marine and coastal areas of Venezuela. Report prepared by the Ministry of the Environment for the Caribbean Action Plan of UNEP Caracas, Venezuela). 113 pp.

Walters, Horace, 1993. Saint Lucia National Essay, in: *El Manejo de Ambientes y Recursos Costeros en América Latina y el Caribe, Volumen II*, Organización de los Estados Americanos, Washington, D.C. (in press).

World Resources Institute, 1992. World Resources 1992-1993, Oxford University Press, 385 pp.

Young, Roy M. 1993. Management and planning of the coastal zone and the Exclusive Economic Zone and aquaculture applications in Jamaica, In El Manejo de ambientes y recursos costeros en América Latina y el Caribe, Volumen II. Washington, D.C.: Organización de los Estados Americanos.

New Concepts in the Governance of Ocean Space

8

Management Issues in Coastal Lagoons The Case of Venice

Francesco Bandarin
Professor of Planning, IUAV
University Institute of Architecture, Venice, Italy

INTRODUCTION

In the last twenty years, a number of legislative measures have been passed in order to solve the problems affecting the City of Venice and its Lagoon. This legislation allowed the development of a design and experimental activity, now completed and ready for implementation. It is calculated that in less then a decade, the major engineering and environmental projects can be completed, providing a full safeguard to the city of Venice and the other urban settlements of the lagoon against the risks of flooding, and allowing the revitalization of the lagoon ecosystem.

The 4th of November of 1966 was a crucial date in Venice's recent history. On that day, an exceptional storm surge hit the Venice area. The force of the sea breached the coastal defences built over the centuries along the littorals, leaving the city flooded for over twenty-four hours, the population exposed to the severe dangers, and properties and technical installations sustaining extensive damage. The 1966 tide, the highest ever recorded in the secular history of Venice, reached 194 cm over average sea level, as compared to a normal spring tide of approximately 50-60 cm.

The fragility of the city and its risky situation became known world-wide, pushing the Italian Parliament to pass the first Special Act in 1973. This legislation allowed the proposal of a number of projects aimed at controlling the tidal action. In spite of the interest raised by the proposals, none of them was judged acceptable from hydraulic, environmental, and economic viewpoints. In the 1980s, a new opportunity for project design and implementation was opened by the second Special Act of 1984, when significant financial resources were allocated to define a new project aimed at defending Venice against the high tides and restoring the lagoon environment.

A Consortium of private enterprises (Consorzio Venezia Nuova) was charged with coordinating the research, design, and implementation activities, under the control of the State Authorities (Magistrato alle Acque). In the course of this effort, the Consortium defined a comprehensive approach to guide the physical transformations of the Lagoon towards a new equilibrium (Bandarin 1993). The design of a system of mobile gates, aimed at preventing the flooding of Venice during excessive high tides, has been viewed as part of an overall action aimed at protecting and maintaining the entire Lagoon ecosystem.

The system of projects that has been defined can be grouped into two main areas (Bandarin, 1991):

1. Projects aimed at defending Venice and other lagoon settlements against the risks of flooding
2. Projects aimed at reversing the degradation of the environment and limiting the risks to it

These projects define the complete set of actions to be developed to ensure the protection of Venice from flood-related risks, and to gradually recover the environmental quality. It must be remembered, nevertheless, that these tasks are only a part of the complex scheme promoted by the government to protect and revitalize Venice (OECD 1990).

The Special Act of 1984 also provided resources for interventions aimed at abating the pollution outflow from the watershed into the Lagoon, and at facilitating the process of building restoration and urban maintenance in the Historic Center.

These areas of intervention are under the jurisdiction of the local authorities; namely the Regional Government, responsible for the pollution control plan of the watershed, and the City of Venice, responsible for the management of interventions in the urban areas.

In January of 1992, the Italian Parliament has appropriated a significant sum for Venice (about $1.5 billion) to be divided into three areas of intervention: (1) the defense of Venice, (2) the control of pollution in the watershed; and (3) the restoration and other maintenance activities in the Historic Center.

The existence of a plurality of responsibilities entails the need of a coordinating body, which has been identified in the Law as a Special Committee for Venice, formed by Ministries and representative of Local governments.

A reform process of the overall governance of the Venice Project aimed at improving the effectiveness of the Committee, has been in the Parliament's agenda since 1991, and is likely to be completed in the near future. Early conclusions stress the need of more effective coordination among the different public bodies involved in the process and of more speedy approval procedures.

THE PROJECTS AIMED AT DEFENDING VENICE AND OTHER LAGOON SETTLEMENTS AGAINST THE RISKS OF FLOODING

Venice is located in the middle of the largest lagoon of the Mediterranean Sea, a body of water of about 550 sq km, communicating with the Adriatic Sea through three inlets (Lido, Malamocco, and Chioggia). The existing islands (mostly artificial landfills built during the centuries) cover about 50 sq km of the total surface; about 50 sq km are covered by marshlands, while the rest constitutes a very intricate hydraulic system formed by shallows and a tree-shaped network of canals of different size.

The major problem Venice and its Lagoon are facing today is the increased frequency of flooding, the so called "acqua alta" which, besides generating inconveniences to the population and damage to economic activities, represents a potential risk to the integrity of the fragile urban structures built throughout the centuries in the middle of a watery environment. Although floodings have been recorded throughout history, in recent years their frequency and levels have increased significantly.

The main reason for that is the relative change of the levels of water and land that have occurred since the beginning of this century: this difference amounts to about 23 cm. In this period of time, the average sea level has risen by about 11 cm, while that of the land has sunk by 12 cm due to a subsidence process mainly linked to natural compaction and to excessive draining of underground water beds (Gatto,1979; Fontes and Bortolami 1972). This phenomenon brings about serious risks if we consider two factors, one at the local and the other at the global scale: the importance which even minimal tide changes have in Venice and the possibility of sea level rise in the future.

In Venice, the average (not maximum) tidal excursion is about 1 m. In many of the world's littoral areas, the difference between high and low tide is far greater: on the Scheldt, in Holland, the range is about 3 m; in London 4 m; in Rainbow, Alaska, 8.40 m; at Saint Malo 10 m; and in the Bay of Fundy, in Canada, it is an even 15 m.

When confronted with figures of this kind, a change in the difference in level between land and sea such as the one that has occurred in Venice, would certainly be serious, yet controllable.

But in Venice, this difference matters, as confirmed by historical data. In the very first decades of this century, St. Mark's Square was flooded seven times per year. Today, flooding occurs on an average of more than forty times a year (with even higher peaks: in 1967 there were sixty-seven floodings).

As the city's altimetry varies, the extension of flooding depends on the level of the tide: when the sea reaches a level of +80 cm over the Punta della Salute datum (which occurs 40 times a year), Venice begins to be partially flooded. When the water level reaches +100 cm (about 7 times a year), approximately 6% of the town is prone to flooding; at +120 cm, about 40% of the town is impracticable; and at +140 cm, about 70% of the surface of the islands in the lagoon is flooded. Events of this magnitude or larger are less frequent, but they may occur, creating risks for the safety and stability of the residential areas in the lagoon. At the beginning of the century, an exceptionally high tide (such as the one that occurred in 1966) had a return period of about 800 years. Now, only a few decades later, that same return period would be about 200 years.

While the processes of subsidence today are relatively under control (prohibiting the draining of water beds has, in fact, eliminated subsidence due to the work of man, and the rate of natural subsidence, estimated at .4 mm per year, has an impact only in the very long term), the rise in sea level still constitutes a potential risk.

In spite of the uncertainties related to the long-term consequences of global warming, a sea level rise of 30 cm is within the realm of possibility in the next 100 years. This increase would have a significant impact on the city of Venice, due to the small difference of level between the emerged lands and the sea: with just a 30 cm increase, St. Mark's Square would be flooded more than 360 times per year.

Aside from to the inconveniences produced by flooding, there are other risks associated to the change in the relative levels of land and sea. The buildings of Venice are subject to degradation processes which can be attributed, in varying degrees, to the nature of the soil, to very old construction techniques, and to wave motion.

Subsidence and sea level rise have brought the water into contact with the brick and masonry with increasing frequency, flooding over the strip of semi-impermeable Istrian stone which has protected Venetian palaces for many centuries. Unlike the stone, the bricks are very porous and permit the capillary rise of salinity, a phenomenon which triggers off the pulverization of the mortar and the scaling of the masonry. All these interacting factors can cause long-term damages that are difficult to predict.

The physical decay of Venice is also connected to the severe demographic crisis the city has suffered in the post-war era. In 1992 the population of Venice dropped under 76,000, down from a maximum of 175,000 inhabitants recorded in 1951. From the point of view of urban degradation, this exodus brings with it the lack of maintenance of buildings and urban resources and the abandonment of many of the small islands of the lagoon, thereby accelerating the process of decay and increasing the threat to the city.

To cope with this problem, many solutions have been studied in the recent past, ranging from the narrowing of the width of the three harbor inlets of the Lagoon to the reduction of the hydraulic cross-section of the communicating channels between the lagoon and the sea. After many years of research and experimentation, it was concluded that the only solution that would provide a definite solution to the problem was the construction of a barrier system, able to stop the tidal flux at the three inlets for the needed period of time.

This solution alone, however, is not sufficient to provide complete safety to the area. This is why the Consorzio has defined a set of four projects to be implemented at the same time, which will provide a complete and long-term protection to the city and all the urban settlements in the lagoon.

THE MOBILE BARRIERS AT THE LAGOON INLETS

The mobile barrier project stems from the design activity that was developed during the 1970s, when the Ministry of Public Works entrusted a group of engineers with the task of designing a defense system. This proposal, completed in 1981, envisaged the construction of a mixed system of fixed barriers and mobile flap gates at the three inlets, so as to leave openings large enough for shipping traffic. The overall width of the openings was 1,070 m, as compared to a present extension of 1,700 m, with a reduction of the hydraulic cross-section of about 35% of the present value.

The rows of flap gates were to be housed in special foundation structures and consisted of cylindrical caissons filled with seawater in rest position and emptied, by injection of compressed air, when they had to reach working position. The project idea was approved, but the overall scheme was criticized from the point of view of its impact on the Lagoon environment, already subject to alterations due to the dredging of large shipping channels and the growing inflow of polluting substances from the watershed.

Following the conclusion of this project and the approval of the second Special Act for Venice (1984), the task of developing further studies and a new project scheme was assigned to the Consortium.

The new project for establishing the defense system was completed in 1992 and proposed considerable innovations compared to the previous one (CVN 1989a; CVN 1992). The fundamental criterion was to leave the hydraulic cross-section of the inlet channels unchanged, creating a system of barriers that covered the whole width of the existing openings.

The total length of the four openings (Chioggia, Malamocco, and two at the Lido inlet along the San Nicoló Canal and the Treporti Canal) is 1,580 m, representing about a 10% reduction of the hydraulic section with respect to its present size. The flap gate too, is different from the 1981 design, and is composed of a single, rectangular caisson, 20 m wide and 4 m thick, with a variable length depending on the depth of the channel (11 m at Lido and Chioggia; 6 m at Treporti; and 15 m at Malamocco).

To check the functionality of the flap gate, and to test the hinges and protections and the removal and maintenance systems, the M.O.S.E. (Electromechanical Experimental Module) was built: a life-size prototype of one element of the system, which has been operational for four years and which successfully concluded its activity in October 1992.

The Project envisages the construction of a foundation structure composed of large prefabricated concrete caissons, to be laid in a line on the bottom of the shipping channel. Besides providing the foundation of the system, this structure contains the passages for the plants (electricity, air, etc.), and it is shaped in such a way as to house the flap gates in rest position. The project contemplates the installation of seventy-nine flap gates, each of which, in working position, will reach an angle of about 50 degrees. Together, they will provide a surface necessary to resist the thrust of the sea, while allowing each flap gate to oscillate independently by about 5 degrees.

This reduces the thrusts affecting the foundations, since part of the energy of the waves is transmitted to the body of water in the Lagoon by means of the oscillating movement of the flap gates.

The barrier system is designed to contain, at the most, a difference of 2 m in the level of the sea and the lagoon, a situation far above the present variations. The length of the barrier has been designed in order to avoid the risk of waves washing over it, but especially to include in the project a variable element that has not yet been defined for certain: the possible increase of the level of the sea due to greenhouse effect. On either side of the inlets are the shoulders, with the prefabricated buildings for the utilities, control, and management of operations. In order to reduce the impact of these buildings on the visual environment, they have been completely housed inside the shoulder and stand only 1.50 m above ground level.

The final part of the work will take about a year and a half to complete. During this time, it will already be possible to start working on some complementary operations, such as experimentation with the foundation trench and the building of the large basins in which the prefabricate elements are to assembled. The time schedule expects work to be completed by the year 2000.

THE REINFORCEMENT OF THE COASTAL DEFENSES

The project of reinforcement of the coastal defences is, as well as the system of mobile barriers, essential to ensure an effective protection of Venice from the force of the sea. The present day coastal defenses along the littorals were built over 200 years ago by the Venetian Republic (the famous "murazzi").

Although work and reparation has continued since then, they now prove to be insufficient to ensure adequate protection to the Lagoon: littoral erosion and the long-term effect of wave action have created the need for a radical improvement of this vital system. A general project that defines the new coastal defense system was completed, and work started in 1991 on the most critical areas, to be gradually extended to almost 50 km of coastline, and completed in approximately five years.

The project foresees the implementation of a number of actions aimed at improving, and in many cases, at completely renewing the existing system of coastal defenses. In particular, the project comprises the following works:

1. On the Cavallino littoral in the northern section, the correction of the direction of the existing groines, the construction of new ones, and a complete beach nourishment program all along the littoral.
2. Along the Lido littoral in the central section, the construction of an artificial beach protected by five large groines, in order to protect the existing rock defenses.
3. Along the Pellestrina littoral, in the southern part, the construction of a suspended beach, protected by a number of groines and by a submerged berm. This will preserve and defend the historical structure of the "murazzi".

THE REINFORCEMENT OF THE EXISTING JETTIES AT THE THREE LAGOON INLETS

The existing jetties were built in a period from 1839 through 1934 in order to protect the harbor entrances from silting and to allow the navigation of modern vessels. They are now in a very deteriorated state, due to erosion and the sinking of the structures provoked by the deepening of the navigation channels, especially at the Malamocco inlet, which was dredged to a depth of -15 m in the 1960s following the construction of a major navigation channel leading to the industrial zone of Porto Marghera.

The aim of the project is to restore the existing structures, in order to improve their technical performance, both as protection of the waterways and as part of the overall system to defend the city against the high tides. The project has studied, in great detail, the characteristics of the wave pattern in the area of the jetties; the geomechanics of the existing structures and of their foundation soil; and the foreseeable evolutionary trends of the structural deformations. Based on this information, the reinforcement and consolidation measures needed have been defined and tested on mathematical and physical models. The goal of the project is to raise a suitably protected filtering structure, to prevent swell and currents from carrying away foundation soils that are characterized by a smaller particle size, thus hindering the progressive subsidence of the jetties. The work on the jetties will start in 1993, and will take about four years to be completed.

LOCAL PROTECTION AGAINST THE HIGH TIDES

The study of the present frequency of the high tides shows that, while tides over the +80 cm level occur about forty times a year, only seven on average, are higher than +100 cm. It would be sufficient, therefore, to protect the urban settlements up to the +100 cm threshold to reduce the number of closures of the mobile barrier system. This is considered a relevant aspect of the overall project, as it limits both the potential damage to the shipping traffic and does not induces significant fluctuations in the water exchange process between the Lagoon and the sea.

A number of interventions in the inhabited centers along the littorals have already been developed; in particular in the villages of Sottomarina, Pellestrina, S. Pietro in Volta, Malamocco, and Treporti. In these areas, the protection against flooding consists of new barriers to avoid the filtering of water due to the difference of pressure between the higher level of water in the lagoon and the lower inhabited centers and new embankments containing the drainage systems. The urban centers are, therefore, completely surrounded by this new safeguard, capable of defending them against tides up to 140-150 cm. Some of these projects have already been completed and are operational: the overall program will be completed in three years.

The applicability of a similar concept has been studied also for the case of Venice and the other historical centers in the Lagoon. Here, the problem is made more complex by the great value of the historic urban structure: a system of protection such as the one envisaged for the littoral areas is not feasible without altering the architectural nature of the City. Nevertheless, solutions able to protect the historical centers up to the +100 cm level have been designed, and proved to be feasible even in the most complex and important areas of Venice, such as Rialto and St. Mark's Square. Most of the solutions involve only minor changes of levels compared to the existing ones and can be implemented with full respect to the historical characters of the city.

THE PROJECTS AIMED AT REVERTING THE TRENDS OF ENVIRONMENTAL DEGRADATION

Throughout the centuries, the size, shape, and quality of the Venice Lagoon have evolved significantly due to natural and human actions (CVN 1988). In its history, three distinct phases can be identified:

1. *The period between 1300 and 1800*, characterized by the diversion of all the major rivers flowing into the Lagoon, the Brenta, Bacchiglione, Piave, and Sile. These early hydraulic works, whose effects are still visible alongside the Lagoon-land border, were essentially motivated by the silting processes of the Lagoon produced by the sediment transport of the rivers.

As the Lagoon was seen as an unique defense to Venice, the Republic tried constantly to prevent alterations of its morphology that would have endangered the security and the economic life of the city. Furthermore, the reduction of freshwater inflow was considered a measure against malaria. These interventions, albeit necessary, reduced significantly the sediment balance of the Lagoon.

2. *From 1880 to 1920*. This period was characterized by the construction of the jetties at the three port mouths, first at Malamocco (1839-1872), then at Lido (1882-1925), and lastly Chioggia (1911-1934). These works were aimed at resolving one of the major problems that the ancient Venetian had always encountered: the accessibility of the harbor. The increased ship size

required deeper channels, as well as easy maintenance of inlets made possible by the jetties. These interventions had, nevertheless, a devastating impact on the process of sediment transport along the coast and inside the Lagoon, further worsening the sediment balance and triggering an acceleration of the erosion processes.

3. *Period from 1920 to today*. This period is characterized by important transformations inside the Lagoon, mostly linked to: the dredging of ship channels (Lido-Marghera 1919-1930; Malamocco-Marghera 1960-1969); land fillings for agricultural and industrial uses (creation of the 1st and 2nd Industrial zones, and filling of the 3rd); and the increase of the surface of the fish farms along the border of the Lagoon, closed to tidal expansion. In recent years, another issue came to the forefront: the rapid deterioration of water quality and the natural environment of the Lagoon. Water pollution from urban, industrial, and agricultural sources is the principal factor of the decay of this environment, and it is severely endangering the Lagoon's self-regenerating capacity.

Recently, significant algal blooms have brought the issue of the decline of environmental quality of the Lagoon to the forefront. This explains why the Special Act of 1984 has added the goal of environmental restoration to the one of safeguarding Venice. Because of the change in its hydrodynamic system, human action, and pollution, one of the most important component of the Lagoon's morphology, the marshlands, have suffered significant reductions. Additionally, increased stream velocity, man-produced waves, and water pollution (death of vegetation) have reduced by 50% (from 100 to 50 sq km), the surface of marshlands in the Lagoon since 1900. This loss is to be regarded as a severe problem, not only because of the biological importance of these structures, but also because of their role in limiting the erosion processes within the entire Lagoon.

As a result of these processes, the present-day Lagoon shows a very different mix of lands and water surfaces as compared to the Lagoon of only 100-150 years ago. A rapid erosion process is characterizing today's evolution of the Lagoon. The loss of sediment transport from the rivers and the sea, the deepening of channels, and the disappearance of marshes, have determined an increase of the average water depths of the shallow areas of the Lagoon of about 30-40 cm during this century. The Lagoon shows today a net loss of about 1 million cubic meters of sediments per year, and is gradually turning into a marine system.

The environmental degradation's most visible symptom is the algal blooms that have reached, in certain areas, remarkable levels (up to 25 kg/cubic meter). The subsequent decay of these algae triggers a complex process of oxygen consumption and hydrogen sulphide release from the sediments which significantly affects the ecosystem. The disappearance of part of the typical fauna of the Lagoon has contributed to the development of a population of a massive population of mosquitoes (chironomidi) that infests the Lagoon and the city of Venice during the summer (Avanzi et al 1985).

Until the 1960s, the responsibility for water pollution was mostly attributed to emissions from the industrial areas and from urban sources. The building of treatment plants during the 1970s has partially reduced the share of pollution due to these activities, while the contribution of agricultural non-point sources of pollution has dramatically increased.

Eutrophication is today characterizing most of the areas of the Lagoon (CVN 1990), with concentrations varying according to the proximity to the source and to the season. The increase of nutrients inflow is not only responsible for the algal blooms, but also for the change of the quality of the sediments in the Lagoon, which triggers a long-lasting transformation of the entire biota systems.

Although scientific investigations have not yet reached final conclusions (CVN 1989b), a general estimate of the amount of nutrients discharged into the Lagoon annually ranges between 10,000 and 12,000 tons (N+P). The estimated distribution of the emission between the different sources is the following:

1. Urban - between 15 and 20%
2. Industrial - between 20 and 25%

3. Agricultural - between 40 and 60%
4. Atmospheric - around 10%
5. Sediments - not yet quantifiable, but likely to be significant

In order to cope with the very complex issues related to the environmental equilibrium of the Lagoon, the Consortium has developed a set of projects and experimental actions which are strictly related one to the other, and constitute a fundamental line of activity for the next decade, parallel to the action aimed at safeguarding Venice from high tides.

THE RESTORATION OF THE LAGOON MORPHOLOGY

The major goal of this project (CVN 1991) is to stop and reverse the erosion process of the Lagoon. In order to achieve this result, the processes affecting the Lagoon's morphology have been carefully studied, such as the sediment losses, the erosion, and the condition of use and exploitation of the basic natural resources in the area, etc.

In order to define the most appropriate and efficient measures, the Lagoon has been subdivided into ten areas with similar development trends, for which a mix of regulatory policies and direct intervention has been planned. The most relevant intervention that will be implemented are:

1. Reconstruction of marshlands, through the reuse of dredging materials that would be otherwise discharged into the Adriatic. Many experiments have already been successfully conducted in the Lagoon and show that the reconstitution of the marshland vegetation is possible within a relatively short period of time.
2. Maintenance of the complex hydraulic network of the Lagoon, presently altered by the erosion process, in order to allow water circulation in all areas and gradually re-establish the natural condition of water exchange.
3. Replanting of aquatic vegetation (phanerogams), which plays an essential role in stabilizing the shallow areas of the Lagoon and which have been severely damaged by man and by pollution.
4. The construction of sand by-pass, in order to recapture into the Lagoon, part of the sediments carried by the marine currents that are presently disrupted by the inlet jetties.

THE WATER AND SEDIMENT QUALITY CONTROL PROJECT

Coping with the problem of water quality decay of the Lagoon has become an urgent issue. As of today, only one action has been implemented; the collection of algae during the blooming season to limit the extent of anoxia produced by the decay of the macrophites during the spring and the summer. This intervention, a relatively costly one has, unfortunately no significant long-term impact and constitutes a limited solution to the problem.

In order to improve the quality of water and sediments in the Lagoon, many activities have been designed and are currently being tested. These activities include modifications of the hydraulic circulation pattern in the Lagoon in order to accelerate the water exchange process with the sea; the construction of areas of sweet water filtration before it is discharged into the Lagoon; and the treatment of sediments with mechanical tools or through artificial oxygenation.

Many of these activities need a careful test on their effectiveness and require accurate cost analyses before they can actually be applied at a larger scale.

A component of this project can be considered the partial reopening to tidal expansion of the fish farms located in the northern and southern part of the Lagoon. These areas, protected by embankments, are presently closed to tidal expansion because the low quality of the Lagoon water endangers the farming. Reopening these areas to tidal expansion in critical moments for the Lagoon (i.e., during the summer, when anoxic crises are more likely), could be beneficial to the conditions of the nearby open areas of the Lagoon. This is why a scheme has been proposed that allows the preservation of the fish farming activity, and at the same time the reopening of these areas to tidal expansion when it's most needed.

Finally, a fundamental aspect of the water quality control is the pollution control plan of the watershed, the responsibility for which belongs to the Regional Government (Regione Veneto 1990). The plan, whose target is the abatement of about 50% of the nutrient inflow in ten years time, is based on the construction of new sewage systems and treatment plants. It is now in the initial phase of development.

THE SUBSTITUTION OF THE PETROLEUM TRAFFIC IN THE LAGOON

The Venice Harbor ranks amongst the most important Italian seaports, because of the development along the Lagoon shores of a large industrial zone, founded in the 1920s and further expanded in the 1960s (CVN 1987; COSES 1990). The present traffic is about 27 million tons per year. Of this tonnage, over 40% is made of petroleum products, such as crude oil, gasoline, virgin naphtha, and chemical products, used by the large chemical, petrochemical and power plants of the area.

Oil traffic represents a severe potential danger for the Venice lagoon, which is a semi-enclosed environment, extremely fragile, and sensitive to water pollution. Recent modeling tests have shown that an accident involving an oil spill in the Lagoon would have a catastrophic impact on the survival of the peculiar biological aspects of this environment.

In order to increase the safety of the environment, a project has been defined to plan all the measures needed to eliminate the risks connected to oil traffic.

The scheme, which will be implemented gradually so as to allow the industry to absorb the extra costs involved, is based on the transfer of the different oil products to other northern Adriatic harbors, such as Ravenna and Trieste, which are better equipped for a safe transit and are already, or can be, connected with new infrastructures to the industrial zone of Porto Marghera. A priority is given, within this scheme, to the substitution of crude oil traffic and gasoline, alone corresponding to 70% of the total. This proposal requires an investment of about 200 billion Lire (around $150 million), for the construction of a pipeline from Trieste to Venice and allows the elimination of the traffic component considered most dangerous because of the size of the oil tankers.

CONCLUSIONS

Investigations and project design activities conducted in recent years have shown the viability of the comprehensive approach to environmental control and management of the Venice Lagoon ecosystem. This approach corresponds not only to a new technical awareness of the interrelationships between physical, environmental, and economic variables, but also to a growing social and economic requirement of integrated environmental control and guidance.

Nevertheless, this approach opens up a new set of issues, most of which are related to the management aspect of the overall process.

As in many other endangered areas in the world, project implementation in the Venice Lagoon is made difficult by social, political, and economic factors: the difficulties associated with them are often overcoming the technical aspects of the problem and concern all the actors, public and private, responsible for policy formulation.

These difficulties need to be faced and properly answered if the gap between policy formulation and implementation which has characterized the recent history of Venice, is to be closed.

REFERENCES

Avanzi et al. 1985. *Ripristino, conservazione ed uso del l'ecosistema lagunare veneziano*. Venezia: Comune di Venezia.

Bandarin, F. 1993. Safeguarding Venice: A challenge for environmental and hydraulic engineering. *Civil Engineering (forthcoming)*.

Bandarin, F. 1991. An integrated project for the Venice Lagoon. In *Proceedings of the Coastal Zone 1991 Conference*. Long Beach, Ca., July 1991.

COSES. 1990. *Porto Marghera. Proposte per un futuro possibile*. Milano: Franco Angeli.

CVN. 1987. *Rapporto sulla situazione attuale dei porti veneziani*. Venezia: Consorzio Venezia Nuova.

CVN. 1988. Venice Lagoon: Environmental aspects and the Venice Project. In *Proceedings of the Environmental Contamination 3rd Conference*. Venice, Sept. 1988.

CVN. 1989a. REA. *Riequilibrio e ambiente. Progetto preliminare di massima delle opere alle bocche*. Venezia: Consorzio Venezia Nuova.

CVN. 1989b. *L'inquinamento di origine agricola nella laguna di Venezia*. Venezia: Consorzio Venezia Nuova.

CVN. 1990. *Rapporto sullo stato attuale dell'ecosistema lagunare*. Venezia: Consorzio Venezia Nuova.

CVN. 1991. *Interventi di recupero morfologico. Progetto generale di massima*. Venezia: Consorzio Venezia Nuova.

CVN. 1992. *Progetto di massima delle opere alle bocche di porto*. Venezia: Consorzio Venezia Nuova.

Fontes, J.Ch. and Bortolami, G. 1972. *Subsidence of the area of Venice during the past 40,000 years*. Venezia: C.N.R., Laboratorio per lo Studio della Dinamica delle Grandi Masse.

Gatto, P. 1979. La Laguna di Venezia. Conservazione dell'ambiente fisico, subsidenza, idrodinamica lagunare e difesa dei litorali. In *Atti dell'XI Convegno della Societa Italiana di Biologia Marina*. Orbetello, 23 maggio 1979.

OECD. (1990). *Integrated management of the Venice Lagoon*. Paris: OECD, Environmental Committee, Group on Natural Resources Management.

Regione Veneto. 1990. *Piano direttore per il disinquinamento e il risanamento della Laguna di Venezia*. Venezia: Giunta Regionale.

9

The Role of the Public-Private Partnership in the Governance of Coastline

S. Thomas Gagliano, Esq.
The Jersey Shore Partnership, Inc.
Giordano, Halleran & Ciesla, PC, Middletown, NJ

ABSTRACT

The public-private partnership—an association of elected officials, business and labor leaders, academics, and others brought together to address a given set of issues—can help to eliminate fragmentation and strengthen the civility and reasonableness that government requires to get things done. This paper offers a case study in the way one such organization—The Jersey Shore Partnership, Inc.—succeeded in helping to obtain enactment of much needed legislation for coastal protection and how it can help plan for the future.

The problems of the governance of coastline are necessarily linked to the usual problems of governance. This paper is about the general problems of governing a particularly fragile coastline, the New Jersey shore, under the fiscally difficult, political circumstances of today and tomorrow.

CASE STUDY: SHORE PRESERVATION AND THE JERSEY SHORE PARTNERSHIP

During my years in the New Jersey Senate, 1978-1989, the issues of shore protection came up many times. They came and went with the storms that hit our shores. No solution was ever achieved because political divisions in New Jersey, reflecting political divisions nationwide, resulted in deadlock. Since the late 1970s, the Republican party has identified itself with the issue of tax reduction. "No new taxes" has been its slogan. The Democratic Party at times echoed this sentiment, which is very popular with the general public, but more often appealed to the perhaps somewhat contradictory desire by the public for more or improved services. Specifically, on the issue of shore protection, New Jersey Democrats had tended to favor a new tax on hotel and motel rentals as a revenue source. New Jersey Republicans tended to prefer the Realty Transfer Tax for shoreline protection. For a decade or more, the issue was deadlocked, as neither party had the votes in the Legislature to impose its preferred solution. The Governor at that time was a Republican who simply said that he would support whatever policy won the consensus of shore area legislators. None ever did.

Soon after the election of a Democratic Governor in 1989, the tax issue became even more acute. The new Governor had said during his election campaign that he did not foresee any tax increases. Once in office and confronted with declining revenues brought on by a severe recession, he reversed his stance. Seconded by a Legislature controlled by his own party, the Governor substantially raised taxes. This action provoked some of the most intense and widespread criticism I have ever seen in New Jersey. In the 1991 legislative elections, the Republic candidates made an unprecedented sweep of the State Assembly and Senate seats, obtaining such a large majority in both houses that they could override any veto by the Governor.

The Opening

By 1991, I was no longer in state government, but I remained interested in the issues of shoreline protection. At this point, forces much larger than political ones took a hand. In late October 1991, almost on the eve of the legislative elections, a major storm hit the New Jersey coast, causing millions of dollars of damage, including severe beach erosion. My feeling was that the time was finally right for effective political action.

The Players

Accordingly, with the full support and approval of my law firm, Giordano, Halleran & Ciesla, I organized a public-private partnership called the Jersey Shore Partnership, Inc. The Partnership's goal was, and still is, to advocate appropriate legislation and regulation before the State and U.S. Federal governments; i.e., laws designed to solve the problems of shoreline protection, beach, and ocean pollution, and then to address the needs of shore transportation, infrastructure, and economic development. Our membership consisted of three mains groups, each of them necessary for the organization's effectiveness. The first group consisted of local and county governments, which often feel shut out of state and federal government decision making. The second group was local businesses; not so much the major national and international corporations, but regional banks, utilities, my law firm, developers, and a variety of service and manufacturing industries. The third was academia, particularly those academics knowledgeable about the coastal engineering issues that are essential to any rational solution of the Jersey shore's many problems. Examples of the three enlisted groups are:

Government

The counties of Monmouth, Ocean, Atlantic, and Cape May, together making up a population census of 1.3 million, as well as several municipal governments formed the core of this component.

Corporate Sector

American Telephone and Telegraph (AT&T); Atlantic Electric; Business Journal of New Jersey; Casino Association of New Jersey; Chemical Bank; First Boston Corporation; First Fidelity Bank; Giordano, Halleran & Ciesla; K. Hovnanian Companies; International Flavor and Fragrances; Investors Savings Bank; Jersey Central Power & Light; Monmouth Ocean Development Council; New Jersey American Water Company; New Jersey Natural Gas Company; New Jersey Bell Telephone Company; PACO Pharmaceuticals, Inc.; Pine Belt Auto Group; Southern New Jersey Development Council; United Jersey Bank Financial; Wakefern Food Corporation; and numerous Chambers of Commerce were all willing to become involved.

Academia

New Jersey Institute of Technology, New Jersey Marine Sciences Consortium, Rutgers - The State University, Stevens Institute of Technology, Stockton State College, and Monmouth College comprised this third of the equation.

In the past, these groups had not been talking to each other in a systematic way, especially about the Jersey shore. A public-private Partnership provided them the opportunity to work together on problems of mutual interest and then together to take their policy recommendations to State and Federal lawmakers and to the bureaucracies. Because the Partnership represented not only expertise, but also institutions of considerable economic and political power, we were then and still assured of a fair hearing in Trenton and Washington, D.C. We are also assured of our share of attention from local news media which, in turn, exert considerable political influence. It is also helpful that in New Jersey, legislative districts representing shore communities now constitute 20% of all legislative districts in the state.

The Ante

For anyone not accustomed to the complexities of New Jersey politics, the Partnership's first major task—the search for a stable long-term, dedicated revenue source for shore protection—would have seemed fairly straightforward. However, as mentioned above, the Republican-controlled Legislature wanted a source, but it wanted it out of existing revenues. The Democratic Governor wanted an entirely new tax, but, because he had already been injured by the backlash against earlier tax increases, was reluctant to actually propose such legislation. Under normal circumstances one might expect a compromise along the lines of a modest surcharge on services, the proceeds to be funneled into a fund dedicated specifically to shore protection.

But, as always, it wasn't that simple. Although all of the shore area legislators in the four, coastal counties were Republican, they did not speak with one voice. Most supported dedicated revenues from the Realty Transfer Tax, but wanted no part of the hotel/motel tax. Others took exactly the opposite position. Still others wanted neither, preferring to raise the money from the general revenue fund, or from liquor licenses, or via a Constitutional amendment, or even from fees generated by the sales of special license plates bearing the phrase, "Jersey Shore." Estimates of the monies actually needed ranged from $15 million to $30 million annually over the next ten years, at least.

There was another complication. As many of you know, in 1988, Republican Party candidate George Bush was elected in part on the strength of a dramatically stated promise not to raise taxes. A couple of years later, he agreed to a budget compromise with Congress, controlled by the Democratic Party, which included such a tax increase. This compromise or reversal of position inspired considerable criticism of the President and probably contributed to his 1992 defeat. It is understandable that most New Jersey Republican legislators, having campaigned against the Democratic Governor's tax increases, were highly reluctant to advocate any new tax of any kind and even regarded the Governor's preference for "a new source of revenue" as a political trap similar to the one President Bush had fallen into not so long before.

Not only the legislature, but the members of the Jersey Shore Partnership were divided. Those representing real estate businesses opposed any increase in the Realty Transfer Tax. Those from the hotel and motel industry naturally opposed a tax on rentals. While a February "Shore Summit" called by the Governor with full participation by the Jersey Shore Partnership, provided a useful forum for airing these disparate views, it did not result in their resolution.

Poll data measuring public opinion showed that New Jersey voters were also divided. Two polls showed that New Jersey voters strongly desired shore protection, but one showed that 64% supported a tax increase for this purpose, while the other showed only 38% supported such an increase. The basic difference between the polls was the wording of the question: The language of the first poll posed a choice between a tax increase and destruction of New Jersey beaches, the others did not (Ackerman 1992, 1992). Support for a dedicated fund versus an annual appropriation was equally divided.

The Bid

Obviously, the time had come for some action to break the impasse and to crystallize opinion around a specific policy option. My own personal view had been that the source of the revenue was secondary in importance to the revenue itself. But, having been a State Legislator, I fully understood the pressures on the elected representatives involved, both the Governor and the legislators.

Accordingly, in May, at the height of the annual budget debate, and after consulting for a final time with my colleagues within the Partnership, it was announced that the Partnership supported a tax on hotel and motel rooms, on seasonal cottage rentals, and on campground space rentals. We endorsed the relatively painless idea (put forth by, among others, a Republican Congressman) of voluntary "Jersey Shore" license plate fees. We also advanced an entirely new, non-partisan idea for the institution of an authority that would issue bonds to raise monies for shore protection.

Not one of these proposals was enacted in 1992 by the new legislature, which as I stated above, was firmly against any new taxes. But our proposal had the intended effect: to accelerate the speed of actions by all parties, which it did in three ways:

First, I had been a Republican legislator. Understandably, many Democrats viewed the Partnership with suspicion, fearing that it might be or become a vehicle for the Republican Party. Our advocacy of tax proposals demonstrated once and for all that we were what we'd always said we were: a strictly non-partisan, public-private partnership with no narrow, political axe to grind. The bona fides of the Partnership now confirmed, we could engage in friendly discussions with both sides.

Second, we were calling the bluff, so to speak, of both the Republican Legislature and the Democratic Governor. To the latter we were, in effect, saying: "You will have serious political support if you publicly back a hotel motel tax, which you have done so far only by hints and in private". To the Republicans, we were, in effect, saying: "You have a veto-proof majority. If you don't want a new tax, then increase or dedicate an existing one to shore protection". To both sides we were saying: "The time for talk is over. Now is the time to act".

Third, we were keeping the attention of the news media. We had already enjoyed excellent support from the newspapers across the state, and in particular, the Asbury Park Press, whose readership is based along the northern, New Jersey shore. In a thoughtful series of articles and editorials, the Press galvanized public support for shore preservation, ensuring that elected officials kept their minds concentrated on this issue. We did our best to provide all interested media representatives with information and arguments for shore protection.

The Raise

Results were not long in coming. According to New Jersey law, a balanced budget must be in place by the end of June. The Governor's staff had negotiated, and the Legislature approved, an agreement with the Port of Authority of New York and New Jersey to pay the State of New Jersey $200 million for relinquishing leasing rights to the World Trade Center. These one-time revenues included $15 million dollars to be dedicated to shore protection in the 1993 fiscal year.

The immediate problem solved, the Legislature, under the outstanding leadership of Senate President Donald DiFrancesco and Assembly Speaker Garabed "Chuck" Haytaian, two non-shore legislators, and shore legislators State Senator Joseph Kyrillos and Assemblyman Steve Corodemus, went to work on the need for a stable, long-term revenue source. By October, a bill dedicating $15 million per year for shore protection from existing Realty Transfer Tax revenues passed both the State Assembly and Senate. Designed by Atlantic County Senator Bill Gormley, the bill had teeth in it: if the proper amount of funds were not set aside for shore protection, the state bureaucracy would lose all revenue from the tax for that year. It is therefore safe to say that the revenues should be relatively secure from accounting chicanery. The Governor gently campaigned against it and even threatened to veto the bill, but in the end he signed it and politically, he made the right decision.

Nature 'Sees'

A few weeks later as I was writing this article, the worst storm in decades hit the Northeastern United States coastline, causing further damage estimated at several hundred million dollars in New Jersey alone.

A Strong Hand

Although the 1992 storm sounded again a call for action, this case history shows the advantages of public-private partnerships as catalysts for obtaining reasonable and timely governmental action.

1. Public-private partnerships are largely immune to charges leveled against conventional lobbying organizations. They do not represent narrow "special interests" or "single-interest groups" which tend toward factions. They can and do include elected officials, business and labor leaders, and members of the scientific and academic community. These groups now speak to each other as equals, probably because the goals of the Partnership are not "owned" by any of them; instead all of them collectively have a stake in the outcome – a stake in the public good.
2. Because they are broad-based organizations, bringing together individuals and institutions of considerable community influence, public-private partnerships can gain the attention of elected and non-elected officials and the news media.
3. Public-private partnerships are issues-oriented without lapsing into ideological extremism. They, therefore, help to build up reasonable consent among the public. Unlike many "citizens groups", we do not depend upon the passions generated by the issues of the moment. We intend to stay in business a long time, offering a forum for responsible, reasoned opinion on issues facing the Jersey shore. As part of its first year activities, and to maintain public attention to our goals, the Jersey Shore Partnership sponsored the first Jersey Shore Economic Conference attended by 300 persons with the Governor, State Senate President, bankers, engineers and economists as speakers and panelists.
4. As non-partisan organizations, public-private partnerships can be more flexible in their policy recommendations than the political parties can ever be. If we change our position with changing circumstances, no one is angered or dismayed, so long as we keep our principal goals clearly in sight. We can also take positions "out in front" of elected representatives who, by necessity, must move very cautiously when public opinion is ambivalent. The penalties for leadership are less severe for us than for them.
5. Public-private partnerships can serve as an extremely useful link among various levels and kinds of government, business, labor, academia, the news media, and the general public. When the devastating December, 1992, storm hit the coast, the Partnership contacted 48 banks and savings and loan associations, requesting that they be part of the Storm Damage Bank Assistance Program. Many responded with discounted interest rates on loans and waiver of fees, while some made charitable contributions to agencies aiding storm victims.

These links foster unity instead of a fragmented approach that goes nowhere. Public-private partnerships also enable, potentially, the kind of long-range planning and implementation needed in an area as fragile and as diverse in needs as any coastal region.

THE FUTURE

Almost exactly one year after our founding, the Jersey Shore Partnership was able to help bring to successful completion, an effort to obtain a $15 million, stable funding source of public revenue for shore protection. We intend to continue to monitor the situation, recommend adjustments, and continue to lobby as needed. In the meantime, there are two, longer-range issues currently on our agenda.

The Master Plan

We have begun to formulate an updated, master plan for the New Jersey shore; a plan that will include environmental protection, shoreline protection, economic development concepts, regulations, revitalization, and transportation. With the participation of the New Jersey State government, all the Shore county governments, and we hope, an increasing number of municipal governments, we should be able to lobby effectively for such a plan. In fact, the Partnership should be a major contributor in drafting a plan, because we have the expertise drawn from a respected cross section of interested business and professional persons.

The Consequences of Storms

A fundamental issue with respect to shore protection has been whether or not the shoreline can be effectively protected at all in the long run. Some environmentalists point to steadily rising sea levels as proof that no human-made, artificial barriers can hold back the Atlantic Ocean for long. According to these individuals, public monies should not be used for beach replenishment, jetties, and seawalls, but for acquisition of properties damaged or destroyed by our periodic storms. Eventually, there would be little or no financial incentives to build or rebuild near the shoreline, and the shoreline would shift naturally with the tides, maintaining itself at no one's expense. In effect they say, the coastline is humanly ungovernable and when it comes to the coast, human governance can only misgovern. We should retreat, they say.

While I understand the argument, I also understand that there is now no major economic or political incentive to support it. Nor will there be in the immediate future. Human beings just don't give up easily, especially when their property and ways of life are at stake. Besides, retreat brings with it special problems of decay that will only be exacerbated if some properties are abandoned while others are not; some are repaired and others are not.

In order to look at each of these issues, the Partnership has formed a Coastal Resources Engineering and Advisory Committee, chaired by Dr. Bob Abel, President, New Jersey Marine Services Consortium and COSU III Co-Chairman, supporting research into improved methods of beach protection. As shore advocates, we non-scientists want to be sure that we have a working understanding of the new ideas in engineering that will enable us to respect environmental quality without walking away from multi-billion dollar investments.

Hurricane Andrew, the most destructive natural disaster in United States history, did little damage to the renourished beaches at Miami Beach and Key Biscayne, and the latest techniques in beach nourishment design have given us hope for our sand-starved Jersey shore beaches.

To us, restored beaches mean added benefits for tourism, greater recreational opportunities, storm protection, lower insurance costs, and higher land values.

The year-long effort of the Jersey Shore Partnership to gain stable funding for the first time in New Jersey was well worth the energy expended. And, as I look back, it stands as a reminder to me that public-private alliances make sense in a democracy and can be very effective political tools to build the support that is needed in today's fiscally difficult times.

On Sunday, January 3, 1993, less than a month after the latest most devastating storm, an article appeared in the state's leading newspaper, the Star Ledger, with the headline, "SENATORS MAY SEND ELECTION FUNDS TO THE BEACH". Again, Senate President Donald DiFrancesco, recognizing the great need, came forward and called for additional shore funding of $10 to $12 million dollars for 1993. He and Senator John Bennett, Deputy Majority Leader, a coastal advocate of long standing, clearly stated their position that the Jersey shore has a "high priority" status and must get more money. This time the discussion surrounding where the money would come from included taking it from the Gubernatorial Finance Fund, an interesting choice in a year when New Jersey will elect a governor and the only year in four where the election fund is called upon to finance the state's gubernatorial campaigns.

As this case study goes to press, the New Jersey Legislature starts its annual battle of the budget. It appears to me that for the first time, the New Jersey Shore not only has the attentions of the Legislature, but also has a broader-based coalition of supporters among the Legislators. Certainly, this is a good sign; a position of some power in the annual ranking of what gets funded and what does not. So for the time being, against the unfortunate backdrop of dramatically damaging coastal storms, the Jersey Shore Partnership has succeeded in doing what it set out to do. Our success to date, in no small measure, came about through the hard work of the public-private alliance established to deal with the prior lack of resolve and failure of leadership.

The coastline is now enjoying legislative priority status. Our goal has to be to maintain and improve upon that status. As we go forward, we need to help New Jersey learn from the storms, not run from them. Only in that way will we protect and preserve this "jewel" of New Jersey; the centerpiece of a multi-billion dollar tourism industry, the Jersey Shore.

REFERENCES

Ackerman, Alyn. "Shore Projects Backed." *Asbury Park Press*, 10 May 1992.

Ackerman, Alyn. "Shore Recapturing the Affection of New Jerseyans." *Newark Star Ledger*, 31 May 1992.

10

Legal Rules, Administrative Planning and Negotiation to Solve Clashing Interests in Coastal Zones: The Italian Model and Perspectives

Nicola Greco
Professor of Public Law of Environment, Scuola Superiore della Pubblica Amministrazione
Scientific Director of Edistudio, Roma, Italia

AN OVERALL LAW: "PROVISIONS FOR THE DEFENSE OF THE SEA"

The Italian coastline measures more than 7,500 kilometers (4,750 miles) and more than 30% of the entire population of the country lives on or near its shores - a total of about 17 million people. The development of towns, holiday villages, and major industries (shipbuilding, petrochemical, and metallurgical) which have found it economically convenient to site themselves along the coasts, has had a negative impact on the territorial and environmental balance of the coastal area and the sea.

In the spring of 1989, the latest, major recurrence of the grave phenomenon of eutrophication of the waters of the Adriatic Sea took place, probably due to the huge deposits of azotic and other chemical substances. In many areas, bathing has been prohibited because the waters were polluted to an extent deemed dangerous to human health. Obviously, this has caused serious difficulties for the tourist industry. Finally, a fierce, political battle broke out over the siting in coastal areas (Gioia Tauro and Brindisi) of major, coal- and gas-burning, electricity generators. While the existence of these facilities may be necessary for the country's economy and its fight against unemployment, they also threaten new and serious environmental pollution.

Recent events in those sectors are a repeat of similar ones which occurred in the past and which led Parliament to pass in 1982, a general law entitled "Provisions for the Defense of the Sea." The application and the administration of this law were entrusted to the Ministry of Mercantile Marine, which in Italy, is responsible for coastal lands belonging to the State, as well as for ports, navigation, and fishing rights. It also has the power to grant authorization for all production and building activities which either take place along the seashore or make use of the sea.

The 1982 law (Provisions for the Defense of the Sea) provides for

1. The drawing up of a "General plan for the defense of the sea and of the seacoast from pollution and the care of the marine environment" (Article 1).

2. The creation of a service to oversee and to prevent activities and occurrences which may be damaging to the marine environment, together with the building and/or purchase of special ships entrusted for this purpose to the national Navy (Articles 2-9).
3. The setting up of a technical "first-aid" service for the defense of the sea and the marine environment in case of accidents, with the provision of severe penalties to prevent the discharge of polluting substances by merchant naval vessels (Articles 10-24).
4. A procedure for instituting "marine reserves" for the protection of the environment and other scientific and cultural elements of value (Articles 25-30). The law includes an initial list of 20 areas (Article 31).

As a consequence of this important innovation, the law of 1982 provides for a far-reaching revision and a logical improvement in the entire central and peripheral administration of which the Ministry of Mercantile Marine is the focal point. A new service, to be known as the Inspectorate for the Defense of the Sea, is to be created within the Ministry. Additionally, the law provides for a staff of 700 persons for this Inspectorate, made up largely of technical and scientific personnel.

THE CRISIS IN CARRYING OUT THE LAW

Between 1983 and 1993, the activities anticipated by the 1982 law were put into effect slowly and with great difficulty. However, it has created a system for monitoring the level of sea pollution and the national "first-aid" service for accidents at sea which cause pollution. The construction phase of the new, coastal patrol fleet also began during this time.

For the most part, actions aimed at drawing up the "General plan for defense of the sea and the coastal area" met with little success, as did the institution of a system of "marine reserves." A technical and scientific body called the 'Sea Council' had also been trying during that time frame to formulate coordinated projects. Unfortunately, it met with methodological difficulties and with the opposition of many Regional and local authorities who actively undermined the Government's attempts to fulfill its responsibilities. To date (1993), only seven marine reserves have been created: Ustica, Egadi, Ciclopi, Capo Rizzuto, Tremiti, Miramare, and Torre Guaceto.

A fundamental point in the crisis of implementing the law has been the lack of revision and updating of all of the services of the Ministry of Mercantile Marine stemming from the institution of the new Inspectorate for the Defense of the Sea. The law required this revision to have been carried out within six months and anticipated the rapid initiation of the Inspectorate. However, at this moment, the service has only 50 employees.

THE MINISTRY OF ENVIRONMENT: A NEW ACTOR IN THE SAFEGUARDING AND PROTECTION OF THE MARINE ENVIRONMENT AND THE COASTS

The failure to implement the 1982 law is certainly due to technical reasons, not unconnected with bureaucratic obstruction, conflicts between the State and the Regions, and lack of funds. But an important political factor has probably weighed heavily on all this as well. Since August of 1983, the Government included in its program the institution of a Ministry of Environment. This goal was finally achieved under law n. 349, effectuated on 8 July 1986 and published in the Official Gazette of 15 July 1986 (162).

The new Ministry has functions of coordination and control in the field of the environment. However, it also has entirely new responsibilities, such as the regulation and application of the Environmental Impact Assessment (EIA) procedures as laid down by the 85/337/EEC directive. But the new Ministry also has taken over functions which were previously ascribed by law to other departments of the State, as in the example of assuming a large part of the responsibility for the safeguarding and protection of the marine environment and the coasts, which the law of 1982 had previously assigned to the Ministry of Mercantile Marine.

Under this new law, the Ministry of Environment is obliged to

1. Promote, preserve, and restore the condition of the environment over the whole territory of the nation (Article 1). Thus the new Ministry takes over the tasks concerning the marine environment and coasts which were assigned in 1982 to the Ministry of Mercantile Marine.
2. Draw up a concrete agreement with the Ministry of Mercantile Marine for the formation of a "General plan for the defense of the sea and the sea coasts" by issuing all necessary provisions for this purpose (Article 2, s. 8).
3. Undertake primary responsibility for the institution and management of the "marine reserves" (Article 2, ss. 9, 11, and 12).
4. Take the place of the Ministry of Mercantile Marine in authorizing discharge at sea by ships and airplanes and in all relations with international authorities over the pollution of the sea (Article 4).

Therefore, many of the procedures which were set in motion with considerable difficulty by the Ministry of Mercantile Marine for effective steps to care for the coastal environment and the sea (and which had not yet been brought to completion) will now have to be set in motion, once again, by the new Ministry of Environment. Furthermore, since the Mercantile Marine has been left with some responsibilities in this field, it will be necessary for the two Ministries to conclude a number of agreements before any concrete action takes place. We can, in fact, look forward to further bureaucratic complications and slow downs in administrative action, even though in principle, the creation in Italy of a Ministry of Environment is certainly a most important political move.

In reality, however, the Inspectorate for the Defense of the Sea, set up by the law of 1982, will continue to develop slowly, and at the same time there will continue to be growth in the technical staff and bureaucratic power of the Ministry of Environment which currently has 400 employees. Part of this staff will be dealing with the coastal and marine environment.

AN EXAMPLE OF CONFLICT BETWEEN PUBLIC AND PRIVATE INTERESTS: AMNESTY FOR ILLEGAL BUILDERS ALONG THE SEASHORE

As soon as he took office in 1986, the new Minister of Environment opened up another area of conflict which directly concerns the field of coastal management.

The recorded exodus to the coasts of large sections of the Italian population has had a serious, degenerative effect over the last few years. Along the seashore, over 1,500,000 buildings, both large and small, have been constructed; none of which were provided for in town development schemes and none of them authorized. Many of these buildings have done serious harm to the structural and environmental balance of the coast and have caused major and constant pollution of offshore waters. In the majority of cases, these buildings are not linked to the official public drainage systems of the local authorities; often such systems have not even been built, and the direct outlet for the discharge of sewage has therefore been the sea.

Law n. 47 of 28 February 1985, approved by Parliament during the discussion for the institution of the Ministry of Environment, provides for a general "amnesty" for all buildings and houses which have been put up illegally. These total more than seven million throughout the whole country and 1,500,000 of them are on the coast. The object of this law was to bring some order into the administration of the towns, but it was also aimed at bringing in a very substantial financial return (estimated at about 6,000 billion liras) through the payment of a fine by the illegal builders who request the "amnesty". A major part of this sum should have gone to the local authorities for the construction of public works which would help to restore the environment and control various forms of pollution. These programs were to be carried out under the auspices of the Ministry of Public Works.

It is the Minister of Public Works, therefore, who was given authority by the 1985 law, to run the major, town planning reform. But the Minister of Environment soon came into conflict with him, for in August of 1986, he ordered the local mayors not to grant the "amnesty" to owners of buildings along the seashore which were not linked up to any effectively functioning drainage system. The Minister of Environment, in a circular addressed to the Mayors, asserted that even after the amnesty, these buildings "will continue to pump sewage into the sea and that they will therefore be unacceptable and in conflict with the duty of the new Minister to safeguard the quality of the environment, including that of the sea and coasts."

The conflict between the Minister of Environment and the Minister of Public Works was resolved by the Government through a compromise: the local authorities were aided financially to speed up their building of services along the coast, while in the meantime, they were able to provide the amnesty for all those owners of buildings and houses along the coast who applied for it.

This is yet another instance of the conflict between the public interest in preserving the environment and other public and private interests. This conflict makes the political and legislative path toward an effective and rational management of the Italian seas and coasts much more difficult.

A SHORT CONCLUSION

Today, March 31, 1993, the Italian coasts and seas are still waiting for the "General plan for the defense of the sea and of the seacoast from pollution and the care of the marine environment" provided by the law in 1982. The Minister of Mercantile Marine and the Minister of Environment are still searching for an agreement on this matter. In other words, conflict, negotiation, and formal compromises among the many and different public and private interests involved seem to be the actual model for the government and administration of the coastal zones. It is not an effective model.

11

The Coastal Use Framework as a Methodological Tool For Coastal Area Management

Adalberto Vallega
Istituto di Scienze Geografiche, Università di Genova,
Genova, Italy

ABSTRACT

According to the concept of *Integrated Coastal Area Management* (ICAM) as introduced in *Agenda 21*, Chapter 17, and agreed upon at the *United Nations Conference on Environment and Development* (UNCED), as well as its supporting, sustainable development principle, the consequent methodological needs are taken into consideration. In this respect, the representation and investigation of the coastal use framework is considered to be the initial stage of the methodological track. The framework is constructed using the following steps: (1) the setting up of a three-level classification whose function will be to provide a taxonomical tool useful for investigations on regional, national, and local levels and (2) conducting a matrix-based analysis of the location and spatial implications of coastal uses. Based on this data, relationships between various uses are considered and an explanation pattern, based on the concepts of spatial differentiation of coastal uses and coordination between them, is presented. The final part of the methodological approach is concerned between uses and the coastal ecosystem.

THE MESSAGE FROM UNCED

As is well known, *Chapter 17* of UNCED's *Agenda 21* provides a detailed code of conduct and guidelines for setting up sustainable development-inspired policies on coastal area management on global, regional, and national levels. Integrated Coastal Area Management (ICAM) is, at the same time, the starting basis and a core objective of this *Agenda* for oceans and seas, not only because it specifically deals with related areas, but also because it could influence policies for other key ocean issues such as the management of living resources in national seas. Based on that approach, it follows that research must first establish methodological tools consistent with ICAM. In this respect, future plans must take into consideration that, historically, integrated management has been the final step in the teleological evolution, in which decision-making centers have passed from one or a few coastal use-based management patterns to multiple, use-based patterns, to, during the 1980s, comprehensive coastal area management, aimed at encompassing all existing uses. *Integrated* is a further step, implying not only that all coastal uses are dealt with, but that: (1) they are developed so as to refrain from altering the coastal ecosystem, and that (2) as great number of uses as possible be reserved for future generations.

At present, the management background which is able to support methodologies consistent with this objective is the general system theory. Here, there is no room to deal with the conceptual endowment of this theory, since attention has to focus on the methodological issue. In this respect, it is appropriate to consider only the first stage of the methodological path which should be followed according to the general system theory; namely, the analysis of coastal uses and their environmental implications.

THE COASTAL USE FRAMEWORK

The basic concept for the creation of the necessary methodological tools is that of *coastal structure*, regarded as consisting of two modules:

1. One or more coastal ecosystems
2. The set of existing coastal uses

As a result, the assessment of the coastal structure is focused on four sets:

1. The set of coastal uses
2. The set of relationships between uses
3. The set of biotic and abiotic components of the coastal ecosystem
4. The set of relationships between coastal uses on the one hand, and the ecosystem on the other.

In order to deal with this subject, first sea uses are to be clustered. As is well known, literature has usually constructed two level-based classifications consisting of categories, which are, in turn, subdivided into kinds of uses. In this respect, the *global marine interaction model* (Couper 1983) has played a leading methodological role. Anyway, in order to provide a more specific tool for sea management, a *coastal use framework* can be set up, based on three levels:

1. upper level - categories of uses
2. intermediate level - subcategories of uses
3. lower level - kinds of uses

In its total extent, this framework includes eighteen categories of uses, sixty-four subcategories of uses, and 250 kinds of uses.

The categories are:

1. Seaports
2. Shipping carriers
3. Shipping routes
4. Shipping navigation aids
5. Sea pipelines
6. Cables
7. Air transportation
8. Biological resources
9. Hydrocarbons
10. Metalliferous resources
11. Renewable energy
12. Defense

13. Recreation
14. Waterfront man-made structures
15. Waste disposal
16. Research
17. Archaeology
18. Environmental protection and preservation.

The whole framework was presented in 1992 (Vallega) and in the *Appendix: The coastal use framework*. While there is no room to describe it in detail in this paper, its complexity can be understood by looking at the category *5.sea pipelines*, which itself can be divided into five subcategories:

1. Slurry pipelines
2. Liquid bulk pipelines
3. Gas pipelines
4. Waste pipelines
5. Water pipelines

Again as an example, the subcategory *5.1 slurry pipelines*, can be divided into these six kinds of uses:

1. Fine coal slurry pipelines
2. Coarse coal slurry pipelines
3. Limestone slurry pipelines
4. Phosphate slurry pipelines
5. Ore slurry pipelines
6. Copper slurry pipelines

Two methodological remarks are appropriate. First, this framework, as well as any other framework supported by the same method, has only an ephemeral role, namely, it is a reference point to be borne in mind in order to help group together the sea uses of a given marine area. Between the *a priori* build-up of the coastal use framework and its application in the investigation and management of given marine areas, a methodological, positive feed-back exists; starting from the *a priori*-built framework, the frameworks relating to specific coastal areas can be sketched in. In turn, the setting up of frameworks designed for specific coastal areas, leads us to adjust and implement the *a priori* coastal uses framework serving as a reference basis, etc., until a circular, methodological path comes to light, similar to those usually developed by any model-based methodology.

Secondly, the coastal use framework is necessarily references to a past or present time. When it is concerned with the future, the expected evolution of the sea structure is to be investigated through prospect-oriented analyses, building up scenarios on man-sea interaction.

LOCATION AND EXTENT OF COASTAL USES

Two issues are worth considering: (1) the location of coastal uses and (2) the extent of the coastal ecosystem subject to the implications generated by those uses.

As far a the first factor is concerned, two kinds of extent are relevant for coastal management; that of the ecosystem and that of the marine national jurisdictional belts.

The *ecosystem* is to be considered through both its abiotic and biotic components. As far as the abiotic components are concerned, it is worth considering those parts of the continental margin in which the coastal use is located; namely, the shoreline, the continental shelf, the continental slope, and the continental rise, as well as the overlying waters. For instance, recreational uses are located on the shoreline and seaports usually involve the shoreline and the continental shelf, while waste disposal pipelines can extend up to the slope and rise. As far as the biotic components - namely food chains -

are concerned, it could be useful for ICAM to distinguish those uses which exploit the food chain as a whole from those which exploit only their demersal or pelagic components.

In determining the *legal extent*, coastal uses are to be clustered according to where they are located within the internal waters, the territorial sea, the continental shelf, the Exclusive Economic Zone, and the Exclusive Fishery Zone. The importance of this classification stems from the circumstance that coastal, island, and archipelagic states that have different rights and subsequent management capabilities, according to the jurisdictional belts in which uses are developed.

Shifting from this aspect to the analysis of the *spatial extent* of the implications generated by coastal uses, four sets can be distinguished.

1. *Uses located onshore which generate no coastal sea impacts*, such as manufacturing plants not related to maritime transportation and not producing riverine or marine discharges
2. *Uses located onshore which generate impacts on the coastal sea*, such as seaports, recreational uses, and sand and gravel dredging
3. *Uses located offshore which do not generate impacts on the shoreline and/or land*, such as navigational aids and research installations
4. *Uses located offshore which generate impacts on the shoreline and/or land*, such as offshore oil unloading port terminals.

THE COASTAL USE-USE RELATIONSHIPS MODEL

According to the current methodological approaches, models of coastal use frameworks, as well as frameworks including both ocean and coastal uses, lead to the establishment of a square matrix (Table 1) in which the relationships between uses are represented. this matrix can be based on various approach levels.

1. Only on the categories of coastal uses, which provides only a general overview of the management content *(upper level of use-use relationships)*
2. On both the categories and subcategories of coastal use, which gives shape to a quite detailed overview *(intermediate level of use-use relationships)*
3. On the categories, subcategories, and kinds of use which provide a very detailed framework, essential to developing planning *(lower level of use-use relationships)*

This matrix could be called *coastal use-use relationships model*. There is no room to display the complete model and, *inter alia*, it does not seem necessary. As an example, the part of the model showing the relationships between *defense* (category 12 of the coastal use framework) and *research* (category 16) as represented in Table 1, includes only the subcategories (intermediate level of the coastal use-use relationships).

It is no use stressing that this coastal use-use relationships model is a characteristic product of a structuralist approach, aiming at identifying coastal structures and assessing the relationships between the elements of the structure. From the methodological point of view it is to be taken into account that, by its nature, it undergoes two constraints.

1. In itself it is not a tool for diachronic analysis so, when the evolution of coastal use framework is to be investigated, complementary methodological tools are to be applied.
2. Only bilateral relationships, use-to-use, can be represented, so complicated networks or relationships which simultaneously involve a set of issues, cannot be displayed.

Table 1

The Coastal Use-Use Relationships Model: Defense and Research

RESEARCH FIELDS Xj interacting with	Water Mass	Seabed and subsoil	Eco-system	External environment interaction	Special areas and particularly sensitive areas	Sea Management
DEFENSE X_i						
Exercise areas	B2	A	B2	A	B2	A
Nuclear test areas	B2	B2	B2	B2	B2	A
Minefields	B1	B2	B1	A	B2	A
Explosive weapons	B2	B2	B2	A	B2	A

Types of relationships

A: No existence of relationships
B: Existence of relationships
B1: Neutral relationships
B2: Conflicting relationships
B3: Reciprocally beneficial relationships
B4: Relationships beneficial to use x_i
B5: Relationships beneficial to use x_j

The use of the model implies that use-use relationships are classified. From the methodological point of view, this is a very delicate concern because there does not exist a consolidated procedure to create a taxonomy for use-use relationships. As an example, the mentioned *global marine interaction model* was based on the following classification:

1. Harmful or conflicting interactions
2. Potentially hazardous interactions
3. Mutually beneficial interactions
4. Harmful to activity at matrix right
5. Beneficial to activity at matrix left

In any event, the assessment of relationships between coastal uses should be consistent with the specific needs of planning, so classifications different from the global marine interaction model can be profitably used. In addition, classifications can be built up according to two, alternative procedures:

1. A *deductive procedure*. This first sets up an *a priori* classification serving as a starting point, then moves from it to assess specific use frameworks and, in its terminal phase, provides elements to implement the *a priori* framework.
2. An *inductive procedure*. This moves from one or more specific use frameworks, set a classification and then implements it through the analysis of other specific use frameworks.

Both procedures give shape to a *circular methodological path*, in which the theoretical approach and the results from the case studies interact.

The following is the suggested, tentative cluster of *coastal use-use relationships* consisting of a square matrix [$CU._i$, $CU._j$]:

1. *No existence of use-use relationships*
2. *Existence of use-use relationships*
   a) Neutral relationships
   b) Conflicting relationships
      1) $CU._i$ giving conflicting impulses to $CU._j$
      2) $CU._j$ giving conflicting impulses to $CU._i$
      3) $CU._i$ giving conflicting impulses to $CJ._j$, and *viceversa*
   c) Beneficial relationships
      1) $CU._i$ giving beneficial impulses to $CU._j$
      2) $CU._j$ giving beneficial impulses to $CU._i$
      3) $CU._i$ giving beneficial impulses to $CU._j$, and *viceversa*

It is self-evident that as far as coastal management is concerned, conflicting relationships have the highest relevance. In this respect, as a tentative approach, the following framework of conflicts between coastal uses can be assumed (Vallega 1992).

1. *Location-generated conflicts*. They occur where two different coastal uses cannot be located in the same place. Conflicts between mercantile navigation and naval exercise areas, as well as between offshore oil and gas exploitation installations and submarine cables, for example.
2. *Organization-generated conflicts*. They occur where a coastal use is organized in a way that conflicts with the organization of another use. For example, maritime transportation serving oil and gas offshore exploitation platforms (crude oil and product carriers, supply vessels, etc.) on the one hand, and yacht racing and cruising on the other.
3. *Environmental implication-generated conflicts*. These take place where a coastal use brings forth environmental impacts that another use cannot tolerate. As an example, the warm water discharges from coastal thermoelectric plants alter the coastal ecosystem preventing the establishment of marine parks.
4. *Aesthetic feature-motivated conflicts*. These occur where one use gives shape to a picture that the other use cannot tolerate. As an example, this incompatibility occurs between coastal manufacturing plants and recreational facilities.

THE DIFFERENTIATION-COORDINATION REFERENT

At this point, methodology should face the evolution of the whole range of coastal uses. In order to facilitate reasoning on this issue, the *differentiation-coordination referent* must be taken into account. As can be seen from Figure 1, this referent consists of a three-axis diagram: differentiation, coordination, and time.

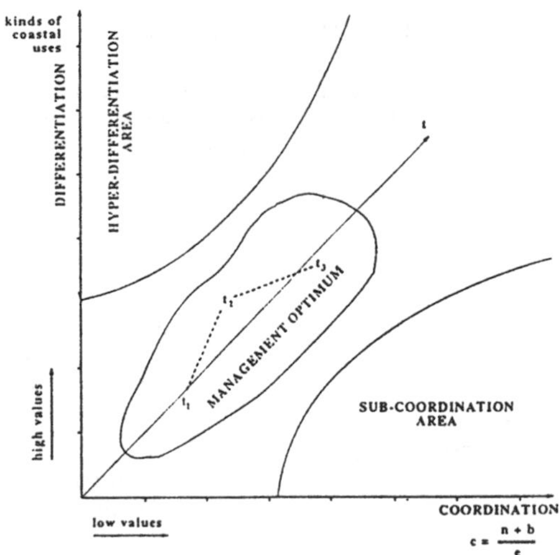

Figure 1. The Coastal Differentiation-Coordination Model

The *differentiation axis* (Y) represents the level of complexity that the set of uses acquires because of the growth of the exploitation of sea resources and the environment. Differentiation depends on three variables:

1. The number of kinds of coastal uses
2. The number of uses pertaining to each kind
3. The technological level on which uses are managed

Simply, the model presented here includes only the kinds of coastal uses. The values increase, moving from the origin of the diagram.

The *coordination axis* (X) shows the degree to which the coastal uses are involved by relationships, bringing about cohesive energy. This implies that conflicting relationships are regarded as reducing the coordination level, and neutral and beneficial relations as acting in the opposite direction. As a first approach, coordination could be expressed by the relation

$$(n + b): c$$

where n is the number of neutral relationships; b is the number of beneficial, and c the number of conflicting ones. The value of this ratio decreases, moving from the origin of the diagram.

This *time* axis represents the evolution of differentiation and coordination. At a given time, which is characterized by a given level of technological advance, organization patterns, and other features:

1. The number of kinds of coastal uses cannot exceed a given threshold because beyond that, the differentiation of the set of uses becomes too high to be managed.
2. The ratio $(n+b):c$ cannot slow down below a given threshold because below that, the coordination between coastal uses becomes too weak to enable decision-making centers to manage uses.

The set of values of differentiation which range above the differentiation threshold form the *hyper-differentiation area*. The set of values of the ratio (n+b):c below the coordination threshold form the *subcoordination area*. Between these two critical areas the *differentiation-coordination optimum* extends. It is a management area within which the number of kinds of coastal uses and the relationships between them-and factors influencing them as well-change without transferring the set of coastal uses into critical situations.

THE SEA USE-ENVIRONMENT RELATIONSHIPS MODEL

The environment is thought of as consisting of (1) biotic and (2) abiotic components of the ecosystem, and (3) the natural context. The latter can be assumed as the reference basis for evaluating if and where environmental impacts take place. Bearing in mind the complexity of the coastal ecosystem and the need to distinguish the coastal area from the ocean, the scheme presented in Table 2 develops.

Table 2

The Components of the Marine Environment

Coastal Area	Ocean Area
Emerged Land Surface	
Periodically Emerged Land Surface	
Sea Surface	**Sea Surface**
Water Column:	**Water Column:**
upper layers	upper layers
intermediate layers	intermediate layers
lower layers	lower layers
Seabed	**Seabed**
Subsoil	**Subsoil**

As a consequence, each kind of use can be evaluated according to this pattern in order to decide whether it is related to ocean and/or the coastal ecosystem and which components of the ecosystem are involved.

Relationships between uses and the coastal ecosystem acquire unusual levels of complexity in coastal management. In this respect, a matrix involving the mentioned components of the coastal ecosystem (lines) and coastal uses (columns) can be imagined. In this, relationships are represented according to their directions:

1. Inputs from the ecosystem to coastal uses
2. Inputs from coastal uses to the coastal ecosystem
3. Inputs moving in both directions

Inputs moving from coastal uses to the ecosystem are the most important for management. In accordance with the implications they are able to generate, the following taxonomic framework can be considered:

1. U_b, inputs beneficial to the coastal ecosystem
2. U_n, inputs neutral for the coastal natural context, because they cause neither damage nor benefits to the coastal ecosystem and/or the geomorphological assessment
3. U_r, inputs harmful to the coastal natural context, calling for protection and preservation
4. U_z, inputs hazardous to the coastal natural context calling for contingency plans or similar tools

Focusing attention on the evolution of coastal use structure, the evolution of a relationship over time is evaluated in order to understand whether it leads the coastal ecosystem only to simple adjustments or the morphogeneses. In this respect, as can be seen from Figure 2, coastal management can be represented by the *coastal exploitation-change* model which shows two areas of concern.

1. The *adjustment area*. In this area the set of neutral and beneficial coastal uses brings about inputs larger than those generated by the set of harmful and hazardous coastal uses. The latter categories of uses not only are less extended but are kept under control also, thanks to the role

played by beneficial uses. The amount of global biological and chemical oxygen demand-which can be regarded as a significant parameter of the pollution of the coastal marine area-grows so slowly that it does not alter the ecosystem. As a final result, the ecosystem only undergoes adjustments.

2. The *collapse area*. The set of harmful and hazardous coastal uses brings about inputs larger than those generated by the set of neutral and beneficial uses. In addition, the latter is not able to reduce the effects produced by harmful and hazardous uses. As a consequence, the global demand for biological and chemical oxygen grows to the point of generating a collapse in the ecosystem. In this case the morphogenesis of coastal management can take place.

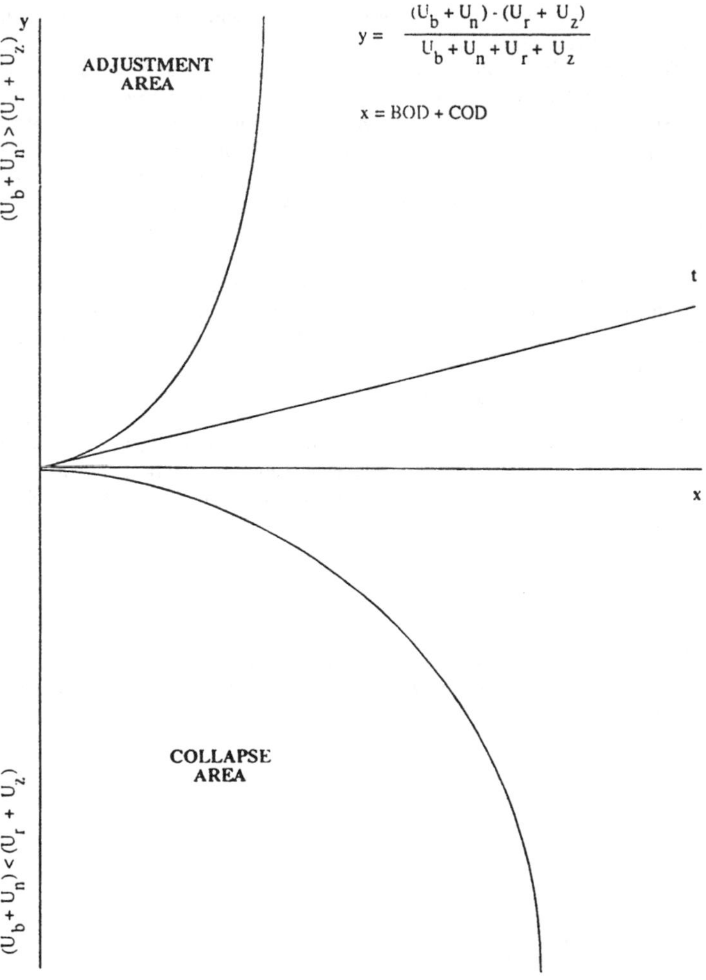

Figure 2. The Coastal Exploitation-Change Model

CONCLUSION

In creating methodologies consistent with ICAM, one needs to deal with concerns quite different from conventional ones for three reasons.

1. Coastal ecosystems and their ruses should be considered through their interaction, which implies the creation of methodologies crossing both natural science- and social science-based approaches.
2. The coastal structure, consisting of the ecosystem and its uses, is to be analyzed in its evolution, which implies the establishment of relevant, diachronic methodologies focused on the changing phases of both the ecosystem and its uses.
3. The goals of management are basically those of preventing human-generated alterations of the ecosystem and maximizing the degrees of freedom with which future generations will be able to use coastal resources.

As is self-evident, UNCED objectives and general guidelines call for great efforts, especially from the methodological point of view. If it is agreed that, at the present state of the art, the most appropriate basis upon which to develop transdisciplinary methodologies consistent with this political turning point is provided by the general-system theory, it implies that:

1. Contrary to conventional approaches, the ecosystem is not regarded as a trivial machine, but one that is capable of giving a range of responses to the same input.
2. The self-regulation and sustaining properties of the ecosystem are considered to their whole extent.
3. Sea uses and the ecosystem are regarded as two modules of a unique system.
4. Adjustments and morphogeneses are the two, main conceptual reference categories used to assess the evolution of the ecosystem, particularly when its implications in sea uses is the core of research.
5. As a consequence, management patterns are based on a non-deterministic view.

This concern has taken shape, not only in scientific thought and research on man-environment relationships *per se*, but it acquires special importance when relationships between human pressure and relevant activities on the one hand, and the coastal ecosystem on the other, are concerned. The prospect of achieving results mainly depends on the interactions that can be set up between scientific thought and management principles and methodologies.

REFERENCES

Couper, A.D. ed 1983, *Atlas of the Oceans*. London: Times Book.

Vallega, A. 1992. *Sea Management. A Theoretical Approach*. (Chapter 7 and Appendix D) London: Elsevier Applied Science.

APPENDICES

Appendix 1
The Coastal Use Framework: General Considerations

The *coastal use framework* derives from a general, *sea use framework*, including both coastal and ocean uses, which is represented in its entirety in Appendix A *Sea Management. A Theoretical Approach*. Since the coastal uses are numbered as they appear in the basis sea use framework, some numbers do not appear in this version because they refer to ocean uses.

The concept of sea use, from which that of coastal use derives, only appears to be easy to formulate and to apply in the investigations of coastal and ocean use structures. To begin with, it is self-evident that this concept embraces two others: facilities and social behavior. Facilities are concerned in the investigation because plants, installations, and man-made structures (seaports, offshore platforms, and pipelines) are considered. While the concept of coastal use could be related only to facilities or only to behavior, it is very difficult for a model, if it is consistent with only one of these criteria, to be useful for managerial purposes.

When a sea use framework includes both facilities and activities:

1. Some sectors of marine resource exploitation are considered only through the kinds of activities of which they consist.
2. Other sectors are considered through the kinds of facilities which they employ.

Including activities or facilities in the coastal use framework depends on the role that the framework has: in some cases it is better to reference activities; in other cases it is more useful to refer to man-made structures located at sea. For instance, in the following coastal use framework, recreational uses in offshore areas (category 13, subcategory 2) are considered through activities such as fishing, sailing, cruising, etc., while the exploration, exploitation, and storage of oil and gas (category 9) are considered through plants and installations.

Nevertheless the real methodological issue does not consist of opting for a framework based only on facilities or only on activities or-as has been done here-contextually on both; in fact all components (facilities and activities) lead to considering *functions of man's presence on the sea*. The real issue is to identify the objectives for which a function-expressed by a facility or an activity-has been set up. At the present time these objectives occur:

1. The implementation of progress in the knowledge of the sea
2. The advance in the exploitation of non-renewable resources
3. The implementation and diffusion of renewable resource exploitation, particularly if these act as substitutes for non-renewable resources
4. The progress in the management of the marine environment

According to the view supporting the policy of the United Nations and these numerous governmental and intergovernmental organizations, as well as the scientific world, coastal uses should be clustered and investigated in such a way as to make understandable:

1. Whether and how the exploitation of non-renewable resources is being rationalized
2. Whether and how the exploitation of renewable resources is advancing, with the aim of minimizing the exploitation of non-renewable ones
3. To what extent management tends to minimize environmental impacts and, in particular, to avoid man-induced morphogeneses in marine ecosystems

Keeping these considerations in mind, *the coastal use framework* is presented next. Since this is not the proper forum to present the framework in its entirety, only *Category I* (seaports) is displayed with its subcategories and kinds of use, while the rest of the categories are displayed only with their subcategories. This is the introductory step in the setting up of the *coastal use-use relationship model*. As can be seen in this framework, coastal uses have been considered *sensu stricto*: only facilities and activities that, by their nature, are functionally tied to the sea have been included in the framework.

Appendix 2
The Coastal Uses Framework: Category 1 - Seaports

Subcategories	Uses
1.1 waterfront commercial structures	1.1.1 liquid bulk terminals
	1.1.2 solid bulk terminals
	1.1.3 multiple bulk terminals
	1.1.4 general cargo terminals
	1.1.5 lo-lo container terminals
	1.1.6 ro-ro container terminals
	1.1.7 lo-lo and ro-ro terminals
	1.1.8 heavy and large cargo terminals
1.2 offshore commercial structures	1.2.1 Catenary Anchor Leg Moorings (CALMs)
	1.2.2 Exposed Location Single Buoy Moorings (ELSBMs)
	1.2.3 Single Buoy Moorings (SBMs)
	1.2.4 Single Anchor Leg Moorings (SALMs)
	1.2.5 Articulated Loading Platforms (ALPs)
	1.2.6 Mooring towers
	1.2.7 Single Buoy Storages (SBSs)
	1.2.8 Single Anchor Leg Storages (SALSs)
	1.2.9 floating transhipment docks
	1.2.10 artificial islands
1.3 dockyards	1.3.1 ship building dockyards
	1.3.2 ship repairing dockyards
	1.3.3 oil and gas platform building dockyards
	1.3.4 oil and gas platform maintenance dockyards
1.4 passenger facilities	1.4.1 cruise ship terminals
	1.4.2 ro-ro carrier terminals
	1.4.3 hydrofoil terminals
1.5 naval facilities	1.5.1 naval dockyards
	1.5.2 surface vessel terminals
	1.5.3 submarine terminals
	1.5.4 weapon storages
	1.5.5 nuclear material deposits
1.6 fishing facilities	1.6.1 vessel terminals
	1.6.2 fishery facilities

1.7 recreational facilities	1.7.1 sail-propelled vessel terminals 1.7.2 engine-propelled vessel terminals 1.7.3 wind surfing facilities 1.7.4 yacht racing facilities 1.7.5 semi-submersible and submarine vessel facilities

Appendix 3
The Coastal Use Framework: Other Categories (abbreviated version)

2. Shipping, Carriers	2.1 bulk vessels 2.2 general cargo vessels 2.3 unitized cargo vessels 2.4 heavy and large cargo vessels 2.5 passenger vessels 2.6 multipurpose vessels
3. Shipping, Routes	3.1 short-sea routes 3.2 passages 3.3 separation lanes
4. Shipping, Navigation Aids	4.1 buoy systems 4.2 lighthouses 4.3 hyperbolic systems 4.4 satellite systems 4.5 inertial systems
5. Sea Pipelines	5.1 slurry pipelines 5.2 liquid bulk pipelines 5.3 gas pipelines 5.4 water pipelines 5.5 waste disposal pipelines
6. Cables	6.1 electric power cables
7. Air Transportation	7.1 airports 7.2 others
8. Biological Resources	8.1 fishing 8.2 gathering 8.3 farming 8.4 extra food products
9. Hydrocarbons	9.1 exploration 9.2 exploitation 9.3 storage
10. Metalliferous Resources	10.1 sand and gravel 10.2 water column minerals\

11. Renewable Energy Sources	11.1 wind
	11.2 water properties
	11.3 water dynamics
	11.4 subsoil
12. Defense	12.1 exercise areas
	12.2 nuclear test areas
	12.3 minefields
	12.4 explosive weapon areas
13. Recreation	13.1 onshore and waterfront
	13.2 offshore
14. Waterfront Man-Made Structures	14.1 onshore and waterfront
	14.2 offshore
15. Waste Disposals	15.1 urban and industrial plants
	15.2 watercourses
	15.3 offshore oil and gas installations
	15.4 dumping
	15.5 navigation
16. Research	16.1 water column
	16.2 seabed and subsoil
	16.3 ecosystems
	16.4 external environment interaction
	16.5 special areas and particularly sensitive areas
	16.6 coastal management
17. Archaeology	17.1 onshore and waterfront
	17.2 offshore
18. Environmental and Protection and Conservation	18.1 onshore and waterfront
	18.2 offshore

12

Improvement in One-Dimensional Mathematical Modeling of Shoreline Evolution: An Application to a Venetian Beach

Piero Ruol, Researcher and Masimo Tondello, Doctoral Candidate
Istituto de Costruzioni Marittime e di Geotecnica, University of Padova, Italy

ABSTRACT

In this paper, a short review on proposed mathematical models for the study of sediment transport and morphology problems in maritime engineering is given. In particular, the useful, 'one-line' models are analyzed in their details and a proposed model of such a type is described, and the problems concerning the choice of the number and sequence of representative incident waves are discussed.

As an area for application of this model, a stretch of coast close to the Venetian Lagoon was considered. First of all, the mathematical model was applied to simulate the beach evolutions occurring over the last decades: the field measurements of 1968, 1980, and 1992 were considered in order to test and calibrate the model. In the end, a forecasting analysis was carried out by the researchers.

INTRODUCTION

The prediction of the induced shoreline evolution related to the construction of a coastal structure (like a harbor, a simple jetty, a series of groins, or detached breakwaters) appears as one of the most interesting problems in coastal management. This is mainly correlated to the engineering and environmental aspects concerning the sudden modification of beach configuration. Useful information can be obtained, for example, to help judge whether local works will be required or to help assess the level of maintenance that may be necessary.

This problem may be studied in a wave laboratory through physical simulations, but the well-known problems related to such an approach (i.e., the scale effects associated with a model study and the impossibility of correctly reproducing inertia forces together with those of gravity and viscosity) rarely lead to accurate results. In addition, large costs and the very long executing times required greatly limit such an approach.

As a consequence, the study of sediment transport and morphology problems is often based on a mathematical approach. The general scheme is found in the equation of motion

$$\Sigma F = \frac{d(mv)}{dt} \tag{1}$$

and on the continuity equation:

$$\frac{dm}{dt} = 0 \tag{2}$$

In previous equations, the general force F usually includes gravity, pressure, wave induced forces, etc.; m denotes mass; and v and t, velocity and time, respectively.

The results obtained through such hydrodynamics models are then used as input data for calculating sediment transport rates and finally, for determining evolving coastal morphology.

MATHEMATICAL MODELS

Proposed mathematical models for the study of sediment transport and morphology problems can be divided into three, general groups.

1. Three-dimensional models (3-D)
2. Two-dimensional models (2-D)
3. One-dimensional models (1-D)

The first group consists of a series of very sophisticated models, able to represent bed-morphology along the three Cartesian coordinates. This complete 3-D approach needs a lot of conditions and assumptions that generally cause large approximations in the results. The difficulties connected to similar studies often lead to serious practical limitations, so as a consequence, this approach is rarely used in practice.

The next lower level of sophistication involves two-dimensional models. Such models can be referred either to a cross-shore section (vertical 2-D) or to a horizontal, Cartesian coordinates system (horizontal 2-D). Vertical 2-D models are able to determine induced modifications of a cross-shore section and are mainly based on the evaluation of hydrodynamics and sand concentration fields (Armanini and Ruol 1988), together with a conservation of mass equation. Through such an approach (Kamphuis 1992), only profile modifications can be evaluated, with the planimetric evolution of the beach, the object of this study, not being addressed.

Horizontal 2-D models on the other hand, take into account morphology modifications of the coastal zone, considering the horizontal coordinate system with one axis parallel to the shoreline and the other perpendicular to it. Such models are based on an in-depth, integrated version of hydrodynamics and continuity equations of sediments (Basco 1983).

Through such an approach, sediment transport is generally not correctly evaluated, because sand entrainment is strictly related to bottom hydraulics conditions and not to the mean velocity used during the mathematical approach. In addition, both a numerical solution based on the Finite Difference Scheme and on the Finite Element Method are often plagued by numerical instabilities. In conclusion, the described 2-D approach is also generally not used for practical applications.

Recently, a "Quasi-three-dimensional" model was proposed (Briand and Kamphuis 1990). Such a model is essentially similar to the horizontal 2-D one, but a non-uniform velocity distribution over the vertical can also be taken into account. Even if nowadays this approach appears to be very interesting, further research is needed in order to allow this model to be used for engineering applications.

The previous, short review of more sophisticated solutions for morphology computations shows the importance of finding a simpler model which limits computation costs and especially, reduces the large approximations of input data. Such conditions are easily satisfied by one-dimensional (1-D) models, generally called 'one-line' models, which represent the site morphology just through the shoreline and the cross-shore profile of the beach. The mathematical formulation is based on a sediment-continuity equation, coupled with a sediment transport formula. For a correct application of this group of models, it is essential to consider a complete, local wave climate and to determine the calibration constant of the latter equation through analyses of the historical shoreline evolutions.

PROPOSED ONE-LINE MODEL

The proposed, mathematical model is based on the sediment continuity equation

$$\frac{\partial Q}{\partial x} + (D+d)\frac{\partial y}{\partial t} = q \qquad (3)$$

in which the x-axis is parallel to the shoreline and the y-axis perpendicular to it, and:

- Q is the long-shore, sediment transport
- D is the closure depth at which beach profile is supposed unchangeable under wave attacks (Hallermaier 1978)
- d is the swash limit, i.e., the closure height on emerging beach
- q is the net, cross-shore, sediment transport rate per unit distance of beach

The continuity equation can be easily understood through Figure 1 in which it can be observed that the profile is supposed to move on or offshore without changing its shape.

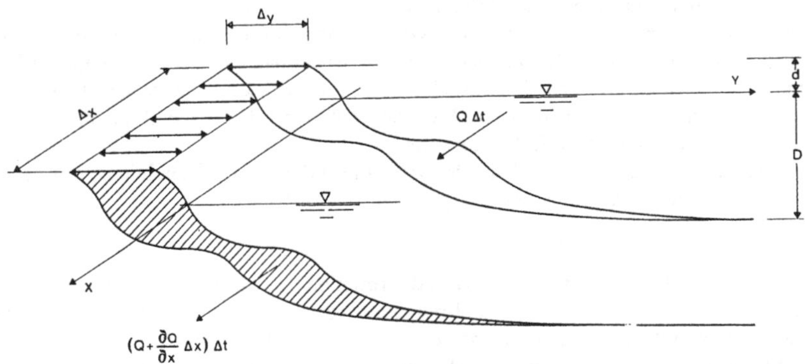

Figure 1. Continuity equation scheme

The conservation of mass equation must be solved, coupled with a sediment transport rate equation (Swart 1976). The formula used for sediment transport computations is based on the well-known Energy Flux Method which considers the hypothesis of a direct relationship between the long-shore component of wave energy flux in the surf zone and long-shore sediment transport (CERC 1984.) In practice, the following equation was used:

$$Q = \frac{K}{g\varrho_s} (\bar{E}\, c_g)_b \quad \sin 2\alpha_b \tag{4}$$

In it

$$\bar{E} = g\,\varrho\, H_{sb}^2 /8 \tag{5}$$

is the energy density, being:

- g acceleration of gravity
- ϱ water density
- c_g group velocity
- α_b angle from breaking wave crests and shoreline
- ϱ_s immersed sand density
- b subscript b denotes breaking values
- K a parameter without dimensions

The K-value depends on local hydrodynamics conditions and must be determined through local calibration iter: generally the actual sediment transport rate is indeed lower than the potential transport rate obtained through the Energy Flux Method. As a consequence, it does appear very important to calibrate such a constant, trying to fit the numerical results with the historical evolutions of the studied beach.

For an accurate, numerical simulation it is very interesting to determine the correct wave climate of the site. Deep-water wave climate could be determined considering a long-term time series of wind data, but this approach is usually not desirable because some approximations (especially in the wave direction parameter) are introduced. More interesting appears to be the approach based on the direct measurements (or observations) followed in this study.

Once the deep-water wave climate was determined, the choice of the number of representative waves to be used as input for the calculations is essential. The usual, one-line models approach is based only on one, single incident wave condition; such an approach appears to be inadequate. In fact, it had been pointed out by the first author (Ruol and Trivellato 1986) that the choice of the number and of the sequence of incident waves largely affects the results.

Since the sediment transport equation is based on the surf zone wave conditions, it is necessary to transform the deep-water wave parameters using:

$$H = H_0\,(K_s\, K_r\, K_d) \tag{6}$$

for the shoaling (K_s coefficient), refraction (K_1), and diffraction (K_d) phenomena must be included.

In the proposed model, the shoaling and refraction coefficients are accounted for and calculated through the small-amplitude, linear theory of waves (assuming all bathymetries parallel to the beach line). Since the shoreline is generally not straight, the combined effect of shoaling and refraction appears to be of great interest because it acts differently on different locations along the shoreline. For this calculation, the following expression was used (Hunt 1979):

$$\frac{c^2}{gh} = [y+(1+0.6522y+0.4622y^2+0.0864y^4+0.0675y^5)^{-1}]^{-1} \qquad (7)$$

with: $y = kd \ \tanh kd$ (k being the wave number)

The model is also able to account for the effects of the diffraction phenomenon present in the regions closer to a structure located along the studied littoral zone: the K_d coefficient was computed through the Goda method (1985):

$$K_d = 0.008\ \Theta + 0.69 \qquad \text{(in the 'shadow zone')} \qquad (8a)$$

$$K_d = 0.370\ \Theta + 0.71 \qquad \text{(outside the 'shadow zone')} \qquad (8b)$$

θ being the angle (in degrees) from the incident wave direction (i.e., the shadow-line) to any point of the littoral zone.

Finally, the breaking wave conditions (essentially wave heights H_b and direction α_b) were determined considering the modified, incident wave characteristics (due to shoaling, refraction, and diffraction), together with the breaking criterion developed for solitary waves:

$$H_b = 0.78\ d_b \qquad \text{(being } d_b \text{ the breaking depth)} \qquad (9)$$

Numerical Solution and Boundary Conditions

The described model was numerically solved using a finite difference technique; the scheme used is basically similar to the Ozasa-Brampton (1979) one and considers the shoreline apportioned into a series of sections of finite length. The Explicit Finite Difference Scheme, based on the next equations, was followed:

$$y(n+1,t+1) = y(n+1,t) - \frac{\Delta t}{2\ (D+d)\ \Delta x}\ [Q(n+2,t) - Q(n,t)] \qquad (10)$$

$$Q(n,t) = \frac{K}{g\ \varrho_s}\ [E(n)\ c_g(n)]_b\ \sin 2\ \alpha_b(n,t) \qquad (11)$$

in which besides the symbols already defined, Δt, Δx are time and space steps, and n and t define the computed section and time of the numerical scheme:

$$\alpha_b(n,t) = \alpha_x(n) - \operatorname{atan}\left[\frac{y(n+1,t) - y(n-1,t)}{2\ \Delta x}\right] \qquad (12)$$

To solve the problem, the boundary conditions at the two ends of the simulated shoreline must be defined. Generally, such conditions are satisfied by imposing constant sediment transport values at both ends of the beach analyzed.

The boundary conditions can also be 'internal,' and in such cases it is possible to define local situations; for example, the presence of an obstacle to long-shore sediment transport (a local by-passing condition must be imposed on parameter Q) or a river delta (sediment discharge q) must be specified. One of the most interesting simulations is the study of the effects induced on the shoreline by a generic obstruction built along a shore: in such case, a constant sediment transport far from the structure

$$Q_{begin} = Q$$

and a null sediment transport at the structure

$$Q_{end} = 0$$

are usually included. As a beach accretes however a different boundary condition should be considered in order to account for the by-passing of sediments beyond the structure. This condition can be expressed by means of beach accretion rates and of the structure length (Kamphuis 1992).

Definition of Time and Space Steps and of Wave Conditions

As already discussed, the choice of time and space steps appears to be of primary interest in solving the problem: the selection of Δt must in fact satisfy the Courant condition, while the first author's remarks can be useful in the selection of Δx (Ruol and Trivellato 1986). It must be noticed that time and space steps are not assumed constant along the shoreline and generally the faster the beach evolution, the smaller the steps. For example, in the regions close to an obstruction, Δx is imposed smaller than when far from it, while Δt is continually decreasing in approaching the structure (such a variation is automatically undertaken by the program itself).

In the end, the definition of the offshore wave module is the object of an interesting discussion: the simplest assumption of one single incident wave condition was observed to be inadequate; hence the problem consists of defining both the number and the sequence of incident waves. This problem was solved by trying to find which incident wave conditions are to be assumed in order to obtain results not so significantly affected by the operator's choice.

The wave climate offshore from the Venetian beaches was considered for such research: in Table 1, the significant wave height percentages as a function of wave directions are reported. Such a table is the result of a statistical analysis of the well-known KNMI visual observations.[(1)]

Referring to such data, a typical Venetian beach (south-east-faced) with an interposed, impermeable structure perpendicular to the shoreline was considered, and some computations in the region closer to the obstacle were performed.

The first considered problem concerned the choice of the sequence of representative incident waves. As already pointed out, the order of assumed waves can affect the numerical results; this questions was solved introducing a 'random-routine' for the waves sequence choice. If the number of waves is great enough, in fact, it was pointed out that the results do not change considerably. Assuming, for example, twelve different wave conditions per year controlled by random routine and a typical simulation period of ten years, it was pointed out that the results appeared to be very similar to each other for every analyzed sequence. These results, referred to in Table 1, are graphically drawn in Figure 2.

[1] The Dutch Meteorologic Institute (KNMI) original data, referred to the period 1961-1990, were kindly furnished by ENEL-CRIS of Mestre-Venezia.

Table 1

Wave Climate Offshore Venetian Beaches

H_s(m) / α	0.5	1.0	1.5	2.0	2.5	3.0	3.5	4.0	4.5	5.0	5.5	6.0	>6.0	TOT
0°	3.48	1.22	0.32	0.24	0	0	0	0	0	0	0	0	0	5.27
30°	5.83	2.59	1.22	0.97	0.16	0	0	0	0	0	0	0	0	10.78
60°	3.00	1.38	1.54	0.49	0.24	0	0	0	0	0	0	0	0	6.65
90°	1.13	0.81	0.41	0	0	0	0.08	0	0	0	0	0	0	2.43
120°	2.92	0.89	0.24	0.08	0	0	0	0.16	0	0	0	0.08	0	4.38
150°	2.19	0.24	0.24	0.57	0.32	0.08	0	0	0	0	0	0	0	3.65
180°	2.03	0.16	0.08	0.16	0.08	0	0	0	0	0	0	0	0	2.51
210°	1.46	0.32	0.16	0.08	0.08	0	0	0	0	0	0	0	0	2.11
240°	0.65	0.65	0.16	0.24	0	0.08	0	0	0	0	0	0	0	1.78
270°	1.05	0	0	0	0	0	0	0	0	0	0	0	0	1.05
300°	1.46	0.32	0.08	0.08	0	0	0	0	0	0	0	0	0	1.94
330°	2.03	0.57	0.16	0.08	0	0	0	0	0	0	0	0	0	2.84
360°														
TOT	27.23	9.16	4.62	3.00	0.89	0.16	0.08	0.16	0	0	0	0.08	0	45.38
CALM PERCENTAGE:														54.62

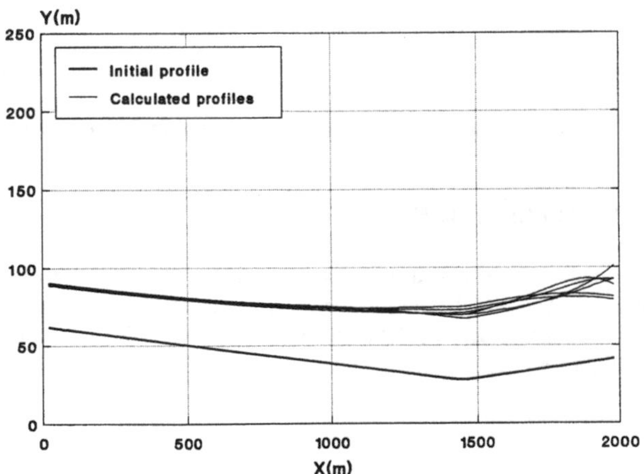

Figure 2. Influence of incident waves sequence (with twelve wave attacks)

The appropriate number of representative waves to be considered in the numerical computation was studied afterwards and it was evaluated that by increasing the number f waves, every time the results The were closer to a specific solution. Some results, again referring to Table 1 and ten years of simulation, are drawn in Figure 3. It can be observed that a number of about twelve different waves (randomly arranged year by year) is enough to reach a successful result.

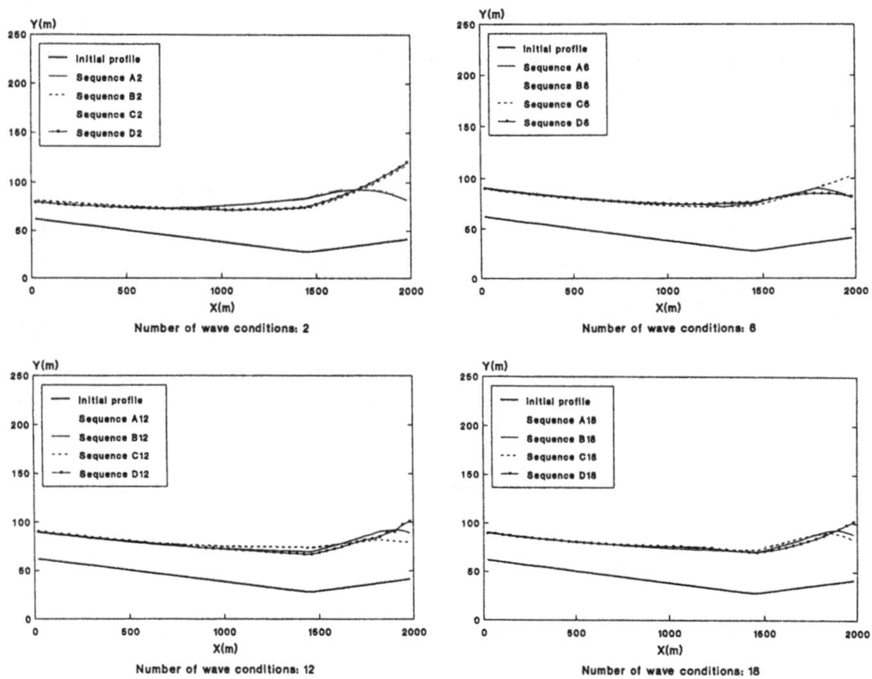

Figure 3. Influence of the number of chosen incident waves

PRACTICAL MODEL APPLICATION

As a typical example, the described model was applied to the Venetian coastline close to Lido Inlet (Cavallino Beach). The 3,600 mile-long, northern jetty of the lagoon inlet, built up in 1888, caused a huge accretion of sandy beach (Matteotti and Ruol 1986). This effect, that is still evident to this day, is shown in Figure 4.

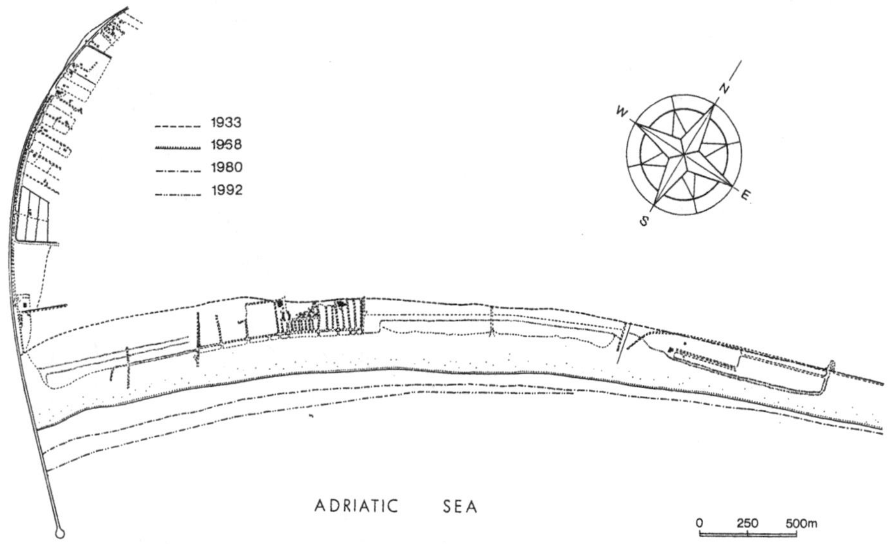

Figure 4. Recent evolution of Cavallino Beach close to Lido Inlet

Wave climate was determined considering the described data measured by the KNMI; the input data concerning both wave heights and directions were calculated through an average procedure; and for each assumed direction of incoming waves, the height value:

$$H = \left[\frac{\Sigma H_i^{5/2}}{N} \right]^{2/5} \tag{13}$$

N being the number of all waves of the same class, was assumed.

This expression was adopted because the sediment transport was assumed to be proportional to the wave height raised to the 5/2 power (in fact, referring to eq. 4, it can be observed that the energy density is proportional to H^2 and the group velocity to $\sqrt{H}$).

The calibration of the model was based on a correct simulation of the beach-line evolutions occurring over the period 1968-1980. It was established (Figure 5) that the value

$$K = 0.19$$

to be introduced in the transport equation was correct for the analyzed beach and for the considered waves (twelve, different wave attacks were considered). This value appears in the range of the K-values determined by other researchers through prototype measurements (Dean et al 1982).

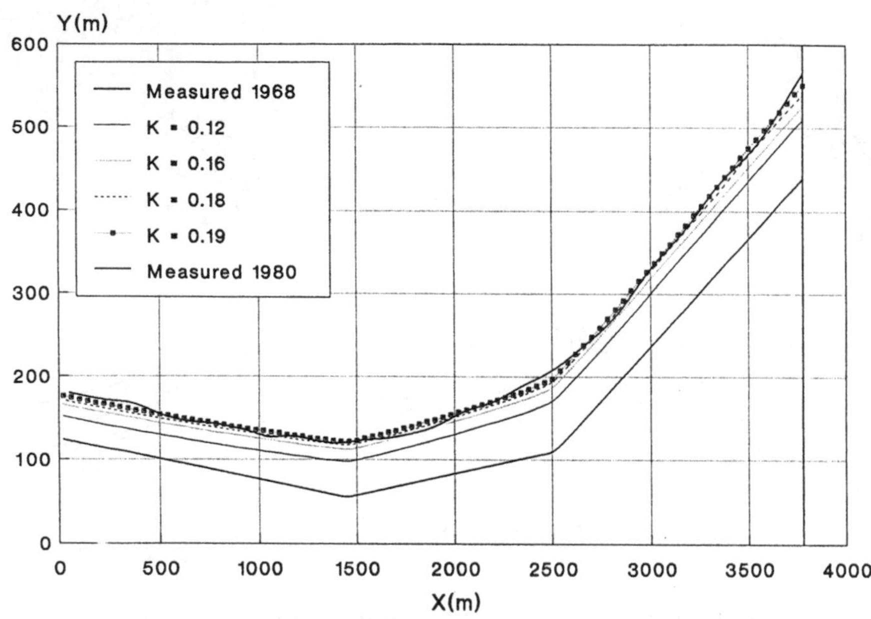

Figure 5. Comparison between measured profiles and calculated ones for different values of the K-coefficient

The determination of the K-coefficient based on the grain-size-distribution of sediments (Figure 6) (Ruol 1988) would have given a higher value (greater than unity) and as a consequence, would have overestimated actual sediment transport (CERC 1984; Dean et al 1982). This means that the potential sediment transport of the analyzed beach is higher than the actual one, being probably (as in most Adriatic beaches) the sand with limited extent of volume.

The model was then applied to the analyzed beach and the results obtained starting from the 1968 condition for subsequent intervals of twelve years were compared continuously to the field measurements (the measured profile of 1968, 1980, and 1992 were available). A general agreement between calculated and measured profiles was pointed out: there does appear to be a beach increase of about 9 m/year in the regions closer to the jetty (Figure 7).

Finally, a beach evolution forecasting was carried out in order to determine the probable beach-line position in the future: the calculated beach-line of the year 2004 is also shown in Figure 7.

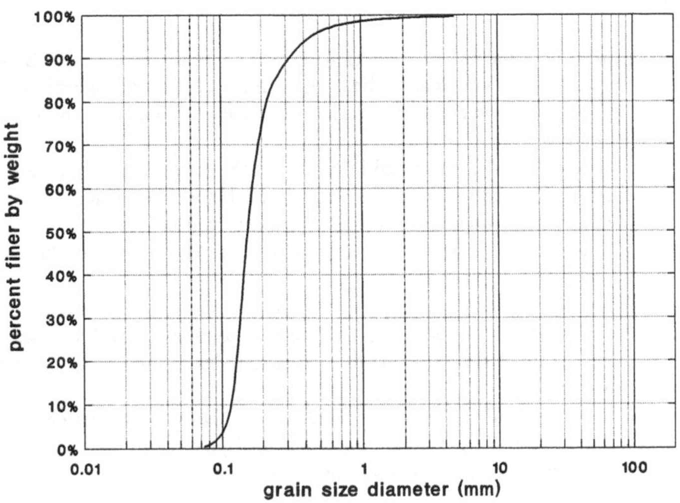

Figure 6. Grain size distribution of Cavallino Beach sediments

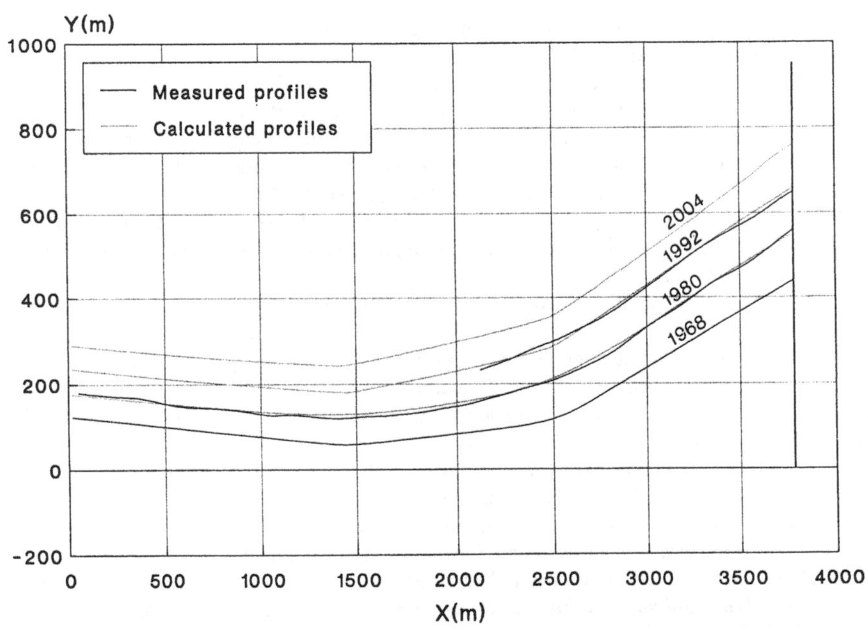

Figure 7. Comparison between measured and calculated beach profiles and a forecasting of plainimetric evolution

CONCLUSION

Among the proposed, mathematical models for studying beach evolutions, the one-dimensional type appears to be very useful for practical applications. In fact, the more sophisticated approaches are generally plagued by numerical problems and need a lot of conditions and assumptions that often lead to great approximations in the results.

The proposed 'one-line' model was tested in order to determine the wave conditions to be used for obtaining the correct results. In particular, it was determined that a result which does not depend on the wave sequence is obtained if the number of assumed incident waves is greater than ten (for practical applications, twelve wave conditions were selected), while a 'casual routine' for the sequence of waves to be applied year-by-year is recommended.

The major importance of the calibration of the model was point out, comparing the K-value obtained through a direct test on measured morphologic evolutions (involving the actual sediment transport) with the one obtained simply from the grain-size-distribution of sediments (implying the potential sediment transport).

This practical application of the model carried out on the Venetian beach close to Lido Inlet showed a close correlation between calculated and measured planimetric evolutions of the beach and allowed a forecasting to be made on future evolutions of the coastline.

REFERENCES

Armanini, A., and P. Ruol. 1988. Non-uniform suspended sediments under waves. In *Proc. XXI Int. Conf. on Coast. Eng.,* 1129-39. ASCE.

Basco, D. R. 1983. Surfzone currents. *Coastal Engineering* 7:331-55.

Briand, M. H. G., and J .W. Kamphuis. 1990. In *Proc. XXII Int. Conf. on Coast. Eng.,* 2159-72. ASCE.

CERC. 1984. Shore protection manual. 4th ed., U.S. Government Printing Office. Washington, D.C.

Dean, R.G., E. P. Berek, C. G. Gable, and R. J. Seymour. 1982. Longshore transport determined by an efficient trap. In *Proc. XVIII Int. Conf. on Coast. Eng.,* 954-68. ASCE.

Goda, Y. 1985. *Random seas and design of maritime structures.* Tokyo; University of Tokyo Press.

Hallermaier, R. J. 1978. Uses for a calculated limit depth to beach erosion. In *Proce. XVI Int. Conf. on Coast. Eng.*, 1493-512. ASCE.

Hunt, J. N. 1979. Direct solution of wave dispersion equation. *Waterways, Port, Coastal and Ocean Engineering* 105: 457-9.

Kamphuis, J. W. 1992. Computation of coastal morphology. In *Proc. of Short Course on Design and Reliability of Coastal Structures.* 211-57 (1-3 Oct.) Venice.

Matteotti, G., and P. Ruol. 1986. Considerazioni in margine all'evoluzione del litorale dell'Alto Adriatico. *Porti, mare, territorio* 7(2): 61-71.

Ozasa, H., and A. H. Brampton. 1979. Models for predicting the shoreline evolution of beaches backed by sea-walls. *Hydr. Res. Station,* Report no. IT 191. Wallingford.

Ruol, P., and F. Trivellato. 1986. Simulazione matematica dell'evoluzione planimetrica di un litorale. In *Proc. XX Conv. Idraulica e Costruzioni Idrauliche*, 354-68. Padova.

Ruol, P. 1988. On the suitable diameters of bed materials to model sediment transport under waves. In *Proc. Conv. Trasp. solido ed evoluz. morfol. dei corsi d'acqua*, 193-203. Trento.

Swart, D. H. 1976. Predictive equations regarding coastal transport. In *Proc. XV Int. Conf. on Coast. Eng.*, 1113-32. ASCE.

13

The ECO^2 Concept of the MED-ARCOBLEU Program

Ing. Alfredo Fanara
Director of Business Development and Marine Resources
Alenia Elsag Sistemi Navali, Genoa, Italy

THE ECO^2 CONCEPT

In 1970 a new concept began to spread around the world as the basic philosophy of UNESCO's "Man and Biosphere" Program. The general objective was:

> To develop, within the natural and social sciences, a base for the rational use and conservation of the resources of the Biosphere, and for the improvement of the relationship between Man and the Environment; to predict the consequences of today's actions on tomorrow's World; thereby to increase man's ability to manage efficiently the natural resources of the Biosphere.

The new approach was to consider that "Man belongs to Nature" and that any human development should be planned as "Sustainable" in the long run, allowing resources exploitation without cumulative and irreversible degradation.

To accomplish this goal, any human activity should be carefully balanced between economic exploitation and the ecological impact that it produces. If the main objective is to protect Man and Nature, one has to consider the modern developments together with and in the light of:

1. Man's safety
2. Man's activities impact
3. Biosphere protection

The first and the latter subobjectives are linked by the intermediate one.

The Sea environment is a particular zone of the Biosphere where these concepts apply in very obvious and tangible ways.

The Coastal Areas and Their Ecological Value

Let us focus our attention on coastal areas, because they support an intense concentration of human activities and as a result, the "sustainable development" is closer to the limits.

From the ecological viewpoint, some marine areas have already been designated as "zones of high Ecological interest" (Figure 1) acknowledging their importance and uniqueness, trying to keep them in the original status, and avoiding further degradation of the local biosphere. Practically, these attitudes become reality, by:

1. Scientific observation
2. Flora and fauna protection
3. Natural resources development
4. Local and nearby biological repopulation

Most of these areas of concern are located in close proximity to the coast; however, in the near future it is logical to assume that some will extend to the pelagic zone.

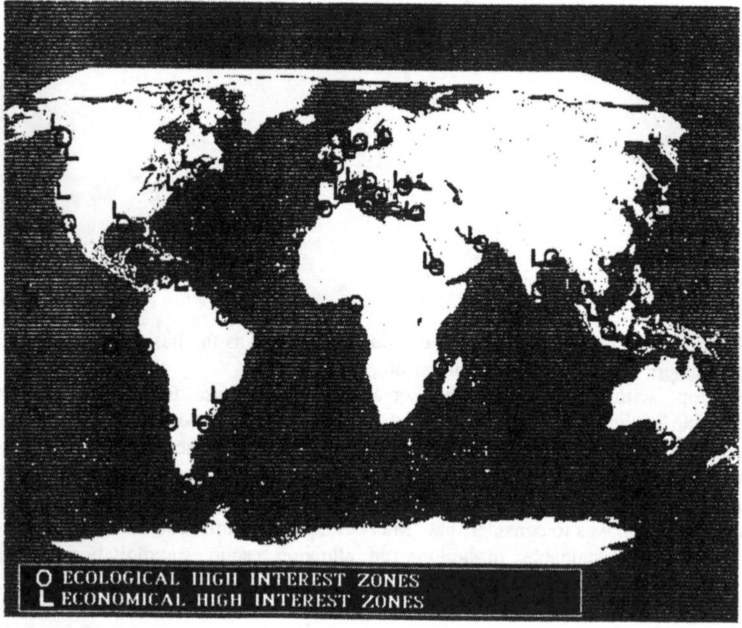

Figure 1. The high interest zones

The Economic Value of Some Coastal Zones

Independent of the ecological zones, around the world some marine coastal regions have assumed an important economic role (Figure 1), due to one or more of the following factors:

1. High urban concentration
2. Heavy shipping traffic, necessitating harbors and intermodal, transport facilities
3. Fishery, aquaculture, or mineralogic exploitation
4. Touristic interest, resorts development recreation-oriented harbors, beaches, etc.
5. Industrial plant location

Due to the high population density and activity level in these areas, the chances are good that they have a formidable impact upon the environment. Industries, transports, and urban centers may cause dangerous pollution. Urban waste waters, agriculture, and aquaculture may also increase the pollution in terms of nutrients and chemical substances flowing into coastal waters. Nutrients in turn, lead to high eutrophication of the sea, bringing worrisome consequences to the marine flora and fauna.

But the fact is that these areas support the human interests and represent a "must" that is related to civilization, industrial progress, and modern urban life.

Zones Where Ecological and Economic Interests Overlap

What happens if the areas of ecological and economic interest overlap each other in the same coastal spot? Of course, their opposite characteristics may give rise to a situation fraught with conflict.

Nonetheless, to try a compatible and sustainable approach is worthwhile. The fact is that these zones are very important and their interaction is critical, therefore they must be cared for more than twice; the complexity of their management is then best represented by an exponential factor. Let us call ECO^2 the zones of high interest, both ECOlogically and ECOnomically. From a certain viewpoint the concept resembles that of the Marine Reserves.

An ECO^2 zone could be considered a particular portion of the marine biosphere that encompasses Nature and Mankind with its complex activities, a sort of *"ecological lab"* where all processes should be harmonized and controlled in order to create a compatible environment. The definition and the study of these zones should be approached scientifically; a possible approach is that proposed by the ARCOBLEU program with the analytical study of the "homogeneous zones".

The ECO^2 Value

In Figure 2 the main human activities that may influence the marine ECO^2 value are shown, together with their positive and negative weights and the relevant underlying forces: technological progress and commerce on one side and geo-biological factors on the other.

There are mutual interactions between economy and ecology; typically, if the ecological status is downgraded, the economic sphere may be influenced as well (as in the case of exploitation through tourism).

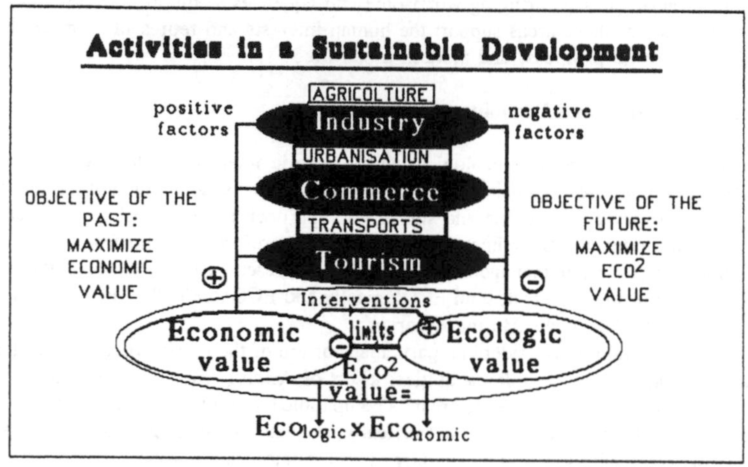

Figure 2. Activities in a sustainable development

The real value of an ECO^2 zone is *"Eco^2"* and it is given by its *"economic value"* multiplied by its *"ecologic value"*: Econ x Ecol. The objective is to maximize that Eco^2 value, keeping in mind that with a null value of one component, the total will also be null. One way to compensate for damages to the ecological sphere is to finance corrective interventions that may restore the original status.

The selected ECO^2 zones are characterized by the fact that they rank very high in *"ECO^2"* values, and must be cared for accordingly. The trend for the future is: *To maximize the ECO^2 value of ECO^2 zones.*

Application of the ECO^2 Concept to Europe and the Mediterranean Sea

When applying the ECO^2 concept to Europe, many interesting zones can be found, some clearly being stand-alone, geo-ecological entities. Incidentally, the most interesting areas have bi- or multinational boundaries.

Europe of today is leaning toward political unity and since the E. C. promotes transnational collaboration for environmental purposes, this so, internationalism should be considered a fostering and positive factor.

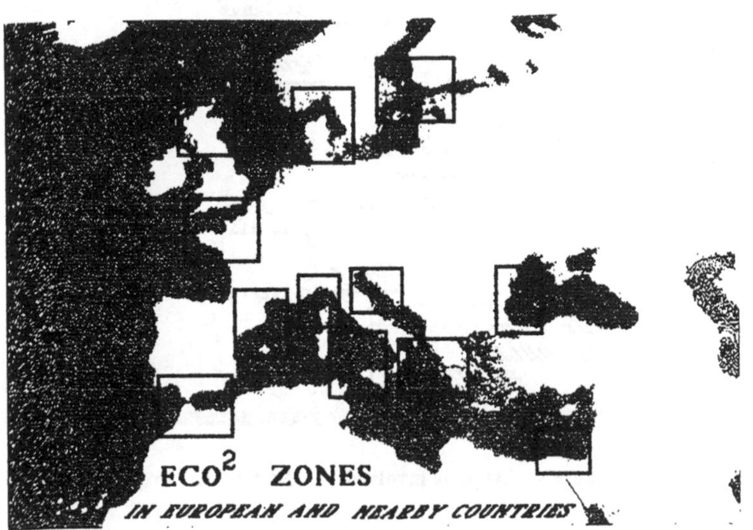

Figure 3. ECO² zones in European and nearby countries

In the next section of this paper, the MED-ARCOBLEU program concerning the Mediterranean Sea will be introduced.

As an "environmental laboratory", the Mediterranean Sea is very challenging, because:

1. It has very peculiar characteristics.
2. It is an enclosed Sea with low water exchange rate.
3. It has recently been the scenario of major accidents.

And, at the same time:

4. It possesses unmatched natural and touristic attractions.
5. It hosts highly valuable fauna and flora species.
6. It is an area of strategic value for the world.
7. It is economically important.

MED-ARCOBLEU: AN INTEGRATED SURVEILLANCE SYSTEM AGAINST CUMULATIVE AND ACCIDENTAL MARINE POLLUTION

The equilibrium of ecosystems leads to consider human activities (agriculture, industry, transport, and tourism) not only from an economic, but also from an ecological point of view. On the Mediterranean coast, the increasing strain from human activities, in conjunction with the characteristics of a closed basin, urgently requires a method of prediction, at any given time, of the impact of such activities on the marine ecosystems.

MED-ARCOBLEU stands for a program of Global Surveillance of and Control System against cumulative and accidental pollution in certain Mediterranean areas called ECO² zones for their economic and ecological high interest.

ARCOBLEU is the pilot project dealing with a large portion of the high Tyrrhenian Sea.

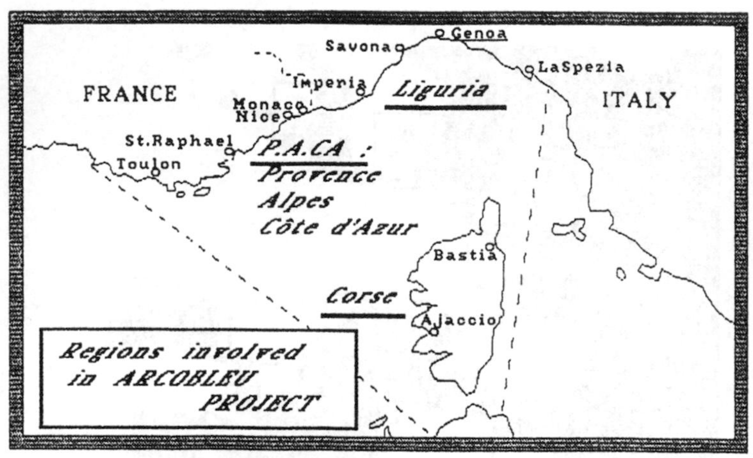

Figure 4. Regions involved in ARCOBLEU Project

The Selected Scenario

The proposed area covers the Corso-Liguro-Provencal basin which presents many homogeneous characteristics with regard to morphology, climate, oceanography, and economy.

Its morphological configuration is characterized by a narrow continental shelf with very high slopes ploughed by deep, submarine valleys down to 3000 meters.

Water masses circulation results from the input of Atlantic water which travels along the African coastline (Algerian stream), before moving northward to Sardinia and Corsica. Here, it creates the Liguro-Provencal stream which then travels along the coastlines of Liguria and France with a cyclonic, counterclockwise movement.

The area is generally oligotrophic, with low or medium primary production, although it is influenced by its eutrophic neighbors with high primary productivity, the Rhone estuary and the Lion Gulf.

The flora and fauna of the area make it a specific zone within the Mediterranean Sea, including near the coast, the Posidonia meadows and some cleanliness indicators, such as the Lithophyllum, and further from the coast, the great diversity of pelagic fishes and the abundance of Cetaceans.

It is worthwhile to remember the governmental, pro-nature initiatives started in order to protect the zone and its biological treasures (RAMOGE, Pelagos, and the Sea Sanctuary proposed by the Environmental Ministries of both countries.)

During the industrialization period of the 60's and 70's, some important industrial/harbor zones were created in the considered area, such as GENOA and FOS/MARSEILLE.

A Vulnerable and Stressed Environment

Since the beginning of the century, the Mediterranean coastal region has had to face an explosive demographic growth. Additionally, the proposed area of coastline, as a part of the so called European "Sunbelt", has to cope with a strong, summertime rush of more than two million people (tourists), increasing the density from 75 to more than 300 inhabitants per sq. km.

This coastline also has large industrial and harbor zones which still handle important maritime trading traffic, despite the oil crisis and the development of new energy sources. The traffic in the Ligurian sea (Savona, Genova, La Spezia) corresponds to 26% of the Italian total traffic, while the French portion represents 35.8% of the national total. The international trade using Mediterranean Sea lanes reaches about 15% of all the world sea exchanges. This has resulted in a large volume of maritime traffic, with important communication lanes through the high Tyrrhenian Sea, from west Sardinia to Liguria or to the Lion Gulf.

All the related activities, navigation and communications, mineral and biological resources harvesting, urban and industrial wastes, tourism, etc., place a considerable stress on the marine ecosystem. The potential, environmental imbalance could cause ecological disasters, either from cumulative or accidental pollution.

THE ANSWER: ARCOBLEU

Particularization of the **MED-ARCOBLEU** concept (extended to all the Mediterranean), **ARCOBLEU** is a "pilot" project, dedicated to the concerned Italo-French area, proposed for Italy, by **ALENIA ELSAG SISTEMI NAVALI** and the **CONSIGLIO NAZIONALE DELLE RICERCHE** and for France, by **IFREMER** (Institut Francais pour l'exploitation de la Mer) and **THOMSON SINTRA-ASM.**

ARCOBLEU: Its first goal is to study the development of an operational and integrated system for the surveillance and control of marine environmental quality. Such a system may be considered the technological vehicle thorough which man can protect coastal environments using general objectives such as the protection of human health and evaluation of intervention efficiency, as well as being the tool for accumulation and retrieving data on exploited resources status, marine life, and pollution levels and trends.

The **ARCOBLEU** project corresponds to an innovative approach requiring:

1. Strong functional and hardware integration
2. Interdisciplinary cooperation
3. International collaboration

Integration

If past and present experimental networks dedicated to environmental surveillance have demonstrated their technical feasibility, they have also clearly shown their intrinsic limits. An overspecialization never allows such networks to be of general purpose, and their objectives are usually just as restrictive. Moreover, by failing to provide all the needed data, they do not allow the extrapolation of really efficient decision-making aids, without a synthesis of the collected information.

ARCOBLEU's main objective is basically a highly integrated one: the surveillance of a risky area must be a global merge of different data and must support several shared needs: e.g., health resources, environment etc.

A functional integration of networks and processes, along with source data fusion (sensors, data bases, etc.) will lead to substantial investment savings.

Interdisciplinary Cooperation

The success of such a program requires expertise in various fields:

1. Scientific (e.g., biology, chemistry, modeling)
2. Technical (e.g., sensors, networks, hardware, software)

3. Industrial (e.g., methodology, projects, planning)
4. Operational (e.g., end-users, operational people, decision makers)

Operationals and end-users will be consulted and, if possible, included in the team at all times during design and development phases.

International law, maritime rules, and all existing intervention means will also have to be taken into account, in a coordinated plan.

The **ARCOBLEU** organization is therefore based, from the very beginning, upon interdisciplinary working groups coordinated by a technical team responsible for all aspects of the program.

International Collaboration

Pollution has no boundaries, therefore, an international collaboration is a prerequisite to any such program.

Since the proposed area is the high Tyrrhenian Sea, the **ARCOBLEU** program starts with French-Italian cooperation, but is open to other E.C. Mediterranean nations, such as Spain and Greece.

The success of this program must be based on the association of competent teams, with both expertise in marine pollution (surveillance and effects) and experience with large and complex systems.

The **ARCOBLEU** program will thus allow a global optimization of such a system in the pilot zone and any other risky ECO^2 area and will provide to both end-users and decision makers, the appropriate tools for management and decision support (impact studies, risk analysis, cost/efficiency, etc.).

An Integrated Surveillance and Control System

Coastal, marine environmental protection requires a well-matched surveillance system continuously evaluating water quality parameters (levels and short-term/long-term trends) in order to detect any change of the medium which could lead to major perturbations of the ecosystem (defined as a set of biological and physico-chemical elements in equilibrium).

The **ARCOBLEU** program will consider a family of systems, taking into account both existing and future components. The already developed and reliable pollution detection networks will be considered as **ARCOBLEU** primary inputs. It will be the same for any other useful source of information, such as meteorological data and maritime traffic data (to corroborate ships to accident pollution).

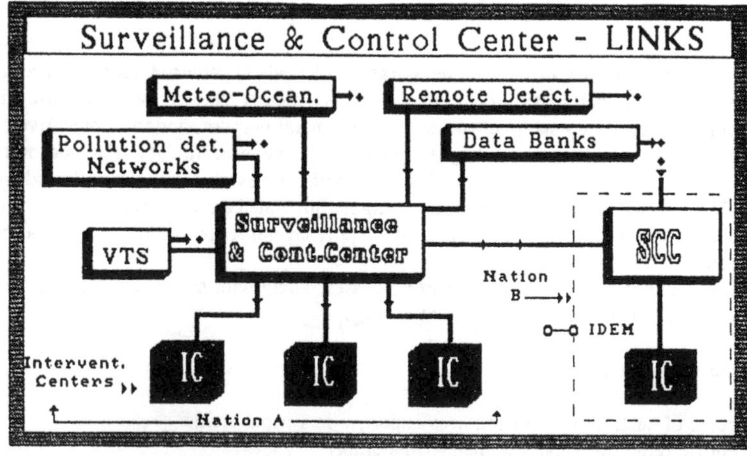

Figure 5. Surveillance and control center: links to sources

A complete synthesis of conventional and modern sensors will be done to evaluate different candidate data sources:

1. Visual and infrared sensors
2. Physico-chemical analyzers
3. Sonars, radars, and lidars
4. Biological analyzers

Selections of sensor platforms and data acquisition strategy will also be determined, taking into account economic constraints.

1. Fixed or mobile stations
2. Satellites
3. Airplanes and helicopters
4. Ships and autonomous vehicles (ROVS)

Potential use of data from such different information sources requires a functional and hardware architecture both flexible and open in order to integrate existing and modern means to obtain a global cumulative and accidental pollution synthesis.

Several, local "peripheral" stations, dedicated to specific sensors or primary networks, will perform raw data acquisition, preprocessing, and local control. Preprocessed data will then be sent to a Control and Supervision Center (CSC) which will manage their long-term storage, as well as merge and integrate them to provide, in real time, synthetic pictures corresponding to the global situations concerning typical areas, such as:

1. Pollution: levels, trends, and thresholds for general water quality parameters and main pollutants
2. Ships traces for correlation to accidental pollution
3. Meteo-oceanographic situation
4. Means available to support intervention.

The CSC will also have to be equipped with the means required for evaluating the risks and supporting emergency operations.

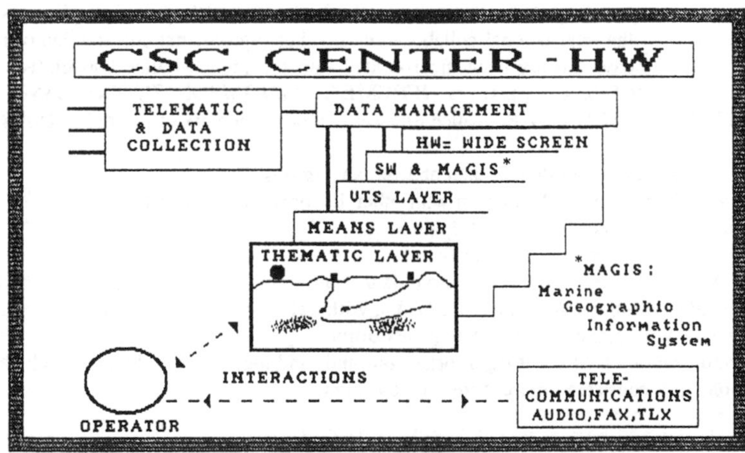

Figure 6. CSC Center-HW

The CSC center has several characteristics in common with modern C^3I (Command, Control, Communication and Information) centers.

1. Automation of data acquisition
2. Automation of the surveillance processing
3. Correlation and fusion of various data
4. Data management and long-term storage (sensor and data banks both national and international)
5. Decision aids and expert systems
6. Reliable communication network (data and voice)

The development of such complex functions requires a proven, system design methodology. The presence of industries (AESN, THOMSON-SINTRA) having a long standing experience in similar systems, guarantees a successful program.

The CSC will also have to be connected to regional intervention centers in order to provide them with all relevant information. The system's modular structure will allow an easy duplication of its core elements (CSC) for any participating nation. This capability will be analyzed in the perspective of an international collaboration, in order to fully match the different existing national procedures. The legal aspects of a multinational control procedure will also be taken into account.

The first phase of the program will be limited to feasibility studies or to risk assessment in the area, defining the precise requirements and the possible options, together with the relevant economic implications.

ARCOBLEU: FUTURE PROSPECTS

The **ARCOBLEU** system is based upon a highly modular structure to allow any adaptation to evolving operational needs. Therefore, it is open to be extended to other areas with different operational needs. This extension and philosophy of openness covers several aspects:

1. Integration of different information sources, including any present or future international standards
2. Progressive upgrading of the system from the basic core to the full-size, pilot system
3. Expansion of the international collaboration to other regions, nations and companies

These future prospects have already been incorporated into the French-Italian cooperative agreement among the current **ARCOBLEU partners (IFREMER - THOMSON SINTRA.ASM - ALENIA ELSAG SISTEMI NAVALI - C.N.R.)** which foresees, in the framework of future **MED-ARCOBLEU** program:

1. A collaboration with other French and Italian regions
2. An already planned collaboration with other European nations of the Mediterranean, like Greece and Spain.
3. A potential enlargement to other nations.
4. A "Mediterranean", full-scale surveillance and control system against cumulative and accidental pollution (**MED-ARCOBLEU**), with a step-by-step development of several national segments with maximum investment flexibility.
5. Harmonization of different legal, organizational, and logistic procedures of each participating nation, in a consistent and coherent global plan.

14

Minera Escondida's Environmental Management in the Coastal Area at Coloso

Andres Camano
Marine Biologist, Natural Resources and Environment Department, Minera Escondida Limitada, Antofagasta, Chile

Eduardo Silva
Geographer, Natural Resources and Environment Department, Minera Escondida Limitada, Antofagasta, Chile

SUMMARY

This paper describes the different uses of the coastal area in Coloso, a bay located in northern Chile near Antofagasta. A historic summary of its development and later decline around 1930 is included, as well as the situation before 1989. Also discussed is the impact generated by the construction and start-up in 1990, of Minera Escondida's filtration plant and shiploading port, emphasizing the environmental program currently in effect. Additionally, the relationship between Minera Escondida and the community of fishermen living near the port facilities is also outlined.

INTRODUCTION

With the growing diversity of human activities and uses of the ocean, a series of conflicts have arisen. This occurs especially in the coastal areas where the increased impact generated by human activity is more easily noticed and emphasized.

Chile has experienced an accelerated increase in the usage of its coastal areas due to urban development of cities and growth of industries in the coastal zone which in turn, have generated a series of problems related to the use of the ocean. Experience has demonstrated that certain uses are compatible, while others are totally or partially incompatible. For example, under certain circumstances, activities related to maritime transport, recreation, housing, and elimination of wastes (with a basic treatment), can be considered compatible. At the same time, the above are almost always incompatible with activities such as ocean farming and natural parks or reservations. It is evident that the use of a coastal area cannot permit a complete destruction of the coastal environment nor the complete prohibition of its development. The answer is to optimize the usage of coastal resources by facing the problem in a sensible and responsible manner. The preservation of the most valuable characteristics of the coastal area and the maximization of economical benefits do not have to be in conflict. On the contrary, both actions should form inseparable elements.

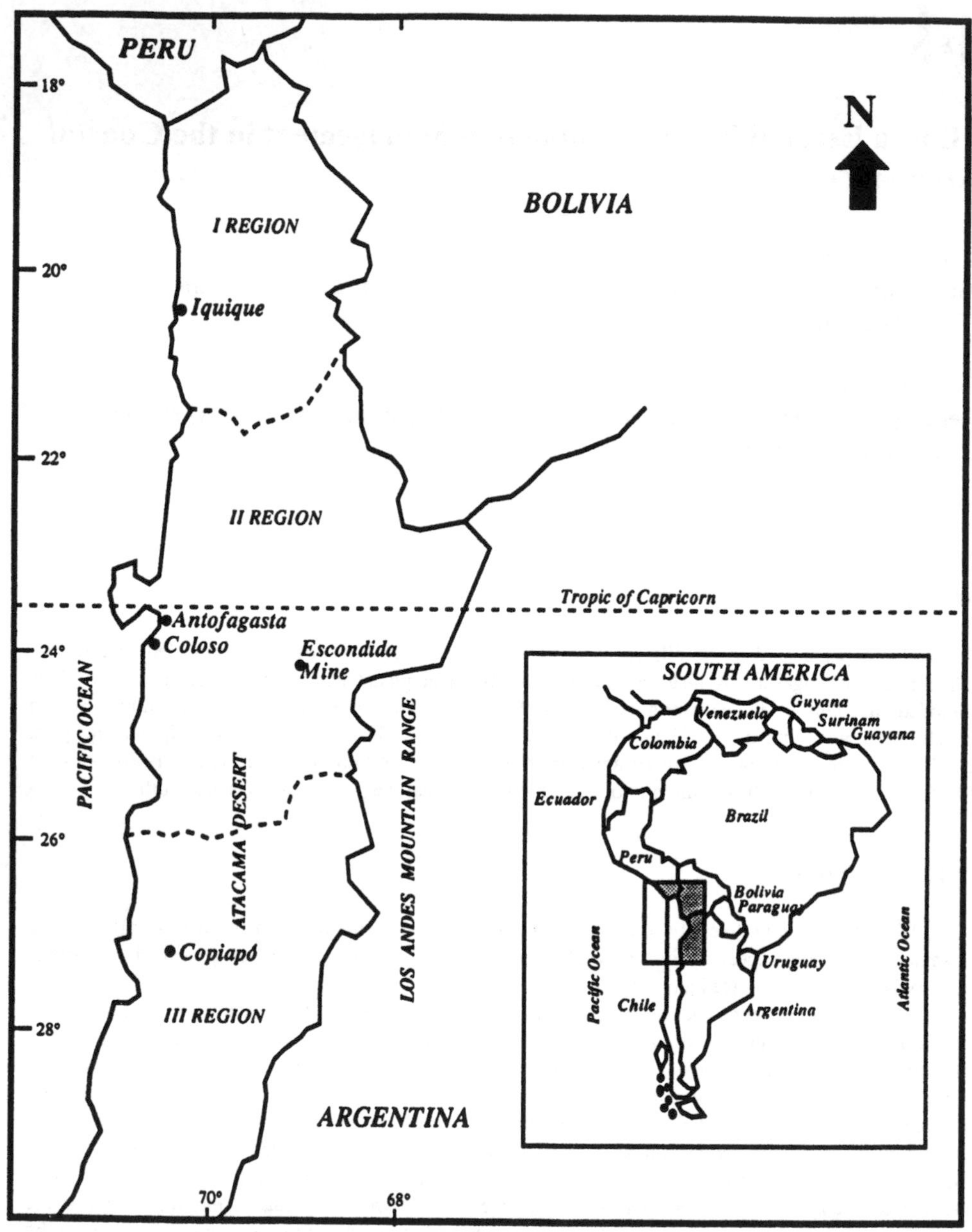

Figure 1. Location of study area. Antofagasta, Chile

MINERA ESCONDIDA

Minera Escondida Limitada is a mining company which currently exploits the Escondida copper ore deposit, located in the Atacama Desert in northern Chile, 160 km southeast of Antofagasta, capital city of the II Region, and 3,100 m above sea level (Figure 1). Minera Escondida is a joint venture of the Australian company, Broken Hill Proprietary (BHP), 57.7%; the British company, Rio Tinto Zinc (RTZ), 30%; the Japanese-Escondida Corporation (JECO), 10%; and the International Finance Corporation of the World Bank (IFC), 2.5%. Escondida is the second largest copper mine in the world, with an annual production of 320,000 tons of fine copper contained in concentrate. The copper concentrate produced at the minesite is transported, via a 170-kilometer slurry pipeline, from the mine to the port of Coloso, located 14 km south of Antofagasta. At Coloso, the concentrate is dewatered and shipped to smelters located throughout the world. Escondida is currently the world's largest producer of custom copper concentrate.

In 1990, Escondida began operation at the Coloso port. While the area historically was an important industrial center, it was currently an artisan, fishing, and tourist area. It was important then, to develop operating practices that would maintain harmony with these activities. In the following pages, a general description is made of the different activities developed to achieve these objectives.

HISTORICAL BACKGROUND

For over half a century, nitrate was Chile's major industry, carrying life to formerly desolate regions and giving employment to thousands spread over hundres of mining operations.

The beginning of industrial activity at Coloso is registered in 1879, when the 'Canton de Aguas Blancas' (mining district), located 80 km southeast of Coloso, started to produce nitrate. Around 1881, six nitrate mines in operation (Central, Oriente, Florencia, Encarnacion, Maria Teresa, and Santa Rosa) produced up to 50,000 bushels of nitrate. That same year, the Chilean government established that the nitrate exported from 'Aguas Blancas,' as well as from other mines, had to pay a tax for each metric bushel produced. This severe blow would result in the shutdown of several mines, with only a slight recovery of the district several years later with the opening of the 'Pepita' operation in 1903.

In 1901, the construction of the Coloso-Aguas Blancas railroad began, with operations starting in 1902. On May 21, 1901, a presidential decree officially recognized Coloso Bay as a minor port, starting January 1, 1992. In 1907, Coloso had two docks that could handle nitrate export shipments and reception of coal and general goods. Along with the construction of the dock and breakwater, a series of other buildings appeared, forming the final structure of Coloso, as shown in Figure 2. The official 1907 census registered a population of 2,032, the highest human concentration in the history of the area.

At the beginning of 1920, Antofagasta and Coloso were finally connected by a road that supported automobiles and other vehicles (Figure 3). The number of vehicles moving between Antofagasta and Coloso was unusually high compared to traffic movement between Antofagasta and other locations (Tables 1 and 2). Due to the short distance to Antofagasta, this road quickly became the main communications media. Coloso, a highly specialized center with basic, public service facilities, soon established a functional relation with the main city. As well as being the exportation port for the nitrate production, Coloso had the railroad which soon converted it into the link between the nitrate operations at Canton de Aguas Blancas and Antofagasta. This fact could well explain the important amount of passenger vehicles moving between Antofagasta and Coloso, which in 1931 was 95% of the total traffic in the region.

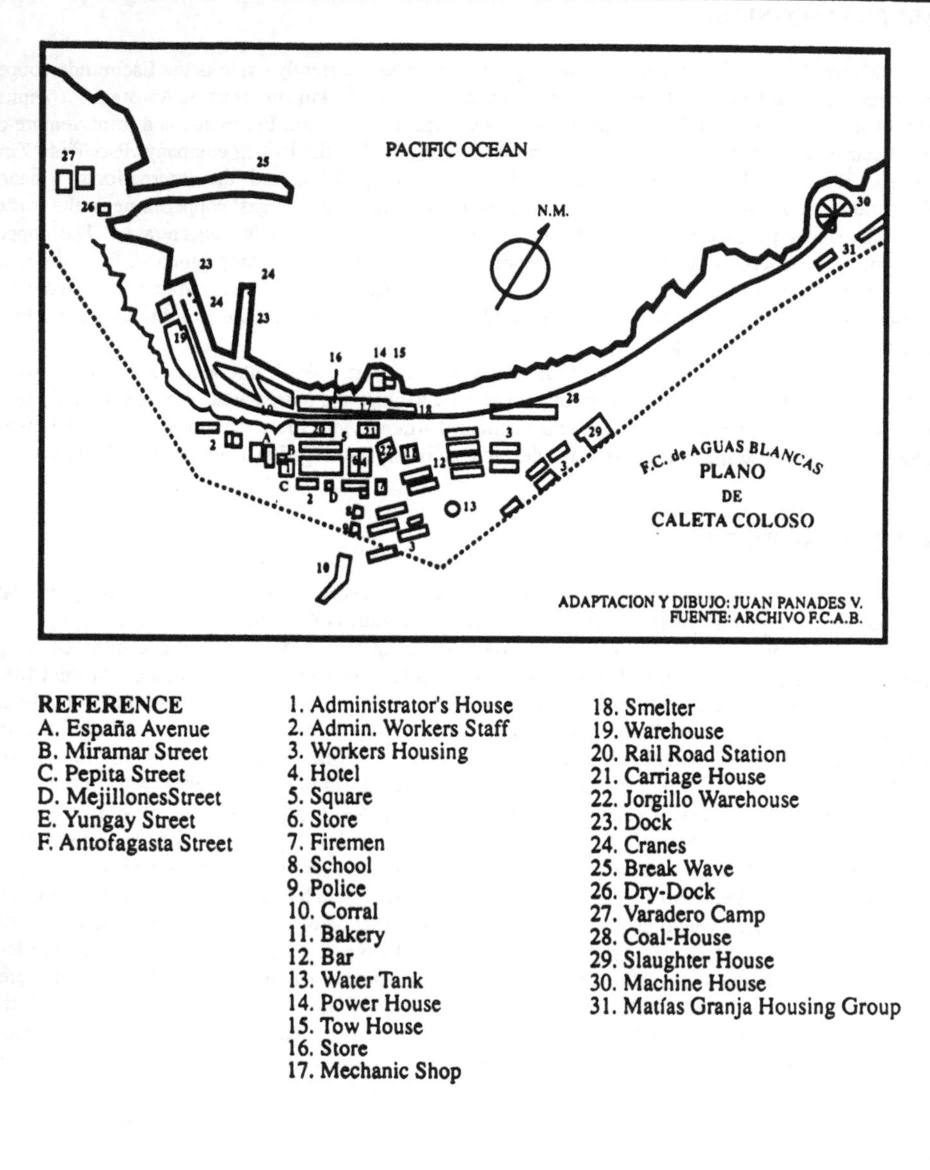
PACIFIC OCEAN
N.M.
F.C. de AGUAS BLANCAS
PLANO
DE
CALETA COLOSO
ADAPTACION Y DIBUJO: JUAN PANADES V.
FUENTE: ARCHIVO F.C.A.B.
REFERENCE
A. España Avenue
B. Miramar Street
C. Pepita Street
D. MejillonesStreet
E. Yungay Street
F. Antofagasta Street
1. Administrator's House
2. Admin. Workers Staff
3. Workers Housing
4. Hotel
5. Square
6. Store
7. Firemen
8. School
9. Police
10. Corral
11. Bakery
12. Bar
13. Water Tank
14. Power House
15. Tow House
16. Store
17. Mechanic Shop
18. Smelter
19. Warehouse
20. Rail Road Station
21. Carriage House
22. Jorgillo Warehouse
23. Dock
24. Cranes
25. Break Wave
26. Dry-Dock
27. Varadero Camp
28. Coal-House
29. Slaughter House
30. Machine House
31. Matías Granja Housing Group

Figure 2. Urban situation of 1931

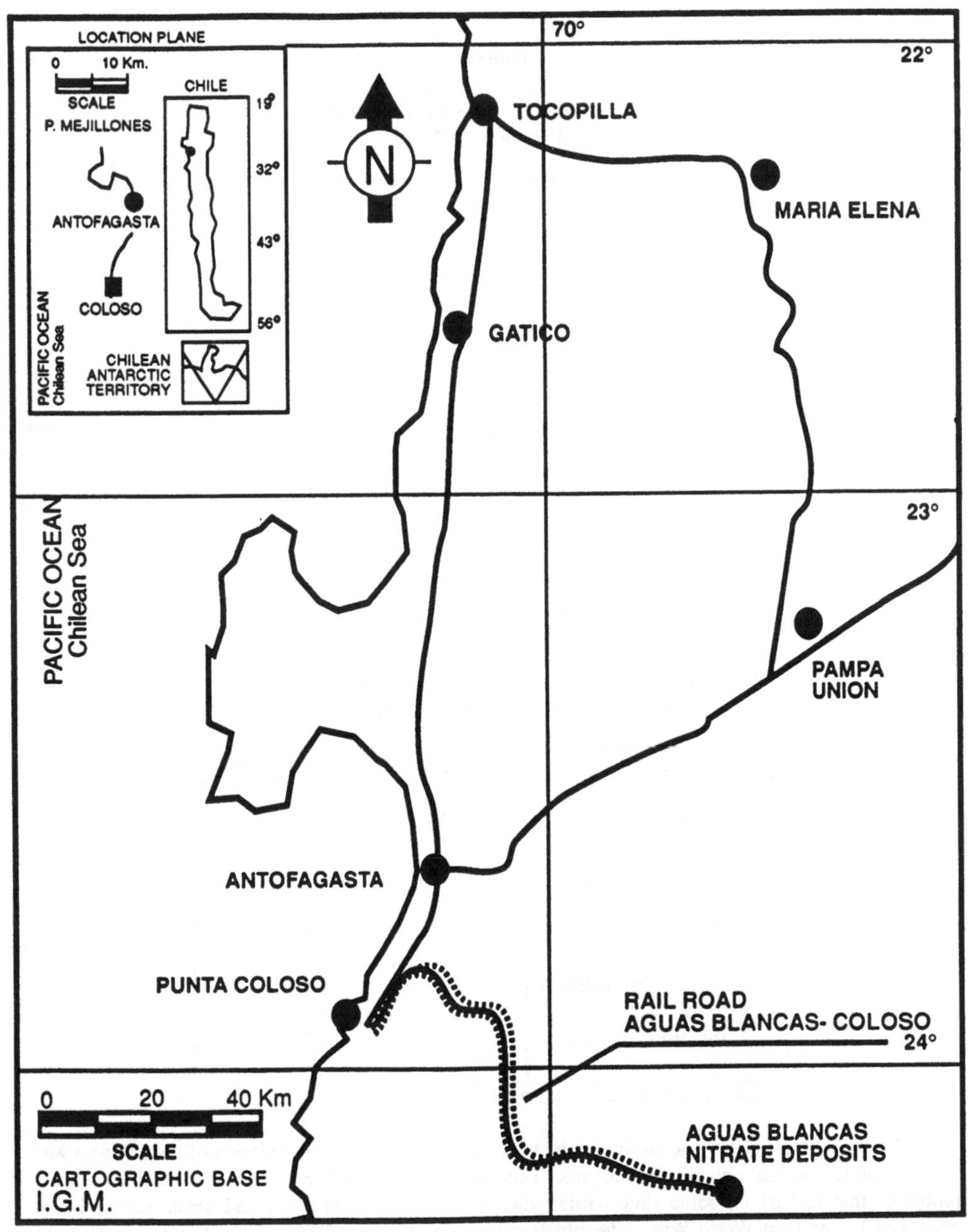

Figure 3. Location main highway flow. 1931.

Table 1

Daily Automobile Traffic on Different Roads
II Region. 1931

Antofagasta - Union	126
Antofagasta - Gatico	95
Tocopilla - Gatico	83
Tocopilla - Maria Elena	98
Antofagasta - Coloso	300
Calama - Chuquicamata	150

Source: Provincial Road Bureau, 1931

Table 2

Classification and Composition of Traffic
in the II Region. 1931 (No. of Vehicles)

ROAD	GOODS	PASSENGERS
Antofagasta - Union	54,871	44,895
Antofagasta - Gatico	9,217	34,675
Tocopilla - Gatico	4,455	20,295
Tocopilla - Maria Elena	16,070	35,770
Antofagasta - Coloso	5,763	109,500

Source: Provincial Road Bureau, 1931

One year before (1930), according to the tenth national census, the population of Antofagasta was 53,591, and Coloso had only 422 inhabitants. This strong decrease in the population of Coloso reflects problems that had begun some time earlier, due to an excess production and stock accumulation generated by the First World War. The situation became critical in 1930, due to the growing use of synthetic nitrate in the world markets. This affected most large, northern Chilean ports, as well as synthetic nitrate in the world markets. This affected most large, northern Chilean ports, as well as Coloso. In the following years, Coloso shipped only the remnant stocks of nitrate. At the same time, the Antofagasta port was built and in 1932, the permits authorizing the operation of Coloso port were terminated. The port was dismantled in 1933 (Recabarren et al. 1989). From 1933 until 1990, Coloso was used as a recreational area by the inhabitants of Antofagasta and as a fishermen settlement, which culminated in the current, Caleta Coloso.

USE OF THE AREA AND CONDITIONS BEFORE 1990

Use of the Area

For evaluation purposes, the area under study (Figure 4) is between El Huascar resort, 3.9 km north of Coloso and El Lenguado Bay, located 1.8 km south of Coloso (Escondida Project 1990).

During 1989, the following human activities were observed in the area:

1. Artisan fishing for commercial purposes
2. Sportfishing
3. Beaches
4. Camping
5. Domestic dwelling

1. *Artisan Fishing for Commercial Purposes.* An important artisan fishery for commercial and subsistence purposes was carried out by fishermen in Caleta Coloso. The most exploited resources were shellfish and to a lesser degree, fish. Within the area under study, the most frequently used places are: El Huascar, El Way, Los Murallones, Paso Malo, Playa Amarilla, Caleta Coloso, Punta Coloso, and El Lenguado (Figure 4).
2. *Sportfishing.* The coastal area near Antofagasta is very rich with a wide variety of species, making it very attractive for sportfishing. Antofagasta inhabitants as well as visitors have made use of the area for recreational fishing for many years. This activity takes place year-round, with a significant increase during the summer months.
3. *Beaches.* In the area under study, it is possible to identify two important beaches. The first, El Huascar, is located 10 km south of central Antofagasta on the road that leads to Coloso. In 1989, El Huascar had approximately twenty summer homes and some recreational infrastructure. Public transportation to the beach was available during the summer months. Playa Amarilla, located 14 km south of downtown Antofagasta, is a very popular beach for the people of Antofagasta, but has virtually no recreational infrastructure. Public transportation reached the area only during summer months (Figure 4).
4. *Camping.* The only authorized campsites in 1989 were Roca Roja and Las Garumas. The first had some basic installations but was used infrequently; the latter had more facilities and was more popular (Figure 4).
5. *Housing.* Two urban sites are located in the area under study. The first is the summer home site at El Huascar, and the second is Caleta Coloso, where in 1989, twenty-three family homes were registered.

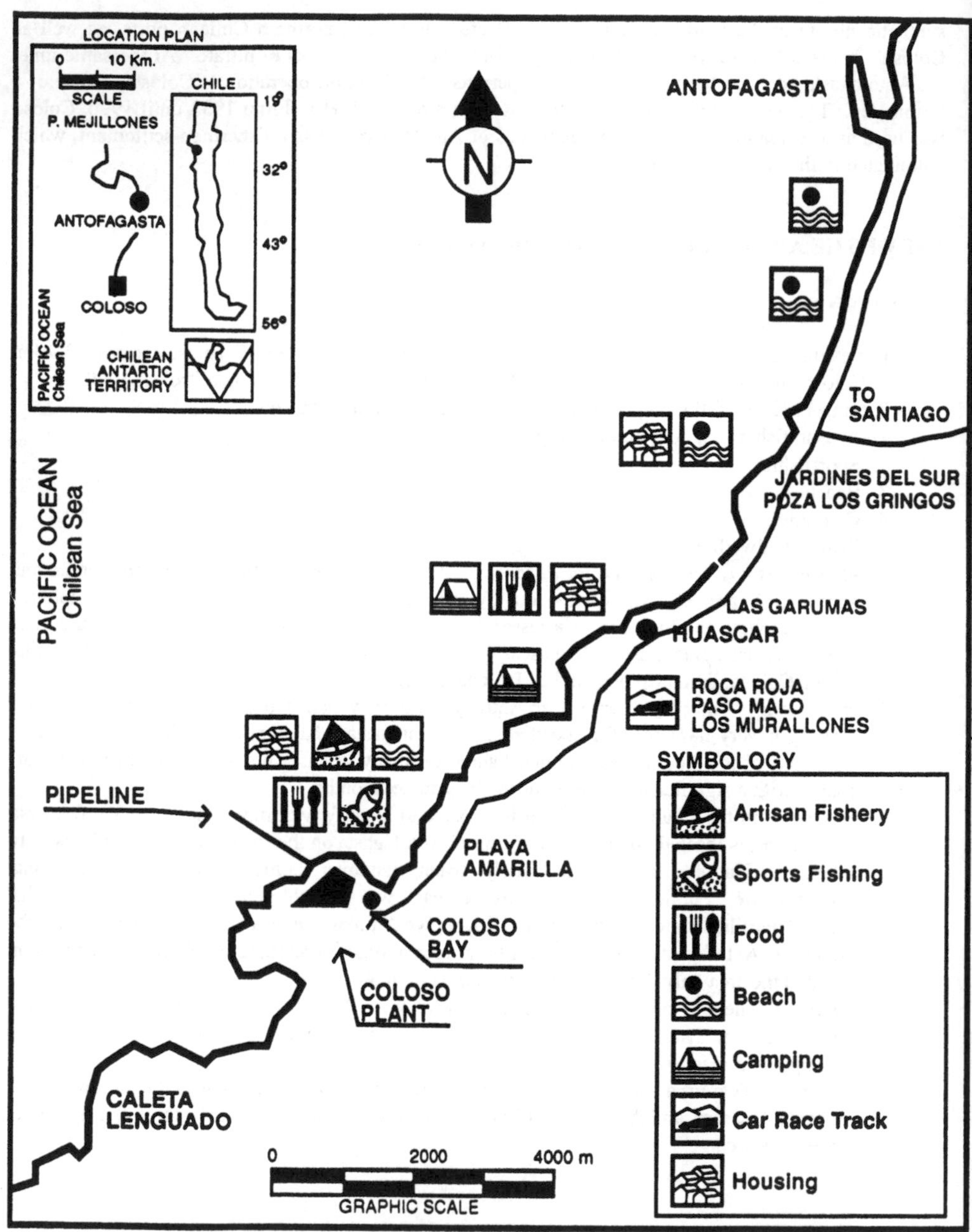

Figure 4. Uses of the area. 1989

Situation of Caleta Coloso Before the Installation of Minera Escondida (1990)

In September 1989, twenty-three fishermen and their families lived in Caleta Coloso, making a total of sixty-five people, most of them living in rudimentary conditions. They did not have electricity and had a serious problem with drinking water, which was purchased from tank trucks at high prices. The restaurant facilities were limited to three stands where visitors could eat seafood, not always prepared under proper, hygienic conditions. The Bay did not have port facilities, so the boats had to anchor offshore. This operation was dangerous under bad weather conditions, and sometimes it was simply impossible to launch or dock the fishing boats. On the other hand, the regular conditions of the access road limited massive visitation to the area. This piece of information was gleaned from personal communication with Mauricio Osses, President of the First Fishermen Union, Calesta Coloso. Under these general conditions, and as part of the Escondida project, construction of the Coloso port and process facilities began on September 15, 1989.

MINERA ESCONDIDA OPERATIONS

The Escondida mineral deposit was discovered in March 1981. Based on study, extensive development, and involved financial arrangements, pre-mine stripping and facilities construction started in August 1988. The concentrator began production in November 1990, treating 35,000 tons/day of copper sulfide ore, using a conventional flotation process to obtain 2,000 tons/day of copper concentrate. Tailings are thickened before being deposited in a natural basin adjacent the mine. The concentrate, in slurry form, is transported to Coloso in a 170-kilometer pipeline. At Coloso, the slurry is received in two agitating tanks, decanted and passed to the filter plant for dewatering. The final product is a fine, copper concentrate powder, with 9% moisture, which is sent via conveyor belt to a closed storage building of 100,000-ton capacity. During the shiploading operations, the concentrate is transported at a 1,000 ton/hour rate, via conveyor belt system to a mechanical shiploader on the dock and loaded into international ships. Currently, Coloso makes two or three shipments per month, with loading requiring between one and three days, depending upon vessel size. This means that on average, the shiploading work does not take more than ten days per month (Minera Escondida Limitada 1991).

MINERA ESCONDIDA'S ACTIONS TO REACH AN HARMONIOUS USE OF THE COASTAL AREA AT COLOSO

The decision to transport the copper concentrate via pipeline from Escondida to Coloso and to construct a plant and shiploading port was based on consideration of such problems as the avoidance of heavy truck movement between Escondida and the central Antofagasta port, reduction of operational problems, avoidance of dust emissions, and the project's cost. A clean operation at Coloso was an important factor in maintaining the different uses of the area, which should not be affected by the industrial activity. A series of actions were implemented. Some were legal requirements, such as those relating to the possible impact on the marine environment by the effluent from concentrate dewatering, while others were at Escondida's own initiative.

One of the relevant legal requirements is contained in resolution Ord. 12600/550, issued by the General Bureau for Maritime Territory and Merchant Marine (DIRECTEMAR) on August 21, 1987, relating to the conservation of marine environments. In accordance with this legislation, construction of a water treatment plant began, along with an environmental impact study, both to be described in some detail.

Water Treatment Plant in Coloso

Water from the concentrate slurry dewatering process is taken to a thickener in which suspended solid particles are recovered. The thickener overflow is then sent to a three-stage decanting pond to remove those particles which are not recovered by the thickener. After this, the water passes through a polishing filter to remove fine particles, passing then to a storage tank. Part of the water is used in a fire system that covers all the installations and for plant cleaning operations. The rest is the effluent, which is discharged in the ocean after a secondary treatment that includes duplex filters; bag filters; five, activated carbon cells to reduce the content of organic material; and four cartridge filters in the final treatment stage to remove particles under 2 microns. Before discharge, a part of the effluent (approximately 16%) is by-passed to a Reverse Osmosis plant to obtain irrigation water used in a special project which is described later on in this document. The current disposal system includes a 1,320-meter long, submarine pipeline whose interior diameter is 185 mm. At the end of the pipeline, a diffuser with sixteen holes is located perpendicular to the pipeline and in line with the strongest currents, controlling the effluent emission to ensure maximum dilution (Figure 5)(Escondida Project 1989).

Environmental Impact Study at Coloso

The slurry dewatering process at Coloso brought forth the need to anticipate the possible effects that the liquid outfall could have on the marine ecosystem. To achieve this, and according to the regulations established by the maritime authority, a program was implemented, including a simulation of the metallurgical process at both Escondida and Coloso in order to design a safe discharge system. The latter involved a detailed, physical oceanographic study; modeling of the outfall's behavior according to the criteria suggested by the U.S. Environmental Protection Agency (EPA); establishment of an Environmental Baseline; a prediction of the impact evaluation; and a verification study with Rhodamine WT of the diffusion effects, performed shortly after the submarine outfall began operation.

The Environmental Baseline Study included various sampling campaigns which were carried out in February and September 1989, and in January and August 1990 (Figure 6). During these baseline studies, considerable information about the area's ecosystem was obtained. The main goal was to characterize the environmental situation before start-up, in order that any inconvenience that might eventually appear could be detected and the necessary mitigation actions taken (Escondida Project 1989).

The following were the main activities considered in the Environmental Baseline Studies:

1. Meterology
2. Oceanography
3. Soft seabed communities
4. Rock communities
5. Artisan fishing statistics
6. Air quality
7. Trace metals in water
8. Trace metals in organisms
9. Trace metals in sediments

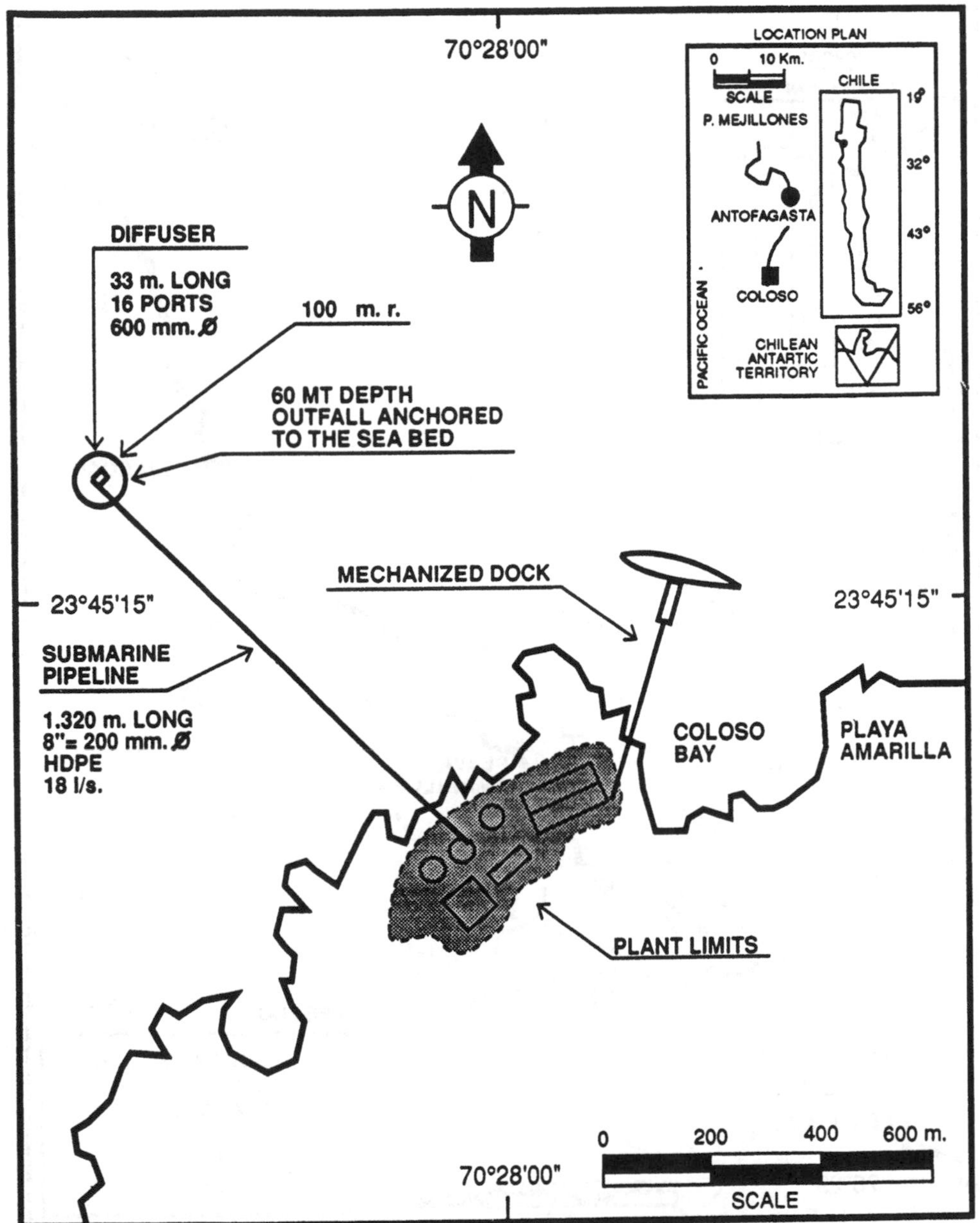

Figure 5. Location of the Coloso submarine outfall system

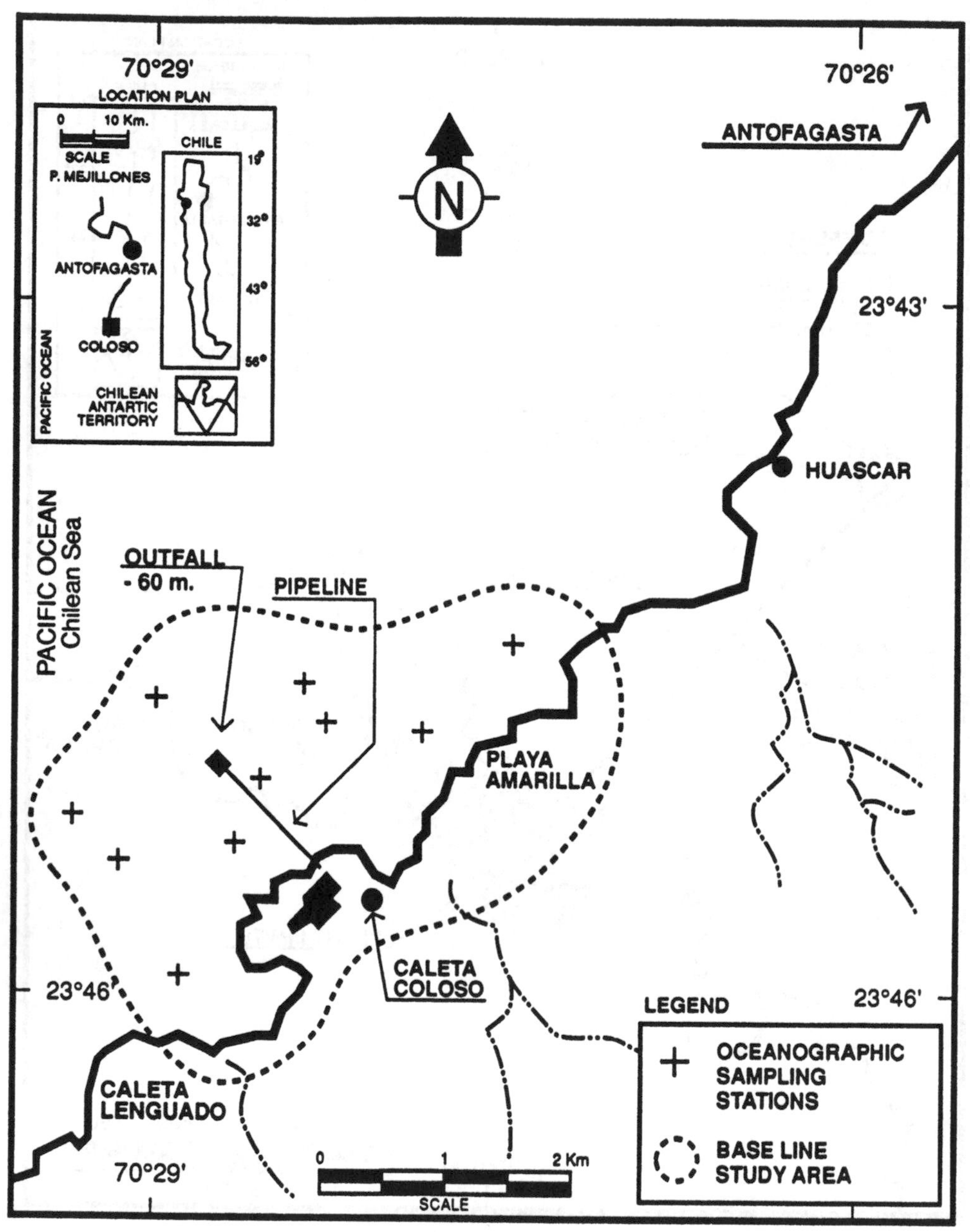

Figure 6. Baseline Study area

Based on the effluent characteristics, the information collected during the Baseline Study, and mathematical models for circulation and dispersion, a study was performed to evaluate the impact of the outfall on the ocean. To do this, the main components that could be affected were considered, such as water quality, biota, inter- and subtidal communities, and artisan fishing. The intensity and extent of the area that eventually could be affected were quantified, as well as the importance and probability of occurrence of an impact. The results of this evaluation process showed a 0.8% probability of impact, which in absolute terms is low. Likewise, the study indicates that no cumulative and irreversible effects will occur and that the ecosystem at Coloso will not be measurably altered.

Follow-Up and Monitoring

A complete monitoring program has been designed to obtain systematic, detailed, and accurate information to control potential effects that the outfall might produce on the environment. This includes the following activities:

Emission Monitoring. The purpose of this program is to establish a control system to monitor the maximum concentration levels in the effluent after treatment in Coloso plant. To do this, a daily control of the head tank is performed, analyzing for the most important elements. Likewise, pH values are registered every day, as well as the amount of effluent discharged. Average of the results and norms for the emission point are shown in Table 3.

Table 3

Maximum Concentrations Accepted by the Specification in the Effluent and Average Concentrations Registered

PARAMETER	MAXIMUM CONCENTRATION OF DISSOLVED ELEMENTS ACCEPTED BY THE SPEC. (mg/1)	AVERAGE DISSOLVED ELEMENTS IN EFFLUENT WATER. PERIOD: 07/Jan/91-19/Apr/92
Arsenic	0.100	< 0.010
Cadmium	0.010	< 0.005
Copper	1.000	0.210
Cyanide	1.000	< 0.100
Mercury	0.002	< 0.001
Lead	0.010	< 0.005
Selenium	0.100	< 0.011
Zinc	0.050	< 0.013

Start-Up Monitoring. Along with the start-up of operations at Coloso port and the first effluent emission to the environment, a short-term monitoring program was carried out in three stages between December 1990 and March 1991. During the program, several activities were developed, most of which indicated results similar to those obtained in the Baseline Study.

Long-Term Monitoring. After the start-up monitoring program terminated, a biannual, long-term monitoring program began, covering the same area as the Baseline Study. This monitoring program measures those ecosystem components that would reveal long-term impacts and considers the same activities developed during the Baseline Study. The results obtained to date confirm the impact predictions. No impact has been detected as a result of effluent discharge through the diffuser. Nevertheless, the experience gained from the port operation indicated the need to perform additional engineering studies to further reduce fugitive concentrate dust generated by the strong winds which occur in the area. The results of the mentioned studies indicated the need to implement corrective actions with an approximate cost of $1.6 million U.S. These actions consisted of the complete closure of the stockpile building and the construction of an extension of the existing building to cover the conveyor belt bin area which was originally exposed.

Bioassay Program

During 1990, as part of the effluent water quality control, acute toxicity assays were performed quarterly, using the ASTM-STPG634.1977 method, with the local species of fish, *Cheirodon pisciculus*. Undiluted effluent water was used during the assays. Results for the first two assays showed 100% survival, the results for the next two were 95% and 90% (Instituto de Investigacion Pesquera 1992).

Concentrate Transport

The use of gravity to transport concentrate as a slurry in a pipeline had operational justifications where environmental considerations were relevant. In the mid- and long-term, the total cost of this system will prove to be lower in comparison to railroad or truck transportation. Also, with this system, problems related to bulk transport are avoided. These include transferring material into trucks or railroad cars with the unavoidable generation of airborne dust and the constant flow of trucks to the port with a noisy, contaminating, and unsafe operation, especially during traffic through Playa Amarilla. This is particularly important, considering the local popularity of Playa Amarilla Beach or central city congestion at the port of Antofagasta (Minera Escondida Limitada 1991).

Green Areas Project

A project was designed to visually improve the area surrounding the plant. To do this, a portion of the treated effluent is passed through a Reverse Osmosis plant to produce a maximum of 200,000 liters per day of irrigation water. This water is currently being used to create a green area at Coloso. This project is developed jointly with two local universities and CONAF (National Forestry Corporation)(Minera Escondida Limitada 1991).

The irrigation water is pumped to a tank located in the hills behind Coloso, and another pipeline conducts water by gravity to the different irrigation areas. More than one thousand trees and plants are irrigated via a time-controlled system.

This project considers three stages.

1. Planting of trees in the coastal area for ornamentation purposes and to create a windbreak in the plant installations.
2. Planting of flower and ornamentation species.
3. Planting certain coastal desert species, with emphasis on those with conservation problems.

Visitor Observation Post

Since the beginning of the environmental programs, special care has been given to inform the community of the scope of the studies and actions taken by the Company to develop a technically- and scientifically-oriented, environmental management program. To achieve this, a visitor observation post was built near the plant, where the community can obtain direct information of the technical aspects of the Escondida operations.

ESCONDIDA'S RELATIONSHIP WITH CALETA COLOSO COMMUNITY

When Escondida officially announced its intention to build a shiploading port at Punta Coloso on November 28, 1988, the fishermen living nearby formed the First Fishermen Union of Caleta Coloso, with the purpose of having a formal organization to negotiate with the Company. Early in 1989, a committee was formed by the Ministry of Housing and Urbanization (MINVU), the Municipality of Antofagasta, the local Maritime Governor, and Minera Escondida, to study the situation of a small number of families that would be affected by the construction of the access road to the port facilities being built by the Company. As a result, the Company built three family homes and presented a project for construction of a seafood and fish market.

At the same time, the Fishermen Union worked with the Municipality and MINVU to formalize the distribution and allocation of lots on the Bay. The Social Development Department of the Municipality was assigned to transfer the families and their belongings to the new lots, an activity eventually carried out by Minera Escondida, which contributed construction materials and work force. It is important to point out that all these activities were performed with the knowledge of the corresponding authorities, and according to the conversations and agreements between the Company and the Fishermen Union of Coloso (Mauricio Osses, personal communication). Other developments included the construction of a fish market, three seafood restaurants, and public washrooms. Additionally, roads were improved, housing slabs were poured, more homes relocated, and a social hall rehabilitated. On a permanent basis, Escondida assists with the busing of the Coloso schoolchildren, garbage removal, and the industrial water supply for the public area of Coloso. Likewise, Escondida has carried out maintenance and cleaning activities in Playa Amarilla, installing access stairs and trash cans.

CURRENT SITUATION OF THE COASTAL AREA AND ITS USES

After two years of operating the facilities at Coloso Port, the situation of the uses of the area has not changed relative to those observed before Escondida's start-up. Most of the uses, in fact, show an increase (Table 4). For example, Playa Amarilla has maintained its importance as a main tourist attraction in Antofagasta, which is reflected in the information obtained in a traffic monitoring activity carried out in the southern area of Antofagasta (Table 5). It can be observed in Table 5, that on the weekend of September 5 and 6, 73% and 68.3% respectively, of the vehicles registered were passenger vehicles, most of which were headed to the coastal area at Coloso. As for Caleta Coloso, currently twenty-six families live there on a permanent basis, making a total of seventy-nine people. This slight increase in the number of inhabitants is also reflected in the activity of the Bay, which has benefited by the infrastructure developed by Minera Escondida in the area.

Table 4

Current Situation of the Different Uses in the Coloso Coastal Area

USES	LOCATION		INFRASTRUCTURE	
	1989	1992	1989	1992
COMMERCIAL FISHING	El Huascar Murallones El Way El Lenguado	El Huascar Murallones El Way El Lenguado	* Road in regular condition	* Dock * Dry dock * Fish/food market * Road in good condition
SPORT FISHING	Coast area El Huascar El Lenguado	Coast area El Huascar El Lenguado	* Road in regular condition	* Road in good condition
CAMPING AREAS	Las Garumas	Las Garumas	* Cabins * Bathrooms	* Cabins * Bathrooms * Elec. power
	Roca Roja	Roca Roja		* Elec. power
		EMPORCHI		* Elec. power
BEACHES	El Huascar	El Huascar	* 1 dine * 2 discos	* 2 diners * 3 discos * Elec. power
	Playa Amarilla	Playa Amarilla	* 3 basic sale stands	* 3 sale stands * Public restrooms * Dock * Dry dock * Access stairways * Daily cleaning
	El Huascar	El Huascar		* Elec. power
HOUSING	El Huascar	El Huascar		* Elec. power
	Caleta Coloso	Caleta Coloso		* Elec. power * Community hall

It is also important to point out that during the community studies performed in 1990, concrete evidence of natural repopulation processes of some coastal species were observed, specifically within the limits of the maritime concession granted to Minera Escondida. The above was due to an important reduction in the extraction of the mentioned species, resulting from the strict protection of the coast

Table 5

Vehicle Count in the Southern Area of Antofagasta

	NUMBER OF VEHICLES							
Day/Category	A	V2R	B	C	CN	BI	OM	TOTAL
Tuesday 01	1347	23	710	151	814	14	9	3068
Wednesday 02	1495	23	701	138	799	13	3	3172
Thursday 03	1425	31	717	182	804	23	8	3190
Friday 04	1575	64	150	147	911	32	7	3486
Saturday 05	2824	79	798	103	1123	30	6	4963
Sunday 06	3298	705	645	35	1127	8	9	5827

	PERCENTAGE						
Day/Category	A	V2R	B	C	CN	BI	OM
Tuesday 01	43.9	0.75	23.14	4.92	26.53	0.46	0.29
Wednesday 02	47.13	0.73	22.10	4.35	25.19	0.41	0.09
Thursday 03	44.67	0.97	22.48	5.71	25.20	0.72	0.25
Friday 04	45.18	1.84	21.51	4.22	26.13	0.92	0.20
Saturday 05	56.90	1.59	16.08	2.08	22.63	0.60	0.12
Sunday 06	56.60	12.10	11.07	0.60	19.34	0.14	0.15

Notes: A : Automobiles
V2R : Two-wheel vehicles
B : Urban buses
CN : Pickup trucks
BI : Institutional or Interurban buses
OM : Other not classified (heavy equipment, etc.)

Source: Jardines del Sur Traffic Monitoring Study, Transport Engineering Department, Pontificia Universidad Catolica de Chile, September 1992

during the construction stage. As a complement of the community study, focal studies of several species with conservation problems were conducted, specifically with abalone and sea urchin. To date, the three studies conducted show that the repopulation process which occurred within Minera

Escondida's concession limits has been significant. For example, the third focal study shows an abalone and urchin capture per unit of effort between 3 and 7 times greater inside the concession than outside. Additionally, the average size of the individuals of both species are significantly larger inside the concession than outside its limits. The evaluation of the reproductive potential shows spectacular results in the abalone population within the concession area compared to similar areas outside. In May 1992, the potential number of abalone larvae liberated to the plankton by the adult population within the concession limits was calculated at 43,548,222. Similar areas outside the concession limits that are exploited by skin divers show values around 680,702 and 13,678 larvae. The overall results of the studies show that the intertidal zone (its biota and specific physical environment) of Minera Escondida's concession area under environmental protection due to safety and customs reasons, has suffered no damaging effects from the plant's operation. In fact, it not only shows strong evidence of having positive effects in terms of repopulation of key organisms, but the large number of larvae found within the protected concession indicate it is acting as a hatchery and a marine reserve for the surrounding area as well (Geotecnica Consultores 1992).

CONCLUSION

The main conclusion obtained from this work is related to the fact that the installation and operation of Escondida's concentrate dewatering and shiploading facilities have not interfered with the uses that the area had previously. Moreover, the existence of the Escondida plant in the Coloso area has improved and reinforced those uses, especially as a resort and tourist attraction, after the initiation of the Company's industrial activity.

Likewise, an equally important fact is the maintenance of the environmental conditions established in the Baseline Study prior to operations start-up, indicating that the Escondida effluent has not produced cumulative or irreversible effects. Nevertheless, it is important to point out that it was necessary to take corrective actions to reduce the loss of airborne concentrate. Further studies have shown that the problem of airborne dust has been controlled.

Finally, considering both items above, it can be concluded that industrial activities like the large-scale processes developed by Minera Escondida, are perfectly compatible with the different uses of the coastal area at Coloso, as long as operations are carefully planned and monitored.

ACKNOWLEDGMENTS

I wish to acknowledge the Company Management for their constant support and encouragement in the development of the environmental programs at Coloso and the minesite. A special recognition is due to Mrs. Besie Harvey for her help in the transcription and translation of this document.

REFERENCES

Geotecnica Consultores. 1990. *Environmental Impact Evaluation.* Escondida Project, Chap. 4, 1-118.

Geotecnica Consultores. 1991. *Third Inter and Sub Tidal Focal Study, Escondida Coastal Area. Playa Amarilla.*

Instituto de Investigacion Pesquera, Environmental Studies Bureau. 1992. *Toxicity bioassay program to evaluate the biological effects of Minera Escondida's effluent on Cheirodon pisciulus.* Tacahuano.

Minera Escondida Limitada. Environmental Impact Study of the Submarine Outfall at Puerto Coloso. 1991. Working document of the Audit, *Workshop on the Environmental Impact Evaluation on the Marine Coastal Environment Produced by a Mining Project Case Study: Minera Escondida.* Organized by CPPS/UNEPA.

Osses, Mauricio (President of First Fishermen Union). Personal communication to author, n.d.

Recabarren, F., A. Obilinovic, and J. Panades. 1989. *Coloso, una Aventura Historica.*

The Coastal Environment: Assessment Standards and Issues

15

Integrated Coastal Ocean Space Management: Challenges for the Next Decade

Charles N. Ehler
Director, Office of Ocean Resources Conservation and Assessment (ORCA)
National Oceanic and Atmospheric Administration
Rockville, Maryland

"The future never just happened. It was created." – Will and Ariel Durant, *The Lessons of History*

THE PROBLEM

One of the greatest challenges that we face over the next decade is restoring, maintaining, and enhancing the quality of our coastal areas under the pressures of projected population growth and coastal development.[1] The difficulties associated with achieving this goal, along with the intrinsic complexities of the natural and anthropogenic systems that converge in coastal areas, is a management challenge as complex as any we face. The challenge is international–applicable to all coastal nations.

The population of the world has doubled between 1950 and 1985. Today it is growing at a rate that adds a billion people every eleven years. The next ten years represent the last chance to stabilize human population at something less than double the current world population of 5.4 billion by the middle of the next century (Camp 1993).

More than 95% of future population growth will occur in the developing countries of Africa, Asia, and Latin America.[2] Most of the developing countries, outside of East Asia, still have annual

[1] "Coastal areas" are defined from a natural systems viewpoint as, at minimum, the coastal watersheds that drain directly into estuaries, other coastal land areas that drain directly into other coastal waters, and the adjacent coastal waters and sediments out to the edge of the continental shelf or 200-mile exclusive economic zone (EEZ). From the viewpoint of political jurisdictions, NOAA has defined "coastal counties" as those with at least 15% of their land area in a coastal watershed. Of the 3,141 counties in the United States, 672 are classified as "coastal" (NOAA 1990a).

[2] Today, developed countries comprise 22% of the global population, with a growth rate of only 0.6% per year. In contrast, the developing regions encompass 78% of the world's total.

population growth rates of between 2.5% and 3.5%. At these rates, populations will double in fewer than thirty years. India and China together accounted for a third of the record 93 million-person increase in 1992. India alone adds 18 million people—equivalent to the population of cities like Mexico City or Sao Paulo—to the world's population every year.

At least half of the world's population will live in cities by the first decade of the next century, most of them in developing countries, and almost all in coastal areas. Of the twenty-three cites that will exceed 10 million people by the year 2000, seventeen are in developing countries, and all but six are located in coastal areas (Table 1). These dramatic increases are due not only to fertility and survival of urban residents, but the migration of people from rural to urban areas, and particularly from "non-coastal" to coastal areas. Imagine the coastal management challenges of areas such as Sao Paulo, Calcutta, Bombay, Jakarta, Dhaka, Cairo, and Bangkok, all of which will have doubled in population from 1970-2000. This doubling of the world's population will be of far greater significance in terms of energy, resource consumption, and stress of the environment, especially the coastal environment, than any previous doubling of worldwide population.

Table 1.

The World's Largest cities, Year 2000 (Millions of people)

Rank	Agglomeration	Country	1970	1985	2000	1970-2000
1	Mexico City	Mexico	9.12	17.30	25.82	16.70
2	**Sao Paulo**	**Brazil**	**8.22**	**15.88**	**23.97**	**15.75**
3	**Tokyo/Yokohama**	**Japan**	**14.91**	**18.82**	**20.22**	**5.31**
4	**Calcutta**	**India**	**7.12**	**10.95**	**16.53**	**9.41**
5	**Greater Bombay**	**India**	**5.98**	**10.07**	**16.00**	**5.98**
6	**New York**	**USA**	**16.29**	**15.64**	**15.78**	**-0.51**
7	**Shanghai**	**China**	**11.41**	**11.96**	**14.30**	**2.89**
8	Seoul	Korea	5.42	10.28	13.77	8.35
9	Tehran	Iran	3.29	7.52	13.58	10.29
10	Rio de Janeiro	Brazil	7.17	10.37	13.26	6.09
11	**Jakarta**	**Indonesia**	**4.48**	**7.94**	**13.25**	**8.77**
12	Delhi	India	3.64	7.40	13.24	9.60
13	**Buenos Aires**	**Argentina**	**8.55**	**10.88**	**13.18**	**4.63**
14	**Karachi**	**Pakistan**	**3.14**	**6.70**	**12.00**	**8.86**
15	Beijing	China	8.29	9.25	11.17	8.29
16	Dhaka	Bangladesh	1.54	4.89	11.16	9.62
17	**Cairo/Giza**	**Egypt**	**5.69**	**7.69**	**11.13**	**5.44**
18	**Manila/Quezon**	**Philippines**	**3.60**	**7.03**	**11.07**	**7.47**
19	**Los Angeles**	**USA**	**8.43**	**10.05**	**10.99**	**2.56**
20	**Bangkok**	**Thailand**	**3.27**	**6.07**	**10.71**	**7.44**
21	**London**	**United Kingdom**	**10.59**	**10.36**	**10.51**	**-0.08**
22	**Osaka/Kobe**	**Japan**	**7.61**	**9.45**	**10.49**	**2.88**
23	Moscow	Russia	7.07	8.97	10.40	**3.33**

Note: Coastal cities are indicated in bold type.

Source: United Nations. 1987. The Prospects of World Urbanization, 1984-1985. New York: UN.

The population problem is less dramatic in developed countries, but serious nonetheless. In 1991, the United States had an estimated 252 million inhabitants (NOAA 1990a). Current Bureau of the Census projections foresee that births and immigration will drive that number to 345 million by 2030 (currently about 70% of U.S. population growth is attributed to natural increase and 30% to net migration). That means 93 million more Americans—more than half of whom will choose to live in only three states—California, Texas, and Florida—and more than half of whom will choose to live in coastal areas.

Today over 135 million people—more than half of our total population—live in coastal areas of the United States. Coastal populations along our Gulf of Mexico and Pacific coasts have more than doubled since 1960. Eastern Florida will have increased more than 200%. Imagine the coastal management challenge to maintain any semblance of coastal environmental quality that most Americans desire.

While the popular impression is that the United States is a sparsely populated country, along its coasts it is among the more densely populated countries in the world (Fox 1992). Population density in coastal areas is four times the U.S. national average. Four states in the northeastern United States have population densities greater than 1,000 persons per square mile.

THE ENVIRONMENTAL EFFECTS OF COASTAL POPULATION GROWTH

The coastal impacts of population growth and related development have been well documented. In 1990, the Joint Group of Experts on the Marine Environment in its report, *The State of the Marine Environment,* noted that:

> The coastal strip, encompassing the shallow-water and intertidal area along with the immediately adjacent land, is clearly the most vulnerable as well as the most abused marine zone. Its sensitivity is directly tied to the diversity and intensity of activities which take place there, and the threat to its future is related to the increasing concentration of the world population in this area. *The consequences of coastal development are thus of the highest concern* (emphasis added) (GESAMP 1990).

Today in the United States, many of our most important coastal water bodies are paying the price of coastal development in the form of degraded water and habitat quality, real losses of habitat, declines in fish and shellfish populations, limitations on commercial harvests of important shellfish, and public beach closures. The findings of recent studies by the National Oceanic and Atmospheric Administration (NOAA) and others continue to paint a dismal picture of the health of our coastal areas:

1. Relatively high levels of toxic contaminants (trace elements and synthetic organic compounds) have been found in urbanized estuaries such as Boston Harbor, western Long Island Sound, the Hudson River-Raritan Bay estuary, Mobile Bay, San Diego Harbor, and San Francisco Bay (NOAA 1990b).
2. Although no national monitoring program to measure nutrient overenrichment exists in the United States, many estuaries that have large loading of nutrients, coupled with poor circulation, have recurring problems of eutrophication. A 1985 NOAA report estimated that thirty-seven of fifty-five major, U.S. estuaries had chronic hypoxia problems in all or part of the estuary (Whitledge 1985).
3. Over a third of the Nation's coastal waters classified for shellfishing had some form of harvest restriction in 1990, largely due to discharges of fecal coliform bacteria from coastal development. If the trend that has been observed over the past twenty-five years continues, wild, natural shellfishing in coastal waters could be wiped out (NOAA 1991a).
4. Over the past three decades, commercial stocks of estuarine and coastal shellfish have continued to decline nationwide, despite restoration efforts such as reef replenishment, increased hatchery operations, and selective breeding (NOAA 1991a).
5. Losses of coastal wetlands between the 1950s and the 1970s—when U.S. population, spurred by an unprecedented domestic "baby boom," grew by over 50 million people—were especially large because of increased pressure of coastal development. By the mid-1970s over half of the original salt marshes and mangrove forests in the United States had been destroyed (Johnston et al. 1992).

6. Populations of virtually all estuarine and inshore species of fish have been reduced to historically low levels of abundance by overfishing, habitat loss, and pollution (NOAA 1991b).
7. In 1991, beaches were closed or advisories issued against swimming on more than 2,000 occasions in coastal states that monitor beach quality. Over two-thirds of these closings occurred along the densely populated coastlines of New York, New Jersey, Connecticut, and California. High levels of bacteria, primarily from raw human sewage, were responsible for the overwhelming majority of these closures and advisories. Only four states (Connecticut, New Jersey, New York, and Hawaii) monitor the entire length of their shorelines and another eight states monitor portions of their shorelines (Natural Resources Defense Council 1992).

Not all of the news is bad. A recent report from NOAA's National Status and Trends Program shows decreases at many sampling sites in levels of toxic contamination between 1986-1990, especially for chlorinated organic contaminants such as DDT and PCBs, whose use has been banned in the United States since the early to mid-1970s. Concentrations of many trace elements also appear to be decreasing (NOAA 1992a). Our investment of billions of dollars in pollution control equipment and absolute prohibitions of certain compounds is working, at least in the short run. However, these improvements could be illusory, as additional sources of contaminants locate in coastal areas.

Some of our most sensitive and valuable coastal areas, such as the Florida Keys and Monterey Bay, have only recently been designated as marine sanctuaries, with the promise of a level of protection commensurate with the value of their natural resources.

Given our love of living, working, and playing in coastal areas, conditions are likely to get worse before they get better. Even in those areas that show evidence of progress toward improved environmental quality, increasing population densities and development could offset short-term improvements. Clearly, if we are going to continue to increase the use of coastal area resources to produce goods and services that we desire, whether it is increased food, recreation, energy, or protected areas, we are going to have to pay more attention to the way we manage the use of coastal resources over the next decade and beyond.

COASTAL MANAGEMENT IN THE UNITED STATES TODAY

For the past 20 years, NOAA has provided about $700 million to coastal states for the development of coastal management programs (NOAA 1992b). This amount has been matched by state resources, so that the total investment is well over $1.0 billion. This year, NOAA will provide almost $50 million to state coastal management programs. Today, twenty-nine of the thirty-five eligible states and territories have NOAA-approved coastal management programs. Several states, such as Georgia, Texas, and Indiana, that are not part of the program today are actively developing management programs. Some of the world's most innovative ideas and important lessons about how to manage coastal areas have come from real-world experiments in states such as California, Hawaii, Oregon, North and South Carolina, and Massachusetts.

Despite this level of investment in state coastal management programs (and an order of magnitude higher public and private investment in coastal pollution control, particularly waste-water treatment plants), something is clearly wrong. Coastal environmental quality conditions continue to deteriorate. In addition to declining environmental quality conditions, the vulnerability of coastal areas to natural hazards—from storms and hurricanes in the short run to long-term sea-level rise—continue to increase as population increases.

In spite of these problems and increasing statutory responsibilities and public expectations, a fact of life in the 1990s is that we are faced with decreasing resources with which to manage these problems throughout the world.

INTEGRATED COASTAL MANAGEMENT: THE MEANS

We face tremendous challenges in coastal nations, in public agencies at all levels of government, in the private sector, and in the academic community about how to collectively use our limited resources to manage these valuable areas. Many would argue to limit government's role and let the market take care of things. Unfortunately, most coastal resources are "common property" with "open" or "free" access, with no price imposed on their use. This situation often, if not always, leads to excessive use of the resource and, eventually, to its degradation or exhaustion. The market cannot perform the management task alone.

Unfortunately, neither can most of our existing public institutions at any level of government perform the management task, particularly if we continue to tackle these problems in a disjointed manner. Almost all decisions about the use of coastal resources continue to be made today on a single use, sector-by-sector basis. Some process is needed to decide what is the total mix of products and services that coastal areas should produce at any particular time, who should produce and pay for them, and who should benefit, and by how much. That process is integrated coastal management (ICM).

The need for an integrated approach to analyzing and managing coastal resources is now widely accepted by many decision makers. It is becoming much more widely understood that ecological systems underpin economic systems and that human activities influence the capacity of ecological systems to maintain such activities in coastal, as well as inland, areas.

Integrated coastal management is a continuous process, comprised of a set of tasks or functions, each of which is necessary to produce desired results. The goal of integrated coastal management is to produce the "optimal" mix of products and services from a coastal region over time, where what constitutes "optimal" is the mix that results in the maximum net social benefits. The mix is defined by the political process and changes over time as a result of changes in:

- The interests and priorities of society as interpreted by political institutions and reflected in legislation, policy statements, principles, rules and regulations
- Scientific knowledge about problems and their nature
- Technologies that help define, analyze, and solve problems
- The national and regional economies that affect the relative priority of coastal management issues
- Trade policies and compliance with international convention. Improved integration can take place
- Across sectors of the coastal economy—fisheries, ports, marine transport, recreation, waste disposal, and energy
- Across management objectives related to economic development, environmental protection, resource use or conservation, habitat restoration
- Across media—water, land, atmosphere
- Across space—from the inland boundary of coastal watersheds (or even further "upstream" to the exclusive economic zone)
- Across institutions—local, state, national, international—as well as across public and private entities
- Across the continuum of coastal management tasks—planning, implementation, operation and maintenance, monitoring, enforcement, and evaluation and
- Across relevant disciplines—physical and biological sciences, economics, engineering, political science, and law

The demand to produce all of the goods and services from coastal regions cannot be satisfied simultaneously in most areas. Therefore, conflicts about the use of these resources continue in many areas and are likely to increase in the future. Integrated coastal management is a means to resolve, or at least to minimize, these conflicts. Through an integrated approach in all of the dimensions just mentioned, trade-offs—across sectors of the economy, across management objectives, among environmental media—can be explicitly identified, evaluated, and determined.

But how do we make it work? Aren't we talking about the same kind of coastal management programs that we've had for the past twenty years? The answer is both yes and no. Clearly, every coastal problem does not require an integrated approach. However, we have to do business differently if we are going to meet the complex challenges of coastal management in the future.

First, it is not enough simply to recognize that we are all in this together. We have to fundamentally change our approach to management, both at the level of the individual and at the level of institutions. As coastal problems increase in scope and scale, as we struggle to deal with issues related to population increase and coastal development with hard choices that involve trade-offs between coastal resource use and protection, one "fact of life" is clear—we will be expected to do more with the same or fewer resources. All of our collective resources, both financial and intellectual, at all levels of government, as well as across the public and private sectors, will be needed to get the job done well.

Second, clearly we are talking about coastal management strategies that are and will continue to be implemented, for the most part, at the "point of attack": regional and local coastal management capabilities will be more important than ever before. But as coastal conflicts become both more frequent and more complex, as more rigorous analysis is expected from decision makers, technical information and services that exist in national agencies like NOAA and academic institutions will be needed to complement the capabilities of local governments. New roles and institutional partnerships will be needed to make this work.

Third, if we're going to take a more integrated approach to coastal management, then we have to pay more attention to all of the components of management, from planning to implementation, to monitoring and evaluation. Typically, we spend too much time producing a "plan" that is frozen in time with too few resources allocated to implementation and monitoring over time. The plan is put on the shelf; day-to-day decisions are carried out on an *ad hoc* basis with little relevance to the plan. Regulations are written for marine sanctuaries with few or no resources available for enforcement. Coastal management must be a continuous process that should be constantly monitoring progress and feeding back information on performance to adjust management strategies to changing conditions. It is a definition straight out of most management textbooks, but rarely applied in the real world of coastal management decisions.

Fourth, we have to recognize that increased participation, not only by the general public, but by the wider scientific and management communities, is not only desirable, but is absolutely critical for long-run success. Not only do these "stake-holders" possess information that is invaluable to the management process, but by explicitly involving them in the process many conflicts can be avoided. For example, is it better to resolve differences and misunderstandings about issues, objectives, and management strategies early in the process, or simply to leave critical management decisions to the legal system? We have to learn how to involve scientists and their knowledge base in the identification of problems, the specification of management objectives, the formulation and evaluation of alternative management strategies, and so on. We have to involve managers with responsibilities for local and regional land use management, for economic development, for water quality management, and for natural resource management in the integrated coastal management process.

UNDERSTANDING INTEGRATED COASTAL MANAGEMENT

Most coastal resource managers and analysts generally agree that current management practice is poorly coordinated, fragmented, and less than effective when measured in terms of results, however they are defined. About every five to ten years the coastal management community makes this rediscovery and then proclaims that "working together" and "better communication" are the answers.

Interests rise in investigating "new" techniques or methods that on the surface appear to support conventional wisdom and make sense even to the lay person. Integrated or comprehensive coastal management has this ring and has been "reinvented" before. What a good idea; let's "integrate!" Who can argue against integration or being comprehensive?

However, we typically begin our reevaluation by asking existing managers and analysts to review their own efforts. Just as expected, the conclusion is that they are already applying these new concepts and all that is needed is a little fine tuning—the desired "quick fix." The reevaluation is completed and business goes on as usual. Little change takes place, except in the actual coastal environment where deteriorating conditions continue and progress remains elusive despite relatively large investments of financial and human resources. Clearly, something is not working, but very little thought is ever given to changing the way in which we do business.

Integrated coastal management is not a new idea or concept. It has been around for many years and some aspects of it have been applied to some degree and called by various names, e.g., comprehensive watershed management, regional environmental quality management, basin-wide planning, area-wide waste treatment management, or special area management. The new idea is actually to *apply* these concepts in a *comprehensive and continuous manner.*

Integrated coastal management is simply a concept that organizes proven planning, systems analysis, economic analysis, ecosystem analysis, and management principles into a more formal and explicit framework. It is a framework that makes the information, analysis, and institutional requirements for integrated management explicit. It enables us to understand in operational terms what actual "integration" means, and provides a yardstick to tell us how adequately or inadequately integration is done. If the patient is suffering from a lack of integration, ICM provides the framework to diagnose the causes and recommend possible cures. One reason most integrated management attempts have failed is because they have not understood how to treat the patient, regardless of its symptoms. Just as a physician must know how the human body works before he can treat it, resource managers must understand the tasks of management and how integration actually works before they can make it happen. Better communication and coordination is not enough. ICM provides a basis for bringing the required knowledge in a systematic manner to the practitioners of coastal management.

Most confusion regarding integrated coastal management revolves around how to actually implement the ideas, concepts, and principles it embodies. Several points are important for practitioners to keep in mind. *First and foremost, the application of ICM is a process that begins with direct participation of the managers, analysts, scientists, and the concerned public of a region or area.* Not all steps in the process or aspects of the framework need always be applied, nor given the same emphasis. Each application is context-specific and depends on the problems of concern, the existing information base, existing institutional arrangements, personnel capabilities, available time, priorities, and resources available. However, regardless of the scope and focus of any ICM application, the framework requires that the decisions about how integration will be addressed be made explicit. All participants must agree and understand the trade-offs to be made and their consequences. Developing an integrated management approach does not take place quickly; it evolves over time based on incremental gains that build one upon another.

A FRAMEWORK FOR ANALYSIS FOR INTEGRATED COASTAL MANAGEMENT

Analysis is a basic task of integrated coastal management. The function of analysis for ICM is to generate information for management decisions, i.e., resource allocation decisions. Because management is a continuous process, analysis must generate information that is delivered at specific points in time within the decision process. Action plans or management strategies should be continually modified based on this new information.

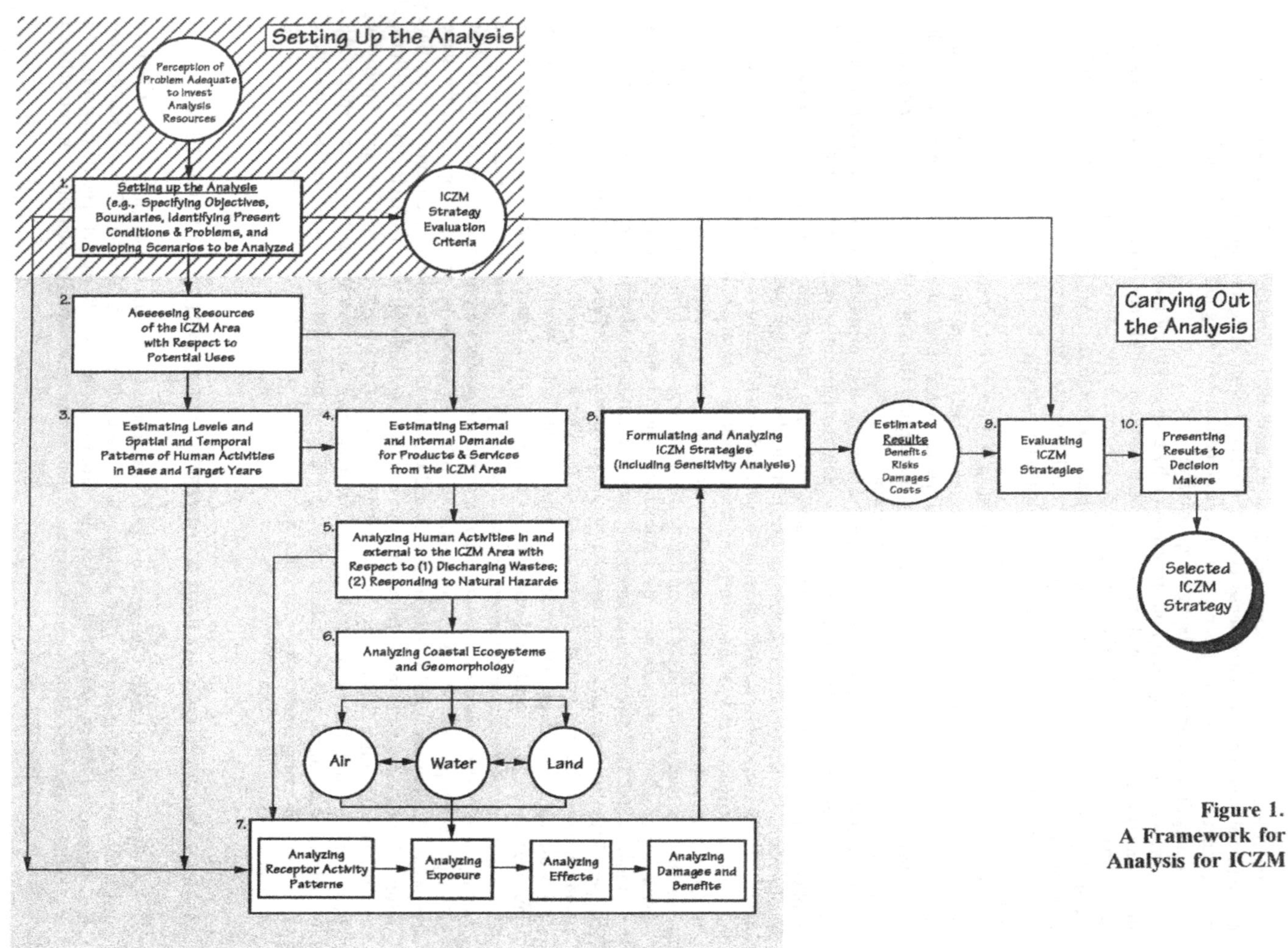

Figure 1.
A Framework for
Analysis for ICZM

A framework for analysis for integrated coastal management is shown in Figure 1. This "framework for analysis" is simply a system or methodology for making quantitative or qualitative estimates relating to problems of, and strategies for, integrated coastal management. The framework systematically links the analysis of the different components of the coastal system, e.g., the human activities, natural systems, effects of changes in ambient environmental quality on receptors, to generate information for resource allocation decisions.

SOME PRINCIPLES RELATED TO MANAGEMENT STRATEGIES

1. The management strategy selected at a given point in time should consist of the next actions to be taken to move toward achieving the objectives. All actions – physical and institutional – take time to put into place and begin operating. Some take very little time, some more, and some multiple years. The next round of analysis should produce the next round of actions to be undertaken, which will enable considering very probable changes in conditions.
2. All of the components of a management strategy are necessary if the desired objective is, or desired objectives are, to be achieved. Too often an analysis for ICM "stops" after identifying some physical measures and estimating costs of those measures. Limiting the analysis to those components is a recipe for inaction.
3. In any given ICM area there are many possible products and services that can be produced, physical measures through which to produce them, implementation incentives to induce action, institutional arrangements, and means of financing. This means that the total number of possible combinations of management strategies is very large. It is impossible, nor is it necessary, to analyze all combinations. Some procedure must be devised to reduce the number of combinations to be analyzed to an operational number in relation to the available analytical resources.

IMPLEMENTING THE FRAMEWORK: THE FLORIDA KEYS NATIONAL MARINE SANCTUARY MANAGEMENT PLAN

Talking about the need for more integrated management is easy and lots of people are doing it. Carrying out integrated activities in the real world is more difficult. Few examples or case studies exist. One example of an integrated approach to a real-world application is NOAA's recent experience in developing the management plan (and continuing management process) for the new Florida Keys National Marine Sanctuary (FKNMS). With the designation of the Florida Keys and Monterey Bay national marine sanctuaries, NOAA has been challenged to move away from managing relatively small, isolated areas such Gray's Reef off Georgia and the Flower Garden Banks in the Gulf of Mexico, to large areas where integrated management is not only possible, but required.

In 1990, the U.S. Congress designated the Florida Keys National Marine Sanctuary, requiring NOAA to develop a comprehensive sanctuary management plan by 1993. Not surprisingly, Congress did not provide new funds to manage the sanctuary, which includes an area extending from just south of Miami to the Dry Tortugas, encompassing almost 2,800 square nautical miles of coastal waters (Figure 2). This large region is one of the most heavily used coral reef tracts in the world, attracting over a million divers a year, and containing a myriad of competing, often conflicting uses and overlapping agency jurisdictions and interests. The scale of the region, the multitude and variety of its uses and users, the diversity and complexity of the ecosystem, the incomplete and fragmentary nature of available data, and the short time frame available to complete the job, have required an innovative approach to applying the available knowledge base to develop the management plan.

Given insufficient resources and time to complete the plan, what did we do? First, we recognized that developing an initial plan for the sanctuary was only the beginning of a continuing management process that would evolve over the period of time. Second, we recognized that to make the most out of what was already known about the Florida Keys, it was necessary to form an integral working relationship with relevant public agencies including the State of Florida (including the Department of Natural Resources and the Department of Environmental Regulation), Monroe County (the local government of the Florida Keys), the U.S. Environmental Protection Agency (the Federal agency responsible for developing a water quality management plan for the Florida Keys), and the U.S. Park Service. Stately simply, we wanted to make the maximum use of local scientists and other experts, managers, and decision makers in the development of the plan. Third, we recognized the need to identify a clear framework and process to develop the management plan. That process is very similar to the generic one just described above. Fourth, we recognized that because of the ecological and economic importance of the Florida Keys, any management strategy would have to incorporate an operational level of detail to be taken seriously and have any hope of implementation. Finally, we recognized that the process had to be a "back-to-front" one that developed a plan first, which would then be used to structure a directed data collection, analysis, and research program, instead of the other way around. We explicitly acknowledged that the detailed analysis and research required to evaluate the efficacy of the management actions would have to take place as part of the continuing management process.

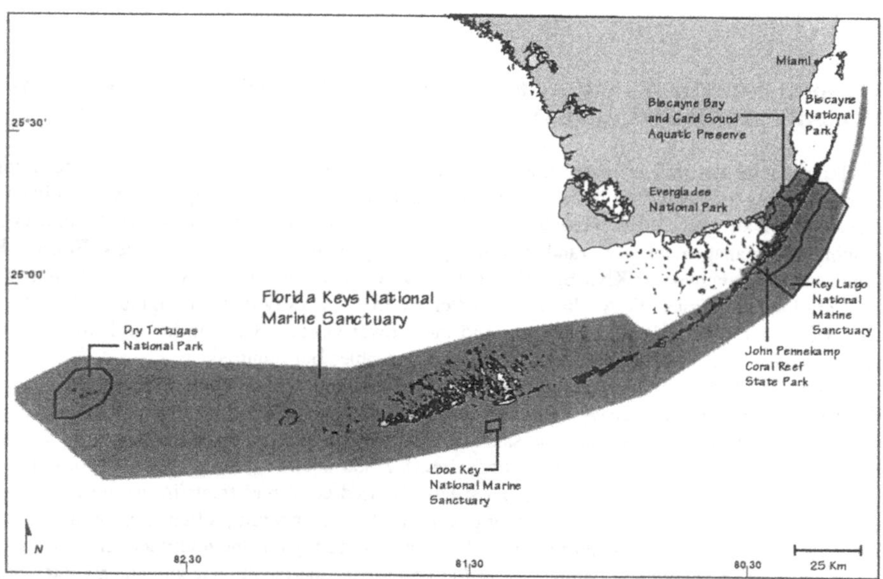

Figure 2. The Florida Keys National Marine Sanctuary

Many carefully organized and structured work sessions have been conducted. For example, in February 1992, an intense, four-day work session was conducted with about fifty Federal, state, and local resource managers with direct working experience in the Florida Keys. A process using predetermined information acquisition and encoding forms and a rotating group format was used to develop the initial information base for formulating the FKNMS management plan. The results of this session appear in the *Management Strategy Identification and Description Workbook,* which identifies, describes, and characterizes the potential impacts of approximately 300 individual management strategies for further consideration in the plan development process (NOAA 1992c). These strategies have been edited and condensed into a package of about 100 strategies. Costs of implementation and institutional arrangements required to implement the strategies have been explicitly identified.

The Florida Keys legislation directed NOAA to consider zoning as a management strategy. Working through a highly participatory process, we have now identified a set of small, well-defined "preservation areas," designed to protect specific reefs, to sustain important marine species, and to reduce conflicts between consumptive and nonconsumptive uses. In addition, "replenishment reserves," designed to encompass large, contiguous habitats, have been proposed to provide spawning, nursery, and permanent residence areas for the protection of marine life. The reserves are intended to protect areas that represent the full range of diversity of resources and habitats found throughout the sanctuary.

Developing the management plan through this integrated process has had positive results in the Florida Keys. An extensive network of experts, relevant management institutions, and affected communities have been formed into a team that continues to support and work together through this integrated process. We are convinced that the results to date are superior to those that would have been produced through a less integrated, less participatory approach. A plan is emerging that can be implemented and will achieve desired results. Perhaps most importantly, the requirement for a continuing management process for the sanctuary has been accepted.

INTERNATIONAL EFFORTS TO IMPLEMENT INTEGRATED COASTAL MANAGEMENT

NOAA is not alone in concluding that better integration is an important starting point for improving the management of coastal areas. A soon to be completed study by the Committee on Waste Management in Coastal Areas of the National Research Council concludes that integrated coastal management is a useful approach for addressing waste management problems beyond our current technology-based approach. While stopping short of specifying how an ICM "model" could be implemented, the Committee clearly endorses many of the same ideas that are included in our approach.

Interest in improving integrated coastal management capabilities is not limited to the United States. One of the principal recommendations of Agenda 21, the "oceans chapter" adopted at the United Nations Conference on Environment and Development in June 1992, was a call for coastal states to commit themselves to integrated management and sustainable development of coastal areas and the marine environment under their national jurisdiction. Integrated coastal management will continue to be an organizing principle for the next several years in the context of upcoming U.N. conferences on sustainable development, land-based sources of marine pollution, and special problems of small island nations. A working group of another U.N. effort, the Intergovernmental Panel on Climate Change (IPCC), has identified integrated coastal management as a primary response capability for dealing effectively with long-term sea-level rise and other effects of climate change. Regional workshops on ICM and a World Coast Conference, scheduled for November 1993 in the Netherlands, are being sponsored under the auspices of the IPCC. The Environment Directorate's Group on Natural Resources Management of the Organization for Economic Cooperation and Development (OECD) has initiated a study of the integration of environmental considerations into coastal zone management in its seventeen member countries, including the United States (OECD 1993). Finally, the World Bank is developing policies and technical guidelines on ICM to guide its investment decisions, particularly through the Global Environmental Facility (GEF).

CONCLUSION

The major challenge of the future to those of us encouraging a move to more integrated coastal management is the paradigm shift from single-sector management to a broader focus that attempts to define the overall interests of countries in their coastal areas, their resources, the quality of their coastal ecosystems, and the balance of economic development into the regional management equation.

Not only will this be difficult, but it is entirely novel. It is against history and human nature. We must adopt the long view and put our efforts into education, training, and changing the perspectives of decision-makers and scientists.

REFERENCES

Camp, Sharon L. 1993. Population: the critical decade. *Foreign Policy* 90 (Spring: 126-144 1993).

Fox, Robert W., and Ira H. Mehlman. 1992. Crowding out the future. *World population growth, U.S. immigration, and pressures on natural resources.* Washington, DC: Federation for American Immigration Reform. 64 p.

GESAMP (IMO/UNESCO/WMO/WHO/IAEA/UN/UNEP Joint Group of Experts on the Scientific Aspects of Marine Pollution). 1990. *The state of the marine environment.* UNEP Regional Seas Reports and Studies No. 115. UNEP. 111 p.

Johnston, James B., et al. 1992. Disappearing coastal wetlands. In *Stemming the tide of coastal fish habitat loss,* ed. Richard H. Stroud, pp. 53-58. Savannah, GA: National Coalition for Marine Conservation.

National Oceanic and Atmospheric Administration. 1990a. *50 years of population change along the nation's coasts, 1960-2010.* Rockville, MD: Office of Ocean Resources Conservation and Assessment, Strategic Environmental Assessments Division. 41 p.

National Oceanic and Atmospheric Administration. 1990b. *Coastal environmental quality in the United States, 1990: chemical contamination in sediment and tissues.* Rockville, MD: Office of Ocean Resources Conservation and Assessment, Coastal Monitoring and Bioeffects Assessment Division. 34 p.

National Oceanic and Atmospheric Administration. 1991a. *The 1990 national shellfish register of classified estuarine waters.* Rockville, MD: Office of Ocean Resources Conservation and Assessment, Strategic Environmental Assessments Division.

National Oceanic and Atmospheric Administration. 1991b. *Our living oceans. The first annual report on the status of U.S. living marine resources.* NOAA Tech. Memo. NMFS-F/SPO-1. Washington, DC: National Marine Fisheries Service. 123 p.

National Oceanic and Atmospheric Administration. 1992a. *Mussel Watch. Recent trends in coastal environmental quality.* Rockville, MD: Office of Ocean Resources Conservation and Assessment, Coastal Monitoring and Bioeffects Assessment Division. 46 p.

National Oceanic and Atmospheric Administration. 1992b. *Biennial report to Congress on coastal zone management.* Washington, DC: Office of Ocean and Coastal Resource Management. 2 vols.

National Oceanic and Atmospheric Administration. 1992c. *Florida Keys national marine sanctuary management plan – management strategy identification and description workbook.* Rockville, MD: Office of Ocean Resources Conservation and Assessment and the Office of Ocean and Coastal Resource Management. Natural Resources Defense Council. 1992. *Testing the waters. A national perspective on beach closings.* New York: NRDC. 67 p.

Organization for Economic Cooperation and Development. 1993. *Coastal zone management: integrated policies.* Paris, France: OECD. 126 p.

Whitledge, Terry E. 1985. *Nationwide review of oxygen depletion and eutrophication in estuarine and coastal waters: executive summary.* Rockville, MD: NOAA Office of Oceanography and Marine Assessment, Ocean Assessments Division (now the Office of Ocean Resources Conservation and Assessment).

16

A Sea Change for Oil Tanker Safety

Charles A. Bookman
Director, Marine Board
National Research Council, Washington, D.C.

ABSTRACT

Oil tanker accidents, from the *Torrey Canyon* in 1967 to the *Braer* in 1993, have compelled intense scrutiny of marine oil transportation systems. Significant changes have been made, which, collectively, have improved safety and environmental protection; but much more can be done. This paper reviews the record of oil spills from marine transportation over a twenty-five year period. Comparative and changing risks are emphasized. Changes in government rules and industry operating practices are briefly reviewed, with special emphasis on recent developments in the United States. Three uses of coastal space that have particular application to preventing tanker accidents are appraised. These are vessel traffic control systems, tanker safety zones, and offshore loading terminals for supertankers. The implementation of these innovative uses of coastal ocean space in the United States is reviewed.

MARINE TRANSPORTATION OIL SPILLS

The amount of petroleum that enters the sea from the marine transportation system has declined substantially over the last twenty-five years. The reduction has been estimated to be on the order of 75 % (National Research Council 1975, 1985; U.S. Coast Guard 1990). Table 1 shows the trends in four categories: tanker operations, drydocking, marine terminal spillage, and tanker accidents.

Several factors contribute to the favorable trend towards less petroleum in the sea from the marine transportation sector. The most prominent is the raised floor of performance of the marine transportation system that has resulted from implementation of the international agreements that have been achieved through the International Maritime Organization.[1] Standards have changed both in terms of operating procedures and in the design and construction of the vessels themselves.

[1] Safety of Life at Sea (1980) and Protocol (1981); Regulations for Preventing Collisions at Sea (1977); Prevention of Pollution from Ships (1983); Facilitation of International Maritime Traffic (1967); Load Lines (1968); Intervention on the High Seas in Cases of Oil Pollution Casualties (1975) and Protocol (1983); Civil Liability for Oil Pollution Damage (1975) and Protocol (1981); Establishment of an International Fund for Compensation for Oil Pollution Damage (1978); Standards of Training, Certification and Watchkeeping for Seafarers (1984).

Table 1

Estimated Oil in Waters from Marine Transportation System (millions of tons)[2]

Category	1973	1981	1989
Tanker Operations	1.08	1.02	.41
Drydocking	.25	.03	.004
Marine Terminals	.50	.04	.03
Accidents	.20	.40	.12
Total	2.03	1.50	.57

Sources: National Research Council 1975, 1985; U.S. Coast Guard 1990

Additional improvements occurred in the United States with the increase in maritime educational institutions that were created in the Vietnam War period. New training institutions extended the number of mariners who were exposed, in a rigorous educational environment, to competent operating practices with the use of state-of-the-art training devices such as real-time simulators.

A heightened awareness of the problems associated with the spillage of oil into the world's waters has also occurred, as well as an awareness of increased costs and liabilities. Traditional operating practices that involved some leaking, spillage, or overboard wash of oil were found increasingly to be unacceptable with the rise of environmental concern and the identification of short- and long-term damage from oil pollution. Moreover, the cost of pollution incidents dramatically increased the exposure of underwriters and, consequently, the cost of insurance. Today, oil tanker operators have billions of dollars of financial exposure in the United States. This was shown in the *Exxon Valdez* incident and is widely acknowledged for potential accidents in the future.

While catastrophic spills such as the *Braer* and the *Aegean Sea* capture headlines, smaller spills account for the majority of the oil spilled into the sea. Moreover, marine accidents account for only approximately 20% of the oil spilled from the marine transportation system, and the number of major accidents has decreased sharply. Figure 1 depicts the record of oil spills in excess of 1,000 barrels in coastal and open seas since 1974 for the United States and the world. In the mid-1970s, an average year saw about forty major spills over 1,000 barrels; by the late 1980s, the number was less than half of that. The volume of oil spilled as a result of these accidents has shown a corresponding decrease. Estimates suggest that total pollution from vessel accidents now is about one-quarter of its high point in the mid-1970s (U.S. Coast Guard 1990).

Despite the encouraging long-term trends, some disturbing developments in the marine transportation system, if un-checked, could lead to more—not less—accidental spillage in the future. The world fleet is aging. More than half of today's tankers and two-thirds of the Very Large Crude Carriers are over fifteen years old. Another disturbing factor is that the industry is currently unable to generate sufficient revenues to man and maintain vessels to the highest standards (Salvarani 1992).

[2] Oil enters the sea from myriad sources (National Research Council 1985). The U.N. Group of Experts on the Scientific Aspects of Marine Pollution (GESAMP) estimated in 1990 that only 10% or less of oil in the sea is estimated to come from ships (GESAMP). Of the fraction that comes from ships, 80% is estimated to come from operating practices; 20% (i.e., 2% of the total) is from shipping accidents. Land-based sources contribute 44%, 33% comes from the atmosphere, 12% from marine transportation and facilities, and the remaining 1% from offshore production.

The result is the use of older vessels with inadequate maintenance and inexpensive crews, with corresponding increase in risk of pollution. Shell Oil has estimated that over 20% of the world's fleet does not meet internationally agreed standards (Shell International Marine Ltd. 1993). Nor do inspection practices meet the need for oversight of the tanker fleet.

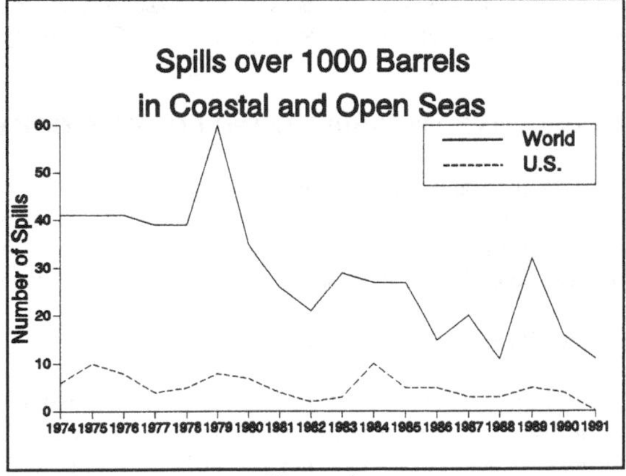

Figure 1.

NEW STATUTES AND RULES

Unfortunately for the industry, changes in government rules affecting it have been affected more by the histrionics associated with catastrophic spills than by systematic analysis of the marine transportation system. The *Exxon Valdez* incident in the United States and other accidents in 1989 led to the Oil Pollution Act of 1990 (P.L. 101-380), which is having profound implications on how the industry will operate in next decade. The *Braer* and the *Aegean Sea* have increased awareness in the European Community, with various changes being discussed (European Economic Commission 1993). Among these are requirements for fixed itineraries and routes, exclusion areas, mandatory inspections, sanctioning of vessels for violating safety standards, and the banning of specific vessels identified as unsafe.

Cargo owners and charterers are becoming increasingly active in determining which vessels they will use. Major oil companies like Shell and Mobil are insisting on vessels passing strict inspection criteria prior to chartering. One major oil company conducted over 3,500 inspections in 1992 (Shell International Marine Ltd. 1993). Such internal industry measures can be a major factor in the safety of the industry; however, they impact only about 25% of the total cargo afloat (that portion owned or controlled by the major oil companies).

The extent to which the marine transportation industry is under scrutiny in the United States is fairly typified by the various studies and recommendations mandated in the United States under the Oil Pollution Act of 1990. Some of the requirements of that Act include:

1. A deepwater ports study
2. Suspension and revocation of licenses, certificates of registry, and merchant mariners' documents for alcohol and drug abuse
3. Manning and crew standards for U.S. and foreign tank vessels
4. Vessel traffic service system studies
5. Overfill devices
6. Tank level or pressure monitoring devices
7. Tanker navigation safety standards
8. Evaluation of double hull tankers and damage stability
9. Establishment of double hull requirements
10. Existing tank vessel requirements
11. Lightering requirements
12. Analysis of alternative vessel designs to double hull designs
13. Requirements for escorts for certain tankers
14. Vessel and facility response plans

As a result of recent accidents, similar measures and assessments have been undertaken in Australia, the United Kingdom, the European Economic Community, and elsewhere.

PREVENTIVE MEASURES USING COASTAL OCEAN SPACE

While significant progress has been made in reducing oil pollution from the marine transportation system, several important measures utilizing coastal ocean space that have significant potential for improving safety and environmental protection have yet to be systematically applied. Three measures that use coastal and ocean space to minimize risk are discussed below.

Marine Traffic Control Systems

Marine traffic control systems have been defined as, "Any service, implemented by a competent authority, which interacts directly with the traffic and in response to that traffic in real time in order to improve safety and efficiency of traffic and to preserve the integrity of the environment" (Cutland et al. 1988).

The rationale for the initiation of vessel traffic control systems (VTS) has evolved from systems to improve the efficiency of traffic for economic advantage (which has been typical in Europe), to the improvement of safety (the primary motive in the United States and Canada), and more recently, to achieve environmental objectives. Initially restricted to a single port area, the VTS concept is expanding to incorporate regional traffic control.

Ship operators traditionally have jealously guarded their independent role in deciding routes and speeds. It is probable that this independence will be modified as nations move singly and together to protect the environment.

In the United States, two separate initiatives were launched in the wake of the *Exxon Valdez* incident. The first was a Port Needs Study to determine the needs of twenty-three U.S. ports for new, expanded, or improved VTS (Maio et al. 1991). This study employed risk analysis to rank twenty-three regions by net benefit to be gained from installation of a VTS. The second initiative, the Vessel Control Study (not yet undertaken), will review the U.S. Coast Guard's existing authority to direct vessel movements and the exercise of that authority. This study could have far-reaching impacts, on the premise that the government may direct the movement of vessels, including potentially ordering the master to follow certain maneuvering orders.

In a related development, all tankers in the Alaska trade transiting Prince William Sound are now required to be outfitted with differential, global positioning system receivers and to transmit positioning information to the Coast Guard at Valdez. The information provides the basis for advisory services and real-time tracking in all weather.

The potential for improving safety and protecting the environment through control of the movement of vessels must be thoroughly considered. Present technologies have the capability to assist in improving the operation significantly. Precise navigation systems, coupled with real-time, interactive transmission of data, and the ability to use real-time simulation to predict vessel transit paths are available and can make a substantial contribution to protection of the environment. While VTS has great promise to improve maritime safety, the implications of altering the traditional independence of the master to direct the vessel's passage have yet to be comprehensively explored (National Research Council 1994 [in press]).

Tanker Safety Zones

The large oil spills off Spain and the Shetland Islands in December 1992 and January 1993 stimulated interest in the manner in which tank vessels transit coastal waters. Various countries, including France, Italy, the Netherlands, and the United Kingdom, either have instituted or are considering regulations that restrict or exclude the movement of tankers in particular waterways or areas.

Over the years, various areas have been prohibited to vessel traffic for a variety of reasons, generally because of conflicting use such as military activities. The currently considered restrictions are not because of conflicting use, however. They are, rather, to eliminate or minimize the danger to the environment of a potential accident by eliminating the cause—in this case, the proximity of the tanker to the shore. The rationale for restricting traffic to patterns farther offshore the Shetland Islands, for instance, is that if the loss of propulsion on the *Braer* had occurred 50 miles offshore rather than 20, then there would have been time to provide a rescue tow before the vessel grounded and subsequently lost its entire cargo.

There is a compelling argument in this approach: if one can keep tankers offshore, then there will be additional time to respond to emergencies, and the likelihood of routine breakdowns becoming total losses will be greatly lessened. The impact of tanker safety zones on the costs of operations could be significant, however. There also are significant concerns about the imposition by a coastal state of unilateral restrictions, particularly beyond the legally recognized contiguous zone.

In the United States, a number of separate and independent actions are being undertaken. As a result of Oil Pollution Act of 1990 (OPA 90), the U.S. Coast Guard has been directed to evaluate the need to limit or prohibit the movement of tankers in certain areas. The results of the evaluations will be available in early 1995. In the interim, a variety of voluntary programs on the part of industry have been instituted. After the *Exxon Valdez* accident, participants in the Alaska/Lower 48 trade agreed to maintain at least a 50-mile standoff from coastal areas. At least one tanker operator is voluntarily standing off 85 miles from shore. In the Florida Straits, major users have advocated keeping at least a 6-mile offset from sensitive coral reefs.

Offshore Loading Terminals for Supertankers

The U.S. East and Gulf Coasts are characterized by wide continental shelves and shallow harbors. The economics of transporting crude oil from foreign sources to the United States necessitates the use of large tankers that, for the most part, are unable to transit the channels of the major U.S. oil importing harbors and ports. The maximum tanker size for most U.S. ports is about 80,000 DWT, limiting the employment of the large, more cost-effective tankers for direct transport of imported oil to those ports. Importers have attempted to overcome this inefficiency by transshipping in Caribbean ports, where large tankers offload into smaller tankers for final delivery to the United States; by lightering directly from large tankers to smaller ones in the Gulf of Mexico and to a much lesser extent on the East and West Coasts; and by the use of the deep water offshore terminal, the LOOP, off New Orleans.

Transshipping and lightering offer economic advantages to the marine transportation system, but they may actually increase the probability of oil pollution. Each requires the entry into port of a tanker, while also requiring double handling of the oil with accompanying risk of spillage while transferring. Transfer at sea, while appearing to have a highly developed procedure and a very good track record, carries with it significant risks. The *Mega Borg* incident, which involved a lightering operation, indicates that the potential of spillage of vast quantities of oil is possible, up to 500,000 tons. This potential cannot be disregarded in assessing the risk from the lightering operation. Lightering presently accounts for about a million barrels a day, or about 13%, of U.S. imports, and almost a thousand port calls annually.

The environmental benefit of the deep water, offshore port alternative is that it eliminates the need for the laden tanker to enter port to offload its cargo. Approximately 31% of U.S. oil spills over a thousand barrels from tankers occur in port areas; these spills would be eliminated with offshore terminals (Lujan and Williamson 1990). The United States currently has just one offshore port, the LOOP, off New Orleans. The LOOP has been an exceptionally safe operation. Similar environmental records can be expected from future offshore terminals.

Two proposals for new offshore ports have surfaced in the United States in the past two years after a hiatus of more than seven years. Despite their environmental advantages, the outlook in the United States for new offshore ports is cloudy. Based on operating costs alone, offshore ports are not economically competitive with lightering, but compelling environmental reasons may already exist to spur government requirements for new offshore ports. Forthcoming U.S. Coast Guard studies on deep water ports and their relative risk versus other modes of operation may kindle additional interest. However, without government action, new construction of these expensive facilities is unlikely.

SUMMARY

There has been a significant reduction in the amount of oil in the sea from marine transportation. The reduction is estimated at annual outflows of 25% of those that occurred in the early 1970s. However, catastrophic accidents continue to occur. Concurrently, public concerns and industry liabilities are increasing and forcing fundamental structural changes in the marine transportation system. Clearly, additional steps need to be taken to eliminate or minimize the impact of oil pollution resulting from the marine transportation system. Innovative uses of coastal ocean space—in particular, vessel traffic services, tanker safety zones, and offshore loading terminals—offer significant environmental and preventive promise. They should be seriously considered for the marine transportation system.

ACKNOWLEDGMENTS

Richard M. Willis provided research assistance for this paper. The views in this paper are the author's and do not necessarily reflect those of the National Research Council or its constituent units.

REFERENCES

Cutland, M. J., C. Deutsch, and C. C. Glansdorp. 1988. The application of VTS concepts and operational functions in the design of a vessel traffic service. In *Proceedings of the Ninth International Harbour Congress Conference*, 3.39-3.62. Antwerp, Belgium: Royal Society of Flemish Engineers.

European Economic Commission. 1993. *A common policy on safe seas*. Brussels, Belgium: European Economic Commission.

Joint Group of Experts on the Scientific Aspects of Marine Pollution (GESAMP). 1990. *The state of the marine environment*. Oxford, U.K.: Blackwell Scientific Publications.

Lujan, M., Jr., and B. A. Williamson. 1990. *Offshore oil terminals: Potential role in U.S. petroleum distribution*. OCS Report MMS 90-0014. Washington, D.C.: U.S. Department of the Interior, Minerals Management Service.

Maio, D., R. Ricci, M. Rosetti, J. Schwenk, and T. Liu. 1991. *Port needs study*. Report No. DOT-CG-N-01-91-1.2. 3 vols. Prepared by John A. Volpe National Transportation Systems Center. Washington, D.C.: U.S. Coast Guard.

National Research Council. 1975. *Petroleum in the marine environment*. Report based on a workshop held by the Ocean Affairs Board, Airlie, Virginia, May 21-25, 1973.

National Research Council. 1985. *Oil in the sea*. Washington, D.C.: National Academy Press.

National Research Council. In press. *Minding the helm: Marine navigation and piloting*. Washington, D.C.: National Academy Press.

Salvarani, R. 1992. Maritime safety and sea pollution prevention policy for the European Community in the field of VTS. Unpublished paper, June.

Shell International Marine Ltd. 1993. Prevention of oil spills from tankers. Unpublished policy paper, 9 January.

U.S. Coast Guard. 1990. Update of inputs of petroleum hydrocarbons into the oceans due to marine transportation activities. Paper submitted to IMO Marine Environment Protection Committee 30, 17 September.

17

Scientific Approach For Evaluating The Sites Of Coastal Thermoelectric Power Stations

Ing. Mario Tomasino
National Electric Board (ENEL) - S.p.A., Centre for Hydraulic and Structural Research
Mestre (Venice), Italy

Dr. R. Ambrogi
National Electric Board (ENEL) - S.p.A., Centre for Environmental and Material Research
Milan, Italy

Dr. E. Ioannilli
National Electric Board (ENEL) - S.p.A., Central Laboratory
Piacenza, Italy

ABSTRACT

Coastal power stations are major industrial settlements that involve careful programming during their design, realization, and operation. The first step, after a site has been selected for the power plant, is to gather all possible information about the characteristics of the environment that can be of use, in order to understand the impact of the plant itself on the environment, and of natural forces on man-made structures.

This aim can be achieved only with a thorough investigation of the oceanographic properties of the coastal area surrounding the proposed site. For this purpose, the National Electricity Board (ENEL) has conducted a series of oceanographic cruises from 1986 to 1989 in nine sites along the Italian peninsula where it had indicated the possibility of installing new, thermoelectric plants.

A common framework was used in all sites, but this was adapted to the peculiar features of each individual location, based on the existing knowledge of the general hydrography of the district.

Seasonal cruises were performed, each of them lasting about fifteen days, in order to collect data on the relevant phenomena of water movement and to have a statistically significant data base. A number of five current meter chains were positioned in a previously determined sea area of variable size and shape, and were left in position for the whole duration of the cruise. Lagrangian current buoys were radar-tracked in two typical situations to obtain a picture of the water movement. At fixed intervals, a grid of stations was occupied for CTD measurements and water sampling. Water quality determinations comprised general indicators, nutrients, and pollutants. The biological survey was centered on zooplankton monitoring during two periods of 24 hours in the station closest to the proposed point of effluent discharge.

This body of information was stored in a data base, and preliminary analyses of the data were published in a cruise report. The data were used for simulation of the warm water effluent dispersion at sea, for designing the intake and discharge works and protection dikes, and to predict the solid transport along the coast. Chemical and biological information were used in the Environmental Impact Assessment (EIA), to describe the baseline situation of water quality and of biomass transport in the area.

INTRODUCTION

The construction and operation of large industrial settlements such as modern, thermoelectric power stations, implies a very complex procedure taking into account a host of technical, environmental, social, and economic consequences.

The particular case of coastal power plants using sea-water in once-through cooling systems, is of special interest in the Italian situation, where coastal areas are densely populated and subject to multiple uses (Bertacchi et al. 1986). In addition, Italy is bordered by a wide variety of sea conditions, from the oligotrophic Western Mediterranean waters, the open situation of the deep waters of the Sicilian Channel and Ionian Sea, to the very eutrophic and shallow water situation of the Northern Adriatic.

It is evident, even through such simple considerations, that very intense public scrutiny must have focused on the problems related to the siting of power stations along the coasts of the peninsula.

The aim of this paper is to illustrate some of the research activities performed by ENEL in the framework of a program of scientific investigation of sea characteristics before the construction of coastal power plants. Of course this program does not fulfill all the requirements from both the engineering and the environmental point of view, but it is considered to be a very important starting point, gathering and organizing a great amount of scientifically sound data, necessary at different levels and times for the management of a substantial coastal space utilization.

Not withstanding the interruption of the nuclear program in 1986, the need for a well-documented programming of plant construction has been a constant characteristic of ENEL's activities in this sector.

ENEL had a previous experience of oceanographic cruises performed as a support for designing new power stations. These were done from 1970 to 1981, but their structure had to be completely revised in order to meet the new objectives and standards for oceanographic knowledge. Besides that, computing and modelization capabilities of the 1980s had to be incorporated in the general plan of the cruises.

Rationale of the Approach

The goals of the program of oceanographic cruises were to:

1. Represent the marine ecosystem surrounding the proposed site (descriptive phase)
2. Study the impact of the plant activity on the ecosystem (predictive phase)
3. Establish correct data for the design of maritime works (project phase)

The methodology of this approach is summarized in Figure 1.

The main objectives of the research are related to the following problematic areas.

1. *Recirculation problems and compliance with the existing legislation*. Once the prevailing current regime has been described, the optimal positioning of the intake and discharge channels can be studied, in order to eliminate or reduce to a minimum the water recirculation. In addition, a prevision can be made on the isotherm pattern in relation to the law requirements, and alternative hypothesis can be tested (surface or submersed diffuser, etc.).
2. *Environmental impact on the ecosystem*. The trophic state of the ecosystem potentially influenced by the power station is described on the basis of the distribution patterns of

suspended and dissolved matter and compounds, and of the assessed transport of zooplankton biomass.

3. *Problems of sizing, stability, and accessibility of the structures*. The statistics of wave energy, by means of direct measures or hindcasting, allows one to apply the theory of extremes to define the project wave, the wave height for a given return time, and the wave-breaking area.
4. *Sediment transportation along shoreline, shore stability*. The shore evolution can be predicted on the basis of data on littoral transport, taking into account the new, maritime works of the power station that could alter the flux of sediments.

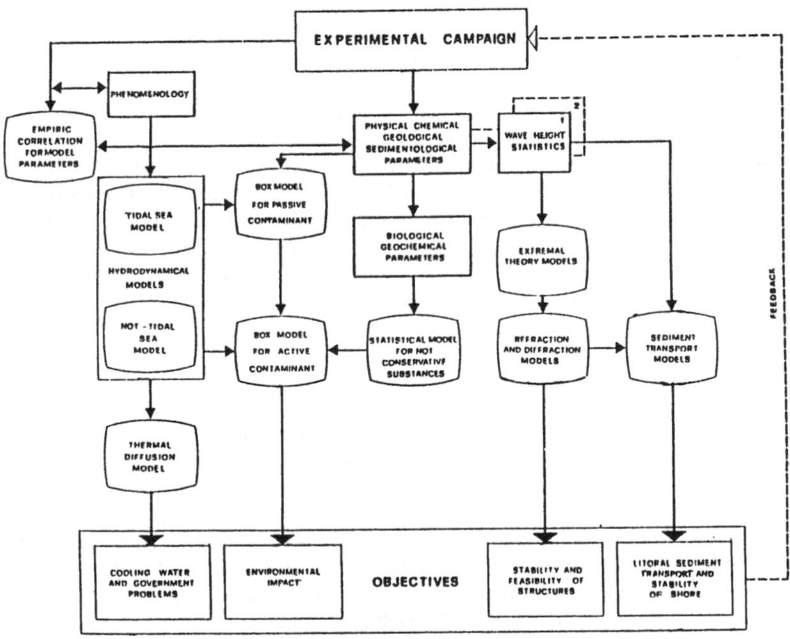

Figure 1. Conceptual scheme for the organization of the oceanographic study.

The only basis upon which to unify such a disparate set of objectives, given the budget limitations of the ship, equipment, and staff involved, has been the development of numerical models for the reproduction of the individual phenomena. In this manner, the models become the means to cope with the multiple objectives. Such means contribute to the comprehension of the system, and hence to the prediction, management, and planning of the same system.

Considering that a "universal" model is not available and that no model, even if very sophisticated, can give a perfect reproduction of reality, we chose some "ad hoc" models from the existing literature for entities that we wanted to describe and whose evolution we wanted to predict (Dazzi and Tomasino 1978).

The following models have been used:

1. Hydrodynamical models were used to describe current patterns through which the evolution of water masses in time and space can be understood and simulated. The intent was to evaluate

the residence time of water masses in the area. Additionally the evolution of the circulation field facilitates the linking of chemical and biological parameters to water movements.

2. Ecological models were implemented for dissolved and suspended matters in the water. Nutrients and pollutant substances can be considered as "passive" when they are transported with the water masses by advection and turbulent diffusion without transformation, whereas the non-conservative processess of a chemical and biological nature produce transformations in the "active" components.
3. Wave refraction models were utilized to convert offshore measurements of wave height and direction into estimates applied to the inshore bathymetry under investigation (Lamberti and Toffolo 1986).
4. Wave refraction and diffraction models were applied to study the behavior of man-made structures (embankments, harbors, dikes) and the wave attenuation inside harbors (Mattioli 1984).
5. Solid transport models used for shoreline environments were linked to refraction models, whose output they used for computation (Caccavella et al. 1987).

To implement such means, the requirements of the experimental campaigns are very strict and demanding in terms of complexity of parameters to be measured and of temporal and spatial extensions of data acquisition. In fact, it is clear that observations made during the sea cruises of phenomena occurring in a given area formed the basis for model elaboration. Conversely, the modelization steps led to the redefinition of specific items of the experimental network used during the cruises in order to better fit into the refined conceptual model.

ORGANIZATION OF THE CRUISES

In the period from 1985 to 1989, thirty-two cruises were completed in eight sites along the Italian coasts (Figure 2). The scheme of the approach followed in each individual site is explained in this section. The cruises have been repeated four times a year for each individual site, in different seasonal conditions to account for the variability of both meteo-climatic events and biological differences.

The scope of the area covered by the cruises has been selected based on the local coastal morphology. As a reference for open sea linear coasts, an area of 200 km2 (20 x 10 km) has been chosen. The scheme of spatial display of measures is shown in Figure 3. As an example of application, Figure 4 shows the area chosen for the site of Montalto di Castro (Tyrrhenian Sea, central Lazio coast), which has a linear coastal development.

The R/V Ciprea of the company "Generale Prospezioni" has completed all the cruises. The ship is 31 m long, weighs 198 tons, and is equipped with all necessary instrumentation; namely, a computer for data acquisition and storage. The navigation system is comprised of Loran C and a hydrographic echosounder.

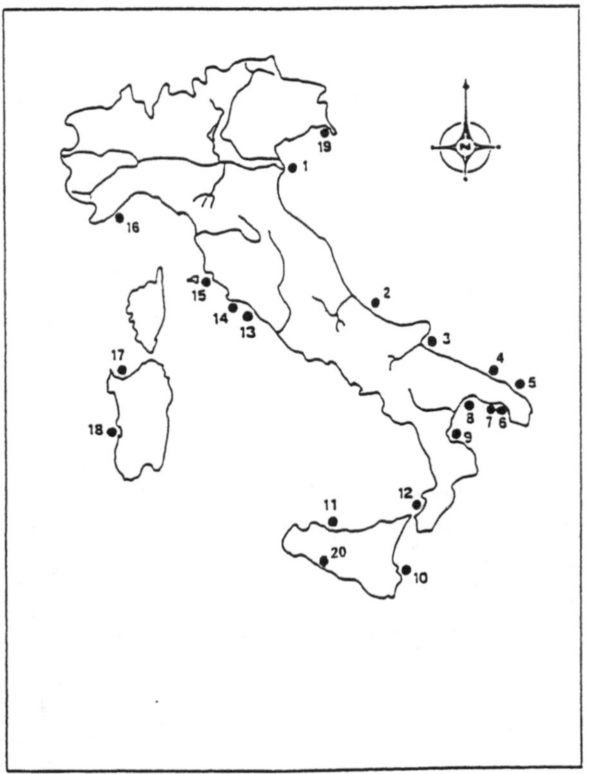

Figure 2. Location of the surveyed sites

4. Salentino-Adriatica
6. Salentino-Ionica Est
7. Salentino-Ionica Ovest
12. Gioia Tauro
14. Montalto di Castro
17. Fiume Santo
19. Friuli
20. Gela

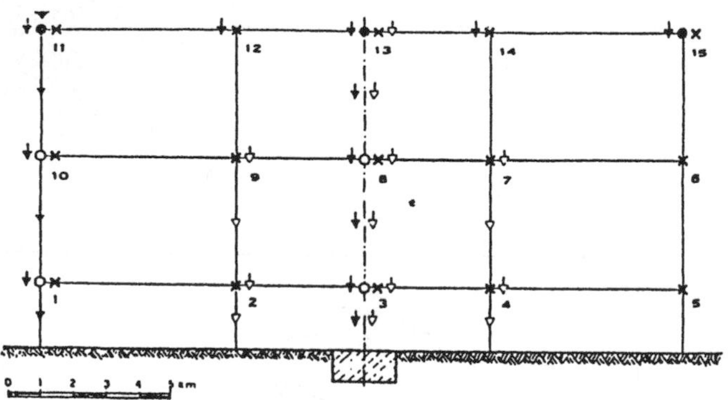

Figure 3. Scheme of the spatial arrangement of measures in an hypothetical site

Notes:

O Current meter chain (One instrument)	▾ Mareograph
● Current meter chain (Two instruments)	↓ Vertical profiles (Grid A)
x Water quality sampling	↑ Vertical profiles (Grid B)

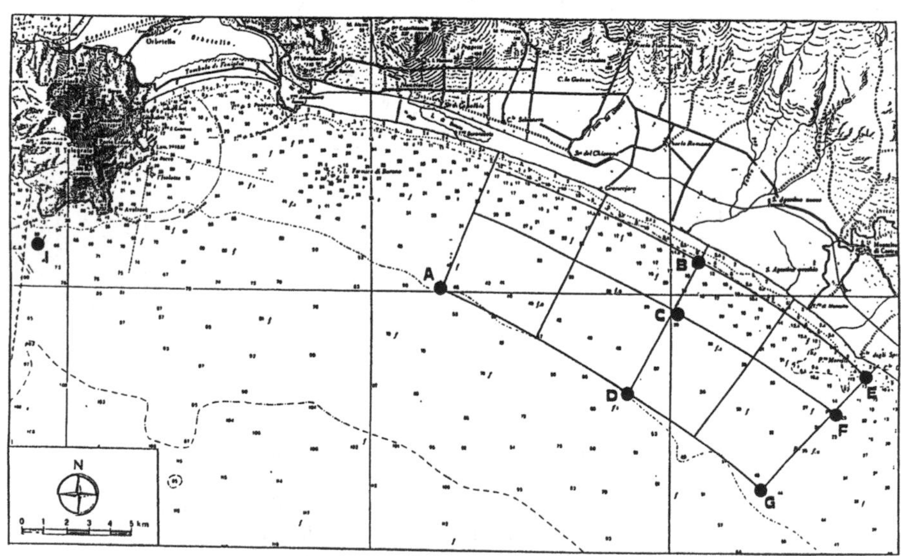

Figure 4. Example of an area of study with the location of the current meter chain (Montalto di Castro)

The following sections of the research have been carried out in each cruise.

Bathymetry and Sedimentology

Bathymetric surveys along five equally spaced transects perpendicular to the coast and continued to the - 20 m isobath were done. The goal of these measurements was to learn about the evolution of the coast and, indirectly, the entity of the littoral solid transport. Sediment samples were taken in a variable number of stations on one occasion.

Hydrodynamics

1. *Current metering with fixed current meter chains*. As a rule, 5 of them are positioned along the perimeter of the study area and three in the internal part. The buoyed chains bore two, Aanderaa RCM4/S current meters, at - 8 m from the mean sea level and at + 10 m from the bottom. In shallow waters, the bottom current meter was not deployed. The instruments also measured temperature and salinity. A ten minute interval of data acquisition was set.
2. *Vertical profiles of water velocity, direction, and temperature*. These were repeated at least twice for each campaign in two sampling grids, a larger one for the general area (Grid A) and a smaller one, onshore (Grid B). A Neil Brown DRCM-2 current meter was used for vertical measures. In the last cruises, a tomographic profiler (RD - Instruments, model ADCP DR 0300) was used.
3. *Lagrangian measures with current buoys*. Groups of three buoys were immersed from the anchored ship at - 0.5 m and at - 8 m and followed with the radar. The measures obtained by 2. and 3. were taken for comparative purposes and are considered as qualitative with reference to the fixed system of current meters described.

Hydrology

In the station points of Grids A and B, vertical profiles of water temperature, salinity, pH, and dissolved oxygen were taken with an OTS probe of Meerestechnich Elektronic GmbH, on three occasions during each cruise.

Water level

A mareograph Aanderaa WLR-5 was positioned at the basis of one of the current meter chains, registering water level throughout the campaign every ten minutes.

Suspended solids

In 10 of the hydrologic grid stations, at the surface and at - 10 m, water samples were taken with a reversing bottle and filtered with 0.45 μm pore filters. The dry weight represented the quantity of solid material per unit volume at the two depths.

Meteo-Marine Observations

A complete meteorological station with the following sensors was deployed on the shore: (a) wind direction and velocity at 10 m from ground level; (b) air temperature and humidity at 2 m; (c) solar radiation; (d) atmospheric pressure; and (e) rain.

Waves (height and period) were measured with a Datawell accelerometric wave recorder anchored offshore (normally 3 nautic miles).

Meteorologic observations were also recorded on board the R/V.

Water quality

Measures of water quality were taken together with hydrological measurements at the same stations of Grids A and B, on two different days.

In the fifteen stations of Grid A a duplicate sample was taken in the surface layer for the analysis of nutrients (different forms of inorganic nitrogen, ortophosphate and soluble silica). In the six stations closer to the coast (Grid B), samples were taken at three depths within the photic zone (surface, - 10 m, and - 20 m). From these samples, nutrients, photosynthetic pigments (Chlorophyll a and c, carotenoids, and phaeopigments), together with primary production, were determined. In the surface sample, the following trace metals were detected: iron, copper, nickel, zinc, lead, and cadmium.

Zooplankton

Sampling was made by a Bongo net (mesh 0.5 mm) equipped with a flowmeter and hauled by the R/V in the surface layer of water at a speed of about 2 knots. Towing was repeated at 3-hour intervals during two, twenty-four hour sampling periods along a transect parallel to the coast, between the three current meter chains closer to the proposed site of the power plant.

The sample of one of the nets was used for biomass evaluation (Lovegrove 1966), the second for the determination and counting of plankton.

DATA PRESENTATION AND ANALYSIS

Before data can be used in numerical models and organized for the different purposes of the study (taking into account the seasonal variations within each site and a comparison between sites), a preliminary, standard treatment was effected for each cruise, to produce a cruise report.

The main steps of such a preliminary elaboration have been as follows:

1. The cruise data produced on magnetic tape have been validated, eliminating all dubious values.
2. The velocity (module and direction) has been graphed, and the average and standard deviations calculated. Frequency values have been represented by means of current roses.
3. Scatterplots of current velocity and direction are plotted in the map for each current meter (Figures 5-8). This representation gives the general pattern of the marine currents in the area during a seasonal cruise.
4. The hourly values of the EW - NS components have been represented in the plane of the odograph, and the prevailing directions have been assessed. This allowed us to define a new reference system for each current meter.
5. The measurements, referred to in the new reference system, are distinguished in two main components: along shore currents and the offshore currents.
6. The temporal series obtained has been analyzed in frequency by means of the Fourier discrete transformer. The energy spectra have been represented by plotting the period (the inverse of frequency) on the abscissa.
7. Every component has been filtered by a low-pass filter with a cutting period of thirty-six hours in order to separate long-period components (currents influenced by meteorological events) from short-period components (tidal and inertial currents).

8. Also tidal level, barometric, and anemometric data have been studied by spectral analysis and filtered to separate long-term components from high-frequency oscillations.
9. Ondametric data have been elaborated, calculating the significative height (the average of one third of the waves registered by the ondameter) and the associated period. These statistics were repeated every three hours.
10. Hydrological and chemical data have been plotted in maps and isopleths traced for the various depths (Figures 9 and 10).

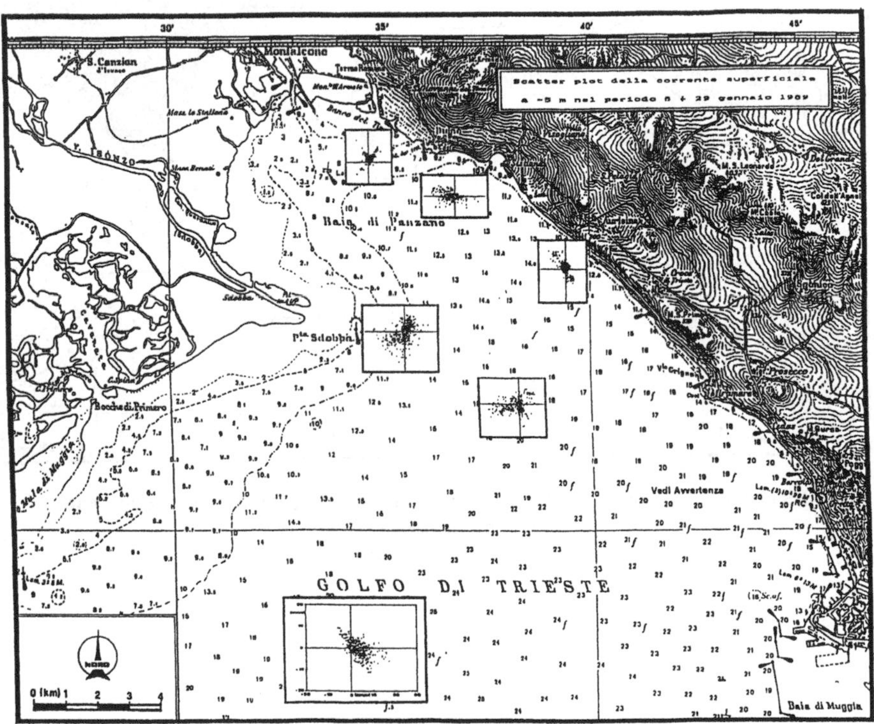

Figure 5. Example of scatter-plot diagrams showing the pattern of current distribution in the Friuli site (Winter situation)

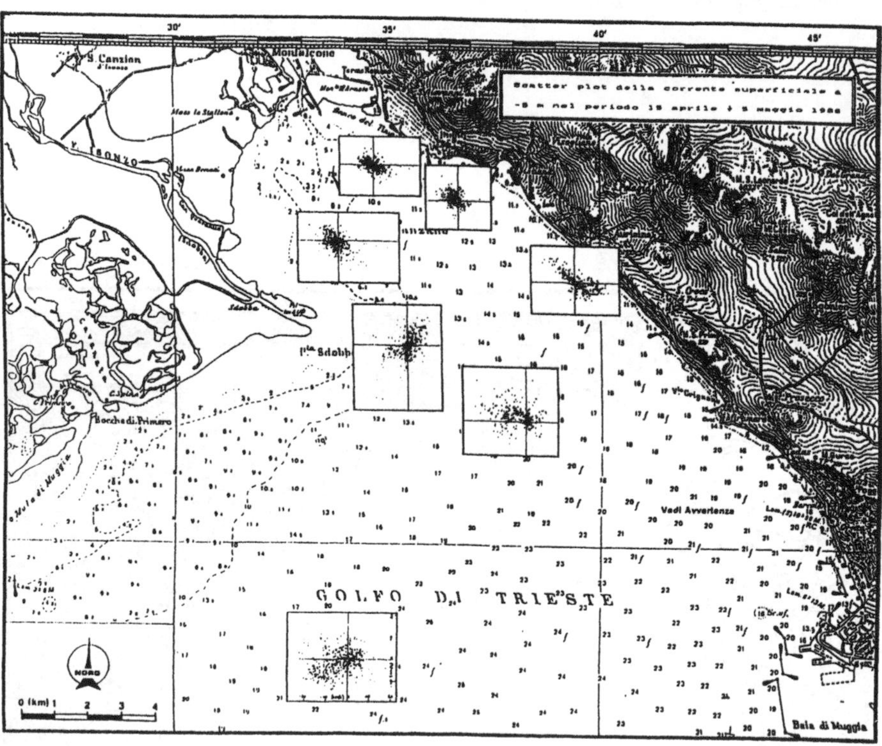

Figure 6. Example of scatter-plot diagrams showing the pattern of current distribution in the Friuli site (Spring situation)

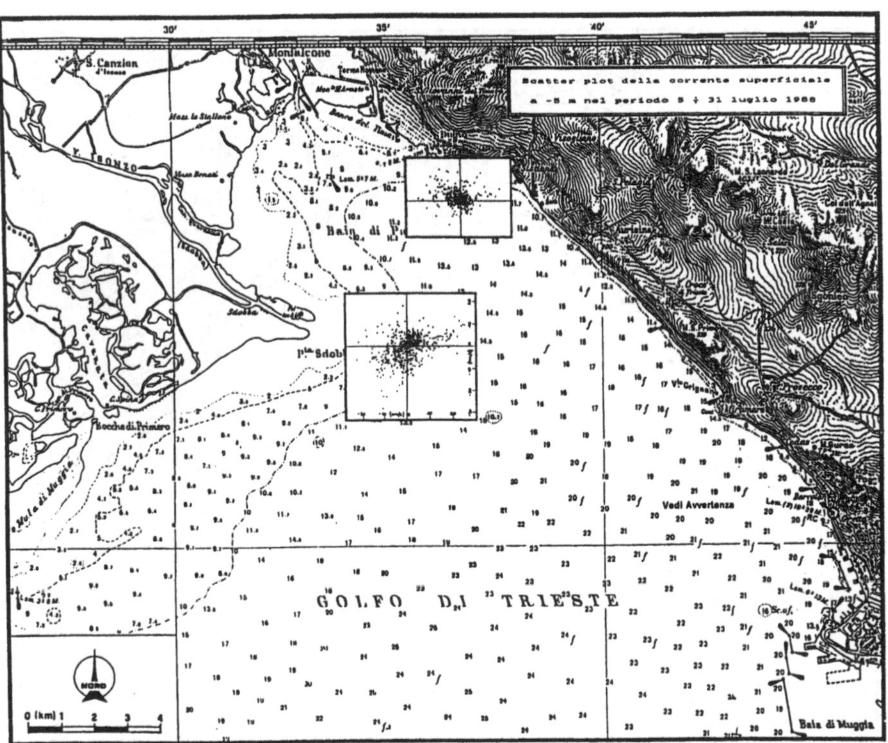

Figure 7. Example of scatter-plot diagrams showing the pattern of current distribution in the Friuli site (Summer situation)

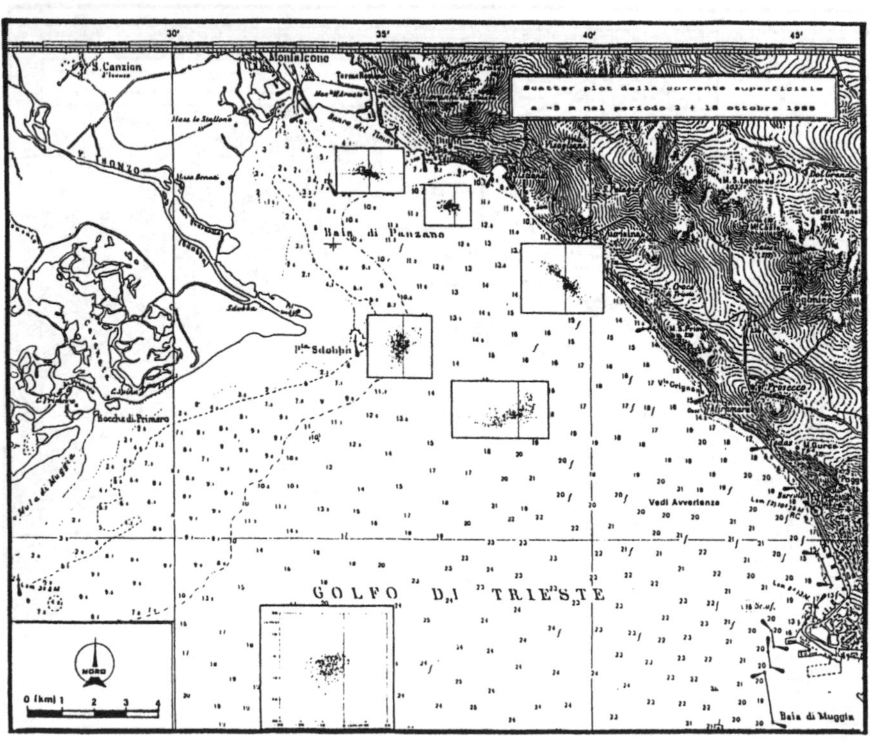

Figure 8. Example of scatter-plot diagrams showing the pattern of current distribution in the Friuli site (Autumn situation)

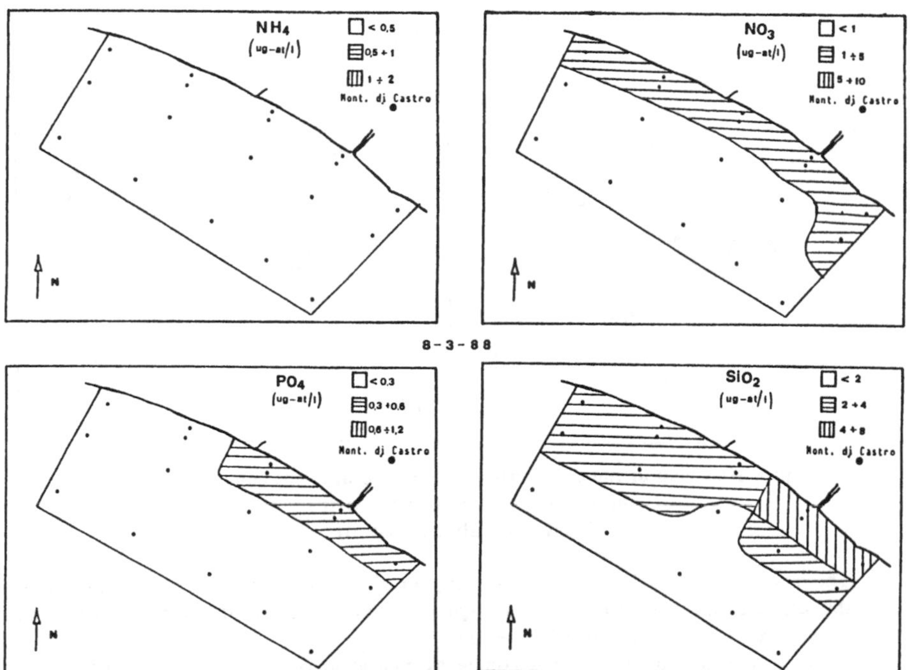

Figure 9. Map of the distribution of surface nutrient concentration in the cruise Montalto di Castro (Winter situation, Grid A)

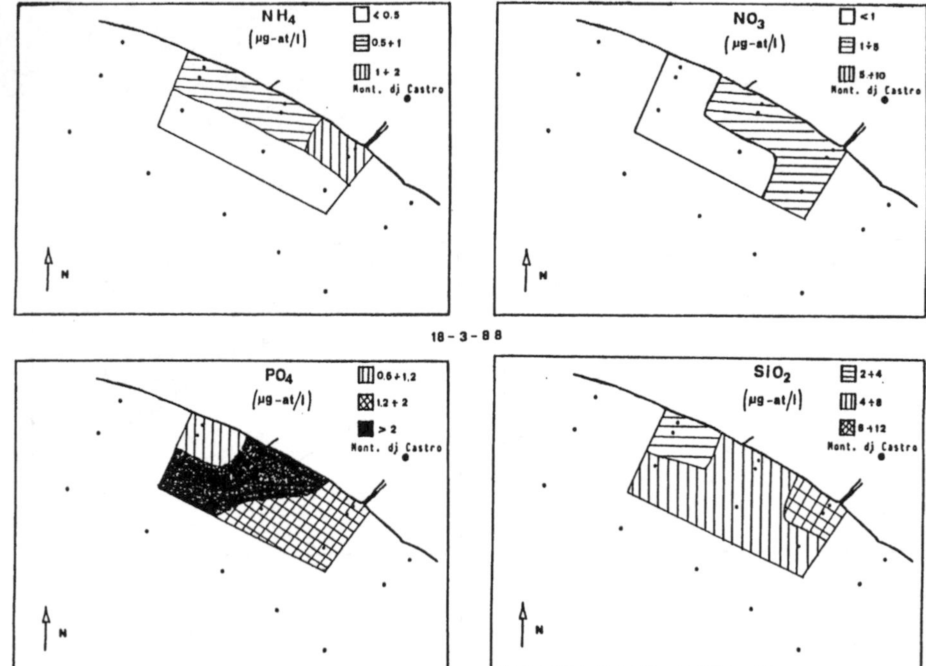

Figure 10. Map of the distribution of surface nutrient concentrations in the cruise Montalto di Castro (Winter situation, Grid B)

11. The range of values and the average of heavy metal concentrations have been compared with the standard values indicated by Water Quality Criteria (1972) for uncontaminated coastal waters, safety threshold, and risk threshold.
12. A quantitative, structural index based on the average abundance of zooplankton (ind. m-3) has been described for each site, both for total densities and for the abundance of common taxa. The qualitative structure has been described by means of the frequency percentage of taxa > 1%.
13. The transport of zooplankton biomass (dry weight) is estimated with reference to the current field determined at the moment of sampling.
14. The analysis of zooplankton data is also performed with reference to the day and night and lunar cycle, which are known to strongly affect the distribution of these organisms.

CONCLUSIONS

At this moment the output of the oceanographic study program has been incorporated in the EIA dossier for the sites of Brindisi (Southern Adriatic Sea); Gioia Tauro (Southern Tyrrhenian Sea); Montalto di Castro (Central Tyrrhenian Sea), where the proposed nuclear power plant has been converted in a project of gas-fueled station, and Fiume Santo (Northern Sardinian coast), where two new units have been added to the existing ones.

The models of thermal plume diffusion (Di Monaco and Leoncini 1984) have been used for designing the cooling circuits of the plants, and other ecological information has been evaluated in order to predict the potential effects of the power plant operation on the marine environment.

A substantial knowledge of local coastal hydrography has been gathered in the process for large sectors of the Italian shoreline. This host of detailed information is available upon request to other users interested in developing industrial or navigational projects in the coastal zone. It has to be stressed that this kind of information is also the basis for any meaningful attempt to describe the marine ecosystem in the areas involved, as demonstrated by the in-depth study on metal distribution in various biotic and abiotic components in the Tyrrhenian area of Montalto di Castro (Cambiaghi et al. 1987).

In the area of NW Sardinia (site of Fiume Santo), an extensive ecological investigation has also been completed under the aegis of these oceanographic cruises.

Some observations on the zooplankton investigations in the Tyrrhenian site of Montalto di Castro has compared with the Northern Adriatic site of Friuli, are reported in the communication of Ambrogi et al. (1993) in this congress.

REFERENCES

Ambrogi, R., G. Queirazza, and T. Zunini Sertorio. 1993. Oceanography and coastal environmental assessment: Two case studies of different areas in the Tyrrhenian and Adriatic Seas. Paper presented at *COSU III Conference*, March 30 - April 2, 1993, Santa Margherita Ligure, Italy.

Bertacchi, P., F. Mioni, and M. Tomasino. 1986. L'oceanografia costiera per la localizzazione delle nuove centrali ENEL. *I Congr. AIOM, Venezia*.

Caccavella, P., M. Venturi, and G. Grancini. 1987. Studio della dinamica costiera e degli effetti in sul litorale dall'opera di scarico a mare del sistema di raffreddamento della centrale nucleare Alto Lazio dell'ENEL a Montalto di Castro. *Conv.Soc. Ital. Navigazione*, Ravenna.

Cambiaghi, M., G. Queirazza, G. Strobino, and M. Tomasino. 1987. Modeling the distribution of radionuclides in a Mediterranean coastal ecosystem. *Environ. Software* 2:207-20.

Dazzi, R. and M. Tomasino. 1978. Limiti e peculiarit*a* di un modello matematico costiero. *XVI Conv. Idraulica e Costruz. Idrauliche*, Torino.

Di Monaco A. and A. Leoncini. 1984. Applicazioni di un modello matematico tridimensionale di diffusione in ambiente idrico. *XIX Conv. Idraulica e Costruz. Idrauliche*, Pavia.

Lamberti, A. and F. Toffolo. 1986. Calcolo dell'onda di progetto in acque basse. *I Congr. AIOM*, Venezia.

Lovegrove, T. 1966. The determination of the dry weight of plankton and the effect of various factors on the values obtained. In *Some contemporary studies in marine science*, ed. H. Barnes, 429-67. London.

Mattioli, F. 1984. Prestazioni numeriche della tecnica di espansione in serie agli elementi finiti per onde marine superficiali lineari. *XIX Conv. Idraulica e Costruz*. Idrauliche, Pavia.

W.Q.C. 1972. Water Quality Criteria. American Association Water Pollution, Washington.

18

Estuarine Dynamics and Global Change

Norbert P. Psuty
Associate Director, Institute of Marine and Coastal Sciences,
Rutgers - The State University of New Jersey, New Brunswick, New Jersey, USA

ABSTRACT

Estuarine systems are the products of inputs from continental, as well as oceanic sources. Spatial gradients and temporal variations within the estuaries are characteristic of the natural system and also are products of the anthropogenic environments. Modern day management is directed toward the maintenance of environmental quality as well as toward commerce and the production of food. Future challenges to the allocation and utilization of estuarine space will be driven by sea level rise and the variety of management approaches and strategies applied to accommodate continuing land-use demands within the spatially dynamic, estuarine systems.

INTRODUCTION

Estuaries are ubiquitous systems found on every inhabited continent and in every portion of each of these continents. Their very abundance is responsible for the great interest and attention directed to them and this same abundance will be responsible for the major changes that will continue to characterize their systems and to demand a flexible and enlightened approach for management.

Estuaries are by definition, the semienclosed embayments found in the transition area between the continental and the oceanic environments. They share some of the characteristics of each of their neighboring environments and respond to changes in each of them, as well as to the changes that are occurring within their own boundaries. The transition zone between the continents and the oceans is extremely varied, showing differences in such basic conditions as geology, geomorphology, climate, vegetation, biologic and mineral resources, and human use and occupation, amongst other characteristics. The estuary may be funnel shaped or it may take on a linear form which is either parallel to the shoreline or transverse to the shoreline, as represented in the Fairbridge classification of estuary morphological types (1980). Each estuary derives some of its character from the flows inputted from the land and from the sea, producing circulation patterns, salinity gradients, and stratification phenomena that offer further opportunities to categorize and differentiate estuarine systems (Pritchard 1955, 1967).

Whereas there has long been recognition of the estuarine region as a special location for natural and cultural systems, it is only recently that the estuary has been the focus of inquiry and special management concern. The lateness of this attention is somewhat a product of semantics because there has always been a general interest in the broad coastal zone that lies at the confluence of the land and water. The newly found focus was in looking at the more restrictive estuarine or transition zone as a location at the continental margin that has specific processes operating and one that has models of interaction. Thus, the growth in interest is in large part, due to the recognition of the unique character of the region, the realization that neither the terrestrial nor the oceanic models could adequately describe the conditions present in these semienclosed aqueous systems, and that the management approaches must also be unique.

Several landmark publications have helped to define and to detail the character of the land/sea transition zone and to identify the management problems. They include the early volume by Lauff (1967) that summarized the first conference proceedings specifically directed toward the status of research in estuaries. This was followed by additional conferences organized by the Estuarine Research Federation, and the production of conference proceedings edited by Cronin (1975a, 1975b), by Wiley (1975, 1976), by Kennedy (1980, 1982, 1984), and by Wolfe (1986), which described the nature of the knowledge regarding estuaries at the time and the areas of future research and concern. Another major effort was the Coastal and Estuarine Lecture Note series published through Springer-Verlag, including three on estuaries (Sündermann and Holz 1980; Tomczak and Cuff 1983; Pequeux, Gilles, and Bolis 1984). The most recent contribution in this arena is an excellent review and summary in a two-volume set by Kennish (1986, 1990) on physical, chemical, and biological aspects of estuaries; and another by Kennish (1992) focusing on the human impacts on estuaries and a wide variety of management issues.

ESTUARIES IN DYNAMIC CHANGE

Because the estuary is positioned in the transition zone, it responds to conditions coming from both directions. Whether it be the changing discharge of the surface streams that flow into the estuaries or the changing composition of the flow, the estuary will be reflecting the abundance or paucity of such variables as nutrients, salinities, changes in water levels, and sedimentation. A similar driving force is causing stresses from the oceanic side as storm surges, salinities, sediments, organisms, and the like are impelled into the lower reaches of the estuaries, and then farther and farther upstream. Estuaries are in constant flux. There is always a forceful battle for domination between the continental processes and the oceanic processes. In some instances, the condition produces an oscillation of boundaries that shift upstream and down in response to the flow dominance from one direction or the other. In other cases there are quasi-permanent shifts, either gradually or step-wise, as the boundary migrates in a net direction. This latter case gives rise to a changing situation in the estuary that may be the quintessential feature of this zone: that of change in response to external processes.

There is no reason to assume that estuaries are fixed in space or that their distribution of natural features, such as wetland communities and aqueous habitats, are invariable. To the contrary, the estuaries give evidence that the ecological boundaries within are in constant flux and that migration, expansion, and loss are part of the characterization of the system. Likewise, it is necessary to realize that the estuarine system is unlike either the adjacent oceanic or terrestrial systems. The processes of wave action and currents flows are not exactly the same in the estuaries as in the ocean. The variations of salinity, temperature, and oxygen are unique to the estuaries. Additionally, many of the human adaptations in the form of resource extraction, in occupation, in land/sea tenure systems, etc., are likewise very specific to estuaries. Nordstrom (1992), in providing a comprehensive view of estuarine beaches, draws our attention to the particular characteristics of these restricted beaches compared to their oceanic counterparts. Others will soon be describing other aspects of the estuary which are as unique.

Physical System

Most of the estuarine systems are the product of the worldwide rise of sea level during the last 10,000 years when oceanic waters were encroaching upon the continental margin and producing semienclosed bays in the irregular topography of the pre-Holocene, subaerial surface. Contemporary sedimentation processes associated with fluvial discharge, coastal transfers alongshore and cross-shore, and barrier island development further molded the estuarine topography. It is likely that the present, general estuarine shoreline configuration, although inherited from earlier times, was initiated on the order of 3000-6000 years ago, when the rate of sea level rise decreased substantially (Walker and Coleman 1987). Following the cessation of the relatively rapid transgression, the estuarine systems became somewhat stationary in location, but began to develop their internal characteristics, such as the distribution of sediment types, the hydrography, the development of barriers at their mouths, the formation of wetlands, and the creation of ecologic habitats and communities. Further, the accumulations of sediments, nutrients, and other components were derived from oceanic sources as well as from the continent (Meade 1969; Clark and Patterson 1985). It is likely that the developmental processes are still occurring, although some investigators suggest that the same processes are now producing an attenuation of some of the forms and features of the estuaries (Kearney and Stevenson 1991; Psuty 1986a).

Human System

Given the large concentration of the world's population at the coastal zone, it is not surprising that the world's estuaries are being subjected to many modifications produced by the human population. Manipulation of the estuarine environment has produced alterations of the natural systems and has introduced completely different circulation patterns, different sedimentation sequences, much different water chemistries and particulate loadings, and different habitats. There are very few estuarine areas that have not been modified to some degree. Even the so-called 'pristine' locations bear the imprint of humans in the aquatic zone, in the surrounding wetlands, and in the sediments. The effects of humans and their modifications are not necessarily derived from the immediate surroundings. The effects may be caused by dams and flood control structures upstream. They may be the product of waste disposal or agricultural fertilizers somewhere in the drainage basin. They may be the product of forest clearing or burning in the upper reaches of the watershed which eventually discharges into the estuary.

Not all estuarine changes are driven by forces from the continent. It has been shown that sediments, toxics, and some associated water chemistry are derived from sources seaward of the estuary as well (Hall, Nadeau and Nicolich 1987). Sediments and/or chemical compounds discharged at one site can be transported in the offshore zone and eventually travel through an inlet to accumulate in an estuarine system far removed from the original oceanic outfall. Some changes are created by additional nutrients being pumped into the system. The result may be the development of extensive vegetative growth but with a corresponding decrease in available oxygen in the system to support other parts of the ecologic community.

Change is part of the dynamics of the estuarine system because the estuary exists at the transition zone. Therefore, as the natural environment is slowly altered by broad global changes or as the human population changes its technology or utilization of the estuarine resources, the components of the estuaries will be affected by the modification of the forcing dynamics and will somehow reflect these conditions in their internal dynamics.

MAJOR CHALLENGES

Environmental Quality

The most pressing problem associated with management of estuaries is environmental quality. This issue extends to the fully developed estuaries with their myriad of industries, harbor features, population centers, and other aspects of human utilization, and to the 'undisturbed' estuaries which continue to retain much of their natural system composed of wetlands, shellfish beds, fishery nursery grounds, and unpolluted waters. Whereas it is impossible to return each and every estuary to the pristine state that existed prior to human occupation, it is within the realm of management to control excessive pollution, to protect designated habitats, to allocate areas for special land uses, and to strike balances among the various competing interests for the physical, biologic, and economic resources of estuarine systems. Progress has been made in some estuarine areas simply by concentrating types of land uses, thereby preserving environmental quality in a portion of the estuary while attempting to areally restrict or delimit degradation.

Because estuaries are systems, it is necessary to approach each and every management objective from the view of the entire system. Estuarine quality is in large part, the product of the quality of the drainage basin leading to the estuary, and thus it is necessary to initiate any strategy in a basin-wide application. Nutrient loadings are accumulative and must be addressed by remedying inputs from point and non-point sources throughout the gamut of inputs to the estuary. Altered systems may have to subscribe to different standards than unaltered systems because it is unrealistic to achieve pristine components in a partial system scenario. Further, some environmental changes can be acceptable to the human populations in the estuarine system at levels other than the original values. Thus, degraded systems may have to be improved, but they may have target values which are above the loadings of the original, ambient levels. Unaltered systems, however, may be managed to retain their characteristics within the range of natural fluctuations.

As Wilson (1988) has indicated, estuarine pollutants consist of a number of groupings: organic matter, petroleum and related products, heavy metals, organochlorines, and radioactivity. Further, the effects of each of the groups relate to their concentrations, their persistence in the environment, and their toxicity. Therefore, effective management and improvement in estuarine quality must direct controls and programs to each of these pollutant groupings individually and in combinations of their concentrations, persistence, and toxicity.

Food Production

Whether in the form of animal protein or grains, the estuaries have historically been an important source of food. Estimates from the United States suggest that the estuarine-dependent species comprised 71% by value and 77% by weight, of the total commercial fisheries landings during 1985 (Chambers 1991). Obviously, this means that this source of animal protein is dependent on the quality of the environmental system to maintain that level of production. Further, estuarine systems have been and are being modified to support crops and animals. In many places in the world, wetlands have been drained, diked, and planted to grain crops, such as rice. Large areas have been converted to ponds to produce shrimp and fish. Other portions produce salt in evaporation pans or provide pasture for livestock. In one sense, this is conversion of food production from a gathering economy to one of sedentary agriculture as the harvesting of products is restricted to confined spaces.

In many estuaries, management decisions will have to be exercised to supervise the modification of the natural system to fields and ponds involved in food production. It is likely that the estuaries will see an increasing demand for space related to aquaculture development, as traditional fishing methods are compromised by environmental changes, competition for space, and poor return on investment. Land and sea tenure systems in the estuaries will also be subject to review and modification as fields and ponds replace tidal flats and open water habitats.

The opportunities for forms of aquaculture in estuaries seems endless, because so many of the commercial finfish and shellfish are estuary-dependent. As natural stocks of fish are depleted, there will be a increased emphasis on generating these same products in controlled settings within the estuaries. Thus, the future fishery species may be those which are cultured in ponds and pens within the estuary (a) taking advantage of the aqueous system but kept within reach for easy harvesting, (b) using enhanced foodstuffs to speed biomass production, (c) concentrating on species which mature to market size quickly, and (d) using genetic engineering to raise organisms which are healthy and resistant to local diseases. Aquaculture will include finfish and shellfish, each with particular niches, and probably will include both saltwater and freshwater species within adjacent artificial habitats. The production of these species will have to be totally integrated within the estuarine system so that wastewater will probably be used to culture algae, and waste products will become the bases of other feeds. The economic and management challenges will be to incorporate the full range of commercial finfish and shellfish products and all of the ancillary steps in the production of these protein sources, as well as their feeds, their wastes, the nutrient rich waters, and their genetic improvements, into an environmental system that will not be degraded; and to make each step of the process contribute to the economic success of the venture.

Estuaries are ripe for management efforts on many different scales and in many different aspects. Management can extend from the basic natural system of circulation, sediment input, water chemistry, and aquatic organisms to the many variables associated with human use and exploitation. Management can be directed toward creating preferred habitats, such as new wetlands or areas of submerged aquatic vegetation. Management may take the form of marshalling the transfers of sediments from dredged areas to sites where sediment is needed; a type of recycling. The driving force in estuarine management is the knowledge that the natural system is a dynamic unit and the physical and biological resources are in constant flux. It will probably not be possible to deter or prevent all of the changes. Therefore, the management goal should be to identify those attributes which are desirable and to create pathways to retain those characteristics within the multiple uses imposed on the system.

Sea Level Rise

It is likely that the most significant new development affecting the estuarine system and its management is the issue of accelerated sea level rise. Because the estuaries are largely the products of the variation of sea level rise which was initiated several thousand years ago, it is expected that the more rapid rise of sea level predicted for the future will produce changes in the present pattern of estuaries and in their components. A rise in sea level will reestablish the transgressive nature of the water/land boundary and will cause shifts in the many kinds of gradients within the estuarine systems. The extent to which the estuaries will adapt to the rise will somewhat depend on the rate of rise, as well as the balance of other variables that describe and define each of the estuaries. For example, if the transgressing estuary were migrating into a developed urban area, it is likely that dikes and levees would be used to limit the inland shifts of the aquatic systems, thereby reducing the extent of the estuarine environment. Also, if the balance of sediment, nutrients, water quality, and other natural factors were disrupted, the composition of the estuary would change. It might become more saline, the vegetation cover would change its community structure, and the depths of the bays might become too great to support submerged aquatic vegetation. Whereas these aforementioned situations might be the product of a changing sea level, they are also challenges for the effective management of both the natural and human modifications of estuarine system dynamics.

One of the predictions regarding sea level rise comes from the Intergovernmental Panel on Climatic Change. This international body has had its constituent committees review the data available and determined that the best estimate regarding future sea level rise is an elevation of the world's ocean level of 0.66 m by the year 2100 (Warrick and Oerlemans 1990). All of the world's estuaries will be affected by this magnitude of rise.

Using the extensive wetland area adjacent to the Great Bay estuary in New Jersey, United States of America as a case study, several scenarios can be developed that point to management problems associated with sea level rise. Interpretation of the data from cores taken throughout the marshes in Great Bay suggest some interesting associations of estuarine development and rate of rise (Psuty 1986a). The basic sea level rise data set from the area (Figure 1) is interpreted to represent two, major rates of rise and encroachment on the continental margin. The older period extends from about 7,000 years ago as a minimum to about 2,500 years ago. During that time, the sea level was rising at the rate of about 2 mm/year (0.2 m/cent.). This calculation is a long-term average and was certainly quite variable during these thousands of years. However, about 2,500 years ago, the rate of sea level rise decreased substantially, now approaching an average of 0.75 mm/year (0.075 m/cent.). Of especial interest are the accompanying events within this estuary which are thought to have existed behind a fronting barrier island during this time.

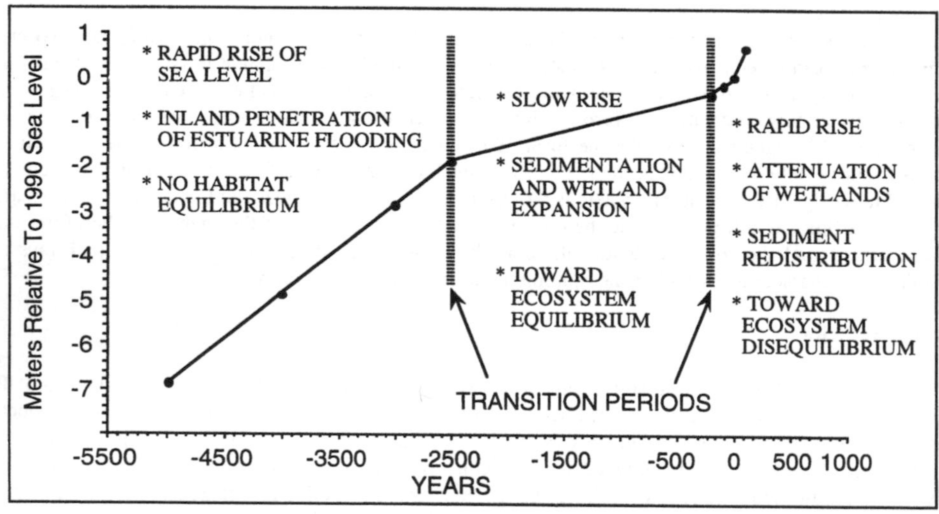

Figure 1. Portrayal of the rates of sea level rise and the expansion or attenuation of the estuarine habitats in a temporal context.

Note: The changing areal dimensions and the spatial associations are related to the balance between the three dimensional volumetric rates of inorganic and organic accumulation in the estuaries and the one dimensional vertical rate of sea level rise. The rates of rise in the past century, and probably the preceding one as well, exceed the rates of the earlier period when the estuaries were largely aquatic environments.

Interpretation of the sequence of stratification units suggests that most of the area of the estuary remained open water during the several thousand years that sea level was rising at the rapid rate (Psuty 1986a). It was only after 2,500 years ago and the reduction of the rate of inundation by the rising sea that the salt marsh became more than a fringing feature and began to broaden horizontally and extend into aquatic habitat. That indicates that the earlier rate of rise was greater than the rate of alluviation and that tidal flats were not developing during the pre-2500 B.P. phase of estuarine history. It was only in the recent millennia that sedimentation was sufficient to prograde the margins of the fringing marsh and begin to extend across the estuarine surface. This sequence is reflected in the stratigraphy of the deposits that show a thin peat zone (former salt marsh), fronted by open bay sediments until 2,500 years ago, when the vertical sequence changed to a horizontal sequence. A similar sequence and temporal span is derived from strata and isotopic dates analyzed in the Sado estuary in Portugal (Psuty and Moreira, manuscript). The interpretation, in both cases, is that slower, sea level rise conditions were accompanied by transfers of sediments into the estuarine environments during the post-2500 B.P. period and that the extensive wetlands developed during the most recent millennia. The interpretation is based on a sediment budget scenario that relates rates of volumetric accumulations in the estuarine environments to rates of vertical displacement of the sea level surface.

There is concern at present that sea level rise is once again upsetting the balance between the rate of volumetric accumulation and the vertical and horizontal continuity of estuarine wetlands (Orson, Panageotou, and Leatherman 1985). If so, it is likely that the estuarine wetlands may be reducing in areal dimension. Several investigators have commented on the attenuation of wetland surface area in estuaries in this century (Kearney and Stevenson 1991; Psuty 1986a; Stevenson, Ward, and Kearney 1986). They suggest that the sedimentary supply to the estuaries is inadequate to maintain the marsh areas so the new, higher sea level rates are causing the marsh to break into smaller units, and that the marsh boundary is shifting to produce more open water area in the estuaries. The net sea level rates of 2-4 mm/year (0.2-0.4 m/cent.) along the east coast of the United States (Hicks, Debaugh, and Hickman 1983; Lyles, Hickman, and Debaugh 1988) would seem to substantiate the claim of an imbalance because it is greater than the rate of the earlier rapid sea level increase and thus, the wetland processes may be leading to fragmentation and overall areal decrease (this net value includes an absolute eustatic sea level rise plus subsidence of the land).

MANAGEMENT IMPLICATIONS

In itself, the reduction of estuarine wetlands is a suitable topic for management policies and strategies because the wetlands are a very important component of the productivity of the estuary. But coupled with the driving force of sea level rise, an inadequate sediment supply, and the anthropogenic effects, the dimensions of the change can be dramatic. However, not all areas are undergoing similar rates of relative sea level rise or similar rates of sediment input. Very basic questions are posed concerning the rates at specific places and of the responses to these rates (Hoffman, Kaye, and Titus 1983) and how this change is manifested in the geomorphological (Psuty 1986b) and biological components of the estuarine system Frank, Perry, and Drinkwater 1990; Kennedy 1990). The questions require an analysis of the events of the past, either the past centuries, or at least the changes within this century. However, it is not always easy to separate the changes produced by the natural variations in the system from those caused by human perturbations. This is a challenging component of the riddle and one that will be difficult to unravel. But the issue may not be so dependent upon identifying the reason for the changes as identifying the changes, the trends of the changes, and the manifold consequences of the changes.

Sea level fluctuation is not a new area of inquiry. What is new is the stress on future sea level projections produced by global warming (Warrick and Oerlemans 1990). These values garner attention because they project elevation rates two-to four-times those of the past century. It is not surprising that the specter of coastal drowning holds great fascination for the coastal scientist and for the managers of this fragile environment. Drowning may lead to greater volumes of water and aquatic habitats in the estuaries. It may lead to many topographic changes, to different salinity gradients, to changes of plant community composition, to different benthic communities, to different circulation patterns, to different nutrient availabilities, and to many other natural variations. Of course, the changes will also be present in the human dimension. Waterfront development, residential communities, infrastructure, agriculture, aquaculture, and other aspects of human utilization of the estuarine environments will undergo changes or will have to develop protective strategies to cope with the changing conditions. According to Klarin and Hershman (1990), some coastal management programs are making strides in acknowledging the issue of sea level rise and are incorporating aspects of changes due to sea level rise in the decision-making process. Yet, as is witnessed in statements from other documents, many of the world's nations are not reacting to the scenarios of changes produced by predictions of near-future sea level elevations (Titus, Wedge, Psuty, and Fancher 1990).

Of especial importance is that most of the world's estuaries have a history of sea level rise in the past several centuries, and there is considerable information that can be derived from a study of these environments that will assist in developing models as to what the future sea level rises will bring. The data are in the charts, maps, aerial photos, in the sediments, in the wetland peats, and in the organisms that have occupied these environments. The record is there, it will have to be sampled, deciphered, and extrapolated to future rates of changes. Equally important, the scientific knowledge that is gained by the realm of investigations into estuarine dynamics associated with sea level changes will have to be translated clearly and unequivocally to the managers in order that decisions can be made based on this information. The managers and the scientists must have a similar data base so that policies, programs, and strategies can be derived from the factual information. In this way, the management decisions can be exercised to enhance the desired objectives and can relate the components of the estuarine system to the planned multiplicity of uses in a proactive style.

CONCLUSIONS

In the short term, the major issue of environmental quality will drive most management decisions. Maintenance of those areas that continue to be productive natural systems will be a prime concern, as will the improvements to those locations whose areas are degraded. Economic development will continue to stress the estuaries because of the concentration of industry, commerce, food production, and urbanization. The short-term challenge will be to develop compromises that benefit both the humans in the estuarine settings and the resource itself.

The long-term challenge (a century) is more daunting. Rising sea levels will continue to cause natural stresses in the estuaries. Boundaries and gradients will shift spatially, and natural ecosystems will adjust to new distributions of water levels, salinities, temperatures, water depths, sediment budgets, and nutrient budgets. Barriers such as dikes and walls will partially limit the extent of the changes in the estuaries, but they will probably redirect the changes and cause increases in the magnitudes elsewhere. Of special importance is that the rise will interact with the cultural patterns of infrastructure and land use. It will be impossible to retain the status quo: management decisions will be required to accommodate the relocation of resources and of land use. Management will mean providing for the multiplicity of uses within a dynamic system. The management practices will have to be equally dynamic and will have to retain a flexible allocation of resources as the zones of the estuary shift inland.

REFERENCES

Chambers, J. R. 1991. Habitat degradation and fishery declines in the U. S. In *Coastal Wetlands*, ed. H. Suzanne Bolton, 46-60. New York: American Society of Civil Engineers.

Clark, J.S., and Patterson, W. A., III 1985. The development of a tidal marsh: Upland and oceanic influences. *Ecological Monographs* 55:189-217.

Cronin, L. E., ed. 1975a. *Estuarine Research. Volume 1, Chemistry, biology, and the estuarine system*. New York: Academic Press.

Cronin, L. E., ed. 1975b. *Estuarine Research. Volume 2, Geology and engineering*. New York: Academic Press.

Fairbridge, R. W. 1980. The estuary: Its definition and geodynamic cycle. In *Chemistry and biochemistry of estuaries*, ed. E. Olausson and I. Cato, 1-16. New York: John Wiley & Sons, Inc.

Frank, K.T., Perry, R. I., and Drinkwater, K. F. 1990. Predicted response of northwest Atlantic invertebrates and fish stocks to CO^2-induced climate change. *Transactions of the American Fisheries Society* 119:353-65.

Hall, M. J., Nadeau, J. E., and Nicolich, M. J. 1987. Sediment transport from Delaware Bay to the New Jersey Inner Shelf. *Journal of Coastal Research* 3:469-74.

Hicks, H.R., Debaugh, H. A., Jr., and Hickman, L. E. 1983. *Sea level variations for the United States, 1855-1980*. Washington, D. C.: US Department of Commerce, National Oceanic and Atmospheric Administration, National Ocean Service.

Hoffman, J.S., Keye, D., and Titus, J. G. 1983. *Projecting future sea level rise: methodology, estimates to the year 2100, and research needs*. Washington, D.C.: U.S. Environmental Protection Agency, Strategic Studies Staff.

Kennedy, V. S., ed. 1980. *Estuarine perspectives*. New York: Academic Press.

Kennedy, V. S., ed. 1982. *Estuarine comparisons*, New York: Academic Press.

Kennedy, V. S., ed. 1984. *The estuary as a filter*. New York: Academic Press.

Kennedy, V.S. 1990. Anticipated effects of climate change on estuarine and coastal fisheries. *Fisheries* 15:16-24.

Kennish, M. J. 1986. *Ecology of estuaries: Volume 1, Physical and chemical aspects*. Boca Raton, Florida: CRC Press.

Kennish, M. J. 1990. *Ecology of estuaries: Volume 2, Biological aspects*. Boca Raton, Florida: CRC Press.

Kennish, M. J. 1992. *Ecology of estuaries: Anthropogenic effects*. Boca Raton, Florida: CRC Press.

Kearney, M. S., and Stevenson, J. C. 1991. Island land loss and marsh vertical accretion rate evidence for historical sea-level changes in Chesapeake Bay. *Journal of Coastal Research* 7:403-15.

Klarin, P. and Hershman, M. 1990. Response of coastal zone management programs to sea level rise in the United States. *Coastal Management* 18:143-65.

Lauff, G. H., ed. 1967. *Estuaries*. American Association for the Advancement of Science, Publication 83, Washington, D.C.

Lyles, S. D., Hickman, L. E., and Debaugh, H. A., Jr. 1988. *Sea level variations for the United States,* Washington, DC.: US Department of Commerce, National Oceanic and Atmospheric Administration, Office of Oceanography and Marine Assessment.

Meade, R.H. 1969. Landward transport of bottom sediments in estuaries of the Atlantic coastal plain. *Journal of Sedimentary Petrology* 39:222-34.

Nordstrom, K. F. 1992. *Estuarine beaches*. London: Elsevier Science Publishers.

Orson, R., Panageotou, W., and Leatherman, S. P., 1985. Response of tidal salt marshes of the U.S. Atlantic and Gulf coasts to rising sea levels. *Journal of Coastal Research* 1:29-37.

Pequeux, A., Gilles, R., and Bolis, L., eds. 1984. *Osmoregulation in estuarine and marine animals*. Volume 9 of Lecture Notes on Coastal and Estuarine Studies. Heidelberg: Springer-Verlag.

Pritchard, D. W. 1955. Estuarine circulation patterns. In *Proc. Amer. Soc. Civ. Eng.*, 81:1-11.

Pritchard, D. W. 1967. Observations of circulation in coastal plain estuaries. In *Estuaries*, ed. G. H. Lauff, 37-44. American Association for the Advancement of Science, Publication 83, Washington, D.C.

Psuty, N. P. 1986a. Holocene sea-level in New Jersey. *Physical Geography* 7:154-65.

Psuty, N. P. 1986b. Impacts of impending sea level rise scenarios: The New Jersey barrier island responses. *Bulletin, New Jersey Academy of Science* 31:29-36.

Psuty, N. P., and Moreira, M. E. prepared manuscript. Holocene sedimentation and a new sea level rise curve, Sado estuary, Portugal.

Stevenson, J.C., Ward, L. G., and Kearney, M. S. 1986. Vertical accretion in marshes with varying rates of sea level rise. In *Estuarine variability*, ed. D. A. Wolfe, 241-259. New York: Academic Press.

Sündermann, J., and Holz, K. P. 1980. *Mathematical modelling of estuarine physics*. Volume 1 of Lecture Notes on Coastal and Estuarine Studies. Heidelberg: Springer-Verlag.

Titus, J. G., Wedge, R., Psuty, N., and Fancher, J. 1990. *Changing Climate and the Coast: Report of the Intergovernmental Panel on Climate Change from the Miami Conference on Adaptive Responses to Sea Level Rise and Other Impacts of Global Climate Change*, Washington, D. C.: US Environmental Protection Agency.

Tomczak, M., Jr., and Cuff, W., eds. 1983. *Synthesis and modelling of intermittent estuaries.* Volume 3 of Lecture Notes on Coastal and Estuarine Studies. Heidelberg: Springer-Verlag.

Walker, H. J., and Coleman, J. M. 1987. Atlantic and Gulf coastal province. In *Geomorphic systems of North America*, ed. W. L. Graf, 51-110. Boulder, Colorado: Geological Society of America, Inc.

Warrick, R. A., and Oerlemans, H. 1990. Sea level rise. In *Climatic change: The IPCC Scientific assessment*, eds. J. T. Houghton, G. J. Jenkins, and J. J. Ephraums, 257-281. Cambridge: Cambridge University Press.

Wiley, M., ed. 1976. *Estuarine processes. Volume 1, Uses, stresses, and adaptation to the estuary.* New York: Academic Press.

Wiley, M., ed. 1977. *Estuarine processes. Volume 2, Circulation, sediments, and transfer of material in the estuary*. New York: Academic Press.

Wilson, J. G. 1988. *The Biology of estuarine management.* London: Croom Helm.

Wolfe, D. A., ed. 1986. *Estuarine Variability*. New York: Academic Press.

19

Oceanography and Coastal Environmental Assessment: Two Case Studies of Different Areas in the Tyrrhenian and Adriatic Seas

Dr. Romano Ambrogi and Dr. Giulio Queirazza
National Electric Board (ENEL) - S.p.A., Center for Environmental and Material Research, Milan, Italy

Prof. Tecla Zunini Sertorio
University of Genova, Institute of Marine Environmental Sciences, Santa Margherita Ligure, Italy

ABSTRACT

Coastal environmental assessment is an important part of Environmental Impact Assessment procedures and comprises the oceanographic survey of the coastal area. In the framework of a nationwide program for studying the sites proposed for thermoelectric power plants, a special interest is devoted to the comparison of two areas, placed in the Tyrrhenian Sea and in the Adriatic Sea, whose oceanographic characteristics are widely different.

The analysis of the differences between the two sites focuses on the data regarding their biological components, namely, zooplankton, taxonomic composition and biomass and its trace metal content.

The different environmental characteristics of the two sites influence, not only the presence and abundance of the major taxa of the zooplankton in the four seasonal samplings, but also the biomass differences and the chemical elemental composition.

The average abundance of total zooplankton was at its maximum during summer in the Adriatic site (average 40,485 ind. m^{-3}), when Cladocera predominated over all other taxa. Total biomass was 159.2 g m^{-3} (dry weight) and was significantly correlated with the abundance of both Cladocera and Copepoda. Copepoda dominated zooplankton in the Tyrrhenian site in every season, but the maximum abundance and biomass was recorded in spring, with much lower values than in the Adriatic (average 7,035 ind. m^{-3} and 23.6 g m^{-3}).

The elemental concentrations in zooplankton were corrected according to observation of possible contamination due to sampling (correction with reference to Nickel content) and to the possible contribution of terrigenous components in the sample itself (correction with reference to Aluminium).

The average value of metal content has been compared between the two sites, taking into account the data of "clean" oceanographic regions. In the majority of cases (Mg, Cr, Ni, Cu, Cd, Ba, and Pb) the metal levels were higher in the Tyrrhenian than in the Adriatic zooplankton. These findings can be interpreted to be the consequence of the higher levels of some elements in the sea water of the Western basin of the Mediterranean, in comparison with the Eastern basin, to which the Adriatic belongs.

In conclusion, oceanographic baseline studies can take advantage of the determination of zooplankton composition and biomass, coupled to some measure of "pollution marker", such as heavy metals, provided that the sampling strategy accounts for the high variability of coastal conditions.

INTRODUCTION

In the course of ecological monitoring in coastal areas, for the purpose of assessing a baseline condition before any individual coastal zone utilization, it is a common experience to encounter very heterogenous ecosystems, different from each other in many respects. On the other hand, the need of designing and using a "standard" approach in the examination and description of the environment is also felt.

Even in the case of a relatively small country such as Italy, surrounded by the Mediterranean Sea, which may seem rather a homogenous situation with respect to coastal characteristics, a nationwide program of oceanographic investigation for the siting of thermoelectric power stations (Tomasino et al. 1993) has had to cope with substantial differences in the general situation of the studied coastal ecosystems.

As an example of such differences we chose to illustrate the results of the cruises carried out at Montalto di Castro (on the Lazio coast, central Tyrrhenian Sea) and in the coastal area of Friuli (Northern Adriatic Sea) in front of the mouth of the river Timavo. The oceanographic, climatic, trophic, and pollution conditions are palpably different in the two sites, so it was felt that a comparison would prove instructive in assessing the capability of a standard program for determining baseline ecological conditions.

With this aim, besides the parameters measured in all the other cruises of the National Electricity Board (ENEL) oceanographic program (Tomasino et al. 1993), in this work we also measured the elemental composition of the zooplankton.

The presence of metals in marine ecosystems can be determined by indicator organisms such as zooplankton, and because the concentration of metals is three or four orders of magnitude greater in the zooplankton than in sea water, the detection of low levels of metal pollution is made easier. Furthermore, the heavy metal concentrations in the aquatic organisms reflect their persistency and availability in the different compartments of the ecosystem.

Until now, few monitoring surveys of metals in zooplankton have been conducted in the Mediterranean Sea (Bernhard and Zattera 1975; Fowler 1990). Nevertheless, plankton constitutes an important reservoir for accumulating and cycling several metals.

In the present study, various zooplankton samples were analyzed to: (1) evaluate if several samples of zooplankton seasonally collected in two sites of the Mediterranean Sea would give a meaningful indication of metal levels; and (2) determine possible differences of the metal levels between the Tyrrhenian Sea (in the Western basin) and the Adriatic Sea (in the Eastern one).

The biological component in the program of oceanographic cruises is thus used as an important indicator of environmental conditions existing before the initiation of siting procedures for large industrial settlements such as power stations.

GEOGRAPHICAL SETTING

The Tyrrhenian Sea is part of the Western Mediterranean (Margaleff 1985) and exhibits many "typical" characteristics of the Mediterranean, being rather oligotrophic, with high salinities, and scarcely influenced by river inputs. The water circulation is driven by the general Mediterranean current, flowing alongshore in a northerly direction and is strongly influenced by the prevailing winds. Information on zooplankton is given, e.g., by Scotto di Carlo and Ianora (1983).

Near the studied site of Montalto di Castro, no local inputs of large towns or industrial settlements are present, nor large river inflow, the area being rather in good condition, as demonstrated by the presence of a large Posidonia oceanica meadow. This site is well-known both from a hydrodynamic point of view and from the chemical and biological standpoint (Queirazza et al. 1987). A detailed modelization approach has been used to describe the behavior of radionuclides in the water, sediments, and biotic compartments (Cambiaghi et al. 1987).

The Northern Adriatic Sea, where the Friuli site is located, is a very peculiar sector of the Mediterranean, being very shallow (maximum depth about forty meters) and receiving a large amount of freshwater input from a number of rivers, among which the most important is the Po River. Another peculiar characteristic of this sea is the presence of tides, that, unlike the rest of the Mediterranean, can reach a maximum range of sea level variation of about one meter.

It has been estimated that, in relation to the area of the basin, the Adriatic receives a specific discharge of river outflow of 39,000 $m^3\ s^{-1}\ km^{-2}$, compared with only 5,300 of the Tyrrhenian Sea. Since the majority of these rivers flow through densely populated and intensively cultivated lands (the Po plain features 30% of the Italian population and 50% of the agricultural crops, not mentioning the intense cattle rearing), the load of nutrient input to the Northern Adriatic is also of great importance. Urban and industrial pollutant loads are also heavier than from any other source to Italian seas (Marchetti 1991).

The biological features of this area are also rather well known, and zooplankton has been investigated in detail (Specchi et al. 1979; Benovic et al. 1987). In particular, the Friuli site is located in the Gulf of Trieste, which is the northeastern reach of the Adriatic. This area is not influenced by the Po inflow, but receives many rivers (Isonzo, Timavo, etc.) that are scarcely polluted.

MATERIAL AND METHODS

The oceanographic cruises were performed in the area shown in Figure 1 (Montalto di Castro) and Figure 2 (Friuli).

Plankton samples were collected with a Bongo net, diameter 20 cm, mesh size 200 μm equipped with a flowmeter. One of the nets collected the samples for taxonomic examination; the other for biomass determination and chemical analysis of the zooplankton. The net was towed at the surface, at three-hour intervals during two, 24-hour cycles, by the R/V Ciprea at a speed of two knots. The two cycles were generally performed at an interval of about five to six days.

Taxonomic determination was performed at the generic level for Copepoda and Cladocera; higher taxa were considered for other groups (Appendicularians, Chaetognaths, Molluscan larvae, etc.). Biomass was measured according to Lovegrove (1966). The dry weight determination references the whole sample, even in the presence of detritus or inorganic particles.

Zooplankton samples were stored by adding 5% (v/v) formaline solution (40% w/v). This procedure assures a minimal metal loss as checked by Queirazza (1979).

For metal determination, zooplankton samples were filtered on 80 μm mesh nylon net, then the recovered material was dried at 60°C for twenty-four hours. About 0.5-1.0 g of dried material were treated with 15 ml of the acid mixture ($H_2O/HNO_3/HCl=2.5/2.5/1.0$) and transferred to a Teflon bomb. To the sample was added 5 ml of H_2O_2, which was then allowed to stand at room temperature for about one hour. When all residual reaction stopped, the bomb was tightened; introduced into the microwave oven and heated at 40% full power (1200 W) for fifteen minutes; allowed to cool for fifteen minutes and, finally, heated at 40% full power (1200 W) for thirty minutes. After cooling, the solution was transferred to a 25 ml polypropylene flask and taken up to volume with deionized water. Samples were analyzed by Atomic Emission Spectroscopy (Varian SpectrAA-10) for Na and K; by Graphite Furnace-Atomic Absorption Spectroscopy (Varian AA-1475) for Cd, Co, Pb, Cr, and Ni; and by

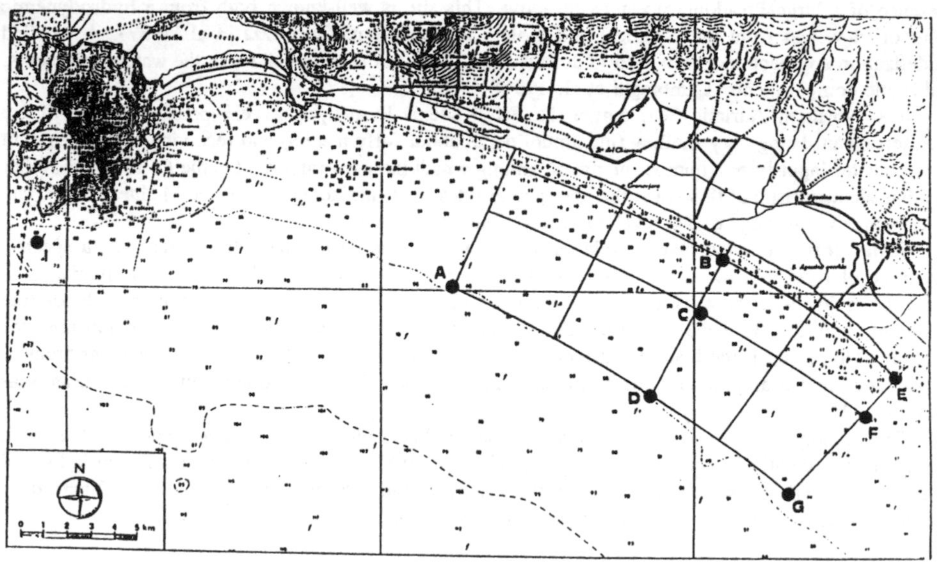

Figure 1. The studies area of Montalto di Castro in the Tyrrhenian Sea (Zooplankton samples collected at station B)

Inductively Coupled Plasma- Atomic Emission Spectroscopy for the other elements (Fe, Mn, Al, Zn, Ba, Mg, Cu, and Ca).

The analytical procedure was checked using standard reference materials. Results were in agreement with the NBS certified values.

RESULTS

Zooplankton Biomass and Composition

The zooplankton assemblages, as expected, are fairly different in the neritic waters of the Friuli site, typically eutrophic, and in the Montalto di Castro site, being rather oligotrophic. The Friuli zooplankton is richer in biomass and total numerical abundance (Figures 3 and 4) but scarcer in number of taxa. The difference is particularly evident during summer, when Cladocerans are massively present in the neritic waters of Friuli.

Also the seasonal abundance cycle is different in the two sites, showing maxima and minima in different seasons. However, the comparison may be biased by the fact that seasonal samplings in the two sites could be performed at one or two months interval.

The important biological load of the eutrophic waters of the Northern Adriatic is apparent from the Copepod abundance (Figure 5). This fact is also evident if one examines other groups, in particular the Cladocerans (Figure 6), that are most numerous in the Adriatic.

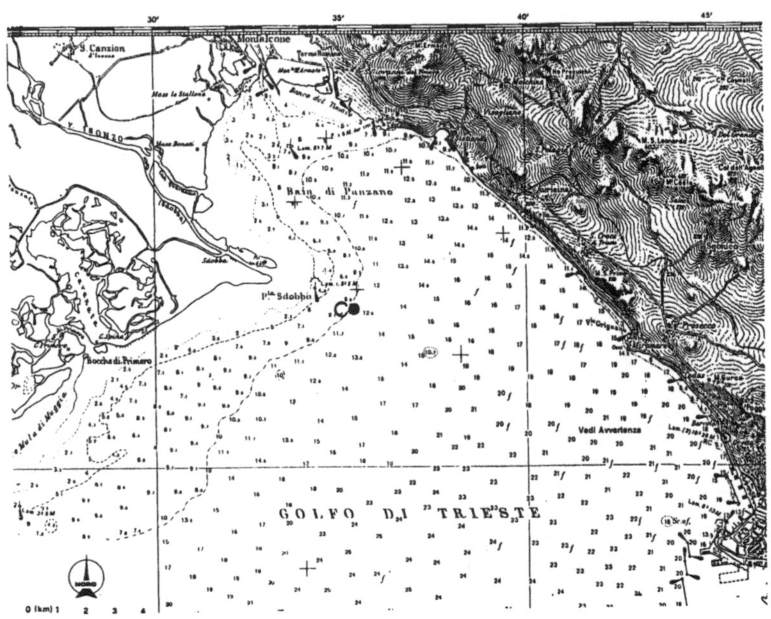

Figure 2. The studied area of Friuli in the Adriatic Sea (Zooplankton samples collected at station C)

The dates of sampling were as follows:

Montalto 1st	: March 6-7, 1988
(Winter)	March 8-9, 1988
Montalto 2nd	: May 24-25, 1988
(Spring)	June 9-10, 1988
Montalto 3rd	: August 24-25, 1988
(Summer)	August 31 - September 1, 1988
Montalto 4th	: November 17-18, 1988 * (biomass not measured)
(Autumn)	December 8-9, 1988
Friuli 1st	: April 23, 1988 * (biomass not measured)
(Spring)	April 27, 1988 * " " "
	April 30 - May 1, 1988
Friuli 2nd	: July 9-10, 1988
(Summer)	July 21-22, 1988
Friuli 3rd	: October 3-4, 1988
(Autumn)	October 9-10, 1988
Friuli 4th	: January 9-10, 1989
(Winter)	January 21-22, 1989

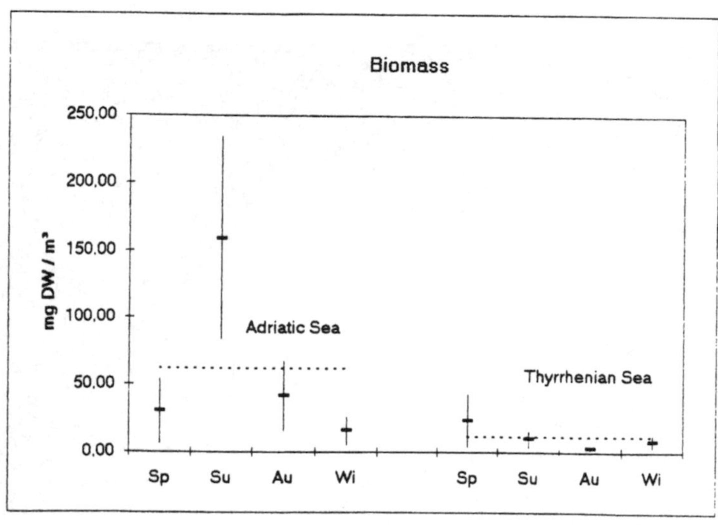

Figure 3. Seasonal zooplankton biomass (mg DW m^{-3} ± 1σ)in the Adriatic Sea (Friuli) and in the Tyrrhenian Sea (Montalto di Castro)

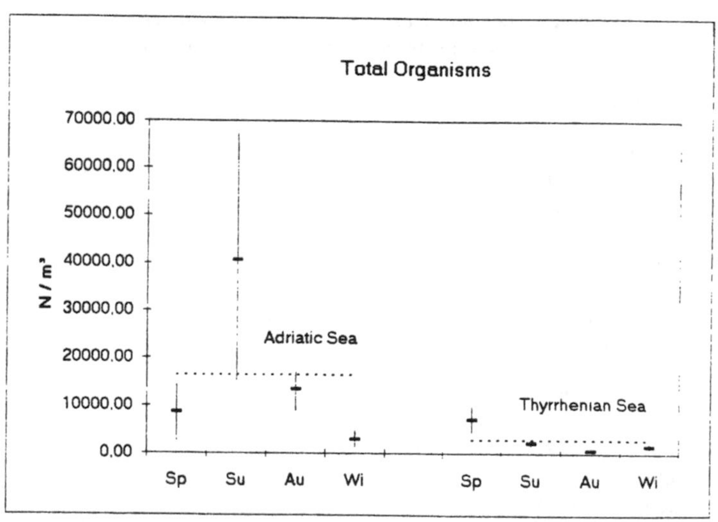

Figure 4. Seasonal abundance of total zooplankton (ind. m^{-3}•1σ)in the Adriatic Sea (Friuli) and in the Tyrrhenian Sea (Montalto di Castro)

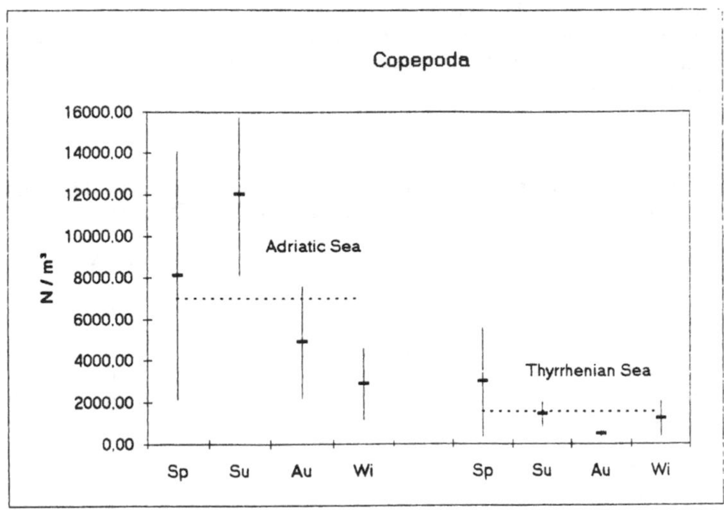

Figure 5. Seasonal abundance of Copecods (ind. m$^{-3}\pm\sigma$) in the Adriatic Sea (Friuli) and in the Tyrrhenian Sea (Montalto di Castro)

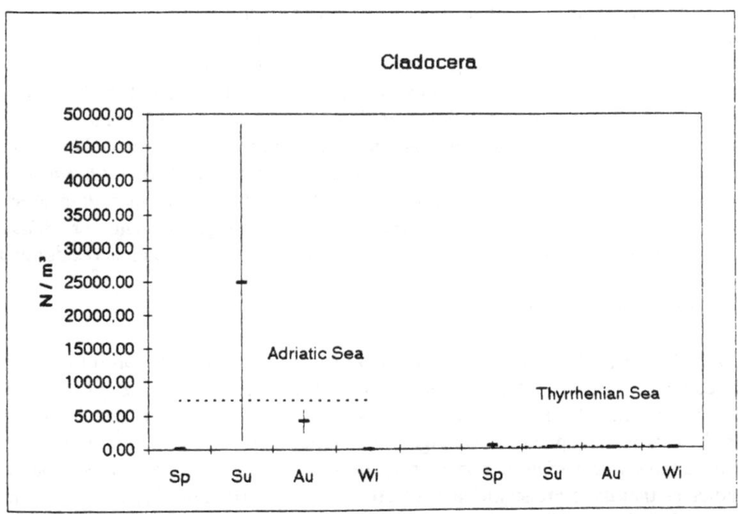

Figure 6. Seasonal abundance of Cladocerans (ind. m$^{-3}\pm1\sigma$)in the Adriatic Sea (Friuli) and in the Tyrrhenian Sea (Montalto di Castro)

The correlation coefficient r indicates that biomass (DW) and total numerical abundance of organims are significantly ($p < 0.05$) correlated in all seasons in both sites, except in two cases. The abundance of Copepods is also highly correlated with biomass at Montalto di Castro and Friuli with the same pattern. The abundance of Cladocerans is correlated only in summer at the Friuli site and in Spring at Montalto (Table 1).

Table 1

Correlation Coefficients (r) Between DW Biomass and Abundance of Copepods, Cladocerans, and Total Zooplankton in Samples of Different Seasons

	ADRIATIC SEA			TYRRHENIAN SEA		
	Cop	Clad	Total	Cop	Clad	Total
Spring	0.414 ns	0.228 ns	0.441 ns	0.978 **	0.837**	0956 **
Summer	0.888 **	0.780 **	0.870 **	0.005 ns	0.103 n	0.008 ns
Autumn	0.775 **	0.449 ns	0.838 **	0.791 **	0.167 ns	0.839 **
Winter	0.860 **	-	0.859 **	0.860 **	-	0.979 **

Note: * significance level 5%
** significance level 1%
ns non significant correlation

Seasonally the zooplankton assemblage is characterized by different groups of organisms in the two sites. The qualitative pattern (Figure 7) clearly shows that Copepods dominate at Montalto di Castro in all the seasonal samplings and that they are overcome by Cladocerans during summer in the Friuli site.

The Copepod assemblage in the two sites has five genera in common (*Paracalanus, Centropages, Acartia, Oithona, and Oncaea*) in all seasons. At the Montalto di Castro site, five more genera are common to all the four seasonal samplings (*Clausocalanus, Corycaeus, Isias, Euterpina, and Temora)*, showing a more diverse assemblage. The number of genera present in each of the four sampling seasons ranges from eleven to fourteen at Friuli and from fifteen to nineteen at Montalto.

As far as the percentage dominance is concerned, the Friuli site is dominated by Acartia in spring, by *Paracalanus* and the Cladoceran, *Penilia*, in summer and autumn, and still by *Paracalanus* in winter. In the Tyrrhenian waters, Acartia, Paracalanus, and Centropages, together with Appendicularians dominate in spring; *Paracalanus, Clausocalanus*, and *Centropages* in summer; *Clausocalanus* in autumn; and *Paracalanus, Clausocalanus*, and *Acartia*, along with Gastropod larvae dominate in winter.

Elemental Composition of Zooplankton

STATGRAPHICS package has been adopted to process data. Distribution fitting procedure (based on Chi-Squared test) has shown a log-normal distribution in nine cases (Al, Ba, Co, Cr, Fe, Mg, Mn, Ni and Zn) and normal distribution in three cases (Cd, Cu, and Pb), for both sampling sites. Non-parametric tests were applied to investigate differences between space and time.

Two-way analysis of variance, considering two sites and four seasons as separate sources of variation, indicates that there are significant differences ($P < 0.01$) both between space and time, with higher concentrations in the Western basin for the majority of the elements, and in summer and autumn periods (summer: Al, Ba, Co, Cr, Fe, Mn and Ni; autumn: Cu, Mg and Pb; winter: Cd; spring: Pb).

The data obtained suggest that some samples were probably contaminated because values for some elements (Ni, Cr, Co, Fe, and Mn) were found to be too high. This contamination may be artificial, due to the sampling system used (Bongo net), hydrowire and other items commonly found aboard ships; and "natural", due to biotic (organic aggregates) and abiotic components (for instance, clay particles) present in the sample, together with zooplankton.

In order to check these two sources of contamination, our data have been validated as follows:

1. Considering only zooplankton samples with a Ni concentration level below 12 µg/gDW. This level has been selected after Martin and Knauer (1973), who reported this value as the maximum of the distribution of "clean" Copepod samples.
2. Considering corrected data for the Montalto site where samples were characterized by a high level of Al. In fact this element can be considered an independent indicator of the insoluble terrigenous marine fall-out.

The average concentration for site and season, for each element (selected for Ni < 12 µg/gDW), are reported in Figures 8 and 9. After this selection, data were reduced by about 30% of the original number of samples for each season.

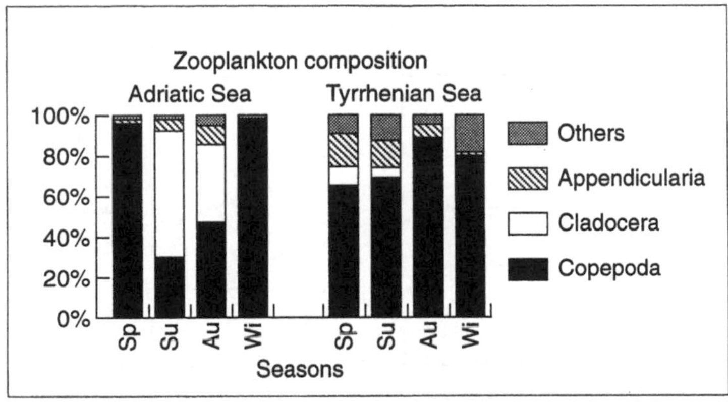

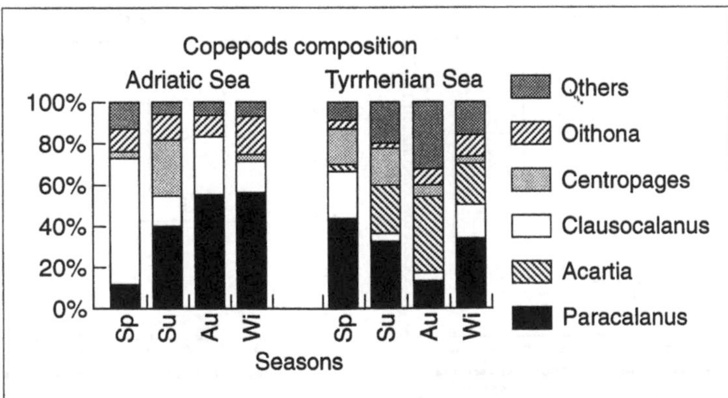

Figure 7. Seasonal zooplankton and Copepods composition in the Adriatic Sea (Friuli) and in the Tyrrhenian Sea (Montalto di Castro)

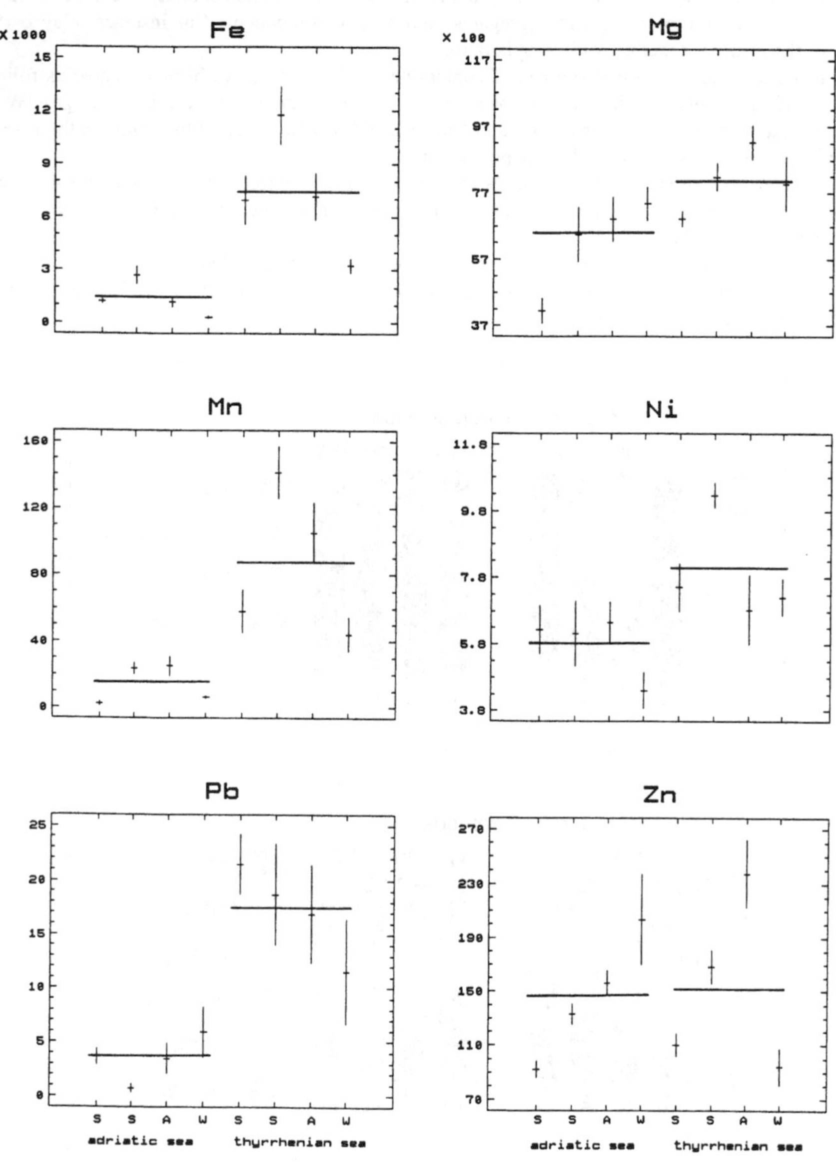

Figure 8. Seasonal concentration of Fe, Mg, Ni, Pb, and Zn in zooplankton samples (μg/g DW ± 1σ) in the Adriatic Sea (Friuli) and in the Tyrrhenian Sea (Montalto di Castro)

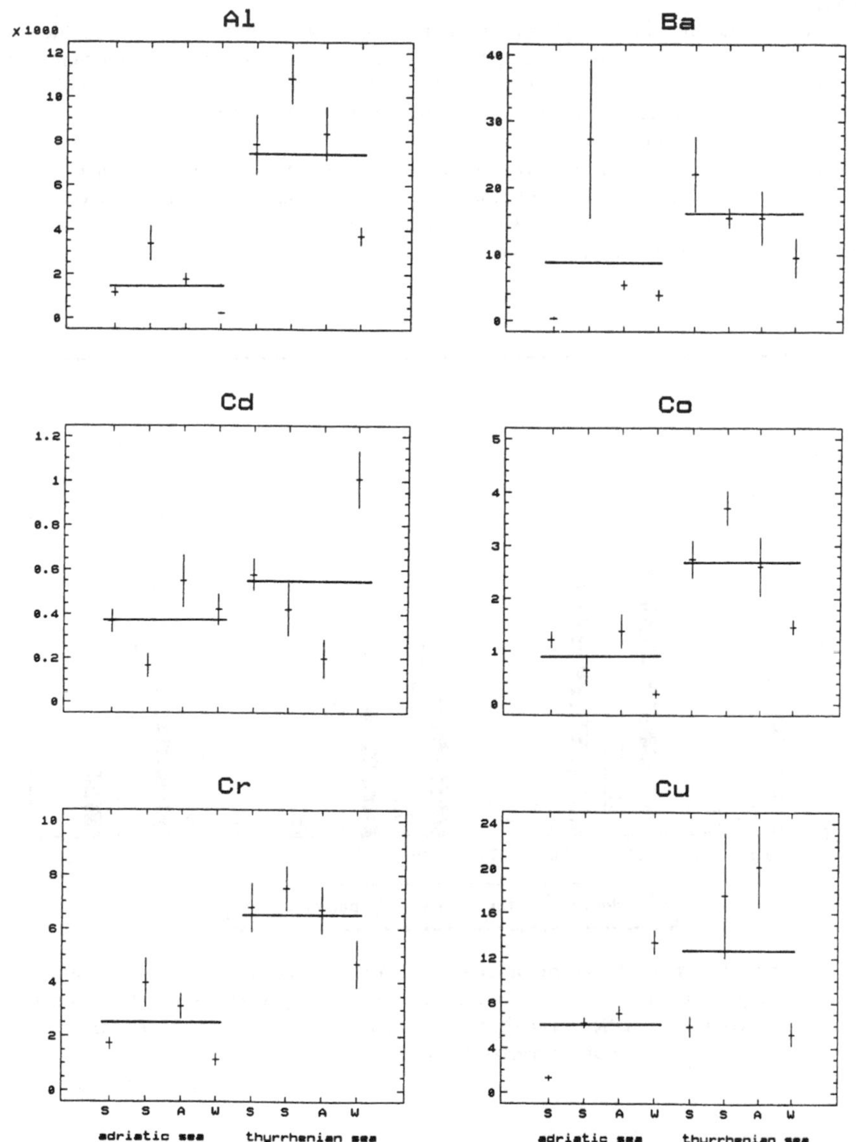

Figure 9. Seasonal concentration of Al,Ba, Cd, Co, Cr, and Cu in zooplankton samples (μg/g DW ± 1σ) in the Adriatic Sea (Friuli) and in the Tyrrhenian Sea (Montalto di Castro)

In order to prove the consequence of this selection, the Spearman rank correlation test has been applied. Positive correlation (99% significance level - $p<0.01$) was observed among Ni and Co, Cr, Fe, and Mn ($r=0.72$, 0.79, 0.67, 0.67). After selection of data, the r values fall to 0.65, 0.65, 0.66, and 0.50 respectively.

In order to quantify the natural contamination, regression analysis between Me (dependent variable) and A1 (independent variable) was applied. The plots obtained show a linear relationship with $r>0.97$ for Cr, Co, and Fe, where r equals 0.88 for Mn and 0.74 for Pb. Taking as A1 natural level the average concentration measured in the Friuli site (1642 μg/gDW), only correlated data were corrected applying the regression equation Me-A1.

Annual mean values and standard deviations of the concentrations of the studied elements in the Adriatic and Tyrrhenian Seas, as well as reference values for zooplankton of the Pacific Ocean (Martin and Knauer 1973) are given in Figure 10. The values of Cr, Co, Fe, Mn, and Pb shown in this Figure derive from the fitting procedure described above.

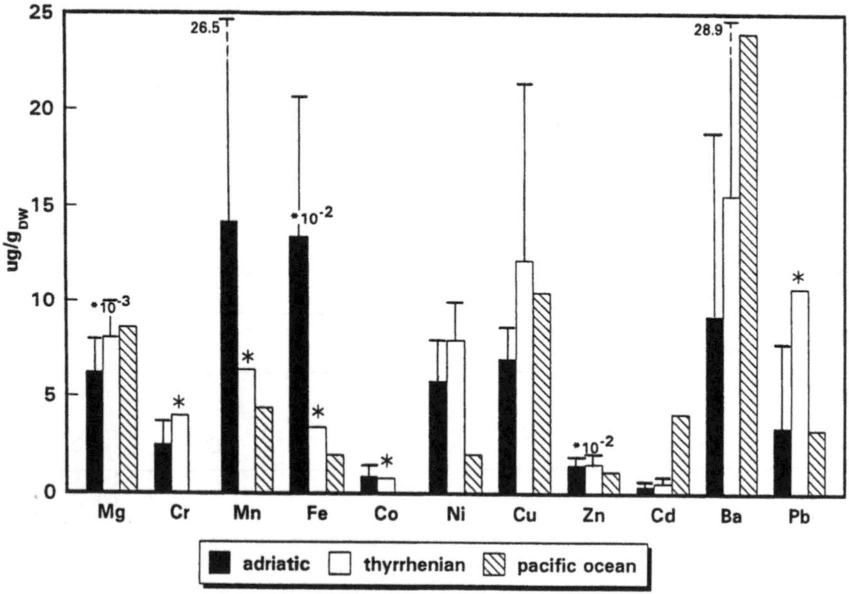

Figure 10. Annual average concentration of Mg, Cr, Mn, Fe, Co, Ni, Cu, Zn, Cd, Ba, and Pb in zooplankton samples (μg/gDW ± 1σ) in the Adriatic and in the Tyrrhenian Sea. Data concerning copepods collected in the Pacific Ocean (Martin and Knauer 1973) are also given for reference. (*Data corrected by A1 levels [see text]).

DISCUSSION

The high biological load of the eutrophic waters of the Northern Adriatic in comparison with the Tyrrhenian Sea, is confirmed by the zooplankton biomass and abundance (Figures 3 and 4). This fact is also evident at the level of particular taxa, and in particular for the Cladocerans (Figure 6), that are most numerous in the Adriatic.

The correlation coefficient r indicates that biomass (DW) and total numerical abundance of organims are significantly ($p < 0.05$) correlated in all seasons in both sites. It can be concluded that, although non-zooplankton material constitutes the biomass sample, the latter is indicative of zooplankton abundance, and in particular of Copepods. Cladocerans are a major component of biomass only in summer at the Friuli site (Table 1).

As can be seen from the results in Figures 8 and 9, the element concentrations in the zooplankton exhibit seasonal variation.

The average concentration levels of the "biologically-essential", macroconstituent elements such as Na, K, Ca, and Mg, were detected with a presence of 6,840; 889; 5,169; and 6,217 μg/gDW for the Friuli site, and of 17,586; 1,680; 13,620; and 8,061 μg/gDW for the Montalto di Castro site, respectively. The data are poorly representative of the reference values (Martin and Knauer 1973) particularly for K and Na. Moreover, a fairly important difference between the two sites is apparent, probably due to a marked "natural" contamination of the Tyrrhenian samples.

Considering the group of the metals (Cr, Cu, Fe, Mn, Ni, Co, and Zn) which can be either beneficial or harmful to biota, depending on their concentration or chemical form, only for Fe, Mn, Co, and Cr is a overall clear seasonal behavior distinguishable (Figures 8 and 9), with a maximum in the summer - autumn period (r values among elements ≥ 0.67). In contrast, for the remaining elements, no similar trends are evident.

During this period, Cu and Zn concentrations in zooplankton appear close to or higher than annual mean values (Figures 8 and 9). The phenomenon of "biological dilution" is not evident for these metals in comparison with the data obtained by Hardstet-Romeo and Laumond (1980). Nickel levels in zooplankton varied only in one season. In particular a low value was found in winter for the Friuli site and a high value in summer was recorded at Montalto di Castro.

Despite being masked by a variety of factors, an overall increase of the element concentrations was registered in the Tyrrhenian Sea with reference to the Adriatic (Figures 8 and 9). Taking into account the contribution of the "natural" contamination (data corrected by Al content), the above statement still holds except for Fe and Mn (Figure 10). This fact can be attributed to zooplankton uptake of more available forms of Fe and Mn, coupled with their adsorption on organic detritus in the Adriatic site, because of the consistently higher biomass production in this area. However, the mechanisms responsible for the control of Fe concentrations are not altogether clear (Wangersky 1986).

In the absence of the precise knowledge of the transfer mechanisms and their rates, the relative ability of organisms to concentrate trace elements is often expressed by a concentration factor, defined as the ratio of the amount of element per unit fresh weight (or dry weight) of tissue to that in an equal weight of sea water (Fowler 1983). For some of the investigated elements (for instance, Mn, Fe, Co, and Zn) values of the dry-weight concentration factors are over three orders of magnitude (Santschi and Honeyman 1989). This fact enables us to use zooplankton as a biological monitoring community of metal levels in different water masses (Western and Eastern basins of the Mediterranean and, as a reference, the Pacific Ocean).

From Figure 10 trace metals in zooplankton organisms follow the expected enrichment of the Mediterranean sea water in respect to ocean water (Ruiz Pino et al. 1990) for Mn, Fe, Zn, Ni, and Cu (for the Western basin only).

Between the waters of the Mediterranean Sea, trace metals, as a whole, were slightly enriched in Western basin vs Eastern ones (Ruiz Pino et al. 1990). Because of the similarity in the concentrations for the zooplankton collected in the two basins, it is possible, given the variability shown in Figure 10, to distinguish only a qualitative agreement (Cr, Ni, Cu, and Zn) in respect to the expected general behavior of trace elements.

The ability of zooplankton to concentrate large quantities of essential elements such as Cd, Pb, and Ba, has been reported (Martin and Knauer 1973; Romeo and Nicolas 1986; Hardstedt-Romeo and Laumond 1980). Concentrations in marine organisms are in the state of dynamic equilibrium and are the net result of both element uptake and elimination processes occurring simultaneously.

The latter process can be important for biologically nonessential elements. In particular, Pb and other elements which are particle-reactive and have a strong propensity for hydrolysis, would show little penetration into algal cytoplasm and negligible retention in zooplankters. Cd is found in the cytoplasm of algal cells and would be, therefore, assimilated by grazers, as shown by Fisher et al. (1991).

A minimum zooplankton concentration of Cd (both for the Adriatic and the Tyrrhenian) and of Pb (for the Adriatic) could be recognized during summer and autumn (Figures 8 and 9). These lowest levels were found during periods of high phytoplankton productivity, which presumably diluted the amount of the metal per unit mass of phytoplankton. The high biomass of zooplankton during these periods (Figure 3) supports such an hypothesis. Such a minimum confirms the cited phenomenon of "biological dilution" which affected Cd levels more than Pb levels in zooplankton ingesting phytoplankton.

In contrast, an uncommonly high level of Ba was found during the summer in the Adriatic zooplankton (Figure 9). Since our collection system does not exclude the presence of zooplankton faecal pellets and other organic detritus with the zooplankton collected, it is consistent with the experimental laboratory observations of Fisher et al. (1991) that Cd (and perhaps Pb) was lost rapidly from faecal pellets due to the strong affinity for chloride in seawater. Whereas Ba could combine largely as barite inclusions to faecal matter (Dehairs et al. 1980).

Concerning Cd and Pb concentrations in zooplankton, our values (Figures 8 and 9) are in agreement with those reported by Romeo and Nicolas (1986), but they are much lower for Cd and slightly higher for Pb than the results reported by Martin and Knauer (1973). This fact appears as a consequence of the different water concentrations of the Mediterranean Sea and Pacific Ocean. In fact Cd concentrations in the Mediterranean Sea are very low (~ 5 ng L^{-1} in Mediterranean, [Laumond et al. 1984]; from 6.7 to 7.2 ng L^{-1} in Western basin and from 6.2 to 7.2 ng L^{-1} in Eastern basin [Ruiz Pino et al. 1990]), whereas Pacific waters are characterized by higher concentrations (from 1 ng L^{-1} in surface to ~ 1000 m [Bruland in Romeo and Nicolas 1986]).

Lead concentrations in the water column of the Mediterranean present higher values (30-50 ng L^{-1}, [Laumond et al. 1984]; 26-62 in Western basin and 21-62 in Eastern basin, [Ruiz Pino et al. 1990]) than the Pacific Ocean (5-12 ng L^{-1}, [Bruland in Romeo and Nicolas 1986]). Moreover, Cd and Pb levels in zooplankton (Figure 10) reflect the slight difference between the two Mediterranean basins.

ACKNOWLEDGMENTS

The technical assistance of N. Belgiovine and D. Portaluppi is deeply appreciated. The statistical treatment of data is due to A. Bozzani and M. Petrillo.

REFERENCES

Benovic, A., S. Fonda Umani, A. Malej, and M. Specchi. 1984. Net zooplankton biomass of the Adriatic Sea. *Mar. Biol.* 70:209-218.

Bernhard, M., and A. Zattera. 1975. Major pollutants in the marine environment. In *Marine pollution and marine waste disposal*, ed. T.H. Pearson and E. Frangipane, 195-300. Oxford and New York.

Cambiaghi, M., G. Queirazza, G. Strobino, and M. Tomasino. 1987. Modeling the distribution of radionuclides in a Mediterranean coastal ecosystem. *Environmental Software* 2:207-220.

Dehairs, F., R. Chesselet, and J. Jedwab. 1980. Discrete suspended particles of barite and the barium cycle in the open ocean. *Earth Planet. Sci. Lett.* 49:528-550.

Fisher, N.S., C. Nolan, and W. Fowler. 1991. Assimilation of metals in marine copepods and its biogeochemical implications. *Marine Ecology Progress Series* 71:37-43.

Fowler, S.W. 1983. Radioecological aspects of deep-sea radioactive waste disposal. In: *Third National Symposium on Radioecology*, 22-23 June 1983, Bologna, Italy, pp. 237-258 Bologna University - AIRP.

Fowler, S.W. 1990. Critical review of selected heavy metal and chlorinated hydrocarbon concentrations in the marine environment. *Marine Environmental Research* 29:1-64.

Hardstedt-Romeo, M., and F. Laumond. 1980. Zinc, copper, and cadmium in zooplankton from the N.W. Mediterranean. *Marine Pollution Bulletin* 11(5): 133-138.

Laumond, F., G. Copin-Montegut, P. Courau, and E. Nicolas. 1984. Cadmium copper and lead in the Western Mediterranean Sea. *Marine Chemistry* 15:251-261.

Lovegrove, T. 1966. The determination of the dry weight of plankton and the effect of various factors on the values obtained. In: *Some contemporary studies in marine sciences*, ed. M. Barnes, pp. 429-467, London.

Marchetti, R. 1991. Quadro complessivo delle condizioni delle acque del fiume Po negli anni '90 e tendenze evolutive. In: *La qualita' delle acque del fiume Po negli anni '90*. Istituto di Ricerca sulle Acque - CNR, Quaderni, 92 : 16.1 - 16.26.

Margaleff, R. 1985. *Key environments: Western Mediterranean*. Pergamon Press.Martin, J.H., and G.A. Knauer. 1973. The elemental composition of plankton. Geochimica et Cosmochimica Acta 37: 1639-1653.

Queirazza, G. 1979. Problematiche analitiche nel dosaggio di tracce di elementi. *Giornale degli Igienisti Industriali* 4 (1-2).

Queirazza, G., L. Guzzi, G. Ciceri, and P. Frigieri. 1987. Environmental studies at pre-operational and operational stages of nuclear power plants in Italy: chemical and radioanalitycal implications. *Science Total Environment*, 64:191-207.

Romeo, M., and E. Nicolas. 1986. Cadmium, copper, lead and zinc in three species of planktonic crustaceans from the east coast of Corsica. *Marine Chemistry* 18:259-367.

Ruiz Pino, D.P., C. Jeandel, J.P. Bethoux, and J.F. Minster. 1990. Are the trace metal cycles balanced in the Mediterranean Sea? *Paleogeography, Paleoclimatology, Paleoecology (Global Planet. Change Sect.)* 82:369-388.

Santschi, P.H., and B.D. Honeyman. 1989. Radionuides in aquatic environments. *Radiation Physics Chemistry*, 34(2) : 213-240.

Scotto di Carlo, B., and A. Ianora. 1983. Standing stocks and species composition of Mediterranean zooplankton. In: *Quantitative analysis and simulation of Mediterranean coastal ecosystems: the Gulf of Naples, a case study*. UNESCO Rep. mar. Sci., 20:59-69.

Specchi, M., S. Fonda Umani, and G. Radini. 1979. Biomassa del plancton nel Golfo di trieste. *Atti Convegno Scientifico Progetto Nazionale Oceanografia CNR*, Roma: 1-12.

Tomasino, M., R. Ambrogi, and E. Ioannilli. 1993. Scientific approach for evaluating the sites of coastal thermoelectricpower station. *COSU III Conference*, March 30 - April 2, 1993, Santa Margherita Ligure, Italy.

Wangersky, P.J. 1986. Biological control of trace metal residence time and speciation: a review and synthesis. *Marine Chemistry*, 18:269-297.

20

Impacts of Marine Recreation in the Americas: How Much is Too Little - How Much is Too Much?

Dr. Don Walsh
President, International Maritime Incorporated
San Pedro, California

ABSTRACT

While many in the ocean industry do not recognize it as an important business sector, marine recreation is the fastest growing segment of ocean industry. There are many forms of marine recreation ranging in cost from simply visiting the beach to being captain of your own yacht. Essentially, all marine recreation takes place within the coastal ocean. This narrow band of the world ocean is also the place where the majority of the world's population lives. And the demographic reality is that this move towards coastal ocean space will continue and intensify. This will result in increased multiple use conflicts among and between the users of this zone. At some point, the use pressures will result in loss of activity quality for the recreational users. That is, the basic idea to have fun and to 're-create' will begin to be lost.

This paper briefly outlines ocean industry and recreation's place within it. Next, mostly using examples from the Americas, it reviews the various kinds of marine recreation activities, where they take place, and by whom. Finally, it discusses how loss of quality can occur and what can be done to ensure that marine-related recreation and leisure can remain a quality experience for the user.

INTRODUCTION AND BACKGROUND

Marine tourism is a relatively new and unknown *marine business* sector. For this reason, it will be helpful to first provide some general introduction and background prior to addressing the main topic of this paper. First, there is the relative size of marine recreation compared to other ocean industry sectors. And, second, there is the definition and scale of what is considered to be marine recreation. With these concepts defined, this paper will then develop its primary theme of how marine recreation can be both positive and negative factors in the uses of the coastal zone.

This conference is concerned with coastal ocean space worldwide. Because of its inherent importance as a 'user' of the coastal ocean, marine recreation must be a major consideration in any coastal planning and management process. In fact, it is often overlooked or is not considered as important as the more customary concerns of coastal industrial development, population expansion, and environmental pollution. The very small number of papers addressing marine recreation and leisure in this symposium are evidence of this assertion. To many planners, managers, and regulators, recreation is not a serious topic. This is too bad. It is hoped that this paper can help put a better focus and emphasis on this under-appreciated use of the sea.

North America, especially the United States and the Caribbean, are the largest marine recreation markets in the world. Thus, many of the examples provided in this paper are from these sectors. While the scale of activity is much larger than most other areas of the world, the operations, opportunities, and problems are universal; only the scale changes.

OCEAN INDUSTRY TODAY: RELATIVE IMPORTANCE OF THE SECTOR

In general, "uses of the sea" and "ocean industry" are the same thing. That is, deriving goods and/or services from the oceans that are of benefit to mankind. In all but the military sector, there is an economic test involved. Can the goods and/or services be produced that are price-competitive with terrestrial resources (i.e., oil and gas) or are they uniquely in demand (i.e., marine fish products)? National security activities at sea (navies, coast guards, marine police, etc.) are services that are for the national well-being, even though they fail the traditional economic test. The following major areas of ocean business are presented in order of overall economic activity.

MILITARY AND NATIONAL DEFENSE

There are over 135 coastal nations in the world, almost all of them have some type of navy, coast guard, or maritime police agency. This sector will always command the largest budgets in the international ocean industry. For example, the prototype of the U.S. Navy's new SEAWOLF class of nuclear submarine will cost about $5 billion. This is about the same as the entire U.S. annual budget for ocean science and technology research.

MARINE TRANSPORTATION

By tonnage, 98% of all the world's trade is carried by ships at some point between the producer and the consumer. As the world's population increases and obtains higher standards of living, the volume of marine transportation will also increase. This category also includes port and harbor developments, since modern marine transportation must be considered a transportation *system* of which the ship is only one element.

OFFSHORE OIL AND GAS

At present, about 22% of the world's energy comes from beneath the sea. Since the majority of the larger finds on land have been located and developed, the importance of subsea sources will become increasingly important. The major offshore developments in the U.S. Gulf of Mexico and the

North Sea have provided the scientific and technical know-how for these developments. The rate of development will be dependent upon the oil commodity market.

FISHERIES

On a global basis, about 14% of man's protein requirements are derived from consumption of seafood. However, it appears that the world catch is nearing its peak limits at about 100 million tons per year. Clearly, marine protein will never supply more than a small part of the world's protein demand.

SHIPBUILDING AND REPAIR

This is the industrial base required to support both the military and marine transportation sectors. However, the majority of shipbuilding activity is limited to about a half dozen nations. For more than a decade, Japan accounted for more than 50% of all commercial shipbuilding tonnage constructed, with Korea being in second place. Gradually the European shipbuilders have improved their percentages in recent years. As of September, 1990, the top six shipbuilding nations were: Japan (25.2), Korea (11.2), Germany (2.3 total of West and East Germany), Denmark (1.7), Yugoslavia (1.6), Italy (1.3), and Spain (1.3). The numbers in parentheses are millions of gross registered tons. It is instructive to note that Japan still accounts for more tonnage than the sum of the five other members.

MARINE RECREATION

As this paper will show, of all of the ocean industry sectors, marine recreation is the fastest growing. Even though it is one of the least organized and understood sectors, it is an area of great interest to investors, entrepreneurs, and coastal planners.

OCEAN MINING

Since the early 1970s, various international consortia and individual companies have spent in excess of one billion dollars to develop the science and technologies for ocean mining. Nevertheless, the international level of effort today is quite small. The reason for decreased levels of investment and effort are associated with the economics of ocean mining. At present, terrestrial sources of most of these minerals are widely available and at much lower costs. Ocean mining will be an important ocean industry sector, but this may not happen for ten to twenty years.

AQUACULTURE

At present, aquaculture accounts for about 10-12% of the production of man's aquatic food supply. This percentage is slowly increasing, however this sector would never be a major economic force in the overall world ocean industry.

MARINE RECREATION: THE FASTEST GROWING OCEAN BUSINESS

In some societies, the idea of its members seeking out recreational opportunities is not highly respected. Recreation is considered to be a waste of time and resources which could be better used in the workplace. Yet, in an increasingly complex world where the day-to-day stresses are compounded by a variety of pressures, the need to have leisure or 'play' time is more important than ever. As a popular phrase says, "you must take some time to smell the flowers".

As societies emerge from their developmental stages, their members gradually achieve increased free time and disposable income. That is, time that individuals can use for their own purposes and discretionary money to spend on those purposes.

Experts who have studied the impact of a balanced recreational regime in a person's life have found that these people are actually more effective and productive workers when on the job, are happier with their lives and workplace, and are ill less often. In other words, they are better workers than those people who rarely do anything else, because recreation provides their minds and bodies with necessary and vital breaks in life's routines. Truly, the term, "re-creation" seems self-defining.

THE WORLDWIDE LEISURE INDUSTRY

According to a 1987 study commissioned by American Express, the travel and leisure industry is the single largest employer in the world. Nearly 6% of the world's population is employed in this two trillion dollar-a-year industry. While these numbers are for both travel and leisure, the latter is by far the largest part.

Despite these very large numbers, the majority of the people in the world have not traveled very much beyond their home-towns. Thus, one can appreciate the future potential of this market as global travel becomes even more universal. Low cost means of mass transportation are opening up every corner of the world to travelers from all levels of the budget spectrum.

MARINE RECREATION DEFINED

Marine recreation is a very broad term which refers to anything from enjoying an ocean sunset/sunrise to a cruise ship voyage, to sport fishing on the sea or SCUBA diving under it. For the purposes of this paper we will define it as having two, broad subcategories: marine sports and marine tourism.

Marine sports represent the more active aspects of marine recreation. Some examples would be fishing, diving, surfing, and sailing. Pursuit of this form of marine leisure requires physical and active participation by the persons involved.

On the other hand, *marine leisure* implies a more passive form of recreational activity such as taking advantage of seaside resorts, cruise ships, marine parks, and tourist submarines.

It is believed that this categorization of the marine recreation market will help facilitate the relative activities in each of these subsectors. By defining marine recreation in terms of these two, broad, activities components, one can clearly prove that marine recreation can be pursued by people of all ages and with various physical capabilities. For example, modern cruise ships now have a number of cabins designed for physically handicapped passengers. Further, physical handicaps do not rule out marine sports, for there are special diver training programs to teach these individuals how to be safe and effective SCUBA divers.

For the investor and the entrepreneur the lesson is clear: marine recreation is a large market and it can be made larger by creative development of facilities and opportunities for the public.

THE EXPANDING LEISURE TIME MARKET

While the United States is currently the world's major marine recreation market, there are perhaps faster rates of growth in other areas of the world. For example, the economic unification of Europe in 1992 should result in the largest economic block in the world. If the predictions of greatly increased prosperity are realized, then there will be even more tourism by Europeans. Furthermore, the eventual political and economic integration of Europe will make all of its areas more accessible to transboundary tourism.

It is not only greater prosperity that will fuel the growth of leisure time, but also the increasing availability and flexibility of time for enjoying recreation. In many societies recreational time is carefully regimented either by work ethic,
national custom, or by business practices. This is often independent of whether an individual has the means to enjoy more extensive holidays. Often, an entire country will have certain weeks or months in which all citizens are encouraged to take their vacations. Offices, factories, and shops all will close. Government and business organize their work year around these national holiday periods. The individual has little flexibility about when he/she can take a vacation.

In addition to scheduling, there is also the problem of the work week. In some countries it is still expected that employees will work more than five-day, forty-hour weeks. A two-week vacation period is the norm and is rarely exceeded, no matter how long one has worked for the employer.

These limitations are gradually changing in the countries where they are still applied. The individual has more disposable income, more leisure time, and the flexibility to choose vacation periods. For example, the Japanese now are developing a leisure time outlook that is truly international. Until recently, this society was the model for the restrictions just discussed. Much of their vacation time was organized around specific national holiday periods and travel was often internal to Japan, since there was not enough time to go very far away.

The evolution of work rules and holiday periods, together with a sustained high value of the yen against most other currencies, allows the Japanese traveler to enjoy significant buying power and more are going abroad to do just that.

An example of their new mobility can be seen in the fact that over 80% of the passengers for the Guam, Saipan, and Rota Island tourist submarines are Japanese. Although two tourist submarine operations have been set up in Japan (Kyushu and Okinawa), travel within Japan tends to be expensive. It is almost cheaper to fly to Guam than to Okinawa, and once there, the yen buys a lot more in its dollar-based economy. Nevertheless, as late as 1987, fewer than 25% of the Japanese had gone abroad.

The other Asian "tigers", or more correctly, "newly industrialized nations" (NIC's) are South Korea, Taiwan, Thailand, and Singapore. All have robust economies which means that there will be an expanding segment of those populations which will have more disposable income and the time to spend it.

In other developed nations, larger numbers of people are joining the search for international leisure time activities. Pacing this evolution is continuing deregulation or rationalization of airline fares. This permits the traveler to move to resort areas throughout the world at a frequency unheard of twenty years ago. Additionally, the international travel business has always been highly competitive in chasing customers, and it is even more so today. For example, with careful shopping, a determined shopper can find a non-stop airline fare from Los Angeles to London which is equivalent to, or, in some instances, less than a fare to New York.

The convergence of all these trends means that patterns of tourist movements are changing, and heretofore remote places are now affordable for large numbers of people.

MARINE RECREATION: THE U.S. EXPERIENCE AS A MODEL

Marine recreation is international in scope, however for illustrative purposes we will use statistics and examples derived from the U.S. marine recreation market, the largest in the world. In the United States, water-oriented (fresh and saltwater) recreation, depending upon how you define the sector, has a gross economic value of $25-30 billion a year. This is a significant figure considering that the estimate for all types of recreation and leisure time activities is in the order of $52 billion a year. In other nations, these numbers may not be quite as large or cover the same range of activities, but the basic principles and issues are universal. Also, the U.S. experience can offer a useful model for marine recreation developers in other nations.

RECREATION AND THE CROWDED COASTAL ZONE

In the United States, approximately 80% of the population lives within 180 kilometers of a coastline (including the Great Lakes). Unlike many nations, the United States has a huge interior so people who live near the water are there because they choose to do so. Furthermore, demographic trends show that the number of people moving to the coastal zone will continue to increase at a rapid rate. Thus it is easy to understand that marine-oriented leisure time activities are a major growth area.

This is not unique to the United States either, for in most nations of the world the major population concentrations are within 200 km of coastlines. And these concentrations continue to grow. In other nations, such as Japan and Great Britain, geography has forced this outcome, since no one can live far from the sea. But it has not been until recently that national governments began to realized the importance of long-range coastal use planning.

The basic idea is that proximity to the sea means this also becomes the 'playground of choice' for recreation. The use of the 'open fields' of coastal ocean space overcomes the increasing population densities at the land-sea interface. It should be noted that this does not define recreational use for only the wealthy or for foreign visitors. In most parts of the world you will find local residents enjoying trips to the beach, using this part of the coastline as people would use inland parks. No great investment of time or money is required to have a family recreational experience by the sea. All that is required is for the local government to provide reasonable access and appropriate facilities for the recreational area.

Thus it is no surprise that multiple uses of coastal lands and the adjacent ocean space are interactive, overlapping, and often in conflict. Of the several sectors comprising marine and ocean industries, there are currently only three significant growth areas: port and harbor development and operations, seagoing national defense, and marine-oriented recreation.

BEACHES: THE FIRST LINE OF MARINE RECREATION

As noted above, going to the beach to enjoy swimming, surfing and sunning is one of the cheapest forms of recreation. However, beach areas are land-extensive, and it is often difficult to find sufficient room to develop proper beach facilities. This is less of a problem in developing nations. Most beaches do require maintenance in the form of trash cleanup, restoring sand lost to wave action, and parking areas close to the site.

An example of the pressure coastal populations can put on a beach is seen in southern California. On a warm, sunny day the number of people on the beaches in the three adjoining counties of Ventura, Los Angeles and Orange will be as many as two million people. If this number of people on the beach were made into a U.S. city, it would rank as the sixth largest in the United States.

DESTINATION RESORTS AND MARINE PARKS

A destination resort is a hotel complex which contains vacation resort amenities such as tennis courts, golf courses, and other leisure activities. It is a destination for the traveler rather than a place to stay while enroute from one point to another. Not all destination resorts are located next to the sea, however some of the most successful are oceanside resorts. These tend to be luxury hotel complexes and are designed to optimize the proximity to the sea. Most offer a wide variety of water sports and there have been some preliminary studies on having smaller (i.e., ten to sixteen passenger) tourist submarines operated by some of the larger resorts. At an average room cost of several hundred dollars a night, the marine-oriented destination resorts market to the upper level of the traveling public.

At the other end of the cost scale are the marine parks. The designs of these parks vary widely. Perhaps the less costly for the visitor is the government-owned and operated aquarium. Next up the scale are the commercial parks such as Sea World Park system in the United States. The five Sea World parks are owned by Anheuser Busch, the U.S. beer company, and the Florida Sea World Park is the third largest United States theme park in terms of public attendance, (4 million visitors a year), ranking after Disneyworld and Disneyland.

Another variant on resort hotels and theme parks are the underwater versions of these tourist facilities. The first underwater park was the John Pennycamp Underwater Park which was developed several years ago by the State of Florida in the Key Largo area. This is a public park and the diver-visitors follow underwater nature trails which are marked with waterproof signs.

The first, and only, operating underwater hotel is called "Jules Undersea Lodge" and it is also located in Key Largo, Florida. Opened in 1986, and located in a water depth of thirty feet, this underwater hotel provides space for eight guests. Large viewing ports provide a view of the ocean surrounding the hotel.

MARINAS: PATHS TO THE SEA

"Ma-ri-na: a small harbor with dockage, supplies, and services for small pleasure craft."

Increasing population densities and multiple uses of the coastal zone result in severe competition for available land. Yet all forms of marine recreation require public access to the shoreline. For many, the availability of an open beach for swimming or a pier for fishing is all that is required. However, for a much larger category of public access, there is a need to have a means to get across the land-sea interface.

Traditional marinas operated as the dictionary definition quoted above indicates. Public access was restricted to persons who had boats located there, boating was the only activity at the marina, and their locations were in parts of harbors that had no commercial shipping value. But this is now changing. Modern marina design not only provides the traditional support for their boating tenants, they are also places with a high degree of public access. People can come to these areas to watch the harbor and boating activities; small shops and restaurants offer a variety of goods and services; and walking and bicycle paths offer exercise opportunities for the visitor. The modern marina is also a place of education where one can learn about boating, sport diving, or take a trip to see marine life such as migrating whales. These facilities provide a true path between the land and sea for the ordinary citizen seeking marine recreation. Along the United States' coastline there are over 8,500 marinas, public dock facilities, and yacht clubs.

PLEASURE BOATING IN THE U.S.: A "POPULATION EXPLOSION"

In 1990, 73.4 million Americans participated in boating activities. The United States has over sixteen million leisure-time boats, and this 'boat population' is growing faster than our 'people population'. But this is not a wealthy man's sport. The average American leisure boat is only 5 meters long and has an outboard motor.

Virtually every marina in the United States has waiting lists for rental dock space. In southern California, marina occupancy rates typically run at 95% of capacity.

However, California is not the largest boating state in the United States. Michigan tops the list with 857,000 boats, and California is second with 753,000. Other major boating states are: Florida (711,000), Minnesota (706,000), Texas (604,000), Wisconsin (482,000), New York (420,000), Ohio (380,000), Illinois (340,000), and South Carolina (302,000). In fact, the top twenty U.S. coastal states each have a boat population exceeding 200,000 crafts.

Thus it's no surprise that recreational boating is a $13.7 billion a year business. This sector also accounts for about $800 million a year in boating-related imports.

With respect to boating imports, one of the fastest growing segments is in "personal watercraft"—the jetskis that were developed in Japan. There are currently over 100,000 in the United States, and the makers and sellers of these watercraft even have their own trade association, the Personal Watercraft Association.

SPORT FISHING FOR FOOD AND FUN

If America's sport fishing fleet were considered part of the nation's commercial fishing fleet, it would rank third in terms of boats and "landed value" of the catch. In 1989 seventeen million recreational fishermen took sixty million fishing trips to catch 469.2 million pounds of fish. Despite all this activity, the average fishing vessel is less than six meters in length, and most fishing effort takes place within six kilometers of the shore. The economic value of this marine 'fishing industry' is in excess of $7.5 billion a year.

CRUISE SHIPS: FAST GROWTH AND DIVERSIFICATION

People like to 'run away to sea', however the perception is that cruise ships are very expensive compared to air travel and land based vacations. This is not correct, but it is an image problem that the cruise ship industry is attempting to address through its advertising programs.

In fact, in 1990 nearly four million people took cruises, producing a revenue of nearly $6 billion for this industry. Industry experts estimate that the full potential of the world cruise ship market to be in the order of $60-70 billion a year.

The at-sea experience of being on a cruise ship usually gives passengers a special feeling of closeness with the sea. To encourage learning more about the marine environment, several of the cruise ship companies carry lecturers on board who tell the passengers about the sea.

Cruising is enjoying a rate of growth in the order of 11% per year, however most of the economic activity is confined to the Caribbean market where the majority of the customers are from North America. The Miami area (including Fort Lauderdale) has the largest passenger terminals in the world. To meet this customer demand there currently are thirty-three cruise ship vessels under construction for delivery by the end of 1994. This will provide a 50% increase in passenger capacity in for the world's cruise ship fleet.

The North American geographic region makes up about 83% of the world cruise ship industry. The truth is, most people who have the disposable time and funds have not taken a cruise. Furthermore, this community of potential customers is growing as more people are able to enjoy upscale marine leisure activities.

An interesting characteristic of this industry is its diversification into related areas. Modern cruise ship operators offer their passengers a large 'menu' of choices associated with their cruise. Shore tours, visits to remote tropical islands, diving expeditions, etc., are some examples of activities that are owned by the ship operators. In the area of tourist submarines, it is known that at least two cruise ship companies have considered operating them as a tie-in with their ship itineraries. This will undoubtedly happen before long. Underwater recreation, whether it is snorkeling, SCUBA, or tourist submarines, is an ideal port activity for the cruise ship passenger who has 'run away to sea'.

UNDERWATER RECREATION: SOMETHING FOR EVERYONE

The sea is an opaque medium; a person standing at the seashore cannot see into it for more than a few meters of depth. But almost all want to know what's there and to see it first- hand. As a consequence, many different means have been developed to permit man to visit the 'inside' of the ocean. Aquariums on land essentially put a cube of the sea in a room where we may look through windows at the marine life and its activities. At some seaside areas, underwater viewing chambers which are accessible from the shore, provide a direct viewing experience. For the more daring tourist, the use of swim fins, facemask, and snorkel permit one to go into the sea for a very limited time and distance. However the ultimate *in situ* experience is through the use of SCUBA. As noted earlier, with this underwater breathing equipment, the sport diver can visit inside the ocean for periods ranging from a few minutes to over an hour, depending upon depth, air capacity, and the divers' experience.

While diving is the best way to experience the inside of the ocean, for safe enjoyment, it requires regular physical conditioning, formal training courses, and a reasonable investment in equipment. In addition, some people are simply uneasy at being 'alone' in the sea; they worry about sharks, snakes, and other hazards (even though all are improbable) that would surround them. Unrealistic fears or not, it is not a casual recreational activity. Only a small fraction of the ocean-visiting tourists will do this as a regular recreational pursuit.

The tourist submarine business will attract leisure time participants who wish to sample to ocean experience without the rigors of having to learn diving or the dangers (largely fictitious) perceived to be part of diving. It is a relatively benign way to take millions of people beneath the sea for the first time and, hopefully, for many more visits. The safety and comfort of modern tourist submarine designs insure a pleasant and safe visit to the world of 'innerspace' for the average person. Also there is considerable public confidence in a passenger carrying system where both the operating personnel and the vessel must meet several sets of strict standards imposed by government, classification societies, and the insurance market.

TOURIST-IN-THE SEA: SPORT DIVING IN THE U.S.

One means to put the tourist into the sea is through use of diving technique and equipment. At virtually all oceanside resorts throughout the world, diving equipment and instruction are available to the tourist. Sport diving includes both snorkeling, which is free (breatholding) diving from the surface and use of SCUBA (self-contained underwater breathing apparatus) equipment. In the first case, diving is limited to how long you can hold your breath and generally to depths of not more than 10 meters. With SCUBA, the diver's theoretical limit is the amount of air in his tanks and time required to decompress. As a practical matter, most responsible authorities and training courses advise the SCUBA-equipped sportdiver to not exceed 40-45 meters.

In the United States, sportdiving has been relatively safe. Furthermore, SCUBA diving has been self-regulated by the diving community. Generally, before equipment can be rented, the renter must produce a certification card showing that he/she has successfully passed one of the nationally-recognized training courses. This is not the case in many foreign countries and increased accident rates are the result.

Sportdiving is a fast growing sector in marine recreation. It is estimated that at present there are about two million SCUBA divers in the United States. This is the largest recreational diver population of any country in the world. The market consequences are formidable. The cost to equip the average SCUBA diver will be in the range of $2,000-3,000. In order to get to good dive sites a person must travel considerable distances to coastal areas where hotel, meals, and dive boat charters add to the divers' recreational costs. It is not a low-budget sport.

Packaged charters, with all-inclusive fees, are another major trend in this area. This is becoming a very big business in sport diving. At the recent SCUBA 90 conference and exhibition in Long Beach (June 1990), at least 15% of the booths in the exhibit area were travel companies offering packaged diver tours to exotic locations. These tie-in, pre-sold packages should be an important part of tourist submarine marketing. In the past few years, specialized, larger cruise ships have been put into service which cater to the avid sport diver.

TOURIST SUBMARINES: A NEW WAY TO GO UNDER THE SEA

Constructed for the 1964 Swiss National Fair, the AUGUSTE PICCARD "mesoscaphe" (PX-8) was the first designed-for-the-purpose tourist submarine. It was built at Monthey, Switzerland in 1963. In one year it carried nearly 34,000 tourists, in groups of forty and a crew of four, on dives into Lake Geneva (Lac Leman) to depths of 220 meters. The maximum rated operating depth of this submarine was 610 meters. This is still the deepest diving of any tourist submarine ever put into service.

Unfortunately, AUGUSTE PICCARD was an idea ahead of its time. After the Fair, quite a bit of effort was devoted to continuing its operations as a tourist submarine, but it never again carried tourist passengers.

In 1983, Sub Aquatics Development Corporation (now Atlantis Submarines International) was founded in Vancouver, Canada, to design, build, and operate the first built-for-the-purpose tourist submarines since the AUGUSTE PICCARD nearly two decades earlier. By mid-1987, the profit potential and customer acceptance of the ATLANTIS submarines was evident and several other companies were organized to enter this new business sector. By the end of 1988, there were eleven, built-for-the-purpose submarines in service. At the end of 1990, there are about twelve companies which were designing and/or building tourist submarines. How many of these companies will succeed is another question. At present there are about 20 tourist submarines in service with another ten under construction/conversion. These submarines are operating in over a dozen countries throughout the world and on every continent except Australia.

By October, 1992, ten ATLANTIS-class submarines had carried two million paying passengers safely beneath the sea at a modest price of $60-85, depending upon the location. By the end of 1991, there were four Atlantis class submarines operating in Hawaii alone. The most active Atlantis location was Guam, where in 1990, nearly 90,000 passengers enjoyed submarine rides at $85 for a one hour trip. Over 80% of these passengers were Japanese tourists visiting Guam.

THE QUESTION OF CARRYING CAPACITY

While tourism is not a resource use in that no removal of a living (i.e., fishing) or non-living (i.e., oil and gas) resource takes place, the very act of observing something can often adversely affect that which is being seen. This a paradox understood by scientists, especially those in life sciences. But the principle also applies to international tourism. That is, how many people at one time can visit an historic or nature site without affecting the thing they have come to see? And at what point is the balance between tourist and locals tipped, so that the quality of life at the site is adversely impacted?

The primary issue is one of "carrying capacity" for any given recreational site. That is, how many visitors can be sustained at a level of activity which provides them with a quality experience while minimizing disruption for the host region?

The carrying capacity problem suggests that the best remedy would be to limit access of tourism to those numbers who can be sustained at the visit site. In recent years, many natural and historical sites have established restricted access, or have been closed entirely, when the presence of too many people has had a damaging effect. The caves of Lascaux in France, with their grand prehistoric paintings, were closed when damage occurred from moisture (i.e., breathing) brought in by visitors. An example of limited access is Ecuador's Galapagos Islands National Park. However, most actions of this sort are not undertaken until actual damage had taken place. There should be a better means to anticipate problems and take preventative actions through effective management of tourism.

Access restrictions have been successfully applied to sharply-defined areas such as a single historical site or a nature park. For larger areas, such as entire coastal regions, the problem is much more difficult, even though the need might be just as great. For many popular coastal areas, tourism is the chief economic factor and primary creator of employment. There will be considerable reluctance to reduce this source of income, even though the quality of the local environment is being damaged by an overpopulation of visitors.

Even if local tourism authorities do not wish to restrict access, the perception of exceeding carrying capacity may come instead from the visitors themselves. From their point of view, if a place is seen as overcrowded and unattractive, then they will not come. This is true even if they are involuntary visitors such as when the cruise ships stops in the port. They simply will not go ashore. Last year the QUEEN ELIZABETH 2 (QE2, with a passenger capacity of over 1,900) stopped in one large port in the Mediterranean. It was a port with a poor reputation as a place to visit and the cruise ship pier was an older, commercial dock in a remote area of the port. Half the passengers did not want to go ashore, so it is unlikely that Cunard, QE 2's owners, will schedule the ship here again. True or not, it was the passenger's perception that this was not a good place to visit.

In the immediate future, it will probably be the companies that sell tourism that will determine the trade for a given area. While there is great reluctance to give up well advertised, traditionally popular tourism locations, this will be done if their customers complain about their experiences. Again, this will be done after problems have arisen, rather than anticipating them.

Changes by the tourism industry can be made relatively quickly. A recent example was the 1991 Mid-East war and its affect on Mediterranean tourism. It was a short war, but almost the entire tourism season of six to eight months was badly damaged by mass cancellations of cruise and packaged tours. If a location gets a bad reputation, then sometimes it will be extremely difficult to get back into the game. There are too many competitive sites waiting to fill the gap.

In general, both host population and visitors can benefit if equitable schemes could be developed to regulate the number of visitors that could be accepted at any one time. In this way, the visitor would have a high quality experience and the local people would be able to keep intact the local amenities that originally made the site attractive for tourism.

EXAMPLE OF TOO MUCH: MARINE RECREATION IN THE CARIBBEAN

For more than three decades, the Caribbean has been the number one marine recreation region in the world. Florida's southeast coast, primarily the Miami area, is the principal gateway to this region. And it is here that the majority of the world's cruise ship trade originates. Whether the visitor arrives by plane or by ship, Caribbean destinations are closely spaced and most have excellent facilities for planes and ships. There is also the advantage of being able to visit several countries while on a brief vacation of three to seven days. In the Caribbean, places are close together and easy to reach. As we will see, this is both an advantage and a problem.

While there are distinct tourism seasons in the Caribbean, there is tourism throughout the year. Compared to other locations in the world, this region has a large, year-round supply of leisure business. This 'steady-state' characteristic is fairly unique and permits more economically efficient operations. This has resulted in consistently better vacation bargains for the tourists.

Modern, jumbo jet aircraft and mega-cruise ships have become efficient delivery systems to bring the tourists into this region. Trends in modern transportation development indicate that larger aircraft and ships will be employed in the future. Travel will be even more affordable and attractive to a broad range of the recreation-seeking public.

It is clear that the Caribbean will maintain its preeminent role for many years to come. However, there are already signs that many of the popular areas are reaching saturation in terms of how visitors can be accommodated. For a variety of reasons, which we will address later in this paper, there are capacity limits for virtually all tourism destinations.

COASTAL TOURISM IN DEVELOPING COUNTRIES

Tourism is especially important to developing economies. Many nations now undergoing economic expansion and development find that they have many historical and natural sites that visitors want to see. Frequently, these sites are relatively unspoiled when located in areas where there is a small local population and previously only a few foreign visitors. Furthermore, tourism is a quick means to create a lot of employment for relatively untrained local people. And once the visitor flow begins, there are added benefits of tourist spending for out-of-hotel meals, shopping, and services.

Development of planning strategies to determine the tradeoff between new economic activity and possible degradation of visitor sites is extremely difficult. To voluntarily cut back on economic activity in the interest of some future goal is perhaps more theoretical than practical. Yet, at some point this must be done, otherwise 'market forces' will begin to work when the quality of the recreational experience begins to deteriorate. Careful planning and tradeoff studies can develop plans and means to permit a sustained level of tourism that will provide optimum economic benefits for a prolonged period of time.

This discussion does not suggest that carrying capacity is not a universal problem. Major urban areas on the coast are, of course, capable of absorbing large quantities of visitors with minimal impact on the local environment. But as marine-related recreation moves to areas away from the major population centers, the question of how many tourists is too few and how many are too many becomes very important. Super-efficient aircraft and ships can literally submerge tourism areas with a sea of visitors. Before-the-fact planning can help anticipate the problems and put plans in place that will result in the least amount of economic dislocation.

Good models exist on how to limit access, provide substitutes for closed sites, and efficiently handle visitor throughput for historic and nature sites. These can be extrapolated to larger coastal regions. This will ensure that such areas can maintain and sustain significant economic activity with minimal disruption of the use of favorite recreational areas. However, planning after the damage has been done can result in chasing the problem with inadequate solutions and the adverse economic impacts may be permanent.

FUTURE DIRECTIONS FOR MARINE RECREATION

Marine recreation is a major growth industry, however, despite its size, the general literature on marine recreation is sparse compared to the scale of its economic activity. Remarkably, when one reviews the abundant literature on coastal zone planning and development, or ports and harbors, there is relatively little said about marine recreation. On paper this sector doesn't seem to exist as a major use activity contributing significant revenues to these areas, one which involves a significant number of the population, and one which often conflicts with other uses.

This is not a trivial problem and there's little evidence to indicate that the traditional ocean planning process is coming to grips with the question. For this reason, it is difficult to find help and expertise to solve problems arising in this area.

Meanwhile the ships and aircraft will continue to arrive, delivering visitors each of whom is hoping to have a quality recreational experience. It is the obligation of the tourism authorities in both business and government to ensure this happens. If it does not, then the tourist will go elsewhere. It is easy to do in today's environment of plentiful, cheap, travel and abundant information on the world's tourism sites.

21

Integrated Coastal Management (ICM) -- An Idea Whose Time Has Come

Dr. J.R. Schubel
Dean and Director, Marine Sciences Research Center
The University at Stony Brook, Stony Brook, NY

Jeanne Gulnick
Marine Sciences Research Center
The University at Stony Brook, Stony Brook, NY

Alessandra Conversi
Marine Sciences Research Center
The University at Stony Brook, Stony Brook, NY

ABSTRACT

The concept of Integrated Coastal Management (ICM) is not new, but it is experiencing a renaissance. From the recent Earth Summit in Rio to the National Research Council's Water Science and Technology Board's report on Wastewater Management in Urban Coastal Areas, scientists, engineers, managers, and environmentalists are extolling the virtues of ICM and calling for its application. We too are firm believers in the potential power of ICM in arresting further degradation of coastal environments and in rehabilitating degraded coastal systems. If we are to exploit the renewed interest in a water quality-based approach to coastal environmental management, we need to work together to demonstrate in a compelling way, the power of ICM. It remains a seductively simple and elegant concept that has little impact on how we do business. We will offer some suggestions for changing that.

BACKGROUND

Roughly 50% of the population in the United States and throughout the world lives within 100 miles of the coast. The numbers and the percentages are increasing, and with them, the impacts of society on the coastal ocean. And the worst is yet to come. Approximately 95% of the world's projected population growth will come in developing countries, much of it in coastal regions where there is virtually no infrastructure to deal with the burgeoning amounts of human, municipal, and industrial wastes. Many of the human wastes will be discharged raw, directly into coastal environments. Valuable natural resources will be damaged or destroyed. According to Ray (1988), "the coastal zone is being altered just as fast as tropical forests." What are the major threats to coastal environments that are causing these undesirable alterations?

A recent report of the Joint Group of Experts on Scientific Aspects of Marine Pollution (GESAMP) (1991) pointed out that "while man's fingerprints are found throughout the World Ocean, the open ocean is still relatively clean, but there are serious problems in the Coastal Ocean." The report adds:

> In contrast to the open ocean, the margins of the sea are affected almost everywhere by man, and encroachment on coastal areas continues worldwide. Irreplaceable habitats are being lost to the construction of harbors and industrial installations, to the development of tourist facilities and mariculture, and to the growth of settlements and cities... If left unchecked, this will soon lead to global deterioration of the marine environment and of its living resources.

GESAMP summarized the major problems of the World Ocean as:

1. Nutrient contamination
2. Microbial contamination of seafood
3. Disposal of debris (particularly plastic debris)
4. Occurrence of synthetic organic compounds in sediments and in predators at the top of the marine food chain
5. Oil in marine systems, mainly the global impact of tar balls on beaches and the effects in local sheltered areas
6. Trace contaminants such as lead, cadmium, and mercury when discharged in high concentrations.

They added that radioactive contamination is a <u>public</u> concern. They did not consider items 5 and 6 to be particularly important globally. In the summary of their findings, they stated:

> We conclude that, at the start of the 1990s, the major causes of immediate concern in the environment on a global basis are coastal development and the attendant destruction of habitats, eutrophication, microbial contamination of seafood and beaches, fouling of the seas by plastic litter, progressive build-up of chlorinated hydrocarbons, especially in the tropics and sub-tropics and accumulation of tar balls on beaches.
>
> ... not enough attention is being given to the consequences of coastal development,
> ... actions on land continue to be taken and executed without regard to consequences in coastal waters.

The Joint Group of Experts on Scientific Aspects of Marine Pollution added:

> The exploitation of the coast is largely a reflection of population increase, accelerating urbanization, greater affluence and faster transport -- trends that will continue throughout the world. Controlling coastal development and protecting habitat will require changes in planning both inland and on the coast, often involving painful social and political choices.

The GESAMP assessment is a global assessment of the entire world ocean, including its coastal component. It's clear that their concern for the future of the world ocean is concentrated on the threats to the margins. The major threats to the world's coastal ocean are also the major threats to United States' coastal ocean.

Each year the twenty-three coastal states, jurisdictions, and interstate commissions of the United States are required to submit to the U. S. Environmental Protection Agency, reports on the condition of their *estuaries* and the reasons for their failure to meet designated uses. In the most recent State Section 305(b) reports to the U.S. Environmental Protection Agency, the twenty-three coastal states, jurisdictions, and interstate commissions reported that

1. Nutrients accounted for 50%[1] of the total impaired[2] area of estuaries.
2. Pathogens accounted for 48% of the total impaired area.
3. Organic enrichment/low dissolved oxygen accounted for 29% of the total impaired area.

It is clear that the problems of the U.S. coastal ocean and the causes of those problems, are similar to those of the coastal ocean of most of the rest of the world. The first order problems are eutrophication, pathogens, and habitat destruction. All are caused primarily by an increasing population, by their waste management practices, and by changing land use patterns. It is also clear that even in the United States, in spite of large expenditures on environmental protection, many of the impacts of society on the coastal ocean are expanding and intensifying. What were once local problems are becoming regional problems.

As the GESAMP report points out, protecting coastal environments requires planning and management not only on the coast, but inland as well. For some estuaries, such as Chesapeake Bay, that planning and management must extend throughout much of the watershed. In some cases it may even have to extend beyond the watershed to the airshed. In other coastal systems, such as Long Island Sound, the area of terrestrial influence is concentrated in the near-coastal zone. For each coastal system, the zone of influence of human activities needs to be identified and to become the basic planning and management unit.

Current wastewater and stormwater management policies and practices in the United States are rooted in the 1972 amendments to the Federal Water Pollution Control Act, reauthorized in 1977 as the Clean Water Act. The 1972 act provided substantial amounts of federal grant support to assist municipalities in improving their wastewater management practices. The Act produced an extensive effort which resulted in significant environmental improvements, particularly in rivers and lakes. In many coastal marine environments, however, the benefits were less clear.

The Clean Water Act required the establishment of uniform, minimum federal standards for municipal and industrial wastewater treatment. The Act did not adequately address regional variations in environmental systems and the interactions of those natural systems with their human systems. In some areas it prescribed a greater level of protection than was required; in others, it prescribed less than was required. The standard for municipal wastewater treatment, secondary treatment, is a technology-based standard. A technology-based standard is a standard based on the capacity of an existing technology as opposed to a performance standard based upon receiving water requirements.

The Clean Water Act of 1972 included a provision to recognize regional differences by allowing coastal dischargers to apply for waivers that would exempt them from the technology-based requirement, if they could demonstrate that their treatment and disposal practices provided adequate protection of the environment. Dischargers who were granted waivers were required to implement source controls and monitoring programs that went far beyond those required for dischargers who chose to go to full secondary treatment.

The period for initial waiver application expired on 31 December 1982. More than a decade later, a number of applications are still pending, and approximately forty dischargers are operating under waivers. When one considers that there are more than 1,400 municipal wastewater treatment plants in the U.S. that serve coastal communities, it's clear that for whatever reasons, the waiver option resulted in few exceptions to the minimum, technology-based standard -- secondary treatment.

There are a number of reasons to move away from technology-based standards to a water quality-based approach to coastal environmental management; and in particular to moving toward an integrated, coastal management strategy. The use of technology-based standards is a simplistic approach to an increasingly complex problem. In many situations it requires too much in the way of protection,

---

[1] The percentages total more than 100% because more than one stressor contributes to impairment of an area.

[2] Impairment of an area is defined as failure of that area to fully support designated activities.

and in many others it requires too little. It is not an efficient use of fiscal resources, short-term or long-term. Specifying a technology in law discourages investment by venture capitalists in environmental technologies and inhibits innovation in developing new environmental technologies. But most important, in the case of secondary treatment, specifying a single technology-based standard has failed to provide consistently a level of protection needed to conserve healthy coastal environments, or to restore degraded coastal environments.

One frequently hears that public policy is a "blunt instrument". Making secondary treatment the law of the land is a good illustration of just how "blunt" it can be. The increase in population in coastal regions, the changing mix of land use activities, increased environmental awareness, demands for cleaner, more diverse coastal environments, and significant advances in scientific understanding of coastal processes and our ability to model them, all converge to argue for a new approach to coastal environmental management. More of the same approach won't even keep pace with increasing pressures on coastal environments and the resulting deterioration, let alone rehabilitate already degraded environments.

In business, they talk about S-curves which are plots of "benefits" on the ordinate against "investment" on the abscissa. So long as one is on the steep part of the S-curve, an additional investment results in a good return. But when one moves up onto the shoulder of the curve, the benefits of additional investment are small, zero, or even negative. An example that is frequently used to illustrate this principal is National Cash Register, the manufacturer of the classic mechanical cash registers. While National continued to make further investments in perfecting their mechanical machines, a number of other manufacturers were developing electronic cash registers. National continued to lose market share. In such situations, the only way to make a significant improvement is to jump to a new S-curve. In coastal marine environmental management, we have moved up onto the shoulder. The only way to achieve a significant benefit from the next round of investments is to jump to a new S-curve. We need a new coastal management paradigm. The paradigm that appears to have the greatest promise is Integrated Coastal Management (ICM).

INTEGRATED COASTAL MANAGEMENT

What is ICM?

Although many of the basic tenets of ICM have existed for well over 30 years, the name is new. The most compelling argument for ICM and the most complete description of the approach is outlined in a National Research Council report "Managing Wastewater In Coastal Urban Areas" (1993). The underlying goal of ICM is to switch from the current technology-based approach to a water quality approach. If successful, this approach should provide greater environmental protection (in some cases, at lower cost) and could stimulate the development and implementation of improved, more effective technologies.

Integrated Coastal Management is a holistic strategy. It starts with the coastal water body and its terrestrial zone of influence (typically the drainage basin) as an integrated, interactive system (Schubel 1992). The cornerstone of the concept is the integration of programs and plans for economic development and environmental quality management; including, integration of cross-sectoral plans for fisheries, energy, transportation, shipping, waste disposal, habitat protection, water quality, and tourism. ICM requires a rigorous cross-disciplinary pursuit. Input from scientists, engineers, economists, sociologists, political scientists, lawyers, and the public is critical. Finally, ICM must integrate the responsibilities for management actions across all relevant levels of jurisdiction, which may include international, national, state, and local. It must also integrate the views of stakeholders from the public and private sectors.

The principle objectives of ICM are two-fold: to maintain, or restore, the integrity of coastal ecosystems and to maintain important human values and uses associated with those coastal resources.

The first step in the integrated coastal management process is to set environmental quality goals for a coastal waterbody. These goals should be expressed in terms of water and sediment quality and other environmentally-based parameters. To accomplish this, existing information and data should be used to identify critical ecosystem values and functions that are to be conserved or, if necessary, restored. Also, human expectations of uses and benefits that are to be derived from the particular water body must be identified explicitly. Broad consultation with stakeholders through a series of public hearings or workshops is an effective method of establishing the uses and values important to society (Schubel et al. 1990). In most coastal areas, there are multiple values and uses important to society, and many are in conflict. As a result, they must be ranked. Identification of important ecosystem values and functions and the environmental qualities needed to ensure them, and identifications of human uses and values important to society, are essential steps in creating goals for the future of a coastal waterbody.

After appropriate goals are set, the next step is to identify the problems and activities, which pose the greatest threat to attaining and sustaining those goals. The identification of the chief threats – their sources and strengths – defines the geographic area of concern, *the zone of influence*. The zone of influence for any waterbody will have aquatic, terrestrial, and atmospheric components. Each component must be realistically defined to take into account all of the threats that have a measurable impact on important societal values and uses and on ecosystem integrity. The zone should be the minimum zone needed to provide an adequate level of protection to ensure important values and uses. If the zone is smaller than necessary, management strategies, even integrated management strategies, will be ineffective. If the zone is larger than necessary, it will lead to an inefficient or perhaps an ineffective use of resources.

Other critical steps in the integrated management process are (1) the assessment and comparison of ecosystem risks and (2) the assessment and comparison of management options to reduce those risks. These steps are required to ensure that whatever fiscal resources are available for management are directed at the causes (sources) of priority problems, i.e., those sources that are amenable to control and which if controlled, will produce predictable and significant reductions in ecosystem risk and in risk to values and uses important to society.

Management alternatives should be based on a combination of the best science and the most appropriate technology. One effective way of bringing new insights into the selection of management strategies to address persistent problems that have eluded solution is to convene a workshop. Not just any workshop, but one which brings together experts on a rich mix of different aspects of a problem – its causes and consequences – along with a few creative problem solvers from entirely different fields. This "view from the balcony" sometimes leads to fresh insights and, as a result, to reformulation and reframing of an issue that make it more amenable to an effective attack.

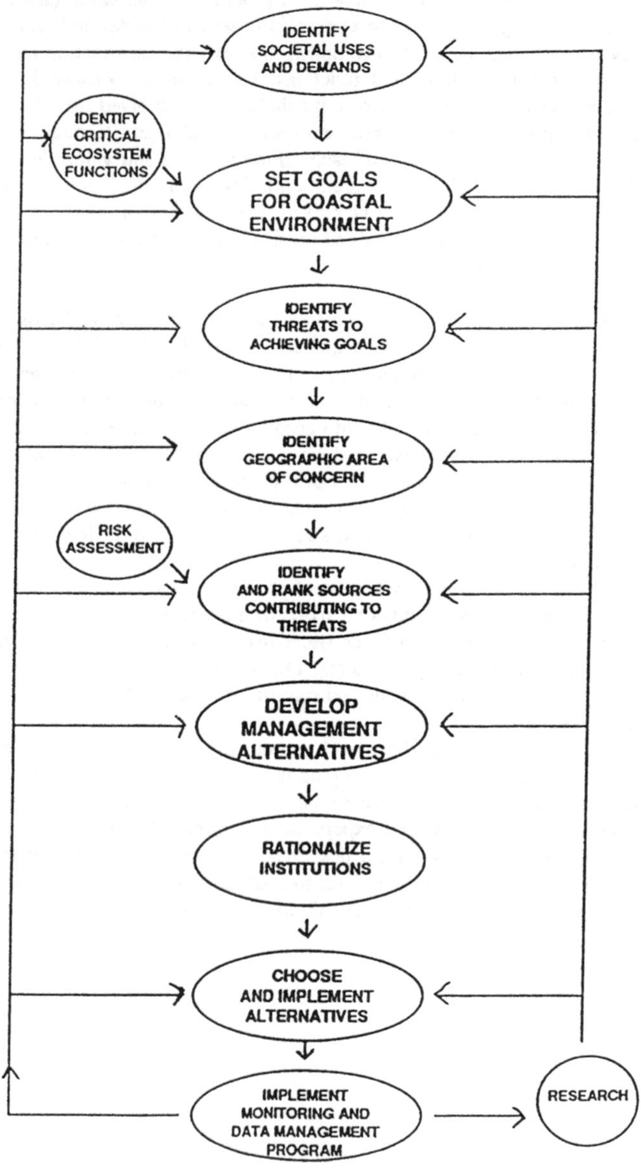

Figure 1. A flowchart to the ICM approach

The ultimate success of ICM depends upon achieving and sustaining a level of institutional coordination and integration that has been achieved only rarely in environmental management. The basic unit of management in ICM is the waterbody and its zone of influence. In most cases, this area will cut across a number of jurisdictional boundaries. In order for management alternatives to be implemented and for feedback from monitoring programs to be used to create new, more effective alternatives, a high level of institutional coordination may be required. One approach is to create a new institution and charge it with responsibility and authority for the integrated coastal management process, providing it with the resources needed. The resistance to such a move may be strong and reliance on existing institutions may be necessary. In this case, an explicit and formal allocation of responsibilities will be required and an interactive network of communication and coordination established among existing institutions to ensure the long-term success of any integrated coastal management plan.

The institutional arrangements must include a long-term, comprehensive, environmental monitoring program to assess the efficacy of the integrated management program. Also, monitoring is useful in identifying new areas of research and in providing information to the public. The results of the monitoring and research programs provide the information needed to refine and enhance the integrated coastal management plan.

ICM is a dynamic process that incorporates the best and most recent science and that generates a steady flow of new information to improve the plan and the program (Figure 1). ICM is a flexible, adaptable, continuous process that changes as societal values and demands on a particular coastal area change, as new threats emerge to reaching societal goals and as new information and technologies develop. Therefore, strong institutional and political support for the planning process is essential. To paraphrase Winston Churchill "In ICM the plan is nothing; planning is everything."

Applications of ICM

ICM is an elegant concept. Will it become more than a concept or will it suffer the same fate as previous water quality-based approaches? Does the National Research Council report describe the process in sufficient detail and with sufficient clarity that the paradigm will be used by groups who are typically charged with the task of managing complex coastal environments? Are the arguments sufficiently compelling to overcome the bureaucratic hysteresis and bureaucratic hysteria even the mention of ICM will provoke?

The only way to answer these questions is to test the paradigm in a number of coastal areas. To date, ICM has not been applied, in all its dimensions to any coastal region. It should be tested on a number of different kinds of coastal environments. A good approach would be to start with existing data and information and to push it to the limit to see how far one can go. This should be done in workshops that bring together representatives of all the stakeholder groups. A second approach is to secure funding for a comprehensive, multiyear study of a coastal system with the objective of developing an integrated coastal management plan. The U.S. Environmental Protection Agency's National Estuary Program may provide such an opportunity. There is another – Mamala Bay.

Mamala Bay extends along the southern shore of Oahu (Figure 2), the most populous of the Hawaiian Islands. Water quality is of utmost importance in Mamala Bay. Tourism, which is responsible for approximately half of all income on the island, hinges on clean, clear, safe coastal waters and a healthy ecosystem. Degradation of existing water quality would be economically devastating. The Bay is used extensively for recreational purposes by both residents and tourists for swimming, surfing, boating, and fishing. If these uses were compromised, tourism would not continue at its current level.

In 1991, the Sierra Club Legal Defense Fund filed suit against the City and County of Honolulu for being in violation of the Clean Water Act of 1972 by failing to provide full, secondary treatment at its Sand Island Treatment Plant which discharges sewage from a deep water outfall into Mamala Bay. The lawsuit resulted in a Consent Decree ordering the formation of the Mamala Bay Commission. One of the three Commissioners is an author of this paper (Schubel). The charter that established the Commission charges it with the responsibility for designing and overseeing a research program for Mamala Bay, a program that will generate the information necessary for the identification and assessment of management alternatives to ensure ecosystem integrity of the Bay. The Commission decided to carry out the study in the context of Integrated Coastal Management.

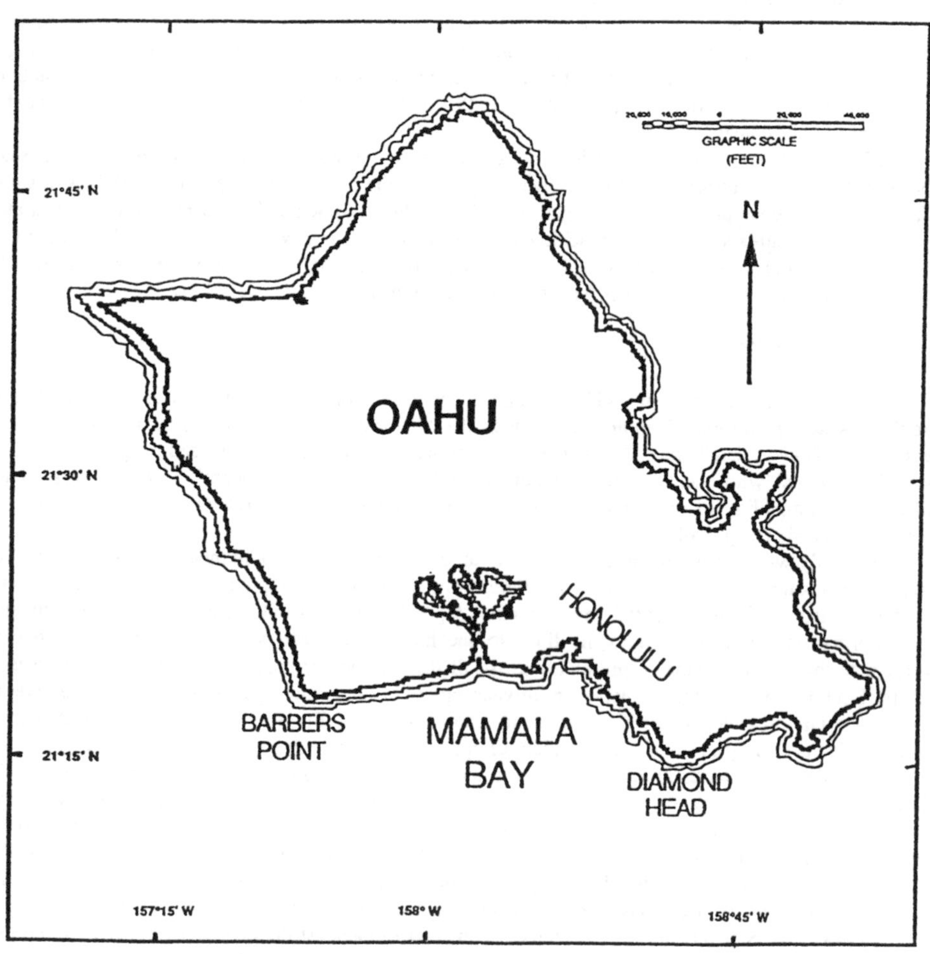

Figure 2. A site to test ICM: Mamala Bay, Hawaii.

A preliminary review of existing information indicates that a principal threat, perhaps the principal threat, to important uses and values of Mamala Bay is contamination of popular bathing areas by pathogens and the assumed human health risk. This threat to one of the most important uses of Mamala Bay has been documented by the episodically high number of indicator organisms off bathing beaches throughout the Bay. A secondary threat may be nutrient enrichment, which potentially could lead to ecosystem degradation, particularly of coral reefs.

The Mamala Bay Study Commission consulted extensively with experts to formulate a comprehensive and flexible study plan that should provide the information needed to develop an integrated coastal management plan. A series of requests for proposals (RFPs) were issued. Each made it clear that any contractor selected would become part of an integrated coastal management team and would be required to take an active part in integrating the results of their projects into a coherent management plan. The first meeting of contractors was held in Honolulu in July 1993. Another critical step that must be taken is the early involvement of local environmental management agencies in an exploration of institutional models to carry out an ICM plan.

The Mamala Bay Study is in its very early stages. Because of its restricted geographic area of influence, its relatively simple jurisdictional regime (the entire area falls within the jurisdiction of one state and one county), and because of the relatively small range and intensity of problems threatening it, Mamala Bay provides an excellent opportunity to test, refine, and enhance the ICM paradigm.

The former, great, U.S. baseball player and homespun philosopher, Yogi Berra, once mused that "when you come to a fork in the road, take it." In management of our coastal marine environments, we have come to a fork in the road. Let's take it.

ACKNOWLEDGMENTS

We thank Joanne Cosgrove, Gina Anzalone, and particularly Maureen Flynn for their patience and help in preparing this paper.

REFERENCES

Joint Group of Experts of the Scientific Aspects of Marine Pollution (GESAMP). 1991. *The state of the marine environment*. Blackwell Scientific Publication, Oxford Press.

National Research Council. 1993. *Managing wastewater in coastal urban areas*. National Academy Press.

Ray, G.C. 1988. Ecological diversity in coastal zones and oceans. In *Biodiversity*, ed. E.O. Wilson. National Academy Press.

Schubel, J.R. 1992. *The need for new research modalities in waste management*, Waste Management Research Report. Vol. 4, No. 3. News from SUNY at Buffalo and Stony Brook, and Cornell University.

Schubel, J. R., W.M. Eichbaum, and S.E. Schubel. 1990. Uses and values for the Delaware Estuary in 2020. Working Paper 43, Reference No, 90-8. Marine Sciences Research Center, SUNY at Stony Brook.

22

IFREMER's Activities in Environmental Monitoring

Jean Jarry
Director
IFREMER Mediterranean Center, Toulon, France

INTRODUCTION

IFREMER, the French Research Institute for the Exploitation of the Oceans, is a public organization created in 1984 by merging CNEXO (The National Center for the Exploitation of the Oceans) and ISTPM (The French Marine Fisheries Institute). IFREMER is sponsored by the French Ministry of Research and also gets directives from two other Ministries: the Ministry of the Sea and the Ministry of the Environment.

IFREMER's goal is to enhance the knowledge about almost all the aspects of the oceans and to help human activities connected with the oceans. IFREMER's organization follows a thematic approach, and has four main directorates:

1. *Oceanic sciences* with laboratories in marine geology, geophysics, geochemistry, physical oceanography, and deep organisms biology.
2. *Marine living resources* with laboratories in fish behavior and stock evaluation, fishing techniques, and fish and shellfish farming.
3. *Ocean engineering* with laboratories in manned and unmanned submersibles, robotics, underwater acoustics, oceanographic instruments, and offshore engineering.
4. *Environmental coastal studies*. This Directorate was created two years ago by combining laboratories which dealt with environmental problems, either in the living resources domain or in oceanic research. About 210 people work in Environmental coastal studies, in research laboratories, coastal laboratories, and services.

COSTAL ENVIRONMENTAL DIRECTORATE

Research Laboratories

Environmental research stimulates multidisciplinary approaches. Thematic laboratories are thus necessary for better efficiency. However, there is no border between them and they undertake many cooperative programs and projects.

The seven research laboratories are:
1. Natural cycles chemistry and modeling laboratory, in Brest
2. Pollutants chemistry and modeling laboratory, in Toulon, Nantes and Brest
3. Hydrodynamics and sedimentology laboratory in Brest
4. Ecology laboratory in Brest, with three smaller teams specializing in benthic ecology, pelagic ecology, and coastal ponds ecology, this last one in Sète, along the Thau Lagoon
5. Microbiology laboratory in Brest
6. Ecotoxicology laboratory in Nantes
7. Phycotoxines and nuisances laboratory in Nantes

Coastal Laboratories

Well-integrated in the coastal areas, coastal laboratories act as regional observatories of the coastal environment. Their main concern is to monitor water and resources quality, then to advise, and transfer information to administrative and local authorities, as well as to shellfish farmers. They also have to answer the questions of local communities regarding water quality, living resources quality, and coastal management.

There are twelve coastal laboratories scattered along the French coasts; three on the British Channel, six on the Atlantic Ocean, and three on the Mediterranean Sea.

Services

Besides research and coastal laboratories, the Coastal Environment Directorate includes three services in charge of collecting information from different geographical sectors and synthetizing it for people or agencies who need it. The three services correspond to three themes: water quality, living resources quality, and local developments and advices.

RESEARCH PROGRAMS

Major National Programs

IFREMER is involved in national programs in close cooperation with other French marine research agencies such as the National Center for Scientific Research (CNRS), Overseas Scientific and Technical Research Office (ORSTOM), National Weather Service, etc.

The two most important national programs in coastal environmental research are:
1. The National Toxic Algae Efflorescence Program. Launched in 1989, it includes two research axes dealing with determinism and proliferation of toxic microalgae, and production mechanisms and activities of toxic substances,
2. The National Coastal Oceanographic Program. Its goal is to answer two questions:
   a) What is the role of coastal ecosystems in the larger biochemical cycles of the oceans?
   b) How is it possible to make compatible the quality of the environment and the management of its resources and space?

Other Research Programs

Other research projects deal with all problems connected to the coastal environment:
- Relations between biological production and materials flow
- Mechanisms of toxic algae

- Hydrodynamics and sedimentology
- Exchanges between sediments and seawater
- Evolution of anthropic pollutants in seawater, particles, and sediments
- Loire Estuary eutrophization
- Seine Estuary pollution
- Dissolved oxygen assessment in Thau lagoon (OXYTHAU program)

A new research program, called "Aquaculture-Environment", is in the funding process. On several defined sites (on Atlantic and Mediterranean coasts as well), this new program will try to understand two, complementary problems:

1. Effect of water pollution on the farmed species
2. Effect of farms on water pollution induced by sea farming: animal feces, food debris, etc.

APPLIED RESEARCH PROGRAMS

Introduction

Coastal laboratories have in their missions, by using the results of research programs, to find solutions to actual local or regional problems.

In this framework, IFREMER is responsible for the animation of three national networks which monitor the quality of coastal waters.

Environmental Quality Networks

National Monitoring Network (RNO)

Established in 1974, the National Monitoring Network (RNO) measures the amount of chemical pollutants in seawater. Along the French shores, there are forty-three sampling sites, fourteen of which are located on the Mediterranean Sea). Monitored chemicals are heavy metals (Cd, Cr, Hg, Pb, Zn), DDT, PCB's, and polyaromatic hydrocarbons. (Figures 1, 2)

Phytoplanktonic Network (REPHY)

The Phytoplanktonic Monitoring Network (REPHY) measures the amount of toxic microalgae in seawater. These species are not dangerous for shellfish but can induce diseases in people who eat them.

REPHY was set in France in 1984 and its main goals are:

1. In terms of environment:
   a) Collect data on phytoplankton populations
   b) Provide very close monitoring of discolored waters and other exceptional blooms
2. In terms of protection of public health:
   a) Detect and monitor the development of species toxic to humans, such as dinophysis, Alexandrium, etc.
3. In terms of protection of marine animals health:
   a) Detect and monitor the development of species toxic to fish and shellfish, such as *Gyrodinium*.

REPHY includes 110 sampling stations, 37 for monitoring, and 73 for warning.

Routine surveys and warning system organization are as follows:

1. Routine surveys on monitoring stations only:
   a) Winter (September to April): seawater sample taken twice a month with systematic counting of all phytoplankton species

b) Summer (May to August): seawater sample taken once a week with systematic counting of all phytoplankton species done twice a month and counts of toxic species done on the remaining samples.

Phytoplanctonic blooms, are unfortunately, quite frequent; between 1975 and 1990, 278 blooms have been observed along the Channel, Atlantic, and Mediterranean French Coasts.

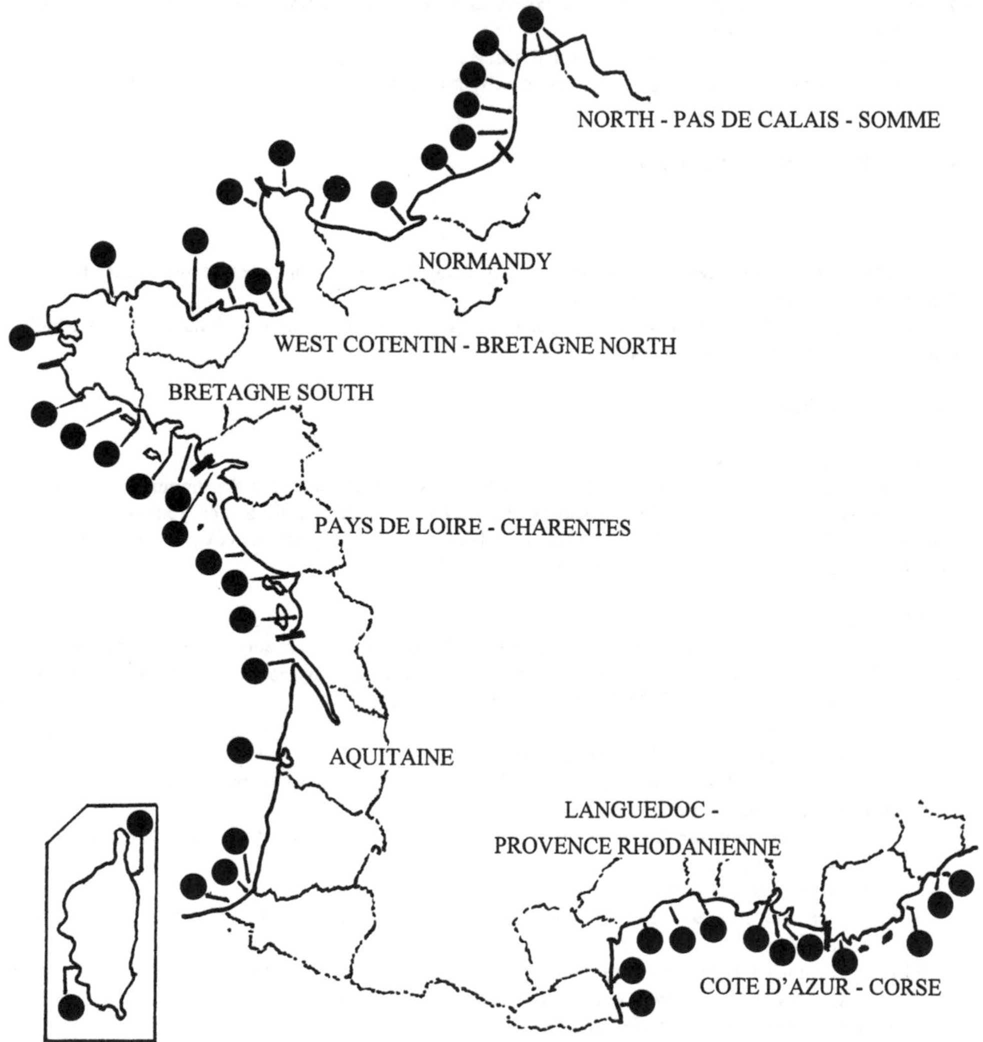

Figure 1. Marine environmental quality National Observation Network Monitoring locations

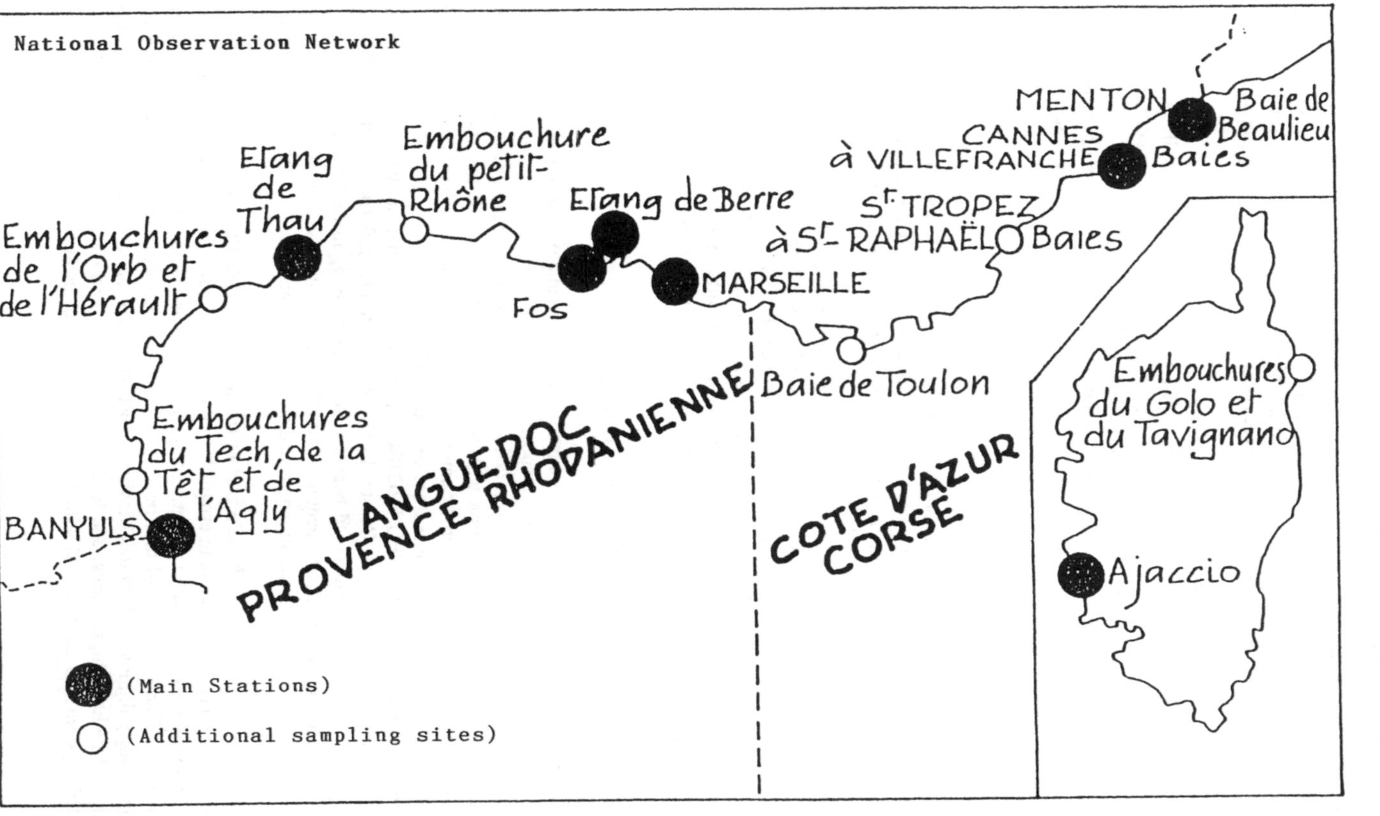

Figure 2. RNO sampling areas along the French Mediterranean coast

Microbiology Network (REMI)

The Microbiology Monitoring Network (REMI) detects pathogenic germs, dangerous for humans when they are ingested. Like REPHY network, REMI is organized in routine survey and warning systems. Set in 1989, REMI includes 345 survey stations and 278 sampling stations.

Economical Consequences

In case of DSP (Diarrheic Shellfish Poisoning) or PDP (Paralyzing Shellfish Poisoning) toxicity, networks measurements are transmitted to local governmental local who must decide whether or not to issue a temporary interdiction on shellfish harvesting and trading.

Economical consequences are important since, each time, it is a loss of income for the infested farms. Between 1984 and 1989, an average of 400 farms have been yearly concerned, with a peak of 1,313 in 1987.

Specific studies

In cooperation with many university laboratories, IFREMER scientists study anomalies in the coastal zones such as proliferation of large algae (ulvae), toxic phytoplankton, and ecology of "new" species, recently introduced in an area. Examples are "*Sargassum muticum*", "*Undaria pinnitifida*" and, since last year, "*Caulerpa taxifolia*".

This last species had such an impact on the media that IFREMER devoted a special effort to it, coordinating the actions of many laboratories. IFREMER itself undertook the cartogaphy of the areas infested by *Caulerpa*, with two campaigns at sea per year to monitor growth and extension of the areas colonized by this algae. IFREMER also undertook the study of several chemical, physical, or mechanical processes to eradicate Caulerpa and protect the other species such as posidoniae.

These actions are a part of IFREMER's "public service" mission.

COASTAL ENVIRONMENTAL TECHNOLOGY

For their research, scientists need sophisticated instrumentation, sometimes available on the shelf, sometimes specially developed to solve a specific problem.

Engineers of the Engineering Directorate are responsible for designing such instrumentation, which will be developed in close cooperation with industry. As examples of specific and new instruments developed for the needs of coastal environmental research, let us cite: a laser nephelometer, sensors for measuring nitrates, chlorophyll, plankton, to name only a few.

The objectives of the RAVEL PROJECT (Reseau Automatis de Veille pour l'Environnement Littoral - Automated Network for Coastal Environment Monitoring) are to design, build, and operate a network of buoys and instrumented beacons for measuring significant parameters of the quality of seawater. Data thus automatically acquired would be transmitted via satellite to a landbased facility where they should be processed, stored, and distributed to the users. In 1993, a prototype buoy will be built and tested.

In the future, RAVEL could be a part of the MEDARCOBLEU project which has been proposed by Italy and France. The goal is to establish in the coming years, an integrated system of information, data transmission storage, and exploitation to give the authorities, in real time, necessary and sufficient information. These leaders will thus be able to make the best decisions to prevent major accidents or to fight major pollutions and protect Mediterranean waters and shores in the most efficacious way. This system will include ships, buoys, and land based receiving and retransmitting stations.

This system will have to deal with chronical as well as accidental pollution.

If this program is funded, the first feasibility studies will start in 1993. The four partners who work in close cooperation are, on the Italian side, CNR and SELENIA, and on the French side, IFREMER and Thomson-Sintra.

In other technological domains, IFREMER's goals can be defined in three words: measurement, modelization, and protection. We have to understand what are the actual risks and corresponding stakes, and thus define technologies necessary to prevent and protect.

Projects underway in 1993 reflect these goals.

1. Currents and waves modelization in coastal areas
2. Conceptual studies of shore protection devices
3. Improvement of used, urban waters sea disposals

CONCLUSION

IFREMER's coastal environmental laboratories play an important role in the study of coastal phenomena. Scientists monitor all the parameters of the marine environment, especially on the Mediterranean coast. Using research results, coastal laboratories help local communities and people to solve local, environmental problems. They thus contribute to the general protection of our beautiful but fragile Mediterranean Sea.

23

Prediction of Aquatic Biotoxin Potential in Fish and Shellfish Harvesting Areas: Ciguatera and Diarrheic Shellfish Poisoning

Douglas L. Park
Professor, Department of Nutrition and Food Science, University of Arizona
Tucson, Arizona
and Hawaii Chemtect International, Pasadena, California

ABSTRACT

Ciguatera (CTX) and diarrheic shellfish poisoning (DSP) toxins are odorless, tasteless, and generally undetectable by any simple chemical test: therefore, bioassays have been used traditionally to monitor suspect fish and shellfish. Programs designed to provide some degree of assurance that foods susceptible to natural toxicant contamination are wholesome and safe to eat require several facets, including marketplace screening of suspect foods for identification of contaminated product; separation of adulterated product to less risk uses; and, where feasible, development of systems or monitoring programs designed to predict potential hazardous food production/harvesting areas.

With respect to public health concerns, prevention of CTX or DSP toxin production would be, of course, the best strategy; however, because of its complexity this approach is not feasible. The next alternative would be a monitoring program for screening marketplace seafoods as well as harvesting areas. In order for an analytical method to be of value in screening marketplace seafoods, it must meet the following criteria: (1) facile to use and interpret; (2) rapid, i.e., able to test a large number of samples in a short period of time; (3) accurately differentiate between toxic and non-toxic product; (4) low cost; (5) available in sufficient quantity to meet private, industrial, and regulatory agency testing demands; and (6) where feasible, provide for a means of confirmation of identity for toxic products.

The solid-phase immunobead assay (S-PIA, Ciguatect™), currently available from HawaiiChemtect International, has the highest potential for application to marketplace screening of fish products for CTX and DSP toxicity. Testing fish or shellfish early in the plan is recommended, since this will minimize the cost expended for the product and potential economic losses to industry. The kit can be used onboard fishing vessels, at receiving docks, processing plants, distribution organizations, retail outlets, consumers, and regulatory agencies. The self-contained assay is available as a single analysis kit designed for non-laboratory use by untrained personnel. Organizations conducting large numbers of analyses would be more inclined to use the laboratory kit which contains sufficient material for fifty tests.

In conclusion, a seafood safety monitoring program would entail a testing program which includes: (1) monitoring fish and shellfish harvesting areas to determine toxic potentials; (2) developing a sampling plan applicable to screening fish and shellfish in the market place for CTX or DSP toxicity potential; (3) screen fish at various points in commercial channels for CTX or DSP toxicity potential, i.e., onboard fishing vessels, at receiving docks, processing plants, distribution organizations, retail outlets, consumers, and regulatory agencies; (4) if desired, re-analysis of products testing positive using alternative extracting and analytical methods; and (5) diverting toxic products to lower risk uses. The S-PIA method (Ciguatect™) can be used to monitor reef fishing areas for ciguatera potential, shellfish beds for DSP toxins, shellfish depuration operations for elimination of DSP toxins, and screen for toxic fish and shellfish in the market place.

INTRODUCTION

Ciguatera fish poisoning is an illness endemic to tropical and subtropical areas which has existed for centuries (World Health Organization 1984). Humans are exposed to the toxins through the consumption of fish which have accumulated toxins produced by microscopic algae, i.e., dinoflagellates. It has been estimated that between 50,000-500,000 cases of ciguatera food poisoning occur each year (Ragelis 1984). Although outbreaks occur primarily in tropical regions of the world, due to interregional shipment of seafoods, outbreaks have been reported in the continental United States and Canada. Symptoms of the disease are gastrointestinal, neurological, and cardiovascular and can persist for weeks and even years (Juranovic and Park 1991). Because of the severity of the ciguatera phenomenon, public health agencies have been striving to implement a seafood safety program for years.

Public health research activities with respect to phycotoxin contamination, i.e. ciguatera (CTX) or DSP, have focused on the protection of human health and the enhancement of commerce of subtropical reef fish. This goal can only be realized, however, by having a control program for effectively removing toxic fish or shellfish from the marketplace. In order to understand how to set up an effective seafood safety monitoring program, it is necessary to understand how the products become toxic and have the analytical tool to identify high risk products. Unfortunately, due to the lack of adequate standards and past absence of rapid screening methods for monitoring the presence or absence of these toxins in fish at various points before they reach the consumer, this goal has not yet been attained. Rapid, reliable analytical methods are crucial to effective seafood safety monitoring programs. Historically, but not today, methods for analysis for CTX and DSP toxins have been labor-intensive and time-consuming, and lack specificity (Juranovic and Park 1991).

TOXIN PRODUCING ALGAE AND PRINCIPAL TOXINS FOR CTX AND DSP

Ciguatera toxins accumulate in fish feeding of macro- and micro-algae and they will eventually pass up the food chain to man. Benthic, toxigenic dinoflagellate species implicated in ciguatera poisoning include *Gambierdiscus toxicus*; *Prorocentrum* sp. (*P. lima, P. concavum, P. emarginatum, P. mexicanum, P. rathynum*); *Amphidinium carterae*; *Ostreopis* sp. (*O. ovata, O. siamensis, O. lenticula*); *Coolia monotis*; *Scrippsiella subsalsa*; and *Thecadinium* sp.

Several toxins may be responsible for ciguatera. The primary toxin, ciguatoxin, has been isolated from large carnivores and in smaller amounts in herbivores. An explanation for this could be that CTX accumulates preferentially in large carnivores due to its greater lipid solubility. Considerable circumstantial evidence has linked *G. toxicus* to this toxin; Murata et al. (1990) reported the structures of ciguatoxin from the moray eel (*Gymnothorax javanicus*) and its likely precursor from the dinoflagellate, *Gambierdiscus toxicus*. The congener was shown to be a less oxygenated analog of ciguatoxin. However, it has not yet been conclusively demonstrated that the toxin produced by the dinoflagellate is the precursor to ciguatoxin(s) accumulating in fish. Until suitable detection methods for these toxins are developed, it will be difficult to determine toxin properties. There are at least five toxins involved in ciguatera named ciguatoxin (CTX), maitotoxin (MTX), scaritoxin (STX), okadaic acid (OA), and a recently named toxin, prorocentrolid (Bagnis et al. 1974; Chungue et al. 1977; Tachibana 1980; Tindall et al. 1984; Yasumoto et al. 1971; Yasumoto and Murata 1988b; Yasumoto and Murata 1988a; Yasumoto and Scheuer 1969). Recent studies suggest that an excess of twenty toxins may be involved in the ciguatera phenomenon (Juranovic et al. 1994; Park et al. 1994; LeGrande 1991).

Diarrheic shellfish poisoning (DSP) has been associated primarily with shellfish harvested from temperate regions of the world such as Japan and Europe (World Health Organization 1984). Again as with ciguatera, the toxins are produced by dinoflagellates; however, different species are involved. These microscopic algae occur naturally in the ocean environment either associated with near-shore environments or free swimming. *Dinophysis* sp. (*D. acuta, D. acuminata, D. caudata, D. fortii, D. mitra, D. rotundata, and D. tripos*) are the principal dinoflagellates associated with blooms responsible for DSP (Moulin et al. 1992).

Okadaic acid, dinophysistoxin-1, and several analogs are the principal toxins responsible for DSP poisoning outbreaks (Murata et al. 1982). Okadaic acid is involved in both DSP and CTX poisoning outbreaks (Gamboa et al. 1992; Park et al. 1992c). However, for CTX, ciguatoxin, maitotoxin, and their analogs are the toxins with the highest toxic potentials. The toxic potential of okadaic acid is several orders of magnitude lower than ciguatoxin. DSP is associated primarily with shellfish (mussels, scallops, etc.) in temperate regions of the world, whereas CTX is associated with finfish from tropical regions.

ANALYTICAL METHODOLOGY

The format of analytical methods for phycotoxins vary according to the application or purpose of the method, i.e., screening, confirmation of identity, reference, etc. For ciguatera, bioassays have been used traditionally to monitor suspect fish. Most earlier methods were based on biological endpoints which had major limitations on levels of detection and specificity. Many native tests for toxicity in fish have been examined, including discoloration of silver coins or copper wire, the repulsion of flies or ants, and rubbing the liver on the gums to ascertain if it causes a tingling feeling (Juranovic and Park 1991). But all of these, with the possible exception of rubbing the liver on the sensitive tissues of the mouth (Lewis 1986), have proven invalid. As more reference material and standards became available, the emergence of chemical and immunochemical methods became apparent.

The mouse assay has been used traditionally; however, it involves a time-consuming process of obtaining the lipid-soluble extracts and it lacks specificity (Yasumoto et al. 1971). Other major disadvantages include costs associated with maintaining a mouse colony, death time is subjective, and the relationship of death time to dose is nonlinear. The method consists of injecting serially diluted, semipurified or crude toxic extracts into mice (usually intraperitoneal [IP]) and observing the symptoms for twenty-four to forty-eight hours. The results are expressed in mouse units where one mouse unit is identified as the amount of toxin that kills a 20 g mouse in twenty-four hours (World Health Organization 1984). This assay is unsuitable as a market test. The mouse assay gave reproducible results following IP injections of toxic fish or dinoflagellate extracts (Hoffman et al. 1983; Gamboa and Park 1985; Park et al. 1994; Sawyer et al. 1984; Gamboa et al. 1992). The measurement of rectal temperature of the animals immediately before the administration of the extracts and periodically for sixteen or forty-eight hours showed a pronounced drop in the body temperature of the mice following exposure to the toxins (Park et al. 1994, Gamboa et al. 1992; Gamboa and Park 1985; Hoffman et al. 1983; McMillan et al. 1980). A symptomatological analysis has been prepared to facilitate the comparison between ciguatera research reports (Hoffman et al. 1983). A dose-response curve was prepared for purified ciguatoxin obtained from toxic blackfin snapper collected from the Virgin Islands. The mouse bioassay has been used extensively in the Pacific and is described in detail by Yasumoto et al. (1984). An in-depth review on the evolution of methodology for assessing aquatic product safety has been prepared (Park, 1994).

An innovative, rapid, solid-phase immunobead assay (S-PIA, Ciguatect™) for the detection of toxins associated with CTX and DSP outbreaks has been developed (Park and Goldsmith 1991; Park et al. 1992a and 1992c). The presence or absence of the toxins is determined by binding the toxins to a membrane attached to a plastic strip and exposing the toxin-ladened membrane to a monoclonal, antibody-colored, latex bead complex which has a high specificity for the toxins of interest. The intensity of the color on the membrane denotes the presence of the toxins. CTX and DSP toxicity potential can be determined directly on edible tissue or following specific extraction procedures. The method has been used to evaluate CTX potential in fish obtained from Hawaii, Australia, and the Caribbean (Park et al. 1992a) and DSP potential in mussels collected from Denmark and France (Park et al. 1992c). This study confirmed the presence of okadaic acid and related DSP toxins in mussels implicated in a DSP poisoning outbreak (Denmark) and mussel depuration operations (France) (Fremy et al. 1994).

For those products where additional testing is desired, possibly for samples testing positive, the University of Arizona and HawaiiChemtect International have developed a rapid extraction method (REM™) capable of extracting and partial purification of toxins associated with ciguatera poisoning in less than thirty minutes (Park et al. 1992b). Toxins are extracted with a chloroform:water:methanol mixture and partitioned into selected phases by varying polarity. When the REM™ is used to extract and purify toxic components, the limit of detection for the Ciguatect™ test kit is <0.05 ng. Also, at this point chemical methods based on thin layer (TLC) and high performance liquid chromatography (HPLC) technology can be used to confirm the presence of individual toxins.

Methods based on TLC and HPLC have been developed for selected, individual toxins associated with CTX and DSP. These methods can be applied as a regulatory tool where sophisticated laboratory facilities are available. HPLC techniques have been applied to the analysis of okadaic acid in fish tissue (Yasumoto 1985; Lee et al. 1987; Dickey et al. 1990). Since okadaic acid is the principal toxin associated with DSP, this methodology has been applied to shellfish (Stabell et al. 1991; Lee et al. 1989). Park and co-workers (personal communication) have developed an TLC method for okadaic acid in fish tissue and dinoflagellate cultures. Specificity of this methodology is obtained following exhaustive purification of toxins extracted from fish tissue. These methods, although not suitable for routine screening programs, would play an important role in confirming the presence of individual toxins in fish products.

HPLC methodology have been reported for ciguatoxin and several analogs (Murata et al. 1990; Lewis et al. 1991). These studies reported four major ciguatoxins. LeGrand and co-workers (LeGrand et al. 1990; LeGrand 1991) used HPLC methodology to isolate multiple ciguatera toxins from wild *Gambierdiscus toxicus* and toxic herbivorous and carnivorous fish.

METHOD VALIDATION

Any method intended to be used in a seafood safety monitoring program must be validated through an interlaboratory study to determine the precision and accuracy parameters of the method. Method validation programs are administered by AOAC International (AOAC) and International Union for Pure and Applied Chemistry (IUPAC). The precision of the solid-phase immunobead assay (Ciguatect™) to detect toxins associated with ciguatera poisoning has been evaluated through analysis of toxic and non-toxic fish fillets (amberjack, surgeon, and parrot fish) and REM™ extracts obtained from fishing areas around the Hawaiian Islands. The precision of the assay has been evaluated through the AOAC/IUPAC interlaboratory validation mechanism (Park et al. 1992a, 1992b). Analysis of toxic and nontoxic amberjack, surgeon, and parrot fish flesh and extracts showed acceptable repeatability and reproducibility parameters (Table 1). University of Arizona, Food and Drug Administration (FDA) and National Marine Fisheries Service (NMFS) laboratories participated in the study. The study confirmed excellent performance of the method and interpretation of the results, and demonstrated acceptable precision parameters (Park et al. 1992a, 1992b).

Table 1

Precision parameters of collaborative data for solid phase immunobead assay (Ciguatect™) determination of ciguatoxins and related polyether compounds.

Species	Mean	S_r	S_R	RSD_r(%)	RSD_R(%)
			Fish Fillets		
Parrot Fish (*Scarus* sp.)	1.2	0.16	0.53	13.5	44.4
Surgeon Fish (*Ctenochaetus* sp.)	1.7	0.15	0.50	9.0	29.7
Amberjack (*Caranx* sp.)	3.6	0.15	0.51	4.3	14.3
			REM Extracts		
Parrot Fish (*Scarus* sp.)	3.1	0.18	0.37	5.8	11.9
Surgeon Fish (*Ctenochaetus* sp.)	3.8	0.18	0.38	4.8	9.9
Amberjack (*Caranx* sp.)	4.9	0.18	0.37	3.7	7.6

SEAFOOD SAFETY MONITORING PROGRAMS

An effective food safety monitoring program is comprised of primarily three components: (1) monitoring fish and shellfish harvesting areas for CTX and DSP potential, (2) establishment of regulatory limits, and (3) a screening program for testing fish products in commercial channels designed to test a large number of samples in a short period of time so that toxic or high-risk products can be separated from healthful foods. Violative or unacceptable product can then, if desired, be subjected to additional testing to confirm the presence and identity of the toxin(s).

Monitoring Fish Harvesting Areas

A key aspect of an effective seafood safety monitoring program, where feasible, is the establishment of a program for predicting and identifying high-risk fishing areas. This is accomplished through the testing of marine specimens endemic to fish harvesting locations. This will vary whether for CTX or DSP. Since seafoods commonly associated with ciguatera poisoning outbreaks come from highly mobile fish, collection and testing the fish alone could provide incorrect information, both false positive and false negative predictions. This possible problem can be corrected by identifying and testing a specimen or bio-marker of limited motility, i.e., invertebrates endemic to the fishing area.

The Ciguatect™ solid-phase immunobead assay has been used to screen thirty-six species of near-shore invertebrates off the coast of the Island of Hawaii for ciguatoxin and related polyether compounds (Figure 1). Specimens included snails, sea urchins, sea cucumbers, crabs, brittle stars, bivalves, and zoanthids. Invertebrates were collected at six "toxic" locations having documented history of ciguatera fish poisoning along the Kona coast, and at three "nontoxic" sites along the Hamakua coast where there had been only one reported case of ciguatera since 1980.

A significant, positive correlation between assay results and site-specific ciguatera history was found for the cowry *Cypraea maculifera* (Figure 2). While assay results for most other species indicated very low or no ciguatoxin present, cone snails (*Conus* spp.), ophiuroids (*Ophiocoma* spp.), and sea cucumbers (*Holothuria* spp.) frequently tested positive. There was no correlation, however, for these three genera between assay results and site history. These results suggest that invertebrates, particularly grazers and deposit feeders, and especially cowries, accumulate ciguatoxins and related polyether compounds at sites known for ciguatera fish poisoning outbreaks and have the potential utility of being a bioindicators of reef toxicity. This marine specimen or another invertebrate native to the area under study, could be an integral part of the ciguatera monitoring program.

This assay has been used to test for DSP toxicity potential in mussels implicated in a DSP poisoning outbreak in Denmark and DSP depuration monitoring studies along the Atlantic coastline of France (Figure 3; Table 2) (Park et al. 1992c; Fremy *et al*. 1994). These study results demonstrate that the assay could identify DSP-contaminated shellfish as well as serve as a useful tool to monitor shellfish on-shore or, during depuration operations, determine at what point shellfish can enter the market following a *Dinophysis* sp. bloom.

Establishment of Regulatory Limits

An important aspect of any food safety monitoring program is the establishment of regulatory limits designed to assure wholesomeness of the food supply. Animal toxicological and human clinical data are crucial information needed for the determination of this value. Since multiple toxins are involved with CTX and DSP poisoning outbreaks, it is not practical to use a single compound for this regulatory limit. Historically, the establishment of a seafood safety monitoring program for ciguatera has been hampered by the lack of reference standards. At the present time, okadaic acid is the only toxin associated with both CTX and DSP poisoning outbreaks in sufficient quantities to serve as a reference standard. Therefore, the term okadaic acid equivalents (OAE) could be used in the establishment of regulatory limits, being aware of the relative potencies of other toxins involved with the CTX or DSP phenomenon. Again, the term OAE is used because multiple toxins are involved in the poisoning outbreaks.

Screen Fish in the Marketplace/Commercial Channels

With respect to public health concerns, prevention of CTX or DSP toxin formation in the environment would be, of course, the best strategy; however, because of the etiology on how toxins accumulate in fish and shellfish in the ocean environment, this approach is not possible. Therefore, screening the product in the marketplace becomes necessary. In order for a method to be of value in screening marketplace seafoods, it must meet the following criteria: (1) facile to use and interpretation; (2) rapid, i.e., able to test a large number of samples in a short period of time; (3) accurately differentiate between toxic and nontoxic samples; (4) low cost; (5) available in sufficient quantity to meet private, industrial, and regulatory agency testing demands; and (6) where feasible, provide for a means of confirmation of identity.

The S-PIA method (Ciguatect™), currently available from HawaiiChemtect International, has the highest potential for application to screening marketplace fish for ciguatera toxicity. The kit can be used at several points along the marketing plan, i.e., harvesting, processing, distribution, retail, etc. Testing fish early in the plan is recommended, since this will minimize the cost expended for the product and potential economic lost to the industry. The kit can be used onboard fishing vessels, at receiving docks, processing plants, distribution organizations, retail outlets, consumers, and regulatory agencies. The self-contained assay is available as a single analysis kit designed for nonlaboratory use by untrained personnel. Organizations conducting large numbers of analyses would be more inclined to use the

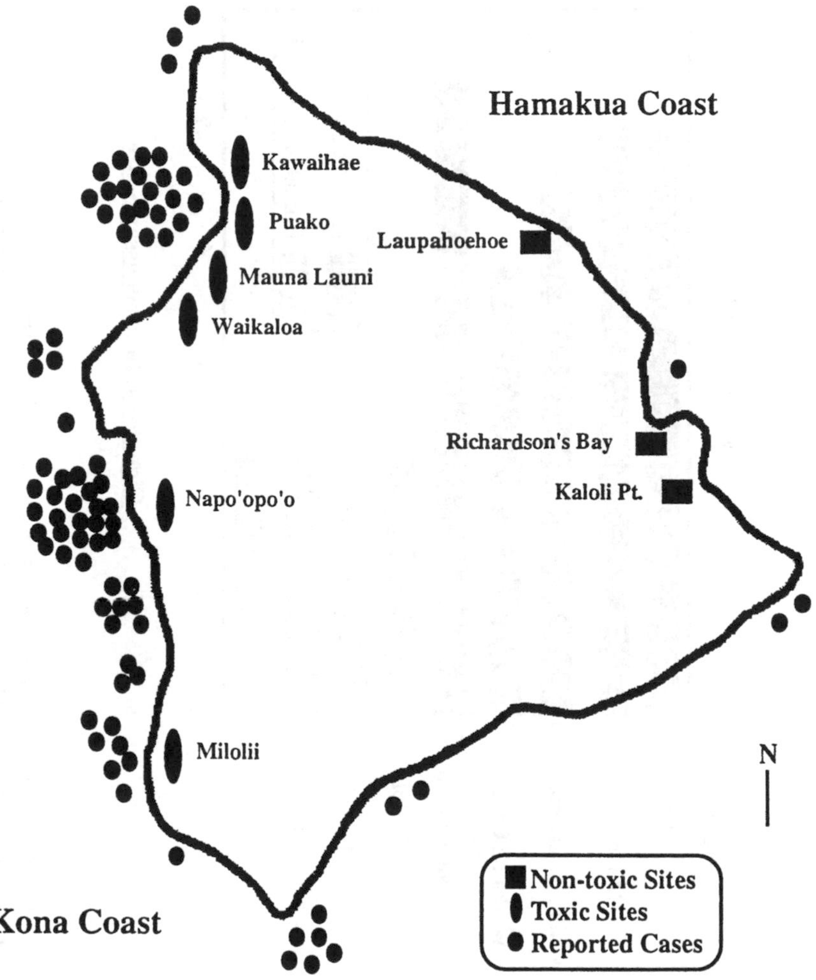

Figure 1. Map showing reported ciguatera outbreaks since 1980. The Kona coast is considered to be toxic and the Hamakua coast nontoxic. Study sites are noted. (R. G. Knitek and D. L. Park, Unpublished Data)
Map courtesy of Hawaii State Department of Health.

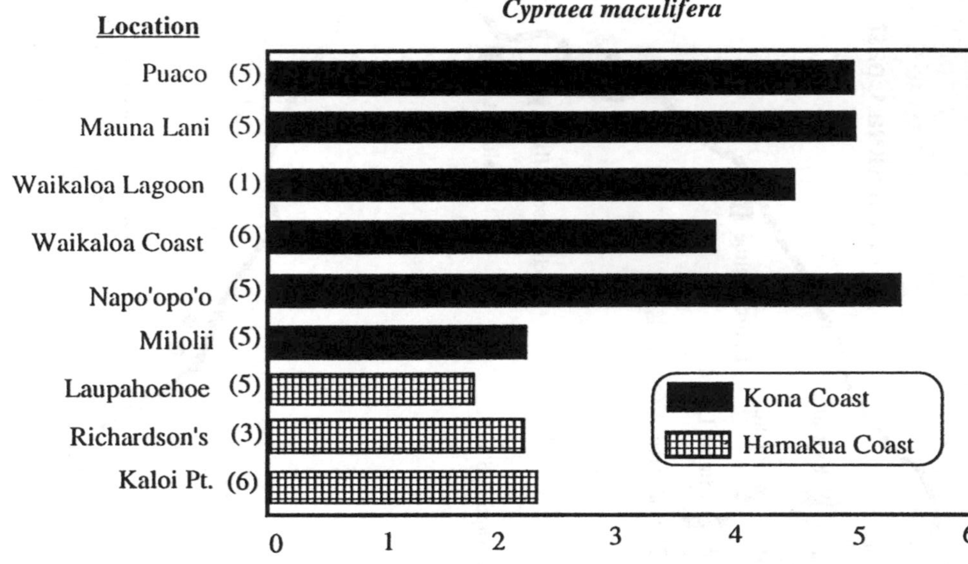

Figure 2. The scores reflected are for specimens that were pooled according to species. The numbers in () indicate the number of individual specimens pools. According to reported cases of illness, the Hawaii Department of Public Health considers the Kona coast to be toxic and the Hamakua coast a non-toxic site. The survey shows the same results.

Table 2.

Comparative results between HPLC and Ciguatect™ for okadaic acid and related polyether compounds in mussels (*Mytilus* sp.) monitored along French coastline.

Sampling location	Date year/month/day	µg OA/g HG[a] HPLC	ng OAC[b] Ciguatect™
Groix	86/6	5.0	4.0
Bois Cise	87/5/21	<0.5	5.0
Douarnenez	88/3/14	<0.5	1.0-2.0
	88/6/12	11.5	>100.0
	88/8/23	<0.5	0.0-1.0
Baie Vilaine	88/3/21	<0.5	2.0
	88/6/06	3.7	100.0
	88/6/27	2.0	10.0
	88/8/17	<0.5	0.0-1.0

Notes:
[a] HG = hepatopancreas gland; Method of Lee *et al.*, 1987.
[b] OAE = Okadaic acid equivalence; color intensity on the test strip assigned a value between 0-5 where 0 = non-detectable and 5 = color intensity equal to 5 ng okadaic acid.

laboratory kit which contains sufficient material for fifty tests.

The Seafood Safety Monitoring Program would involve large-scale testing of fish according to an acceptable sampling plan. Fish or lots testing negative to the screening procedure would be allowed to proceed in commercial channels. Each point identified above would be a quality control point. Product testing positive for toxic potential would be diverted to lower risk uses or retested to confirm toxic potential. This can be done by using the REM procedure which concentrates the toxins and retesting or by using alternative test methods for specific toxins.

SUMMARY

In conclusion, the seafood safety monitoring program would entail a testing program which includes: (1) monitoring fish harvesting areas to determine ciguatera potential; (2) developing a sampling plan applicable to screening fish in the market place for ciguatera potential; (3) screening fish at various points in commercial channels for ciguatera potential, i.e., on-board fishing vessels, at receiving docks, processing plants, distribution organizations, retail outlets, consumers, and regulatory agencies; (4) if desired, reanalyzing fish testing positive using alternative extracting and analytical methods; and (5) diverting toxic fish to lower risk uses.

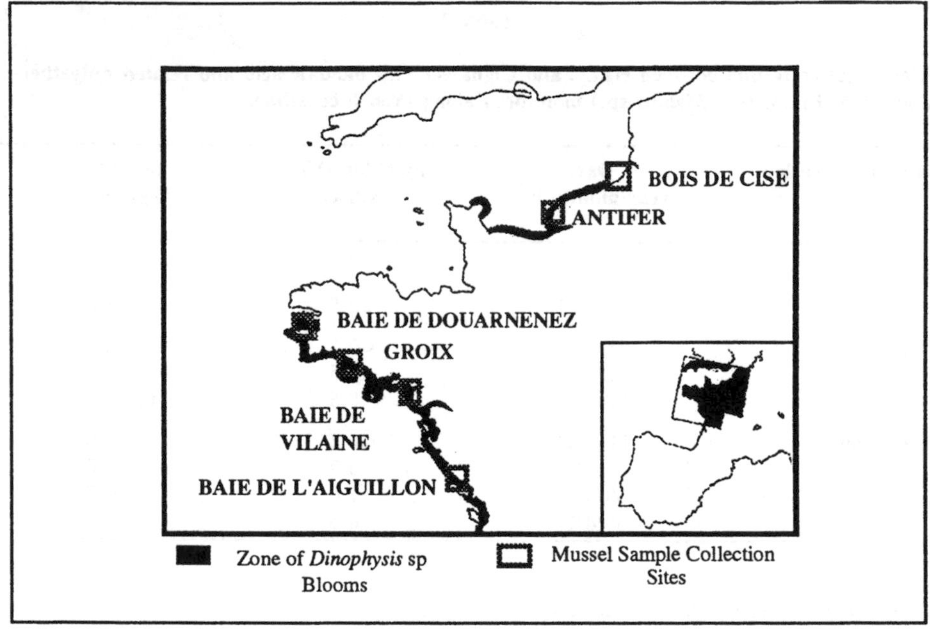

Figure 3. Mussel (*mytilus* sp) collection sites and *Dinophysis* sp. bloom locations long French coastline

REFERENCES

Bagnis, R., M. E. Luossan, and S. Thevenin. 1974. Les intoxications par poisons perroquets aux Iles Gambier. *Med Trop.* 34:523-527.

Chungue, E., R. Bagnis, N. Fusetani, and Y. Hashimoto. 1977. Isolation of two toxins from parrotfish *Scarus gibus*. Toxicon 15:89-93.

Dickey, R. W., S. C. Bobzin, D. J. Faulkner, F. A. Bencsath, and D. Andrzejewski. 1990. Identification of okadaic acid from a Caribbean dinoflagellate, *Prorocentrum concavum*. *Toxicon* 28(4): 371-377.

Gamboa, P. M., and D. L. Park. 1985. Fractionation of extracts of *Prorocentrum lima*, *Gambierdiscus toxicus*, and ciguatoxic fish using counter current chromatography. Paper presented at 2nd Conference on Ciguatera, 23-25 April, San Juan, Puerto Rico.

Gamboa, P. M., D. L. Park, and J. M. Fremy. 1992. Extraction and purification of toxic fractions from barracuda (*Sphyraena barracuda*) implicated in ciguatera poisoning. In *Proceedings of the 3rd International Conference on Ciguatera Fish Poisoning*, ed. T. R. Tosteson, 13-24. Morin Heights, Quebec, Canada: Polyscience Publishers, Inc.

Fremy, J. M., D. L. Park, S. K. Mohapatra, H. M. Sikorska, and E. Gleizes. 1994. Application of immunochemical methods for the detection of okadaic acid in mussels. in press. *J. Natural Toxins*.

Hoffman, P. A., H. R. Granade, and J. P. McMillan. 1983. The mouse ciguatoxin bioassay: A dose response curve and symptomatology analysis. *Toxicon* 21(3): 363-369.

REFERENCES

Bagnis, R., M. E. Luossan, and S. Thevenin. 1974. Les intoxications par poisons perroquets aux Iles Gambier. *Med Trop.* 34:523-527.

Chungue, E., R. Bagnis, N. Fusetani, and Y. Hashimoto. 1977. Isolation of two toxins from parrotfish *Scarus gibus*. Toxicon 15:89-93.

Dickey, R. W., S. C. Bobzin, D. J. Faulkner, F. A. Bencsath, and D. Andrzejewski. 1990. Identification of okadaic acid from a Caribbean dinoflagellate, *Prorocentrum concavum*. *Toxicon* 28(4): 371-377.

Gamboa, P. M., and D. L. Park. 1985. Fractionation of extracts of *Prorocentrum lima*, *Gambierdiscus toxicus*, and ciguatoxic fish using counter current chromatography. Paper presented at 2nd Conference on Ciguatera, 23-25 April, San Juan, Puerto Rico.

Gamboa, P. M., D. L. Park, and J. M. Fremy. 1992. Extraction and purification of toxic fractions from barracuda (*Sphyraena barracuda*) implicated in ciguatera poisoning. In *Proceedings of the 3rd International Conference on Ciguatera Fish Poisoning*, ed. T. R. Tosteson, 13-24. Morin Heights, Quebec, Canada: Polyscience Publishers, Inc.

Fremy, J. M., D. L. Park, S. K. Mohapatra, H. M. Sikorska, and E. Gleizes. 1994. Application of immunochemical methods for the detection of okadaic acid in mussels. in press. *J. Natural Toxins*.

Hoffman, P. A., H. R. Granade, and J. P. McMillan. 1983. The mouse ciguatoxin bioassay: A dose response curve and symptomatology analysis. *Toxicon* 21(3): 363-369.

Juranovic, L. R., and Park, D. L. 1991. Food borne toxins of marine origin: Ciguatera. *Rev. Environ. Contam. Toxicol.* 117:51-94.

Juranovic, L. R., D. L. Park, J. M. Fremy, and C. A. Neilsen. 1994. Isolation/separation of toxins produced by *Gambierdiscus toxicus* and *Prorocentrum concavum*. submitted for publication *J. Aquatic Food Product Technology*.

Lee, J. S., M. Murata, and T. Yasumoto. 1989. Analytical methods for the determination of diarrhetic shellfish toxin. In *Mycotoxins and Phycotoxins*, ed. S. Natori, K. Hasimoto and Y. Ueno, 327-334. Elsevier Science Publishers.

Lee, J. S., T. Yanagi, R. Kenma, and T. Yasumoto. 1987. Fluorometric determination of diarrhetic shellfish toxins by high performance liquid chromatography. *Agric. Biol. Chem.* 51(3): 877-881.

LeGrand, A. M. 1991. Les toxines de la ciguatera. In *Proceedings of Symposium on Marine Biotoxins*, 30-31 January, Paris, France.

LeGrand, A. M., M. Fukui, P. Cruchet, Y. Ishibashi, and T. Yasumoto. 1990. Characterization of toxins from different fish species and wild *G. toxicus*. In *Proceedings 3rd International Conference on Ciguatera*, ed. T. R. Tosteson, 25-32. Morin Heights, Quebec, Canada: Polyscience Publishers, Inc.

Lewis, N. D. 1986. Diseast and development: Ciguatera fish poisoning. *Soc. Sci. Med.* 23(10): 986-993.

Lewis, R. J., M. Sellin, M. A. Poli, R. S. Norton, J. K. MacLeod, and M. M. Sheil. 1991. Purification and characterization of ciguatoxins from Moray Eel (*Lycodontis javanicus*, Muraenidae). *Toxicon* 29(9): 1115-1127.

McMillan, J., H. Ray, and P. Hoffman. 1980. Ciguatera fish poisoning in the U.S. Virgin Islands. Preliminary studies. *J. Coll. Virgin Is.* 6:84-107.

Moulin, F., J. P. Vernoux, J. M. Fremy, and M. Ledoux. 1992 Dinoflagellate toxins involved in marine foodborne intoxication. CNEVA, Laboratoire Central d'Hygiene Alimentaire. Paris, France.

Murata, M., A. M. LeGrand, Y. Ishibashi, M. Fukui, T. Yasumoto. 1990. Structures of ciguatoxin and its congener. *J. Am. Chem. Soc.* 112:4380-4386.

Murata, M., M. Shimatani, H. Sugitani, Y. Oshima, and T. Yasumoto. 1982. Isolation and structural elucidation of the causative toxin of diarrhetic shellfish poisoning. *Bull. Japan Soc. Scient. Fisheries* 48(4): 549-552.

Park, D. L. Evolution of methods for assessing ciguatera toxins in fish. in press. *Rev. Environ. Contam. Toxicol.*

Park, D. L., and C. H. Goldsmith. 1991. Inter-laboratory validation of the solid-phase immunobead assay for the detection of toxins associated with ciguatera poisoning. Paper presented at the 5th International Conference on Toxic Marine Phytoplankton, 28 October-1 November, Newport, Rhode Island.

Park, D. L., P. M. Gamboa, and C. H. Goldsmith. 1992a. Rapid facile solid-phase immunobead assay for screening ciguatoxic fish in the market place. Paper presented at the 4th International Conference on Ciguatera, 4-8 May, Papeete, Tahiti, French Polynesia.

Park, D. L., P. M. Gamboa, and C. H. Goldsmith. 1992b. Validation of the solid-phase immunobead assay (Ciguatect™) for toxins associated with ciguatera poisoning. Paper presented at the 106th International AOAC Annual Meeting, 31 August-3 September, Cincinnati, Ohio.

Park, D. L., J. M. Fremy, C. Marcaillou-Lebaut, P. M. Gamboa, E. Gleizes, P. Masselin, and C. H. Goldsmith. 1992c. Innovative rapid solid-phase immunobead assay for the detection of okadaic acid and related DSP toxins in shellfish. Paper presented at 2nd International Conference on Shellfish Depuration, 6-8 April, Rennes, France.

Park, D. L., L. R. Juranovic, S. M. Rua, Jr., and C. A. Nielsen. 1994. Toxic/mutagenic potential of toxins produced by *Gambierdiscus toxicus* and *Prorocentrum concavum*. submitted for publication. *J. Aquatic Food Product Technology.*

Ragelis, E. P. 1984. Ciguatera seafood poisoning overview. In *Seafood Toxins,* ed. E. P. Ragelism 22-36. Washington D.C.: American Chemical Society.

Sawyer, P., D. Jallow, P. Scheuer, R. York, J. McMillian, N. Withers, H. Fudenberg, and T. Higerd. 1984. Effect of ciguatera-associated toxins on body temperature in mice. In *Seafood toxins*, ed. E. Ragelis, 321-329. Washington, D.C.: American Chemical Society.

Stabell, O. B., V. Hormazabal, I. Steffennak, and Pedersen. 1991. Diarrhetic shellfish toxins: Improvement of sample clean-up for HPLC determination. *Toxicon* 29(1): 21-19.

Tachibana, K. 1980. *Structural studies on marine toxins*. Ph.D. Diss., University of Hawaii.

Tindall, D. R., R.W. Dickey, R.D. Carlson, and G. Morey-Gaines. 1984. Ciguatoxic dinoflagellates from the Caribbean Sea. In: *Seafood Toxins.*, ed. E.P. Ragelis, 225-240. American Chemical Society. Washington D.C.

World Health Organization. 1984. Aquatic (marine and freshwater) biotoxins. *Env. Health Criteria*. 37. Geneva, Switzerland.

Yasumoto, T. 1985. Recent progress in the chemistry of dinoflagellates. In: *Toxic Dinoflagellates*, ed. D. M. Anderson, A. W. White, and D. G. Baden, 259. Elsevier, New York. p259.

Yasumoto, T., Y. Hashimoto, R. Bagnis, J. E. Randall, and A. H. Banner. 1971. Toxicity of the surgeonfishes. *Bull Jpn Soc Sci Fish.* 37:724-734.

Yasumoto, T., and M. Murata. 1988a. Polyether toxins implicated in ciguatera and seafood poisoning. Faculty of Agric., Tohoku Univ., Tsumidori, Sendai 980, Japan.

Yasumoto, T., and M. Murata. 1988b. Polyether toxins produced by dinoflagellates. Faculty of Agric., Tohoku Univ., Tsumidori, Sendai 980, Japan.

Yasumoto, T., U. Raj, and R. Bagnis. 1984. Seafood poisoning in tropical regions. Lab of Food Hyg, Fac of Agric., Tohoku Univ., Japan.

Yasumoto, T., and P. J. Scheuer. 1969. Marine toxins from the Pacific-VIII ciguatoxin from moray eel livers. *Toxicon* 7:273-276.

24

Large Marine Ecosystems: A New Concept in Ocean Management

Thomas L. Laughlin
Office of International Affairs
NOAA, Washington, D.C.

Kenneth Sherman
NOAA, National Marine Fisheries Service, Narragansett, Rhode Island

INTRODUCTION

Recently the United Nations Conference on the Environment and Development (UNCED) focused attention on the possible harmful effects of climate change on the long-term sustainability of marine resources and related linkages to sustained economic growth. It appears that consensus was reached among the participants, underscoring the observation that human activity has reached a level where the global environment is at risk. Short-term economic gains appear to impose significant risk to long-term sustainability of marine resources. Among the issues raised at UNCED that require attention by marine scientists are the observations that: (1) unrestricted use of chlorofluorocarbons appear to be reducing the atmospheric ozone levels, and (2) the levels of CO_2 emissions appear to be enhancing the greenhouse effect, thereby accelerating the global warming trend. Against this background of global climate change is an increasing trend toward coastal habitat loss, pollution, and overexploitation of living marine resources, resulting from the increasing needs of a growing coastal, urban population.

Nearly 95% of the usable global biomass yield of living marine resources is produced within and adjacent to the boundaries of coastal nations. It is the coastal ecosystems that are being stressed from habitat degradation, pollution, and overexploitation of marine resources. The major biomass of fish populations is caught within the geographic limits of 49 large marine ecosystems (LMEs). The LMEs are extensive areas of ocean space of approximately 200,000 km^2 or greater. They are characterized by distinct bathymetry, hydrography, productivity, and trophically dependent populations (Sherman and Alexander 1986). Several occupy semienclosed seas, including the Black Sea, Mediterranean Sea, Baltic Sea, and the Caribbean Sea. Some of the LMEs can be divided into domains or subsystems; for example, the Adriatic Sea as a subsystem of the Mediterranean. In other LMEs, geographic limits are defined by the scope of the continental margins. Among these are the U.S. Northeast Continental Shelf, the East Greenland Sea, and the Northwestern Australian Shelf. The seaward limit of the LMEs extends beyond the physical, outer limits of the shelves, to include all or a portion of the continental Slopes. Care was taken to limit the seaward boundaries to the areas affected by ocean currents rather than relying on the 200-mile Exclusive Economic Zone (EEZ), or fisheries zone limits. Among the ocean current LMEs are the Humboldt Current, Canary Current, California Current, and Kuroshio Current (Figure 1).

For nearly seventy-five years since the turn of the century, biological oceanographers did not achieve any great success in predicting fish yield based on food chain studies. As a result, through the mid-1970s, the predictions of the levels of biomass yields for different regions of the world ocean were open to disagreement (Ryther 1969; Alverson et al. 1970; Lasker 1988). It is clear that "experts" have been off the mark in earlier estimates of global yield of fisheries biomass. Projections given in *The Global 2000 Report* (U.S. Council on Environmental Quality 1980) indicated that the world annual yield was expected to rise little, if at all, by the year 2000 from the 60 million metric tons (mmt) reached in the 1970s. In contrast, estimates given in *The Resourceful Earth* (Wise 1984) argue for an annual yield of 100-120 mmt by the year 2000. The trend is upward; the 1989 level of marine, global fishery yields reached 86.5 mmt (FAO 1992). The lack of a clear definition of actual and/or potential global yield is not unexpected, given the limited efforts presently underway to improve the global information base on living marine resource yields. A milestone in fishery science was achieved in 1975 with the convening of a symposium by the International Council for the Exploration of the Sea (ICES) that focused on changes in the fish stocks of the North Sea and their causes. The symposium, which dealt with the North Sea as an ecosystem, following the lead of Steele (1974), Cushing (1975), Andersen and Ursin (1977), and others, was prompted by a rather dramatic shift in the finfish community of the North Sea from a balance between pelagic and demersal finfish prior to 1960 to demersal domination from the mid-1960s through the mid-1970s. Although no consensus on cause and effect was reached by the participants, it was suggested by the convener (Hempel 1978) that the previous studies of seven-and-a-half decades may have been to narrowly focused, and that future studies should take into consideration fish stocks, their competitors, predators and prey, and interactions of the fish stocks with their environments, the fisheries, habitat change, and pollution from an ecosystems perspective.

The LMEs that together produce nearly 95% of the annual global fisheries biomass yield are listed in Table 1. Although the United Nations has shown an upward trend in annual biomass yields for the past three decades, it is largely the clupeoids (herring-like fish) that are increasing in abundance (FAO 1992). A large number of stocks have been and continue to be fished at levels above long-term sustainability. The variations in abundance levels among the species constituting the annual global biomass yields are indicative of changing LME states caused by natural environmental perturbations, overexploitation, and pollution as either principal; secondary, or tertiary driving forces affecting biomass yields. Although the spatial dimensions of LMEs preclude a strictly controlled experimental approach to this study, they are perfectly amenable to the comparative method of science as described by Bakun (1993). since 1984, results have been reported for twenty-nine case studies investigating the major causes of large-scale perturbations in biomass yields of LMEs (Table 2).

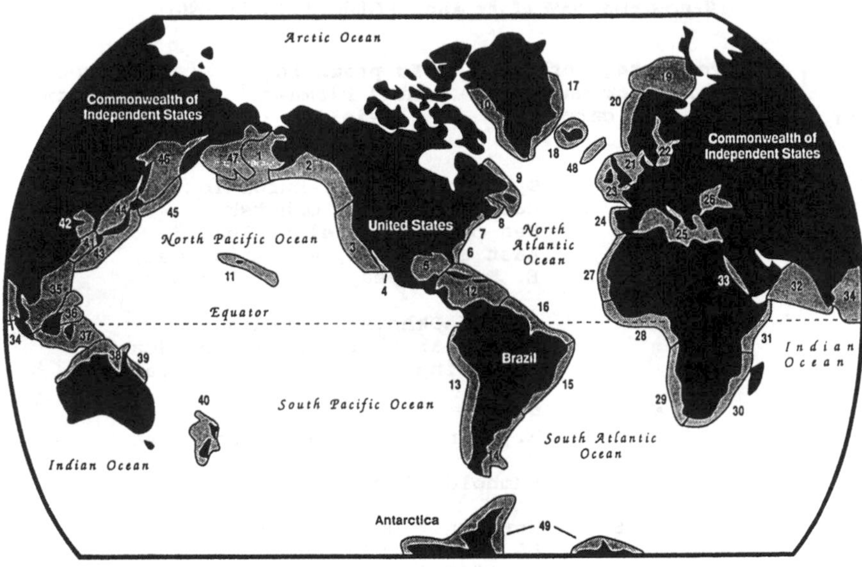

Figure 1. Provisional designation for 49 Large Marine Ecosystems. Criteria used in defining the geographical limits of LMEs include the distinct and coherent character of regional bathymetry, hydrography, productivity, and trophic relationships. In relation to hydrography, care was taken here to limit the seaward boundaries to the areas affected by the continental shelves and the currents, rather than relying simply on the 200-mile exclusive economic or fisheries zones limits. Among the ocean current LMEs are the Humboldt Current, Benguela Current, Canary Current, and Kuroshio Current.

1. Eastern Bering Sea; 2. Gulf of Alaska; 3. California Current; 4. Gulf of California; 5. Gulf of Mexico; 6. Southeast U.S. Continental Shelf; 7. Northeast U.S. Continental Shelf; 8. Scotian Shelf; 9. Newfoundland Shelf; 10. West Greenland Shelf; 11. Insular Pacific--Hawaiian; 12. Caribbean Sea; 13. Humboldt Current; 14. Patagonian Shelf; 15. Brazil Current; 16. Northeast Brazil Shelf; 17. East Greenland Shelf; 18. Iceland Shelf; 19. Barents Sea; 20. Norwegian Shelf; 21. North Sea; 22. Baltic Sea; 23. Celtic-Biscay Shelf; 24. Iberian Coastal; 25. Mediterranean Sea; 26. Black Sea; 27. Canary Current; 28. Guinea Current; 29. Benguela Current; 30. Agulhas Current; 31. Somali Coastal Current; 32. Arabian Sea; 33. Red Sea; 34. Bay of Bengal; 35. South China Sea; 36. Sulu-Celebes Seas; 37. Indonesian Seas; 38. Northern Australian Shelf; 39. Great Barrier Reef; 40. New Zealand Shelf; 41. East China Sea; 42. Yellow Sea; 43. Kuroshio Current; 44. Sea of Japan; 45. Oyashio Current; 46. Sea of Okhotsk; 47. West Bering Sea; 48. Faroe Plateau; 49. Antarctic.

Note: This global map is a provisional depiction of the limits of identifiable LMEs. Initial emphasis has been placed on the 49 LMEs that together produce 95.8% of the annual global fishery yield.

Table 1

Contributions by Country and Large Marine Ecosystem (LME) (Representing 95% of the Annual Global Catch in 1990)

Country	Percentage of[a] world marine nominal catch	LMEs producing annual biomass yield	Cumulative percentages
Japan	12.25	Oyashio Current, Kuroshio Current; Sea of Okhotsk, Sea of Japan, Yellow Sea, East China Sea, W. Bering Sea, E. Bering Sea, and Scotia Sea	
USSR	11.37	Sea of Okhotsk, Barents Sea, Norwegian Shelf, W. Bering Sea, E. Bering Sea, and Scotia Sea	
China	8.28	W. Bering Sea, Yellow Sea, E. China Sea, and S. China Sea	
Peru	8.27	Humboldt Current	
USA	6.76	Northeast US Shelf, Southeast US Shelf, Gulf of Mexico, California Current, Gulf of Alaska, and E. Bering Sea	
Chile	5.98	Humboldt Current	52.91
Korea Republic	3.28F[b]	Yellow Sea, Sea of Japan, E. China Sea, and Kuroshio Current	
Thailand	2.96F	South China Sea, and Indonesian Seas	
India	2.78	Bay of Bengal and Arabian Sea	
Indonesia	2.76	Indonesian Seas	
Norway	2.11	Norwegian Shelf and Barents Sea	
Korea D. P. Rep.	1.98F	Sea of Japan and Yellow Sea	
Philippines	1.96	S. China Sea, Sulu-Celebes Sea	
Canada	1.90	Scotian Shelf, Northeast U.S. Shelf, Newfoundland Shelf	

Notes: a - Percentages based on fish catch statistics *Source: FAO 1990 Yearbook, v. 70, FAO, 1992.*
b - Percentage calculated using FAO estimate from available sources of information.

Table 1-Continued

Country	Percentage of[a] world marine nominal catch	LMEs producing annual biomass yield	Cumulative percentages
Iceland	1.82	Icelandic Shelf	
Denmark	2.07	Baltic Sea and North Sea	76.25
Spain	1.73	Iberian Coastal Current and Canary Current	
Mexico	1.46	Gulf of California, Gulf of Mexico, and California Current	
France	1.03F	North Sea, Biscay-Celtic Shelf, Mediterranean Sea	80.47
Viet Nam	0.74	South China Sea	
Myanmar	0.72	Bay of Bengal, Andaman Sea	
Brazil	0.71F	Patagonian Shelf and Brazil Current	
Malaysia	0.71F	Gulf of Thailand, Andaman Sea, Indonesian Seas, and S. China Sea	
UK-Scotland	0.70	North Sea	
New Zealand	0.68	New Zealand Shelf Ecosystem	
Morocco	0.68	Canary Current	
Argentina	0.66	Patagonian Shelf	
Italy	0.57	Mediterranean Sea	
Netherlands	0.52	North Sea	
Poland	0.52	Baltic Sea	
Ecuador	0.47	Humboldt Current	
Pakistan	0.44	Bay of Bengal	
Turkey	0.41	Black Sea, Mediterranean Sea	
Germany (F.R. and N.L.)	0.41	Baltic Sea and Scotia Sea	
Ghana	0.40	Gulf of Guinea	

Notes: a - Percentages based on fish catch statistics *Source: FAO 1990 Yearbook, v. 70, FAO, 1992.*

Table 1-Continued

Country	Percentage of[a] world marine nominal catch	LMEs producing annual biomass yield	Cumulative percentages
Portugal	0.39	Iberian Shelf and Canary Current	90.20
Venezuela	0.38	Caribbean Sea	
Namibia	0.35	Benguela Current	
Faeroe Islands	0.34	Faeroe Plateau	
Senegal	0.34	Gulf of Guinea and Canary Current	
Sweden	0.31	Baltic Sea	
Bangladesh	0.31	Bay of Bengal	
Ireland	0.28	Biscay-Celtic Shelf	
Hong Kong	0.28	S. China Sea	
Nigeria	0.26	Gulf of Guinea	
Australia	0.25	N. Australian Shelf and Great Barrier Reef	
Iran, I.R.	0.24F	Arabian Sea	
UK Eng., Wales	0.21	North Sea	
Cuba	0.20	Caribbean Sea	
Panama	0.19	California Current and Caribbean Sea	
Greenland	0.17	East Greenland Shelf, West Greenland Shelf	
Sri Lanka	0.19	Bay of Bengal	
Greece	0.16	Mediterranean Sea	
Oman	0.15	Arabian Sea	
Angola	0.12	Guinea Current, Angola Basin	
United Arab Em.	0.11	Arabian Sea	95.01

Notes: a - Percentages based on fish catch statistics *Source: FAO 1990 Yearbook, v. 70, FAO, 1992.*

Table 2

List of Large Marine Ecosystems (LMEs)
(Reported Biomass Yield Data)

Large Marine Ecosystem	Volume No.*	Authors
U.S. Northeast Continental Shelf	1	M. Sissenwine
	4	P. Falkowski
U.S. Southeast Continental Shelf	4	J. Yoder
Gulf of Mexico	2	W. J. Richards and M. F. McGowan
	4	B. E. Brown et al.
California Current	1	A. MacCall
	4	M. Mullin
	5	D. Bottom
Eastern Bering Shelf	1	L. Incze and J. D. Schumacher
West Greenland Shelf	3	H. Hovgaard and E. Buch
Norwegian Sea	3	B. Ellertsen et al.
Barents Sea	2	H. R. Skjoldal and F. Rey
	4	V. Borisov
North Sea	1	N. Daan
Baltic Sea	1&5	G. Kullenberg
Iberian Coastal	2	T. Wyatt and G. Perez-Gandaras
Mediterranean-Adriatic Sea	5	G. Bombace
Canary Current	5	C. Bas
Gulf of Guinea	5	D. Binet and E. Marchal
Benguela Current	2	R.J.M. Crawford et al.
Patagonian Shelf	5	A. Bakun
Caribbean Sea	3	W. J. Richards and J. A. Bohnsack
South China Sea-Gulf of Thailand	2	T. Piyakarnchana
Yellow Sea	2	Q. Tang
Sea of Okhotsk	5	V. V. Kusnetsov
Humboldt Current	5	J. Alheit and P. Bernal
Indonesia Seas-Banda Sea	3	J. J. Zijlstra and M. A. Baars
Bay of Bengal	5	S. N. Dwivedi
Antarctic Marine	1&5	R. T. Scully et al.
Weddell Sea	3	G. Hempel
Kuroshio Current	2	M. Terazaki
Oyashio Current	2	T. Minoda
Great Barrier Reef	2	R. H. Bradbury and C. N. Mundy
	5	G. Kelleher
South China Sea	5	D. Pauly and V. Christensen

Sources:

* *Vol. 1, Sherman, K., and L. M. Alexander (Editors), 1986. Variability and Management of Large Marine Ecosystems, AAAS Selected Symposium 99, Westview Press, Boulder, CO.*

* *Vol. 2, Sherman, K., and L. M. Alexander (Editors), 1989. Biomass Yields and Geography of large Marine Ecosystems, AAAS Selected Symposium 111, Westview Press, Boulder, CO.*

* *Vol. 3, Sherman, K., L. M. Alexander, and B. D. Gold (Editors), 1990. Large Marine Ecosystems: Patterns, Processes, and Yields, AAAS Symposium, AAAS Press, Washington, DC.*

* *Vol. 4, Sherman, K., L. M. Alexander, and B. D. Gold (Editors), 1991. Food Chains, Yields, Models, and Management of Large Marine Ecosystems, AAAS Symposium, Westview Press, Boulder, CO.*

* *Vol. 5, Sherman, K., L. M. Alexander, and B. d. gold (Editors), 1993. Stress, Mitigation, and sustainability of Large Marine Ecosystems, AAAS Press, Washington, DC.*

With a minimum of expense and effort, ongoing Food and Agriculture Organization (FAO) fisheries programs can be strengthened by refocusing them around the natural boundaries of regional LMEs. The United Nations Environment Program (UNEP) Regional Seas programs can be enhanced by taking a more holistic ecosystems approach to pollution issues as part of an overall effort to improve the health of the oceans. The LE research and monitoring strategies are compatible with the proposed Global Ocean Observing System (GOOS), and will, in fact, strengthen GOOS by adding an ecosystem module to the existing physical and meteorological modules (IOC 1992a). The LME approach will complement the Global Ocean Ecosystem Dynamics Studies (GLOBEC) of the U.S. National Science Foundation and provide useful data inputs to the Joint Global Ocean Flux Study (JGOFS). The concept has been discussed at the International Council for the Exploration of the Sea (ICES), International Council for the Scientific Exploration of the Mediterranean Sea (ICEM), FAO, International Oceanographic Commission (IOC), World Conservation Union (IUCN), and the United Nations Environment Program (UNEP) with generally favorable responses for developing the concept more fully and implementing it more widely within the United Nations framework of ongoing programs.

The concept is wholly compatible with the FAO interest in studying "catchment basins" and quantifying their impact on enclosed and semienclosed seas (e.g., LMEs). The observations presently underway under the IOC/OSLR (Ocean Studies Related to Living Resources) program, now operating within the California Current, Humboldt Current, and Iberian Coastal Ecosystems, provide an important framework for expanded LME studies of these systems in relation to not only fisheries issues, but also problems of pollution and coastal zone management.

The regional seas at the borders of the ocean adjacent to the continental land masses continue to be degraded form pollution, habitat loss, and overexploitation of resources. The U.N. Conference on Environment and Development provided an important opportunity to take the necessary steps to reduce pollution, reclaim lost habitat, and promote long-term sustainable development of the coastal zones and ocean resources. Concerns are growing over ocean pollution, the depleted state of living resources, the degradation of coastal zone habitats, and the need for improved strategies to monitor the health of the oceans in a systematic manner consistent with the objective of reducing human-induced stress through scientifically based mitigating actions. The principal inhibitor to rapid progress has been the sectorization of ocean research, monitoring, and management. The various U.N. Agencies engaged in ocean activity are specialized around specific problem areas (e.g., FAO-Fisheries; UNEP-Pollution; IOC-Science). We believe that while the specialization is important for dealing with specific areas of ocean activities, it would be desirable to broaden the scope of the U.N. approach to include support of a regional approach to research, monitoring, and management of entire marine ecosystems, including consideration of coastal zone management, pollution reduction, fisheries sustainability, and habitat protection and restoration.

As a positive step toward ocean development and sustainability, the LMEs, which are ocean regions wholly compatible with FAO interest in catchment basins, serve as ecologically based units for tracking the health of the oceans. Significant progress has already been made in developing monitoring strategies for LMEs (IOC 1992b). On an international level, Knauss (1993) has suggested the organization of a set of regional programs, each designed for a specific LME. Each nation or set of nations bordering an LME would be responsible for the design and prosecution of the program. Such programs would have as a goal the monitoring of the ecosystem, understanding how it works, and how humans are perturbing the system. Those responsible for the program of each LME could meet locally, on a regular basis. Every few years, representatives from each region could come together internationally to compare notes and report on the health of all LMEs. In effect, they would be reporting on the health of the ocean.

APPLICATION OF THE LME CONCEPT TO OCEAN MANAGEMENT

The principal attraction of the LME concept as an organizing tool for ocean management is that it is science-based. Management areas are defined on the basis of ecology, rather than by legal/political parameters.

The benefit of this approach is that it provides a factual basis on which all interested parties can agree. Such agreement does not automatically lead to consensus on a management strategy which crosses political borders. However, an agreed factual basis facilitates discussion of management alternatives by narrowing the range of debate. This was the experience with the negotiation of the Montreal Protocol, wherein agreement on remedial steps was considerably easier due to scientific information about the causes of ozone depletion.

A good example of an area where the LME approach is useful is the Gulf of Guinea ecosystem. This LME includes the EEZ's of eleven West African coastal states, as well as areas of the high seas. Some of the "slices" of EEZ jurisdiction are no more than 40-50 miles wide. These thin slivers of ocean space are not viable management units since their health can be significantly affected by activities in adjacent EEZ segments. An understanding of the ecosystem can form the basis of multistate cooperation. Cooperation is necessary to sustainable use of this ocean space.

The report of the World Commission on Environment and Development recognized the need for new departures in the management of ocean space. The report states that "sustainable development, if not survival itself, depends on significant advances in the management of the oceans." [and] "An international ecosystem approach is required for the management of [ocean] resources for sustained use." (World Commission on Environment and Development 1987).

The report's recommendation is a combination of distillation of experience and prescription for the future. Lets look at both United States and international examples of ecosystem management.

In the United States, laws refer to the need "...to maintain the health and stability of the marine ecosystem": (MMPA 1992) [and] "...to safeguard the water quality and ecosystem health of each region...") MPRSA 1972). in some regions, state governments have based research designs related to natural resource decision making on the LME approach. For example, a research plan for the Washington and Oregon continental shelf refers to the need to "Designate the ecosystem and the primary unit of environmental planning and management" (Bottom et al. 1989). In Hawaii, legislation has been introduced which calls for the holistic management of "...the total Hawaiian archipelago as a single ecosystem...." (Hawaii 1992).

Internationally, perhaps the best example of an ecosystem approach is the Convention on the Conservation of Antarctic Marine Living Resources (CCAMLR). This Convention calls for "maintenance of the ecological relationships between harvested, dependent and related populations...and prevention of changes or minimization of the risk of changes in the marine ecosystem...." (CCAMLR 1980). CCAMLR also calls for consideration of associated activities and environmental changes.

The focus on the interrelationship of species and the consideration of other effects on the ecosystem with the aim of sustained conservation (defined to include rational use), make CCAMLR a leading example of an ecosystem approach to multilateral, ocean resource management.

Other international agreements, while not so comprehensive as CCAMLR, demonstrate a growing global awareness of the need for a cross-sector, interdisciplinary approach to ocean management, based on regional understanding of relevant ecosystems. For example, the Regional Convention on Fisheries Cooperation Among African States Bordering on the Atlantic Ocean calls for "...the protection and preservation of the marine environment as well as the management of coastal areas of the Region" (Regional Convention on Fisheries Cooperation Among African States Bordering on the Atlantic 1991). The Convention on the Protection of the Marine Environment of the Baltic Sea Area, 1992 calls upon the Contracting Parties to "promote the ecological restoration of the Baltic Sea Area and the preservation of its ecological balance" (Convention on the Protection of the Marine Environment of the Baltic Sea Area 1992).

In the Pacific, the newly created North Pacific Marine Science Organization's initial, scientific, preparatory meeting conclude that a "basin-scale and interdisciplinary approach to thinking about the subarctic Pacific" is needed. Such an approach would address the central scientific issues of the region in the context of weather and climate changes, fisheries, and overall environmental quality (PICES 1992).

There is not always a congruence between scientific and management thinking, on the one hand, and that in donor organizations on the other. In this case, however, the encouraging news is that the Global Environment Facility has funded two large marine ecosystem projects; one in the Gulf of Guinea and one in the Yellow Sea. In the Yellow Sea, on ramification of this funding is that the Peoples Republic of China (PRC) and the Public of Korea are working together to implement the program. Another is that the PRC will be incorporating the LME concept into their ocean research and management programs for all adjacent marine areas.

Turning to the U.N. conference on Environment and Development (UNCED), we find that the Oceans Chapter of Agenda 21 endorses a multispecies, regional, cross-sectoral approach to sustainable development of the oceans. The Chapter covers fisheries, pollution, coastal zone management, research, data requirements, training, and needed institutions. When considered in toto, it is clear that UNCED recognized the need for regional, ecosystem-based approaches to ocean management.

In closing, let us suggest eight aspects of the LME approach to understanding and managing ocean space.

1. Interdisciplinary understanding is required.
2. An intersectoral approach is needed.
3. Requires long-term data collection, research, and modeling.
4. Requires cooperation among numerous agencies, national, and international.
5. Often requires regional cooperation.
6. Provides a mesoscale unit, defined on the basis of ecology in which the above can occur.
7. Lends itself to periodic assessments of the health of the Earth's oceans.

In your deliberations as to future ocean management alternatives, we hope you will favorably consider this promising avenue to regional cooperation.

REFERENCES

Alverson, D. L., A. R. Longhurst, and J. A. Gulland. 1970. How much food from the sea? *Science* 168:503-505.

Andersen, K. P., and E. Ursin. 1977. A multispecies extension to the Beverton and Holt Theory of Fishing with accounts of phosphorus circulation and primary production. *Meddelser fra Danmarks Fiskeri-og Havundersogelser N.S. 7:319-435.*

Bakun, A. 1993. The California Current, Benguela Current, and Southwestern Atlantic Shelf ecosystems: A comparative approach to identifying factors regulating biomass yields. *In: Stress, Mitigation, and Sustainability of Large Marine Ecosystems,* ed.

Sherman, K., L. M. Alexander, and B. D. Gold. Washington, DC: AAAS Press.

Bottom, D. L., K. K. Jones, J. D. Rodgers, and R. F. Brown. 1989. *Management of living resources: A research plan for the Washington and Oregon continental margin.* National Coastal Resources Research and Development Institute, Newport, Oregon. NCRI-T-89-004, 80 pp.

CCAMLR (Convention on the Conservation of Antarctic Marine Living Resources). 1980. Article II. [Complete copy of treaty may be obtained by writing: CCAMLR, 25 Olde Wharf, Hobart, Tasmania 7000, Australia.]

Convention on the Protection of the Marine Environment of the Baltic Sea Area. 1992. Article 3.

Cushing, D. H. 1975. *Marine ecology and fisheries*. London: Cambridge University Press.

FAO (Food and Agriculture Organization of the UN). 1992. *FAO Yearbook of Fishery Statistics*. Vol. 70 Rome: FAO.

Hawaii, State of. House Bill No. 3443. 16th Leg. 1992. Section 4.

Hempel, G., ed. 1978. Symposium on North Sea fish stocks -- Recent changes and their causes. *Rapports Proces-verbaux Reunions Conseil international por la Exploration de la Mer* 172:449.

IOC (International Oceanographic Commission of UNESCO). 1992a. *GOOS. Global Ocean Observing System, An initiative of the Intergovernmental Oceanographic Commission (of UNESCO)*, Paris: IOC/UNESCO, Paris.

__________________________. 1992b. *The use of large marine ecosystem concept in the Global Ocean Observing System (GOOS)*, Twenty-fifth Session of the IOC Executive Council, Paris, 10-18 March 1992, IOC/EC-XXV/Inf.7.

Knauss, J. 1993. Preface. *In: Stress, mitigation, and sustainability of large marine ecosystems*, ed. K. Sherman, L. M. Alexander, and B. D. Gold. Washington, DC: AAAS Press.

Lasker, R. 1988. Food chains and fisheries: An assessment after 20 years. *In: Toward a theory on biological-physical interactions in the world ocean*, ed. B. J. Rothschild. 173-82. NATO ASI Series. Series C: Mathematical and Physical Sciences, Vol. 239, Dordrecht, The Netherlands: Kluwer Academic Publishers.

MMPA (Marine Mammal Protection Act). Section 2(6), 16 USC §1361 (*et. seq.*). 1992.

MPRSA (Marine Protection, Research and Sanctuaries Act). Section 401(1), 33 USC §1401 (*et. seq.*). 1972.

PICES (Pacific International Commission for the Exploration of the Sea). 1992. Summary Report of a scientific workshop. 10-13 December 1992.

Regional Convention on Fisheries Cooperation Among African States Bordering on the Atlantic. Article 12. 5 July 1991.

Ryther, J. H. 1969. Relationship of photosynthesis to fish production in the sea. *Science* 166:72-6.

Sherman, K. and L. M. Alexander, eds. 1986. *Variability and management of large marine ecosystems*. AAAS Selective Symposium 99, Boulder, CO: Westview Press, Inc.

Steele, J. H. 1974. *The structure of marine ecosystems*. Cambridge: Harvard University Press. U.S. Council on Environmental Quality and the Department of State. 1980. *The global 2000 report to the President: Entering the twenty-first century*. Vols. 1-3. Washington, DC: U.S. Government Printing office.

Wise, J. P. 1984. The future of food from the sea. *In: The resourceful earth*, eds. J. L. Simon and H. Kahn, 113-27, New York: Basil Blackwell, Inc.

World Commission on Environment and Development. 1987. Our common future. Oxford/New York: Oxford University Press.

25

Xenobiotics In Mytilus Galloprovincialis As Biondicator In Tyrrhenian Marine Coastal Zones

Isabella Buttino, Domenica Fierro, and Daniele Merola
Researchers
Department of Fisiologia Generale ed Ambientale Università degli Studi di Napoli Federico II, Napoli

Giovanni Sansone
Senior Researcher
Department of Fisiologia Generale ed Ambientale Università degli Studi di Napoli Federico II, Napoli

ABSTRACT

Mytilus galloprovincialis was used to monitor xenobiotic presence in two, shallow marine areas of differing water quality. One-year old mussels were placed in two sampling stations along the Tyrrhenian Campania Coast: Pozzuoli Bay (polluted area) and Cilento Coast (unpolluted area). One hundred and twenty days after the placement of these biomonitoring stations, the mussels were collected and analyzed to determine organochlorine compounds. Results showed xenobiotic elements present in the mussels collected from the Pozzuoli Bay Station: Lindane, Chlorothalonil, Aldrin, Endrin and pp'DDD ranged from 30-100 ng/g (w/w). In the mussels coming from the Cilento Coast biomonitoring station, only the presence of Chlorothalonil was noted, with a maximum concentration of 35 ng/g (w/w). Data confirmed that mussels can be used as bioindicators for organochlorine compound pollution and showed that pesticide pollution in Pozzuoli Bay is higher than along Cilento Coast.

INTRODUCTION

Shallow marine areas house a great number of marine organisms, and the preservation of an ecological balance in this habitat is critical for the survival of endemic species. Technical and scientific advances made within the last century have brought with them chemicals, the quantity and toxicity of which, have had a decidedly negative impact upon the environment. While the impact of agricultural chemicals on aquatic environments have been quite devastating, especially for inland waters such as rivers and lakes, they have also had a significant impact upon neighboring seas.

The heightened sensitivity to environmental problems and the growing awareness that changes in the biological balance are tipping towards the deterioration of marine coastal environment, have promoted research projects to control pollution. The strategies developed to control pollution were designed to determine the level of contaminants in the different environmental niches and to establish the dose-effect relationship of xenobiotics on monitored ecosystems by "sentinel" organisms (Ade *et al.* 1984; Payne 1984; Viarengo and Canesi 1991).

Mussels have often been used in monitoring programs as bioindicators of chemical marine pollution (Goldberg *et al.* 1978). In fact, mussels are benthic and filter-feeding and may accumulate many xenobiotics in their tissues, such as chlorinated pesticides present in the habitat (Ernst 1977; Renberg 1985; Van der Oost *et al.* 1988). This study is part of a biological, marine monitoring program of the mussel *Mytilus galloprovincialis*, along the Campania Tyrrhenian Coast (Figure 1).

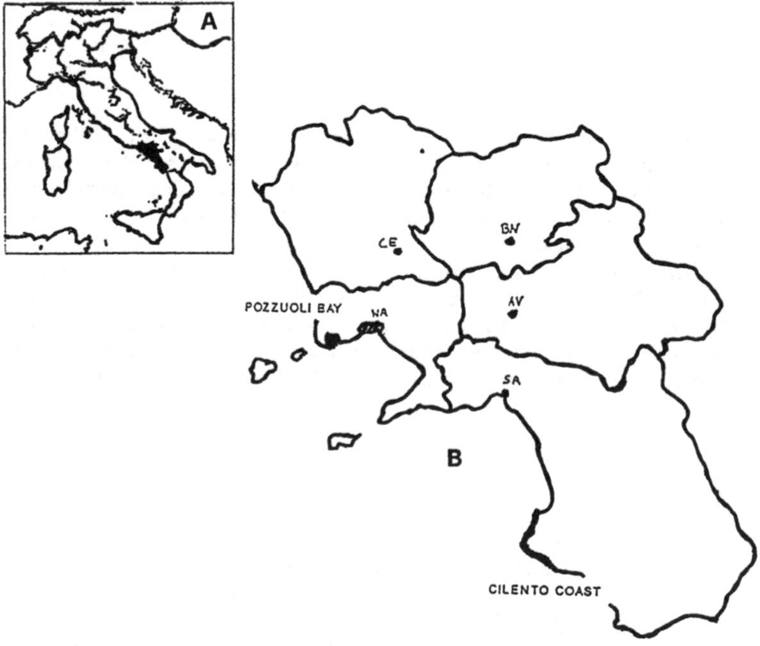

Figure 1. (A) Tyrrheninan Campania Coast. (B) Biomonitored Coasts: Pozzuoli Bay and Cilento Coast.

The aim of the present work has been to investigate the levels of bioaccumulated xenobiotics, such as chlorinated pesticides, in mussels used as bioindicators in two, different marine coastal areas: Pozzuoli Bay and Cilento Coast (Figure 1). The coastal zone of Pozzuoli Bay is a densely populated region with burgeoning Agriculture and Industry, while the Cilento Coast is an area with little agricultural or industrial. Both areas are crossed by rivers with estuaries located close to the two zones investigated. Xenobiotics, including Aldrin, Endrin, Dieldrin, and pp'DDD, that are normally forbidden, were used more frequently in the agricultural activities of the more inland of the two considered zones (Sansone 1990).

The hypotheses of this work were that xenobiotic elements present in inland waters (Amodio et al. 1988), were transferred into the shallow areas and probably accumulated in mussels coming from the Pozzuoli Bay (polluted area) and would decrease after repositioning in the Cilento marine areas (unpolluted zone), because of the passive mechanisms of bioaccumulation of chlorinated xenobiotics. Finally, the Cilento Coast's, biomonitored mussels should be used as the controls with respect to mussel population positioned in Pozzuoli Bay.

MATERIAL AND METHODS

One-year old mussels, coming from a mussel farm in Pozzuoli Bay, were placed in two biomonitoring stations along the Campania Coast: Pozzuoli Bay and Cilento Coast (Figure 1). The stations were placed 8 m depth and 500 m off-shore. At positioning time (0 time) and 120 days after placing the stations, mussels were collected and analyzed to determine xenobiotics, such as organochlorine compounds. Fifty grams (wet weight soft body) of mussels were homogenyzed in acethone (1:15 w/v), centrifugated, and concentrated by rotavapor. The organic phase was fractioned on a PL gel-column (500A x 5μm x 60cm) by an HPLC-UV detector (Shimadzu LC10), and the fraction with organochlorine compounds was analyzed in a GC-ECD detector (Dany 8510). Pesticides analyzed were: Lindane, Chlorothalonil, Endosulfan, and Captan. These pesticides were selected because they were the most diffused in the practical agriculture in Campania. In addition, organochlorine compounds such as Aldrin, Dieldrin, Endrin, and pp' DDD were analyzed.

RESULTS AND DISCUSSION

The analyses showed that in *Mytilus* coming from a mussel farm on Pozzuoli Bay, beforepositioning in two monitoring stations at time zero, were present only five compounds with these concentrations (ng/g): Lindane (42$\pm$14), Chlorothalonil (41$\pm$16), Aldrin (60$\pm$19), Endrin (36$\pm$18),pp' DDD (78$\pm$21) (Figures 2 and 3). After 120 days, the mussels collected from Pozzuoli Bay monitoring station were found to contain the same five compounds with similar concentrations at zero time: respectively, 50$\pm$12; 38$\pm$17; 55$\pm$22; 35$\pm$13; and 104$\pm$45. Any significant differences were recorded (Figure 2) However, after 120 days of biomonitoring, in mussels coming from the Cilento Coast , were found to contain only Chlorothalonil (17$\pm$9 ng/g), but with a lower concentration than that recorded at zero time ($0.05 > p > 0.02$) (see Figure 3). Data are the results of two replicate analyses (n=4).

Results confirm the hypotesis of the work. In fact, bioaccumulated, xenobioticsin Pozzuoli Bay mussels were fully released after 120 days of stabulation in Cilento Coast. Only Chlorothalonil remained present, but in minimal concentrations.

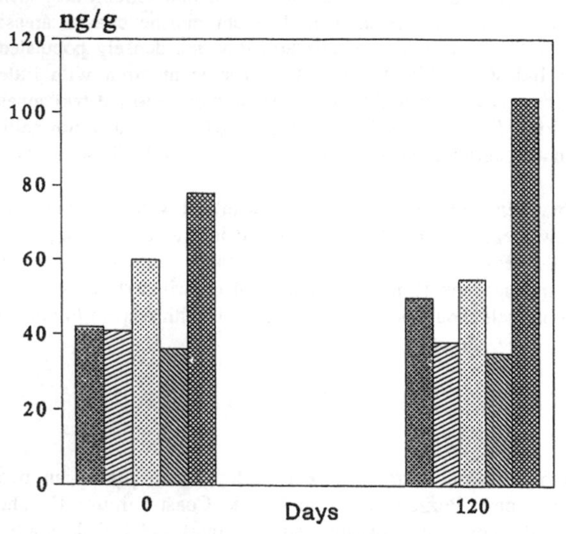

Figure 2. Xenobiotic concentrations at zero time and 120 days of biondicator mussels in Pozzuoli Bay.

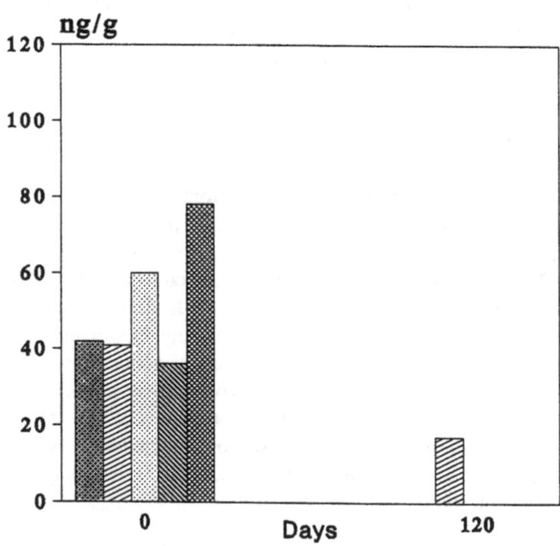

Figure 3. Xenobiotic concentrations at zero time and 120 days of biomonitor mussels in Cilento Coast.

Xenobiotic concentrations in Pozzuoli Bay mussel tissues, instead, remained constant after 120 days. This situation might be attributed to the probable, continual turnover of xenobiotics.

Regarding Chlorothalonil behaviour, the authors suggest two hypotheses. First, Chlorothalonil is present in the sea water of the Cilento coastal area, and the established concentration is the result of xenobiotic uptake. It is used in this zone for agricultural activity in extremely small quantities (Sansone 1990). Secondly, the bioaffinity of Chlorothalonil for mussel tissues could be higher than other xenobiotics and the release is probably slower.

The recorded levels of xenobiotics in mussels coming from the monitoring station at Pozzuoli Bay clearly demostrate that this zone is polluted by organochlorine compounds.

In conclusion, our data confirm the negative impact of practical, agricultural chemicals of Pozzuoli Bay's inland area on this shallow marine area. Moreover, the constant presence of xenobiotics could pose a very real problem in the maintenance of a biological balance among the many organisms endemic to this zone.

ACKNOWLEDGEMENT

The research was supported by financial Grant of Agensud Project B-2197/88.

REFERENCES

Ade, P., M.G. Bonchelli Soldaini, M.G. Castelli, E. Chiesara, F. Clementi, R. Fanelli, E. Funari, G. Ignesti, A. Marabini, M. Orunesu, S. Palmero, R. Pirisino, A. Ramundo Orlando, R. Riezzi, V. Silano, A. Viarengo, and L. Vittozzi. 1984. Biochemical and morphological comparison of microsomal preparations from rat, quail, trout, mussel and water flea. *Ecotoxicol. Environ. Safety* 8:423-46.

Amodio, R., R. Cocchieri, and A. Arnese. 1988. Organochlorine pesticide residues in fish from southern Italian rivers. *Bull. Environ. Contam. Toxicol.* 40:233-39.

Ernst, W. 1977. Determination of the bioconcentration potential of marine organisms. A steady state approach. I. Bioconcentration data for seven chlorinated pesticide in mussel (*Mytilus edulis*) and their relation to solubility data. *Chemosphere* 11:731-40.

Goldberg, E.D., V.T. Bowen, J.W. Farrington, G. Harvey, J.H. Martin, P.L. Parker, R.W. Risebrouger, W. Robertson, W. Schneider, and E. Gamble. 1978. The mussel watch. *Environ. Conserv.* 5:101-25.

Payne, J.F. 1984. Mixed-function oxidases in biological monitoring programs: Review of potential usage in different phyla of aquatic animals. In *Ecotoxicological testing for the marine environment*, ed. G. Personne, E. Jaspers, and C. Claus, Vol 1, 625-655. Bredene, Belgium: State University Ghent and Institute of Marine Scientific Research.

Renberg, L., M. Tarkpea, and E. Lindén. 1985. The use of the bivalve *Mytilus edulis* as a test organism for bioconcentration studies. *Ecotoxicol. Environ. Safety* 9:171-78.

Sansone, G. 1990. Rischi derivanti dall'uso continuo e prolungato di prodotti chimici in agricoltura sulla qualità della vita delle popolazioni ittiche delle acque interne. I and II Rapporto Regione Campania. Unpublished data.

Van der Oost, R., H. Heida, and A. Opperhuizen. 1988. Polychlorinated biphenyl congeners in sediments, plankton, molluscs, crustaceans, and eel in a freshwater lake: Implications of using reference chemicals and indicator organisms in bioaccumulation studies. *Arch. Environ. Contam. Toxicol*. 17:721-29.

Viarengo, A., and Canesi, L. 1991. Mussels as biological indicators of pollution. *Aquaculture* 94:225-43.

26

Institutional Bases for the Planning and Management of the Coastal Development in the Biobío Region, Chile

Gonzalo A. Cid
Coastal Management Group
Center EULA-CHILE
University of Concepción, P.O.Box 156-C, Concepción, Chile.

Víctor A. Gallardo
Head of Ph.D. Program on Environmental Sciences
Center EULA-CHILE
University of Concepción, P.O.Box 156-C, Concepción, Chile.

INTRODUCTION

While the cause of conflicts and loss of opportunities in the coastal zone appears well identified, i.e., sectorial planning and management (Sorensen and McCreary 1990) and the options to overcome the shortcomings of this outlook to environmental management, i.e., integrated coastal zone management (ICZM) (Sorensen and McCreary 1990) is also well defined, the way to reach this goal is not clear. Probably there are many ways depending on the country involved.

In recent years, the coastal area degradation in the Biobío Region (Chilean Eighth Region), especially in the enclosed bays and estuaries, has been notorious and at least partially attributable to the growth of industrial activities. These problems are rooted in bad planning and management of its rich, diverse, and complex coastal resources which have been exploited under an ineffective administration by the local governments (Municipalities). At this time, the progressive loss of resources, habitats, biodiversity, and quality of life in the region forms the principal issue in coastal affairs.

Geographically and economically speaking, the Biobío Region is predominantly maritime. This is shown by the great industrial and fishery development taking place in its coastal area. Also, due to industrial development, public and social affairs are important in the human context, especially in two bays of the region: Concepción Bay and San Vicente Bay, as well as for the adjacent Gulf of Arauco to the south, with its own embayments of Coronel and Lota.

A special case to consider in this coastal system are these two latter embayments, where general agreements are being sought for the implementation of a Sectional Plan which will allow for the siting of a major commercial port (Bay of Coronel) and the installation of an industrial park in its waterfront (Bay of Lota) (Figures 1, 2, and 3).

So far, regional (and national) management of coastal and oceanic resources has followed a single sector approach with all the limitations of multiple-agency responsibility; overlapping jurisdictions, profuse and confused regulations, and the prevalence of short-term economic returns versus total evaluation of coastal resources, causing conflicts and meager macroeconomic results.

We find it necessary to create chances for integrated coastal resources management in Chile, as is the experience in so many countries, both developed and developing, starting with the United States of America (Coastal Zone Management Act, 1972) (Olsen and Seavey 1990). The aim would be to develop a hierarchical governmental administration for coastal zone management, which will have both a "top down" (general policies and minimum standards) and a "bottom up" (local councils for coastal zone management) structure, the latter starting at the **Comunas** (Chilean geopolitical division in charge of a Municipality) with coastal territory. Each comuna would look after its coastal resources to obtain a managed and sustainable development in its area.

Just as in so many countries around the world that have established coastal management legislations in their national policies, the socioeconomic importance of the coastal areas and its character must be recognized as worthy of special attention and management.

As a predominantly maritime country, Chile must become aware of this international trend and avoid ineffective coastal resource administration and conservation. Drawing from its experience in the Biobío River watershed study, the Europe-Latinamerica Center for Research and Education in Environmental Sciences (EULA-CHILE Center) at the University of Concepción (established with the assistance from the Italian Government), is setting the bases for the planning and management of coastal resources in the Eighth Region of Chile, and hopefully, influencing national government to the establishment of a coastal zone management national policy (Gallardo 1992).

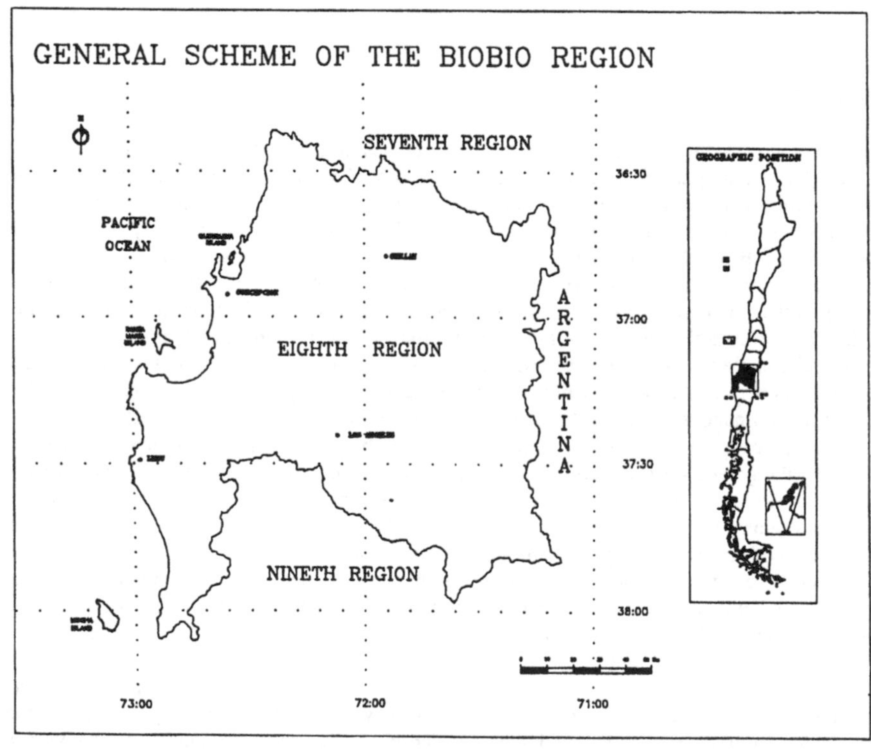

Fig. 1. General view of Biobio Region

DESCRIPTION OF THE AREA

In this paper, we are giving special attention to the Bays of Concepción and San Vicente, because they are the most important, regional coastal development poles at this time. On the other hand, Coronel and Lota Bays are examples of the limitations of present approaches to the management of coastal resources in Chile, and the Eighth Region in particular.

The Bay of Concepción and the Bay of San Vicente are template bays located between 36° and 37° S lat. and ca. 73° W long. in the Chilean central zone. The most remarkable differences between them are their mouths' orientation and their relation to the water dynamics of the area which is strongly influenced by seasonal upwelling processes.

The Bay of Concepción

This Bay, with a surface of 167.4 km², an average depth of 18.5 m, and a volume of 3.09×10^9 m³ is north-oriented. Twenty-five percent of its volume is under the depth of 30 m. The Quiriquina Island, at its mouth, plays an important role in the Bay of Concepción. This island divides the Bay's mouth into two channels which creates a special current system within the Bay.

The natural bottom sediments in the Bay are highly enriched, black, reduced muds with high hydrogen sulfide production. According to the kinds of sediments, the Bay can be divided into:

1. An external zone with fine gray sediments
2. A bordering middle zone, with a depth of no more than 10-15 m, with sandy and a mixture of mud with fine sands substrata
3. A central zone with anoxic, fine, clayey, black sediments. This zone forms the major part of the Bay bottom.

Oceanographic studies in the Bay of Concepción reveal three, well-defined water masses influencing the Bay:

1. Subantartic waters
2. Equatorial-subsurface waters
3. Intermediate antartic waters

Subantartic waters have been characterized with a salinity of 33.8×10^{-3}, temperatures up to 11.5 °C, and dissolved oxygen content of 3-6 ml O_2/l (Ahumada 1976).

The San Vicente Bay

This is located in an area with seasonal coastal upwelling, which mainly occurs during the months of September to March.

The northwest oriented San Vicente Bay has a surface of 13.2 km², an average depth of 20 m, and a volume of 260.5×10^6 m³. It is a relatively shallow bay with an approximate water residence time of twenty-four hours. Despite that, its mouth opens further to the west than the Bay of Concepción's. Water dynamics are more important in this Bay than in the Bay of Concepción, due to the proportions of its mouth width in relation to its length. The result is that the Bay of Concepción is more protected. As far as we know, there has never been a large vessel stranding due to storms in the latter bay as it has happened in the San Vicente Bay (i.e., the stranding and oilspill produced by the tanker CABO TAMAR in 1976).

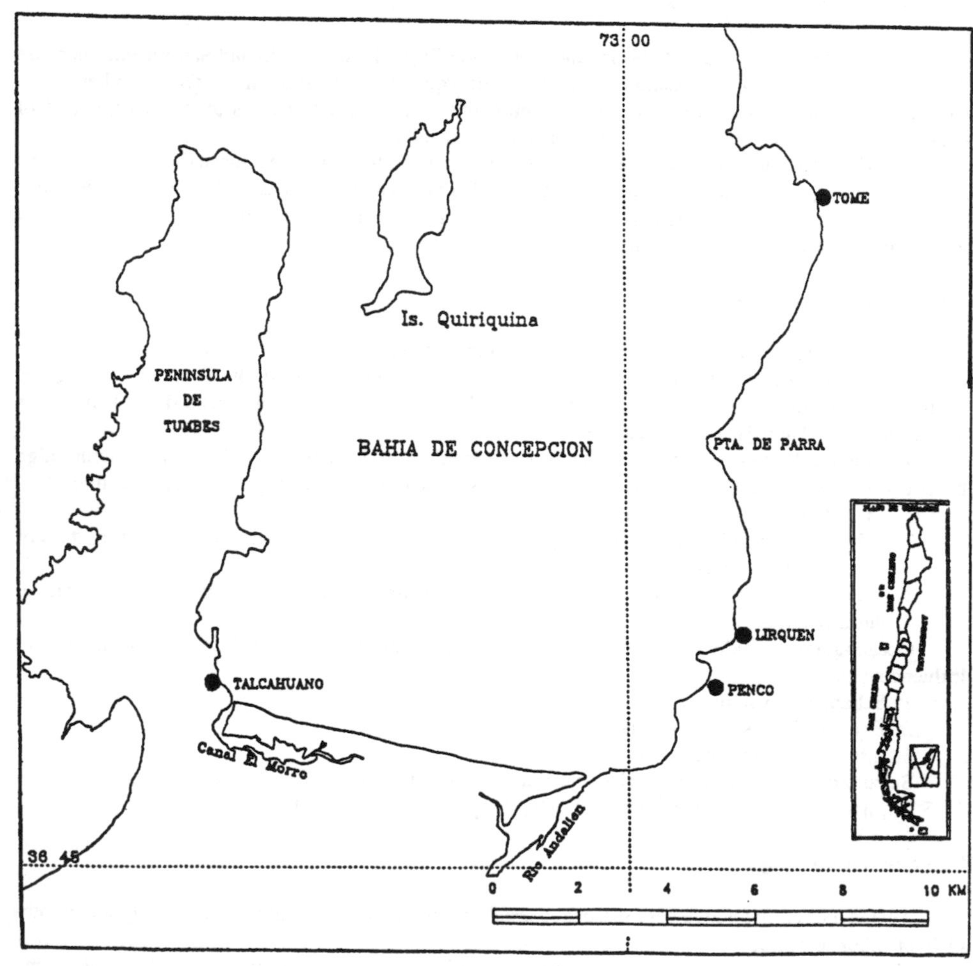

Figure 2. Bay of Concepcion

This Bay receives discharge of the Andalién River at its southeastern corner and of several other smaller flows; it is also bordered in its southern reaches by an extensive area of wetlands and tidal areas which empty into the Bay at its southwestern corner (Canal El Morro) (Ahumada and Rudolph 1988).

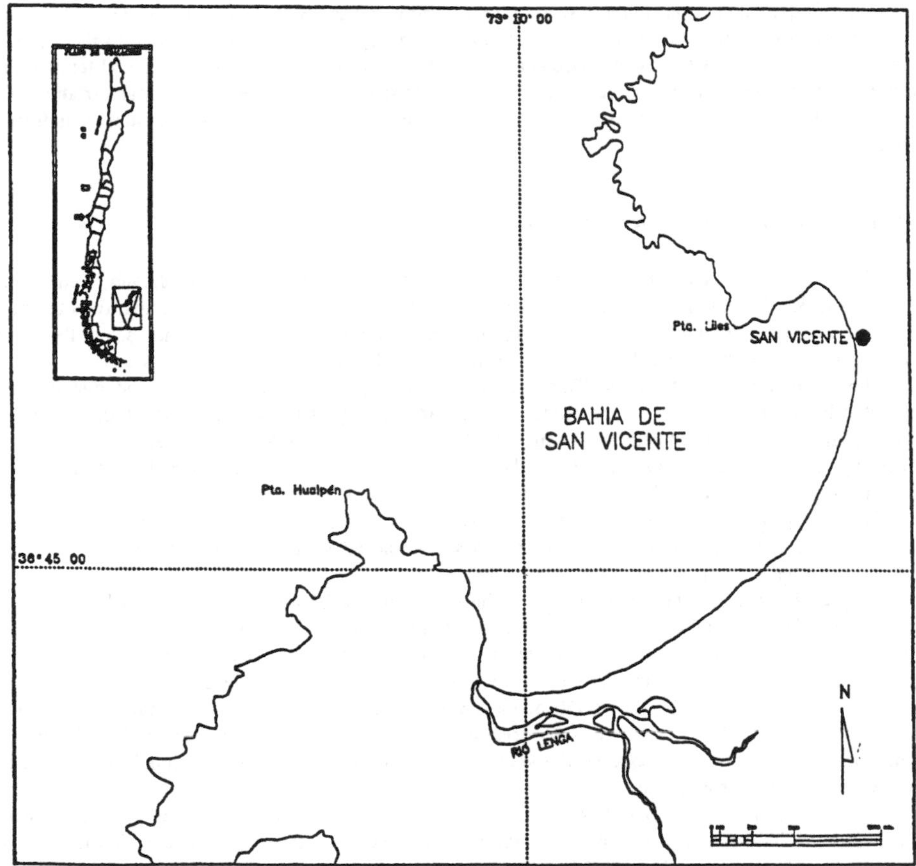

Figure 3. San Vincente Bay

Before the 1940s, the beaches of both bays were used by fishermen as settling areas and by the populations of Talcahuano and Concepción, the two major human settlements of the area, for recreation. The present bay use (industrial siting) began with the first installation of Huachipato Siderurgical Plant in 1945.

The coastal upwelling processes in this area give rise to important industrial fisheries, with landings of more than 2 million tons in 1992 in the ports of the Eighth Region. The unloading of this fish, which is done by pumping in an open system, plus the production of fish-meal, create the main environmental problems in the bays under discussion, contributing in 1988 some 6.36 ton/day BOD_5 in 1988.

The coastal waters are used to discharge other liquid industrial wastes as well, among these: acids, ammonia, iron salt solutions, cyanides, oils, lubricants, phenols, etc. The principal industries in this zone are: siderurgical, petrochemical, industrial plastics, oil refining, fisheries (principally fish meal), cement, chemicals, and wires (Ahumada 1992). Conflicting activities are marine cultures and recreation.

Hydrographic conditions in the Bay are rather homogeneous. Salinity is between 33.5-34.8 x 10^{-3} in winter and rises during the summer. Great changes in depth are not detected, due to the effect of the rain and the Lenga estuary which opens at the southern corner of the Bay. Superficial temperature presents small changes, with a range of 12.3-14.5 °C. Dissolved oxygen has a seasonal variation with higher values in winter. The low oxygen concentration is due to the high organic matter of industrial wastes.

LAND USE AND TENURE IN THE BAYS

Three Municipalities have jurisdiction over the Bay of Concepción: Talcahuano, Penco, and Tomé, while San Vicente Bay is within the Talcahuano Commune only. This Municipality is the most important administrative unit in relation to coastal environments, their resources and uses in the whole area, due to the intensive coastal utilization within its jurisdiction.

The industrial development in Talcahuano has taken place in two different and distant areas (separated by 2 km). The first is the Rocuant sector in the Bay of Concepción, where there are nine large fishing industries (1988 data base) which discharge wastewaters from fish unloading and fishmeal processing onto the Rocuant mud flat, inside the Bay. This causes great environmental deterioration and has won the reputation for Talcahuano of being one of the most polluted city/ports of the world.

The second one is San Vicente Port in San Vicente Bay, where eight fishing industries are based; three of which discharge wastewaters into the coastal enclosed waters of the Bay and four, into the municipal sewage system.

Despite the fact that the human effect in the nature of the Bíobio region coastal zone is evident (wastewaters discharge, fetid odors, ugly scenic views, recreational place and nautical sports loss, etc.), just recently, efforts and money have been spent into quantifying the environmental impact instead of enforcing the law to create an integrated coastal resources management program.

The recent resolution N°12600/550 of the Dirección General del Territorio Marítimo y Marina Mercante (DGTM & MM), a Chilean Navy organization that regulates and enforces trade and activities in the coastal zone, has made possible the identification and localization of the global impact and the effect of the coastal resources use.

South of San Vicente Bay lies the Arauco Gulf, a very important oceanographic system, influencing the industrial and economic activities of the Bíobio Region, which contains the Bay of Coronel, mainly administered by the Municipality of Coronel and Lota Bay, administered by the Lota Municipality. The first has two development areas: industry and local recreation. Mainly represented by fishing industries, the first activity is located in the Bay's northern area. Recreation and local tourism activities are located in the central-south area at the popular "Playa Blanca" (White Beach) site, one of the few important, publicly accessible beaches in the region. The effect of the industrial impact has been a loss of the quality of the environment, of the scenic views, and of the natural coastal resources. This fact has not only been bad for recreation and local tourism, but also has produced confrontations to obtain administration and Bay use among the different interested groups. Nowadays, this can be seen in the plan for the building of a port and in a correct definition and use of the seaport conception.

Governmental Administration

Considering the Comunas as an administrative division of the country governed by Municipalities, and having the quality of an autonomous public rights corporations as defined in the National Law N°18,695 dated on March 31, 1988 (the "Ley Orgánica Constitucional de Municipalidades"), these organizations have the capability of defining the development areas within their geographic borders, taking into account the national, regional and, obviously, "comunal" (from Comuna) interests.

A first point to consider is the coastal zone territory definition and its differentiation from the rural and urban development areas. Thus, in the case of Chile, we could have three possible national development areas: the urban, the rural, and the coastal areas, which have not been well-defined or set aside by the law.

Chilean legislation does not define the coastal zone, even though, this is needed for an adequate planning and management of the coastal zone. Actually the concept of a coastal zone as understood in other legislations, i.e., U.S. legislation, is still not understood in Chile. In fact, the problems associated with the lack of a national coastal zone management policy have not reached the national environmental agenda as yet, except for the mentioning of particular extreme cases, such as the Canal El Morro situation in Talcahuano. Such a definition is necessary, among other things, for the solution of agency jurisdictional conflicts.

According to Chilean climatic and physical-geographic characteristics, the coastal zone may have influences reaching, in some parts of the territory, to the Andes mountains. This influence from the ocean to the land makes is difficult to define a coastal zone for the whole country. This would have to be a local decision, perhaps based in the Regional divisions.

Nonetheless, this is a major problem, because it affects the development projects as far as the building and urbanization law of the Ministry of Housing and Urbanism ("Ley General de Urbanismo y Construcciones") issued on February, 1991. Since this legislation does not define the coastal zone: neither do the National Laws. In conclusion, the coastal zone is neither an urban nor a rural area, and its exploitation depends on the nearness of human settlements. Perhaps, the Exclusive Reserve Zone of 5 nautical miles created for the artesanal fishermen (as defined in the present Chilean Fishery Law), or limits among tides, or waste discharge limits, or biological resources limits, etc., can serve as patterns to define a coastal zone. As it is now, enforcement becomes a difficult problem for the Municipal administrative body.

An appropriate and modern definition of "coastal zone" is necessary as the basis for an adequate plan of coastal development and protection of the environment within the "Regulatory Plan". This plan is the only management instrument available at present from which a coastal zone program could be derived. This "Regulatory Plan" is defined as an instrument formed by the set of norms concerning the adequate conditions of hygiene and safety in buildings and urban spaces, and comfort in the functional relations between areas for living, work, services, and entertainment. This is needed for each coastal commune as a first step.

We can see that the definition of "Regulatory Plan" is legislatively poor, is at the root of many conflicts, and becomes at times, a source for arbitrary or inadequate administration measures, such as the situation in Lota, where the population was in conflict with the Mayor because he had made irregular changes in the Regulatory Plan in order to authorize the setting up of a fishery plant in the communal beach. Although the revision of all matters concerning the "Regulatory Plan" is urgent for the good management of coastal resources, it is a difficult matter because the law N°18,695, Article 4, gives the communal development responsibility to each Municipality. This can be done independently or with another national administrative organization. We propose that, having environmental protection functions, the "Maritime Municipalities" must define their coastal limits in order to achieve the best possible development options. According to Chilean Building and Urbanism General Law, Article 41 (1992), "Communal Urban Planning" is defined "as the promoter of the harmonic development within the communal territory, especially in populated places, and according to social and economic regional development aims."

It is important to consider that this definition involves only urban actors and does not consider rural and coastal areas unless these have human or industrial settlements. This law attributes zoning and use of the territory, the priority in land urbanization for city growth, etc., to the "Regulatory Plan."

If we now consider the hierarchy that starts in a Regional Plan supervised by a Regional Intendancy, follows for an Intercomunal Plan (a plan for common interests and resources), and finishes in a communal plan for the possible coastal areas' protection and management; the law N°18,695, Article 5 establishes that the Municipalities as autonomous bodies will be able to:

1. Execute the Development Comunal Plan and the programs needed to obtain it
2. Create workshops called "Unidades Vecinales" (associations of neighbors with common interests), to obtain a balanced development and to channel a public participation.

One nonessential municipal duty in Chilean legislation is the general enforcement of environmental protection inside the municipal jurisdiction. The duty should be redefined by the government in order to give a higher priority to the coastal management and the environmental protection.

According to the municipal and urban structure of the national legislative body, and without considering the new general and specific provisions of the National Law, called "Ley de Bases del Medio Ambiente" (Chilean Environmental Law), we think that there is a way to guide a coastal management initiative in Chile. The only barrier for this is to define the needs, the uses, and the limits of the coastal zone.

In Chile, the administrative, governmental organizations are created at the municipal level. Thus, a bottom-up policy is hierarchically considered. But, in this case, perhaps the best policy, according to our own national reality may be a kind of top-down one (from the national or regional government to the smaller governmental divisions), with an enclosed seas and coastal zones management policy. For this, it would be necessary for the actual regional autonomy to define their own development plans (as defined in Chilean Regional Government Law).

The Special Case of the Port of Coronel

In the case of Coronel Bay, the Ministery of Housing and Urbanism has started a study for the "Plan Seccional del Puerto de Coronel" (a plan to build a seaport in Coronel Bay). In this plan, the seaport concept is a priority. Here, the problem is to define the Coronel Port. The Navy Hydrographic Institute defines its 'port', in this case, as the Bay's waters and adjacent lands. This is also called "the urban sector around the port - Entorno Urbano del Puerto". On the other hand, this "Plan Seccional's" goals are to assist in the Coronel Port's development, its associated service-providing system, and in the free access to its shore; it also takes into account the zoning of the urban lands and marine waters uses (Secretaria Regional Ministerial de Vivienda y Urbanismo, VIII Region 1992).

The plan to follow includes the modification of the pre-existing "Regulatory Plan". This clashes with the views of different interested groups present in the area and produces an arbitrary port area zoning, and a directional development which mainly satisfy particular interests. From this it becomes obvious that it is necessary to unify criteria on the basis of objective analysis in order to obtain an optimal, multi-objective mix. Thus the Coronel case is a very good study point undertaken by Center EULA-CHILE.

The last point to close the coastal-associated organizations' political-administrative vision, is to mention that each municipality must have a Common Development Unit which will be able to propose and create some actions for its environmental protection. It also must have a Municipal Works Unit, leading to the creation and enforcement of a "Regulatory Plan" and to prevent environmental deterioration (Ley Organica Constitucional de Municipalidades 1988).

CONCLUSION

Despite the lack of a Coastal Management Program in Chile, in some coastal areas, especially in the Bays considered in this work, we can see there is plenty of opportunities to overcome this planning shortcoming. With the present legislative framework which gives autonomy to regional and municipal government organizations, environmental planning opportunities must be created considering watershed, flora, fauna, and coastal zone management.

In the same way, in the case of Chile, we think that a national coastal resource management policy must consider all the needs and actors involved in this system, which include environmental, economic, and social areas connected between them. This makes the coastal zone one of the most complex environmental systems (Gallardo 1976).

Considering Concepción and San Vicente Bays' case, where industrial development is sometimes more important than social welfare, their conflicting uses are:

1. Industries
2. Fisheries raw material discharge and processing
3. Coastal tourism and amusement
4. Coastal forestry activities
5. Housing building
6. Ports and piers

For this reason, the planning focus must integrate all the activities from a multisectorial viewpoint so that all the organizations having area jurisdiction can participate. In order to accomplish this, the municipalities of Talcahuano, Penco, and Tomé must consider a "Plan Regulador Intercomunal" for Concepción Bay which is a common good, as soon as possible, in order to avoid basing environmental planning on an underdeveloped, country policy. This means foreseeing the damage before it is too late.

Finally, and according to our experience, the Municipality of Talcahuano can be the most important organization for a coastal zone development plan in the Eighth Chilean region.

In Coronel Bay, there is much to do. Nonetheless, because of its plan for future development, it soon will become a possible "study" area for the other regions and for the rest of the country.

Some of the general goals of a national coastal management policy include (Olsen and Seavey 1990):

1. Identification of problems and needs
2. Establishment of hierarchy
3. Initiation of study programs with the experience of NGOs (non-governmental organizations) and GOs (governmental organizations)
4. Organization for public participation
5. Enforcement
6. Financial resources
7. Competent and qualified personnel participation

Important steps to establish the basis of a Land and Coastal Zones Global Program and a conceptual framework for a social and economic resources evaluation are: (a) the creation of communal teams, (b) the implementation of the coastal zone concept, and (c) tenure, use, and planning of natural resources. In this way it will be possible to achieve the adequate institutionalization needed to coordinate national, regional, and communal actions in the coastal zone.

Our local experience tends to support the hypothesis that, with the goal of ICZM in mind, supporting universities is a good investment. In fact, the role of universities in the formulation of environmental policy has been often stressed.

Although Center EULA-CHILE from the University of Concepción, because of its emphasis on developing management proposals for the Biobío River Basin, initiated efforts on the coastal zone management problems in this, one of the most coastally-dependent and coastally-oriented regions of Chile, the elaboration of an ICZM program is still ahead of us. Elements towards this goal are already in place, i.e., physical infrastructure, trained staff, and strong connections with the political and social web. We only need and increment in our data bases and specially focused effort on coastal issues in order to achieve major goals in the subject.

REFERENCES

Ahumada, R.B. 1976. "Contribución al conocimiento de las condiciones hidrográficas de la Bahía Concepción y áreas adyacentes (CHILE)". Thesis to obtain the Licence degree in Biology, Universidad de Concepción, Chile.

Ahumada, R.B., 1992. "Distribución espacial de metales traza (Cd, Cr, Cu, Ni, Pb y Zn), en una bahía de uso múltiple: San Vicente, Chile". Tesis to obtain Ph.D. in Environmental Sciences. Centro EULA-CHILE, Universidad de Concepción. 105p.

Ahumada, R.B. & Rudolph A.J. 1988. " Estudio de la selección de un sitio de descarga de RIL, provenientes de la Industria Pesquera. Technical Project, Final Work. Pontificia Universidad Católica de Chile, sede regional Talcahuano.

Gallardo, V.A. 1976. "Hacia una administración moderna de la zona costera de Chile". In: Preservación del Medio Ambiente Marino (págs. 270-281). Edited by Fco. Orrego Vicuña. Inst. de Estudios Internacionales. Universidad de Chile. Edited by Universidad Técnica del Estado-Chile.

Gallardo, V.A. 1992. "Descripción de la situación ambiental de la zona costera del gran Concepción. Impacto social de la contaminación marino-costera". Informe final Seinario Internacional sobre Planificación Territorial. Centro EULA-CHILE, Universidad de Concepción.

Ley General de Urbanismo y Construcciones. 1991. Ministerio de Vivienda y Urbanismo. División de Desarrollo Urbano. República de Chile. Versión no oficial revisada, 1991.

Ley Orgánica Constitucional de Municipalidades, Ley N°18.695 de la República de Chile, publicada en el Diario Oficial del 31 de Marzo de 1988.

Olsen, S., and G.L. Seavey, 1990. The State of Rhode Island. Coastal resources management program, (as amended). Coastal Resources Management Council. Narragansett, Rhode Island: Graduate School of Oceanography, U.R.I.

Ordenanza General de Urbanismo y Construcciones. Ministerio de Vivienda y Urbanismo, División de Desarrollo Urbano. Versión Oficial. Santiago de Chile, Mayo de 1992.

Secretaría Regional Ministerial de Vivienda y Urbanismo, VIII Región. 1992. Términos de Referencia para Contratar la Elaboración del "Estudio para el Plan Seccional Puerto de Coronel". Concepción, Chile.

Sorensen, J.C., and S.T. McCreary. 1990. Institutional arreangements for managing coastal resources and environments. National Park Service, U.S. Department of Interior and U.S. Agency for International Development, 194 p.

27

Assessment of Effects of Coastal Power Plants on Marine Biological Resources in Italy

Romeo Cironi
Senior Scientist, ENEL S.p.A.,
Electric Energy Joint Stock Company, Piacenza, Italy

Edmondo Ioannilli
Manager, ENEL S.p.A.,
Electric Energy Joint Stock Company, Piacenza, Italy

Roberto Vitali
Senior Scientist, ENEL S.p.A.,
Electric Energy Joint Stock Company, Piacenza, Italy

ABSTRACT

The concerns about the environmental effects of the operation of coastal power plants on marine resources put forward in the early seventies, triggered in all industrialized countries, the issue of stringent limitations on thermal discharges, which severely impacted the design and the operation of such plants. Consequently, in the following years, many experimental studies have been performed to assess these worrisome effects in real life conditions. In Italy, very complex and long-term investigations have been performed in several coastal sites hosting large power plants (Piombino, La Spezia, Torre Valdaliga, and Fiume Santo). These locations, representative of different ecological situations, were investigated by independent study groups belonging to the Italian universities. The aim of these studies was to detect possible modifications of the structural and dynamic characteristics of several ecosystem components (namely phyto- and zooplankton, benthos, and fishes, together with physical and chemical parameters) associated with thermal discharges.

In this paper, the methodologies of the quoted studies are delineated, together with their results. In general, these studies failed to detect any biologically important effect, even in the vicinity of the outfalls. In light of these results, we have drawn the conclusion that the early concerns about the effects of thermal discharges were largely overestimated and, consequently, that it would be wise to mitigate the present limitations which have proven more stringent than necessary to protect the biological resources of coastal sites.

INTRODUCTION

The concerns about the environmental effects of the operation of coastal power plants on marine resources put forward in the early seventies, triggered in all industrialized countries, the issue of stringent limitations on thermal discharges which severely impacted the design and the operation of such plants.

Consequently, in the following years, many experimental studies have been performed to assess these worrisome effects in real life conditions. In Italy very complex and long-term investigations have been performed in several coastal sites hosting large power plants (Piombino, La Spezia, Torre Valdaliga, and Fiume Santo) (Figure 1). These locations, representative of different ecological situations, were investigated by independent study groups belonging to the Italian universities. The aim of these studies was to detect possible modifications of structural and dynamic characteristics of several ecosystem components (namely phyto- and zooplankton, benthos, and fishes, together with physical and chemical parameters) associated with thermal discharges.

The experimental approach was generally based on statistical comparisons between preoperational and operational phases and among zones differently impacted by temperature rise (Ioannilli and Franco 1988). This approach, if rigorously followed, seems adequate to detect possible ecosystem modifications, independent from the actual mechanisms of action, including impingement and entrainment, as well as synergisms between the heat discharge and the pollutants independently existing in the water body.

In this paper, the methodologies of the quoted studies are delineated, together with their results.

Figure 1. Study area

EXPERIMENTAL APPROACHES AND RESULTS

Power Plant of Piombino (Central Tuscany)

In the sea area in front of the Thermoelectric Power Plant between the Port of Piombino and Torre Mozza on the Gulf of Follonica (Figure 2), thorough ecological research projects were carried out from 1975 to 1981 at three different times by research groups from the Universities of Modena and Pisa.

The first survey was made in 1975, before the start-up of the power plant, to assess the situation of an area already subject to the discharge of the nearby steel industry. After the start-up, a second survey was made from September 1977 to September 1978, and a third from October 1980 to September 1981. The most significant studies regarded the pattern of the benthic communities' structure and the heavy metals incorporated into the mussels.

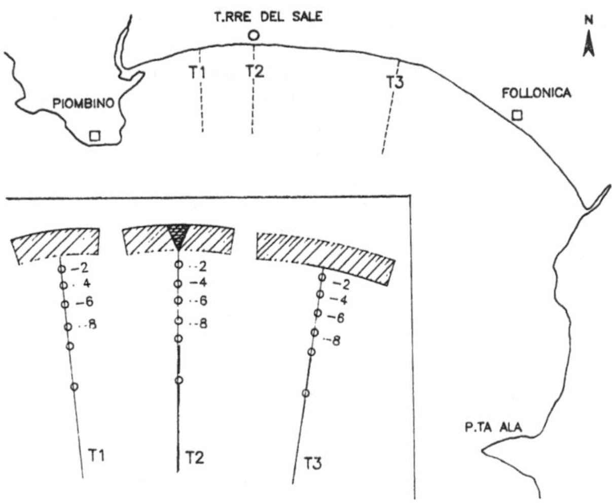

Figure 2. Study area

Briefly, the research has shown that the benthic communities' structure in the Gulf of Follonica is regulated by natural factors such as the depth, the sediment features (connected to the terrigenous input of Cornia River), the Posidonia prairie, and the local hydrodynamic regime (Figure 3).

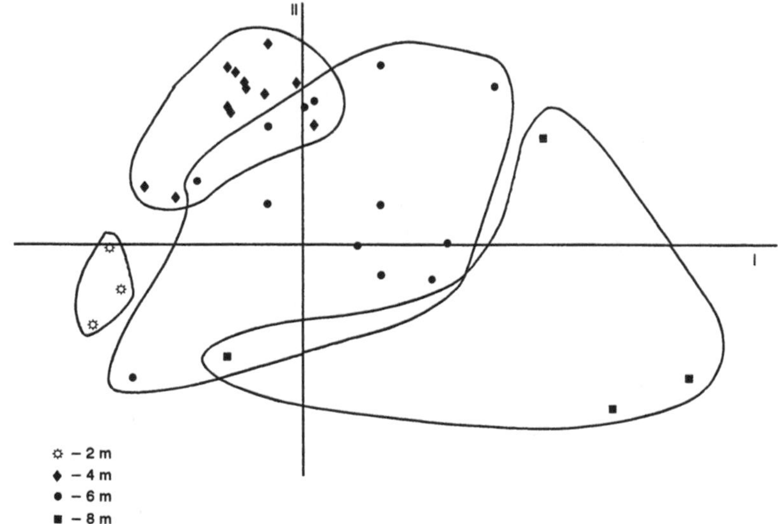

Figure 3. Principal Components Analysis: sampling points in the subspace defined by I and II principal components

Interestingly enough, no influence connected to the thermal discharge of the power plant was detected, even using reasonable statistical power (Bonvicini Pagliai et al. 1980; Bonvicini Pagliai et al. 1982).

As regards the mussels, it was found that those sampled in the discharge area (Torre del Sale) presented metal concentrations (especially of lead) lower than those sampled in nearby areas, even if higher than those sampled in cleaner areas (Punta Ala and Gulf of Baratti). Certainly the temperature rise did not increase the heavy metal accumulation by mussels.

These results are detailed in the following Reports:

1. Univ. Pisa, Ist Zoologia; Univ. Modena, Ist. Zoologia; ENEL-DCO/ULP (1983): "Centrale Termoelettrica di Piombino: Risultati della terza indagine ecologica nel tratto di mare prossimo all'impianto (Ottobre 1980 - Giugno 1981)".
2. Univ. Pisa, Ist. Zoologia; ENEL-DCO/ULP (1989): "Centrale Termoelettrica di Piombino: Risultati della quarta indagine ecologica nel tratto di mare prossimo all'impianto (Ottobre 1987 - Giugno 1988)".

Power Plants of Civitavecchia (Northern Latium)

The study goals have been extended to a global assessment of the ecosystem quality affected, not only by power plant discharges, but by the urban effluent of Civitavecchia and by harbor activities as well (Figure 4).

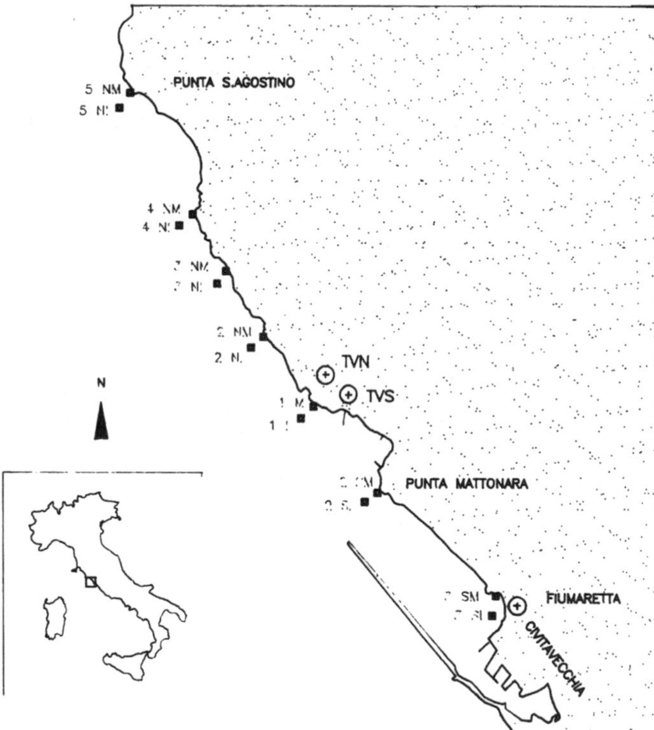

Figure 4. Study area

The research carried out examined the following environmental components:

1. 1977 - Water quality and plankton and the nature and morphology of the depth and benthos
2. 1983-84 - Water quality, meso- and infralittoral benthic biocenosis, and the epiphyte community of *Posidonia oceanica*
3. 1988-89 - Water quality and zooplankton, infralittoral benthic community, *Posidonia oceanica*, and the epiphyte community.

Water Quality

The water quality in the study area shows evidence of eutrophication, with nutrients more or less abundant, depending on the oceanographic condition. The eutrophication seems to increase with time in the study period, even if the deleterious effects usually connected to this condition, i.e., lack of oxygen, low water transparency, and death of benthic species, were not observed until the date of the last research (December 1989). In fact, because of the strong oxygen production on the bottom due to the presence of Posidonia and the active water currents in the area, good water quality is seemingly assured (Figure 5).

The concentration of dissolved oxygen decreases slightly just in front of the power plants discharges, but only when it is near or higher than saturation level, without effects on the water quality (Crema et al. 1979).

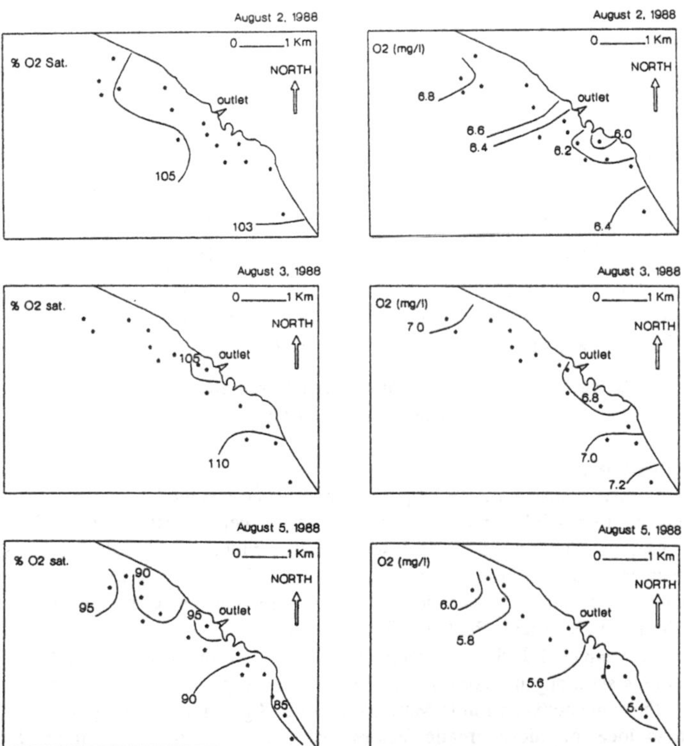

Figure 5. Patterns of oxygen distribution near surface

Plankton

In the first period (1977), the study covered phyto-, zoo- and ichthyoplankton. The survey, carried out in Spring and in Summer, took into consideration ten stations placed in five transects along the coast, at two depths (-8 and -16 m) in the tract of sea between Punta Mattonara and Capo S. Agostino. For all the three components, the functioning of the power plants that were operational at that time (Fiumaretta and TVS) did not show any considerable, direct or indirect influence on the checked population structure; whereas variations connected to the dilution of the water coming from the Port of Civitavecchia with the open sea water were more or less evident (Ioannilli et al. 1979).

In the years 1988-89, the research on plankton concentrated only on the zooplanktonic component which was sampled every fifteen days at four stations set between Capo Linaro and Punta S. Agostino, about 1 mile from the coast.

The results of the research showed that this component is completely controlled by the seasonal cycles, typical of the principal taxa, and that there was no significant difference among the historic series taken in stations set at different distance from the thermal discharge (Figure 6).

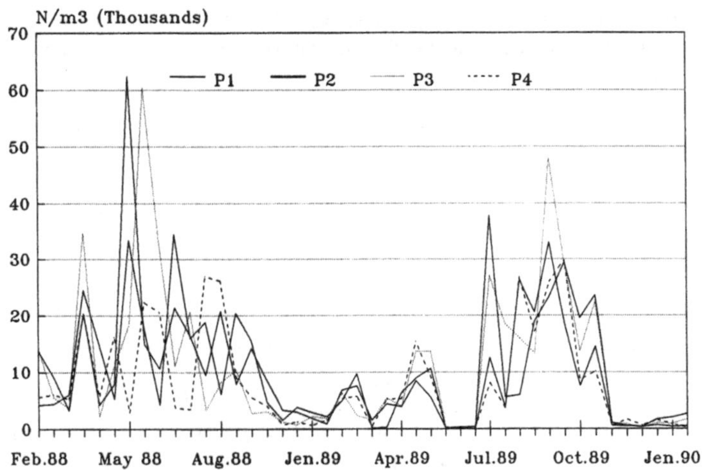

Figure 6. Historic series of Copepods abundances at different distance from discharge

Benthic Community

The experimental plans of the three periods have been changed to make the research methods increasingly sensitive to possible direct or indirect effects of thermal effluent on the benthic population.

The first research (Summer 1977) dealt with a rather wide, spatial scale (twelve stations set at -4 -8 and -16 m between Punta Mattonara and S. Agostino). The pattern of the population structure was found to be determined mainly by the depth, without showing any effect correlated to the thermal perturbation (Figure 7) (Bonvicini Pagliai 1979).

The research conducted during the years 1983-84 (before the start-up of the TVN Power Plant) regarded the stations set along the coast in the mesolittoral strip (depth between 0.5-0.3 m); the study area was always the same between Punta Mattonara and S. Agostino. The variations among the stations were attributed to local or microclimatic factors. Even with this new experimental plan, particular effects such as the reduction of the specific richness, or numeric abundance, or the settlement of thermophile species, were not noticed (Taramelli et al. 1985; Pelusi et al. 1985).

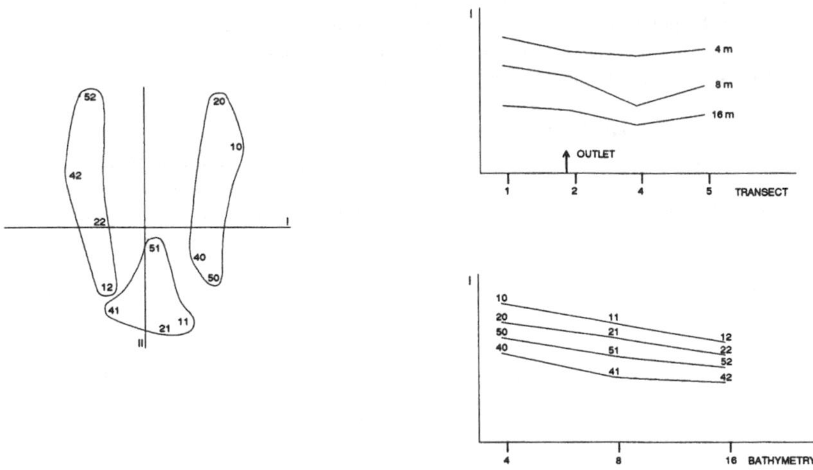

Figure 7. Principal Components Analysis: Benthos 1978

In the last survey period (1988-89), the study was limited to the benthic community of *Corallina facies*, examined at thirty stations, set in a range of about 1,000 m from the thermal effluent of the TVN Power Plant; whereas the benthic community on hard substratum in shallow water was sampled at thirty stations, set along the coast within a range of about 2,000 m from the thermal effluent. Both research samplings were repeated four times in two years. Also the results of these last studies, designed in order to enhance the survey's sensitivity, did not show any striking variations of the benthic community associated with the thermal perturbation (Figure 8).

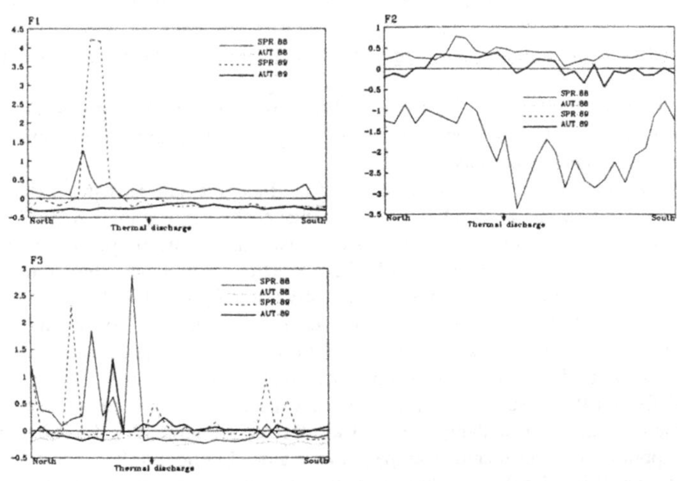

Figure 8. Factor Analysis: Benthos

Some elements would suggest that the spatial structure is established mainly by the presence of urban and port discharges from South. In conclusion, the thermal impact on the benthic biocenosis, if present, is restricted at a distance shorter than 250 m from the point of discharge; that is to say in an area subject to a substantial thermal perturbation (Chimenz et al. 1985).

Posidonia Oceanica and Epiphyte Community

In terms of phenologic parameters, the prairie of *Posidonia oceanica* is characterized by a condition of productivity and epiphyte community structure quite normal for the Tyrrhenian coast of the Italian Peninsula. The fluctuations of the epiphyte community structure (Figure 9) and productivity values (Figure 10) are not dependent on the distance from the point of the thermal discharge intake.

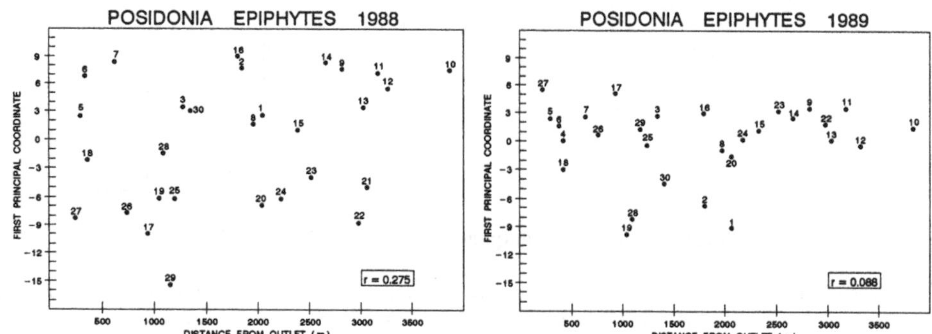

Figure 9. Posidonia epiphytes: regression between distance from outlet and sampling points coordinates on I axis

The cited results are detailed in the following works:

1. ENEL-DCO/ULP (1978): "Valutazione degli effetti degli scarichi termici di una grande Centrale Termoelettrica costiera. Indagine presso la Centrale di Torrevaldaliga".
2. ENEL-DCO/ULP (1983): "Indagine sull'ecosistema marino prospiciente le Centrali Termoelettriche di Civitavecchia. Primo rapporto di avanzamento. Indagini termiche, chimiche e biologiche dell'estate 1982".
3. ENEL-DCO/ULP (1983): "Indagine sull'ecosistema marino prospiciente le Centrali Termoelettriche di Civitavecchia. Secondo rapporto di avanzamento. Indagini termiche, chimiche e biologiche del marzo 1983".
4. Univ. Roma La Spienza; Ist. Zoologia (1985): "Indagine sull'ecosistema del tratto di mare prospiciente le Centrali Termoelettriche di Civitavecchia. Indagine biologica".
5. ENEL-DCO/ULP (1989): "Indagine sull'ecosistema marino prospiciente la Centrale Termoelettrica di Torrevaldaliga Nord. Caratteristiche chimiche e biochimiche dell'acqua. Primo rapporto di avanzamento. Campagna estiva dell'Agosto 1988".
6. ENEL-DCO/ULP (1989): "Indagine sull'ecosistema marino prospiciente la Centrale Termoelettrica di Torrevaldaliga Nord. Caratteristiche chimiche e biochimiche dell'acqua. Secondo rapporto di avanzamento. Campagna invernale del Dicembre 1988".
7. ENEL-DCO/ULP (1990): "Indagine sull'ecosistema marino prospiciente la Centrale Termoelettrica di Torrevandaliga Nord. Caratteristiche chimiche e biochimiche dell'acqua. Terzo rapporto di avanzamento. Campagna estiva del Luglio 1989".
8. ENEL-DCO/ULP (1990): "Indagine sull'ecosistema marino prospiciente la Centrale Termoelettrica di Torrevandaliga Nord. Caratteristiche chimiche e biochimiche dell'acqua.

Quarto rapporto di avanzamento. Campagna tardo-autunnale del Dicembre 1990".

9. Commissione Tecnico-Scientifica Comune di Civitavecchia-ENEL (1991): "Indagine sull'ambiente naturale nel tratto di mare antistante la Centrale di Torrevaldaliga Nord".

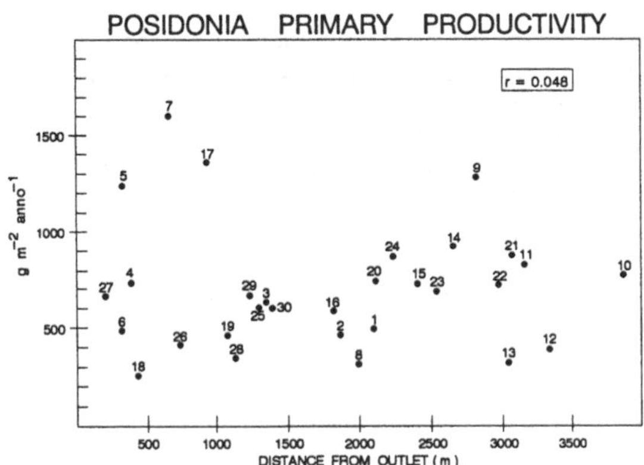

Figure 10. Posidonia productivity: regression between distance from outlet and sampling points coordinates on I axis

Power Plant of Fiume Santo (Northern Sardinia)

The study was carried out in the area shown in Figure 11, in the periods before and after the power station start-up (respectively both Spring and Autumn 1981 and Spring and Autumn 1985). It covered the pattern of the thermal perturbation of the water quality parameters and of the structure of planktonic and benthic populations. The studies on the water quality and on the planktonic component were carried out either in a coastal strip of 6x3 km in front of the thermal discharge (mesoscale), or in the thermal plume itself (microscale).

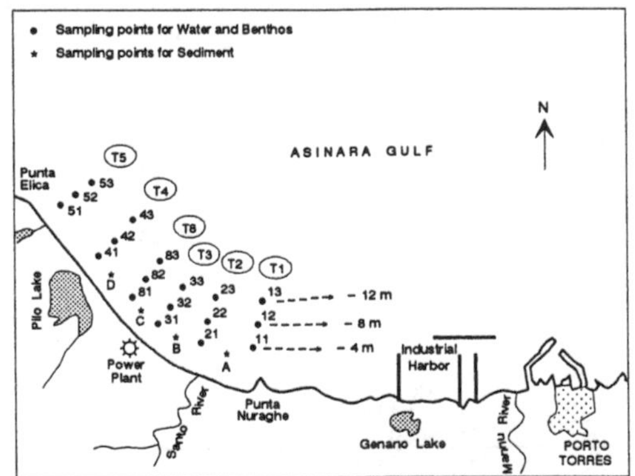

Figure 11. Study area

Water Quality

In both preoperational and operational periods, the water quality pattern in the mesoscale turned out to be determined by the same influencing factors: the urban and industrial discharges (coming from the industrial harbor, a petrochemical complex and the town of Porto Torres); the oceanographic situation; the depth; and the season. No effect associated with the thermal discharge was detected in this spatial frame.

Effects connected to the power plant were noticed on some parameters in the microscale investigations (Figure 12). The main effects of the passage through the condenser (station 61 vs station 62-63) were: (a) a temperature rise of about 8.2 °C; (b) a decrease of 0.5 ppm of DO; (c) a decrease of Chl<u>a</u> (from 0.34 to 0.11 mg/m^3) and of the primary productivity (from 6.4 to 0.9 mg/m^{3}h), with a consequent increase of pheopigments; and (d) small increases of P.tot, P-PO4, and TOC with the N-NH4 showing an irregular pattern. All parameters recovered their natural values a short distance from the discharge structure (Franco et al. 1988).

The apparent increase of the dissolved oxygen along the thermal plume is largely due to the lower oxygen content of the intake water. The east side of the structure appeared to be more affected by the cited discharges than the west side.

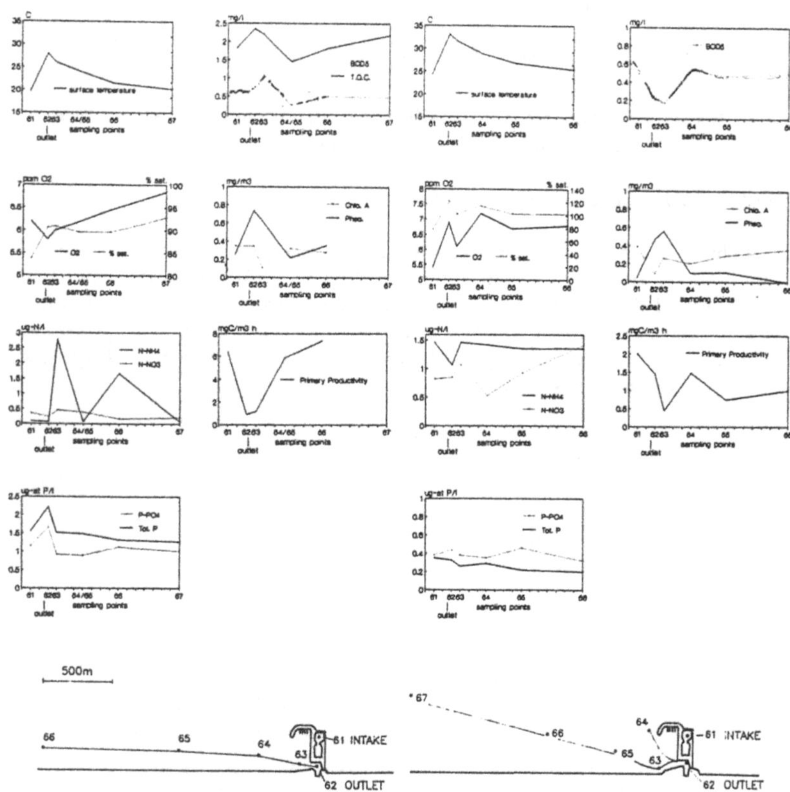

Figure 12. Spatial trend of some environmental variables along thermal plume

Planktonic Population

The analysis of the spatial distribution in the mesoscale of the planktonic population showed that the same factors influenced, directly or indirectly, the abiotic components and the plankton. The study of the spatial distribution of the population structure did not show any impact by the thermal perturbation, except the localized effect identified in the microscale survey (Figure 13). This showed that the passage through the cooling circuit reduced significantly the biomass of the planktonic population; however, out from the station following the discharge (at about 200 m), a gradual increase of abundances was noticed: the original values (and the population structure) were recovered at a distance of about 750 m (Arru et al. 1988).

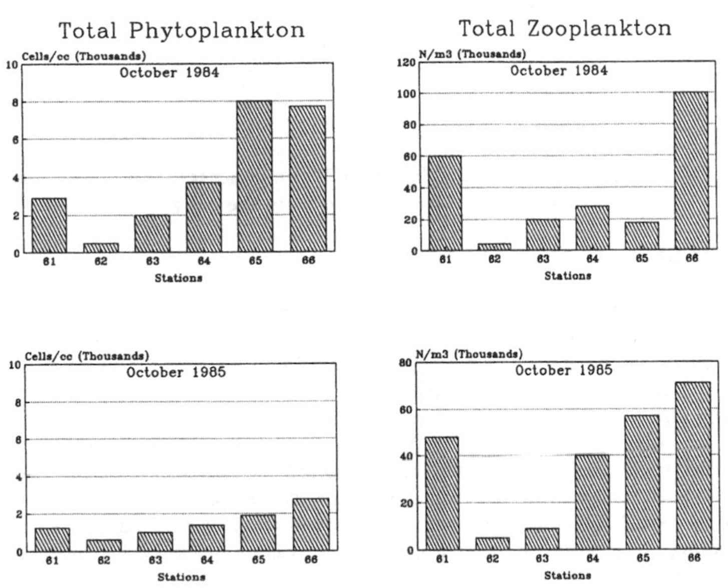

Figure 13. Numerical abundance, station by station

Benthic Component

A prairie of *Posidonia oceanica* represented the biocenosis at vegetable dominance in the studied area. The prairie structure showed the typical aspect of the infralittoral plan and the vegetative bloom of Posidonia was normal. The analysis of the spatial distribution of the benthic population showed the depth as the main influencing factor. In the stations nearer to the coast, the population structure was influenced by many local factors whereas, going toward greater depths, the communities located in all the stations were found to be more uniform, so that significant differences among the various zones under the influence of the thermal impact were not evident (Figure 14) (Arru et al. 1988).

The results above-mentioned are reported in:

1. Univ. Sassari: Centro Studi Territorio, Ist. Chimica Analitica, Ist. Antropologia (Cattedra Biologia Generale); ENEL-DCO/ULP (1986): "Controllo dell'ambiente marino antistante la Centrale Termoelettrica di Fiume Santo".

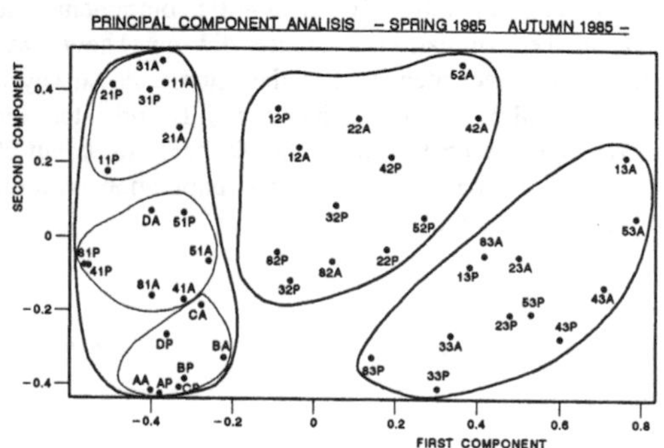

Figure 14. Sampling defined by I and II components. Note: Bold line depth; thin line, distance from the outlet.

Power Plant of La Spezia (Eastern Ligury)

The research was planned to establish the compatibility among the competing uses of water, e.g., for cooling of the power plant and other available environmental resources. This would allow local authorities to document assessment on the social acceptability of the present utilization of the gulf, and it would supply technical information to modify it if necessary. The environmental resources that can be theoretically affected by the discharge, directly or indirectly, are the following (Figure 15):

1. In the areas heavily utilized for the harbor in front of the town (area A) the aesthetic quality of the water could theoretically be considered an alterable resource, but it is already considered (unjustifiably) an unalterable biological resource.
2. In the coastal area toward the South, in the open areas inside the dike (area B) the alterable resources, besides the aesthetic quality, are the water's fitness for the mussel culture and for recreational use (bathing, fishing, etc.)
3. In the areas outside the dike (area C) the most exigent resource (whose protection safeguards all the other ones) is certainly the water body's capacity to support an indigenous biological community.

To estimate the influence of the power plant's operation over the port waters and their inhabitants, it was necessary to describe their condition in all the interested areas and to link it with the degree of the perturbation induced by the power plant, taking the other influencing factors into consideration.

The hydrodynamic modeling of the whole basin was the basis upon which we estimated the flows among the different areas for the interpretation of the observed patterns of various parameters.

The following components were chosen, considering their biological meaning and suitability of their description:

1. General indicators of the water quality (including heavy metals and metallorganics)
2. Indicators of biological activity (chlorophyl, primary and heterotrophic productivity)
3. Mussels as bioindicators
4. Zooplankton
5. Macrobenthos

The research was carried out in Summer 1989, in Winter 1989-90, and in Summer 1990.

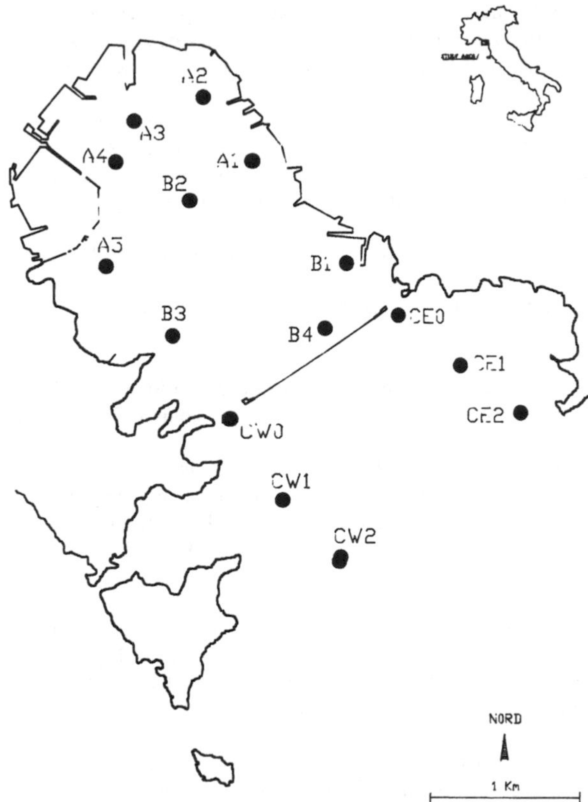

Figure 15. Study area

Water Quality

The water of the La Spezia Gulf showed the effects of urban waste and run off. However, the pollution levels were unexpectedly low for a harbor zone because of the sustained water dynamics in the Gulf, mostly induced by the power plant discharge. Outside the dike, the pollution parameters were always at low levels and the oxygenation degree was good inside and outside, either in Summer or in Winter.

The concentrations of trace metals (Cu, Mn, Zn, Fe, Cd, Pb, As, Sb, total Cr, and exavalent Cr) were in the same range normally found in coastal environments. The IPA, PCB, and organotin compounds were in low concentration compared with those found in similar environments. The analysis of the chlorophyl a in Winter, showed values of concentration of 2.5 mg/l inside the dike, decreasing to about 0.8 mg/l in the outside area.

The concentration of total and fecal coliforms and of fecal streptococci showed variable values depending on the zone; the three bacteriologic components were correlated with the indicators of cloacal discharges, such as salinity and nutrients.

The heterotrophic bacterial activity (AET) was measured to estimate the self-purifying power of the water body. The AET values were similar in both seasons (mean values 11.6 mg C/m^3/hour in Summer and 13.7 mg C/m^3/hour in Winter) and relatively higher than the ones found in other coastal areas of the Northern Tyrrhenin affected by a smaller quantity of urban discharges (Cironi and Ferrari 1991). Any direct influence of the temperature increase caused by the thermal discharge of the ENEL Power Plant was not evident.

The "mussel watch", that is the utilization of mussels as bioindicators for a temporally integrated evaluation of the environmental condition, was applied. Regarding trace metals (Hg, Cd, Pb, Cu, As, Zn, Sb, and Se), the bioaccumulation were similar in Winter and Summer, except for Pb, that in Winter was ten times higher than in Summer. The concentration of Hg and Pb are definitely lower than the limit values set by Italian law.

The preliminary conclusion drawn from this part of the study was that the area's acceptable water quality, quite unexpected, given the burden of urban and industrial discharge, could be largely due to the positive effect of the power plant discharge on the water turnover in the internal zone of the Gulf.

Zooplankton

The study of the zooplanktonic components carried out in Summer 1989 and in Winter 1989-90 identified the natural factors controlling the biomass and the composition of the population in the Gulf of La Spezia.

The spatial trends were interpreted by considering the particular water circulation prevailing in the Bay and the influence of the Magra River water, which was probably the principal factor effecting the structure of the zooplanktonic population in the checked area (Figure 16).

The observed pattern did not show any effect from the thermal discharge.

Benthos

The study allowed the description of the composition and, partially, the structure of the macrobenthic population in the superior infralittoral at the stations of Figure 17 (hard substratum). The population structure in the various stations were clearly interpretable on the basis of the differences of the principal natural factors. Important differences were generally associated with the seasonal cycle and water dynamics.

The existence of an influence on the population in the study area from thermal input can be excluded, since the species composition (341 species were identified) does not show important differences "ceteris paribus" from that of other sites with no thermal discharge.

The above-mentioned results are detailed in the following Reports

1. ENEA/CREA, ENEL-DCO/ULP (1990): "Indagini sulle caratteristiche ambientali delle acque del Golfo di La Spezia in attuazione del protocollo di intesa EELL-ENEL. Primo rapporto di avanzamento".
2. ENEA/CREA, ENEL-DCO/ULP (1991): "Indagini sulle caratteristiche ambientali delle acque del Golfo di La Spezia in attuazione del protocollo di intesa EELL-ENEL. Secondo rapporto di avanzamento".
3. ENEA/CREA, ENEL-DCO/ULP (1992): "Indagini sulle caratteristiche ambientali delle acque del Golfo di La Spezia in attuazione del protocollo di intesa EELL-ENEL. Terzo rapporto di avanzamento".

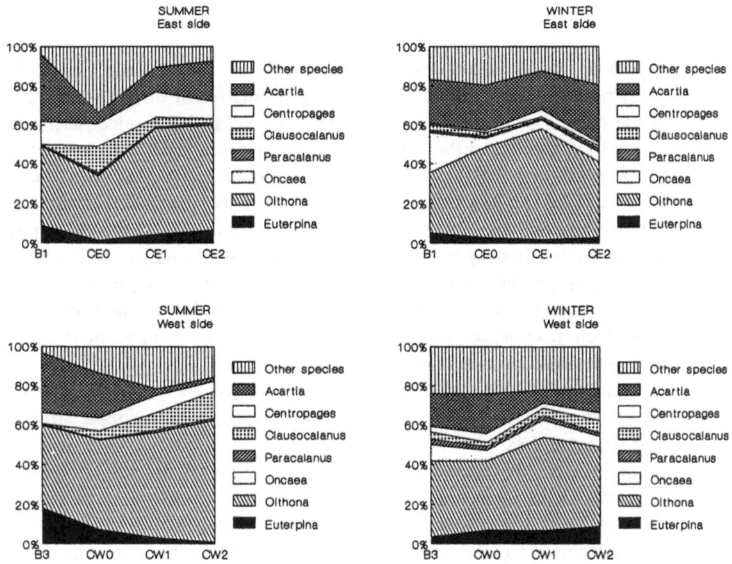

Figure 16. Spatial trend of principal genera of Copepods

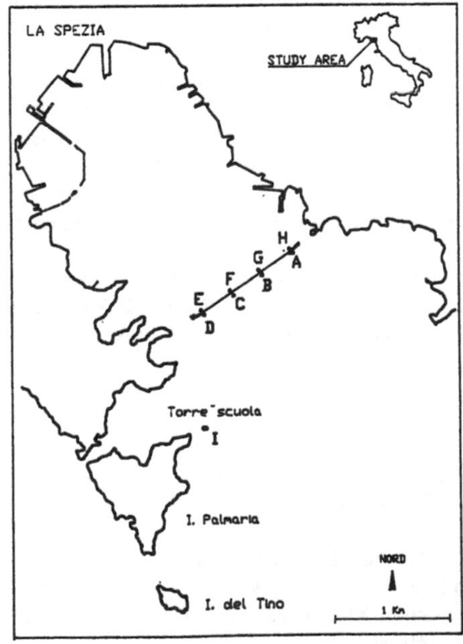

Figure 17. Sampling stations

CONCLUSIONS

This paper reports a review of the studies made with the collaboration of Italian university research groups at many power station sites which were representative of different ecological situations, in order to point out the effects of the thermal discharges of coastal thermoelectric plants.

The methodological approach and the essential results of these studies have been systematically reported, quoting the Reports and papers (available upon request) in which they are described and supported extensively. The general sense of the results is that the thermal discharges do not affect significantly the structural and dynamic features of the main components of the examined ecosystems. In fact, some effects are noticeable only at short distance from the discharge, in the inner body of the thermal plume. Concerning any effects upon the dissolved oxygen, primary productivity, biomass, and plankton composition; they quickly vanish with the progressive dilution of the plume in the receiving water without permanent environmental alterations. Benthic population appears virtually unaffected, even at short distances (hundreds of meters) from the discharge.

Therefore, we can conclude that the concerns put forward in the early seventies about the environmental impact of the thermal discharges of power plants, which caused stringent legislative limitations, were actually excessive.

These experimental results should suggest a reconsideration of the limitations that have been responsible for the grave consequences on the investment and operating costs of power plants, particularly in Italy and in The United States. It is, moreover, evident that it is impossible to establish an unique, suitable regulation for all situations, which could assure, at the same time, the protection of the environment and its productive use. For example, it seems unreasonable to apply to coastal areas which are used exclusively as industrial harbors, the same limits suitable for a natural body of water, which must sustain a balanced, indigenous biological community. In fact, in these areas, the aesthetic water quality (heat, odor, etc.) can be the only resource possibly influenced by the thermal discharge and as a biological resource, it is not identifiable.

In general, legislation relevant to thermal discharges in water bodies issued from most of the OCSE Countries (England, France, Germany, and Spain) often provides a "case by case" approach. In The United States, even if there exist stringent regulations issued from several States, it is provided (Federal Register, Oct. 8, 1974, par. 122.9) that the Regional Administrator may mitigate restrictions if the plant owner proves "that the effluent limitations otherwise required are more stringent than necessary to assure the protection and propagation of a balanced, indigenous community of shellfish, fish, and wildlife in and on the body of water into which the discharge is made, and such demonstration has not been rebutted".

It is urgently necessary that this pragmatic approach, in harmony with the Environmental Impact Assessment Procedure, be introduced in Italy too, as CNR/IRSA had already authoritatively suggested at the Congress "Ten years since the Merli Italian law" (IRSA 1986).

REFERENCES

Arru, A., M.A. Franco, V. Tupponi, E. Ioannilli, M. Bertonati, R. Cironi, and M. Miserocchi. 1989. Valutazione degli effetti di uno scarico termico sull'ecosistema costiero del Golfo dell'Asinara. Parte I. Aspetti biologici. *Inquinamento* 31 n. 12.

Bonvicini Pagliai, A.M., R. Crema, E. Ioannilli, M. Bertonati, R. Cironi, and R. Vitali. 1979. Caratteristiche strutturali del macrobenthos della fascia infralitorale antistante la Centrale di Torrevaldaligha (Civitavecchia).XI Congr. Soc. Ital. Biol. Mar.; *Atti Soc.Tosc. Sci. Nat. Mem., Ses. B, 86, suppl. 1979*, 160-67.

Bonvicini Pagliai, A.M., R. Crema, E. Ioannilli, and R. Vitali. 1980. Spatial and temporal patterns of sandy bottom communities near a coastal power plant. *V Jurnes Etud. Pollutions*. 815-22. Cagliari: C.I.E.S.M.

Bonvicini Pagliai, A.M., R. Crema, and R. Vitali. 1982. Notes on the analysis of the environmental data from pollution surveys by multivariate techniques. In *Environmental systems analysis and management*, ed. S. Rinaldi, 467-76. Amsterdam: North-Holland Publishing Company.

Chimenz., C., E. Taramelli, and R. Cironi. 1985. Preliminary study on animal population of the prairie of Torrevaldaliga's *Posidonia oceanica*. Second international Workshop on *Posidonia oceanica* beds; Ischia 7-11 October.

Cironi, R., and G. Ferrari. 1991. Microbiological study of a coastal water body influenced by the thermal discharge of a power plant. In *Second International Symposium on Microbial Ecology of the Mediterranean Sea*. (15-18 May) Taormina.

Crema, R., E. Ioannilli, A.M. Bonvicini Pagliai, M. Bertonati, R. Cironi, and R. Vitali. 1979. Chimica fisica delle acque e produttività primaria nel tratto di mare antistante la Centrale termoelettrica di Torrevaldaliga. XI Congr. Soc. Ital. Biol. Mar.; *Atti Soc. Tosc. Sci. Nat. Mem., Ses. B, 86, suppl. 1979*, 182-89.

Franco, M.A., V. Tupponi, A. Arru, E. Ioannilli, M. Bertonati, R. Cironi, and M. Miserocchi. 1988. Valutazione degli effetti di uno scarico termico sull'ecosistema costiero del Golfo dell'Asinara. Parte I. Aspetti chimico-ambientali. *Inquinamento* 30 n. 5.

Ioannilli, E., R. Crema, A.M. Bonvicini Pagliai, M. Bertonati, R. Cironi, and R. Vitali. 1979. Qualità dell'acqua e comunità fitoplanctoniche in rapporto allo carico termico della Centrale Termoelettrica di Torrevaldaliga (Civitavecchia). XI Congr. Soc. Ital. Biol. Mar.; *Atti Soc. Tosc. Sci. Nat. Mem., Ses. B, 86, suppl. 1979*, 168-81.

Ioannilli, E., and M.A. Franco. 1988. Considerazioni metodologiche degli effetti di scarichi termici sugli ecosistemi costieri. *Inquinamento* 30 n. 5.

Pelusi, P., E. Taramelli, C. Perticaroli, and R. Cironi. 1985. Crostacei decapodi diTorrevaldaliga (Civitvecchia, Roma). *Oebalia* 11 n. 3.

Taramelli, E., L. Venanzangeli, and R. Cironi. 1985. Amphipoda of Torrevaldaliga Posidonia bed. Second international Workshop on Posidonia oceanica beds; Ischia 7-11 October.

28

The Adriatic Sea and Coastal Resources: A Management and Pollution Control Study

Andrea Chiappori
Environmental Systems Manager
Ansaldo Industria, Genoa, Italy

Roberto Balostro
Planning and Studies Unite Coordinator
Ansaldo Industria, Genoa, Italy

Brenda Thake, Professor
David Santillo, Doctor
Daniel Thornton and Dipen Patel, Researchers
School of Biological Sciences
Queen Mary and Westfield College, London, United Kingdom

ABSTRACT

The Northern Adriatic Sea, a semienclosed water mass situated between eastern Italy and northwestern Yugoslavia, is experiencing severe, water quality problems at present. There is evidence that these problems have become worse in recent years, and the natural assimilative capacity of the water mass, as a whole, for pollutants has apparently been exceeded. The most obvious effect of this is an increase in algal growth, believed to be caused principally by excessive nutrient loads to the system. This has given rise to substantial problems with both fisheries and tourism, and it is estimated that the total costs of the pollution-related impacts exceed 3,000 billion Lire ($2,500 million U.S.) annually.

While it is possible to design antipollution measures to deal with such problems, their success or otherwise in preventing further occurrences of algal blooms cannot be estimated without a reliable, predictive capacity. Thus, any measures taken to control pollution at present cannot be guaranteed to be successful, and significant expenditure may occur without any noticeable improvements to water quality in the area. Similarly, any over-regulation of discharges may succeed in improving water quality, but will involve massive and unnecessary expenditure on capital works. Cost-effective controls can only be put in place if a reliable, predictive capability is developed to match pollutant loads to their impacts on coastal waters.

It has been proposed that a project be undertaken to develop mathematical models of the northern Adriatic Sea and its catchment to provide such a predictive capability. These models must include elements dealing with local and regional meteorology (which affects water movements), pollutant loading from the major riverine sources in Italy and any other significant sources, and the hydraulics and pollution conditions in the coastal waters themselves.

THE PROBLEM: POLLUTION OF THE NORTHERN ADRIATIC SEA

The Northern Adriatic Sea is a partially landlocked body of water with an area of 19,000 km^2 and a mean water depth of 35 m, bounded to the west and north by Italy and to the east by Yugoslavia. This water mass receives the inflows from several rivers, by far the largest of which is the Po River. The latter contributes an annually averaged flow of about 1,500 m^3 sec^{-1} from a catchment area of some 75,000 km^2. The population of the Po River catchment is over fifteen million, and industrial sources of pollutants contribute substantial, additional loads of contaminants. Other rivers of significance in the area are the Adige, the Tagliamento, the Piave, the Reno, the Brenta, and the Isonzo, although none of these approaches the Po River in terms of pollutant loading.

Water quality in the Northern Adriatic Sea has deteriorated significantly during the last three decades, and recent evidence suggests that this deterioration is accelerating. While there is some evidence for localized pollution by trace metals and other conservative contaminants (Albani et al. 1989), the discharge of nutrients constitutes the principal cause for concern.

Nutrients enter the marine receiving waters of the Northern Adriatic Sea, either directly or indirectly, from a variety of sources in both Italy and Yugoslavia; domestic, agricultural, and industrial wastes are all significant in terms of nutrient loading. Italian sources are considered to predominate, and the Po River catchment alone is thought to contribute about 50% of the overall nutrient load to these waters (Degobbis 1989), although reliable loading data are sparse.

The effect of this input of nutrients is to increase primary productivity in the coastal waters of the Northern Adriatic Sea. Recent investigations have shown that the assimilative capacity of these waters is being exceeded, and that an accelerating trend towards eutrophication (overenrichment) is present. Algal growth has gradually increased since the 1950s and has become particularly severe in the last two years (Justic 1987; Degobbis 1989; Cognetti 1989).

This increase in primary productivity, seen mostly as an increase in phytoplankton (microalgae) as opposed to macroalgal species, has given rise to several problems. First, algal blooms cause a marked reduction in dissolved oxygen levels in the near-bottom waters of the Northern Adriatic, and commercial fisheries (especially for benthic species) have suffered considerably. This impact has been observed over a large area of coastal waters, stretching from the Gulf of Trieste to some distance south of the Po River estuary. However, effects are most evident in the region offshore from the Po River.

The second problem concerns the potential from dinoflagellate blooms in this same area of nutrient enrichment. This is of particular concern, as several of these species produce toxins which cause paralytic shellfish poisoning (PSP) and other maladies in humans. Such toxins are passed to humans through marine foods.

Thirdly, certain species of the algae which tend to bloom in the region produce mucilage or slime. When blooms of these species (which are benthic diatoms) occur, the slime produced forms floating patches of considerable size (up to tens or even hundreds of kilometers in length). These foul fishing nets in offshore waters and are also carried towards beaches on the western seaboard of the Northern Adriatic, severely affecting the local tourist industry (Cognetti 1989).

Despite the gravity and geographical scale of the situation, no strategic studies have been undertaken to date on preventative measures. The present proposal thus represents a first initiative to provide a scientifically sound basis for pollution control and management in the Northern Adriatic Sea as a whole.

THE PROPOSED PROJECT

To address this problem adequately demands the development of a thorough understanding of several factors which have a significant influence on water quality in the Northern Adriatic. These factors are

1. *The Meteorology of the Region:.* Algal blooms are thought to be most frequent and intense when the marine receiving waters are stratified, and this depends on meteorological conditions in the area of the Northern Adriatic (rainfall, temperature, and winds).
2. *The Hydraulics of Rivers and Coastal Waters:.* Nutrients are carried to the marine waters by inflowing rivers and are diluted and dispersed by coastal and offshore currents. The hydraulics of the major rivers and of the marine waters are thus of critical importance in determining the incidence and severity of algal blooms.
3. *Pollutant Loads:.* The loads of nutrients entering the rivers and coastal waters of the Northern Adriatic must be clearly defined in space and time if any control strategy for algal growth is to be produced.
4. *Biology of the System:.* To reduce or eliminate the impacts of algal blooms, the biology of the coastal and offshore waters of the Northern Adriatic must be defined. Studies are required, not just on algae themselves and their response to nutrients and other parameters, but also on secondary consumers (as these assist in controlling algal biomass). In addition, the impacts of reduced oxygen levels on key benthic species must be understood, such that the effects of algal blooms on fisheries may be elucidated.

All of the topics listed above are amenable to mathematical modeling, and only through the development and use of such models can a reliable, predictive capability be attained, such that the present pollution may be controlled. This proposal is therefore based fundamentally on the development of such models, supporting data for which will be produced by a variety of ancillary studies. The models to be developed will thus constitute a sound basis for the design of pollution control measures in the catchment as a whole. Such 'intervention' measures cannot be designed or assessed in a cost-effective manner in the absence of these predictive models. It is envisaged that the present project will include the preliminary use of the models to assess alternative pollution control strategies in the Northern Adriatic Sea, this to be undertaken largely in the final year of the studies.

PROJECT FUNDING

Funding for certain portions of the project has already been sought. Thus, the European Commission has funded 0.75 M ECU from the Science and Technology for Environmental Protection (STEP) Program, in order to support the meteorological, ecosystem, and field studies portions of the work. Figure 1 shows the relationship of the STEP-funded part of the work to the remainder of the project., In the last section of this report, a description of the preliminary results of the project is presented.

THE PROJECT AND ITS COMPONENTS

Overview of the Project

The study area is defined as the entire Adriatic Sea, with an emphasis on the northern half of this water mass (Figure 2). The southern portion of the Adriatic is included to provide essential information to drive the hydraulic and water quality models, including boundary data in particular.

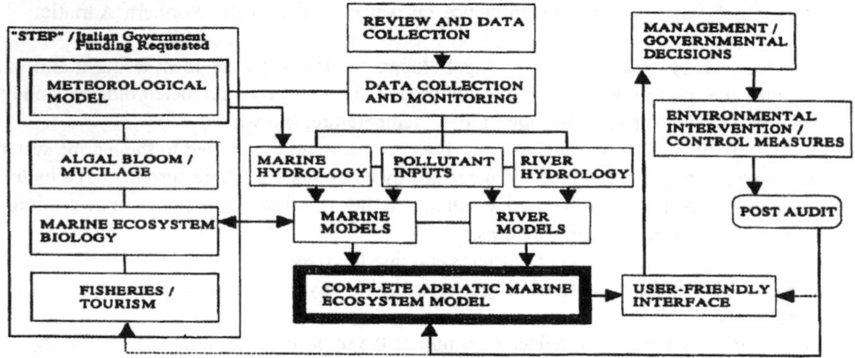

Figure 1. Schematic design specification for the development of an integrated management tool to control pollution and eutrophication problems in the Adriatic Sea.

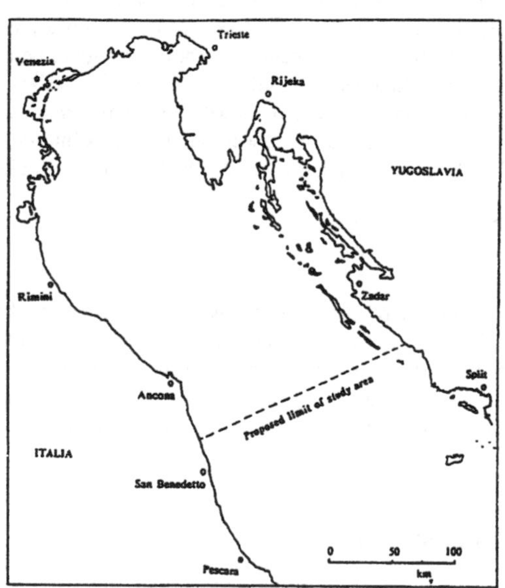

Figure 2. The proposed study area

The project seeks to provide an understanding of the following:

1. *Pollution loading to the Northern Adriatic Sea from the various land-based sources in both Italy and Yugoslavia:* The Emphasis of the project is on the Italian catchments, as these contribute the greatest loads of contaminants to the marine receiving waters of the study area. Nutrient loads will be accorded priority in these studies, as these contaminants are the principal cause of the algal bloom problems in the study area.
2. *The delivery of both freshwater and pollutants to the marine receiving waters from the major catchment rivers:* Simple, mathematical box models of the seven, most important rivers will be developed, and will provide data to drive the models of coastal waters for both hydraulics and water quality.
3. *The hydraulics of the estuarine and marine receiving waters of the Northern Adriatic Sea::* A mathematical model will be developed to describe and predict the hydraulics of these waters under various meteorological conditions and this will be used to drive the water quality model to be discussed further on in this paper.
4. *A water quality model for the Northern Adriatic Sea:* This model will be capable of predicting various water quality parameters in the study area an will provide forecasts of algal growth in different combinations of ambient conditions and nutrient loads.
5. *An ecosystem model for the study area:.* This model will employ data from the water quality model described above to predict the impacts of algal blooms on key benthic species and on other parts of the ecosystem in the Northern Adriatic Sea. Such impacts will include, in particular, those due to hypoxia or anoxia (Lowered oxygen, or lack of oxygen) in the water column.
6. *The potential benefits of possible pollution control or 'intervention' measures in the catchment of the study area:* As noted previously, such intervention measures will be assessed in a preliminary fashion only in the later stages of the present study.

Hydraulic and Pollution Modeling of Rivers

Seven rivers of significance exist in the study area, and of these, the River Po is by far the largest. Three stages are envisaged for the work on hydraulic and pollution modeling of these rivers. These stages are Strategic Planning, Hydraulic and Water Quality Surveys and Pollution Load Surveys, and Mathematical Modeling.

The *Strategic Planning* stage requires an evaluation of existing data source and the definition of the requirement for new data on either river flows or pollutant loads within the seven catchment areas identified.

In the *Hydraulic and Water Quality Survey* stage, local laboratories will be set up for field analysis and data collection and collation. Mobilization of hydraulic survey and water quality investigation equipment will take place, along with the organization of complex analyses of micropollutants, which will take place in the United Kingdom.

The *Pollution Load Survey and Mathematical Modeling* stage of the work will involve the identification of existing and planned point and non-point sources of pollution, the examination of existing and new data, and an evaluation of their significance with respect to potential or planned interventions. The verification of the models developed will be continuous, and the interpretation of a series of planning scenarios will come from the basis of the recommendations to be included in the final report.

Hydraulic and Pollution Modeling of Marine Water

An important component of the study will be review and assessment of existing data and analysis of the dynamics and eutrophication of the Northern Adriatic Sea. The hydraulic and pollution models covering marine waters will be set up to simulate the essential features of the observed phenomena. It will be essential to be able to simulate and predict the day-to-day, seasonal variations in the three-dimensional temperature, salinity, and flow fields in the Northern Adriatic Sea as a basis for driving the pollution (nutrient balance) model, and hence the algal bloom and ecosystem model. The matching set of three-dimensional, segmented and layered hydraulic, pollution, and ecosystem models will have a variable grid to enable them to cover the whole study area and to resolve the relatively thin photic zone and narrow coastal strip. The area of the coastal strip off eastern Italy is of particular importance, as it is here that the greatest impacts occur on tourism through the effects of algal-produced mucilage on the beaches.

The tide-averaged, three-dimensional hydraulic model to be produced for the Northern Adriatic Sea in the present studies will need to accept inputs from the meteorological model in terms of daily need insulation, surface cooling, evaporation, rainfall, wind speed, and meteorological pressure. The hydraulic model will also accept data on freshwater discharges from the river models.

In view of the overlapping time scales for production of the various models (describing meteorological processes, riverine inputs, and marine hydraulics and pollution), it will be necessary initially to construct a pilot model for marine waters, based on less accurate inputs, which will be refined during the course of the project. The results from the pilot model should assist in planning the field surveys. The hydraulic model for marine waters will be retrospectively calibrated using past events (over particular years) and by comparing the model results with observed, seasonal, three-dimensional temperature and salinity distributions and drift velocities.

The pollution model will include the full oxygen balance and nutrient cycles and will therefore be directly linked to the algal bloom and ecosystem model developed by Queen Mary and Westfield (QMW). The pollution loadings will be provided by the river model data. The modeling method allows for the effects of light penetration, photosynthesis, settling of detritus and the resulting decay, benthic oxygen demand, and the recycling of nutrients. The predicted water quality will be sensitive to the seasonal pattern of primary productivity and the succession of the various algal species. The pollution model will be calibrated by hindcasting a measured event and comparing the results with observations of dissolved oxygen, pollutant, and nutrient concentrations.

Meteorological Modeling

The meteorological events in defining stratification of the water mass in the Northern Adriatic are an important facet of this study.

Vertical stratification of the waters has been found by field studies to occur in certain years of either heavy rainfall or high temperatures and calm weather and is created by either salinity or temperature differences in the water column. The occurrence or otherwise of stratification is believed to be critical in determining the timing and extent of algal growth in the Northern Adriatic Sea.

The specific objectives of the meteorological study team are to identify and model the climatic regimes of the Northern Adriatic Sea based upon 'most probable scenarios'. This differs from simpler climatic information processes which identify only averaged meteorological data. Individual days belonging to different climatic regimes may be selected by the former procedure, thus optimizing the design of sensitivity analyses. A limited area model with fine-mesh resolution will provide a hindcast facility for the meteorological field of interest.

Ecosystem Biology and Modeling

Because existing data on pelagic processes are inadequate as a basis for predictions of algal bloom formation, additional research is a prerequisite for the development of an ecosystem model for the Adriatic Sea. This research will cover the following discrete but related areas of concern:

1. *Phytoplankton and Nutrients:* It is not clear why only certain species of phytoplankton bloom under particular combinations of conditions. This problem will be investigated using continuous cultures which maintain populations in a steady state, thus permitting the study of competition between species and the effects of pulsed supply of nutrients. Limiting factors to growth to be studies will include phosphate, nitrate (or ammonia-nitrogen), light, and essential trace elements.
2. *Phytoplankton and Grazers:* Blooms may be at least partly due to the failure of grazers to control phytoplankton growth. The response of zooplankton such as copepods to increasing phytoplankton biomass will be studies to elucidate the response time of grazers to an increased food abundance. The hypothesis that phytoplankton form toxins to reduce grazing will be studies, and the relative importance of benthic and pelagic grazers in the study area will be quantified.
3. *Benthic Community Effects:* Stratified conditions and algal blooms may result in hypoxia or anoxia in bottom waters, affecting benthic communities. This will not only reduce the productivity of the system, but also encourage further algal growth as benthic grazers are removed from the system. Oxygen thresholds for local benthic species will be determined experimentally, and changes in the field (both in meiobenthic and macrobenthic communities) will be elucidated.
4. *Ecosystem Modeling:* Results from the above research will be employed in the development of an ecosystem model of the Adriatic Sea. A biological model capable of simulating the key processes presently thought to be implicated in the formation of blooms will be developed. This model will interface with the hydraulic models produced in the portion of the project funded outside STEP, which will in turn, be dependent in part on the meteorological predictions.

The specific objectives of the work of the Queen May laboratory are to establish the physiological characteristics of blooming microalagl species and the impacts of nutrient upon these; to elucidate the roles of grazers and toxins in algal bloom formation and control; to identify the effects of bloom formation and decay on hypoxia and benthic communities; and to incorporate these findings and the present knowledge of such processes elsewhere into an ecosystem model capable of predicting the incidence and impacts of algal blooms in the Adriatic Sea.

PRELIMINARY RESULTS OF THE BIOLOGICAL RESEARCH IN THE INITIAL, SMALL-SCALE PROJECT IN THE NORTHERN ADRIATIC

Aims

The aims of the present, small-scale study are two-fold:

1. To determine the physical, chemical, and biological factors which stimulate of trigger mucilage production by algae, bacteria, or by consortia of the two, either in the water column or on the sea bed.

2. To produce a mathematical ecosystem model which in this preliminary study incorporates vertical transport functions, but not horizontal ones. The wider scope of the horizontal spatial dimension is addressed in the Global Study.

The present investigation is directed specifically towards depth-dependent variables and to interactions between the water column and the benthos. The study is of two years duration and is at the halfway stage. The approach and some of the preliminary results are discussed.

Approach and Results

Research activities have been focused on five areas.

1. Field monitoring
2. Field experiments in large enclosures
3. Field experiments in small volumes
4. Laboratory experiments in batch culture
5. Model development

A sixth activity, experiments on algal growth dynamics under highly controlled conditions in chemostat culture, are currently being initiated.

1. *Field monitoring* was carried out mainly in the extensive, outer harbor at Marina di Ravenna and additionally on three, short cruises in inshore waters between Porto Garibaldi and Po della Pila, in May, July, and September 1992. Large and diverse algal populations were present at all stations in the May and July sampling periods, but a far lower phytoplankton biomass was present in September. The algal biomass continued to be dominated by diatoms until mid-July, probably owing to an unusually cold period in late May and June. The transition from diatom to dinoflagellate domination occurred in inshore waters in late July. No severe mucilage events were evident in 1992 in the study areas.

   Dissolved nutrient concentrations in the water column were low in early July and even lower towards the end s of the month (Table 1). The euphotic zone extended to little more than 2 m at most sampling sites, and thus it seems likely that light limitation interacted with nutrient limitation in sub-surface waters.

   Experiments with Clarke-type oxygen electrodes of 5-10 ml working volume are routinely used on phytoplankton cultures at QMW for measurements of photosynthesis and respiration. Field populations from the Adriatic did not perform net photosynthesis in the electrode chambers, probably because prominent genera like *Chaetoceros* with long spines and flagellates like *Gyrodinium* did not perform net photosynthesis when stirred at the rate required for the successful operation of these electrodes. After a number of attempts at laboratory measurements at various photon flux densities, primary productivity measurements were carried out in the Harbor itself using bottles placed at one meter depth intervals and incubated over a three-hour period. Charge in oxygen concentration was monitored successfully with an oxygen electrode prove, and was related to chlorophyll-specific phytoplankon growth rates.

   During the cruise in coastal waters towards the river Po Delta during May, a pronounced plume front was observed about 2 km from the shore (Figure 3).

Table 1

Concentration (μmol l^{-1}) of Dissolved Inorganic Nutrients in the Harbor at the Beginning and End of the Field Work Period July 1992

Depth	Sample	NO_3^-	NO_{2-}	NH_4^+	PO_4^{2-}	SiO_2
08/07/92						
	1	1.86	0.43	1.04	0.13	10.2
Surface	2	12.0	0.42	0.84	0.22	10.2
	3	1.74	0.19	0.91	0.39	9.50
	1	1.30	0.35	0.45	0.30	10.1
3 m	2	0.84	0.33	0.45	0.25	10.6
	3	0.84	0.38	0.26	0.38	8.42
	4	1.53	0.21	0.32	0.39	10.8
27/07/92						
	1	0.10	0.39	0.52	<0.1	8.36
Surface	2	0.17	0.67	1.49	0.20	8.51
	3	0.12	0.33	3.25	1.15	3.95
	4	0.19	0.19	1.30	<0.1	7.96

Figure 3. Plume from 2 km from the shore

Examples of hydrographic probe data from the mixed water on the landward side of the front are given in Figure 4 and on the seaward side in Figure 5.
Primary productivity, as shown by dissolved oxygen concentrations, was clearly higher on the landward side, and this is linked with the sharper pycnocline there compared with that on the seaward side. Nutrient concentrations were far higher on the low salinity, landward side of the front (Table 2).

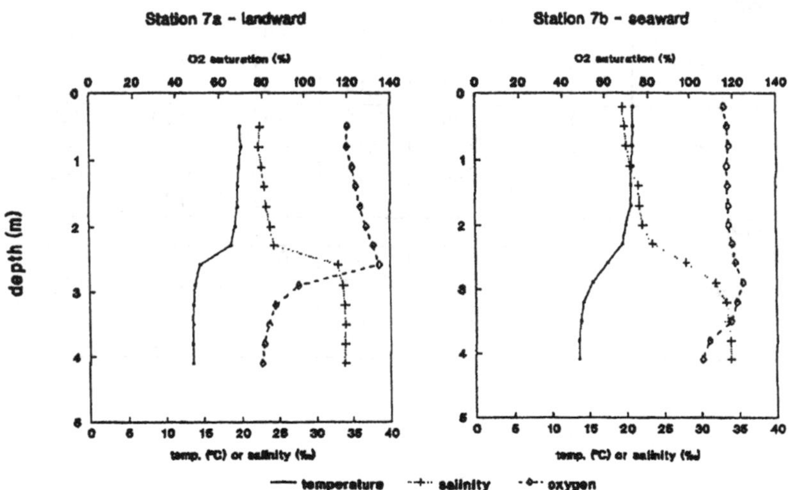

Figures 4 and 5. Hydrographic profiles from the landward and seaward sides of the Po plume front in May 1992

Table 2

Concentrations (μmol^{-1}) of Dissolved Inorganic Nutrients on the Landward and Seaward Sides of the Po plume front at the Beginning of July 1992

Station	Sample	NO_3^-	NO_2^-	NH_4^+	PO_4^{3-}	SiO_2
08/07/92						
Landward	1	81.6	0.82	0.78	0.99	60.1
	2	83.6	0.82	0.78	1.23	44.5
Seaward	1	25.4	0.70	0.45	0.22	29.4
	2	21.2	0.43	0.39	0.52	22.6

This frontal system is clearly an accumulation zone, and surface foams similar to those covering large areas of the sea in 1989 were evident at the frontal boundary (Figure 6).

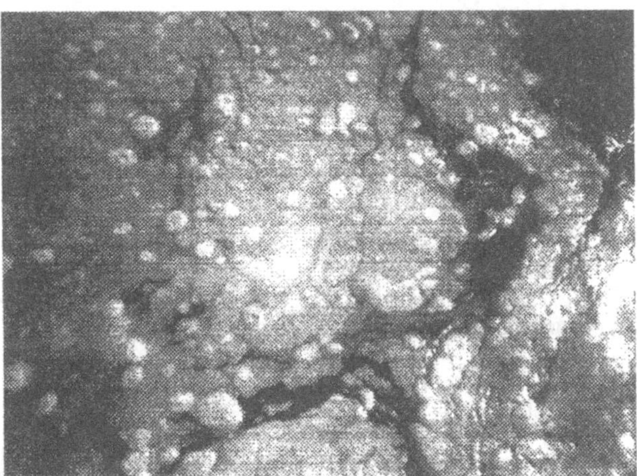

Figure 6. Surface foams

In the first July cruise, the frontal boundary was far less clear, and in the second July cruise, the front was undetectable. It may be the case that the development of plume fronts can be used as indicators or predictors of mucilage events, and in the 1993 season, this issue will be addressed in more detail.

2. *The field experiments in large enclosures* were initiated in early July. An aluminum and heavy-duty polyethylene structure was designed and manufactured at QMW in London, transported in pieces by road to Marina di Ravenna, and reassembled onshore before launching into 4 m of water in the harbor. The structure (mesocosm) was hexagonal in cross-section, 3 m deep, and was divided vertically into six chambers, each of 1,300 liters volume (Figure 8). The mesocosm was left in the water for six days before nutrient additions were made to four of the chambers.

   The principal reasons for using mesocosms were

   a) They provide an opportunity for manipulating water column chemistry while maintaining a natural light and temperature regime.
   b) They allow the hydrographic regime to be simplified by the reduction or the removal of the influence of horizontal water movement and nutrient input.
   c) Water chemistry amendments can be made in sufficiently large volumes so that inoculum effects are obviated. Enclosure of the water columns caused a rapid increase in algal biomass measured as chlorophyll concentration and caused an accelerated species succession from diatom-dominated to dinoflagellate-dominated populations. The chlorophyll concentrations rose inside the mesocosm during the month from about 20 μg/1 to 60 μg/1, but by the end of July, the chlorophyll concentrations outside the chambers had reached those of the inside. The nutrient addition experiments inside the mesocosm

produced inconclusive results, probably owing to the rather variable amount of wash entering the chambers from the harbor when medium-sized boats passed through. In 1993, the mesocosm experiment will be repeated using two sets of chambers built with taller rims to prevent inflow from the harbor.

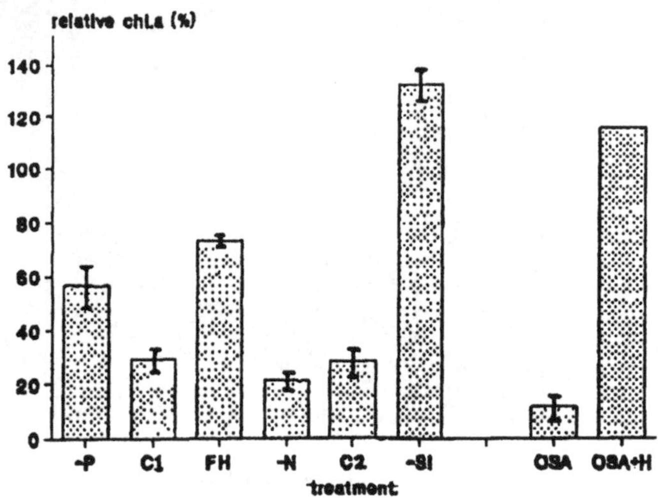

Figure 7. relative phytoplankton biomass, expressed as a percentage of the initial concentration of chlorophyll α, from the nutrient addition experiment in 1.5 liter bottles incubated in situ.

Notes: -P *minus phosphate*; Cl control 1; FH full nutrient addition; -N *minus nitrate;* C_2 control 2; -Si *minus silicate*; OSA harbor water control: OSA+H harbor water plus full nutrient additions

3. Since the amount of dilution of mesocosm chambers was evidently variable, a further nutrient addition experiment was designed, using water from the inside of each chamber as the initial inoculum. Similar concentrations of nutrient were added to closed, polyethylene bottles of 1.5 liters working volume, and these were placed in an incubation crate anchored to the same buoy as the mesocosms. Figure 7 shows that the smallest relative growth was caused by the treatment from which nitrate had been omitted, followed by the minus phosphate treatment, indicating that nitrogen was the primary limiting nutrient in coastal waters at that time of year. Other large-bottle bioassays were also conducted in July in the field laboratory using two, experimental protocols.
   a) A wide variety of manipulations were made in the conventional manner of addition of one nutrient to each vessel and observing the one which gave the greatest growth stimulation.
   b) The second protocol was the same as the one used in the field; that of adding full medium minus a single solute, and identifying the limiting nutrient by lack of growth.

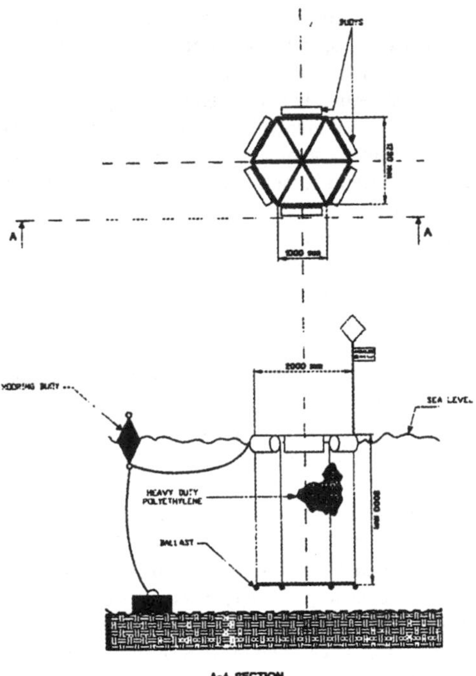

Figure 8. Plan and transverse diagrams of the mesocosms deployed in Marina di Ravenna harbor in July 1992

Other large-bottle bioassays were also conducted in July in the field laboratory using two, experimental protocols.

a) A wide variety of manipulations were made in the conventional manner of addition of one nutrient to each vessel and observing the one which gave the greatest growth stimulation.
b) The second protocol was the same as the one used in the field; that of adding full medium minus a single solute, and identifying the limiting nutrient by lack of growth.

   The results from this series of experiments were highly variable. Nitrogen, phosphorus, and dissolved inorganic carbon as well as changes in salinity, all affected final biomass as measured by chlorophyll concentration. The variability was most likely caused by differences in initial inocula. Trace metals and vitamins were consistently shown not to be growth limiting.

4. *Laboratory experiments in batch culture* have focused upon testing some of the current hypotheses which attempt to identify the triggering mechanisms for mucilage production by algae and bacteria. Algae have been isolated from the Adriatic itself for this purpose, and these cultures have been supplemented with those obtained from other phycologists and from culture collections. The effects of N:P ration, salinity, alkalinity, a range of organic compounds, and the degree of bacterial contamination have been tested. some of the results appear encouraging, but the work as yet is at too preliminary a stage to report here.

One problem is that of finding a satisfactory measure of the amount of mucilage produced. Detailed analyses of specific polysaccharides are inappropriate, as mucilage contains a wide range of these which contribute differing proportions to the total amount produced. On the other hand, total carbon is not particularly helpful because of the large fraction of cell carbon which is not contained in mucilage. At present, two measures of mucilage are being employed in this study; a histochemical one involving a broad-spectrum polysaccharide stain, and a fluorescence spectrophotometric technique. It is likely that at least one of these will be usable in field situations.

5. The development of the ecosystem model is at a preliminary stage since it was intended that the results from the initial field observations and the laboratory experiments would be taken into account from the outset in the modeling work. The following aspects are considered central in the development of a realistic phytoplankton growth model:
   a) The degree of coupling between photosynthesis and growth
   b) The effects of different chlorophylls and the predominance of carotenoids on chlorophyll-normalized rates of photosynthesis and self-shading
   c) Settling and motility in phytoplankton
   d) Vertical dispersion of algal cells and the effect of water column stability on phytoplankton growth.

   In the final ecosystem model, bacterial respiration, which appears to account for a relatively large proportion of the dissolved oxygen depletion, may have to be differentiated from that due to phytoplankton, with the role of bacteria in the water column carbon budget to be considered separately. Hence, specific field and laboratory measurements will be designed to provide coefficients for model calibration.

In progress then, is an interactive process of model development and biological research. In the Algal Research Unit, the close collaboration between algal physiologists, biochemists, ecologists, and mathematicians will prove central to the successful realization of the project objectives.

REFERENCES

Albani, A. D., P. C. Rickwood, V. D. Favero, and R. S. Barbero. 1989. The geochemical anomalies in sediments on the shelf near the Lagoon of Venice, Italy. *Marine Pollution Bulletin* 20:438-442.

Cognetti, G. 1989. SOS from the Adriatic. *Marine Pollution Bulletin* 20:578-588.

Degobbis, D. 1989. Increase eutrophication of the Northern Adriatic Sea. Second act. *Marine Pollution Bulletin* 20:452-457.

Justic, D. 1987. Long-term eutrophication of the Northern Adriatic Sea. *Marine Pollution Bulletin* 18:281-284.

29

Land Utilization Development of the Ocean Coastal Areas of Lagos Metropolis Between Years 1980 and 2000

O. A. Oyediran
Department of Architecture, University of Lagos, Akoka, Nigeria

INTRODUCTION

Lagos State (Figure 1) is one of the smallest states in area that make up the Federal Republic of Nigeria. Situated at the southwest corner of the country, it has a population of over five million people in the 1992 census figures and a land area of about 3,500 sq. mtrs. The entire southern boundary from the west to east side is bounded by the Atlantic Ocean. Within the aegis of Lagos State is the Lagos metropolis itself, which accounts for about 37% of the land surface of the state, 1,200 sq.mtr. of which is mostly bounded by the Lagoon and the Atlantic Ocean. Lagos Port is the major Nigerian outlet for the country's import and export products. According to the official statistics from the Nigerian Port Authority, it handles over 80% of all the country's trade transactions. According to information from the Ministry of Economic Planning and Land Matters (1981) in Lagos, "the city as the main Port and Industrial centre of Nigeria accounts for over 1/5 of total production in terms of value added. The primary sector is insignificant, while manufacturing and Public Administration are the largest activity in metropolitan Lagos. Lagos accounts for 56% of value added in utilities, 49% of manufacturing and crafts, 48% other services and, 23% Distribution." With the above statistics, the author strongly believes that the healthy industrial and commercial development is strongly tied to proper planning and land utilization along the coastal areas.

OBSERVATIONS ON LAND UTILIZATION

This author therefore looked at the Lagos metropolis in terms of its land utilization at present and the proposals as published by the government for the years 1980-2000, and tried to establish the use made of its coastal area where it opens to the outside world. In doing this, a critical look at the government's policy on land use, as well as the state Regional Plan (1980-2000), both published by the

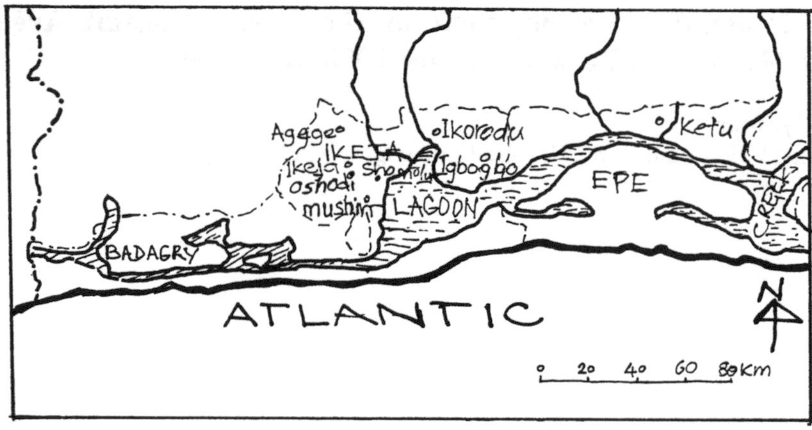

Figure 1. Lagos state

state government in 1981 for public consumption, was undertaken. Furthermore, personal interviews of users and other professional involved in planning the environment were conducted. The findings of both interviews, and the government document were then assessed for a critical conclusion.

The Lagos State Regional Plan (1980-2000) stipulated in its land utilization policy that the southern sector (coastal area) be reserved for "Hotel Development, Holiday Resort, Recreation, and Tourism," while a major center of population is envisaged on Lekki Peninsula.

To achieve the proposals made for the coastal areas, the planning office established a "Residential Estate" along the coastal areas and allotted plots of land accordingly in areas around Lekki Peninsula, Victoria Island, and Victoria Island Annex. This action of the authority has left residents in an awkward and unhealthy situation. Private owners along the coast (e.g., Okun Aja and Ibeju-Alagutan Villages on Atlantic Ocean Coast) started indiscriminate and unplanned development of the land area for mostly residential buildings, which in most cases did not receive the approval of planning offices. Unfortunately, most of the areas on this side of the Lagos metropolis have none of the necessary infrastructures (water, electricity, and good access roads). The only motorable road passing through the area links Lagos metropolis with the next town, Epe, while the road on the waterfront itself is sandy and unmotorable. Generally, coastal frontage (Atlantic Ocean Front) in this area is barren, unplanned, and full of unharvested coconut palms and sands.

Towards the most developed area of the metropolis coastal area is Victoria Island, Maroko, and Victoria Island Extensions (Figure 2). These areas are well planned by the government exclusively for residential purposes. As a result, many of the buildings erected in the area are detached, single family houses with few, interspersed, high-rise residential blocks. However, and like its less developed neighbor, Okun-Aja, many vital infrastructures have not yet been provided. Many buildings have no pipe-borne water for household uses, while constant flooding of most access roads occurs each rainy season.

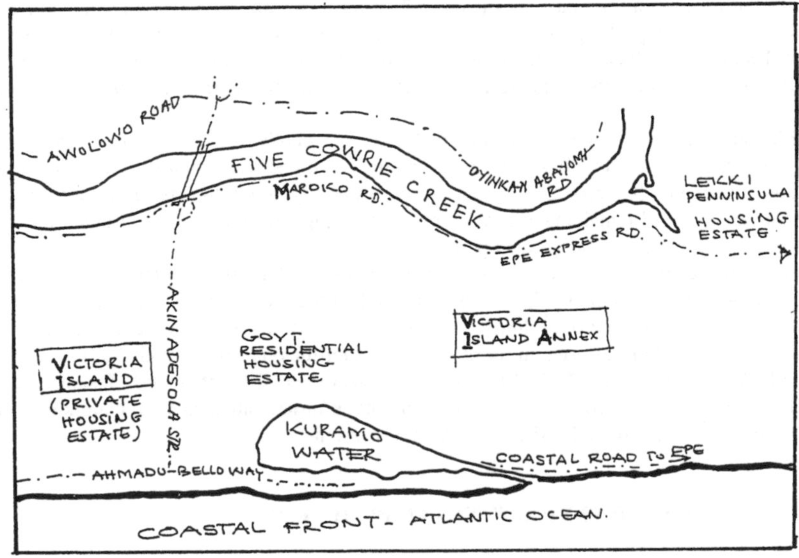

Figure 2. Coastal front - Atlantic Ocean

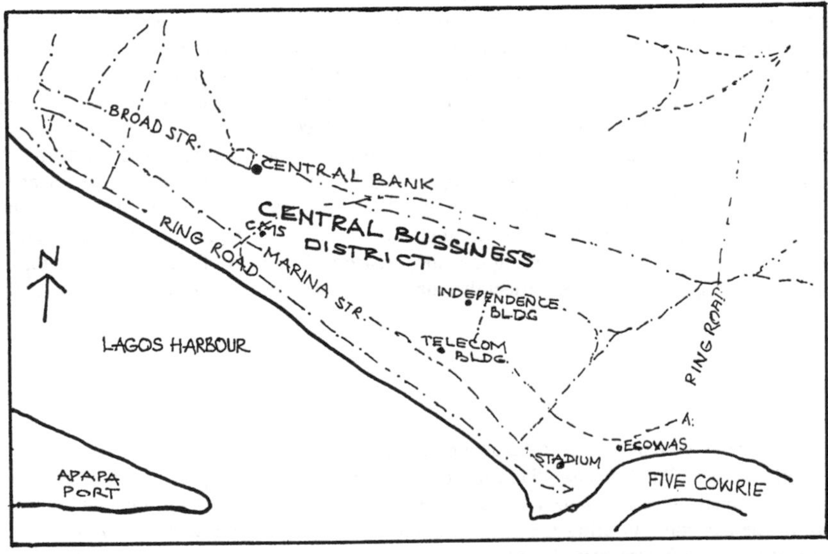

Figure 3. Metropolitan waterfront (partial)

The rest of the metropolis waterfront (Figure 3) is no better off in its planning and land utilization. The major roads running along the waterfront (Marina and Ring Road), host low-rise to high-rise commercial and office blocks. At the same time, local ferry ports dot the main ring road (Marina), making fast and non-stop driving along the waterfront a bit impossible with the constant pedestrian crossing.

Looking at the picture of the coastal front of Lagos metropolis from the ocean, gives an impression of a city front that is adorned with only chains of commercial and residential structures tied to the coast with strings of concrete fly-overs. The human aspect of the city in this area has been sacrificed for some commercial ventures. All remaining available areas at the Lagoon front, as well as under the many city-front fly-overs, are reserved and used as parking lots.

There are however a few places that are being developed along the waterfront as recreation areas, according to the Land Use Policy. These areas are, unfortunately, not well planned in terms of access and safety, and are frequented only during festivals and occasionally on weekends. This is due to the fact that they have nothing to offer in terms of beach facilities like those available in other countries (i.e. the Florida beaches in the United States), where *leisure* is the watchword. The sum total of our observation is that the whole coastal land frontage has been turned into a chain of commercial structures with little concern shown for related environmental problems.

PHYSICAL AND OTHER ENVIRONMENTAL PROBLEMS

As a result of unplanned development along the Lagoon and coastal area, many unforeseen environmental problems have surfaced. Such problems include the constant flooding of the entire frontal area during the rainy season. There are no means of draining the rain water into the Lagoon as a result of the low level land difference. Efforts made in the past included opening up draining passages, but unfortunately, the land area is level in many areas with the low lying sea level.

Apart from this natural phenomenon, a lot of "Sand-Fillings" had been undertaken in the past by the authorities to such an extent that the required volume of space to empty the city drainage is not available. Furthermore, available information shows that the indiscriminate establishment of residential areas along the coast reflect, unfortunately, their lack of community planning. In addition to the problems of the flooding regularly experienced, the inability of the government to establish a planning policy and enforce the same, made it possible for indiscriminate establishment of jetties for private uses along the coast. It is also observed that due to the high water level on the coast, a lot of unwanted articles and oil pillage make the water so unhealthy that local commuters on the waterway are sometimes forced out of their jobs during the rainy season. This is as of result a fluctuating sea level and fear of boats capsizing. All these affect the commercial activities on the coast, especially the fishing industries.

CONCLUSION

From the research undertaken, it became clear that the Lagos metropolis coastal area has not been put to maximum use. The Regional Plan for the years 1980-2000, had a proposal to establish the much needed recreational areas and hotels for commercial activities and tourism along the coast, but the inability of the executors to bring the plan to fruition made it look like a mere writing exercise. The problems of lack of infrastructure along the coast, the flooding and sometimes over-flooding of the land area from the ocean, and the uncoordinated development of private and public lands along the coast, are serious enough to warrant a need for urgent solutions. The authorities should organize the services of professional planners, architects, engineers, and other environmentalists to improve the health and environment on the coast. A solution to the flooding should be undertaken immediately and legislation passed to stem the constant tide of pollution assaulting the waterfront from chemical drains and oil spillage. It is also advisable for the authorities to start enforcing all planning laws regarding new developments while discouraging further encroachment of the coastal areas. There is need also for improvement on existing insufficient facilities along the coast while urgent efforts are recommended for reorganizing and providing leisure areas along the newly opened coastal lands in Okun-Aja, Ibeju, and other settlements along the Atlantic Ocean frontage.

Finally, it would be necessary to undertake a more comprehensive appraisal of the proposed plan with an eye to modifying the same to meet present day needs. A replanning of the city's waterfront is also needed to provide a better environment and to check flooding.

REFERENCES

Ministry of Economic Planning, Ikeja 1981. *Lagos State regional plan (1980-2000).*

Land Policy of the new order - A broadcast by Governor L.K. Jakande. Publicity Dept., Governor's Office, Ikeja 1981.

30

Climate Change and Coastal Vulnerability

J. v.d. Weide
Delft Hydraulics, Emmeloord, The Netherlands

INTRODUCTION

The possibility of a global climate change, as a result of increasing concentrations of greenhouse gases, is now widely accepted by the scientific community. In the Proceedings of the Second World Climate Conference (1990), held in Geneva, the state of the art in climate modelling was reviewed with a view to predicting the resultant temperature rise and marine-meteo conditions, such as an accelerated sea level rise (ASLR) and changing storm conditions, with the related storm surges and waves. The results, in terms of anticipated temperature rise and ASLR for a "business as usual" scenario, is shown on Figure 1. The coastal zone will be most sensitive to such changes. First, a large part of the world coastline consists of loose, alluvial deposits. Such a coast will easily adapt itself to changing water levels and waves. Moreover, large estuaries and lagoons are found here, which experience a delicate balance between tidal forcing and estuarine response. This balance is easily disrupted by ASLR.

Secondly, coastal plains are concentrations of socioeconomic activities and contain many important food producing areas. ASLR, in combination with an increased frequency of storm surges and high waves, may be detrimental for many of these areas due to increased risk of flooding. Finally, intertidal areas and marginal seas are valuable ecosystems, vital for biomass production and indispensable for the survival of many species. In those areas where the ecosystem cannot adapt to these changes, loss of lowlands and biodiversity will be the result.

Numerous authors have described these effects in greater detail for specific regions or nations. In the Proceedings of the Climate Conference, a review is given of the effects of ASLR on coastal zone management (CZM), in the paper of Eid and Hulsbergen (pp. 301-310). A more detailed review of effects for the Mediterranean is given by Sestini (1992). He describes both the present autonomous developments as a baseline scenario, together with possible effects of ASLR. Most authors however list possible effects of ASLR, without quantifying the significance of these effects in physical, economic, or ecological terms. For that reason, The Coastal Management Subgroup of Working Group III on

Response Strategies of Intergovernmental Panel on Climate Change has performed a number of studies to assess these effects in a quantitative way. Although these studies were of a general nature, they emphasized the importance of ASLR and underlined the necessity for a global response.

The first studies included an inventory of possible impacts. Subsequently, methods were developed to describe and quantify these impacts, and to develop appropriate response strategies. Finally, ways and means are being developed to integrate these strategies within a general CZM policy.

In the present paper, the various methods of Vulnerability Assessment are reviewed and some practical results are given.

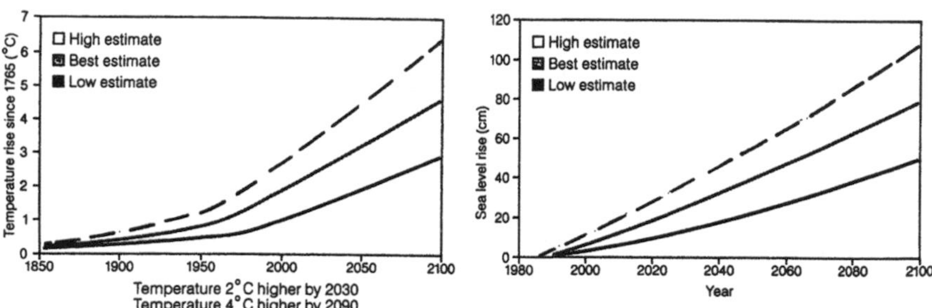

Figure 1. Temperature rise and sealevel change for IPPCC business as usual scenario

COASTAL VULNERABILITY

Climate change will induce secondary effects, such as sea level rise and changed wind and rainfall patterns. These effects will directly influence the coastal system. The extent of the impact depends on the type of impact and the response of the coastal system for that particular type of impact. In this paper, the vulnerability of a coastal system is defined as *the disruption of specified coastal functions as a result of a given climate change scenario.*

According to this definition, vulnerability is a function of the following parameters:

1. The coastal function under consideration
2. The response of that function to changes
3. The intensity of climate change and the corresponding effects in terms of ASLR and changing wind and rainfall patterns

Coastal Functions and Impact Categories

The coast is an area where many economic and social functions are concentrated. Moreover, the coastal area is often of a high economic, ecologic, and cultural value. For a Vulnerability Assessment the following functions should be considered:

1. Safety
2. Economic values and activities
3. Social and public activities
4. Ecological values
5. Cultural values

Those functions and values may be affected in different degrees. It is common to identify the following impact categories:

1. Values at change
2. Values at risk
3. Values at loss

In most Vulnerability Assessment studies, the *safety* of life and property is related to the risk of flooding. The incremental risk of flooding due to ASLR is used to quantify the vulnerability of the area with respect to safety. Most *social* and *economic* activities are of a long-term nature. The vulnerability of these activities are therefore best characterized by changes in the output over a longer period of thirty years. This is normally the basis for the Vulnerability Assessment for this class of function.

Damage to *ecological* and *cultural values*, due to the effects of an advancing sea is often irreversible. This is also the case when coastal areas are being eroded or permanently inundated. For these types of impacts, the vulnerability is best defined by assessing the net *losses* over a period of thirty years.

The following vulnerability profile can, therefore, be composed:

Table 1

Vulnerability profile

Impact category Coastal function or value	Values at risk	Values at change	Values at loss
Safety	*		
Economic values and	*	*	
activities		*	
Social and public activities			*
Cultural values			*
Ecologic values			

Thus, it has not been possible to define a common denominator to express the value all of these functions because different parameters are used, such as number of people at risk, changes in BNP, or areas lost. For each of these values, critical threshold values have to be defined. If these values are exceeded response strategies have to be defined.

Response Strategies

Response strategies may include:

1. Retreat
2. Accomodation
3. Protection

Retreat involves no effort to protect the land. *Accomodation* implies that people continue to use the land without actively protecting it against the effect of climate change. However, the damage is minimized in a passive way, for instance by erecting flood shelters, elevated buildings, and by using flow and salt tolerant crops. *Protection* involves the construction of mitigating measures, such as sea walls, dikes and dunes.

As stated before such strategies should be formulated, taking into account autonomous changes in the external conditions due to economic developments and geophysical processes (e.g. tsunamis and subsidence).

METHODOLOGIES FOR VULNERABILITY ASSESSMENT

In 1990 a worldwide cost estimate was made to assess the cost associated with the protection of the world coastlines against a sea level rise of 1 meter per century. The results of this study are given in the IPCC Summary Report (1990). Although the study was based upon rather crude schematizations of the coastal system, the result was sufficiently accurate to obtain a first estimate of the economic consequences of the greenhouse effect. The study indicated that an average annual investment in the order of $5 billion U.S. was needed to cope with the ASLR of 1 m per century.

Obviously, not all areas are equally vulnerable, due to differences in the natural coastal features and the socioeconomic characteristics of the coastal zone. In order to obtain a more reliable vulnerability profile of the various coastal sections, IPCC developed a methodology for Vulnerability Assessment, commonly referred to as the Seven Steps (1991). These seven steps include

1. *Delineation of the case study area and specification of ASLR boundary conditions.*
2. *Inventory of study area characteristics.* This refers to coherent description of the natural system and the use of the coastal area.
3. *Identification of relevant development factors.* In this step, autonomous developments of both the natural system and its usage are analyzed as a base-line scenario for further Impact Assessment studies.
4. *Assessment of physical changes and natural system response.* In this stage, the impact of ASLR on the natural system is analyzed, taking into account shoreline degradation, risks of inundation, and effects of increased salinity intrusion. These changes may result in direct loss of life and property, production losses, and degradation of ecosystems.
5. *Formulation of response strategies and assessment of their cost and effects.* IPCC considers three types of response strategies:
   a) Retreat, the option without any protection
   b) Defense, the option with full protection
   c) Accommodate, the intermediate option to alleviate the adverse effects of the ASLR
6. *The assessment of the vulnerability profile and interpretation of results.* The IPCC report considers three categories of affected values:
   a) Values at loss
   b) Values at risk
   c) Values at change

   These values have to be assessed for the various selected ASLR scenarios and formulated response strategies. Values may be expressed in geometrical parameters (areas), economic quantities (costs; changes in BNP), or ecological values (habitat and species).
7. *Identification of actions to develop a long-term, coastal zone management planning.*

Meanwhile, a number of regional and national Vulnerability Assessment (VA) studies are underway to test the proposed methodology. Concurrent with these studies, a global vulnerability assessment was made by DELFT HYDRAULICS and The Tidal Waters Division of Rijkswaterstaat (The Netherlands), focusing on the flood prone areas, the number of people at risk, and the area of coastal wetland at loss. Results of this study were published in 1992 prior to the UNCED conference.

The more important results of the study are shown in Figure 2. The study indicated that some 100 to 200 million people are living in flood prone coastal areas defined as the area below the once-per-year sea level. The number will increase by about 50% for an ASLR of 1 m and will be almost doubled in 2020, when the effect of sea level rise and population growth are combined. The study further indicated that an ASLR of 1 m would halve the area of coastal wetlands, now totalling some 910,000 square kilometers.

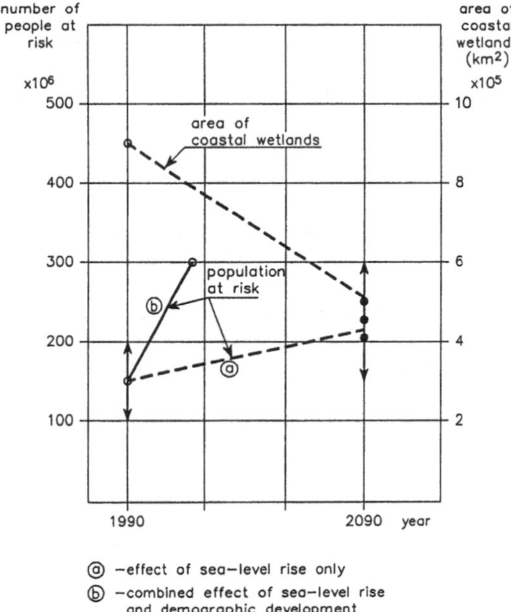

Figure 2. Impact of a sea level rise of 1 meter on wetland losses and number of people at risk in the coastal zone

The above figures emphasize the importance of ASLR for the coastal zone and underline the necessity for a more sophisticated approach. In view of the complexity of the physical and socio-economic aspects, the uncertainty in the definition of the input scenarios, and the large variation in response strategies, a systems analysis is most appropriate for such an approach.

SYSTEMS ANALYSIS AS A TOOL FOR VULNERABILITY ASSESSMENT

Systems analysis was introduced in the early fifties as a decision support tool for policymakers and has since then been developed into an expedient for strategic studies. Its efficiency has been strongly enhanced by the introduction of the personal computer. According to Boulding (1956) system modeling is a level of theoretical model building which lies somewhere between the highly generalized constructions of pure mathematics and the specific theories of the specialized disciplines.

In literature, various definitions are used to describe a system. In this paper the following definition is used: *A system is a schematization of reality by means of a set of elements and their interactions.*

System *elements* are the basic entities of the system which contain all properties (attributes) relevant for the problem to be addressed. *State variables* have to be defined to quantify the value of these attributes for a given point in time.

Interactions are defined by relationships between these elements, which may induce changes in these state variables. Interactions may be described verbally or by means of emperical or theoretical models.

It should be noted that in general, the system under consideration contains only a part of all the elements of the real world (the Universe). Therefore, *system boundaries* are needed which define the system domain, and *input* and *output* variables are formulated to describe the interaction with the system environment.

In order to make the system tangible, *subsystems* may be defined which contain a subset of elements and interactions.

The Coast Area System

If this approach is applied to the domain of coastal zone problems, the coast is schematized as an interrelated set of subsystems. At the highest level of abstraction, the following subsystems are identified (Figure 3):

1. *Natural subsystem*, which may be further divided into its biotic and abiotic components
2. *Socioeconomic subsystems*, including the users of the coastal area, the supporting physical infrastructure for these activities, and the institutional setting (social infrastructure)

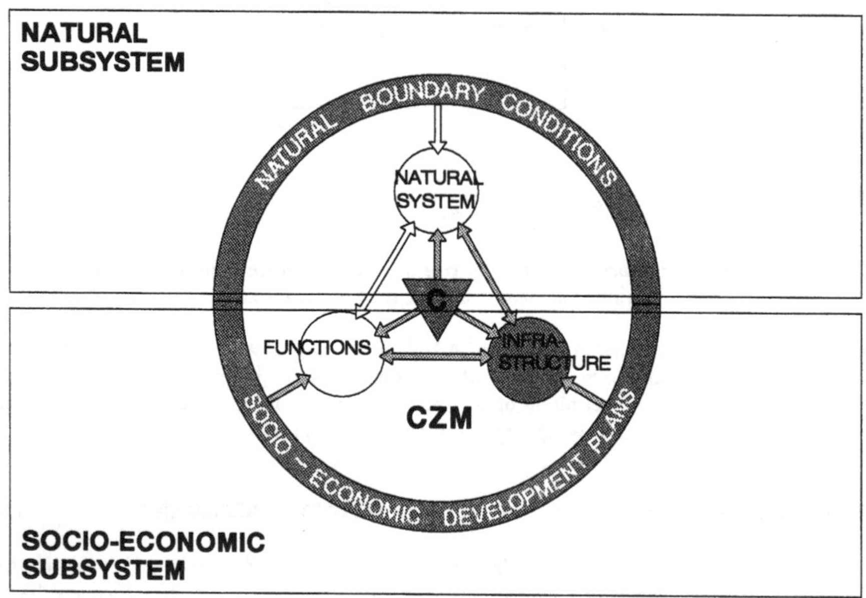

Figure 3. The coast as a system

The system may represent a specific area, but it can also be extended to include the total coastline of the world. The aggregated system description presented can be further refined to represent relevant details of the project area.

Spatial- and time-dependent boundary conditions have to be formulated to describe the interaction with the outside world. Autonomous developments have to be defined in terms of a number of development scenarios.

With the help of such an approach, the effect of changing input scenarios can be evaluated and the efficiency of remedial strategies can be assessed for a selected number of aspects. The method is essentially a more elaborated version of the Seven Steps, more suitable for use on a computer.

The accuracy of the method is determined by the adequacy of the system description and the accuracy of the representation of the interrelation between system elements. This aspect will be discussed in the next section.

System Modeling

Models

Boulding gives an intuitive hierarchy of real world complexity, using nine levels of complexity: the first three levels include *physical* systems; the second three levels include *biological* systems; and the final three levels encompass *social* systems, including man as a specific system entity. All systems have been described verbally, and full scale observations and laboratory experiments have been used to obtain an empirical description of system behavior. Such a description may be used to predict system behavior in the future, provided system parameters and system inputs do not change with time.

Obviously, this method is no longer reliable when transient processes are involved, causing a change in input conditions and/or system parameters. In that case more rigorous mathematical descriptions are needed to model the system. According to Boulding, this is only feasible for the first three levels of system, eg., those described by equations of physics.

In the book, the Art of Modelling Dynamic Systems, Foster Morrison (1991) gives a review of models which can be used to describe these first three types of systems. According to that approach, these systems may be classified as follows:

Zero type: *Constant state* (static models).

Type I: *Solvable, dynamic systems.* For these systems, the system behaviour can be described mathematically. The equation can be solved analytically or numerically.

Type II: *Dynamic system, amenable to perturbation theory.* For these systems, a solution is found in terms of a power series using the solution of a first order, solvable system as a starting point.

Type III: *Chaotic dynamic systems.* These systems show both deterministic and stochastic behavior. A solution for these systems is found by using time-averaged parameters to describe the systems behavior. These time-averaged parameters are then used to identify and model regularities in such a system.

Type IV: *Stochastic systems.* A solution of these systems is found in terms of time-averaged quantities.

It should be noted that all models are a schematization of the real world system. Depending on the sophistication of the system description, first or higher approximation of the real world system can be obtained. Model validation is, therefore, required to check the conformity between model behaviour and the behavior of real world systems.

Often, such a validation has to be followed by a model calibration to establish the most appropriate values for model parameters. For practical purposes, only validated and calibrated models should be used. The order of approximation to be selected is a compromise between *required accuracy, technological possibilities*, and *related costs*. As the outcome of this evaluation differs from one project to the other, a large variety of modeling concepts are being used in coastal sciences.

Models range from validated and calibrated mathematical models to prototypes, which are used for R & D projects only. In this paper, only the first type of models have been considered as representative for the state of the art in coastal engineering.

Inputs and Outputs

System modeling is possible when system parameters and input parameters are properly defined. The efficiency of the system is greatly enhanced when system outputs are presented in a suitable format, which facilitates the comparison of various strategic options. Modern GIS and Data Base techniques are used, therefore, to support modeling and pre/post-processing activities.

Such a modeling system is shown in Figure 4. This approach was followed, when modeling the effect of ASLR for the Netherlands. This is described in greater detail in Peerbolte and Wind (1991).

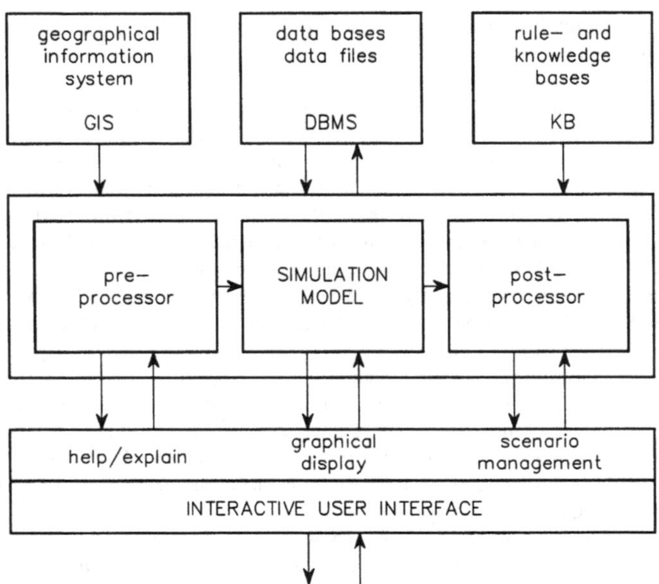

Figure 4. An integrated frame work for interactive modeling

A GLOBAL VULNERABILITY ASSESSMENT FOR THE COASTAL ZONE

The coastlines of the world are the result of morphological changes, governed by climatological and geological processes. On a global scale, this can be schematized by a system description using three, interrelated subsystems.

1. The atmosphere and the governing climatologic processes such as wind, rainfall, etc.
2. The lithosphere and the governing geologic processes, such as regression, transgression of coasts, and erosion of watersheds
3. The hydrosphere, acting as a coupling between the two previous subsystems, through the marine, littoral, estuarine, and fluvial processes, acting in both the aquatic surface and groundwater systems.

Both changes in the atmosphere due to climate changes and changes in the lithosphere due to geological processes, will have an impact on the coastal zone. In this respect, the coastal zone is defined in a broad sense and incorporates the land/sea interface and part of the adjacent marine and terrestrial area.

In order to model the effect of these changes on the coastal zone, a true reproduction of the physical processes in the hydrosphere should be pursued. This simulation should encompass relevant subsystems such as *the seas*, *the coastal waters and estuaries*, and *river systems*; their interaction with the atmosphere and the underlying lithosphere; and should be based upon a reliable, physical description of the governing processes in these subsystems.

A global modeling concept can be developed for the land/sea interaction with only the air, the sea, and the land as system elements. Although too crude to be used for actual modeling of local areas, this model makes it possible to identify and classify specific coastal zones by their geomorphological features and to assess their vulnerability for the effect of climate change. Moreover, this classification may be a basis for a further, in-depth description of the relevant aquatic systems in the hydrosphere for particular types of coasts.

In this paper, only relevant processes in the atmosphere and the lithosphere are reviewed will be reviewed, along with their resultant characteristics of the coastal zone.

The Atmosphere

The relevant atmospheric processes, important for climate change are

- Winds
- Temperature
- Rainfall
- Evaporation

Using these parameters, various climate zones may be identified as shown on Figure 5.

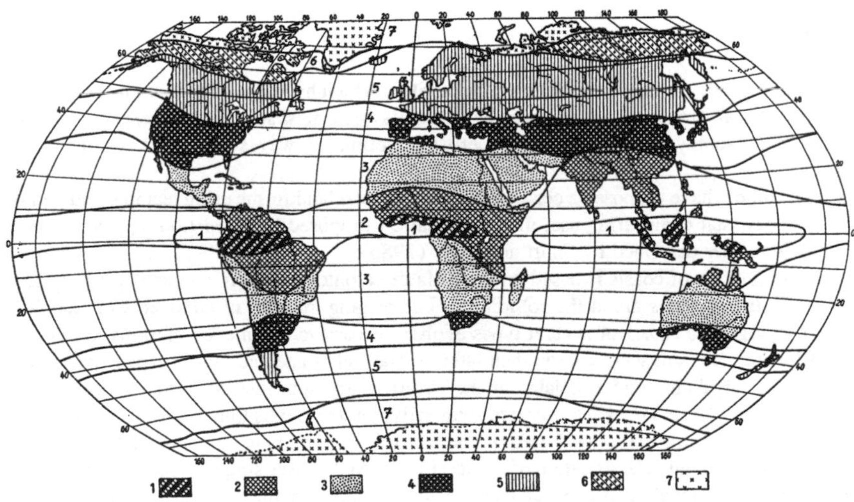

1 = equatorial
2 = sub equatorial
3 = tropical
4 = sub tropical
5 = temperature
6 = sub arctic
7 = arctic

Figure 5. World climate zones

Based upon this classification, the following climate zones, with their characteristic, marine meteorologic conditions may be identified (Table 2).

A somewhat similar climatologic division has been accepted by the International Geosphere Biosphere Program, to setup a network of regional research centers, System for Analysis, Research, and Training (START).

The present contribution is dealing with equatorial, tropical and temperate zones. Arctic regions are not considered.

Table 2

Classification of climate types and values of characteristic climatologic parameters

Climate type	Temperature	Rainfall	Wind intensity
1. Equatorial	high	high	low
2. Subequatorial (0°-10°)	high	high	moderate, monsoons and trade winds
3. Tropical (20°-30°)	high	low	moderate, trade winds*
4. Subtropical (30°-40°)	moderate	moderate	moderate**
5. Temperate (40°-60°)	moderate	moderate	high
6. Subarctic (60°-70°)	low	moderate	high
7. Arctic (70°-90°)	low	low	moderate

* occasional tropical storms and hurricanes
** occasional tropical storms

The Lithosphere

The coastal zones are the result of tectonic activities which have laid the basis for the coastal land forms and geological processes which give the coast its present shape. In this respect, the balance between sediment supply by rivers and the erosive action of waves and tides is of paramount importance.

Various authors have proposed a classification for coastal land forms using tectonic and geological parameters. For coastal engineering use, the classification proposed by Nordstrom appears to be quite helpful. This is further described by Short and Blair (1986):

> This classification of coasts is based on their plate/tectonic setting (Inman and Nordstrom, 1971). *Collision coasts* are at the colliding or converging margins of continents or islands arcs. *Trailing-edge coasts* of several types evolve on passive continental margins, initially as rift-boundaries fault scarp coasts, then later as maturely dissected fault scarps fronted by narrow coastal plains, and still later in the evolutionary sequence as broad sedimentary coastal plains. *Marginal sea coasts* are primarily on the depositional edges of shallow marginal or epicontinental seas. Based on the tectonic environment, Inman and Nordstrom devised a morphologic classification of coasts (Figure 6) that uses simple descriptive phrases for the gross regional landscape of each tectonic type. On the passive trailing coastal margins of continents, rivers drop their loads of sediments, which are then shaped into deltas or a variety of progradational landforms. Mud may be trapped by vegetation to form extensive coastal salt marshes or tropical mangrove swamps. Sand is more likely than mud to move laterally along coasts to build a variety of beach landforms. Approximately 13 per cent of the world's coasts are said to be sandy barrier beaches, primarily constructional in nature (Zenkovich, 1967, pp. 288 and 390).

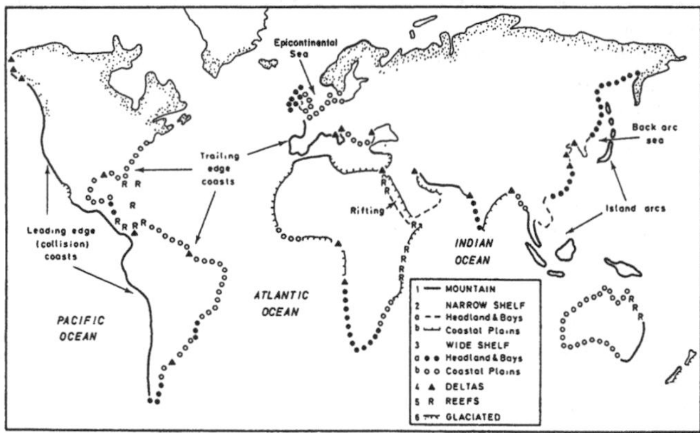

Figure 6. Coastal landforms of the world

An extremely abundant but nevertheless remarkable coastal landform type is that built by biogenic deposits of shallow-water marine animals such as corals. Entire island archipelagos and a very large percentage of the world's coast have been constructed of coralling limestone (Davies, 1980, p. 5), forming reefs that fringe or protect other coastal landforms. The largest of these, the Great Barrier Reef of Australia, extends for 2300 km along the tropical northeast coast of Australia, inhibiting wave erosion of the mainland coast and creating a huge but unique coastal landform assemblage.

The recent geomorphodynamic evolution is determined by the geological composition of the shore material (gravel, sand, mud) and by the local meteo-marine conditions, such as waves, tides, currents, wind, rainfall, and temperatures. In this respect, quarternary geologic formations are of parmount interest for the coastal engineer, as they are most affected by natural forces and human interference.

Coastal Landforms

A large variety in coastal landforms exist resulting from the combination of the above tectonic, geological, and climatologic processes. For coastal engineering applications, the global description presented by Short and Blair is not sufficient. Various authors have proposed a classification of coastal land forms. As various earth sciences are involved, none of these descriptions includes the full spectrum of parameters. Based upon a system analytical approach, the following model for coastal classification is proposed.

The coast is a physical system, which derives its *resistance* against impacts mainly from its geological features. These geological features should be considered constant for time horizons common in coastal engineering, say 50 to 100 years. Long-term processes, stemming from tectonic activities and resulting in transgression or regression, are excluded therefore in the *classification,* but are essential for understanding the *genesis* of the coast. Moreover, autonomous geological developments should be incorporated as exogenic boundary conditions for any coastal study.

In many instances, the physical resistance is enhanced by biological factors such as salt marsh vegetation or mangroves. In a few cases, biological processes are fully determinant for the resistance. This is the case where reef formations are being considered. Freezing may also affect the resistance. As arctic areas are not considered in this paper, this aspect is excluded. Obviously, the natural resistance of each of the above classes of coasts may be affected by human interference, which can lead to artificial reinforcement of the natural resistance.

Based upon physical resistance, the following types of coasts may be distinguished:

1. Natural coasts
   a) Geologic parameters determinant
      (1) Rocky coast
      (2) Cliffs
      (3) Gravel beaches
      (4) Sand beaches
      (5) Mud flats
   b) Geologic and biologic parameters determinant
      (1) Salt marshes
      (2) Mangroves
   c) Biologic parameters determinant
      (1) Reef formations
2. Artificial coasts
   (a) Coastal protection
   (b) Groines
   (c) Offshore breakwaters

The coastal system is exposed to forces which tend to reshape the coastline. These forces may originate from the atmosphere or the lithosphere and may be modified in the hydrosphere. Essentially, the shape of the coast is determined by the following processes:

1. Sediment supply by rivers and surface run-off
2. Sediment diffusion by waves

Although this model is a crude description of the coastal system, it may serve as a basis for a global classification (Figure 7).

Sediment supplied by rivers and surface run-off is given on the vertical axis, whereas sediment diffusion is shown on the horizontal axis. Figures given on the horizontal axis are indicative for low- moderate- and high-energy coasts. Figures given on the vertical axis indicate average sediment yields commonly found in high moderate and low rainfall areas.

If the sediment yield exceeds the wave induced transport, accretion will occur. Depending on the wave intensity (low to moderate), the resultant coast will consist of mud or sand. Foreshore slopes will be mild and the elevation of the coastal plains and deltas will be low. For reduced transports and increased wave intensity (moderate to high), sand and gravel beaches are formed with steeper foreshores and occasional barrier reefs and dunes, especially in areas with high aeolean transport. If sediment supply is small and wave intensity is high, the old rock formulation will determine the coastal shape. Erosion of rock by waves will result in short and shallow abrasion planes with erosion products such as gravel and stones deposited at the beach.

Using the climatologic parameters shown in Table 2, it may be inferred that mud will prevail in the lower latitudes due to the abundant supply of sediment as a result of heavy rainfall combined with low wave intensity due to the mild wind climate. For intermediate latitudes, rainfall intensity drops and wave heights increase, resulting in a increase in sand concentration along the beach. For higher latitudes, a rock and gravel coast will predominate because of their reduced sediment yield as a result of a low runoff combined with severe storms and waves.

The above considerations are restricted to geologic processes: reefs and other biologic formations are not considered. Further details are given in Davies (1977), from which Figure 8 is obtained.

Global Vulnerability

Obviously not all of these landforms are equally vulnerable for the effects of climate change. Their vulnerability is determined by the following factors:

1. The relative change in atmospheric or related hydrodynamic input as a result of climate change.
2. The response of the particular coastal system to change. In this respect, the various impact categories previously discussed should be considered.

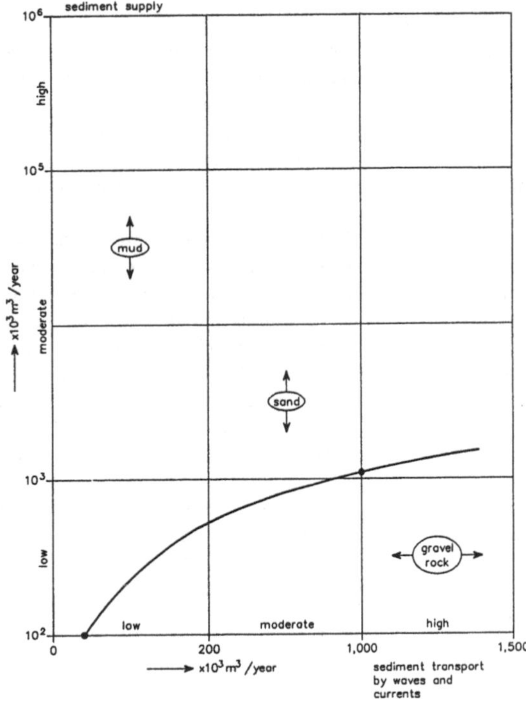

Figure 7. Coastal landforms as a function of sediment inflow and wave intensity

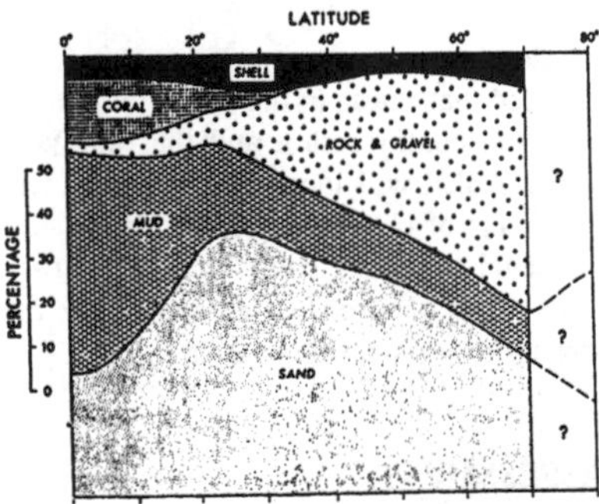

Figure 8. Relative frequency of occurrence of sediment types by latitude on the continental shelves of the world *Source: (After Hayes 1967a), From (Davis 1977)*

Atmospheric and Hydrodynamic Input

Possible changes in climatic conditions will include changing rainfall patterns and changing wind. The latter will be responsible for increased storm surges and higher wave intensity. This will be reflected in higher, averaged annual conditions and higher extremes. Moreover, an accelerated rise of the average mean sea level (ASLR) may be expected.

Response of the Coastal System

As may be inferred from Figure 7, changes in *wave intensity* and *sediment yields* will change the large-scale development of the coast. Higher waves will tend to remove fine material, resulting in a steepening of the foreshore. Moreover, risk of flooding will increase due to extreme waves. It should be noted that this effect is relatively more important in calm areas, e.g., the lower latitudes. Reduced run-off and related sediment yield will cause erosion of the coast. This effect is more important in areas with low to moderate rainfall and will be less pronounced in tropical areas with abundant supplies of water.

An *accelerated sea-level rise (ASLR)* will be felt first in areas where the coast is flat. In these areas, a small increase in sea level rise will result in inundation of large areas. This is again found in tropical areas. Subsequently, all coastal landforms will start to erode, as the cross-shore profile needs to be adjusted to changing hydrodynamic conditions. In this respect, gravel and sand coasts will experience larger deformations. In particular, the gravel beaches will respond quickly to changing conditions due to their steep profile. Mud coasts will respond much slower. The actual erosion rates are determined by a delicate balance between longshore and cross-shore transport. More detailed modeling of these phenomena is needed for a quantitative assessment.

Finally, increased levels at sea may affect flood levels in rivers and may cause higher salt intrusion in surface and groundwater.

The effect of these changes on the socioeconomic system cannot be quantified without knowing the socioeconomic consequences of the changes in the physical system. In general, the following aspects are taken into account when quantifying the aspects shown in Table 1:

- Safety - flood levels along the coast and in the rivers
- Economic and social activities
  - flood levels
  - salinity intrusion
  - erosion
- Ecological values
  - flood levels
  - salinity intrusion
  - loss of land due to erosion

In summary, it may be concluded that coasts in the lower latitudes are highly vulnerable to climate change. Due to their particular coastal configuration and the relatively mild, meteo-marine climate, changes in storm surge intensity and ASLR will increase the risk of flooding. Coasts in the temperate zones will be affected mainly by effects of changing rainfall and resultant changes in sediment yield. Combined with an increase in sea level, this may cause increased erosion. Coasts in higher latitudes are only marginally affected, as the effect on storm surge and waves intensity will be relatively small, and the rock and gravel coasts are more resistant to erosion.

LOCAL VULNERABILITY ASSESSMENT

The previous section a global system description was given which was used to characterize the coastal system and its sensitivity to change. In order to quantify the effect of the impact categories (previously described) for various coastal systems, a more refined system description is needed. This description is given in Figure 9. The left part of this Figure represents the global description previously given, while the right part summarizes the main actors and impacts resulting from climate change. The diagram in the center is a refined system description of the hydrosphere in terms of system elements and their interactions. In order to describe and quantify the impact of climate change, system parameters should be known and interaction processes should be described by means of appropriate input-output relations.

Actors and Primary Impacts

Climate change will result in temperature change, a changing wind climate (in terms of average annual values and extreme events), and changing rainfall patterns. Although climate modeling is still in its early stage of development, a state-of-the-art review presented at the World Climate Conference in Geneva in 1990, showed promising results. For this paper, no review was made of tools and techniques for climate modeling. The paper is restricted to tools and techniques to analyize and predict their impact on the aquatic system.

In this respect the following primary impacts have been distinguished:

1. Sea level rise due to thermal expansion
2. Changes in wave climate and statistics of extreme waves
3. Changing run-off pattern of river basins

These impacts may be considered as the global response of the system environment to climate changes. For the system simulation of the specific coastal system, values should be considered as input scenarios at the seaward and landward boundaries respectively.

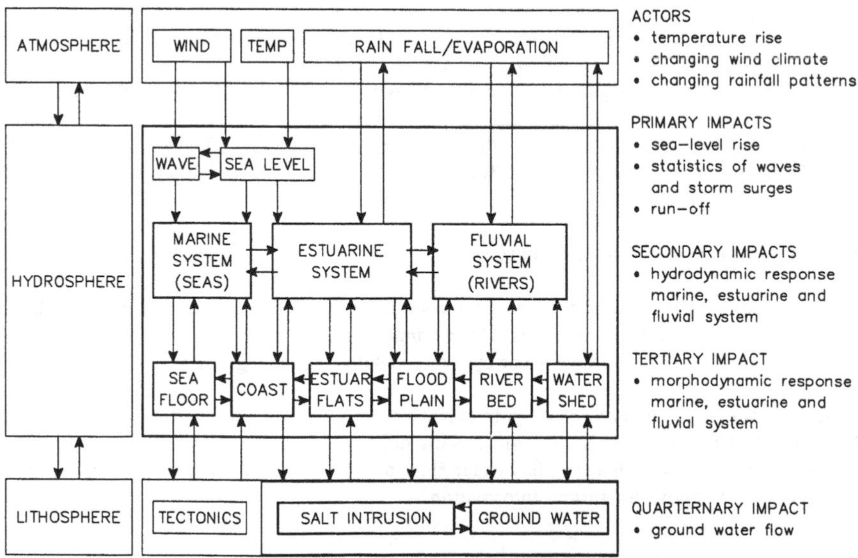

Figure 9. Generalized system diagram of the coastal zone with aquatic subsystem (and main interactions with atmosphere and lithosphere)

Secondary Impacts - Hydrodynamic Response

The sea, the estuaries, and the lower stretches of rivers will respond to the changing inputs. At sea, waterlevels, tides, wind induced currents, and wave conditions will change. Changing tidal elevation and a changing river regime will affect salinity intrusion in estuaries and lower rivers.

Finally, these changes will affect backwater curves and will consequently change high water levels in the estuary and along the river. This will imply an increased risk of flooding. It can be stated that the response of hydrodynamic systems will be almost instantaneous. This means that given the inherent time scale of the primary impact, the resultant changes in the hydrodynamic system will be almost simultaneous. Presently, numerical models are available to compute these changes. Reference is made to Goda (1985), Ippen (1966), and Horikawa (1988) for a state of the art review.

Tertiary Impacts - Morphodynamic Response

Changing hydrodynamic conditions will induce a morphodynamic response of the entire marine, estuarine, and fluvial system.

Flow-induced transport patterns at sea will change, resulting in a dynamic response of the seabed. Wave-induced transports, in both longshore and cross-shore directions, will be affected, resulting in changes in the plan shape and the cross-sectional shape of the beach.

The response of tidal basins is more complex. A changing sea level will affect the delicate equilibrium between tidal volume and cross-sectional area. This will result in sediment transport fluxes to restore the equilibrium. Erosion of tidal flats, outer deltas, and neighboring coasts will be the result.

In estuaries, the situation is even more complex, as the interaction between fresh and saltwater is an important factor in the sedimentation process. The shape and location of the inner deltas in such areas may, therefore, be affected.

Further upstream, the river will respond to changing flow conditions and, possibly, a changing sediment inflow from the catchment area. Erosion and deposition patterns will change, resulting in changes in the alignment and cross-sectional area of the river.

The response of morphodynamic systems to changes is slow. The effects of sudden changes, such as the closure of an estuary or the damming of a river, will still be felt after decades. The effects of transient processes such as a gradual rise of the sea level, will be felt even longer. Models are available or under development for these processes. Reference is made to Teisson (1991) and de Vriend (1991) for a state of the art review.

Quarternary Impacts - Aquifer Response

Groundwater reserves are part of the total volume of water resources stored in the hydrosphere. Seepage and percolation are the processes which connect surface and subsurface resources. Both will be affected by changes in the surface water systems, resulting in changes in water table and increased salt content.

The response of aquifers to changes in surface water systems will be slow. Periods of decades will be needed to arrive at equilibrium conditions for sudden changes; even longer periods will be found for transient processes. Models for groundwater flow are used to quantify these aspects. Reference is made to general text books for further information.

REFERENCES

Boulding, K.E. 1956. General system theory, the skeleton of science. *Management Science* 197-208.

Davies, J.L. 1977. *Geographical variations in coastal development*. London: Longman Group Ltd.

DELFT HYDRAULICS/Tidal Water Division Rijkswaterstaat. 1992 (May). Sealevel rise - A global vulnerability assessment. The Hague. Delft.

de Vriend, H. 1991. Mathematical modeling and large scale coastal behaviour. Part I - Physical processes. Part II - Predictive models. *Journal of Hydraulic Research* 29(6): 727-53.

Goda, Yoshimi 1985. *Random seas and design of maritime structures*. University of Tokyo Press.

Horikawa, Kiyoshi 1988. *Nearshore dynamics and coastal processes*. University of Tokyo Press.

IPCC - Response Strategies Working Group The Seven Steps to the Assessment of the Vulnerability of Coastal Areas to Sea-Level Rise. 1991 (September).

IPCC - Response Strategies Working Group Summary Report. 1990 (April).

Ippen, A.T. ed. 1966. *Estuary and coastline hydrodynamics*. New York: McGraw Hill Book Comp. Inc.

Morrison, Foster 1991. The art of modeling dynamic systems. New York: Joha Wiley and Sons Luc.

Peerbolte, E.B., and H.G. Wind 1991. Policy implications of sealevel rise in the Netherlands. In *Proceedings International Conference on climate impacts on the environment and society (CIES)*. University of Tsukuba, Ibaraki, Japan.

Proceedings Second World Climate Conference 1990. *Climate change: Impact and policy.* Cambridge: Cambridge University Press, 1991.

Sestini, A. 1992. Sealevel rise in the Mediteranean Region - Likely consequences and response options. *Proceedings of the Workshop "Regional Programmes and Environmental Protection."* Genova, Italy. 12-14 February, 1992.

Short, Nicolas M., and Robert W. Blair Jr. eds. 1986. Coastal landforms. In *Geomorphology from space*, NASA Scientific and Technical Information Branch. Washington DC.

Teisson, C. 1991. Cohesive suspended sediment transport, feasibility and limitations of numerical modeling. *Journal of Hydraulic Research*. 29(6): 754-69.

Ocean Resources and Sustainable Development

31

Keynote

An Innovative Power Generation System from Sea Currents in the Messina Strait

Ing. Dario Berti and Ing. Emanuele Garbuglia
Tecnomare S.p.A., Venezia, Italy

ABSTRACT

This paper describes the results of a feasibility study regarding the exploitation of marine currents in the Strait of Messina, Italy. The high kinetic energy contents of these currents suggest the development of some exploitation system. The technical solution proposed in this study consists of seabed-installed, vertical axis turbines, which are able to optimize energy extraction. The work performed at this stage has included a basic design of the system and a preliminary assessment of the environmental impact.

A costs/benefits evaluation showed that electricity produced by the proposed system is competitive with that based on other energy sources; conventional but renewable ones. Moreover, fresh water, considered an alternative to electricity production, could be produced at costs comparable with present supplies by shuttle tankers.

INTRODUCTION

The growing interest in the exploitation of ocean energy, such as marine currents, thermal and density gradients, and wave or tide does not result from the consideration of mere economical advantages, but has its origin in the concurrent trends of the present socioeconomical context.

This aspect was the subject of a recent study carried out by the writers, with the aim of defining a picture of the future demand for marine technology in the medium- and long-term (five to fifteen years) through the analysis of the future marine markets and technology needs. The study entitled

"Marine Technologies 2000", was developed starting from the analysis of the relationship between man and sea in three main areas:

1. Knowledge and safeguard of the marine environment
2. Exploitation of ocean resources
3. Utilization of ocean space

Within these areas, the study focused on some sectors as they related to a general scenario to represent the economic, social, political, technological, and cultural development of society. The prediction of future advances will be based on the opinions of an international panel of experts, consulted by means of questionnaires. The marine current was included in the area of ocean resources that was broken down in the tree-like scheme reported in Figure 1.

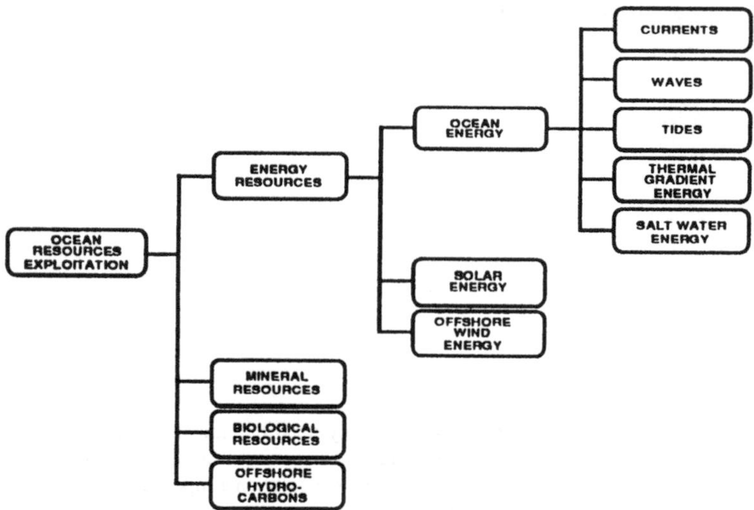

Figure 1. Tree-like scheme of the ocean resources exploitation

The results of the study pertaining to the area of the exploitation of ocean energy resources are briefly summarized, highlighting the socioeconomic factors which influence, both in a positive and negative way, the development of that sector.

A positive impulse to such growth can be given by the worldwide consciousness of real, long-term threats to the environment from thermal energy plants and the uncertain future of safe nuclear energy, especially after the recent Chernobyl disaster. Additionally, considering the fact that the main oil and gas reserves are located in politically unstable areas, there is a call for the creation of adequate strategic reserves and security of supply, which also means exploitation of internal energy resources. That leads the governments, in particular in industrialized countries, to promote new development schemes for renewable energy systems and to plan for a certain amount of non-fossil-fuel power generation. Public opinion also contributes to this impulse, because of the increased level of the "quality of life" and the growing attention given to environmental safeguards and living in a cleaner world.

On the other hand, the development of that sector can also be motivated by the foreseen increase in population and the continuous concentration of people in coastal areas. Logically following would be the future expansion of the towns seaward, first by reclaiming areas and then by the creation of "ocean towns," fully using the ocean space area for urbanization, leisure, industrial activities, and the like. Because of these factors, there is an increased importance placed upon the role of the oceans as a source of materials, food, and energy. The consequent increase in energy consumption which is bound to both the increase in world population and industrial production, cannot be met by a supply of energy based on combustion, but requires a different approach in which the ocean energy resources can play an important role.

It is quite evident that none of these factors alone is able to substantially change the economics of the renewable energy exploitation. The energy market will continue to be oil-dependent for some years. As in all the other renewable and non-conventional energy resources, the energy extracted from marine currents can hardly exceed a few percent of the global energy consumption, even if in several cases it can be very helpful in counteracting energy shortage while coping with environmental constraints in coastal areas or isles by integrating or substituting energy systems based on conventional resources, like oil, gas, or coal.

The incentives are, nevertheless, counteracted by actual and real technological problems. They concern applicability limited to specific areas, usually far from the final users; high cost if compared to the production; and plant maintenance and durability. Most of these aspects are not peculiar to marine energy, but are common to all renewable energy resources. The need is therefore to realistically assess the size of ocean resources and to identify cost-effective solutions to extract the energy in a usable form.

MARINE CURRENTS

With reference to Figure 1, marine currents, representing a large, renewable energy resource, can significantly contribute to the fulfillment of the energy demand, with emphasis put on local supplies.

Among the ocean and renewable energy resources, marine currents have certainly not been given enough consideration by researchers and energy authorities. The reason can probably be found in the fully marine characteristic of the systems suitable for exploiting marine currents, if compared with either OTEC, wave, or tidal energy systems, which are essentially land-based. This problem can be overcome, thanks to the technology transfer from the offshore industry. The technology level reached in the offshore industry, gained in more than thirty years of activities (especially in the field of offshore hydrocarbon resources) provides know-how and operative experience to transfer the advanced land technologies to the subsea environment. The technologies we refer to are those utilized for exploiting wind energy. In fact, the fluido-dynamic analogy between a wind and a marine current suggests the use of the technological and operative experiences gained by the wind energy industry in recent years. Corrosion resistant materials, waterproof components, remote control, marine operations, and subsea maintenance are some of the factors that can actually make the utilization of underwater turbines (derived from the wind energy sector) to exploit marine currents, suitable and effective.

Attention in the following discussion is focused on how and why the marine current energy resource can effectively be exploited, referencing a feasibility study carried out by the writers. The study investigates the exploitation of marine currents in the Strait of Messina (South of Italy), which are characterized by a high, kinetic energy content. While this study is targeted at a particular application, it is aimed at proposing general guidelines and technological solutions for dealing with analogous situations of exploiting marine currents. The size of the proposed hydro-generator (1 MW electric) and the modular system approach make this solution suitable in various contexts.

THE SITE

Geography

The Messina Strait (Figure 2) is a small sea channel about 30 km long with a minimum width of 3,200 m, that separates the Italian mainland from the island of Sicily and connects the Ionian and Tyrrhenian Seas. It presents a variable bottom profile with a depth of up to 1,500 m. The profile is characterized by the presence of a transversal sill in the northern part of the strait which is also the narrowest. In correspondence with this sill, the depth rapidly reaches the minimum of about 80 m over a length of about 7 km. This point constitutes a net separation between the two seas and a "bottle neck" for the water flow. The Strait is considered one of the more singular sites in the world in terms of its oceanographic features.

From the rather limited hydrological data available in literature, the current regimes of the Strait are characterized by high speeds, high turbulence, and instability. A complete oceanographic survey of the Strait was carried out between 1979-80 to support a study aimed at checking the feasibility of crossing the Strait with a stable railroad link. It followed some previous campaigns, limited in extension, to support specific operations such as maintenance of underwater cables or pipeline installation. The results of those campaigns confirmed expectations: current speed of up to 5 m/s was monitored, although seldom and in limited areas, while average speeds of 2-3 m/s were frequently measured.

The oceanographic data describes the current as resulting from components of different origins, among which the tide is the most significant one. Very large values of current velocity are related to the semidiurnal tidal component, which reverses its direction about every 6 hours and has average values over 2.2 m/s. This large tidal component is due to a phase delay of about 180 degrees in the level variation between the Ionian and the Tyrrhenian Seas, which creates a high flow rate along the short length of the Strait. The remaining tidal components are less important, but concur to generate overall current speeds in the order of 3 m/s.

Other components of thermo-haline and meteorological origins are also present. An important component is the current generated by the difference in density (salinity and temperature) between the Tyrrhenian and Ionian Seas. Two approximately stable and opposite flows are present: one at the surface and one near to the bottom. Going from the Ionian Sea to the Tyrrhenian Sea, the bottom one is significant for our application and has a speed in the order of 0.3 m/s. Other transversal and vertical components, mostly from meteorological origin, are present and cause evident turbulence of the current and local increases in its speed.

A local increase in current speed is also observed in correspondence of the sill. The presence of the obstacle, constituted by the rapid variation of water depth, causes a variation of the flow regimen resulting in an increase in speed near the top of the sill.

In conclusion, the current regimen of the Strait, while requiring further and more specific investigations due to its complexity and singularity, shows very interesting levels of current velocity. A quantitative evaluation based on collected oceanographic data results in a total energy content of about 2900 GWh/year. This value fully justifies the interest in exploiting that potential energy.

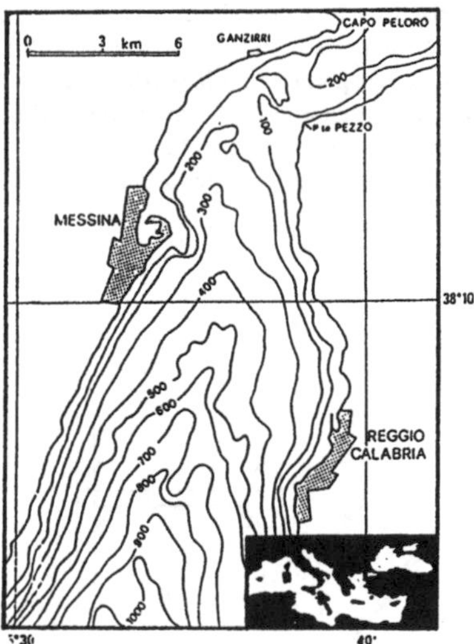

Figure 2. Strait of Messina and installation site

Local Socioeconomic Context

The main activities of the regions facing the Strait are fishery, maritime transport, and tourism. Being periodically crossed by fish schools of high commercial interest, the area of the Strait draws a large number of people employed in fisheries. Moreover, it is characterized by an intense maritime traffic of tankers, passenger, and cargo ships, and it is frequently crossed by ferryboats and private boats linking the opposite coasts.

Two main needs emerge from the present situation and perspective analysis. The first results from a heavy shortage of fresh water, which is heightened in summer. The second refers to the electricity supply. Although at present the balance on electricity import is positive, the future demand forecast calls for an increase not covered by the actual potential.

The National Electricity Plan (PEN), on the other hand, pushes in the direction of increasing local production and exploiting renewable resources. The exploitation of the local energy resource, constituted by the marine currents, seems to be a possible solution to partially meet both needs.

EXPLOITATION TECHNOLOGIES

The research for a technical solution to exploit marine currents must drawn up depending both on the wind generator industry and the offshore industry. Marine currents, interacting with a rotating machine (hydrogenerator), are fluid-dynamically comparable to the wind when passing through a wind generator. This suggests, as a first step, to carefully evaluate the features of those technical solutions currently adopted by the wind industry. Compared to a wind, a marine current generally features lower speeds (in the order of 25%) but higher density (about 800 times greater), and consequently offers greater specific power (about 10 times).

The variability of marine currents and the fact that they are generally not well canalized, does not fit in well with the configurations of ducted axial turbines commonly used in hydroelectric plants to exploit high flow rate characterized by low prevalence, as is the case with rivers or tides.

The solution is to be looked for in the category of free field turbines, like wind generators. The technological and operative experiences gained in this sector in recent years can be profitably transferred to the new category of free field hydrogenerators. Vertical axis turbines (VAT or Darrieus turbine) and mostly horizontal axis turbines (HAT) are used in wind exploitation. In Figure 3, the main features of the two configurations are briefly illustrated to allow a better comprehension of the choices made.

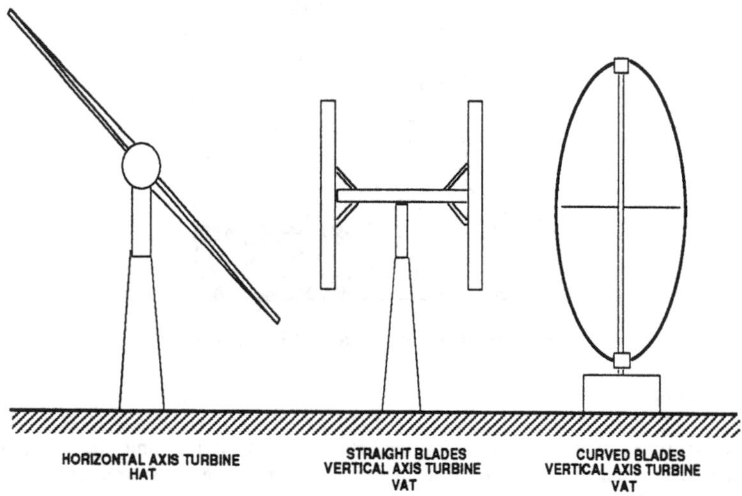

Figure 3. Wind turbines

The VAT does not require repositioning to follow the wind direction and is preferred when the wind flows from varying directions on a wide range. In addition, this system offers a favorable configuration with respect to the gravity loads which are better balanced around the rotation axis.

The flexibility of VAT is, however, overshadowed by limitations resulting from the structural resistance of the blades. In VAT, they are loaded by bending torque, while in HAT, main loads are centrifugal axial forces. This aspect affects the geometry and hydrodynamics of the turbine, limiting maximum blade length, rotating speed, or requiring solutions with curved blades, constraining their tips to the rotation axis (proposed for the bigger VATs). All these solutions are palliative however, because the turbine efficiency decreases. Another disadvantage of the VAT solution is the complexity in implementing the variable pitch feature to the blades, which is theoretically necessary if the current speed varies. On the contrary, the HAT solution make this implementation easier, and variable pitch blades are commonly used.

The heaviest fluids such as water, offering higher energy potentials, make it possible to keep the rotating speed of the turbine low for a constant output power, and make the VAT solutions more advantageous. The inertial centrifugal loads on the blade are reduced to give way to hydrodynamic loads originated by the high specific power of the fluid. The aforementioned advantage of this configuration on the gravity loads is more evident if the electric generator is also considered. Low rotating speeds require large diameter generators (multipole generators) or specially built, large gearboxes to increase the speed (low speeds imply high torque to be transmitted from turbine to generator shaft). It is evident that a VAT configuration is preferred, because it avoids locating the multipole generator on the top of a tower, as in the HAT configuration. Examples of big wind VATs (up to 4MW) using a multipole generator are presently working. For the aforesaid reasons, the VAT solution seems to be preferred for application with marine currents.

The technology level reached in the offshore industry, and mainly in the oil sector, provides know-how and operative experience to transfer the technologies from the wind industry to the subsea environment. Corrosion resistant materials, waterproof components, remote control, marine operations, and subsea maintenance are some of the aspects that can actually make the utilization of underwater turbines to exploit marine currents, suitable and effective.

THE EXPLOITATION SYSTEM

Overall system

The configuration of the system proposed to exploit the Messina Strait currents is based on a hundred power modules to be installed on the sea bottom in correspondence with the Strait sill (Figure 4).

Figure 4. The exploitation system: Artistic view

The choice of the installation site is preliminary and the final choice can only be determined after specific oceanographic surveys have been made. At present, the sill seems to offer the best features of high current speed (up to 3 m/s) and low water depth.

Each power module is designed to produce either electricity or, in alternative, high-pressure water to export to an onshore desalination plant (based on reverse osmosis technique). Globally, the system can produce about 400 GWh/year of electricity (or 170.000 m3/day of high-pressure water).

The configuration of the whole plant is founded on the concept of modularity which fits in well with the irregular morphology of the installation site and allows a great amount of flexibility. At present, each power module is considered with its own link to common land facilities, but within a detailed analysis, the use of underwater collectors directly tied back to the shore will be evaluated. A simplified drawing of the power module is given in Figure 5.

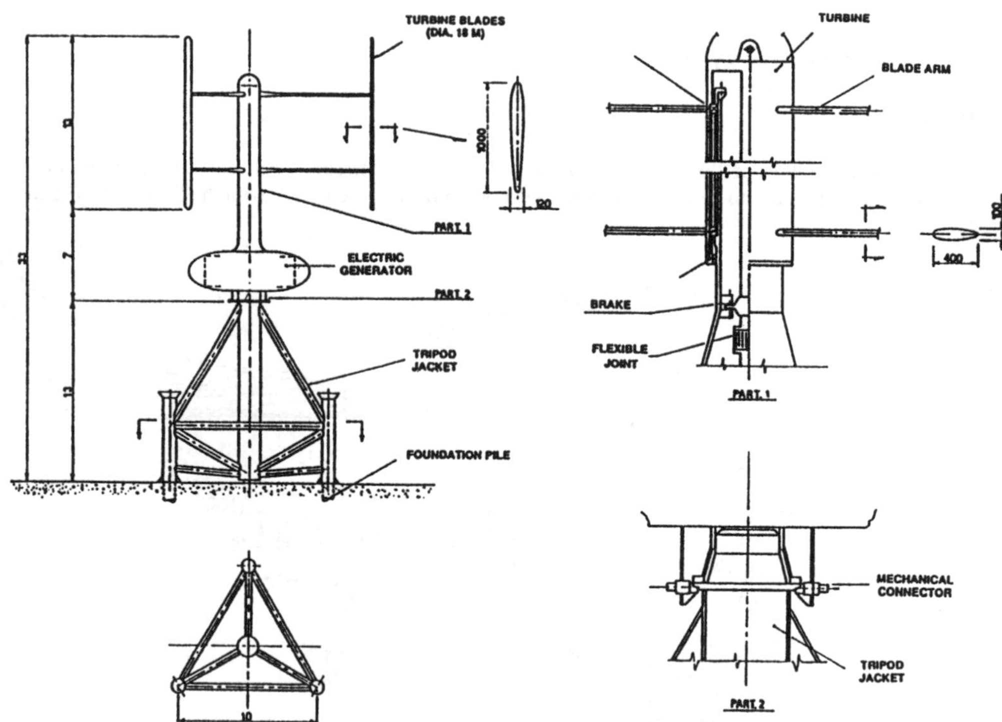

Figure 5. The exploitation system: the power module

Each module is basically composed of:

- The rotor, made up of the turbine, the drive train, and power generator (electrical generator or seawater pump)
- The support structure
- The underwater line (electrical cable or pipeline) to transfer the product (electric energy or high-pressure water) to shore
- The land facilities.

Turbine

The turbulence and variable directions of the Strait currents direct the choice towards a vertical axis, straight blades configuration. The geometry definition and the preliminary turbine sizing were drawn from the analysis of the state-of-the-art wind VATs and were based on the turbomachine and hydrofoil theory. The ability of a turbine to capture the kinetic energy of the flow, depends essentially on its ability to decrease the flow speed and deviate its momentum, and is related to the geometry of the blades' transverse sections. When designed according to foil sections, a blade is submitted to hydrodynamic lift and drag force. When the former prevails, it is transmitted through the support arm, to the turbine shaft on which it generates a rotating torque (Figure 6.)

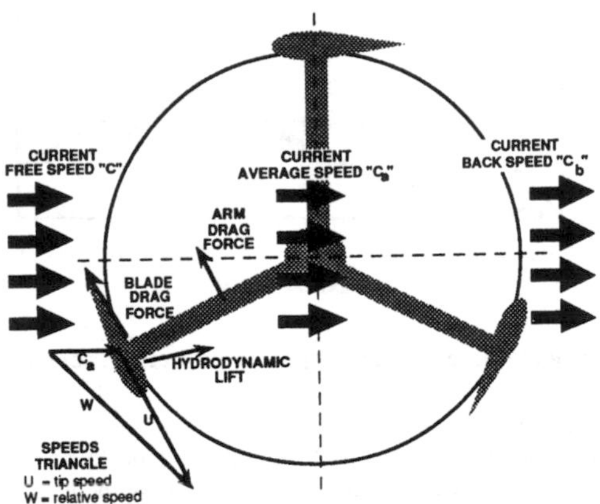

Figure 6. Theorical scheme of the turbine

A short sensibility analysis, addressed to a power module of "square configuration" (turbine diameter comparable to blade length) with reference to about 1 MW output power and a stream free speed of 3 m/s, was carried out to select optimal turbine geometrical features. A three-blade configuration, each supported by one arm, was chosen and the main geometrical characteristics were as follows:

- Turbine diameter - 18 m
- Blade length - 13 m
- Solidity - 0,05
- Blade chord - 1 m

NACA profiles were considered both for the blades and for the support arms. A nominal output power of 1100 kW corresponds to these characteristics.

The global efficiency of the machine reaches its higher values (more that 30%) with a ratio between peripheral speed U (or blade tip speed) and stream free speed C greater than 4, which is assumed as the reference value in maximum current speed conditions. Consequently the rotating speed is 13 rpm and the peripheral speed U is 12 m/s.

Turbine geometry and dynamics made it possible to evaluate the output torque and the overall loads on the structures, taking into account hydrodynamic lift and drag forces on blades and arms. The turbine efficiency was also analyzed by varying the current speeds in two different conditions:

1. Ratio *U/C* constant (equal to 4)
2. *U* constant (equal to 12 m/s)

As shown in Figure 7, with current speed between 2 and 3 m/s, the machine works with high efficiency. Out of this range (which happens during tidal inversions), a regulation at *U/C* constant or a machine stop are required.

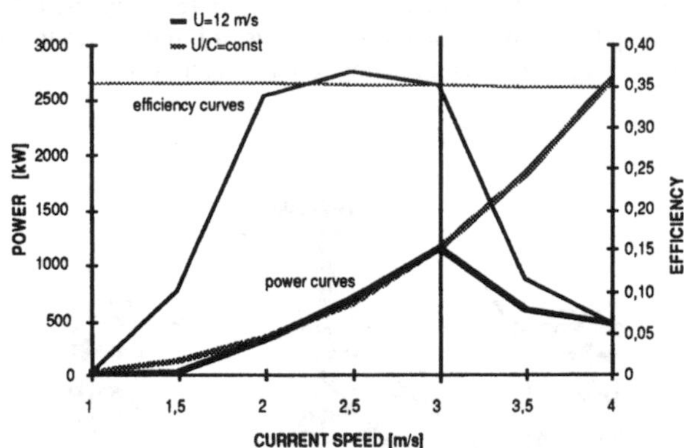

Figure 7. Turbine performance (output power and efficiency) vs free stream speed

Drive Train and Electricity Generator

The low rotating speed (under 13 rpm) and the high torque to be transferred to the generator suggested avoiding the multiplication gearbox and directly coupling the turbine to the generator through a flexible joint. The generator will be a multipole, 3-phase unit (about 130 poles), producing a voltage in the order of 3 kV with a variable frequency up to 20 Hz. Solutions of this type are in use in "Darrieus" wind turbines (VAT type). A mechanical brake, which consists of a thick steel disc, is incorporated into the drive train.

Underwater Cable and Land Facilities

The electricity is transported to the onshore electric station by an underwater cable. The land electric station will host the control room and will be provided with static frequency converters and transformers to convert the electricity into a supply suitable for connection to the electrical grid. The control system is similar to those used to control the wind turbine. A control link, included in the underwater cable, exchanges data between the subsea system and the control room.

Support Structure

The position of the rotor with reference to the sea bottom affects the possibility of uniform utilization of the whole turbine height because of the progressive reduction of the current speed while approaching the sea bottom (boundary layer effect). In order to avoid or at least reduce this negative influence, the rotor is positioned on a support, with an average elevation of about 30 m above the sea bottom. This makes it possible to contain the stream speed reduction under 20% and to rely on a current speed centered around about 2.5 m/s.

The support structure consists of a tripod jacket of about 70 tons, supported by three piles (Figure 5). The rotor is connected to the structure by means of a mechanical connector similar to those used in the offshore industry for structural applications. It makes installation and recovery operations easier, separating the mechanical part from the structural one.

Alternative: Water Desalination

The lack of water for potable uses such as agriculture and industry, affects the installation site areas and suggests the consideration of the production of fresh water as an alternative to electricity production. The generation of a storable product can overcome the problem which results from daily and seasonal energy fluctuation and the consequent need to vary production.

The idea is to pump seawater to an onshore desalination plant using the reverse osmosis technique which requires high-pressure water (50 to 60 bars) in order to work. To avoid multiple energy transformations, (mechanical to electrical and electrical to mechanical) the turbines are directly coupled to the pumping unit. The water is transferred to the onshore desalination plant by means of underwater pipelines. A daily production of 170,000 m^3/day is foreseen.

The load factor of the desalination facilities can be improved by the introduction of electrically driven pumping units (hybrid solution) to cover the daily variation in production; the result of the tidal regimen. While doubling the freshwater output (340,000 m^3/day) and increasing the capital expenditure, the hybrid solution assures the continuity of the supply independent of the tidal variations. Compared to conventional desalination plants, the solution (hybrid or not) allows electric energy recovery and the use of renewable energy resource.

CONSIDERATIONS ON "EIA"

A forecasting study on possible environmental and socioeconomic impacts was carried out. This preliminary study, based on a critical items analysis, considered the most significant environmental and socioeconomic components: marine environment, coastal environment, fisheries, maritime transport, energy, and freshwater supply. The conclusion of this preliminary study are quite revealing.

The presence of underwater bodies could be a reason for fish restoration, but the noise and the disturbance to the current hydrodynamics produced by the rotating machines could modify the underwater habitat and the migration fluxes of some fish species, with possible impacts on the fishing activity. The modification of the currents regimen could also modify the morphology of the sea bottom and perhaps, the coastal profiles.

No significant impacts on coasts are foreseen due to the presence of the land facilities. No significant impacts are foreseen on maritime transport during the installation and maintenance of the underwater systems, as seen from analogous activities being carried out currently.

A positive impact is certainly given by the production of electric power (about 400 GWh/year). On a local scale, this supply covers 5% to 10% of the consumption. The system supplying fresh water also has a positive impact, both for the strategic importance of this resource and for the entity of the supply (170.000 to 340.000 m^3/day), which could satisfy the demand of up to one million people or irrigate up to 8,000 ha of agricultural land.

The confirmation of these preliminary conclusions can only be achieved after a complete EIA, including model testing, forecasting modeling, and risk analysis, since the site is a seismic risk area.

COSTS/BENEFITS EVALUATION

Costs Estimating

The system's capital expenditure (capex), including design, construction, installation, commissioning, and start-up, has been roughly estimated at about $ 250 million U.S. These costs are broken down in Table 1. The most expensive items are the rotors (47%) and the marine operations (20 %), which include site survey and preparation, system transport and installation. The total cost, referred to a system made up of one hundred modules, results from the cost of one module (unit cost) reduced by a quantity (% reduction) based on the "scale effect".

Table 1

Capital Expenditure

CAPITAL EXPENDITURE [1000 US$]	unit cost	% scale reduction	total cost
Rotor	1670	30%	116900
turbine & frame	310		
electric generator	870		
instrumentation	70		
other mechanical comp.	420		
Support structure	270	10%	24300
Marine operations for transport & installation	990	50%	49500
Underwater cable	410	60%	16400
Land equipment & facilities	220	60%	8800
Onshore civil works	400	70%	12000
Engineering	400	80%	8000
Testing, certification & assurance	200	60%	8000
TOTAL CAPEX [1000 US$]	10200		243900

Simultaneously the annual operative costs (opex), covering the management, maintenance, and operational support of the system have been estimated at about 16 million U.S.$ per year, as detailed in Table 2. The costs refer to a utilization factor of 50%, which takes into account the tide fluctuation and an annual production of 430 GWh.

If the alternative of seawater desalination with the hybrid solution is considered, the capital and perative expenditures are respectively increased by about 400 million U.S.$ and 90 million U.S.$/year. These amounts are only relevant to the desalination plants, the flowlines, and other related facilities, while the cost of the power modules and their operation is considered similar to the previous cost estimate. Contingencies have not been explicated because a sensibility analysis was carried out by varying capex and opex.

Economics and Benefits

The costs were then compared to the possible income derived from the energy sale. The reference price of the electricity was fixed at 10 U.S cents /kWh. The internal rate of return (IRR) of the investment and the technical cost of the electricity produced were then evaluated on the basis of a twenty-year operative life. A sensibility analysis of these two parameters is summarized in Figures 8 and 9. The basic case shows an acceptable level of return on the investment (6%), while the technical unit cost is 5 U.S. cents/kWh.

Table 2

Operative Expenditure

OPERATIVE EXPENDITURE [1000 US$/year]	unit cost	factor	total cost
Personnel	75	15	1130
Maintenance (% on plants capex)	166800	3,0%	5000
Assurances (% of capex)	243900	0,5%	1220
Capital costs (5%; 20 years)			7300
General Costs (% of personnel + maintenance costs)	6130	25,5%	1530
TOTAL OPEX [1000 US$/year]			16200

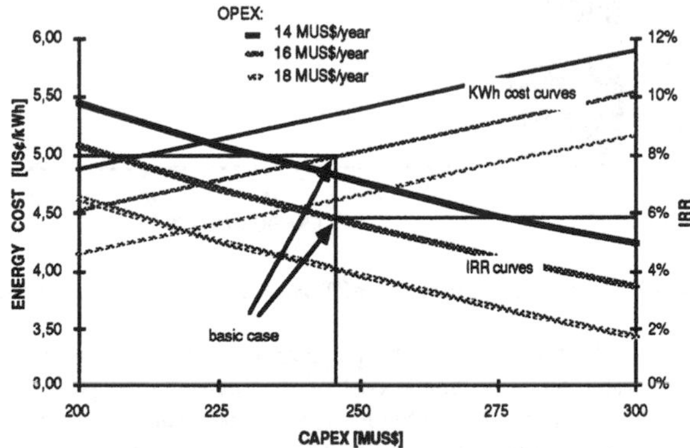

Figure 8. System economics: unit energy costs (US cents/kWh) and internal rate of return (IRR) vs capital and operative expenditure (CAPEX/OPEX)

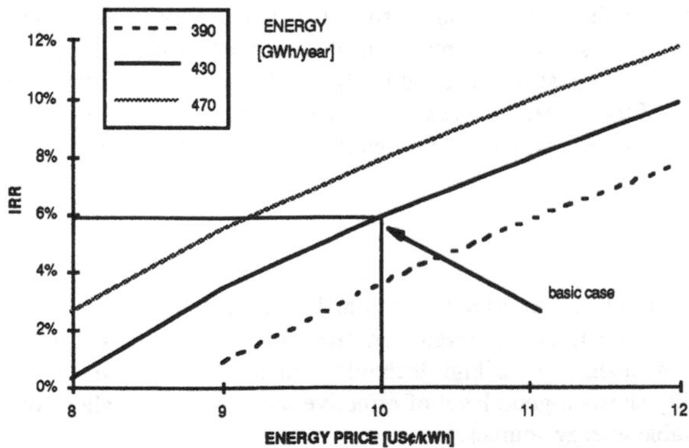

Figure 9. System economics: internal rate of return (IRR) vs unit energy price (US cents/kWh) and annual energy production (GWh/year)

This value was then compared to the technical costs relative to other energy resources and generation systems, as shown in Figure 10.

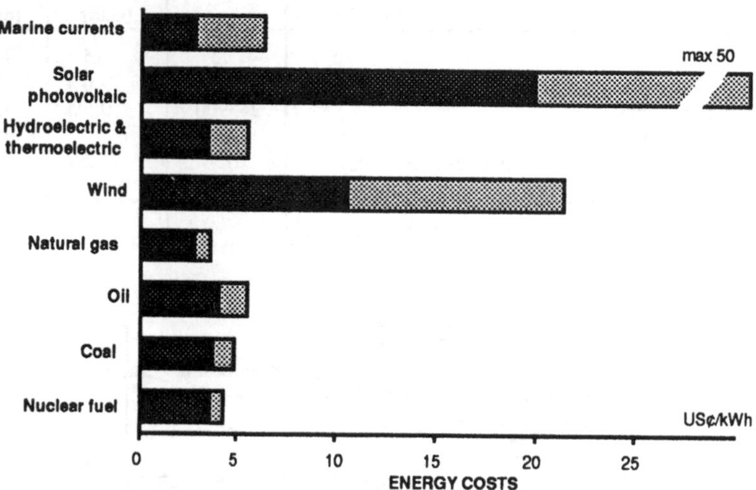

Figure 10. Unit energy costs for different energy resources (min/max)

The energy produced by the system is cheaper than that produced by exploiting other renewable resources, and is comparable to the technical cost of the conventional power plants fed by fossil or nuclear fuels. The benefits offered by the system are the respect for the environment, the exploitation of renewable energy sources, and the advantage taken of local resources.

On the other hand the fresh water supplied by the hybrid desalination plants (340.000 m^3/day), reaches technical costs of about 1.7 U.S.$/$m^3$, which is comparable with the costs obtained in commercial desalination plants. These costs are greater than the actual water price (not bound to market law!) but are smaller than the costs of transporting fresh water by tanker, which is usually done in emergency to satisfy the fresh water demand of the regions around the Strait. The fresh water production, though not fully justified economically, is however very interesting, considering the socioeconomic context of the areas. The scarcity of fresh water and its irregular supply calls for more attention to the proposed system which guarantees the supply continuity of this strategic resource.

CONCLUSION

The technical feasibility of an effective system for the exploitation of marine currents has been demonstrated. The technical solution, derived from the wind industry, can be successfully implemented in a subsea environment thanks to the high technological level reached in the offshore industry. The production of electricity shows a good level of effectiveness, especially when compared to production based on other renewable energy sources.

Moreover, beyond the technical and economical feasibility, other advantages to this solution should be considered, such as to its high production flexibility and its characteristics as a mean of exploiting a local and renewable energy source. The size of the proposed generator (1 MW electrical) and the modular approach upon which the system has been designed, make this solution suitable for diverse scenarios. In fact, other geographical areas show local, marine current conditions that could be effectively exploited.

The system is even suitable for small local application, such as small islands or coastal lands that are difficult to reach or are reachable only a great with cost by the national electric grid. It is also viable when environmental features do not suggest the use of fossil fuels or when landscape constraints do not allow the installation of visible facilities. In conclusion, even if there needs to be more in-depth evaluations done, the proposed solution and systems for marine current exploitation as a whole, are calling for more attention from the energy industry and the energy authorities.

ACKNOWLEDGMENT

We wish to thank the Italian companies TECNOMARE SUD S.p.A., SAIPEM S.p.A., and "Ponte di Archimede nello Stretto di Messina" for the contribution given to the completion of this study.

REFERENCES

Bertacchi P. and M. Tomasino. 1978. Energia dalle correnti marine, *World Offshore*.

Mosetti F. Some news on the currents in the Straits of Messina, *Bollettino di Oceanologia Teorica ed Applicata* July 1988.

Osservatorio Geofisico Sperimentale di Trieste, "Ricerche idrodinamiche nello Stretto di Messina in vista dell'istituzione di un colleamento stabile viario tra Sicilia e Calabria" - rieste 1980.

Ventrone, G. 1981. *The Turbomachines, ed. Libreria Cortina,* Padova.

Vercelli F., "Il regime delle correnti e delle maree nello Stretto di Messina". Deleg. Ital. Comm. Intern. Mediter. Off. Graf. Ferrari, Venezia 1925.

32

United States Ocean Resources and Technology Development

Paul Yuen, Kelton McKinley, and Patrick Takahashi
University of Hawaii, Honolulu, Hawaii

Joseph R. Vadus
National Oceanic and Atmospheric Administration
Washington, D.C.

ABSTRACT

The United States has the largest Exclusive Economic Zone (EEZ) in the world, with an estimated value of over one trillion dollars. Yet, there is no National Ocean Policy or Strategic Plan to pursue the development and utilization of the oceans and their resources in a systematic and sustainable manner for social and economic benefit. Numerous conferences, workshops, and studies have addressed the importance and need for ocean resources, and while proceedings and reports were generated, development and implementation did not materialize. "U.S. Ocean Resources 2000" was a workshop jointly supported by the National Science Foundation (NSF) and National Oceanic and Atmospheric Administration (NOAA) held in Hawaii in June 1992. The primary purpose of the gathering, which drew specialists from the Atlantic, Gulf, and Pacific communities, was to extend the process one step further; that is, to develop a proactive, strategic plan for implementation. The workshop compiled information on the availability of ocean resources and the attendant needs for ocean resource technology development to efficiently carry out a program in development and utilization. Several Mega Projects were proposed, primarily dealing with floating platform applications. There was strong consensus on (1) a call for congressional oversight hearings to focus on creating a national strategy through public-private partnerships aimed at developing U.S. ocean resources to stimulate the economy and create new jobs, and (2) establishment of a cooperative industry-university entity, to be called the National Ocean Resource Technology Development Corporation, to promote resource development proposals and recommend higher-risk, high-payoff projects for federal start-up funding. A key recommendation was for establishment of working partnerships among industry, government, and academia to restore the United States' eminence in the development, utilization, and protection of the oceans and their resources.

INTRODUCTION

World economic powers are increasingly looking to the sea for expanded resources and open space. Nearly half of the U.S. population now live within the coastal zone. This fraction will continue to increase, and with every expansion, more pressure will be placed on our limited coastal areas and surrounding waters (National Oceanic and Atmospheric Administration 1990). Competition for space, food, mineral resources, and the need for improved and expanded waste disposal, will only become more severe.

The Proclamation of 1983, creating the U.S. Exclusive Economic Zone (EEZ), provided a tremendous opportunity for the revitalization of the U.S. economy. While the United States has not yet capitalized on that initiative, workshops sponsored by the National Science Foundation and National Oceanic and Atmospheric Administration showed promise for stimulating a progressive program towards intelligent utilization of the EEZ.

A national effort is needed to reenergize industry, universities, and the government. Creative solutions are needed to enhance growth and development in this important economic sector. At a time when the United States is facing increasing economic challenges at home and increased economic competition abroad, there is much that can be done by government, industry, and academia. Teaming together can result in a strong, viable ocean industry and a revitalized national economy. The rich resources of the ocean must be tapped for food, energy, and materials, while at the same time continuing to utilize our ocean space for recreation and waste disposal in an environmentally responsible manner.

"U.S. OCEAN RESOURCES 2000"

In June 1992, at "U.S. Ocean Resources 2000," the National Science Foundation and the National Oceanic and Atmospheric Administration brought together a diverse group of ocean specialists to develop a strategic plan for the utilization and management of our nation's ocean resources. Thirty-five, invited participants from industry, government, and academia representing the Atlantic, Gulf, and Pacific communities met for three days to discuss the problems and opportunities presented by the ocean. A national effort to revitalize the U.S. economy and reestablish its global competitiveness through demonstrated new leadership in the sustainable development of ocean resources and their related technologies was proposed (Carey and Yuen 1992). Compelling reasons for a strong, national effort included the need to

1. Stimulate the U.S. economy, strengthen the ocean industry, and create new jobs.
2. Recapture the United States' early, leadership position by establishing public-private partnerships and enterprises in this vital area.
3. Promote expanded U.S. exports of ocean resource products and technologies.
4. Foster the necessary advances in technology toward effective and sustainable ocean resource applications via support of research, development, technology transfer, an pre-commercial applications.
5. Use a portion of the reduction in defense spending to invest in the potential wealth of the oceans, and through use of existing expertise and related technologies, redirect military technology developments for societal benefit.
6. Provide direction to reorient U.S. posture towards economic competitiveness and security.

When former President Ronald Reagan proclaimed our EEZ in 1983, he almost doubled the jurisdictional area of the United States and opened a new frontier for exploration, development, and management. The United States now has the largest EEZ in the world with an estimated resource base of over one trillion dollars. The U.S. EEZ can provide:

1. Seafood for our increasing needs, which have been stimulated by changing, nutritional and consumption patterns. The United States currently has a *per capita* seafood consumption level one-half that of the world (and one-tenth that of Japan), and continues to import one-half the seafood consumed.
2. Stategic industrial minerals, nearly all of which the United States presently imports from countries like Zaire and South Africa.
3. Major sources of environmentally-preferable renewable energy which can reduce U.S. dependence on imported oil while minimizing the amount of gases released to the atmosphere through combustion processes.
4. Valuable, coastal ocean space for expanded development, i.e., artificial islands, floating cities, and offshore facilities for high-quality living and recreation and environmentally-compatible industrial activities.

While under 1% of the resources now consumed annually in the Untied States come from the sea, the EEZ proclamation has provided a new, marine opportunity. The "U.S. Ocean Resources 2000" gathering set the stage for developing a national strategic plan to marshal the scientific and engineering resources of the United States toward more productive use of the oceans and their resources.

HISTORY OF U.S. OCEAN INDUSTRIES

"In the United States today, less than 1 percent of the resources consumed annually come from the sea...a tragic underutilization of our ocean resources, which have been estimated at over a trillion dollars." This view, expressed by Senator Daniel Inouye (1992b), is reflective of the general perception that we have failed to take advantage of our rich resources for the benefit of society. The United States as a nation with strong maritime roots, has never developed a strong, marine industrial policy. The enactment of the Marine Resources and Engineering Development Act of 1966, the recommendations of the Stratton Commission, and the Proclamation of a 200-mile EEZ in 1983, provided a basis for pursuing ocean development opportunities.

In the Marine Resources and Engineering Development Act of 1966, the Congress declared that it was the policy of the United States to "develop, encourage, and maintain a coordinated, comprehensive, and long-range national program in marine science for the benefit of mankind to assist in protection of health and property, enhancement of commerce, transportation, and national security, rehabilitation of our commercial fisheries, and increased utilization of these and other resources." This important, visionary legislation led to the creation of a national Marine Sciences Council and more importantly, to the establishment of the Stratton Commission, which in turn fostered the creation of the NOAA.

In 1983, President Ronald Reagan proclaimed the U.S. EEZ. The proclamation was greeted with a great deal of enthusiasm and renewed interest in the potential of ocean resources including: offshore oil and gas, fisheries, non-energy-related mineral resources, renewable (nonfossil) energy resources, waste disposal, and pharmaceuticals from the sea. In May 1984, the National Advisory Committee on Oceans and Atmosphere (NACOA) further emphasized the importance of the development of the EEZ and its resources (1988).

The "U.S. Ocean Resources 2000" conferees noted that U.S. efforts in ocean resource and ocean technology development have been quite limited. The United States continues to map the territorial waters of the nation; the U.S. Navy has developed the most sophisticated ocean robotics and military related hardware; some pioneering research has been conducted on ocean energy conversion; offshore oil and gas development has continued, but with an increasingly multinational face; overall fisheries management has improved, but requires a better understanding of wild stocks and the need for stock enhancement; and limited, basic research has been done on the unique features of the seabed, such as hydrothermal vent systems and deep ocean resource recovery. It is clear that the time has come for a revitalized, concerted effort, expanding on our more fundamental efforts and transferring defense capability to civilian applications.

Looking ahead to meet the challenges of the coming decades, there is need for: a stable, expand source of protein; cost-competitive, environmentally sound sources of energy to fuel industrial production and meet ever-increasing societal demands; new materials to supplement shrinking or environmentally restricted, traditional sources; sound waste management options; and large-scale marine ecosystem and comprehensive coastal zone management programs. Marine biotechnology, including biopharmaceuticals, could usher in a new age for the World. The time is right to lay the foundation for sustainable, marine economic development

In the United States, there are already hundreds offcompanies that design, build, and sell a wide variety of ocean products (e.g., instruments and measuring devices) that support off- and shore-based operations, providing construction and engineering services with an estimated revenue productivity of $54 billion annually in terms of goods and services sold (Covey 1991). These industries hold the key to an improved U.S. posture in ocean resources development, but are struggling in the face of growing foreign competition and unified foreign government/industry policies and partnerships. The conferees suggested a review of philosophy and attitudes regarding government-industry cooperation.

UNITED STATES OCEAN RESOURCE POTENTIAL

The declared, 200-nautical mile limit of the EEZ gives the United States the largest, territorial waters of any nation on Earth. These waters measure 2.3 million square nautical miles and have almost doubled the size of the nation. This area includes the waters off the continental mainland, Alaska, the islands of the Hawaiian chain, and the U.S. Trust Territories in the Pacific.

Approximately 85% of these waters are in the Pacific. It is a territory diverse in its ecology and rich in natural resources. "U.S. Ocean Resources 2000" participants reviewed and summarized U.S. ocean resources as follows.

The Environment and Living Marine Resources

"The rich life and relatively pristine marine environment within [U.S. territorial waters] is perhaps its most valuable resource...Only by maintaining and improving the environmental quality of the [ocean] is it possible to maximize the value of its other resource areas: fisheries, marine wildlife, oxygen production, pharmaceuticals, recreation, and the basic security of a health, rich community of marine life." (Coast States Association 1987). Significant, environmental problems will persist during the accelerated future development of the world's oceans, especially the coastal ocean and its associated resources. Developers must always consider and weigh the biological implications of man's impact.

The coastal ocean is particularly affected. Impacts include sewerage outfalls, runoff from land, ocean, dumping, sand and gravel mining, dredging, oil and gas production, hazardous materials spills, disposal of waste, and contaminants from chemical and industrial product streams. Increased use of coastal ocean space will require long-term monitoring and assessment to ensure adequate protection. Some of the problems that must be averted or ameliorated include eutrophication (over-enrichment) and oxygen depletion in coastal waters; physical modification of seabeds and their associated, bottom-dwelling organisms; damage to fish and shellfish; and natural hazards and coastal erosion. Over the next decade there will be an increasing need for new, high-technology instrumentation to provide near real time measures of the changes in the quality of our marine environment and systems to deliver this information to decisionmakers at all levels.

Fisheries continue to be one of the most important, living marine resources. While yields from the oceans appear to have leveled off, and in some cases declined, there is still tremendous opportunity to develop underutilized species and to apply new technology to areas such as mariculture or open ocean ranching. "Coastal fisheries are an important source of nutrition and recreation, and contribute significantly to the economy, health, and quality of life. They are enormous, yielding about 10 billion pounds of food each year, or nearly 50 pounds for each person in the United States. Added to this is another 750 million pounds caught each year by recreational fishermen. Counting all subsidiary effects, U.S. coastal fisheries contribute more than $23 billion to the economy each year and provide employment for more than a million people. The United States has the greatest abundance and diversity of fish and shellfish off its coasts; fully 15% of the world's living resources are within the U.S. EEZ." (Gordon and Gutting 1984/85; Takahashi and Vadus 1992). These projections do not begin to express the potential for the economic contribution of coastal mariculture and aquaculture to future food (protein) production and for additional uses to be discussed below. The development of these resources should continue to be a high priority over the next decade.

New products from the sea stand as one of the most exciting and potentially significant uses of biological ocean resources. In the past, biotechnology relied primarily upon the exploitation of terrestrial organisms or their components. Since marine organisms are vast in number and are largely unexploited, their potential for biotechnology development is great. Rapid advances in molecular biology, genetics, microbiology, cellular physiology, and biochemistry have made it possible to produce a number of economically important substances of biological origin. Research linking the experience of engineering and the life sciences is crucial to exploratory development.

Marine biotechnology can provide substantial benefits in applications in the food, pharmaceutical, and chemical industries. New techniques to minimize marine fouling can improve the operation and efficiency of marine transportation systems, marine instrumentation, and seawater hear exchangers. New materials derived through research and possessing special properties may lead to exciting breakthroughs in the applied sciences. There are also many applications in aquaculture and seawater development that would benefit from advances in molecular biology and genetic alteration, since controlling the growth, quality, and health of marine life is important inn the seafood industry. Deep-ocean microorganisms associated with hydrothermal vents have unique characteristics associated with their survival in this high-temperature, high-pressure, low-oxygen environment. Their properties could be the basis for new pharmaceuticals (Vadus 1990).

Oil and Gas Reserves

Offshore oil and gas development continues to provide an important, although declining, source of energy for U.S. industry and consumers, providing a cushion against fluctuations in the world supply and a strategic reserve. The current value of oil and gas recovered from federal and state waters approaches $26 billion annually. The United States currently produces $40 billion/year ($20/barrel) of oil and oil equivalents of which $8 billion comes from the U.S. offshore. The United States imports $51 billion of oil annually, which represents over 1/3 of the U.S. annual imbalance of payments. Ninety percent of the oil and virtually all gas produced in federal waters have been from the Gulf of Mexico. The U.S. Geological Survey has estimated that between 26% and 41% of the undiscovered oil and between 25% and 30% of the undiscovered gas reserves in the United States lie offshore (Champ et al. 1984/85).

To what extent these reserves will be, or can be, profitably extracted is not well known. Currently, oil and gas production platforms are capable of operating in depths of 350 meters (Vadus 1990). In the years ahead, new technology to extract and transport these resources in an environmentally safe manner will be needed, especially in the deep ocean and frontier areas such as the Arctic.

Seabed Minerals

Hard mineral resources in the ocean, while yet to be significantly developed, also represent a tremendous, strategic reserve for the United States (Olson and Woolsey 1989). Most deposits lie in deep water and will require investments in high-technology mining approaches to make the extraction of these materials cost-effective as compared to traditional sources. These resources include sand, gravel, placer deposits, phosphorite, manganese nodules, cobalt crusts, and polymetallic sulfides. There are vast quantities of minerals on the seabed, as well as in the form of dissolved elements within the seawater itself. Uncertainties still remain about the distribution and abundance of minerals of economic interest.

For the U.S. territorial waters, seabed gravel resources on the continental margins are estimated to be in the hundreds of billions of cubic meters; placer deposits on the Pacific continental margins are estimated at over 2 billion cubic meters; the Blake plateau off Florida is estimated to have more than 2 billion tons of phosphorites, with the areas off Georgia and the Carolinas perhaps having even larger deposits. The seabeds in the vicinity of the Hawaiian Island chain have potential yields of more than 2.5 million tons of cobalt and 1.6 million tons of nickel at an estimated projected value of $50 billion and $3.2 billion, respectively (State of Hawaii 1987).

Marine mining has remained relatively undeveloped because of the high cost of offshore production relative to land production. However, political implications such as the uncertainties placed on the price of oil, could well determine the fate of this potentially major industry. The strategic metals security factor must be entered into decision making.

Ocean Energy and Deep Ocean Water

Nontraditional sources of energy from wave, tidal action, temperature, or salinity differentials remain one of the more promising areas for development (Vadus, Bregman, and Takahashi 1991). Ocean thermal energy conversion can yield base-load electricity, while the deep ocean water effluent can be a source of nutrients for mariculture and cold for air conditioning. The mounting need for freshwater for terrestrial agricultural and industrial purposes opens up a new area of potential use of ocean resources through desalination and cold water agriculture.

The oceans occupy almost three-quarters of the earth's surface and represent an enormous source of nonpolluting, renewable energy providing an alternative energy source that can offset our reliance on fossil fuels and their attendant environmental problems (i.e., global warming, acid rain, and urban air pollution) with projected, available ocean power far exceeding estimates for the ultimate energy consumption by mankind. While many of the major developed nations have conducted exploratory research and even installed a few commercial facilities, the total operational power available from alternative technologies, with the exception of one French tidal power plant, is less than 100 megawatts.

Much of the development (e.g., tidal, wave, salinity gradient, current, and at-sea wind energy) thus far has focused on the production of a single commodity, electricity. Of these programs, relatively little progress has been made with salinity gradients and currents. Tidal, wave, and wind power have shown good potential for further refinement and within certain niches, are currently within an acceptable range for economic competitiveness. Two other options, ocean thermal energy conversion (OTEC) and biomass production, promise multiple end-products as well as electricity and liquid transportation fuels. The multiplicity of co-products from OTEC and marine biomass production improves the cost competitiveness of these two alternatives. Small-scale OTEC systems providing electric power, nutrients for aquaculture, and fresh water may be ideal for expanding the economic potential of many small island communities and its application in this limited market will enable OTEC technology to mature, proving the technology in use before scaling-up to larger-sized systems (Vadus, Bregman, and Takahashi 1991).

The use of cold, deep water, a resource rich in nutrients, has been successfully demonstrated in aquaculture and biomass production. The natural phenomenon of high biological productivity in upwelling areas, while representing only 0.1% of the ocean's surface areas, accounts for more than 40% of the world's fish catch. Pumping large amounts of nutrient-rich, deep, cold water would be analogous to natural upwelling. The national Science Foundation (NSF) workshop on "Engineering Research Needs for Off-shore Mariculture Systems" described the potential and activity level for this concept (Takahashi, Bardach, Champ., and Welder 1991).

Marine Transportation

Private sector transportation of goods and people is a growth industry, and the volume and dollar value has been steadily increasing. However, the face of the industry has changed dramatically. Thirty years ago, break-bulk ships carried most nonbulk cargo; today, this same cargo is containerized, carried in technologically advanced ships, and transported via a unified, intermodal transportation system. Ports or harbors have also changed in response to this new technology and those that have not are falling behind.

Over the last thirty years, the number of major, U.S. shipping lines has dropped from dozens to only a handful. Construction and operational costs have favored the emerging nations to our detriment, and it is now almost certain that commercial shipbuilding will not develop on a major scale in the United States in the next few decades unless there is a significant global conflict. Another look at the Jones Act makes sense.

Marine Recreation

Marine recreation is the second largest area of spending in the United States; only expenditures for the military are greater. There is a huge public constituency. In the United States, water-oriented recreation (including both fresh and saltwater) has a gross economic value of $20 to $25 billion per year. This is a significant figure considering that the estimate for all recreation and leisure time activities in the Untied States is on the order of $52 billion per year. Americans own more than 13 million boats for leisure, and the boating population is expanding at a rate greater than our nation's population as a whole. Recreational boating is a $9 billion per year business. America's sport fishing fleet ranks third in terms of boats and 'landed value' of the catch, and the economic value of the 'fishing industry' is in excess of $8 billion per year. The U.S.-based cruise ship industry carries over 2 million passengers per year for annual sales of more than $2 billion. This is a rapidly growing part of the $200 billion per year U.S. travel and tourism industry (Walsh 1989).

Ocean Space and Offshore Facilities

Use of U.S. territorial ocean space for both industrial and commercial purposes has been limited, and other countries now lead the United States in this type of development. However, as coastal population pressures mount, this alternative will become increasingly attractive. Applications will demand advances in offshore engineering technology for the construction of artificial islands and for the deployment of offshore

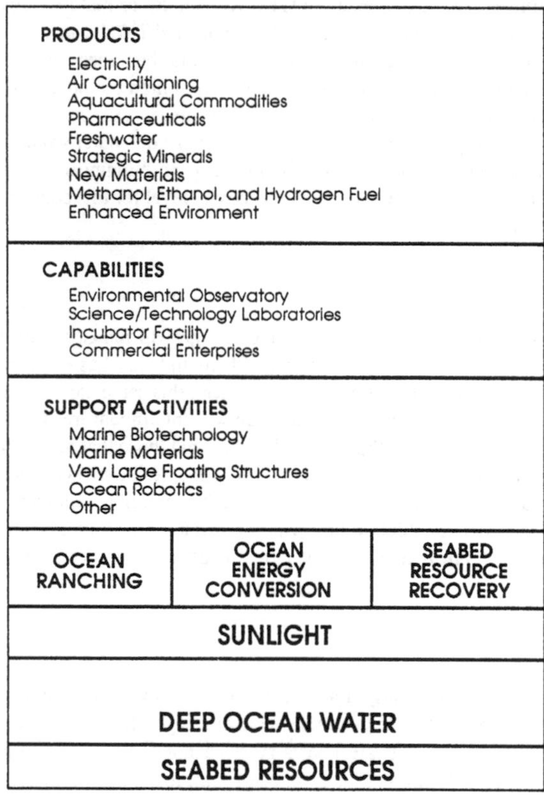

PRODUCTS
Electricity
Air Conditioning
Aquacultural Commodities
Pharmaceuticals
Freshwater
Strategic Minerals
New Materials
Methanol, Ethanol, and Hydrogen Fuel
Enhanced Environment
CAPABILITIES
Environmental Observatory
Science/Technology Laboratories
Incubator Facility
Commercial Enterprises
SUPPORT ACTIVITIES
Marine Biotechnology
Marine Materials
Very Large Floating Structures
Ocean Robotics
Other
OCEAN RANCHING
OCEAN ENERGY CONVERSION
SEABED RESOURCE RECOVERY
SUNLIGHT
DEEP OCEAN WATER
SEABED RESOURCES

Figure 1. Project Blue Revolution Mission

structures or floating platforms to support: fishing, mariculture, ocean ranching, recreation, energy production, manufacturing, environmental monitoring, and mining.

In the United States, about half the population now lives within fifty miles of the coastline; by the year 2010, that proportion could be as high as 60%. Development of ocean space in estuaries and coastal areas, including revitalization of ports, harbors, and communities is gaining momentum (National Oceanic and Atmospheric Administration 1990). A number of major port and harbor communities such as Los Angeles, Long Beach, Baltimore, and New York have begun, or are currently planning large-scale, ocean space revitalization projects.

Over fifty participants from the United States, Japan, France, South Korea, and other nations, representing government, industry, and academia, participated at a workshop during PACON '92 to discuss the potential of Project Blue Revolution. Proposed was a one hectare (about 100,000 square feet) integrated, ocean resource development and management floating platform for operation around the turn of the century at a cost of $500 million. Figure 1 is a systems representation of the concept. The conferees argued that in consideration of the potential economic and environmental benefits to mankind, cost of this incubator plantship can be well-justified compared to current and planned mega-space and military projects. The potential for international cooperation is excellent, with a proposed next step to be a jointly funded development of a strategic plan (Takahashi and Vadus 1991; Inouye 1992a; Takahashi and Matsuura 1991).

Waste Management

As societal opposition and the cost of land-based options have increased, interest in using the ocean space and the seabed for waste management has been rekindled. Growing international restrictions on deep sea disposal and state governmental opposition to near-coastal dumping presently leave only the offshore seabeds of the U.S. Exclusive Economic Zone (EEZ) for the possible disposal of municipal waste and dredged material. The seabed of U.S. waters should be considered along with land-based alternatives with a view to selecting the most environmentally sound solution. Surveys of the U.S. territorial waters, including information on seabed geology, habitat, and living resources, are needed to locate suitable sites for specific classes of wastes and efforts to develop safe and cost-effective alternatives must be continued (Vadus 1990).

INTERNATIONAL COOPERATION

As one example of collaboration on an international scale, the Project Blue Revolution workshop explored:

1. *The major applications (commercial foci) for Project Blue Revolution?* There was a strong sense that useful products or services must be delivered, and that the private sector was a key component. The model served by the Floating Structures Association of Japan, with more than 100 companies and five banks, might be considered in developing an international team. Among the applications discussed were an integrated incubator plantship, waste management facility, energy generation platform, seabed minerals refinery, ocean ranching homeport, and an observatory for ocean ranching. The R&D capability, however, has to be a secondary consideration to complement the commercial product or utility service.
2. *The socio-enviro-political justifications for the project?* A key selling feature of Project Blue Revolution related to the environmental benefits that can accrue. Early public education of the concept could be key to long-term success. Justification points for operation on the open ocean include
   a) Reduction of stress in near coastal waters
   b) Removal of certain industries and processes away from populated regions so that any wastes can be recycled
   c) Good potential for total systems development in harmony with nature
   d) Creation of new fisheries and biomass plantations increase the abundance of the world food supply and provide alternative sources of cleaner energy

e) Possibility of easier and more workable permitting and regulatory approvals, thus minimizing delays
f) The prospect for enhancement of the environment should the various, proposed options for global warming remediation prove to be successful

3. *Effective examples for establishing a consortium for government, industry, and academic partnerships?* Japan's Government-industry cooperative efforts, Apollo, and the space shuttle are some examples which have succeeded because of these partnerships.
4. *The potential for international partnership?* There was a unanimity of opinion that the potential was good. The end of the Cold War has reduced military expenditures and opens to question the major investments in big science and space projects. The ocean is a last frontier, ideal for economic development with a commons that needs to be protected for the world. There was a sense that cooperation can be developed if mutual benefits can be shown.
5. *Where should be the locus of operation?* The location of the initial platform will not doubt be driven by funding. However, site selection criteria need to be established to pick a best international site or one that is ideally mission-oriented.
6. *What is the optimal size?* Again, to a good degree, the amount of dollars will determine the scope of the program and design. One concept is to start small and modularize; another is to pick a mission and design to an operational need. From an engineering standpoint, ultimate scale-up requires a practical, modular size. A 100,000 square foot (or one hectare) platform was deemed as a good design point for incubator applications.
7. *What are some possible strategies for funding an internally cooperative project?* Before international funding can succeed, there must be a credible feasibility and marketing plan. While a B-2 bomber might cost nearly a billion dollars or more, and the Space Experiment could cost up to $50 billion, the military-aerospace industry already exists to propose and lobby for such expenditures. In Japan there are industrial alliances such as the Floating Structures Association, capable of managing such a project. On the surface, in the United States, there is no equivalent organization of Federal agency equipped to start and implement this type of enterprise. However, the recent reduction in defense needs, coupled with a call for dual applications -- current civilian and long-term military -- might stimulate a hybrid organization where Federal funds can be earmarked for specific economic development projects in the spirit of the transcontinental railway system, which opened up the western regions of the United States.
8. *Is the use of Japan's 'Aquapolis' applicable? Others?* There was a general feeling that past generation platforms or even naval craft, were too costly from a maintenance standpoint. There are new materials, equipment, and designs that can be marshaled for the 21st Century.
9. *What should be the source of energy?* All options should be initially studied, but OTEC has particular advantages because of the nutrient-rich fluids useful for co-product development. Wave energy conversion can be added to produce energy while absorbing wave forces to ease stress on the platform.
10. *Who should be the lead organization?* A cooperative, feasibility study group should be established. The lead organization will thus depend on the specific application, location, and funding sources. A multinational consortium or organization with international contacts might be best for this purpose.
11. *What are some other factors requiring consideration?* Little, if any, consideration has been given to the regulatory and policy environment within which the floating platform would operate when ready for deployment. The implications of State (3-12 miles), Federal (EEZ to 200 miles), and International territorial sea conflicts need to be addressed. Regulatory consideration are also raised in regard to whether the platform would be bottom moored or surface propelled. These

matters are essential to business planning and risk assessment as they entail potential 'hidden' costs and uncertainty that need to be identified as soon as possible in the strategic planning process.

12. *Future activities?* Plan for and hold a more comprehensive and structured workshop with greater industrial involvement. This meeting should also involve representatives from the environmental groups. The session co-leaders, Vadus and Takahashi, were charged with a task to organize the follow-up on these activities.
13. *Conclusions*
    a) An optimal size is about one hectare, which can be built, tested, and operated for about $500 million, and serve as a module for larger facilities.
    b) A target date of the year 2000 is reasonable.
    c) There are huge food, energy, materials, and ocean space benefits, that enable sustainable development in harmony with the environment.
    d) International cooperation will facilitate progress.

STEPS TOWARD A STRATEGIC PLAN

Interests in ocean resources and technology development span a broad range of economic and policy concerns. Effective management of the oceans and the marine environment will become a major issue as world population increasingly moves toward the coast. By looking seaward, we can access new avenues to sustain economic development. These resources can provide new sources of food, energy, water, and materials to meet societal needs for present and future generations. They will become even more valuable as the need to find alternatives to traditional, land-based sources of food, energy, water, and materials mounts.

By making an investment now, we can revitalize our ocean industries and devise new approaches to meet growing needs: i.e., develop new sources of protein through fish farming and ranching, and by restoring or enhancing wild stocks; introduce new sources of cost-effective energy to coastal, inland, and island communities; provide fresh water in drought-prone areas for human consumption, manufacturing, and agricultural uses; develop new materials to support industrial development and our quality of life; and create new applications and job categories using ocean-developed technologies.

All of this translates into general economic redevelopment, business growth, and jobs in new an expanded industries. The products and services that are a part of today's maritime community are no longer restricted to the category of raw materials; they are profitable, value-added commodities with 'spin-offs' to other sectors and broad markets. An investment in ocean resources development is an investment that will contribute to the diversification of the domestic industrial base and to the modernization of the economic infrastructure.

Ocean resources and their associated ocean technology have tremendous international trade potential. Two-thirds of the world's population live in the countries bordering the Pacific rim. Many of these nations rely heavily on the oceans.

What are the opportunities for strengthening our existing ocean resources and technology sectors? Where are the new areas for investment likely to emerge over the course of the next decade? There are a number of possibilities. In the traditional resource areas (e.g., fishing, oil and gas, and transportation) there are opportunities for applying new technology toward improving our overall economic productivity. In newly emerging areas (i.e., biotechnology, ocean energy, and ocean space utilization) there are tremendous opportunities to develop new applications that can be targeted for solving major, societal problems.

A Call for a National Program

"U.S. Ocean Resources 2000" created a blueprint for strengthening our nation's ocean industries and making the best use of our ocean resources. A partnership of industry, government, and academia must establish long-term support for an aggressive research and development effort, coupled to an effective technology transfer program within a stable, regulatory environment.

Some of the most noteworthy issues identified by participants include

1. Assigning an oceans and technology resources development role more explicitly to a single Federal Agency to foster accountability and consistency and providing a focal point for the diverse interests that comprise the marine resources community
2. Reviewing current marine resources -- related provisions with a view to reviving current acts and policies and/or pursuing new legislative action consistent with our times
3. Developing a clear, national Plan with priority areas explicitly delineated for substantial, long-term commitment and investment
4. Working out a mechanism for cost-sharing by industries through cooperative research and development agreements with federal laboratories and universities
5. Proposing and funding specific pilot projects executed by these industry, government, and university partnerships; projects promoting and demonstrating new ocean resources end-uses

A key finding was the observation that proactive federal participation will be necessary, even with substantial industry interest and cooperation. Participants acknowledged that the risks associated with ocean resource development remain greater than can likely be assumed by the private sector alone. Like the earlier, U.S. cross-continental railway system and the U.S. space program, government leadership and financing will be required to bridge the gap between the promise of the future and the economic reality of today. With that important acknowledgment before them, participants converged on two plans with parallel paths.

One approach was to build recognition of the importance and opportunities for national growth that are inherent in the wise use of our ocean resources. It was observed by participants that this would require Congressional oversight hearings focused on the importance of ocean resources and technology development and health of the U.S. ocean industry. It was further recommended that a desirable product would be the assignment of ocean resources development and technology to a single federal agency, perhaps an office of Ocean Technology Development within the U.S. Department of Commerce. This office would serve as the focal point for all nonmilitary, oceanic research, technology development, assessment, management, and environmental monitoring. A second and fund-rich source is the Deportment of Defense. As the so-called 'Peace Dividend' begins to materialize, an important transition emphasis will be on dual applications where programs would have both military and civilian value.

The second parallel approach recommended the establishment of a private, not-for-profit, partially governmentally funded organization to serve as a focus for ocean resources programs. This organization would provide the leadership and management oversight for ocean resource development and assist in the establishment and maintenance of industry, university, and government cooperatives.

Participants concluded that whether one or both of these pathways were followed, the best way to convince the American public and Congress was in measurable progress. The participants recommended that one or more visionary mega-projects be initiated. Among the more popular ideas were

1. A multipurpose incubator or large-scale, industrial floating platform for the production of energy, food, and other commodities providing real goods and services to the American public
2. A network of multielement systems for long-term, environmental or climatological observation
3. A seabed observatory permitting measurement and sampling of the ocean environment at depth, accessing the actual 'last frontier' for this planet

4. A large-scale, managed, living marine system, including the concept of ocean ranching and the use of artificially upwelled waters for enhancing the natural productivity of infertile areas in the ocean

The possible integration of all of these concepts into a single, three-dimensional (surface, water column, and bottom accessing) system for ocean resources development, research, and applications were judged by participants as having great potential.

Participants emphasized the importance of a national effort for restoring U.S. economic and technological competitiveness and felt that that action should be taken as soon as possible. It was recommended that Congressional oversight hearings should occur by the Spring of 1993 with the creation of a division or program within an existing federal organization within eighteen months. In parallel with this main effort and to enable an early start, a non-profit organization for ocean resources development is expected to be formed early in 1993. It was also recommended that planning efforts addressing one or more of the proposed mega-projects also begin immediately to enable the United States to launch one or more such projects by the year 2000.

REFERENCES

Carey, J., and P. Yuen. 1992. U.S. Ocean Resources 2000: planning for development and management. Report for National Science Foundation and National Oceanic and Atmospheric Administration, 9-11 June, Honolulu, Hawaii.

Champ, Michael A., et al. 1984/85. Non-living EEZ resources: minerals, oil, and gas. *Oceanus* 27(4): Winter.

Coast States Organization. 1987. *Coastal states and the U.S. Exclusive Econimic Zone*, April 1987.

Covey, Charles W. 1991. The ocean/marine industry - the "other" resources. *Sea Technology*. 32(1).

Gordon, William G., and Richard E. Gutting, Jr. 1984/85. The coastal fishing industry and the EEZ. *Oceanus* 27(4): Winter.

Inouye, Daniel K. 1922a. The American blue revolution: A solution for the 21st century. *Sea Technology* 33(9): 23-26.

marine Resources and Engineering Development Act of 1966. (P.L. 89-454), Section 110(a).

national Advisory Committee on Oceans and Atmosphere. 1986. *The need for a national plan of scientific exploration for the Exclusive Economic Zone*. A special report to the President and the Congress, June 1986.

National Oceanic and Atmospheric Administration. 1990. *The second report of a coastal trens series: 50 years of population change along the nation's coasts*, 1960-2010. April 1990.

Olson, Harry J., and J. R. Woolsey. 1989. Ongoing minerals research at the Marine Minerals Technology Center. Presented at the Oceans '89 Conference.

State of Hawaii, Department of Business and Economic Development, Business Development and Marketing Division ,Ocean Resources Division. 1987. *Mining development scenario for cobalt-rich manganese crusts in the Exclusive Economic Zones of the Hawaiian Archipelago and Johnston Island.* Honolulu, Hawaii.

Takahashi, P. K., J. E. Bardach, M. A. Champ, and J. Weidler, Jr. 1991. *Engineering research needs for off-shore mariculture systems workshop*. National Science Foundation, 26-28 September. Honolulu, Hawaii.

Takahashi, P. K., and R. M. Matsuura. 1991. Blue Revolution 2000. In *Proceedings of the first international workshop on very large floating structures*. National Science Foundation, 24-26 April. Honolulu, Hawaii.

Takahashi, P. K., and J. R. Vadus. 1992. Ocean space utilization: The blue revolution. Presented at PACON '92, 1-5 June. Kona, Hawaii.

Vadus, Joseph R. 1990. Ocean technology in the United States: Recent advances, future needs and international collaboration. *MTS Journal* 24(1).

Vadus, J. R., R. Bregman, and P. K. Takahashi. 1991. The potential of ocean energy conversion systems and their impact on the environment. In *Use and misuse of the seafloor proceedings,* ed. K. J. Hsu and J. Thiede, 373-402. Environmental Sciences Research Report ES 11. 17-22 mrch. Berlin, Germany.

Walsh, Don. 1989. Proceedings of the First International Symposium on Coastal Ocean Space Utilization, 8-10 May. New York, New York.

33

Wealth From The Oceans: A United Kingdom Program

Terry Veness
Department of Trade and Industry, United Kingdom

Cliff Funnell
Consultant to the Wealth From The Oceans Program

ABSTRACT

This paper charts the development of the WEALTH FROM THE OCEANS PROGRAM, a UK national collaborative R&D program, from its inception to date. It sets the Program in its historical context and provides some indication of its aims and objectives from the policymaker's perspective. It also reviews some of the individual projects that have been supported within the Program to demonstrate the range of technologies and disciplines that have come within its area of operation. Although the Program does not finish until April 1993 and many projects will continue beyond this date, an attempt has been made to draw some provisional conclusions from our experience of managing the Program thus far.

THE HISTORICAL PERSPECTIVE

The United Kingdom has a long and enviable tradition of marine science and technology, as well as a long track record in developing its ocean resources. The size of the offshore hydrocarbon industry is well documented, however, marine resource exploitation off the coast of the United Kingdom is not limited to oil and gas. Indeed, fish farming is an important industry, especially in the Highlands and Islands of Scotland. Production of farmed salmon alone rose to 40,000 tons by 1990. Furthermore, marine-dredged sand and gravel production rose from 7 million tons per annum in 1965, to approximately 18 million tons by 1990, representing some 16% of the total annual requirement of England and Wales.

Global interest in the oceans and coastal zones has grown significantly, partly in response to the growing practice of countries to declare Exclusive Economic Zones establishing sovereign rights for the exploration, exploitation, conservation, and management of natural resources, living or nonliving, of the waters above the seabed and of the seabed and subsoil out to 200 nautical miles. This had a major impact on the market for new and improved technology to maximize the potential benefits from this new resource.

In theory, the United Kingdom was well placed to respond to the needs of the growing national and international market by building on its solid foundation of existing expertise. In practice, however, the technological challenges of these emerging markets were and still are daunting, and action was required to kick-start the essential technologies for the future.

In order to exploit significant UK technological expertise, the Advisory Committee on Resources From The Sea was set up by the Department of Trade and Industry (DTI) in 1983. Its aim was to review the then current, worldwide activities directed towards the eventual exploitation of the diverse resources with likely commercial potential available from the seas and oceans. The Committee subsequently recommended that the United Kingdom should undertake a coordinated program of research, exploration, and development, with a view to deriving the maximum, immediate advantage for the nation from its existing significant capabilities in this area and building upon those capabilities to maintain and increase the availability of the relevant technologies in the United Kingdom.

DEVELOPMENT OF A GOVERNMENT-ASSISTED PROGRAM

The program which the Committee proposed was intended to lead towards resource exploitation which would probably not take place until the next century. Industrial benefits were, however, foreseeable throughout the program.

The Committee concentrated its attention on resources other than the well-established areas of fishing, whaling, sand and gravel, and offshore oil and gas. The other principal, marine resources identified as being of potential, future interest were

1. Biological
    a) Plankton
    b) Seaweed
2. Mineral
    a) Manganese nodules
    b) Polymetallic sulfides
    c) Minerals in solution and in seabed deposits
3. Energy
    a) Tides
    b) Currents
    c) Wind
    d) Waves
    e) Thermal

These were regarded by the Committee as an illustrative list only and they anticipated that there would be other resources yet to be discovered and as yet unknown; an example of this is polymetallic sulfide deposits which were discovered only recently. It was also felt that new approaches to existing resources, such as fish, might also emerge which might be relevant to the work and experience of the Committee.

To assist UK industry to identify the key marine technologies that would be required in the future and to help frame an appropriate Government Program to assist the process, the DTI commissioned a number of separate Technology/Market Feasibility Studies.

1. Marine environmental measurement technologies
2. Seabed surveying technology
3. Manned submersibles
4. Surface support for subsea operations
5. Enabling technologies for exploitation of resources
6. Mariculture

On the basis of their recommendations, the Government, through the DTI, established the Resources From The Sea Program in 1986, with the aim of encouraging a major, profitable, UK service and manufacturing industry for the surveying, exploration, development, and management of the ocean and its resources.

As a results of a review of DTI innovation policy at that time, there was a move to support only collaborative precompetitive research. It was decided, therefore, to replace the single-company assistance program (Resources From the Sea Program) with a new ocean technology program that reflected more closely this change in emphasis. To this end, the WEALTH FROM THE OCEANS PROGRAM was launched in April of 1989, with an initial budget of £9 million to be committed over a four-year period. It was intended that this financial commitment would be matched by industry on at least a 50-50 basis and would thus mean an additional investment in ocean technology in the United Kingdom of approximately £20 million.

AIMS AND OBJECTIVES

The main aim of the WEALTH FROM THE OCEANS PROGRAM was to encourage precompetitive research in the advanced technologies associated with the exploration, surveying, development, and management of the oceans and their resources. We take the term ‘resources’ to be much broader than purely living and nonliving, such as minerals and energy. The ocean space and the seabed below it are also resources, and utilization of that space for leisure, artificial islands, as a route for pipelines and telecommunications cables, and hazardous waste, will also require the necessary equipment and services. Salvage or recovery of cargo from the deep ocean, be it military or commercial, valuable or hazardous, presents a technological challenge that could also be addressed by research encouraged by this Program.

Last, but certainly not least, was the technology to manage our oceans. By ‘manage,’ we did not mean just economically, but also through the application of good husbandry. Assessing the environmental impact of our past, present, and especially future use of the oceans will depend more and more on our technological capability in such areas as sensor design and information engineering, e.g., data collection, analysis, and transmission.

The WEALTH FROM THE OCEANS PROGRAM concentrated on the following, key, interdependent areas of technology:

1. Exploration, surveying, and measurement technology, including research and evaluation of instrumentation and sensors, deployment systems, improved survey techniques, mapping, modeling, data management, and analysis

2. Subsea operations and associated support technology, ranging from supporting activities which enable subsea operations to be undertaken, to ocean resource development and management, including marine bioresources, renewable energy, ocean mineral technology, and seabed utilization

Technology specifically intended for the hydrocarbon sector was not included in the Program as similar initiatives were, and continue to be, carried out by the DTI's Offshore Supplies Office. However, many of the technologies to be developed would have application to this sector and research projects with multiple applications were not excluded.

In many cases, technology has been developed for other marine sectors, e.g., oil and gas, fishing, etc. The marine science community already had vast experience working in the deep ocean. The intention was not to duplicate research but to build upon it and where necessary, enhance it to satisfy alternative, new applications.

There was also seen to be scope for technology transfer from the military sector. In order to ensure that there was a free flow of information and real potential for technology transfer, the Admiralty Research Establishment (now Defense Research Agency) was represented on the Technical Advisory Committee that advised the Department on the technical aspects of projects submitted under the WEALTH FROM THE OCEANS PROGRAM. Close contact was also maintained with other Government departments, especially Energy, Agriculture, Fisheries, and Food (MAFF), Scottish Office Agriculture and Fisheries Department (SOAFD), Natural Environmental Research Council (NERC), and the Science and Engineering Research Council (SERC), with whom we shared the Technical Advisory Committee, thus ensuring close coordination of DTI and SERC support of marine technology.

Another promising avenue was seen to be the exploration of the potential for subsea application of technologies such as artificial intelligence, information technology, fiber optics, and advanced materials.

The WEALTH FROM THE OCEANS PROGRAM is not an ocean resources program. It was not the intention to support specific subsea operations, such as deep seabed mining, but to support research into the tools that would enable this and other operations to become commercially viable in the long run. Examples of such enabling technologies are autonomous vehicles, subsea acoustic communications, navigation and positioning, and self-contained power sources.

The aim of the Program, therefore, was to strengthen the UK technology base and hence strengthen UK-based companies in the marketplace through support for the precompetitive, collaborative research into ocean technologies. The Program would support longer-term research projects that were too far from the marketplace to survive without government support. The basic rule of thumb is that projects must involve three companies or organizations (subsequently reduced to two), of which two must be able to exploit the results commercially. Such projects might attract financial support of up to 50% of the total cost.

MANAGING THE PROGRAM

It was decided very early on that the management approach would need to be proactive to ensure the widest possible awareness of the opportunities for collaboration that the Program provided. In addition, we were very keen to encourage industry/university collaboration and assist pull-through of academic, scientific research into the marketplace.

Technology transfer was seen to be a major plank of the Program and our proactive approach enabled us to explore the prospects for research clubs in various market or technology sectors. We believed such clubs might provide an opportunity for small- and medium-sized enterprises to become involved in research which hitherto would have been beyond them for reasons of cost, manpower limitations, or lack of facilities. Interest Groups, as they were more accurately called later, were set up in the following areas:

1. Advanced underwater vehicles, including UUVs
2. Ocean renewable energies
   a) Offshore wind
   b) Wave
   c) Tidal
   d) Ocean current
   e) OTEC
3. Marine environmental
4. Materials

The main aim was to provide a forum within which companies would have the opportunity to discuss and share technological information and consider the technological needs likely to be necessary to satisfy the markets of the future.

Sixteen projects were subsequently funded under the program in the areas of UUV technology; marine environmental measurement, monitoring, and modeling; ocean renewable energy; mariculture; and marine biotechnology. The following are brief descriptions of some of these projects.

Deep Ocean Project

The aim of this project, now complete, was to apply generic engineering research to certain key areas of technology crucial to the exploitation of markets in depths down to full-ocean depth. The areas of research were grouped under the three headings of data and power transmission, handling of heavy loads on the seabed, and preparation of subsea work sites. The work involved comparative studies, pure research and modeling, and testing on both land and at sea. Significant progress was made in the performance of both umbilical and heavy lift systems, in the composition and performance of explosives for accurate cutting at 6,000 meters, and in materials selection and design.

Marine Bioresource Program on Shellfish Hatchery Research

This project is being undertaken by a consortium of the Institute of Offshore Engineering, Orkney Water Test Center, Sea Fish Industry Authority and Scallop King PLC. It is focusing on the transfer of water management technology from the oil and gas sector to the mariculture industry. This is a particularly good example of both university/industry collaboration and technology transfer across market sectors.

Autonomous Underwater Vehicle Research Program

This is a collaboration between Marconi Underwater Systems, Chelsea Instruments, Moog Controls, and Alcan International. The research is aimed at the key areas of energy systems, navigation, communications, and sensors. A feature of the program has been the building of a test-bed Autonomous Underwater Vehicle (AUV). This vehicle is presently undergoing trials in confined waters, but will be involved in under-ice sea trials later this year. It is also hoped that the test-bed vehicle will eventually become available to the wider United Kingdom UUV community for research and evaluation work.

PORPOISE - Power Optimization for Remotely Piloted Offshore In-Sea Environment

This project involves collaboration between Marconi Command and Control Systems, Ruston Gas Turbines, Engineering Research Center, Carlton Deep Sea Systems, Hydrogen Engineering Applications, and J. Marr Technical Services. Its aim is to research the possible use of carbon dioxide scrubbing systems with fuel cells for the provision of power sources to Autonomous Underwater Vehicles (AUVs), and to establish the feasibility of using mathematical modeling techniques to assess the performance of such systems.

A Study of the Application of Low Cost Heat Exchanger Technology to OTEC

This research is being undertaken by a consortium of Marconi Electronics Alcan International, and Newcastle Polytechnic. The project aims to study the application of roll bonding technology for constructing multipanel heat exchangers for applications such as Ocean Thermal Energy Conversion. These modules, based on low cost, corrosion-resistant alloys, represent the first significant move away from the use of noneconomic 'exotic' metals in the evaporator and condenser stages of the system. To replicate the temperature differential of the ocean thermal gradients, the system is being installed in the steam condenser, cooling water inflow and outfall at Alcan's Lynemouth thermal power station. This allows the critical elements of the system to take advantage of existing, closely monitored, high-volume water flows.

FLOAT - Floating Support Structures for Offshore Wind Turbines

The consortium of BMT Fluid Mechanics, Tecnomare UK, and Garrad Hassan are undertaking research to define optimal floating support structures for wind turbines, thereby enabling the economic generation of electricity from windpower in offshore locations.

Mapping and Modeling of Coastal Waters

This project aims to extend the directional current mapping capability of Ocean Surface Current Radar (OSCR) to both a higher resolution system and one that can measure directional wave fields. Another aim is to integrate the data with that of ACDPs to derive 3-D models to describe and predict coastal sea behavior. The eventual deliverables will be shore-based, computerized, environmental monitoring systems for applications such as coastal engineering, ship navigation, and the prediction of the fate of pollutants. The participants in this project are BMT Ceemaid, Marex Technology, and the GEC-Marconi Research Center.

Geochemical Cycling of Pollutants in Estuarine Systems

This project aims to develop a quantitative geochemical model that accurately predicts the movement of pollutants in estuarine and coastal waters. The participants are BMT Ceemaid, Rechem Environmental Research, Plymouth Marine Laboratory, and Polytechnic South West.

The Cultivation of Bioluminescent marine Organisms in a Land-Based System

The aim of this project is to cultivate, on an industrial scale, the bioluminescent, marine, bivalve mollusk, *Pholas dactylus*, in order to harvest bioluminescent materials. The collaborators are Knight Scientific, Seasalter Shellfish (Whitstable), and Pall Process Filtration.

THE EUROPEAN DIMENSION

In parallel with the United Kingdom's WEALTH FROM THE OCEANS PROGRAM, there are two European marine/ocean technology programs which have also been promoted to UK industry under the WEALTH FROM THE OCEANS umbrella. These are the EUROMAR project, within the EUREKA initiative, and MAST (Marine Science and Technology), within the EC Framework Program.

The aims of EUROMAR are the development, application, and successful exploitation of Europe's advanced marine technology with worldwide potential. Its full terms of reference are to:

1. Foster technological progress for integrated ecological management of the marine environment
2. Promote cooperation between industry and science in developing marine instrumentation, and operational systems
3. Improve the productivity and competitiveness in the European marine industry for worldwide application

EUROMAR acts as an 'umbrella' project, generating 'daughter' projects in the following fields of operation:

1. Instruments and carrier systems
2. Bottom systems and mesocosms
3. Remote sensing
4. Data systems
5. Models
6. Atmospheric input

As a result of this promotion through the use of newsletters, seminars, and workshops, UK companies are now becoming increasingly active in European collaboration.

Funding for EUROMAR participants comes from the participating national governments; for example, UK companies seek funding form the DTI. Within MAST however, funding is obtained centrally from the European Commission. The aim of MAST is to contribute to establishing a scientific basis for the exploration, exploitation, management, and protection of European coastal waters and the seas surrounding the EC Member states. Although much of this 103 MECU program is devoted to science, there is a marine technology area that has previously focused on development of instrumentation for undertaking the science. It is to be hoped that any future MAST program will include environmental monitoring technology in its broadest sense and will thus have greater commercial potential, thereby attracting greater industrial interest.

AN EARLY ASSESSMENT OF THE PROGRAM

It is too early to judge the success of the Program or the policy that underpins it. It is however possible and necessary to make some early assessment of the continuing need for a program of this kind.

We are now in the process of making this assessment in order to decide on the need for and the possible content of any new program to replace wealth from the oceans.

In general, we are satisfied that the individual projects supported under the Program have made significant technical breakthroughs at the national or international level. We are also satisfied that they are unlikely to have gone ahead without Government support or that they have displaced other potential United Kingdom developments. However, it is clearly impossible at this stage, to make the most important judgment of how far these projects will lead to substantial commercial developments.

We have some doubts on the underlying rationale for a technology assistance program that tries to cover all of the technologies related to the oceans. It is possible that any future general program should put more emphasis on technology transfer through the promotion of special interest groups and should not dissipate limited project resources over too wide an area. It is certainly the case that any new program should probably put some emphasis on acting as a catalyst to promoting the greater use of European-wide programs. Clearly, any new program will also need to reflect any current Government policies and financial constraints on technology issues.

Our current thinking tends towards the idea that any new program should focus more specifically on those technologies related to ocean monitoring and associated deployment systems. This would seem to offer a better opportunity to maximize the return on Government investment in an area that offers both commercial opportunity and public benefits. It also closely relates to existing European programs.

In summary, therefore, WEALTH FROM THE OCEANS and its predecessors, have provided valuable experience in managing relatively small, but widely-drawn Government technology programs. It is clearly too early to judge the success of a Program that has concentrated on precompetitive collaborative research. Our experience indicates that in the future, Government funding should place more emphasis on technology transfer, while any new project-related program should be more clearly targeted to maximize commercial advantage and public benefit.

34

Coastal Zone Development in the Republic of Korea

Seoung-Yong Hong
Director, Marine Policy Center
Korea Ocean Research and Development Institute

Hyung-Tack Huh
Senior Research Fellow, Marine Biotechnology Laboratories
Korea Ocean Research and Development Institute

ABSTRACT

This paper includes a brief description of the important characteristics of Korea's coastal zone resources and development activities. The poor natural resources inland have had a profound impact on the uses of the coastal zone, reinforcing the nation's economic development strategy. The main issues for coastal zone management in Korea are: coastal fisheries, space creation by reclamation, port development, marine recreation, tidal and nuclear power plants, environment preservation, and governance that assures the enforcement of laws for coastal zone management. Increased use of coastal space has led to competition and conflict. An increased number of critical, coastal zone issues and interactions have been reflected in the creation of new governmental structures that deal with coastal zone resources and environment.

INTRODUCTION

As a country embarks on its development path, it is faced with both constraints and possibilities, depending upon certain initial features of its environment such as natural resources, national ethos, geographic size, and geopolitical location, etc., (Ranis 1989).

Bounded on the south by the Korea Strait, on the north by North Korea, on the east by the East Sea (Sea of Japan), and on the west by the Yellow Sea: South Korea is a geopolitical island and a land bridge connecting the Pacific Ocean and the mainland of Asia (Figure 1). Its population of 43.27 million occupies an area of 99,274 sq.km.q.km.[1] Only about 23% of the total land area consists of lowlands and plains, most of which lie along the coasts and the major rivers.

[1] Korea is about the size of Holland, Hungary, or Guatemala. Korea represents a rare case in that its total land area has expanded from 98,429 sq. km. 1969 to 99,274 sq. km in 1991 by reclamation on the west and the south coasts.

Although Korea's general coastline is estimated at 1,318 km, a measurement taking finer account of its manifold indentations would be closer to 6,200 km (Worldmark Press 1984). About 3,200 islands, most of them off the southern and western coasts, add another 8,600 km of coastline.[2]

The west coast of Korea is low and deeply indented and has long stretches of tidal land due to a high tidal range of more than 5 m. The eastern coastline tends to be straight. The east coast is a coast of emergence, while the west and south coasts are coasts of submergence (Park and Eisma 1985). The south coast, especially the south-southwestern part, is strongly indented and divided into numerous small bays, islands, and peninsulas.

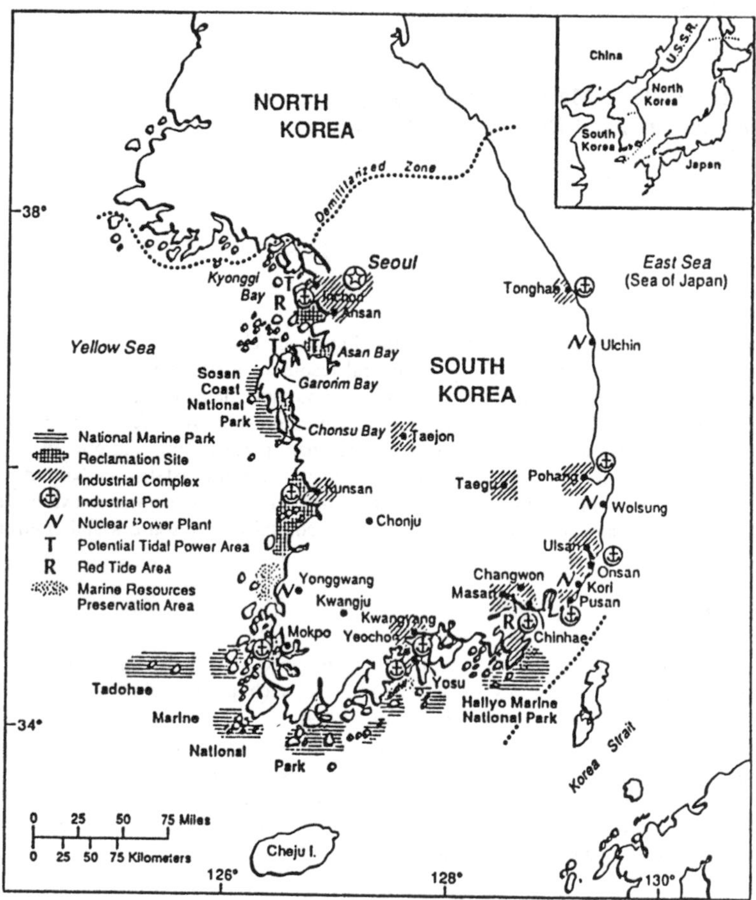

Figure 1. South Korea in its Asian setting and principal coastal zone activities
Source: Hong, S.Y. 1991. Coastam Management 19:392.

[2] A number of islands have been decreasing due to reclamation.

The physical characteristics of Korea with little arable, inhabitable land and poor natural resources, have had a profound impact on the uses of the coastal zone that have reinforced the nation's development strategy. The rapid growth of the Korean economy created an ever-increasing demand for industrial sites. Thus, a variety of development activities such as reclamation, port and tanker terminal construction, shipyards, and power plants have taken place competitively within the coastal zone.

The landward boundary of the coastal zone in this study was defined by using existing administrative subdivisions such as cities and towns adjacent to coastal water, mainly because most basic data for the statistical analysis have been collected and compiled by administrative units. The seaward area of coastal zone extends 3 nautical miles, where most of coastal activities occur. This seaward boundary covers coastal waters of 53,000 sq.km, with 56% in the west coast, 36% in the south coast, and 8% in the east coast.

This paper focuses on the present uses and trends of coastal activities in Korea. Governance arrangements and some prominent issues, such as the West Coast Development Policy and the Pusan Artificial Island Construction Plan, are also discussed.

COASTAL RESOURCES AND PRESENT USES IN KOREA

Since the 1960s, many coastal areas in Korea have been industrialized and urbanized. The urbanization rate in the coastal zone has been higher than that of the whole country. The density of coastal zone population, 608 people/sq. km, is 1.4 times larger than that of the whole country. Gross regional product in the coastal zone accounted for 34.2% of the GNP. The current status of coastal activities in Korea is concerned especially with seven fields: coastal fisheries, space creation by reclamation, port development, marine recreation, tidal power, oil and gas, and nuclear power plants.

Coastal Fisheries

Although the value-added from the fishing industry contributes about 1.1% of GNP (Fisheries Administration 1992), Korea's fisheries have made a significant contribution, both to the diet of Korean people (48% of protein source) and to the nation's export earnings ($1.6 billion US in 1991).

In recent years the distant water fishery of Korea, one of the largest in the world, has suffered from several external factors including the declaration of exclusive economic zones (EEZ) by coastal states, the fluctuation of oil prices, and a fall in fish prices. Consequently, the relative contribution by coastal fishing and mariculture to total fishery production has increased in the 1990s compared to the 1970s (Park 1988).

The production of fisheries in Korea amounted to 2,983 thousand tons in 1991, of which 874 thousand tons were produced in distant waters, 1,304 thousand tons in the coastal waters, and 775 thousand tons from mariculture (Fisheries Administration 1992). During the past decade, the value-added showed a 378% increase, while areas in mariculture farming increased by 39%. Major cultivated species were shellfish and seaweeds, which account for 39.8% and 57.5% of the total culturing area, respectively (Fisheries Administration 1992). It is notable that fish mariculture such as yellow tail, bastard halibuts, rock fish, and parrot fish, increased 70 times from 38 tons in 1980 to 2,656 tons in 1990. The potential area for viable mariculture production was estimated to be 208,000 *ha* and among this, 113,000 *ha* has already been developed.

With such a high utilization rate already achieved, it is increasingly important to manage marine living resources more efficiently, taking into account environmental problems caused by massive land reclamation and coastal industrial complexes built in the last three decades.[3] For this purpose, 413,000 *ha* have been designated as the Marine Fishery Resources Preservation Areas.

Space Creation by Reclamation

Food self-sufficiency has become an elusive goal since the early 1960s, when Korea was a food deficit country (Kuznets 1977). Therefore, the increase of agricultural productivity, which is constrained by the limited area of arable land, has long been one of the top priority items on the national agenda.

Since the beginning of the 1970s, the government has allocated about 18% of its investment funds to agriculture, concentrating on large-scale projects of tidal lands reclamation and improvement of irrigation systems (Cho 1981). A number of large-scale reclamation projects, mainly on the west coast, have been undertaken to exploit their great potential and are recently being revitalized by the West Coast Development Policies (Table 1).

Reclamation of tidal lands and public waters is conducted on the grounds of 'The Reclamation Act of Public Waters of 1962' (Law No. 986, approved on Jan. 30, 1962), and its amendment of 1986 (amended by Law No. 3901, approved on Dec. 31, 1987). However, governmental economic incentive programs and development projects have seriously undercut the wetland preservation goals of the 'Reclamation Act of Public Waters'.

The total reclaimed area is now about 1,100 sq. km. This represents 15% of the total potential area, which is estimated to be 7,270 sq.km (Korea Development Institute n.d.). The reclamation of total potential area would enlarge new arable land corresponding to about 24% of the total existing arable land area.

The purpose of reclamation so far has been mostly for agriculture (about 92% of total reclaimed area), while the areas zoned for the industrial and residential purposes are only 7% and 1% of the total, respectively. In the next decade, however, the area for industrial and residential use is expected to increase (about 30% of total area to be reclaimed) to meet the growing demand. The coastal areas to be reclaimed are particularly subject to multiple use, resulting in various demands by the different interest groups and user agencies. Basic conflicts arise between traditional fishermen and reclaimers. In the following section, two, major, national plans are discussed.

The West Coast Development Policy

Since the mid-1980s, the Korean government has heralded a popular slogan of "The Era of West Coast". In 1989, the government established the Standing Committee for the West Coastal Zone Development under the Prime Minister's Office (Cho 1989).

The internal motivation is to boost more balanced regional development for the West Coast region (provinces adjacent to the Yellow Sea). Due to adverse natural conditions for industrial port development and water resources, the West Coast region has been excluded from the mainstream of buoyant economic development during the past three decades.

The external motivation for the West Coast initiative is primarily concerned with Korea's response to expanding Chinese linkages in terms of trade and resource development and also to help position the nation for the expected coming "era" of economic prominence for the Pan-Pacific and North-East Asian countries.

---

[3] As one response, Paragraph 8 of Article 6 of 'the Land Use and Management Act (Law No. 2048, Dec. 30, 1972, revised Law No. 3910, Dec. 31, 1987)' and Article 6 of 'the Fishery Promotion Act (Law No. 1814, Aug. 3, 1966, revised law No. 3239, Jan. 4, 1980)' provided for the designation and protection of especially sensitive areas to the destructive forces caused by pollution and other man-made changes. Since 1975, pursuant to the above articles, ten sensitive areas in coastal sea water have been designated as the Marine Fishery Resources Preservation Areas. These total areas comprise 413,000 *ha*.

Table 1

Major Reclamation Projects in the West Coast

Region	Area (*ha*)	Project Peroid
Gyewhado	2,500	1962 - 1979
Pyungtaek	18,419	1970 - 1977
Sapgyocheon I	24,700	1976 - 1986
Kimpo	4,900	1981 - 1986
Seosan A	5,930	1982 - 1988
Seosan B	9,664	1983 - 1989
Daeho[a]	7,700	1980 - 1992
Siwha I[a]	2,303	1986 - 1991
Siwha II[a]	9,000	1991 - 1997
Asan[a]	1,132	1989 - 2011
Kunsan[a]	690	1989 - 1992
Kun-Chang[a]	3,092	1989 - 2021
Daebul[a]	1,370	1989 - 1996
Saemankeum[a]	42,000	1987 - 1998
Yongsangang II[a]	20,700	1987 - 1991
Yongsangang III-1[a]	12,200	1985 - 1994
Yongsangang III-2[a]	6,800	1989 - 1999
TOTAL	173,100	

Note: Projects denoted by [a] are included in the West Coast Development Policy.

Source: Hankuk Kyungje Shinmun, Oct. 11, 1989, p. 2 and S.K. Park. 1988. 21 Seki-rul Hyanghan Susanop Paljeon-kwa Bada Yiyong Cheollyak [Development of Fisheries and Strategy for Utilization of the Ocean in 21st Century]. Seoul: Korea Rural Economic Institute. pp. 18-19.

According to China's coastal development strategies, each coastal region is to be matched with a particular foreign economy or region in its trade and investment relations (Zou and Jun 1989). For instance, the Fujian Delta is to be linked with Taiwan; the Zhujiang Delta with Hong Kong and Macao; the Changjiang Delta with Japan; the Hainan Island with the ASEAN countries; and finally, the Liaoning-Shandong Peninsular with Korea. The Chinese open policy has thus spurred Korea's external motivation for the West Coast Development Policy.

The main components of the West Coast Development Policy include the following:

1. *Construction of Industrial Bases.* In order to provide job opportunities in the manufacturing sector in the western part of the country, as well as to promote knowledge-intensive, high technology industries, the tentative plan for a few, large-scale industrial bases along the west coast is currently being prepared.
2. *Expansion of Transportation Networks.* In order to facilitate industrial development and strengthen domestic and foreign trade, expressway, railway, port, and airport facilities will be expanded.
3. *Promotion of Hinterlands.* Major, less-developed urban areas, particularly in the southwestern region, will be developed to play a vital role as industrial centers (Cho 1989).

The Korean government is also planning to construct a new, large international airport on Yongjong Island about 40 km west of Seoul by 1997. The projected Yongjong airport, which will become one of East

Asia's largest airports, having an area of 51 sq. km, will be capable of handling up to 240,000 outgoing and incoming flights and 40 million passengers a year.

The Pusan Artificial Island Construction Plan

Pusan, the second largest city on the southeast coast, has been developed rapidly since the Korean War. However, the city is hemmed in by narrow, inland valleys and thus faced with some problems such as land shortage, high real estate values, wastes disposal, etc..

Hence, the city of Pusan is planning to construct an artificial island similar to the Kobe Port Island in Japan. The size of the Pusan artificial island will be 6.1 sq. km in total, larger than the 4.36 sq.km of the Kobe Port Island (Okamura 1989).[4] The average water depth at this site is about 15 m, and the island will be connected with the land by two bridges of 3.8 km length. The artificial island will be designed for multiple use, aiming at urban development combined with harbor facilities, such as a container terminal, an information-related teleport, conference facilities, and commercial and residential districts. The total investment is estimated at 720 billion won ($1,029 million US) in the municipal sector and 590 billion won ($843 million US) in the private sector (Kang 1989). The completion of this project is scheduled for 1998. The city of Pusan has conducted a scientific investigation on the environmental effects of artificial island construction in the coastal bay.

Port Development

The long coastline of Korea has been favored by the development of marine transportation. There are, in general, no extreme natural navigational hazards except for the high tidal range along the west coast of the Korean Peninsula. Maritime transportation and ports have played a strategic role in the country's export-led trade, economy, and defense. The value of exports grew at an average annual rate of 28% between 1962 and 1991 (Economic Planning Board 1989/1992). Imports also expanded remarkably during the same period, rising from $421 million US to $81.5 billion US.

Korean policy has emphasized the expansion of national shipping capacity to complement a rapidly growing foreign trade. Owing to technological innovations, quality improvements, and market development, South Korea's shipbuilding industry has become one of the largest in the world, second only to that of Japan (Korea Shipbuilder's Association 1992).[5] The total shipping capacity of Korea in 1991 expanded to well over 7.8 million tons, fourteenth in the world in terms of tonnage.

Korea has about 48 commercial ports and 394 harbors in various sites. In 1988, the total volume of port-handled cargo of Korea was 292,909,000 R/T, and the total amount of economic activity generated by the port industry was $112.5 billion US (Economic Planning Board 1989/1992). This means that the total impact of Korean ports on the economy averaged about $308 million US a day. Most of Korea's seaborne cargo is carried from the ports of Pusan, Pohang, and Ulsan on the southeastern coast and Inchon on the west coast. These four ports represent over 75% of Korea's total foreign trade by tonnage.

The following three facts will substantially influence decision making for port development: (1) shifts in maritime trade and transportation due to economies of scale and deregulation within a highly competitive environment, (2) the scarcity and higher cost of capital to build new facilities, and (3) increased pressure from local environmental organizations and public agencies seeking more benefits and better accountability from public ports (Hershman 1988). To meet these changes and the future growing demand for port

---

[4] During several years in the 1980s, Japan completed a number of large-scale, artificial islands in nearshore, such as Kobe Port Island (4.36 sq. km), Osaka South Port (9.37 sq. km), Yokohama Daikoku (3.21 sq. km), Rokko Island (5.83 sq. km), and Yokkaichi (3.87 sq. km). The Kansai International Airport, an area of 5.11 sq. km, will also be constructed.

[5] The Korean shipbuilding volume in 1988 accounted for 29.1% of the world total, while Japan stood at 37.0%.

facilities, the Korean government has made decisions to invest in some expansion and new development projects for Pusan, Inchon, Kunsan, Mokpo, and Kwangyang.

Marine Recreation

Advanced technologies, increases in disposable income, mobility, education, and labor laws all have contributed to increased time available for leisure pursuits (Anderson 1980). About 70% of Korea's population lives within 50 miles from the coast, and marine recreation on the coast and in the nearshore waters is in growing demand by both long- and short-term users.

South Korea, in emphasizing the conservation of unique historical, archaeological, and scenic sites, has established national and provincial marine parks along the coasts. The marine recreational areas account for 40.7% of the total nationwide recreational areas. Plans to develop more marine recreation sites of 250 sq. km are underway for marinas, resorts, beaches, and parking lots to meet the increasing needs.

While the unique environment of the Korean coastal zone provides especially rich resources for marine recreation, there are various difficulties in conserving and protecting the coastal environment. Oil spills have recently had significant impacts on coastal recreation areas. Beach erosion is also a significant problem. The sand of some beaches, like Mallipo on the west coast and Haeundae in Pusan, is undergoing serious reduction because of erosion and the local sea level rise. Shoreline protection policies in Korea that rely on seawalls, bulkheads, and other structures, however, are losing favor because of their high construction costs.

Tidal Power

Planning for tide-derived power in Korea dates back to as early as the 1930s. In recent years, very favorable sites for harnessing tidal power have been reported along the 160 km stretch of the west coast between the north shore of Kangwha Island and the north end of Cheonsu Bay (Korea Electric Company 1978). At least ten suitable sites were identified which could provide a capacity in excess of 30 million kw. Feasibility studies were conducted for Asan Bay, Cheonsu Bay, and Garorim Bay.

The feasibility study on the basis of the price of oil of $13 US per barrel in 1978 by the Korea Ocean Research and Development Institute (KORDI) and the Shawinigan Engineering Company of Canada concluded that none of the sites were economically viable with a differential escalation in the oil price of 2% from the year 1990 to 2010. Owing to a recent, nationwide antinuclear movement in South Korea, however, the idea of establishing a tidal power plant is being revived.

The economic scale of a tidal power plant raises a controversial issue. Kyonggi Bay, with a 7 m average tidal range and a 4,500 sq. km. basin area, can theoretically provide a capacity of 44.5 million kw (annual production of 100,000 Gwh) which accounts for 2.1 times the nation's total electrical need in 1989 (Ahn 1990). The design scale of this Kyonggi Bay scheme is over 130 times larger than the earlier design for Garorim Bay of about 330 Mw. The difficulty in long distance transmission for regions of considerable distance from the coast was the main obstacle to the earlier tidal power pan. The economic potential of the tidal power scheme increases considerably when it is used in conjunction with other power sources in a regional grid or even a national grid as it is in France. Besides, multipurpose development schemes mixed with resorts, mariculture farms, and reclamation areas around a tidal power plant, may enhance its economic viability.

Oil and Gas

As Korea's economy has grown toward energy-intensive industry, the nation's energy consumption has been heavily dependent upon foreign supplies because there is no domestic production of oil. The production of domestic anthracite meets less than 18% of total energy demands. Total import of crude oil was about $1,500 million US in 1976 and rose to about $8,133 million US in 1991, at an average growth rate of over 10% per year since the mid-1960s (Economic Planning Board 1989/1992).

To ease this heavy dependence on imported crude oil, the nation's oil exploration activity has been enhanced in the past two decades. It was not until the late 1960s, when a series of geophysical surveys indicated the likelihood of oil deposits under the seabed of Korea's adjacent seas, that the oil hungry government began to show increased interest in seabed oil development. The U.N. Economic Commission for Asia and the Far East report (ECAFE; renamed in 1974 as ESCAP: Economic and Social Commission for Asia and the Pacific) publicized its optimistic judgment that the sediments beneath the continental shelf in the Yellow Sea were believed to have great potential as oil and gas reservoirs (CCOP/ECAFE). Since 1970, oil development contracts have been signed for seven, seabed oil development zones between the South Korean government and Western oil companies, including agreement between Korea and Japan concerning joint development of the southern part of the continental shelf, designated as Block VII by Korea.[6]

Despite increased attention by the government, Korea's offshore oil exploration was hampered by considerable problems such as a lack of both technological experience and capital availability and international conflicts surrounding seabed oil development zones (Park 1983; Harrison 1977).

During the 1970s, with the promulgation of 'the Submarine Mineral Resources Development Law (Presidential Decree No. 5020, May 30, 1970)', Korea's oil exploration was mainly carried out by Western oil companies. During the 1980s however the government initiated efforts toward an independent drilling capability through the inauguration of the Korea Petroleum Development Corporation (PEDCO), the government-sponsored corporation to manage both downstream and upstream processes of oil development.

Results in offshore oil exploration of Korea since 1969 while they have not yet been appreciable, are not discouraging. Up until now, nineteen holes have been drilled by the PEDCO or foreign concessions, and geophysical surveys have examined 73,722 km (Kang 1989). Although offshore drilling has found no commercial quantities of oil or gas, there remains the possibility of 128 reservoir beds in the continental shelf.

Nuclear Power Plants

Until 1981, oil was the principal source of electric power, and South Korea's 118, oil-burning power plants produced about 75% of its production of electricity totalling 40.2 billion kwh (Economic Planning Board 1989/1992).[7] The government's energy plan for the 1980s sought to deemphasize oil through efforts to achieve a drastic increase in nuclear energy production.

Korea has constructed eight nuclear power plants since the end of the 1970s, and all are located along the coast. Five, additional coastal nuclear plants are underway. Nuclear power plants accounted for 46.9% of the nation's total electrical supply in 1988. Nuclear power's share in Korea's energy supply is expected to stay at the 40% level through the first quarter of the 21st century. Korea is projecting a need for 55 more nuclear power plants by the early 2000s to maintain that level.

---

[6] 'Agreement between Japan and the Repubic of Korea Concerning Joint Development of the Southern Part of the Continental Shelf Adjacent to the Two Countries' was signed on Jan. 30, 1974 and came into force on June 22, 1978.

[7] In 1981, coal-fired thermal power plant and hydroelectric generators produced another 11% and 7%, and one nuclear power plant supplied 7%.

MARINE POLLUTION AND ENVIRONMENT PRESERVATION

The problem of marine pollution in Korea has emerged as an important social issue because of several, sporadic oil spill accidents, increased industrial activities, and land reclamation.

First, the growing tanker traffic due to Korea's increased consumption of imported oil poses a constant threat of oil spills. Crude oil is the largest, single import item consisting of more than 44 million tons annually or one-tenth of Korea's imports (Economic Planning Board 1989/1992). The Ministry of Environment in 1992 reported 2,470 incidents of marine oil spill between 1979 and 1991, with the compensations for damages in fishery and mariculture reaching over $33.5 million US (Table 2).

Second, inflows of urban and industrial wastewater are damaging the coastal ocean environment. Over 10 million tons of wastewater are dumped into the four major rivers daily; the Han River, Kum River, Yongsan River, and Naktong River. The current treatment capacities of industrial effluents and municipal sewage are 34% and 25%, respectively. As a result, growing coastal industrialization, as well as coastal urbanization, have created serious marine pollution problems. Table 3 shows some indication of coastal water pollution in Korea.

The overall Chemical Oxygen Demand (COD) values in coastal areas already show unacceptable levels for the first class seawater quality.[8] Heavy metals such as copper (Cu) and zinc (Zn) are as yet lower than the general criteria for seawater quality.

Total concentration of nitrogenous nutrients exceeds the standard values for the third class seawater quality. Red tides at Chinhae and Masan Bays in the southern coast pose great damage to many sea farms almost every year. The damage, caused by red tides, reached 1.8 billion won ($2.16 million US) in 1981 and 77 million won ($ 0.11 million US) in 1985 (Environment Administration 1992). To remedy chronic red tides, the Environment Administration in 1982 designated 934 sq. km. of coastal waters in the vicinities of Ulsan, Pusan, Chinhae, and Kwangyang Bay, as Special Preservation Sites.[9] In these special sites, some industrial development activities causing marine contamination are limited. To regulate the wastewater from industries, 'effluent standards' have been established and factories are fined or closed if they cannot meet the standards.

Finally, large-scale coastal reclamations along the west and southern coasts have caused great environmental changes. As a result, the various benefits provided by coastal wetlands are being lost and fishing grounds are damaged by reclamation activities which eventually led to conflicts between traditional users and new industrial users. There are also damaging externalities in terms of foregone benefits associated with storm protection, waste assimilation, and wildlife habitat which are related directly to the amount of wetlands loss (Edwards 1987).

GOVERNANCE FOR COASTAL ZONE DEVELOPMENT

In recent years, a significant environmental stress is being placed on Korea's coastal zones due to competing and irresponsible uses. Important physical and ecological processes, as well as social and aesthetic values are being irreversibly damaged and lost. This deterioration is partly attributable to inadequate policy guidelines for coastal zone management, as well as inadequate strategies for optimal planning and development of the coastal environment.

---

[8] The Ministry of Environment classifies seawater quality into I, and II and III. Class I (below COD 1 *ppm*) means seawater quality is suitable for fishing and mariculture. Class II (below COD 2 *ppm*) means the seawater quality is suitable for recreational activity such as swimming. Class III (below COD 4 *ppm*) means the seawater quality is suitable for industrial activity and ship anchoring.

[9] The Environment Administration Order No. 82-6 (Oct. 21, 1982).

Table 2

Oil Spills from Shipwreck Accident

Year	No. of Accidents	Size of Oil Spills (ton)	Compensation for Damages ($1,000 US)	Recovery Cost ($1,000 US)
1981	185	983	2,677	1,418
1985	166	2,204	659	502
1990	248	2,421	9,741	9,043
1991	240	1,257	4,896	5,155
TOTAL ('79-91)	2,470	12,979	33,532	22,852

Source: Environment Administration, 1992. Whankyung Pojun [Environment Preservation]. Seoul.

Table 3

COD Values in Coastal Areas of Korea

(unit: *mg/l*)

Areas	1987	1988	1989	1990	1991
Panwol	2.6	2.6	1.8	2.7	3.3
Kwangyang	2.8	2.9	2.6	2.6	2.5
Masan	6.1	5.1	6.3	4.1	4.3
Chinhae	2.6	2.3	2.6	2.5	2.4
Sokcho	-	7.9	12.3	10.0	7.3
Chumunjin	2.8	4.9	4.6	5.3	4.0

Source: Environment Administration. 1992.

From an institutional perspective, coastal zone resources and environment were simple issues with few functional linkages in Korea before the 1960s. However, since the early 1970s, some major socio-environmental changes, such as rapid industrialization, increased energy needs, growing population density, increased foreign trade, and technological advances, have challenged the traditional hierarchy of national interests and complicated coastal zone management. An increased number of important coastal zone issues and interactions were reflected in the creation of new governmental organizations and new laws which deal with coastal zone resources and environment.[10] The Korean government thus has created over twelve institutions with their subordinate agencies specifically to address major problems in coastal zones. Figure

[10] Until the end of the 1950s, only a few government agencies, such as Ministries of Agriculture and Fisheries, Home Affairs, and National Defense, were associated with coastal zone issues. Since the 1960s, several new national institutions were added to deal with new issues: The Ministry of Construction was established in 1962, Ministry of Energy & Resources in 1978, and the Environment Administration in 1980. In 1990, the Environment Administraiton was upgraded as the Ministry of Environment.

2 shows the ministries that govern the coastal zone and the functional areas they cover.

These governmental organizations tended to set their own policies and priorities, neither conforming to nor being responsible for creating and enforcing a national master plan for efficient coastal zone management. The Economic Planning Board (EPB) is responsible for producing the economic development plans as well as for presenting the Yearly Overall Resource Budgets and Interim Plan Reports. Other relevant Ministries such as Construction, Trade, Industry and Energy, Agriculture and Fisheries, Home Affairs, Transportation, and Science and Technology, are dealing with discretion in the application of rules and regulations. As a result, these organizations have to make hard choices about priorities and performance in order to adjust to the difficulties of implementation.

Government policymakers have solicited the opinions of academics and particularly of government-supported research institutes. The government-supported research institutes represented in coastal policy research are the Korea Ocean Research and Development Institute under the Ministry of Science and Technology, the Korea Research Institute for Human Settlements (KRIHS) under the Ministry of Construction, the Korea Rural Economics Institute under the Ministry of Agriculture and Fisheries, the National Fisheries Research and Development Agency under the Fisheries Administration, and the Korea Maritime Institute under the Maritime and Port Administration.

Article 5 of the Environmental Conservation Act of 1977 (Law No. 3078, approved Dec. 31, 1977) and its Amendment of 1986 (last amended by Law No. 3903, Dec. 31, 1986) can provide room for a compromise settlement measure on the basis of scientific data. It requires would-be-developers to prepare environmental impact statements (EIS) or supply data concerning their coastal zone activities such as industrial complex construction, port development, reclamation, and water resource development. While private or public developers are often in a unique position to generate sophisticated data due to available expertise and funds, there can be some argument that the quality of data generated also depends on the objectivity of the developers. Effective regulation is possible if the standards are linked to a process of scientific assessment of the total effects of pollutants (Miles 1989) and some of the spatial problems, such as area displacement of fishermen, may be mitigated through compensation agreements. However, the impact to coastal fisheries identified in the environmental impact statements prepared for reclamation projects may have been underestimated. This is partially due to the fact that environmental impact statements, in general, have focused on impacts of fisheries in a limited area, without considering broader, second-and third-order effects (Cicin-Sain and Tiddens 1989).

The Minister of Construction, under Article 3(2) of 'the Reclamation Act of Public Waters of 1986', is to establish "A Basic Masterplan for Reclamation of Public Waters" so that public waters are rendered suitable for the nation's total land functions and uses. The established Basic Masterplan is subject to be reviewed every ten years. While Article 14 of its amendment of 1986 articulated the positive participation of commercial construction firms in reclamation, subject to governmental permission for acquiring ownership of the reclaimed area, it strengthened the compensation system to control reclamation activities more tightly than ever before, in order to mitigate the conflicts caused by imprudent projects.[11]

[11] Article 16 of the Amended Reclamation Act of Public Waters aimed at two objectives. The first objective was that of reducing the burden imposed on construction firms that were doing business in the Middle East. South Korean construction firms garnered $8.3 billion US of overseas construction contracts in 1980. However, after the construction boom in the Middle East in the early 1980s, Korea shifted the enormous construction machineries to the coastal zone reclamation. The second goal concerned conflicts between coastal residents and owners in reclamation areas. The government permits the commercial construction firms to acquire the ownership of reclaimed area corresponding to total investment costs, while it requires an obligation for the firm to compensate or establish some measures to mitigate the damages to the previous proprietorship before the commencement of reclamation projects.

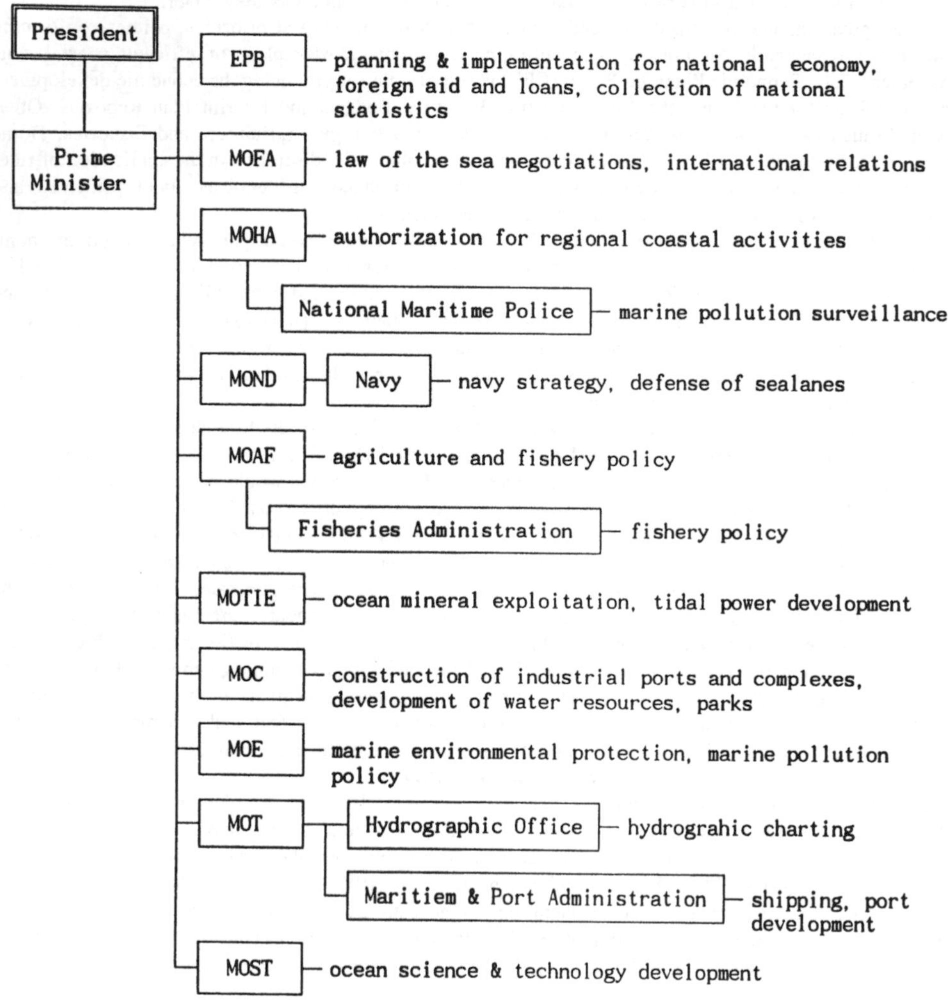

Figure 2. Coastal zone management in Korean government

Note: EPB (Economic Planning Board), MOFA (Ministry of Foreign Affairs), MOHA (Ministry of Home Affairs), MOND (Ministry of National Defense), MOAF (Ministry of Agriculture and Fisheries), MOTIE (Ministry of Trade, Industry and Energy), MOC (Ministry of Construction), MOE (Ministry of Environment), MOT (Ministry of Transportation), MOST (Ministry of Science & Technology)

Coastal zone management in Korea is still embodied in separate laws, each typically addressing one class of resource or environmental matter. Since the early 1950s, the Korean National Assembly has enacted over fifty-four laws relating to coastal zone management (Hong et al 1989). The mechanisms for resolving interministerial conflicts over the management of these resources and environments vary widely from law to law. Furthermore, the government operates seventeen Committees or Commissions under various Ministries for the purpose of resolving interministerial conflicts or improving efficiency of coastal zone management policies.

Some conservancy zoning programs have been applied in some regions to protect areas of special flora, fauna, and geologic or ecologic interest.[12] While local zoning is a broad tool for sensitive area management, it is also subject to many limitations, including a limited data base, lack of administrative expertise, inadequate geographical perspective, and failure to take into account the unique features of each site (Kusler 1980).

The enactment of 'The Marine Development Basic Act (MDBA: Law No. 3983, Dec. 4, 1987)' was thus a welcomed response to the urgent need for a more visionary approach to planning for nationwide coastal zone management, and even further, to coordinating interministerial conflicts rationally through the deliberation of the Marine Development Committee chaired by the Prime Minister. Article 11 of the MDBA stresses that the government shall, for harmonious ocean development, adopt necessary measures and arrangements for rational coordination between marine environment preservation and marine development. The definition of ocean development in this context includes coastal zone activities.

In order to reduce interministerial and central vs. local conflicts and to enhance some grassroot planning for coastal zone management from regional areas,[13] Korea urgently needs to promulgate a special law such as the U.S. Coastal Zone Management (CZM) Act of 1972 and its Amendment of 1990. Under the CZM Act, the States exercise their full legal authority to manage and protect their coastal resources. The federal government offers financial and technical assistance as an inducement to the States and commits itself to act consistently with such programs (Archer 1989). This legally-enforceable commitment by the U.S. federal government constitutes the "federal consistency doctrine" by which States and federal agencies resolve their conflicts.

To date, Korea has had a strong central governance system. From the beginning of the 1990s, however, decentralization in favor of regional government has been decreed.

Upon undertaking this new decentralized governance, Korea would seem well-advised to be equipped with arrangements for assessing trade-offs among competing interests in an objective, comprehensive, and theoretically sound manner.

The basic pattern of coastal zone uses in Korea has changed from the linear expansion of the coastal zone to one of integrated coastal uses. Therefore, an integrated coastal zone management policy, using strategies such as multiple-use planning (Miles 1989), may well have better success if they are begun on a regional basis and then expanded to national land use planning.

CONCLUSION

Coastal zone management in Korea is a new and rapidly emerging policy area. Until the end of 1970, coastal zone activities might not have caused major conflicts. However, various industrial projects for utilizing the coastal areas have recently caused many conflicts which raised social and environmental issues. Coastal development activities, such as coastal reclamation, construction of industrial complexes, and nuclear power plants are thus becoming battlegrounds between new developers and local residents. These issues are manifested at the central government level as conflicts between the Ministry of

---

[12] Article 6 of 'the Land Use and Management Act of 1972', Article 6 of 'the Fishery Promotion Act of 1966', and Article 44(3) of 'the Marine Pollution Control Act of 1977'.

[13] Korea has fourteen administrative units including nine provinces and five special status cities (Seoul, Pusan, Taegu, Inchon, and Kwangju). The provinces are divided into cities (*Shi*), counties (*Gun*), townships (*Up*), and villages (*Myon*).

Construction (MOC) and the Ministry of Agriculture & Fisheries (MOAF) or between MOC and Ministry of Environment (MOE).

Furthermore, the nation's previous economic policy, focusing on the development of some specific coastal zones, caused sectoral and regional imbalances. Recognizing the importance of regional balance, the Korean government is now paying serious attention to regional development and spatial integration on the West Coast. The West Coast Development Policy aims also to expand the linkages with China in terms of trade and resource development.

As a result of sporadic oil spills, increasing industrialization, and coastal urbanization, the problems of marine pollution and ecosystem alterations have emerged as critical social issues. The deterioration of the country's coastal environment can be attributable to inefficient governance by over twelve governmental institutions, each seeking to serve its own policy guidelines which stress sectoral economic development, with little attention to environmental protection. Korea's coastal zone management thus seems to be pushed and pulled by the dynamics of each individual issue, compounded by the urgent demands for economic growth and industrial development. The heightened public concerns on environmental problems, however, are expected to boost more active countermeasures on the part of the government and industries against coastal pollution.

In addition, Korea would be well-advised to be equipped with governance measures similar to those of the U.S. CZM Act of 1972 and also measures of several European countries to assess tradeoffs among competing interests in a theoretically sound manner. For the time being, the Land Use and Management Act of 1972 and the Marine Development Basic Act of 1987 can play major roles in establishing a harmonious, coastal zone management plan and facilitating coordination among various governmental agencies and other involved parties. Finally, multiple-use planning may enhance integrated coastal zone management if it is begun on a grassroots basis, and then expanded to national consensus.

REFERENCES

Ahn, S.H. 1990. *Kyonggiman-e choreok paljeonso-rul seuja (Establishing a tidal power plant in Kyonggi bay)*. In *Wolgan Chosun*. *May*: 504-11.

Anderson, S.H. 1980. The role of recreation in the marine environment. In *Ocean Yearbook* 2, ed. E. Mann Borgese and N. Ginsberg, 183-98. Chicago: The University of Chicago Press.

Archer, J.H. 1989. Resolving intergovernmental conflicts in marine resource management: The U.S. experience. *Ocean and Shoreline Management* 12(3): 253-69.

CCOP/ECAFE. 1969. Geological structure and some water characteristics of the East China Sea and Yellow Sea. *Technical Bulletin* 2: Bangkok.

Cho, D.S. 1981. *Kukje chawonron (Study of international resources)*. Seoul: Pakyongsa.

Choe, S.C. 1989. Korean economic policy and the coastal zone management of the yellow sea. In *Proceedings of an International Conference on the Regime of the yellow Sea*. Seoul, Korea: The SLOC Study Group-Korea and Institute of East and West Studies, Yonsei University.

Cicin-Sain, B., and A. Tiddens. 1989. Private and public approaches to solving oil fishing conflicts offshore California. *Ocean and Shoreline Management* 12(3): 233-51.

Economic Planning Board (EPB). 1989. 1992. *Major statistics of Korean economy*. Seoul.

Edwards, S.F. 1987. *An introduction to coastal zone economics: Concepts, methods, and case studies*. New York: Taylor & Francis.

Environment Administration. 1992. *Whankyung pojun (Environment preservation)*. Seoul.

Fisheries Administration. 1992. *Susanop donghyang-e kwanhan yeoncha pogoso (Annual report on the trends in fishing industry)*. Seoul.

Harrison, S.S. 1977. *China, oil, and Asia: Conflict ahead*? New York: Columbia University Press.

Hershman, M.J. 1988. A new role for the public port. *In Urban ports and harbor management*. ed. 3-25.

M.J.Hershman. New York: Taylor & Francis.

Hong, S.Y, Lee, Won-Kap. Chang, Hak-Bong and Kim, Sung-Kwi. 1989. *Haeyang keabal chujin kiban cheongbi-e kwanhan yeongu (A study on the establishment of the basis for the ocean development)*. Seoul: KORDI.

Kang, C.D. 1989. *Inkon-eun Pusan Simin-eui Ggum (The artificial island in Pusan is Pasan citizens' dream)*. *Wolgan Chosun*. *September*:394-403.

Kang, C.S. 1989. *Yeukkwangku Cheonyeon Gascheung (Potential of natural gas in Block VI)*. *Wolgan Chosun*. *November*: 386-95.

Korea Development Institute (KDI). Various years. *KDI quarterly economic outlook*. Seoul.

Korea Electric Company. 1978. *Korea tidal power study: Phase I*. Seoul: KORDI and Shawinigan Engineering Co.

Korea Shipbuilder's Association (KSA). 1992. *Chosun charyozip (Shipbuilding data bank)*. Seoul.

Kusler, J.A. 1980. *Regulating sentive land*. Cambridge: Ballinger Publishing Co.

Kuznets, P. 1977. *Economic growth and structure in the Republic of Korea*. New Haven: Yale University Press.

Miles, E.L. 1989. Concepts, approaches, and applications in sea use planning and management. *Ocean Development and International Law* 20: 213-38.

Okamura, K. 1989. Future ocean space utilization in Japan. In *Proceedings of an International Conference on EEZ Reousrces: Technology assessment*. 2-28. Honolulu: Internaitonal Ocean Technology Congress.

Park, C.H. 1983. Oil under troubled waters: The Northeast Asia seabed controversy. In *East Asia and the Law of the Sea*. 1-51. Seoul: Seoul National University Press.

Park. D.W., and D. Eisma. 1985. North Korea and Korea. In *The world's coastline*. ed. E.C.F. Bird and M.L. Schwartz. 833-40. New York: Van Norstrand Reinhold Co.

Park, S.K. 1988. *21 Seki-rul hyanghan susanop paljeon-kwa bada yiyong cheollyak (Development of fisheries and strategy for utilization of the ocean in 21st century)*. Seoul: Korea Rural Economic Institute.

anis, G. 1989. *The political economy of development policy change: A comparative study of Taiwan and Korea*. Seoul: Korea Development Institute.

Worldmark Press. 1984. *Worldmark encyclopedia of the nations: Asia & Oceania*. 193-200. New York: John Wiley & Sons, Inc.

Zou, Gang, and Ma Jun. 1989. *Tuijin zhongguoyanhaidiqu yu dongya jingide quanmianhezhuo (The promotion of economic cooperation between the Chinese coastal zones and the East Asia)*. In *Guanlishijie (Management World)* 6:73-81.

35

Lobster Ranching in Coastal Waters

Jens G. Balchen
Department of Engineering Cybernetics, The Norwegian Institute of Technology
University of Trondheim, Norway

ABSTRACT

The dramatic reduction in the population of the European lobster (*Homarus Gammarus)* has created an interest in large-scale industrial production of lobster juveniles for the release under control conditions in coastal waters. A major program in this field has been functioning in Norway for many years with both private and governmental sponsorship. Plans have been devised for a large-scale production facility for 1.2 million, one-year old lobster juveniles per year using warm cooling water from a large oil refinery on the west coast of Norway. Some biological, technological, and organizational aspects of this program are described in the paper and some perspectives for future development are outlined.

INTRODUCTION

Lobsters of many kinds live in most oceans and have been a popular culinary treat for man for ages. In the northern hemisphere, the European Lobster *(Homarus Gammarus*) and the American lobster (*Homarus americanus*) are the most popular. These two kinds of lobsters are very similar and are often treated as one species.

In the last fifty years there has been a dramatic decline in population of the European lobster and the market for the American lobster has increased so fast that the natural production is in danger. More than 100 years ago, attempts were made to hatch lobster larvae and release them into the coastal waters of Norway in order to enhance the population. However, the results of this activity were not encouraging. Similarly, in the United States, a large number of fourth-stage lobster juveniles where released on the east coast around 1950 without noticeable results impact.

Because of the meager results of stocking coastal waters with small lobster juveniles, a number of projects were started in the 1960-80s to develop artificial lobster-rearing facilities starting with fertilized lobster females and ending up with marketable, full-grown lobsters in an aquaculture system (van Olst, Carlberg, and Ford 1977). Major investments were made in this sector but none of the projects seem to have become economically successful.

Around 1975, a project was started in Norway as a cooperative effort between the industrial organization, Tiedemanns, and the author of this paper, with the goal of developing a large-scale industrial production facility for one-year old lobster juveniles for release in specifically controlled areas along the Norwegian coast. After a number of trials, a facility was built near an electrometallurgical factory near Trondheim with the capacity of producing 120,000 juveniles per year. The idea was that a one-year old lobster, raised in this favorably controlled environment (water 20°C), will have a very high probability of survival in the ocean, and would only need three to four years in the ocean before reaching marketable size.

A large number of young lobsters produced at this facility have been released along the Norwegian coast during the last ten years and observations have been made of the influence upon the catch of marketable lobsters. In an attempt to record the recatch rate, agreements were made with local fishermen and lobster receiving stations to keep record of the catch (Karlsen 1992). In the beginning no tagging of the lobsters were done, but more recently, biologists from the Institute of Marine Research, Bergen, Norway have been involved in tagging experiments (van der Meeren et al 1990). Similar investigations have been made in the United Kingdom (Burton 1992).

There is great uncertainty about the probable recatch percentage of released lobster juveniles. At some locations survival rates have been very high (more than 50%), whereas in other places, hardly any of the released lobsters were caught at marketable size. There is general agreement however, that "a small lobster is an aggressive and well-armed adversary, and when given a chance to adapt to the wild by being placed directly on to the substrate, they show a good ability to survive" (Burton 1992). Consequently, the most important factor when releasing young lobsters is to secure a high-quality substrate where the lobster can hide, make its burrows, and find adequate food without being subject to excessive predation.

SCALING UP THE PRODUCTION OF LOBSTER JUVENILES

Since around 1985, concrete planning of a larger production facility for one-year old lobster juveniles has been made in Norway. Since large amounts of warm sea water are needed for a plant producing 1.2 million juveniles per year, the possibility of utilizing cooling water from a very large oil refinery operated by the Norwegian State Oil Company (Statoil) at Mongstad, Norway, has been investigated. A rather detailed plan has been made for an upscaled facility based on the same principles as the first Tiedemann plant. However, some new technological developments had to be made in order to increase the production by a factor of 10 without a similar increase in the demand for factory space.

The development of a total system plan for a Norwegian "lobster industry" encompassing all steps, from berried lobster females to the final marketing, has been under way during the same period. In 1990 the Norwegian government established a program for the development and encouragement of sea-ranching of a number of species including salmon, arctic char, cod, and lobsters (PUSH Program). The planning of the "lobster industry" has been supported by this Program.

THE ELEMENTS OF THE NORWEGIAN "LOBSTER INDUSTRY"

The Norwegian "lobster industry" will consist of the following individual functions, some of which are organized as separate companies:

1. Collection, purchase, and transportation of live, berried female lobsters from local fishermen.
2. Production of 1.2 million one year old lobster juveniles at "Mongstad Lobster" facility.
3. Tranportation of lobster juveniles to individual customers (local companies) along the Norwegian coast.
4. Establishment of about ten, locally owned companies (Utsira Lobster etc.) which purchase approximately 100,000 lobsters per year for release at controlled locations owned by them.
5. The development of new technological means for
   a) The computerized registration of substrate quality with resulting detailed maps of areas suitable for lobster release
   b) Computer controlled release of one-year old lobster juveniles
   c) Systematic surveying of conditions during three to four years of growth
   d) Computer controlled harvesting and registration of lobsters
   e) Efficient technological means for packing and transportation of live lobsters
6. Establishment of a marketing organization owned by the lobster producers for the international marketing of live lobsters, primarily in Europe.

One important aspect of the establishment of a Norwegian "lobster industry" is that the local companies which will purchase and release lobster juveniles must be given the sole right of recatching the lobsters when they have reached marketable size. This poses a legal problem because according to present legislation, everybody has the right to fish for lobsters anywhere during the fishing season. A commission established by the Norwegian government is presently working on new legislation in this respect and hopefully all problems will be eliminated before the first major release takes place. These are difficult and controversial problems which it has taken many years to resolve.

Figure 1. Circular pool

THE PRODUCTION FACILITY FOR ONE-YEAR OLD LOBSTER JUVENILES

The original Tiedemann plant which is still in operation (1993) consists of a 60 m diameter pool divided into eleven concentric rings (width 180 cm, water depth 24 cm) in which are floating 1,000 plastic frames consisting of 120 small separate boxes. In each box lives a juvenile lobster (Grimsen, Jaques, Erenst, and Balchen 1987) (Figure 1).

Preceding the circular pool comes a larval hatching system containing a large number of tanks in which first stage lobster larvae are kept for about one month, moulting four times until they reach the first bottom dwelling stage.

When sorting more than two million bottom stage lobster young, it is necessary to have an automatic CCD camera/image processing system which can select the acceptable individuals. During the motion of the lobster juveniles in the concentric rings, they are being fed from overhead feeders and sorted in different ways for nearly one year.

The new plant which is ten times larger than the original, differs in that it has floating stacks of ten frames each containing 144 individual boxes. Consequently, the floating units contain each 1,440 small lobsters. Also the eleven concentric rings are replaced by a spiral in which each stack is floating from the center to the largest diameter and returning to the center via a mechanical device.

After one year of growth, each lobster weighs about 10 grams. They are packed in moist and cooled boxes, where they can live a rather long time before they are released in the ocean.

SELECTING PROPER PLACES FOR RELEASE OF JUVENILE LOBSTERS

The most important factor for the success of raising lobsters to marketable size in a controlled environment is the quality of the bottom substrate. It should contain a proper mixture of sand, gravel, stones, and boulders with an abundance of seaweeds, benthos, mussels, and sea urchins. Since lobsters are very much subject to predation when moulting, it is of utmost importance that the substrate contains adequate space for hiding in burrows.

It has been observed that lobsters rarely migrate deeper than 40 m. This phenomenon is related to the lack of food at higher depths. This fact can be used as a natural fence against the migration out of the area by choosing shallow waters (an island) surrounded by waters deeper than 40 m. Many islands in Norway satisfy this criterion and are thus candidates for further substrate quality assessment.

A technical system for surveying and recording the quality of the substrate has been suggested and designed. It consists of the following functional units:

1. A surface vessel, preferably of catamaran type (10-15 m long) equipped with multiple thrusters for position control based on precise radio navigation (radar transponders). This system will locate the vessel at any point in the actual area to within an accuracy of about 2 m. Each pixel will have an area of 4 m^2. An actual area (the island of Utsira, Norway) will be 10^6 pixels or 4 km^2.
2. Behind the vessel is hauled a device extending to the bottom which consists of a telescopic tube (length up to about 60 m) having an instrument container at the end (Figure 2). This container has a number of subsystems:
   a) Remotely operated, electrical thrusters for vertical and horizontal movement of the container.
   b) Underwater color camera with pan and tilt unit and light source for inspection and operator based classification of the substrate
   c) A sampling system for collecting physical samples of the bottom substrate through the telescopic tube

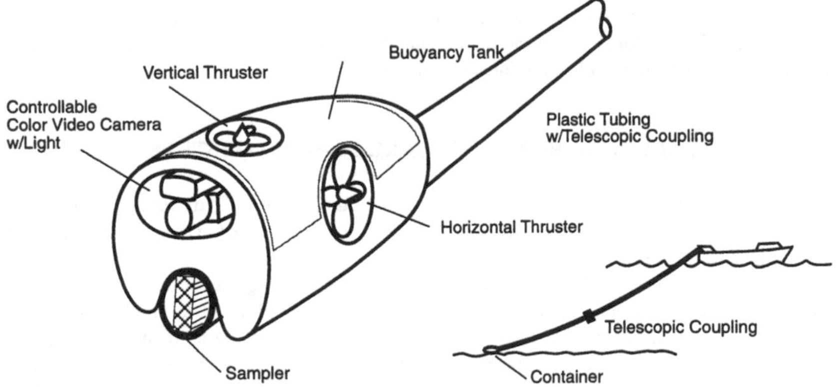

Figure 2. Instrument container hauled by surface vessel

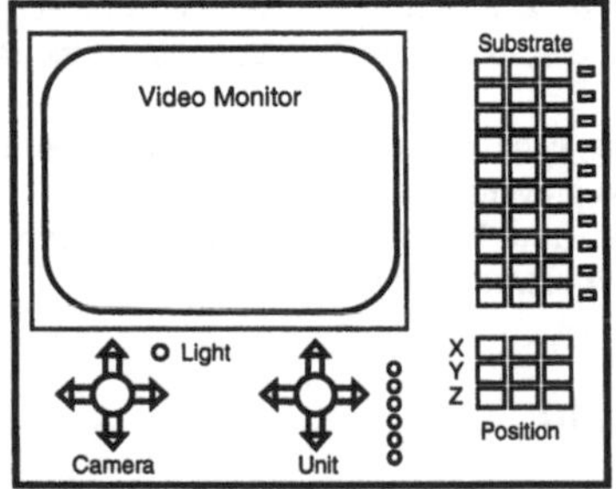

Figure 3. Operator control station

d) An operator control station in the surface vessel consisting of two joysticks for thruster control and camera control, a video monitor, a computer based keyboard for the semiautomatic classification and recording of the bottom substrate in up to twenty categories and three levels (low, medium, high)(Figure 3).

3. A computerized system for the generation of maps describing the bottom substrate in terms of quality parameters from which the acceptable location for release f lobsters can be extracted.
4. A system for computer controlled release of lobster juveniles in locations with proper quality bottom substrate under operator surveillance.

SURVEILLANCE OF RELEASE AREA DURING LOBSTER GROWTH

The system described above, consisting of the surface vessel and the instrument container, is suitable for surveillance of the released area during the growth period of three to four years because it is relatively easy to discover whether or not there is lobster activity in the area. The use of divers for this purpose, as well as in the release function, has been suggested and tried but is expected to be much too expensive for practical use. One important limiting factor in the use of divers is that they can not go deeper that about 20 m and only for very limited time.

TECHNICAL SYSTEM FOR HARVESTING MARKET SIZE LOBSTERS

After the release of one-year old lobster juveniles, it takes three to four years to reach marketable size. If more time is given, larger size will result. It becomes a matter of economic evaluation whether it pays to use longer growing time.

A number of alternative techniques have been suggested for large-scale harvesting of the lobsters which will have a density of about 1 lobster per 10 m^2. One technique is to employ ordinary lobster pots attached to a line at a distance of about 10 m and lines separated about 10 m. In order to deploy and handle such a system with a very large number of lobster pots, a surface vessel with dynamic positioning is necessary. For this purpose the vessel described above is ideally suited. The lobster pots are deployed under dynamic positioning according to the computerized map described above and the catch of lobsters can be semiautomatically recorded, yielding very important data about the catch of lobsters in relation to the release of the juveniles three to four years earlier. This information allows for the updating of the maps in order to help improve the yield over longer periods.

Since the production of marketable lobsters is expected to be very high, it is necessary to have an efficient system for harvesting as described above, in contrast to the standard method employed under normal lobster fishing.

After removal from the pots, each lobster has to be secured with rubber bands around each claw (or similar technique) in order to prevent damage to the animals.

Market-size lobsters are known to be very robust and can survive the rather long transport time from the harvesting facility to the market place.

INTERNATIONAL MARKETING

The European lobster is a very highly priced product which should be presented live to the market in the major population centers in Europe. A special distribution system should be established consisting of means for transportation, storing, and distribution of live lobsters.

CONCLUSIONS

The market in Europe for the European lobster (*Homarus Gammarus*) is increasing at the same time as the natural production is being markedly reduced. Therefore, a very strong interest in "helping" nature by sea ranching of lobsters exists in Norway. Plans have been made for a large-scale production of 1.2 million lobster juveniles per year for the controlled release in special areas under private ownership. Also publicly financed stock enhancement is possible from this source of lobster juveniles. An interesting side effect of this "lobster industry" in Norway is the manufacture of special equipment to make the surveillance of the lobster grounds and the harvest more efficient.

ACKNOWLEDGMENT

The author acknowledges the support of the Norwegian government sponsored "Program for Development and Encouragement of Sea-Ranching" - The PUSH Program.

REFERENCES

Burton, C.A. 1992. Techniques of lobster stock enhancement. *Sea Fish Report*, no. 396. UK: Sea Fish Industry Authority

Grimsen, S., R.N. Jaques, V. Erenst, and J.G. Balchen 1987. Aspects of automation in a lobster farming plant. *Modeling, Identification and Control, Norway*, 8(1): 61-68.

Karlsen, M. 1992. Catch data from Bulandet. Unpublished records in Norwegian.

van der Meeren, G.I., T. Svåsand, S. Grimsen, A. Kristiansen, and Farestveit, E. 1990. Large scale release experiment of juvenile lobsters, *Homarus Gammarus,* in Norway. ICES CM 1990/K2. 9 pp. Mimeo.

van Olst, J.C., J.M. Carlberg, and R.F. Ford 1977. A description of intensive culture systems for the American lobster *(Homarus americanus)* and other cannibalistic crustaceans. In *Proc. 8th annual meeting world aquaculture society*, 271-292. San Jose, Costa Rica.

36

Derivation of β-Carotene from Marine Organisms

Ami Ben-Amotz
National Institute of Oceanography
Israel Oceanographic and Limnological Research
Haifa, Israel

ABSTRACT

Species of *Dunaliella*, a green halotolerant microalga, thrive in media with a very broad range of salt concentrations and accumulate large amounts of the commercially valuable chemicals, glycerol and β-carotene. Glycerol serves as an intracellular osmoticum, protecting the alga under high, extracellular salt concentrations, and β-carotene shields the cell against damage by high light intensity.

Dunaliella is cultivated in coastal ocean areas in large outdoor ponds in regions of high solar irradiation and moderate temperatures. The algae are harvested to produce high β-carotene, dry algal powder for use by the health food industry and for food coloring. The algal β-carotene differs from the available synthetic all-trans β-carotene in its stereoisomeric composition, providing mono-cis and di-cis stereoisomers of β-carotene of unique properties for use as a pharmaceutical drug.

DUNALIELLA AND THE MARINE ENVIRONMENT

The chlorophyte *Dunaliella* is classified under the order Volvocales, which includes a variety of ill-defined, unicellular species. Members of the genus *Dunaliella* are all motile, ovoid bioflagellates with a cell volume ranging from 50 to 1,000 m3 (Butcher 1959). The alga contains one large chloroplast with a single pyrenoid and many starch granules in the basal portion of the chloroplast. *Dunaliella* lacks a rigid cell wall, and the cell is enclosed by a thin, elastic plasma membrane covered by a mucus "surface coat". The lack of a rigid cell wall permits rapid cell-volume responses to the extracellular osmotic changes.

Dunaliella predominates in many aquatic marine habitats and in salt water bodies which contain more than 10% salt. Typical examples are the Dead Sea in Israel, the Pink Lake in Australia, and the Great Salt Lake in the United States. *Dunaliella* is probably the most halotolerant eukaryotic organism known, showing a remarkable adaptation to a variety of salt concentrations from as low as 0.2% to salt saturation of about 35% (Borowitzka 1981).

Dunaliella osmoregulates by varying the intracellular concentration of the photosynthetic glycerol in response to the extracellular osmotic pressure. On growth in media containing different salt concentrations, the intracellular glycerol concentration is directly proportional to the extracellular salt concentration and maintains the cell water volume and the required cellular osmotic pressure. Lacking a rigid cell wall, *Dunaliella* shrinks or swells rapidly when exposed to hypertonic and hypotonic conditions, respectively. Within minutes of the osmotic transition, the cell responds by synthesis or elimination of glycerol in a time kinetic process which ceases when the cell volume returns to its original extent. The mechanism of controlling the synthesis and elimination of glycerol in *Dunaliella* is believed to involve a unique "glycerol cycle" activated by a few novel enzymes (Ben-Amotz and Avron 1981, 1990; Borowitzka and Borowitzka 1988).

β-CAROTENE BIOSYNTHESIS

Dunaliella is the most enriched β-carotene, eukaryotic organism known (Ben-Amotz and Avron 1990; Borowitzka and Borowitzka 1988). Under appropriate cultivation, more than 10% of the dry weight of *Dunaliella* is β-carotene. The accumulation of β-carotene is species-specific and is related physiologically to high light intensity, nitrate limitation, and to most environmental stress conditions such as high salt, high and low temperatures, nutrient deficiencies, etc. β-carotene in *Dunaliella* accumulates within oily globules in the interthylakoid spaces of the chloroplast and is composed mainly of two stereoisomers: 9-cis and all-trans. Both the total amount of the accumulated β-carotene and the 9-cis to all-trans ratio depend on the light absorbed by the cell during one division cycle - the higher the light intensity and the lower the growth rate of the alga, the higher the cellular β-carotene content and the 9-cis to all-trans ratio. Recent studies showed that the isomerization reaction which eventually causes the accumulation of 9-cis β-carotene occurs early in the pathway of carotene synthesis and allows the production of the different β-carotene stereoisomeric intermediates to an equivalent level of the all-trans β-carotene (Shaish, Avron, and Ben-Amotz 1990). The physicochemical properties of 9-cis β-carotene differ from those of all-trans β-carotene: all-trans β-carotene is practically insoluble in oil and is easily crystallized, while 9-cis β-carotene is much more soluble in hydrophobic solvents, very difficult to crystallize, and generally an oil in its concentrated form. The high proportion of 9-cis β-carotene probably accounts for the very high concentration of soluble β-carotene within the lipoidal chloroplastic globule of *Dunaliella*.

The most probable function of the β-carotene globules in *Dunaliella* is to protect the cell against the high-intensity irradiation to which it is exposed in the natural habitat by absorbing energy in the blue region of the spectrum. Strains unable to accumulate β-carotene die when exposed to high irradiation, while the β-carotene-rich *Dunaliella* strains flourish (Shaish, Ben-Amotz, and Avron 1991).

BIOTECHNOLOGY OF DUNALIELLA

Dunaliella is highly suitable for biotechnology. The ability of the halotolerant alga to thrive in media with high salt concentrations allows outdoor cultivation in relatively pure cultures with a low presence of predators and grazers. In addition, its high β-carotene content protects it from the intense solar irradiation in the areas where such cultivation is practical, i.e., arid or desert areas with access to brackish water or seawater. *Dunaliella* is mass cultivated in autotrophic medium at the expense of inorganic nutrients in marine medium. Mass production of *Dunaliella* can be activated by techniques resembling those commonly used for the large-scale production of other algae. However, the commercial production of *Dunaliella* in outdoor cultivation is aimed at maximizing β-carotene production rather than biomass as in *Chlorella* or *Spirulina* culture. As indicated above, β-carotene accumulation in *Dunaliella* increases with increasing light intensity and slower growth rate. However, under these conditions, productivity is rather low, due to the slow rate of growth. Thus, in practice, maximal productivity of β-carotene is attained in areas having high light intensity where the growth rate and the nitrogen supply are controlled. The growth rate of the alga is limited synergistically by the salt concentration in the medium and by controlling the medium content of an essential nutrient such as nitrate or sulfate. Theoretical maximal productivity calculations for natural, outdoor conditions set an upper limit to the light conversion efficiency of approximately 3%, taking into consideration the photosynthetic machinery, solar irradiation, nutrient supply, temperature, and the design of the bioreactor (Avron 1989). A light conversion efficiency of 3% is equivalent to 25 g biomass m-2 day-1. However, intensive culturing attempts to produce high β-carotene *Dunaliella* containing about 30 pg β-carotene per cell, with a β-carotene to chlorophyll ratio exceeding 7, limit the maximal algal outdoor productivity to 5-10 g *Dunaliella* m-2 day-1 of dry biomass containing 4% β-carotene. A wide range of *Dunaliella* biotechnologically commercial approaches, from intensive small-bore plastic tubes through open raceways to extensive, unmixed, large-body, open ponds, allows the production of the algae at different categories of economics.

β-carotene is currently the most important product of *Dunaliella* cultivation. It is used as a food-coloring agent and as provitamin A in human food and animal feed. It is added to different commercial products such as cosmetics, vitamin preparations, and pharmaceuticals. The worldwide market is dominated by synthetic, all-trans β-carotene, a product of complex chemical synthesis developed by researchers at Hoffmann La Roche around 1950 (Isler 1971). Epidemiological and oncological studies suggest that normal to high levels of β-carotene in the body may protect it against cancer (Krinsky 1989a). It has been suggested that the antioxidant function and the ability to quench various radical species may explain the effect of β-carotene as a chemopreventive agent (Krinsky 1989b). Therefore, humans and animals fed on a diet high in carotenoid-rich vegetables and fruits and who maintain higher than average levels of serum β-carotene have a lower incidence of several types of cancer.

The interest in a natural source of β-carotene is increasing with the build-up of information relating carotenoids to preventive medicine. Of special interest in this regard is the observation that natural β-carotene as found in *Dunaliella* and in most fruits and vegetables contains a mixture of stereoisomers of β-carotene of much higher fat solubility and lower crystallization property than the fat-insoluble, crystallizable, all-trans β-carotene. The antioxygenic requirement for higher, absorbed β-carotene in disease prevention may create a new, expanded market for the natural β-carotene stereoisomeric mixture. Most products of *Dunaliella* in the commercial market are extracts of β-carotene in edible oil containing between 1.5 and 30% *Dunaliella* β-carotene. The others are dried *Dunaliella* powder in capsules or tablets of about 5% β-carotene, used predominantly in the health food market under the label of natural β-carotene. The increasing production of *Dunaliella* and its introduction into margarine and other food products as a natural food coloring may gradually replace the synthetic all-trans β-carotene in "all-natural" food items. The natural β-carotene sold in 1992 for approximately 2,000 U.S.$ kg-1 at a total world market of around 15 tons.

No other products of *Dunaliella* have reached the market stage based on feasible economics. Nevertheless, the production of *Dunaliella* biomass for β-carotene allows the extraction and purification of many other biochemical agents, including glycerol, enzymes, vitamins, amino acids, fatty acids, and growth regulators.

REFERENCES

Avron, M. 1989. The efficiency of biosolar energy conversion by aquatic photosynthetic organisms. In *Microbial mats*, ed. Y. Cohen and E. Rosenberg, 385-7. Baltimore: ASM Press.

Ben-Amotz, A., and M. Avron. 1981. Glycerol and β-carotene metabolism in the halotolerant alga *Dunaliella*: A model system for biosolar energy conversion. *Trends Biochem. Sci.* 6: 297-9.

Ben-Amotz, A., and M. Avron. 1990. The biotechnology of cultivating the halotolerant alga *Dunaliella*. *Trends Biotechnol.* 8: 121-6.

Borowitzka, L. J. 1981. The microflora: Adaptation to life in extremely saline lakes. *Hydrobiologia* 81: 33-46.

Borowitzka, L. J., and M. A. Borowitzka. 1988. *Dunaliella*. In *Microalgal biotechnology*, ed. M. A. Borowitzka and L. J. Borowitzka, 27-58. Cambridge: Cambridge University Press.

Butcher, R. W. 1959. An introductory account of the smaller algae of British coastal waters. I. Introduction and Chlorophyceae. *Fish. Invest. Ser.* 31: 175-91.

Isler, O., ed. 1971. *Carotenoids*. Basel: Birkhauser.

Krinsky, N. I. 1989a. Antioxidants fractions of carotenoids. *Free Radical Biol. Med.* 7: 617-35.

Krinsky, N. I. 1989b. Carotenoids as chemopreventive agents. *Preventive Med.* 18: 592-602.

Shaish, A., M. Avron, and A. Ben-Amotz. 1990. Effect of inhibitors on the formation of stereoisomers in the biosynthesis of β-carotene in *Dunaliella bardawil*. *Plant Cell Physiol.* 31: 689-96.

Shaish, A., A. Ben-Amotz, and M. Avron. 1991. Production and selection of high β-carotene mutants of *Dunaliella bardawil*. *J. Phycol.* 27: 652-6.

37

Potential Developments in Marine Biotechnology

Dr. Rita R. Colwell
President
University of Maryland Biotechnology Institute, College Park, MD

Dr. Russell T. Hill
Research Assistant Professor
Center of Marine Biotechnology, Baltimore, MD

MARINE AQUACULTURE AND BIOTECHNOLOGY

The farming of marine finfish, shellfish, crustaceans, and seaweed is growing in economic importance worldwide, although marine aquaculture in the United States remains limited in scope. Total world fish production by aquaculture is ca. 14 million metric tons (mmt) of which only 0.3 mmt is produced in the United States, with nearly three-quarters comprised of freshwater organisms. Marine aquaculture in the United States is dominated by oyster culture, followed in order of importance by clams, mussels, salmon, and shrimp. The U.S. trade deficit in edible fish products was $3.2 billion in 1989 ($5.5 billion if nonfood fishery products are included [National Research Council 1992]). Expansion of marine aquaculture in the United States could contribute greatly to reducing this deficit and meeting the growing demand for fresh seafood. In addition, aquaculture of seaweed and phytoplankton may yield several high-value products which would reduce the component of this deficit-comprising import of algal-derived polysaccharides and chemicals.

Many of the problems that must be resolved for the marine aquaculture industry to grow substantially in the United States may be only peripherially influenced by marine biotechnology. Husbandry of important species can be advanced by further research, and species can be improved by conventional breeding techniques. A regulatory and funding framework that encourages the industry's growth and facilitates the resolution of a number of policy issues at the federal level is important (National Research Council 1992). There are several areas, however, where marine biotechnology is a vital component in the success of marine aquaculture. Great strides have been made in the genetic manipulation of marine organisms and in the understanding of molecular and biochemical processes that influence successful exploitation of these organisms. Advances in research on the hormonal control of reproduction of finfish and shellfish will be increasingly important in fish farming. Marine biotechnology will also play an essential role in the treatment of disease in organisms grown in aquaculture facilities. Aquaculture facilities can have negative environmental impacts, which may be alleviated by biotechnology.

Many of the contributions of biotechnology to aquaculture will benefit freshwater and marine aquaculture. In some cases, this is a rather artificial distinction, since organisms may have both freshwater and marine stages in their life cycles. In general, the following discussion applies to marine organisms, although some fundamental advances are mentioned that have been made in freshwater organisms but which have relevance to marine aquaculture.

GENETIC MANIPULATION OF MARINE ORGANISMS

Selective breeding of fish is a well-established technique that has an important role to play in the aquaculture industry (Hershberger 1990). Direct genetic manipulation of fish by using recombinant DNA techniques has revolutionized this conventional approach in less than a decade. Fish are particularly amenable to genetic manipulation because they characteristically have large eggs which can be externally fertilized through microinjection of DNA construct. Major research efforts in genetic manipulation have been directed at growth enhancement and development of fish with superior cold resistance. Development of disease-resistant fish is an increasingly important objective.

Growth enhancement by foreign growth hormone (GH) genes was first demonstrated in 1982 in mice that developed from eggs microinjected with rat GH gene constructs (Palmiter et al. 1982). Growth enhancement of fish species that are important in aquaculture has since been attempted. The first successful growth hormone transfer into fish was that of the transfer into goldfish of the human GH gene fused to a mouse metallothionein promoter (Zhu et al. 1985) resulting in offspring of these transgenic fish that were considerably larger than untreated fish (Zhu 1992). Since then, several vertebrate genes fused to a variety of promoters have been introduced into fish species. Examples include the expression of a mouse metallothionein human growth hormone fusion gene in Atlantic salmon (Rokkones et al. 1989) and bovine growth hormone (expressed from the Rous sarcoma virus long terminal repeat and carp ß-actin gene promoters) in walleye (Moav et al. 1992).

Growth enhancement has also been achieved by using fish GH. Dr. Chen, at the Center of Marine Biotechnology, Maryland Biotechnology Institute and Dr. Powers, at the Johns Hopkins University, demonstrated that growth hormone in rainbow trout is encoded by two separate genes (Agellon et al. 1988a). A large amount of biologically active rainbow trout GH was prepared by expressing one of these genes in the bacterium, *Escherichia coli*. The administration of this hormone to the rainbow trout by injection or dipping resulted in enhanced growth (Agellon et al. 1988b). Exogenous GH application may not be practical for large-scale aquaculture requirements, but this necessity can, of course, be circumvented by generation of transgenic fish. The GH gene, fused to the long terminal repeat of Rous sarcoma virus was transferred to common carp and channel catfish (Chen et al. 1992).

Growth enhancement of fish has also been accomplished by using "all-fish" gene constructs. Dr. Hackett and colleagues developed expression vectors which contained the proximal promoter and enhancer regulatory elements of the carp-ß actin gene and the polyadenylation signal from the salmon growth hormone gene (Liu et al. 1990). Growth enhancement has been obtained in transgenic Atlantic salmon by using an all-fish gene construct comprised of an antifreeze protein gene promoter linked to a chinook salmon GH gene (Du et al. 1992).

The expression and inheritance of GH genes in transgenic fish is complex. Important considerations are tissue specificity and developmental stage specificity of transcriptional control elements (Moav et al. 1992), as well as stable inheritance and expression of GH genes by offspring. Use of commercial strains of fast-growing transgenic fish in aquaculture must await further research progress. Areas which require further research include determination of physiological, nutritional, and environmental factors that will maximize performance of transgenic individuals and the resolution of safety and environmental impact issues (Chen and Powers 1990).

Genetic manipulation of Atlantic salmon has been done to attempt to increase the cold resistance of this species. Many marine fish inhabiting cold waters produce proteins which act as antifreezes, protecting them by inhibiting the formation of ice crystals in their serum. These proteins are termed antifreeze glycoproteins and antifreeze polyproteins/polypeptides. Atlantic salmon lack genes coding for these proteins and therefore cannot survive in icy waters. The mechanism whereby antifreeze proteins bind ice crystals and inhibit ice formation has been described (Raymond et al. 1989). Genes coding for antifreeze proteins has been transferred, expressed, and inherited in Atlantic salmon. Expression of adequate levels of antifreeze proteins in salmon would enable Atlantic salmon to be grown in aquaculture facilities in Canada and would extend the potential natural range of this fish. Antifreeze proteins from fish may also prove to be valuable for the hypothermic preservation of mammalian organs (Lee et al. 1992).

Genetic manipulation may be useful in enhancing growth of shellfish. GH has been shown to enhance growth rates of California red abalone (Morse 1984), and similar results have been reported for exogenous application of GH to juvenile oysters (Paynter and Chen, unpublished results, cited in [Chen et al. 1992]).

MARINE NATURAL PRODUCTS

Marine organisms are sources of a wide range of natural products that have biomedical, biotechnological, agricultural, and industrial applications. Natural products produced by marine organisms include primary and secondary products as well as enzymes and waste products such as chitin. Marine biotechnology has an essential role to play in the discovery and large-scale production of marine natural products, and genetic manipulation will become increasingly important in this regard. Ironically, enzymes resulting from marine biotechnology have made possible some important new techniques in biotechnology.

Bioactive Compounds From Marine Invertebrates

The marine environment is characterized by physical and chemical properties that are markedly different from those of the terrestrial environment, and these properties have produced complex ecosystems which include many sessile organisms. The production of bioactive chemicals is a common means of defense, especially in sessile organisms and in vulnerable, soft-bodied organisms. Groups of organisms that have been found to produce bioactive natural products include marine bacteria; dinoflagellates; algae; coelenterates (in particular corals); echinoderms such as sea cucumbers and starfish; bryozoans; sponges; soft-bodied molluscs such as sea hares and nudibranchs; and tunicates. The chemical basis of some marine ecological phenomena has been discussed by Scheuer (Scheuer 1990), who has also edited two comprehensive review volumes on organic chemicals of biological marine origin (Scheuer 1987; Scheuer 1988). Bioactive substances from marine organisms have been studied for several decades and thousands of these chemicals have been described. Recent discoveries of marine natural products with interesting biological and pharmaceutical properties have been the subject of a series of comprehensive reviews by Faulkner (1984, 1986, 1987, 1988, 1990), the most recent of which contains 438 references. The majority refer to marine natural products of biological origin described in the previous year. While it is beyond our present scope to review the huge range of natural products that have been discovered in marine organisms, it is useful to list some of the major laboratories in the United States that are involved in this important aspect of marine biotechnology. Examples of significant products that have emerged from this research are given, and approaches that are likely to yield significant results in the future are outlined.

Scripps Institution of Oceanography, La Jolla, California, has played a leading role in the isolation of bioactive substances from marine organisms. Work in Faulkner's laboratory at this institution has resulted in the discovery of numerous bioactive compounds from a wide range of organisms including sponges (James et al. 1991; Kushlan and Faulkner 1991; Stierle and Faulkner 1991) and algae (Trimurtulu et al. 1992). By adopting a similar approach of screening organisms that on the basis of field observations, appear to be using chemical defenses against predation, Fenical and co-workers have isolated a number of bioactive compounds, including anti-inflammatories and antiviral agents from corals (Groweiss et al. 1988; Roussis et al. 1990) and antifungal disulphides from ascidians (Lindquist and Fenical 1990).

Sponges have been a leading source of bioactive compounds. Dysinin-type sesquiterpenes with antihelmintic activity (Horton et al. 1990) and several cytotoxic heterocycles (Quinoa et al. 1986) are among the compounds isolated by researchers at the University of California, Santa Cruz. Researchers in Scheuer's laboratory at the University of Hawaii in Manoa, have isolated a number of cytotoxic compounds from marine sponges (Akee et al. 1990). The Harbor Branch Oceanographic Institution, Ft. Pierce, Florida, has isolated many bioactive compounds from sponges, including an antitumor compound (Sakai et al. 1986) and many cytotoxic and antifungal substances (Gunasekera et al. 1990; Gunasekera et al. 1990).

The wide range of bioactive compounds produced by marine microorganisms emphasizes the great potential of these compounds in biomedical applications, and this has encouraged further large-scale, systematic screening of marine organisms. For example, the US National Cancer Institute (NCI) has established a screening system consisting of 60 *in vitro* cell lines representing seven cancer sites: blood cells, brain, colon, kidney, lung, ovary, and skin (Ansley 1990). Extracts from many marine organisms are tested for their cytotoxic activity. Additional tests are performed to detect anti-HIV activity using a human lymphoblastic cell line infected with the AIDS virus. Some pharmaceutical companies screen marine isolates for anti-inflammatory, insecticidal, and herbicidal activities, in addition to cytotoxic and antiviral screening (Cardellina 1986).

Bioactive Compounds From Marine Algae

Marine alga have been one subject of investigation by Gerwick's group at Oregon State University, Corvallis, Oregon. A mammalian Insulin release modulator (Moghaddam et al. 1990) and a potent mammalian immunohormone (Bernart and Gerwick 1988) are among the biomedically important compounds isolated from these algae. Red marine algae were shown to be a rich source of eicosanoid-type biochemicals (Gerwick et al. 1990), and numerous other compounds have been isolated from algae and cyanobacteria by this group. The Harbor Branch Oceanographic Institution, Ft. Pierce, Florida, has isolated a compound showing antiviral activity from the marine alga, *Halimeda tuna* (Koehn et al. 1991).

Bioactive Compounds From Marine Bacteria

Marine bacteria have great potential in the production of bioactive compounds and pharmaceuticals, but have not been intensively investigated. It is possible that many of the compounds isolated from marine organisms such as sponges, are produced by bacteria associated with those sponges. For example, several diketopiperazines previously ascribed to the sponge, *Tedania ignis,* have been shown to be produced by a marine *Micrococcus* sp. associated with this sponge (Stierle et al. 1988). Isolation of marine bacteria, particularly from sponges that can be screened for production of bioactive substances, is in progress at the Center of Marine Biotechnology, Maryland Biotechnology Institute. There are several practical advantages in the use of bacteria as sources of natural products. Many bacteria can be readily grown in fermenters under controlled conditions for consistent production of compounds of interest. Marine invertebrates, on the other hand, need to be collected from natural ecosystems where they may be inaccessible or present only in low numbers. Large-scale collection for natural product production may, in such cases, threaten endangered populations. Alternatively, specialized conditions need to be established to grow invertebrates in captivity. Production of compounds by bacteria can be enhanced, often by several orders of magnitude, by optimization of fermentation conditions and by selection of high-producing mutants. In addition, genetic manipulation of bacteria is relatively easy, and genes of interest can be cloned into expression vectors and transferred by fermentation processes to bacterial species that are well-adapted for production of compounds. For these reasons, it is advantageous to investigate whether natural products ascribed to marine invertebrates may, in fact, be produced by bacteria associated with those invertebrates and to specifically investigate marine bacteria (in particular, those bacteria closely associated with invertebrates) as potential sources of natural products. In some cases, bacteria may be present in extracellular associations and readily cultured. Such is the case of those found with isolates from the Caribbean sclerosponge, *Ceratoporella nicholsoni* (Santavy et al. 1990), some of which produced compounds with antibacterial and antineoplastic activity (Colwell et al. 1989). However, there is a range of interdependence between bacterium and host, and there is evidence that some bacterial-invertebrate symbioses may date from the Precambrian era (Wilkinson 1984). In cases where true symbiotic relationships exist between host and bacterium, it is likely to be extremely difficult, if not impossible, to isolate and maintain the bacterium in pure culture. Molecular approaches may be useful in these cases. Total DNA can be extracted from the invertebrate (and its resident microbial population) and a composite gene library of the total DNA can be produced in an appropriate host. The genes coding for useful products can be isolated from the composite library by using appropriate screening procedures. This "genetic fishing" procedure obviates the need for prior determination of the cellular source of the products of interest and has the potential of facilitating rapid, large-scale production of marine natural products.

Actinomycetes are a group of bacteria which have been little investigated in the marine environment, but which are known to be metabolically versatile and to produce many bioactive compounds, particularly antibiotics. Terrestrial actinomycetes produce over two-thirds of naturally occurring antibiotics, including many of medical importance (Okami and Hotta 1988). Although not commonly regarded as an important group in marine environments which are typically dominated by gram-negative bacteria, there are several recent reports which indicate that marine environments may be an important new source of actinomycetes. A coryneform- or actinomycete-like group was reported in the Caribbean sclerosponge, *Ceratoporella nicholsoni* (Santavy et al. 1990). Scheuer (1990) raises the interesting possibility that a series of isoquinolenequinones, among them the antibiotic mimosamycin, isolated during an investigation of a nudibranch predator and its sponge prey and found independently in a sponge, *Reniera* sp., might be produced by a *Streptomyces* sp. Jensen et al. (1991) reported that actinomycetes are widespread in tropical marine sediments. Work at the Center of Marine Biotechnology, Maryland Biotechnology Institute has resulted in an efficient method for isolation of actinomycetes from marine samples and demonstrated that a wide range of unusual actinomycetes, different from those typically found in terrestrial samples, are present in sediments from the Chesapeake Bay (Takizawa et al. 1993). Intensive isolation and screening of actinomycetes from marine environments is warranted in view of the enormous range of important compounds that have been isolated from terrestrial actinomycetes in previous decades. Actinomycetes isolated from marine environments are likely to prove equally valuable.

Marine Toxins

Marine toxins are defined here as marine natural products that have specific pharmacological activities that produce adverse effects in animals, generally at very low concentrations. Many marine toxins are produced by dinoflagellates and may be retained or concentrated through several trophic levels before exerting adverse effects on predators higher in the food chain (including man). Examples of toxins from dinoflagellates which may result in fatal poisoning in man are ciguatoxin and saxotoxin. An important toxin which is found in many marine animals is the potent sodium channel blocker, tetrodotoxin, also known as puffer fish toxin. This toxin has been found in a wide range of marine bacteria (Simidu et al. 1988), suggesting that the presence of this neurotoxin in many distantly related animal genera may indicate production of the toxin by bacteria associated with these animals. The techniques of marine biotechnology will be useful in understanding toxin production and in devising methods for detection of toxins, such as the method recently described for the detection of tetrodotoxin (Raybould et al. 1992).

Toxins are of interest in the context of marine natural products because they may have useful medical applications given the appropriate dosages and delivery systems and also have application as research tools, particularly in studies on neuromuscular systems. It has been speculated that toxins from a single genus of predatory cone snails may prove to have pharmaceutical potential comparable to that of plant alkaloids or the fermentation products of microorganisms. *Conus* spp. (approximately 500 in number) produce a vast range of pharmacologically active small peptides, the targets of which include calcium channels, sodium channels, *N*-methyl-D-aspartate receptors, acetylcholine receptors, and vasopressin receptors (Olivera et al. 1990). The study of marine toxins promises to be a particularly productive area of marine biotechnology.

Enzymes

Enzymes from marine bacteria are of interest because they are likely to be salt-resistant, a characteristic often advantageous in industrial processes. Proteases are of particular importance (Kalisz 1988) and have application in detergents and as components of membrane-cleaning formulations (Marshall et al. 1991). *Vibrio* spp. have been found to produce many proteases. The marine bacterium *Vibrio alginolyticus*, in addition to several proteases, produces collagenase, an enzyme with a variety of industrial and commercial applications.

A particularly important group of marine organisms for enzyme production are the hyperthermophilic archaea that have been isolated from hot water seeps and hydrothermal vents. The Archaea form one of the three domains of organisms defined by Woese et al. (1990). The other two domains, Bacteria and Eucarya, are typically found in extreme environments. Hyperthermophilic archaea grow at temperatures over 100°C and therefore require enzyme systems that are stable at high temperatures. Thermostable enzymes may have advantages in industrial processes. In addition, thermostable DNA-modifying enzymes such as polymerases, ligases, and restriction endonucleases have important applications in molecular biology. The use of thermostable DNA polmerases in the polymerase chain reaction (PCR) (Saiki et al. 1988), which is a powerful technique that selectively amplifies a specific DNA sequence of interest from complex mixtures of nucleic acids, has been instrumental in the rapid increase in use of this powerful technique. A thermostable DNA polymerase from the hyperthermophilic archaea, *Pyrococcus furiosus,* has both polymerase and proofreading capabilities, giving the advantage of high fidelity PCR products (Lundberg et al. 1991). A recent development is the emergence of the ligase chain reaction (LCR) as a new technique for the detection of mutations in DNA (Murray 1989). The ligase chain reaction uses a thermostable DNA ligase to detect, amplify, and distinguish specific DNA sequences. Thermostable DNA ligases are therefore important in this regard.

Metabolic enzymes isolated from thermophilic bacteria and archaea are typically markedly thermostable. Glutamate dehydrogenase, a key enzyme in nitrogen metabolism, isolated from *P. furiosus*, has an optimal temperature for enzyme activity of 95°C and a half-life of over 3.5 hours at 100°C (Klump et al. 1992). Glutamate dehydrogenase and glutamine synthetase (also important in nitrogen metabolism) from *P. furiosus* represent the most thermostable versions of these enzymes described to date (Robb et al. 1992). Similarly, the majority of enzymes involved in the primary metabolic pathways of *P. furiosus* and the most thermophilic bacterium, *Thermotoga maritima*, are dramatically more thermostable than their counterparts from mesophilic organisms (Adams et al. 1992). Further study of enzymes from these thermophilic marine microorganisms will contribute to understanding the mechanisms of enzyme thermostability and may enable directed modification of industrially important enzymes to enhance their thermostability. An alternative approach is the isolation from these microorganisms of enzymes suitable for particular industrial applications.

Abundant Marine Natural Products

There are several marine products that are available in very large quantities which may increase in importance as valuable resources through the application of techniques of marine biotechnology. Three of these—chitin, polysaccharides, and kelp— will be considered here.

Millions of pounds of chitin are generated annually as waste products by the seafood industry. This abundant biopolymer and its by-products, chitosan and N-acetylglucosamine, may be useful in several industrial and medical applications. Current mechanical and chemical methods for isolation of chitin from shellfish wastes are relatively harsh and inefficient, and enzymatic methods may be preferable in the purification of chitin and production of chitin by-products. Many bacteria, including marine bacteria of the genera *Vibrio* (Wortman et al. 1986) and *Streptomyces* (Pisano et al. 1992), produce chitinases which may be useful in this regard. Cloning of chitinase enzymes from marine bacteria (Wortman et al. 1986) is one approach toward developing enzymatic systems for the production of useful chitin derivatives. Applications for these products include their use as paper additives (Muzzarelli 1986), pharmaceuticals (Nagai et al. 1984), and absorbable sutures (Nakajima et al. 1984). The addition of chitin to soils can have an inhibitory effect on plant pathogenic fungi (Mitchell and Alexander 1962), so the use of chitin compounds could therefore be beneficial in agriculture.

Many marine bacteria produce abundant polysaccharides which may have commercial application as viscosity-increasing agents, gelling agents, or adhesives (Colwell et al. 1985). Marine polysaccharides and their importance in biotechnology were reviewed by Colwell et al. (1985)., as were the many commercial applications of marine algae and seaweeds (Section XX-Aquaculture). One aspect more appropriately discussed here is the use of kelp as a source of methane production. The Giant Pacific Kelp, *Macrocystis pyrifera*, is particularly suitable as a substrate in view of its high growth rate, high ratio of readily degradable organic compounds, and the soluble nature of its major organic constituents. A marine methanogenic consortium of bacteria capable of methane production from kelp has been characterized (Sowers and Ferry 1984). Methane production from kelp is a potential, renewable energy source.

Biofilms and Bioadhesion in the Marine Environment

The importance of biofilms in the marine environment is widely recognized. Biofilms rapidly form on clean surfaces in seawater. Composed initially of organic molecules and bacteria, the films are subsequently bolstered by other microorganisms and larger animals such as oysters and barnacles. The study of biofilms is aimed at devising strategies to control biofouling of surfaces, understanding the mechanisms of bioadhesion, and investigating ecological relationships within biofilms and between biofilms and organisms in the surrounding water.

The formation of biofilms on ships leads to biofouling, greatly reducing fuel efficiency. It has been estimated that a 200 μm-thick layer on a ship's hull can decrease speed by 20% (Curtin 1985). Application of antifouling paint to the hulls of ships decreases biofilm formation, but these paints frequently contain toxic compounds such as heavy metals that pollute the marine environment. Research on factors controlling formation of the initial film of marine bacteria on surfaces may lead to the development of nontoxic methods to prevent biofouling. Possibilities include use of alternative materials to inhibit biofilm formation and "biological control" of biofouling using thin, bacterial biofilms which resist further colonization.

Biofilms have been implicated in corrosion processes and control of biofilms may therefore be important in preventing metal corrosion in seawater. Microorganisms may be involved in hydrogen embrittlement which results in loss of ductility and tensile strength of susceptible metals. The generation of sufficient hydrogen to affect susceptible metals has been demonstrated with pure-culture bacterial films (Walsh et al. 1989), although the importance of this process with complex biofilms in natural marine environments is unclear. Bacterial biofilms have been implicated as factors in corrosion of copper and nickel in seawater (Little et al. 1990). Corrosion of bare steel in seawater was increased twofold by a mixed culture of marine bacteria, but interestingly, was decreased sevenfold by a different mixed culture (Walsh and Jones 1990). Interactions between biofilms, metals, and protective coatings applied to metals are complex. Increased understanding of these interactions may result in improved methods to prevent corrosion.

Biofilms play an important ecological role in the settlement and metamorphosis of marine invertebrates. This interaction has been studied in detail with larvae of the commercially important, U.S. East Coast oyster, *Crassostrea virginica* and the Pacific oyster, *Crassostrea gigas*, at the Center of Marine Biotechnology, Maryland Biotechnology Institute. Biofilms of the marine bacterium, *Shewanella colwelliana,* were shown to be beneficial to "set" of oyster larvae [Weiner et al. 1989]. "Set" is a general term which refers to both settlement and metamorphosis of larvae, although these processes may be triggered by different cues. Settlement behavior appears to be initiated by L-DOPA and may also be triggered by bacterially produced ammonia (Bonar et al. 1990). *S. colwelliana* synthesizes two tyrosinase enzymes, MelA and MelB. MelB is a conventional tyrosinase which catalyzes the hydroxylation of tyrosine to L-DOPA, which appears to be important in the induction of settlement behavior by *S. colwelliana* biofilms. The MelA gene has been sequenced (Weiner et al. 1991), and codes for a unique enzyme which mediates melanogenesis and which may provide a positive cue influencing larvae to permanently cement to a surface. An acidic exopolysaccharide which is produced by *S. colwelliana* also appears to be an important cue for oyster set (Weiner et al. 1991). This detailed elucidation of factors which are important in oyster set, may have application in enhancing the same process in natural waters and in aquaculture facilities by provision of surfaces which enhance oyster set.

Bioadhesion of bacteria, other microorganisms, and invertebrates is characteristic of biofilm formation. Mechanisms of bioadhesion have been investigated and may give valuable clues for methods of production of water-resistant adhesives. Work at the University of Maryland on the marine bacterium, *Alteromonas colwelliensis,* has indicated that an exopolymer designated "polysaccharide adhesive viscous exopolymer" (PAVE), produced by this bacterium, has great potential as an adhesive (Abu et al. 1991). The adhesive strategies of marine mussels have been explored by Waite and co-workers at the University of Delaware. Formation of byssal threads, whereby mussels attach to solid surfaces, is an exceedingly complex process (Waite 1983). Threads comprise a core of collagen and elastin surrounded by a tough, durable varnish derived from a polyphenolic protein and catecholoxidase. The polyphenolic protein has a highly repetitive structure, rich in L-DOPA (Waite 1991). Analogues of this protein have been produced by recombinant DNA technology (Filpula et al. 1990) and may be useful as protein-based, medical adhesives [Strausberg and Link 1990]. Cell-Tak™, a crude preparation of the polyphenolic protein extracted directly from mussels, has been used to enhance attachment of cells and tissues in culture (Notter 1988).

Although some progress has been made in the study of certain aspects of biofilms and bioadhesion, many basic questions remain about natural biofilms. Some of the complex factors involved in the interactions between bacteria and surfaces were reviewed by Fletcher (1987, 1990). Interesting scientific questions pertaining to biofilms were listed by Walsh et al. (1989), and include what organisms are present in natural biofilms, their metabolic and genetic interactions, and the role of extracellular polymers in biofilm structure. Advances in the use of species specific DNA probes will be important in the investigation of the community structure of biofilms. The application of novel microscopic techniques to study bacterial adhesion to surfaces (Fletcher 1987), as well as the application of confocal scanning laser microscopy and environmental scanning electron microscopy to examine intact biofilms, are providing new insights. Addressing these questions requires a multidisciplinary approach, and this basic research is likely to produce many benefits in marine biotechnology.

BIOREMEDIATION

Bioremediation of pollutants in the marine environment is a field of marine biotechnology that is still in its infancy, but one that has great potential in dealing with pollutants that may be extremely difficult or impossible to remove by other approaches. The aspect of bioremediation in the marine environment that has received most attention is degradation of hydrocarbons, in particular petroleum products, which enter the marine environment from oil drilling, loading of tankers, catastrophic oil spills, and by natural seepage from oil-bearing sediments. Environmental parameters which affect microbial degradation of hydrocarbons in the environment and metabolic and genetic factors that are important in this process have been extensively reviewed (Colwell and Walker 1977; Atlas 1981, 1984; Leahy and Colwell 1990).

Biodegradation of petroleum in the marine environment is a complex process. Physical and chemical factors that affect this process include chemical composition, physical state and concentration of the oil or hydrocarbon, temperature, salinity, oxygen and nutrient availability, and water activity. Important biological factors influencing the rate of biodegradation include the suite of bacteria, fungi, and other microorganisms that are present, and the adaptation of these microorganisms by prior exposure to hydrocarbons (Leahy and Colwell 1990).

In view of the complexity of microbial degradation of hydrocarbons, it is perhaps not surprising that the effectiveness of attempts to enhance natural degradation processes is not at all clear. Bioremediation attempts to treat oil spills have included nutrient addition to enhance growth of the indigenous bacterial community, addition of laboratory-grown inocula of oil-degrading bacteria, and the use of biological surfactants derived from bacteria.

Nitrogen and phosphorus are frequently limiting nutrients in the marine environment, and it has been demonstrated experimentally that the supply of these nutrients can limit microbial degradation of hydrocarbons in seawater (Atlas and Bartha 1972) and in estuarine water and sediment (Walker and Colwell 1974). Encouraging results were obtained by using oleophilic and slow-release fertilizers (Atlas and Bartha 1973; Olivieri et al. 1976). However, one potential effect of fertilizer addition is a direct toxic effect on susceptible marine life; ecological and toxicological effects should therefore be monitored (Clark et al. 1991). Seeding of oil spills with active, hydrocarbon-degrading microorganisms has the advantage of reducing the initial lag period before the indigenous community responds to the oil spill (Atlas 1991) or to nutrient addition. In a trial following an oil spill which came ashore in marshes in Galveston, Texas, portions of contaminated marshland were treated with the bacterial bioremediation agent, Alpha BioSea. The effectiveness of this treatment was not clear (Mearns 1991). The use of bioremediation for destroying oil on the beach in Alaska after the Exxon Valdez oil spill was concluded to be effective, but not at all sites that were treated (Alexander and Loehr 1992). Use of fertilizers appeared to have been useful in this bioremediation effort (Fox 1989). One approach that enhanced oil removal from Alaskan gravel after this spill was the use of surfactants of microbial origin which naturally disperse oil (Harvey et al. 1990). These surfactants are superior to chemical dispersants which are toxic and cause pollution problems. Treatment with dispersants may have disadvantages under some circumstances by resulting in the contamination of larger areas by the dispersed oil. Dispersants have been useful in the emulsification and removal of residual oil in oil tankers (Rosenberg et al. 1975).

Bioremediation attempts have clearly met with mixed success in the treatment of oil spills, but do appear to be effective under some circumstances and warrant further investigation. An interesting, recent proposal is the use of hollow glass beads coated with the catalyst, titanium dioxide, which in the presence of light, initiates oxidation of large organic molecules such as hydrocarbons. Natural microbial degradation may proceed more rapidly on the more soluble, partially oxidized molecules (Rosenberg et al. 1992). Another approach which has great potential is the use of bacterial strains which have been manipulated by recombinant DNA techniques to improve capabilities for hydrocarbon degradation and suitability as seed organisms. However, use of genetically engineered microorganisms in the environment remains a contended issue (Leahy and Colwell 1990).

Degradation of hydrocarbons in marine and estuarine sediments is likely to be dependent on different degradative pathways to those occurring in the water column, because conditions are anaerobic below a thin surface layer of sediment. Preexposure to polycyclic aromatic hydrocarbons was shown to enhance subsequent rates of hydrocarbon degradation in organic-rich aerobic marine sediments (Bauer and Capone 1988). Anaerobic degradation of oxidized aromatic hydrocarbons has been shown to occur under anaerobic conditions. Microbial consortia can also metabolize certain other hydrocarbons. These anaerobic degradative processes were reviewed by Leahy and Colwell (1990).

Bioremediation of pollutants in the marine environment (other than hydrocarbons) has been little studied. Pollutants frequently become dispersed over very wide areas and are greatly diluted. Pollutants such as heavy metals and pesticide residues may however retain toxic effects on susceptible marine organisms even at very low concentrations. While there is meager information on the fate of pesticide residues in the marine environment, biodegradation of pesticides in terrestrial and freshwater ecosystems is better understood (MacRae 1989). Bioremediation may be the only practicable method for removal of these pollutants. Application of techniques such as the use of gene probes to detect and monitor organisms with specific biodegradative capabilities will be useful tools in this endeavor. It may be possible to monitor expression of specific biodegradative genes by detection of messenger RNA transcripts of those genes. An interesting approach that has been proposed is the use of starvation promoters, which are highly expressed in low-nutrient conditions such as those typically found in seawater, to give selective expression of desired genes in metabolically sluggish populations of bacteria (Matin 1991).

In cases where high concentrations of pollutants are confined in a relatively small volume of seawater or sediment, it may be advantageous to employ a closed system where degradative processes are easier to study and control. This approach may be particularly useful in treatment of dredge spoils from harbors which sediments are likely to be among the most highly contaminated marine ecosystems. A disadvantage of closed systems is the requirement for specialized bioreactors or other sophisticated equipment. A novel soil treatment method was developed by Kaake et al., which avoided the use of expensive equipment while retaining some advantages of a closed system. Bioremediation of herbicide-contaminated soils was achieved by nutrient pretreatment. This process stimulated oxygen consumption that lead to anaerobic conditions which established an anaerobic, microbial consortium capable of complete degradation of the herbicide (Kaake et al. 1992). Effectiveness of in situ (or open system) bioremediation of organic pollutants has been difficult to demonstrate in many cases. Convincing, indirect evidence for microbial degradation of polyaromatic hydrocarbons in a contaminated aquifer was obtained by monitoring microbial adaptation to the pollutant, and demonstrating pollutant-stimulated, in situ bacterial growth (Madsen et al. 1991).

Research on bioremediation and biodegradation processes in soil and groundwater in the United States may yield information that can be applied to bioremediation of contaminated marine sites. In a study on biodegradation of creosote and pentachlorophenol in contaminated groundwater, it was found that indigenous microorganisms could degrade the majority of organic contaminants. However, toxicity and teratogenicity of the biotreated groundwater only decreased slightly, indicating that toxicity and teratogenicity were associated with compounds that were difficult to treat with indigenous microorganisms (Mueller et al. 1991). One approach may be to develop genetically engineered microorganisms with specific capabilities against the most toxic, recalcitrant components. Accumulation of creosote pollutants has been reported in a nearshore estuarine environment in Florida (Elder and Dresler, 1988), and it would be interesting to assess the degradative capabilities of microorganisms from this environment.

Heavy metal pollution of seawater and sediments may be particularly damaging to marine ecosystems because they may persist for a long period of time. High mercury levels were found in marine sediments and in crabs, shrimps, and oysters twenty-one years after a mercury release from a chloralkali plant into Lavaca Bay, Texas (Palmer et al. 1992). High contamination in the coastal marine environment by many metals (and pesticide residues and hydrocarbons) has been found associated with dense urban areas on the East and West coasts of the U.S. (Valette-Silver and O'Conner 1992). High heavy metal concentrations were reported in stranded Atlantic bottlenosed dolphins (Haubold and Tarpley 1992), demonstrating entry and concentration of these toxic pollutants in the food chain. Bioremediation of heavy metals in marine ecosystems may be a future beneficial application of marine biotechnology. Bacteria capable of concentrating silver (Goddard and Bull 1989) and copper (Dunn and Bull 1983) have been reported, and an actinomycete was recently isolated that accumulated uranium and lead (Golab et al. 1992). Metal-resistant actinomycetes have been isolated from heavily polluted sediments in Baltimore's Inner Harbor (Amoroso et al. 1993). Bacteria which tolerate and accumulate metals or which convert metals to less toxic forms have great potential in bioremediation of heavy metal pollution.

Application of bioremediation for the effective treatment of pollutants in the marine environment requires further research on the metabolic capabilities of marine and estuarine microorganisms. Progress in bioremediation of soils and groundwater is likely to spur progress in marine systems. Development of molecular approaches to monitor microorganisms in the environment will be important. The issue of release of genetically engineered microorganisms into the environment will also influence marine bioremediation.

These are only a few examples of the potential of marine biotechnology, a major review of which is underway (Hill and Colwell, in preparation).

REFERENCES

Abu, G. O., R. M. Weiner, J. Rice, and R. R. Colwell. 1991. Properties of an extracellular adhesive polymer from the marine bacterium, *Shewanella colwelliana*. *Biofouling* 3:69-84.

Adams, M. W. W., J.-B. Park, S. Mukund, J. Blamey, and R. M. Kelly. 1992. Metabolic enzymes from sulfur-dependent, extremely thermophilic organisms. In *Biocatalysis at extreme temperatures: enzyme systems near and above 100°C.* ed. M. W. W. Adams and R. M. Kelly, 4-22. Washington, DC.: American Chemical Society.

Agellon, L. B., S. L. Davies, C. M. Lin, and T. T. Chen. 1988a. Growth hormone in rainbow trout is encoded by two separate genes. *Mol. Reproduc. Develop.* 1:11-17.

Agellon, L. B., C. J. Emery, J. M. Jones, S. L. Davies, A. D. Dingle, and T. T. Chen. 1988b. Promotion of rapid growth of rainbow trout (*Salmo gairdneri*) by a recombinant fish growth hormone. *Can. J. Fish. Aquat. Sci.* 45:146-151.

Akee, R. K., T. R. Carroll, W. Y. Yoshida, and P. J. Scheuer. 1990. Two imidazole alkaloids from a sponge. *J. Org. Chem.* 55:1944-1946.

Alexander, M., and R. C. Loehr. 1992. Bioremediation review (letter). *Science* 258:874.

Amoroso, M. J., R. R. Colwell, and R. T. Hill. 1993. Isolation of metal-resistant actinomycetes from Chesapeake Bay. 93rd Annual Meeting of the American Society for Microbiology. Atlanta, Georgia.

Ansley, D. 1990. Cancer institute turns to cell line screening. *The Scientist* 4:3.

Atlas, R. M. 1981. Microbial degradation of petroleum hydrocarbons: An environmental perspective. *Microbiol. Rev*. 45:180-209.

Atlas, R. M. 1984. *Petroleum microbiology*. Macmillan Publishing Co. New York.

Atlas, R. M. 1991. Bioremediation of oil spills: promise and perils. International Marine Biotechnology Conference. Baltimore, Maryland.

Atlas, R. M., and R. Bartha. 1972. Degradation and mineralization of petroleum in seawater: Limitation by nitrogen and phosphorus. *Biotechnol. Bioeng.* 14:309-317.

Atlas, R. M., and R. Bartha. 1973. Stimulated biodegradation of oil slicks using oleophilic fertilizers. *Environ. Sci. Technol.* 7:538-541.

Bauer, J. E., and D. G. Capone. 1988. Effects of co-occurring aromatic hydrocarbons on degradation of individual polycyclic aromatic hydrocarbons in marine sediment slurries. *Appl. Environ. Microbiol.* 54:1649-1655.

Bernart, M., and W. H. Gerwick. 1988. Isolation of 12-(S)-HEPE from the red marine alga *Murrayella periclados* and revision of structure of an acyclic icosanoid from *Laurencia hybrida*. Implications to the biosynthesis of the marine prostanoid hybridalactone. *Tetrahedron Lett*. 29:2015-2018.

Bonar, D. B., S. L. Coon, M. Walsh, R. M. Weiner, and W. Fitt. 1990. Control of oyster settlement and metamorphosis by endogenous and exogenous chemical cues. *Bulletin Marine Science* 46:484-498.

Cardellina, J. H. I. 1986. Marine natural products as leads to new pharmaceutical and agrochemical agents. *Pure and Applied Chemistry* 58:365-374.

Carroll, A. R., and P. J. Scheuer. 1990. Four beta-alkylpyridines from a sponge. *Tetrahedron Lett*. 46:6637-6644.

Chen, T. T., C.-M. Lin, R. A. Dunham, and D. A. Powers. 1992. Integration, expression, and inheritance of foreign fish growth hormone gene in transgenic fish, In *Transgenic fish*, ed. C. L. Hew and G. L. Fletcher, 164-175. Singapore: World Scientific Publishing Co.

Chen, T. T., and D. A. Powers. 1990. Transgenic fish. *Trends Biotechnol*. 8:209-218.

Clark, J. R., R. C. Prince, and J. E. Lindstrom. 1991. Monitoring ecological/toxicological effects of oil spill bioremediation treatments. International Marine Biotechnology Conference. Baltimore, Maryland.

Colwell, R. R., E. R. Pariser, and A. J. Sinskey. 1985. Biotechnology of marine polysaccharides. Washington, D.C.: Hemisphere Publishing Corp.

Colwell, R. R., D. Santavy, F. Singleton, T. Breschel, and T. Davidson. 1989. Marine bioactive metabolites. First International Marine Biotechnology Conference. Tokyo, Japan.

Colwell, R. R., and J. D. Walker. 1977. Ecological aspects of microbial degradation of petroleum in the marine environment. *Crit. Rev. Microbiol* 5:423-445.

Curtin, M. E. 1985. Trying to solve the biofouling problem. *Bio/Technol.* 3:38.

Du, S. J., Z. Y. Gong, G. L. Fletcher, M. A. Shears, M. J. King, D. R. Idler, and C. L. Hew. 1992. Growth enhancement in transgenic Atlantic salmon by the use of an all-fish chimeric growth hormone gene construct. *Bio/Technol*. 10:176-181.

Dunn, G. M., and A. T. Bull. 1983. Bioaccumulation of copper by a defined community of activated sludge bacteria. *European J. Appl. Microbiol. Biotechnol* 17:30-34.

Elder, J. F., and Dresler, P. V. 1988. Accumulation and bioconcentration of polycyclic aromatic hydrocarbons in a nearshore estuarine environment near a Pensacola (Florida) creosote contamination site. *Environ. Pollut*. 49:117-132

Faulkner, D. J. 1984. Marine natural products: metabolites of marine algae and herbivorous marine molluscs. *Nat. Prod. Rep.* 1:251-280.

__________ 1984. Marine natural products:Metabolites of marine invertebrates. *Nat. Prod. Rep.* 1:551-598.

__________ 1986. Marine natural products. *Nat. Prod. Rep.* 3:1-33.

__________ 1987. Marine natural products. *Nat. Prod. Rep.* 4:539-576.

__________ 1988. Marine natural products. *Nat. Prod. Rep.* 5:613-650.

__________ 1990. Marine natural products. *Nat. Prod. Rep.* 7:269-309.

__________ 1992. Marine natural products. *Nat. Prod. Rep.* 9:323-364.

Filpula, D. R., S.-M. Lee, R. P. Link, S. L. Strausberg, and R. L. Strausberg. 1990. Structural and functional repetition in a marine mussel adhesive protein. *Biotechnol. Prog.* 6:171-177.

Fletcher, M. 1987. How do bacteria attach to solid surfaces? *Microbiol. Sci.* 4:133-136.

Fox, J. L. 1989. Native microbes' role in Alaskan clean-up. *Bio/Technol.* 7:852.

Gerwick, W. H., M. W. Bernart, M. F. Moghaddam, Z. D. Jiang, M. L. Solem, and D. G. Nagle. 1990. Eicosanoids from the Rhodophyta: new metabolism in the algae. *Hydrobiologia* 204/205:621-628.

Goddard, P. A., and A. T. Bull. 1989. The isolation and characterization of bacteria capable of accumulationg silver. *Appl. Microbiol. Biotechnol.* 31:308-313.

Golab, Z., B. Orlowska, M. Glubiak, and K. Glejnik. 1992. Uranium and lead accumulation in cells of *Streptomyces* sp. *Acta Microbiologica Polonica* 39:177-188.

Groweiss, A., S. A. Look, and W. Fenical. 1988. Solenolides, new antiinflammatory and antiviral diterpenoids from a marine octocoral of the genus *Solenopodium*. *J. Org. Chem.* 53:2401-2406.

Gunasekera, S. P., M. Gunasekera, G. P. Gunawardana, P. McCarthy, and N. Burres. 1990. Two new bioactive cyclic peroxides from the marine sponge *Plakortis angulospiculatus*. *J. Nat. Prod.* 53:669-674.

Gunasekera, S. P., M. Gunasekera, and R. E. Longley. 1990. Discodermolide: A new bioactive polyhydroxylated lactone from the marine sponge *Discodermia dissoluta*. *J. Org. Chem.* 55:4912-4915.

Harvey, S., I. Elashvili, J. J. Valdes, D. Kamely, and A. M. Chakrabarty. 1990. Enhanced removal of Exxon Valdez spilled oil from Alaskan gravel by a microbial surfactant. *Bio/Technol.* 8.

Haubold, E. M., and R. J. Tarpley. 1992. Heavy metal concentrations in stranded Atlantic bottlenose dolphins. American Geophysical Union 1992 Ocean Sciences Meeting. New Orleans, Louisiana.

Hershberger, W. K. 1990. Selective breeding in aquaculture. Food Rev. Int. 6:359-372.

Horton, P., W. D. Inman, and P. Crews. 1990. Enantiomeric relationships and antihelmintic activity of dysinin derivatives from *Dysidea* marine sponges. *J. Nat. Prod.* 53:143-151.

James, D. M., H. B. Kunze, and D. J. Faulkner. 1991. Two new brominated tyrosine derivatives from the sponge *Druinella (=Psammaplysilla) purpurea*. J. Nat. Prod. 54:1137-1140.

Jensen, P. R., R. Dwight, and W. Fenical. 1991. Distribution of actinomycetes in near-shore tropical marine sediments. *Appl. Environ. Microbiol.* 57:1102-1108.

Kaake, R. H., D. J. Roberts, T. O. Stevens, R. L. Crawford, and D. L. Crawford. 1992. Bioremediation of soils contaminated with the herbicide 2-*sec*-butyl-4,6-dinitrophenol (Dinoseb). *Appl. Environ. Microbiol.* 58:1683-1689.

Kalisz, H. M. 1988. Microbial proteinases. *Advances in Biochemistry and Engineering Biotechnology* 36:1-65.

Klump, H., J. Di Ruggiero, M. Kessel, J.-B. Park, M. W. W. Adams, and F. T. Robb. 1992. Glutamate dehydrogenase from the hyperthermophile *Pyrococcus furiosus*. *J. Biol. Chem.* 267:22681-22685.

Koehn, F. E., S. P. Gunasekera, D. N. Niel, and S. S. Cross. 1991. Halitunal, an unusual diterpene aldehyde from the marine alga *Halimeda tuna*. *Tetrahedron Lett.* 32:169-172.

Kushlan, D. M., and D. J. Faulkner. 1991. A novel perlactone from the Caribbean sponge *Plakortis angulospiculatus*. *J. Nat. Prod.* 54:1451-1454.

Leahy, J. G., and R. R. Colwell. 1990. Microbial degradation of hydrocarbons in the environment. *Microbiol. Rev.* 54:305-315.

Lee, C. Y., B. Rubinsky, and G. L. Fletcher. 1992. Hypothermic preservation of whole mammalian organs with antifreeze proteins. *Cryo-letters* 13:59-66.

Lindquist, N., and W. Fenical. 1990. Polycarpamines A-E, antifungal disulfides from the marine ascidian *Polycarpa auzata*. *Tetrahedron Lett.* 31:2389-2392.

Little, B., P. Wagner, R. Ray, and M. McNeil. 1990. Microbiologically influenced corrosion in copper and nickel seawater piping systems. *Mar. Technol. Soc. J.* 24:10-17.

Liu, Z., B. Moav, A. J. Faras, K. S. Guise, A. R. Kapuscinski, and P. B. Hackett. 1990. Development of expression vectors for transgenic fish. *Bio/Technol.* 8:1268-1272.

Lundberg, K. S., D. D. Shoemaker, M. W. W. Adams, J. M. Short, J. A. Sorge, and E. J. Mathur. 1991. High-filelity amplification using a thermostable DNA polymerase isolated from *Pyrococcus furiosus*. *Gene* 108:1-6.

MacRae, I. C. 1989. Microbial metabolism of pesticides and structurally related compounds. *Rev. Environ. Contam. Toxicol.* 109:1-87.

Madsen, E. L., J. L. Sinclair, and W. C. Ghiorse. 1991. In situ biodegradation: microbial patterns in a contaminated aquifer. *Science* 252:830-833.

Marshall, L. C., F. T. Robb, F. L. Singleton, J. Eyal, and D. R. Durham. 1991. Assessment of proteases derived from marine microorganisms to augment the performance of industrial membrane cleaning compositions. International Marine Biotechnology Conference. Baltimore, Maryland.

Matin, A. 1991. The molecular basis of carbon-starvation-induced general resistance in Escherichia coli. *Mol. Microbiol.* 51:3-10.

Mearns, A. J. 1991. Bioremediation attempts in Galveston Bay, Texas: Observations and lessons. International Marine Biotechnology Conference. Baltimore, Maryland.

Mitchell, R., and M. Alexander. 1962. Microbiological processes associated with the use of chitin for biological control. *Proc. Soil Sci.* Soc. 26:556-558.

Moav, B., Z. Liu, N. L. Moav, M. L. Gross, A. R. Kapuscinski, A. J. Faras, K. Guise, and P. B. Hackett. 1992. Expression of heterologous genes in transgenic fish, In *Transgenic fish*, ed. C. L. Hew and G. L. Fletcher, 120-141. Singapore: World Scientific Publishing Co.

Moghaddam, M. F., W. H. Gerwick, and D. L. Ballantine. 1990. Discovery of the mammalian insulin release modulator, Hepoxilin B3, from the tropical red algae *Platysiphonia miniata* and *Cottoniella filamentosa*. *J. Biol. Chem.* 265:6126-6130.

Morse, D. E. 1984. Biochemical and genetic engineering for improved production of abalone and other valuable molluscs. *Aquaculture* 39:263-282.

Mueller, J. G., D. P. Middaugh, S. E. Lantz, and P. J. Chapman. 1991. Biodegradation of creosote and pentachlorophenol in contaminated groundwater: chemical and biological assessment. *Appl. Environ. Microbiol.* 57:1277-1285.

Murray, V. 1989. Improved double-stranded DNA sequencing using the linear polymerase chain reaction. *Nucleic Acids Res.* 17:8889.

Muzzarelli, R. A. A. 1986. Chitin. Oxford: Pergamon Press.

Nagai, T., Y. Sawayanagi, and N. Nambu. 1984. Applications of chitin and chitosan to pharmaceutical preparations. In *Chitin, chitosan and related enzymes*, ed. J. P. Zikakis, 21-39. Orlando: Academic Press, Inc.

Nakajima, M., K. Atsumi, and K. Kifune. 1984. Development of absorbable sutures from chitin. In *Chitin, chitosan and related enzymes*, 407-410. J. P. Zikakis ed., Orlando: Academic Press, Inc.

National Research Council. 1992. Marine aquaculture, opportunities for growth. *National Academy Press*, Washington, D.C.

Notter, M. F. D. 1988. Selective attachment of neural cells to specific substrates including Cell-Tak, a new cellular adhesive. *Exp. Cell Res.* 177:237-246.

Okami, Y., and K. Hotta. 1988. Search and discovery of new antibiotics. *In Actinomycetes in biotechnology*. ed. M. Goodfellow, S. T. Williams and M. Mordarski, 33-67. San Diego: Academic Press Inc.

Olivera, B. M., J. Rivier, C. Clark, C. A. Ramilo, G. P. Corpuz, F. C. Abogadie, E. E. Mena, S. R. Woodward, D. R. Hillyard, and L. J. Cruz. 1990. Diversity of *Conus* neuropeptides. *Science* 249:257-263.

Olivieri, R., P. Bacchin, A. Robertiello, N. Oddo, L. Degan, and A. Tonolo. 1976. Microbial degradation of oil spills enhanced by a slow-release fertilizer. *Appl. Environ. Microbiol.* 31:629-634.

Palmiter, R. D., R. L. Brinster, R. E. Hammer, M. E. Trumbauer, M. G. Rosenfeld, N. C. Brinberg, and R. M. Evans. 1982. Dramatic growth of mice that develop from eggs microinjected with metallothioneine-growth hormone fusion genes. *Nature* (London)

Pisano, M. A., M. J. Sommer, and L. Tars. 1992. Bioactivity of chitinolytic actinomycetes of marine origin. *Appl. Microbiol. Biotechnol.* 36:553-555.

Quinoa, E., E. Kho, L. V. Manes, and P. Crews. 1986. Heterocycles from the marine sponge *Xestospongia* sp. *J. Org. Chem.* 51:4260-4264.

Raybould, T. J. G., G. S. Bignami, L. K. Inouye, S. B. Simpson, J. B. Byrnes, P. G. Grothaus, and D. C. Vann. 1992. A monoclonal antibody-based immunoassay for detecting tetrodotoxin in biological samples. *J. Clin. Lab. Anal.* 6:65-72.

Raymond, J. A., W. Radding, and A. L. DeVries. 1989. Inhibition of growth of nonbasal planes in ice by fish antifreezes. *Proc. Natl. Acad. Sci.* USA 86:881.

Robb, F. T., Y. Masuchi, J.-B. Park, and M. W. W. Adams. 1992. Key enzymes in the primary nitrogen metabolism of a hyperthermophile. In *Biocatalysis at extreme temperatures: Enzyme systems near and above 100°C*, ed. M. W. W. Adams and R. M. Kelly, 74-85. Washington, DC: American Chemical Society.

Rokkones, E., P. Alestrom, H. Skjervold, and K. M. Gautvik. 1989. Microinjection and expression of a mouse metallothionein human growth hormone fusion gene in fertilized salmonid eggs. *J. Comp. Physiol.* B 158:751-758.

Rosenberg, E., E. Englander, A. Horowitz, and D. Gutnick. 1975. Bacterial growth and dispersion of crude oil in an oil tanker during its bassalt voyage. In *Proceedings of the impact of the use of microorganisms on the aquatic environment*, ed. A. W. Bourquin, D. G. Ahearn and S. P. Meyers, 157-167. EPA 660-3-75-001. Corvallis, Oregon: U.S. Environmental Protection Agency.

Rosenberg, I., J. R. Brock, and A. Heller. 1992. Collection optics of TiO2 photocatalyst on hollow glass microbeads floating on oil slicks. *J. Phys. Chem.* 96:3423-3428.

Roussis, V., Z. Wu, and W. Fenical. 1990. New anti-inflammatory pseudopterosins from the marine octocoral *Pseudopterogorgia elisabethae*. *J. Org. Chem.* 55:4916-4922.

Saiki, R. K., D. H. Gelfand, S. Stoffel, S. J. Scharf, R. Higuchi, G. T. Horn, K. B. Mullis, and H. A. Erlich. 1988. Primer-directed enzymatic amplification of DNA with a thermostable DNA polymerase. *Science* 239:487-491.

Sakai, R., T. Higa, and Y. Kashman. 1986. Misakinolide-A, an antitumor macrolide from the marine sponge *Theonalla* Sp. *Chem. Lett.* 1:1499-1502.

Santavy, D. L., P. Willenz, and R. R. Colwell. 1990. Phenotypic study of bacteria associated with the Caribbean sclerosponge, *Ceratoporella nicholsoni*. *Appl. Environ. Microbiol.* 56:1750-1762.

Scheuer, P. J. 1987. *Bio-organic marine chemistry*. Springer-Verlag. Berlin.

___________ 1988. *Bio-organic marine chemistry*. Springer-Verlag. Berlin.

___________ 1990. Some marine ecological phenomena: Chemical basis and biomedical potential. *Science* 248:173-177.

Shears, M. A., G. L. Fletcher, C. L. Hew, S. Gauthier, and P. L. Davies. 1991. Transfer, expression and stable inheritance of antifreeze protein genes in Atlantic salmon (*Salmo salar*). *Mol. Mar. Biol. Biotechnol.* 1:58-63.

Simidu, U., T. Noguchi, D. F. Hwang, Y. Shida, and K. Hashimoto. 1988. Marine bacteria which produce tetrodotoxin. *Appl. Environ. Microbiol.* 53:1714-1715.

Sowers, K. R., and J. G. Ferry. 1984. Characterization of a marine methanogenic consortium. 1984 International Gas Research Conference. Rockville, Maryland.

Stierle, A. C., J. H. I. Cardellina, and F. L. Singleton. 1988. A marine *Micrococcus* produces metabolites ascribed to the sponge *Tedania ignis*. *Experientia* 44:1021.

Stierle, D. B., and D. J. Faulkner. 1991. Antimicrobial N-methylpyridinium salts related to the xestamines from the Caribbean sponge *Calyx podatypa*. *J. Nat. Prod.* 54:1134-1136.

Strausberg, R. L., and R. P. Link. 1990. *Trends in biotechnology*. 8:53-57.

Takizawa, M., R. R. Colwell, and H. R. T. 1993. Isolation and diversity of actinomycetes in the Chesapeake Bay. *Appl. Environ. Microbiol.* In Press:

Trimurtulu, G., D. M. Kushlan, D. J. Faulkner, and C. B. Rao. 1992. Divarinone, a novel diterpene from the brown alga *Dictyota divaricata* of the Indian Ocean. *Tetrahedron Lett.* 33:729-732.

Valette-Silver, N. J., and T. P. O'Conner. 1992. Evolution of the chemical contamination along the coasts of the United States, since 1984. American Geophysical Union 1992 Ocean Sciences Meeting. New Orleans, Louisiana.

Waite, J. H. 1983. Adhesion in byssally attached bivalves. *Biol. Rev.* 58:209-231.

Waite, J. H. 1991. Adhesive stategies of marine mussels. International Marine Biotechnology Conference. Baltimore, MD.

Walker, J. D., and R. R. Colwell. 1974. Microbial degradation of model petroleum at low temperatures. *Microb. Ecol.* 1:63-95.

Walsh, M., T. E. Ford, and R. Mitchell. 1989. Influence of hydrogen-producing bacteria on hydrogen uptake by steel. *Corrosion* 45:705-709.

Walsh, M., and J. M. Jones. 1990. Microbiologically influenced corrosion of epoxy- and nylon-coated steel by mixed microbial communities. *Corrosion/90*. Las Vegas, Nevada.

Weiner, R. M., M. Walsh, M. P. Labare, D. B. Bonar, and R. R. Colwell. 1989. Effect of biofilms of the marine bacterium Alteromonas colwelliana (LST) on set of the oysters Crassostrea gigas (Thunberg, 1793) and C. virginica (Gmelin, 1791). *J. Shellfish Res.* 8:117-123.

Weiner, R. M., W. C. Fuqua, S. L. Coon, D. Sledjeski, and R. R. Colwell. 1991. Tyrosinases, biofilms and oyster set. International Marine Biotechnology Conference. Baltimore, Maryland.

Wilkinson, C. R. 1984. Immunological evidence for the Precambrian origin of bacterial symbioses in marine sponges. *Proc. R. Soc. Lond.* B 220:509-517.

Woese, C. R., O. Kandler, and M. L. Wheelis. 1990. Towards a natural system of organisms: Proposal for the domains Archaea, Bacteria, and Eucarya. *Proc. Natl. Acad. Sci. USA* 87:4576-4579.

Wortman, A. T., C. C. Somerville, and R. R. Colwell. 1986. Chitinase determinants of *Vibrio vulnificus*: Gene cloning and applications of a chitinase probe. *Appl. Environ. Microbiol.* 52:142-145.

Wright, A. E., O. J. McConnell, S. Kohmoto, M. S. Lui, W. Thompson, and K. M. Snader. 1987. Duryne, a new cytotoxic agent from the marine sponge *Cribochalina dura*. *Tetrahedron Lett.* 28:1377-1380.

Wright, A. E., S. A. Pomponi, O. J. McConnell, S. Kohmoto, and M. P. J. 1987. (+)-Curcuphenol and (+)-curcudiol, sesquiterpene phenols from shallow and deep water collections of the marine sponge *Didiscus flavus*. *J. Nat. Prod.* 50:976-978.

Zhu, Z. 1992. Generation of fast growing transgenic fish: Methods and mechanisms. In *Transgenic fish*, ed. C. L. Hew and G. L. Fletcher, 92-119. Singapore: World Scientific Publishing Co.

Zhu, Z., G. Li, L. He, and S. Chen. 1985. Novel gene transfer into the fertilized eggs of goldfish (*Carassius auratus* L. 1758). *Z. Angew. Ichthyol*. 1:31-34.

38

Artificial Habitats for Rearing Slow-Growing Marine Invertebrates

Prof. Giancarlo Albertelli, Dr. Giorgio Bavestrello, Dr. Riccardo Cattaneo-Vietti, Dr. Enrico Olivari, and Dr. Mario Petrillo
Applied Marine Studies Center - (Istituto Scienze Ambientali Marine) - University of Genoa - Italy

INTRODUCTION

The damage or the overfishing caused to several populations of marine invertebrates have recently raised many worries. It would be interesting to verify the possibility of better management regarding these species which are of great economic value. Excluding the sea-farming species reared under restricted conditions, the revitalization of marine overexploited or damaged populations has been successfully carried out with some benthic organisms such as *Posidonia oceanica* (Jeudy de Grissac 1984; Meinesz et al. 1992), giant clams and sea urchins (Trinidad-Roa 1988), *Pinna nobilis* (Hignette 1983; De Gaulejac and Vicente 1990), and hermatypic corals to rebuild damaged reefs (Hadisubroto 1988).

In the Mediterranean Sea, the precious red coral, *Corallium rubrum* (L.) and the date mussel, *Lithophaga lithophaga* (L.), are two economically important invertebrates that are both heavily harvested (FAO 1983, 1988; Hrs-Brenko et al. 1990; Russo and Cicogna 1991, 1992a, 1992b).

A correct management of these resources seems very difficult, as these species show a slow development, and today in many areas, overfishing is certainly a palpable reality. Moreover, their harvesting provokes severe damage to their biocenoses, because the collecting methods are highly disruptive. For example, the exploitation of the date mussel, an endolithic species, causes great destruction to the littoral communities (Marano et al. 1982; Boero et al. 1990).

The main aim of this study is to check the growing efficiency of these organisms in artificial habitats, either in the sea or the laboratory. It would be most useful to verify whether the reproductive capacities of transplanted specimens could support processes of reforming new populations using artificial habitats as larvae spreaders or cause reproduction and larval settling under controlled conditions.

Both of these species seems to be "strong", and consequently could be easily manipulated. The date mussel is, in fact, widely dispersed throughout the Mediterranean infralittoral communities where calcareous rocks are available, and both species seem to survive under polluted conditions (Harmelin et al. 1987; Cattaneo-Vietti [personal observation]). Moreover, the red coral can live detached from its natural substrate (Weinberg 1979; Cattaneo-Vietti et al. 1992a, 1992b; Pais et al. 1992) and settle on iron (Palmulli 1988) or glass (Bianconi et al. 1988; Giacomelli et al. 1988). In the case of the date mussel, no studies were made about its artificial rearing.

MATERIALS AND METHODS

Red Coral Rearing

In January 1989, four reinforced concrete, artificial grottos (Fig. 1A), kindly built by the Association Monégasque pour la Protection de la Nature (AMPN) and weighing 8 t each, were submerged in the Marine Reserve of Monaco, where their continuous surveillance made this an ideal

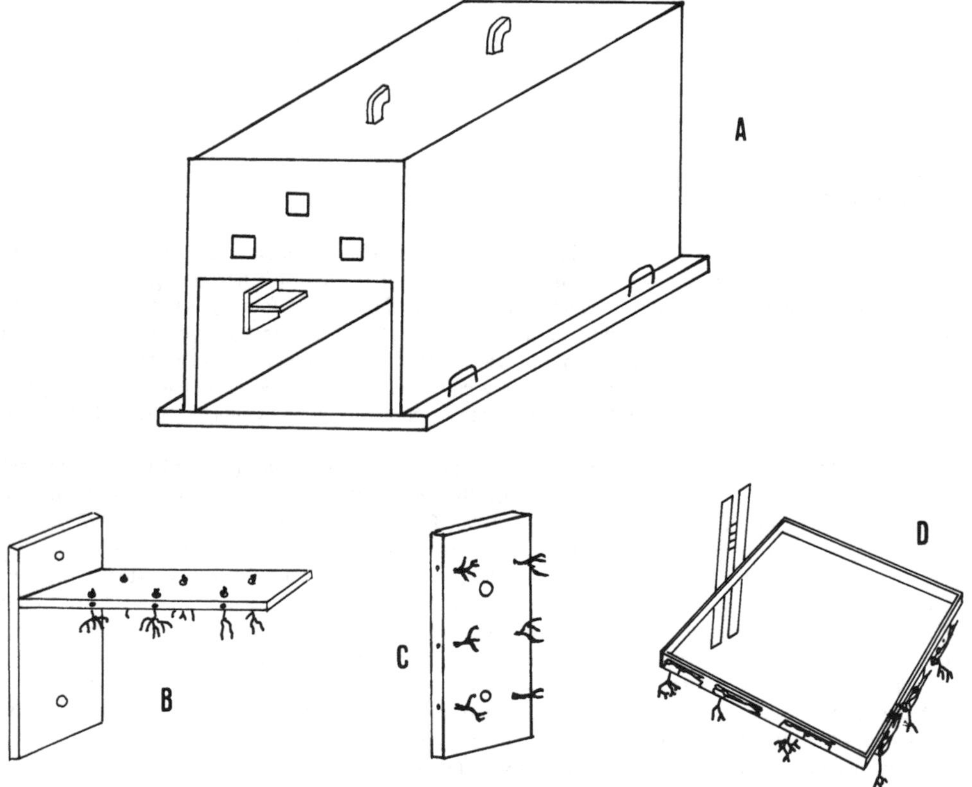

Figure 1. Red coral; Artificial grottos (A), colonies fixed to polypropylene panels (B and C), and to iron frames with clothespegs (D).

spot to manage long-term experiments. The grottos were placed in two different areas: a couple of them were located at the foot of the Loew's Cliff at a depth of 38 m, while the other two were placed at the external limit of the Larvotto Reserve at a depth of 27 m. The grottos were perpendicularly positioned to check the possible influence of both perpendicular and parallel currents along the cliff.

In May 1989, seventy, red coral colonies collected along the Loew's cliff, were fixed with bolts to polypropylene panels (Figure 1B) in a horizontal and vertical position. Each panel (60 x 19 x 3 cm) had six colonies, and its surface was roughened to facilitate larval settlement.

Six months later (October 1989), four iron frames, 40 cm square, similar to those used by Weinberg (1979), with plastic clothespegs (Figure 1C) to fix a new series of colonies, were placed into the caves. Finally, in May 1991, other colonies were fixed using the Devcon resin to porphyry bars.

All the colonies were of noncommercial size (base diameter 8 mm) and their average age, according to Santangelo (1990), varied from four to six years old. Consequently, all were sexually mature. Finally, each colony was photographed in situ with a 1:3 ratio.

Date Mussel Rearing

For a year, beginning in April of 1985, fifty-five specimens of *Lithophaga lithophaga* originating from the Gulf of Taranto, were introduced into a 100 liters aquarium with an areator and a closed filter system. An apparatus kept the seawater at its natural temperature. The specimens were placed in glass modular boxes built for clear vision (Figure 2A).

Biometric measurements (wet weight, shell length, height, and thickness) were taken every quarter, and to evaluate stress conditions, adenilic nucleotides (ATP, ADP, AMP) were measured according to Wiisman (1976) and Viarengo et al. (1985), from July to September 1986. Artificial food was used (1.5 cc of Liquifry Marine, plus 0.8 g of dried powdered meat) and administered at the same time every day for fifteen minutes. Sea biometric experiments were carried out for a year, beginning from November 1988, using forty-two specimens of date mussels put in two VCR (Vinyl Chloride Resin) structures expressely built to prevent predation (Figure 2B).

These structures were placed at a 7 m depth on a sandy bottom, close to a rocky shore near La Spezia.

RESULTS

Red Coral Rearing

In January 1993, four years after the beginning of the experiment, roughly 50 % of the transplanted colonies are alive and appear generally healthy and active.

The biggest loss occurred during the first six months (June-October 1989), owing to our inadequate locking system. The use of bolts, in fact, has often caused the rupture of the calcareous axis and, consequently, the falling off of the colony, even a few hours after transplantation. The fact of not having found dead colonies attached to the panels confirms the thesis of the accidental loss. The clothespegs and the Devcon resin produced better results and no colonies died.

Generally speaking, all the colonies tolerated the transplantation, with an active action of inhibition towards the development of fast growing organisms which occurred a year later (May 1990). Fouling grew particularly fast on the iron frames. In spite of this, the transplanted colonies remained active and were able to cope with this development at their base. Some colonies, slightly infested by bryozoans and serpulids, appeared in good condition and capable of inhibiting their development. Other colonies

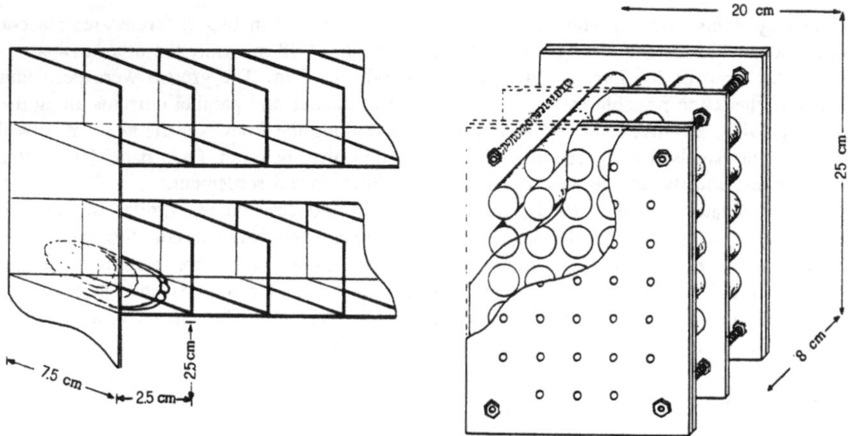

Figure 2. Date mussel; Glass modular boxes (A) and VCR structures (B) used to place the specimens

presenting broken branches were able to repair the damages with their coenenchyma. Two years after the transplantation (April 1991), it was possible to observe in some cases, a development of the coenenchyma on the artificial supports, but it is difficult to determine a growth rate of the colonies without disruptive methods.

At the same time, numerous small colonies of red coral (10-15 colonies/m^2) appeared to have settled only on the roof of the caves. Some of them already have a match-like shape and are 12-16 mm long, but most of them are shorter (5-10 mm), botton-like, and very flat on the substrate. The diameter of the axis varies from 3 to 4 mm, and this depends on the thickness of the coenenchyma. The number of polyps varies from ten to fifteen in the smallest colonies, and from thirty to thirty-five in the largest. All the newborn colonies appeared in good condition and showed expanded polyps.

In January 1993, practically two years after the settling, these colonies are on average, 25 mm long, with two or three branches containing approximately forty polyps. The average diameter of the base is about 4 mm, excluding the coenenchyma thickness.

The density of the settled population is at this time, about 20-30 ind./m^2 in the cave placed parallel to the cliff and it is slightly less in the other one. The increase in the 1992 density compared to that of 1991 suggests a new settling during the last year.

Date Mussel Rearing

The date mussel are very adaptable to artificial conditions: after a year and a half, the largest part was alive and the mortality rate was very low, with a slight increase only in the last quarter of the observation period (5.45%), during the winter.

During the whole period, the annual percent growing increases were very low, sometimes provoking difficulties in the reading of the values: 2.7% ponderal increase, 0.2% length, 0.8% height, and 1.0% thickness. The highest increases were recorded during the summer. All this reveals almost no growth in the lab specimens, even taking the natural slow growth of this species into consideration (Pierotti et al. 1966; Kleemann 1973).

To compare the growth rate in "lab" and "natural conditions", in November 1988, another annual experiment took place, using forty-two specimens put inside VCR structures, immersed at sea in the

Gulf of La Spezia. While after a year, the biometric data of this population confirmed its low growth rate, it nevertheless did reflect a much higher overall increase of more than double that of the natural populations (Figure 3). Also in this case, the highest biometric increase occurred during the summer, and particularly in the specimens with the lowest dimensional and ponderal classes. Finally, to verify the natural growth speed of this species in nature, a population living inside calcareous blocks which form the Genoa harbor pier has been studied.

These blocks positioned in 1970, now host date mussels 70-80 mm long. All this suggests a growth rate at least three times faster in the Ligurian Sea than that reported by Kleeman (1973).

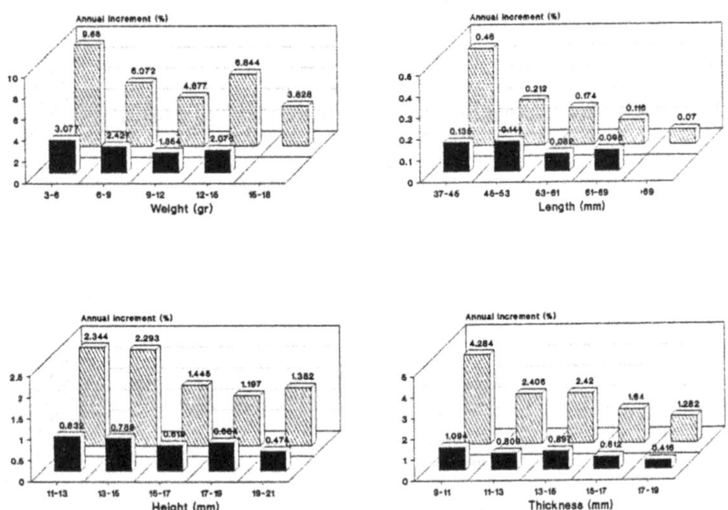

Figure 3. Date mussel; Annual biometric increment.

To determine if the experimental conditions or the kind of food used could have stressed the lab specimens, analysis of adenilic nucleotides were carried out to evaluate their "energetic charge" from July to September 1986. This analysis showed a general decrease during this time, but no significant differences between lab specimens and others collected at sea (Table 1). This decrease can most likely be related to the gametogenesis which occurred in the same period in both populations.

CONCLUSIONS

Red coral and date mussel, although within the limits of their intrinsic slowness of growth, showed an amazing tolerance living out of their natural habitats.

The red coral transferred to artificial substrata has lived for four years, revealing a good adaptation capacity when light, hydrodynamism, and sedimentation rates are optimal. The high mortality in the transplanted colonies during the first period was caused by bolts, while the use of clothespegs and resin was successful. Also, regarding the date mussel, the mortality values were always very low in the artificial structures, demonstrating their adaptability.

Both the species maintained a good physiological condition during the whole period. Generally, the red coral appeared with expanded polyps, able to reproduce, to limit the surrounding fouling, and with their coenenchyma, managing to repair damaged parts. The date mussels reproduced both in lab and sea conditions, showing no differences in stress during the gametogenesis. The reproductive capacities of both the species have been good. The presence of new born colonies of red coral in the caves only two years after the transplantation confirmed the good adaptability of the planula, as it has already been highlighted in the laboratory (Giacomelli et al. 1988; Bianconi et al. 1988). The red coral is not a pioneer species, but if its planula is placed in choice conditions, it will settle onto a well-structured, natural substratum: for example, in natural caves, the red coral settled onto artificial panels after just seven to eight years (Harmelin 1980). This difficulty in settling could be related not only to the smoothness of the proposed substratum, but also to the small vibrations generated by the

Table 1

Date Mussel Rearing for Adenilic Nucleotides and Energy Charge

	July		August		September	
Analysis	CN	CS	CN	CS	CN	CS
ATP (μmoles/g)	0.97	0.76	0.21	0.29	trace	trace
ADT (μmoles/g)	1.02	0.94	0.97	0.47	trace	trace
AMP (μmoles/g)	0.35	0.23	1.40	1.32	trace	trace
Σ neucleotides (moles/g)	2.34	1.93	2.58	2.08	trace	trace
Energetic charge	0.63	0.64	0.27	0.25	-	-

Notes: (CS) "lab" specimens; (CN) "at sea" specimens.

hydrodynamics, thereby, creating inadequate settlement conditions. This hypothesis is supported by the consideration that, on fouling experimental panels, a large part of the settled bryozoans are epiphytic species, i.e., indifferent to the substratum movement. Inside the used artificial caves, these planulae have certainly found an inadequate substratum, but being in a confined space, they were forced to settle, exactly as occurs under lab conditions.

If this hypothesis is true, the type and the condition of the substratum would appear less important than previously thought. This would certainly help the red coral for if it could be reared on a noncalcareous substrata, it would avoid the boring sponge, whose attack generally occurs through the calcareous substratum and causes great losses in the commercial value of the red coral.

To verify all these hypotheses, a new structure has been placed on the red coral communities which is able to collect planulae, as shown in Figure 4. Inside this kind of hood, it is possible to test different substrates which, after the planulae settle, could be transferred to adequate habitats for the colonies' development and repopulation.

Regarding the date mussel's reproductive activity, both the natural and lab specimens have shown the same energetic change in their metabolic rate during spawning, suggesting that it has not been influenced by lab conditions. This may allow for future opportunities to manage a great quantity of larvae produced in a laboratory, settling on adequate substrates.

The slow growth is confirmed in both the species and its evaluation is complex and tedious unless destructive methods are used. Regarding the red coral, the underwater temporal series of photographs have not helped in giving precise measurements, since some parallax mistakes arose and direct measurements on settled colonies were practically impossible. All the same, the data on the new born colonies suggested a growth slightly quicker than that generally known: during the first two years (1991-92), in the Monaco grottos, the colonies showed an average growing length of about 8-12 mm/y and a diameter growth of about 2-3 mm/y.

Measurements of the growth rate of the date mussel in the lab were not easy to evaluate due to their small weight increases and the intervalve water which could invalidate this ponderal data. However it showed growing difficulties in lab conditions and future experiments must be carried out directly at sea, where better growing rates will result. In the lab, only the gametes' laying and the harvesting of the larvae and their metamorphosis on adequate substrata could be followed.

For both the species, seafarming techniques could be applicable, improving the harvesting and at same time, protecting their natural habitats.

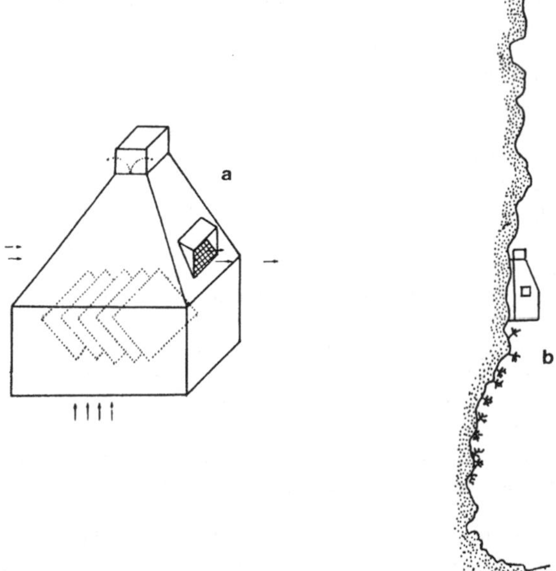

Figure 4. Red coral; Collecting planulae structure

In conclusion, it is too early to assert that these kinds of experiments could be useful for recolonization programs, but the good results obtained regarding the "manipulation" of these species, allows us to feel optimistic for the future utilization of these methods in farming, or at least in reintroduction of certain species in areas where they have disappeared.

For both species, genetic studies which isolate peculiar races, and the application of biotechnological techniques will, perhaps, allow the improvement of their growth speed and establish a new methodology in this field of applied marine biology.

ACKNOWLEDGEMENTS

The authors would like to thank Eugène Debernardi, President de l'Association Monégasque pour la Protection de la Nature, for his help in the field work. Thanks, also to the Marina Militare (Comsubin - Varignano) for the use of a protected area at sea during our experiments on the date mussels.

REFERENCES

Bianconi, C. H., G. Rivoire, A. Stiller, and C.-F. Boudouresque. 1988. Le Corail rouge, *Corallium rubrum* L., dans la réserve naturelle de Scandola (Corse).*Trav. Scient. Parc Nat. Rég. Res. Nat. Corse*16:1-80.

Boero, F., E. Cecere, G. Fanelli, S. Geraci, A. Giangrande, C. Gravili, G. Grilli, R. Piccinni, M. Imperatrice, S. Piraino, D. Saracino, and F. Tucci. 1990. Impact of *Lithophaga lithophaga* Mollusca) fisheries on hard bottom substrata along the Apulian coast (Ionian Sea). *25th European mar. biol. sym.* Abstract.

Cattaneo-Vietti, R., G. Bavestrello, M. Barbieri and L. Senes. 1992a. Premieres experiences d'elevage de corail rouge dans la Reserve sous-marine de Monaco. In Experience de coralliculture en milieu naturel. Ass. Monègasque Prot. Nat., Monaco. 11-17.

Cattaneo-Vietti, R., F. Cicogna and L. Senes. 1992b. Il corallo rosso, una specie in pericolo? *Boll. Mus. Ist. Biol* 56- 57:87-98. Univ. Genova.

De Gaulejac B., and N. Vicente. 1990. Ecologie de *Pinna nobilis* (L.) mollusque bivalve sur le cotes de Corse. Essais de transplantation et experiences en milieu controle. *Haliotis* 10:83-100.

FAO. 1983. Technical consultation on red coral resources of the Western Mediterranean and their rational exploitation. *FAO Fish. Rep.* 306:1-142.

FAO. 1988. GFCM technical consultation on red coral of the Mediterranean. *FAO Fish. Rep.* 413:1-162.

Foster, A.B. 1979. Phenotypic plasticity in the reef corals *Montastraea annularis* (Ellis & Solander) and *Siderastrea siderea* (Ellis & Solander). *J. Exp. Mar. Biol. Ecol.* 39:25- 54.

Giacomelli, S., G. Bavestrello and F. Cicogna. 1988. Experience in rearing *Corallium rubrum*. *FAO Fish. Rep.* 413:57-58.

Hadisubroto, I. 1988. A trial improvement on coral reef in Jepara. In *Regional workshop on artificial reefs development and management* (93-96). Penang, Malaysia ASEAN/UNDP/FAO Manila, Philippines.

Harmelin, J.G. 1980. Etablissement des communautes de substrats durs en milieu obscur. Resultats preliminaires d'une experience a long terme en Méditerranée. *Mem. Biol. Marina e Oceanogr.* 10 (suppl.):29-52.

Harmelin, J.G., J. Vacelet and C. Petron. 1987. Méditerranée vivante. Ed. Glénat. 259 pp.

Hignette, M. 1983. Croissance de *Pinna nobilis* L. (Mollusque Eulamellibranche) après implantation dans le Réserve sous- marine de Monaco, (201-202) Comp.-Rend. 28 Congr. CIESM, Cannes.

Hrs-Brenko, M., D. Zavodnik and E. Zahtila. 1990. The date shell *Lithophaga lithophaga* Linnaeus, and its habitat calls for protection in the Adriatic Sea. In *Les especes marines à protéger en Méditerranée,* ed. C.F. Boudouresque, M. Avon and V. Gravez, 151-158. GIS Posidonie.

Jeudy de Grissac, A. 1984. Essais d'implantations d'especes vegetales marines: les especes pionneres, le posidonies. In *Int. Workshop Posidonia Oceanica Beds*, ed. C.F. Boudouresque, A. Jeudy de Grissac and J. Olivier, 431-436. GIS Posidonie.

Kleemann, K.H. 1973. Der Gesteinsabbau durch Atzmuscheln an Kalkkusten. *Oecologia* 13:377-395.

Marano, G., N. Casavola, C. Saracino and E. Rizzi. 1990. Ciclo riproduttivo di *Lithophaga lithophaga*

(L.) (Mytilidae) nell'Adriatico pugliese. *Oebalia* 16(2 suppl.): 697-699.

Meinesz, A., H. Molenaar, E. Bellone and F. Loques. 1992. Vegetative reproduction in *Posidonia oceanica*. I. Effects of rhizome length and trasplantation season in orthotropic shoots. *Marine Ecology PSZN I* 13(2): 163-174.

Pais, A., L.A. Chessa, and S. Serra. 1992. A new technique for transplantation of red coral *Corallium rubrum* (L.) in laboratory and on artificial reefs. *Rapp. Comm. int. Mer Medit.* 33:46.

Palmulli, D. 1988. Situation de *Corallium rubrum* dans les eaux qui entourent le promontoire de Portofino (GE). *FAO Fish. Rep.* 413:79-81.

Pierotti, P., R. Lo Russo, and S. Sivieri-Buggiani. 1966. Il dattero di mare, *Lithophaga lithophaga*, nel Golfo di La Spezia. *Ann. Fac. Med. Vet. Univ. Pisa* 18:157-174.

Russo, G. F., and F. Cicogna. 1991. The date mussel (*Lithophaga lithophaga*), a case in the Gulf of Naples. In *Les especes marines à protéger en Méditerranée,* ed. C.F. Boudouresque, M. Avon and V. Gravez, 141-50. GIS Posidonie.

Russo G. F.. and F. Cicogna. 1992a. Date mussel (*Lithophaga lithophaga*) harvesting: Evaluation of damage along the Sorrentine-Amalfitane peninsula (Bay of Naples). *Rapp. Comm. Int. Mer Medit.* 33:51.

Russo, G. F. and F. Cicogna. 1992b. Il dattero di mare, *Lithophaga lithophaga* e gli effetti distruttivi della sua pesca sull'ambiente marino costiero: problemi e prospettive. *Boll. Mus. Ist. Biol.* Univ. Genova. 56-57:165-194.

Santangelo, G. 1990. Caratterizzazione della struttura di età di un popolamento di *Corallium rubrum* (L., 1758). 53 Congresso UZI, Palermo, Simposi e Tavole Rotonde. 19-20.

Trinidad-Roa M.J. 1988. Mariculture potential of giant clams and sea urchins in the Lingayen Gulf area. In *Towards sustainable development of the coastal resources of Lingayen Gulf, Philippines*, (133-137) Proc. ASEAN-UN Coast. Res. Man. Project Workshop, Baijang, Philippines.

Viarengo, A., Secondini A., Scoppa P., and Orunesu M. 1985. Metodo rapido per la determinazione della carica energetica adenilica negli organismi marini. Atti del XXXI Congresso Nazionale della Società Italiana di Biochimica. Rimini.

Weinberg, S. 1979. Transplantation experiments with Mediterranean gorgonians. *Bijdr. Tot. de Dierk.* 49(1):31-41.

Wiisman, T.M.C. 1976. Adenosine phosphates and energy charge in different tissues of *Mytilus edulis* under aerobic and anaerobic conditions. *J. Comp. Physiol.* 107:129-40.

Wilkinson C.R., and J. Vacelet. 1979. Transplantation of marine sponges to different conditions of light and current. *J. Exp. Mar. Biol. Ecol.* 37:91-104.

39

Integrated Coastal Policy via Building with Nature

Ronald E. Waterman
Member Government Province of South-Holland
Adviser Ministry of Transport, Public Works and Water Management of The Netherlands

INTRODUCTION

Many civilizations found their origin in delta's and coastal regions. In the year 2000 approximately 80% of the cities (regions) with the largest populations will be found in those areas. Striking examples can be found in nearly all parts of the world. There we have to deal with many existing and future problems that need solutions, but also with challenging opportunities.

Such a densely populated area is situated in the western part of The Netherlands. It is a part that lies below the level of the sea and rivers, so an effective system of water control is needed to keep the land dry and habitable. Natural defenses like sand-dunes and man-made defenses like (reinforced) dikes and strong, solid, sea-wall elements are used for the protection of the low-lying part of The Netherlands. Protection against seawater is needed, which would flood the land via estuaries and inlets, as well as against infiltration by river water, groundwater, and rain. Modern pumping stations work day and night to drain off excess water.

Integrated, multifunctional, sustainable coastal zone development, based on careful analysis of these regions and their climate, their soil and subsoil characteristics, their river systems, the bordering sea, flora and fauna, and present use, gives an answer to the question of how we can solve these problems in relation to each other and in relation to the hinterland on one hand and to the bordering sea on the other.

An important element of integrated coastal policy - apart from, but also including coastal protection and water resources management - is land reclamation using as much as possible the principle of building with nature. Existing and future problems in coastal zone and hinterland can be solved and new opportunities can be found. Learning from mistakes and using the achievements of the past, the challenge of the future can be met, including sustainable development.

In order to assure a future for The Netherlands' coastal zone which is environmentally sound, technically feasible, and economically attractive, two aspects are essential (Figure 1):

1. An integrated approach to coastal zone and hinterland, including old/new land-sea. Many functions have to be considered carefully, while using many different disciplines.
2. Possible realization of new land where nature allows us to do so, using the principle of building with nature. The essence of this principle is flexible integration of land in sea and of water in the new land, making use of materials and forces present in nature, taking into account existing and potential nature values.

This approach is of vital importance in many coastal and deltaic regions of the world. The final development should be such that the overall economy is strengthened and the environment improved.

A PARTLY NEW COASTLINE FOR THE NETHERLANDS

The Netherlands used to be threatened by two 'sea-claws'. One reaches in from the North via the former Zuyder Zee (now Yssel Lake) aiming at the heart of The Netherlands. The other is multipronged and reaches in from the South via the estuaries of South-Holland and Zeeland, also aiming at the same heart.

The present coastline of The Netherlands is the result of natural forces and the action of Man. The country is situated in the delta of the rivers Rhine, Meuse, and Scheldt (Figure 2). Its subsoil consists mainly of sand and clay deposited by the rivers and the sea in the past millennia (Figure 3). Its present coastline is the uneasy border between the North Sea and the alluvial deposits which, through the ages, have been subjected to slow growth and sudden catastrophic setbacks. Storms at high tide in the sea and floods from the rivers caused breakthroughs in the sand barriers which had formed naturally along the beaches. From the earliest times, the inhabitants of these low-laying lands have struggled to save the land from the onslaught of the sea. In 1953, the unfortunate coincidence of an extreme high tide and strong northwesterly gales caused the uneasy equilibrium to be disturbed: the dikes were breached at many places, large areas were inundated, and 1,835 people were drowned. This calamity gave rise to the institution of a special branch of the Ministry of Transport and Public Works which set into motion the realization of the Delta Plan. With all the possibilities of modern technical know-how, radical changes in the existing coastline of the southwestern part of The Netherlands were achieved. The most daring part of the Plan, the storm surge barrier across the Eastern Scheldt, was completed in 1987.

The Delta Plan was the second, recent, spectacular Dutch achievement in this field; the first being the closure of the Zuyder Zee between the First and Second World Wars, followed by the reclamation in phases of 165,000 hectares of new land in the fresh-water lake that came into existence behind the new barrier (Figure 4).

THE DELTA PROJECT

The last occasion on which the sea made major inroads into the land was on February 1, 1952, when large areas in the southwestern part of the country were flooded. The disaster cost 1,835 lives and brought home the need to carry out the Delta Project to close off the estuaries as quickly as possible.

All the estuaries have now been closed, with the exception of the New Waterway and the Western Scheldt which remain open to allow shipping access to the ports of Rotterdam and Antwerp. The Eastern Scheldt has been closed by means of a storm surge barrier which is 3,200 meters long, made up of sixty-five concrete piers between which sixty-two steel gates are suspended. Under normal conditions, the gates remain open and permit the sea to flow in and out

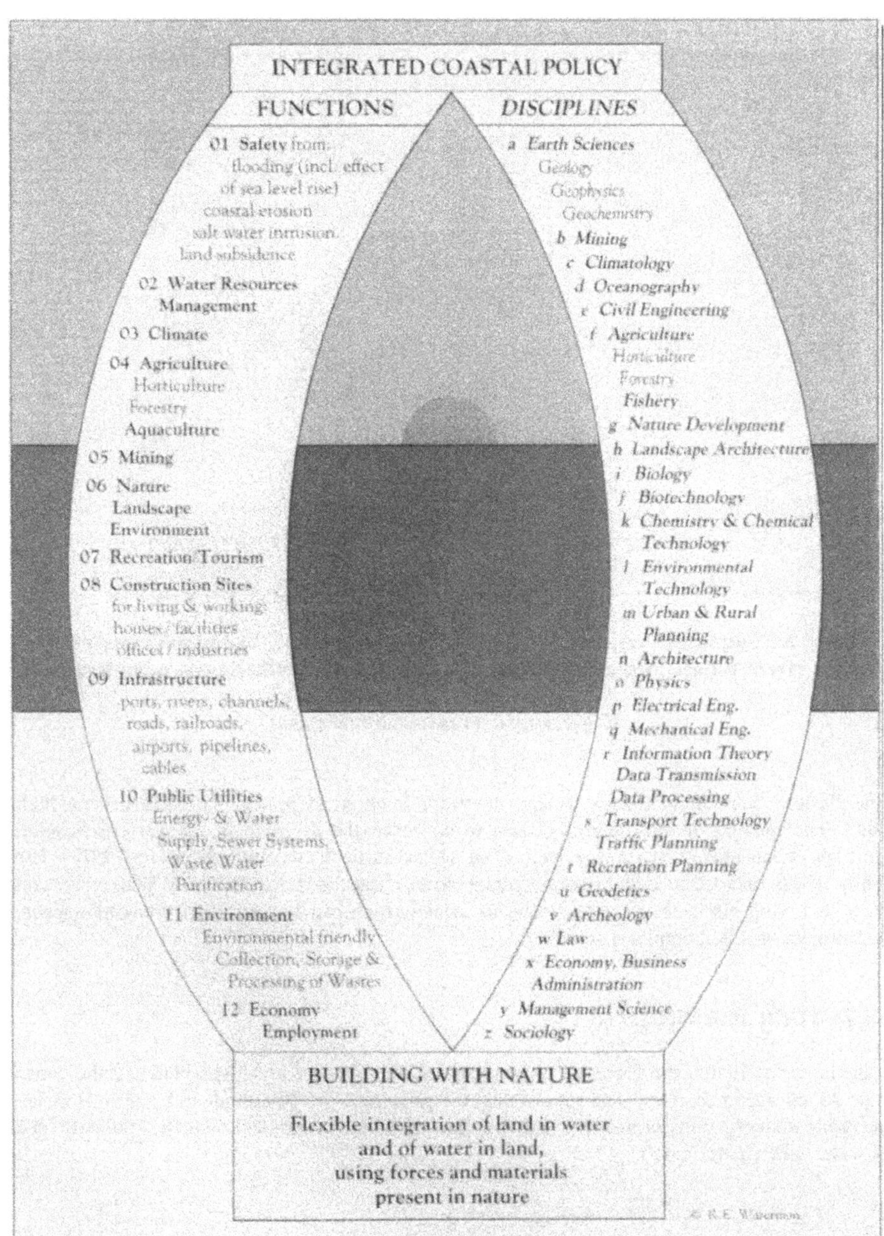

Figure 1. Integrated coastal policy via building with nature

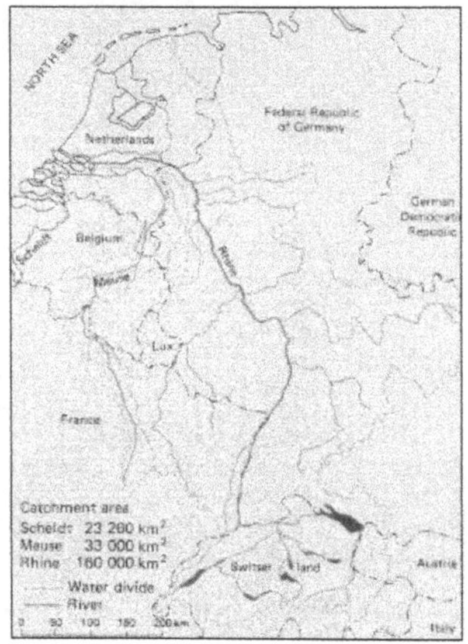

Figure 2. Catchment areas of the rivers Rhine, Meuse & Scheldt

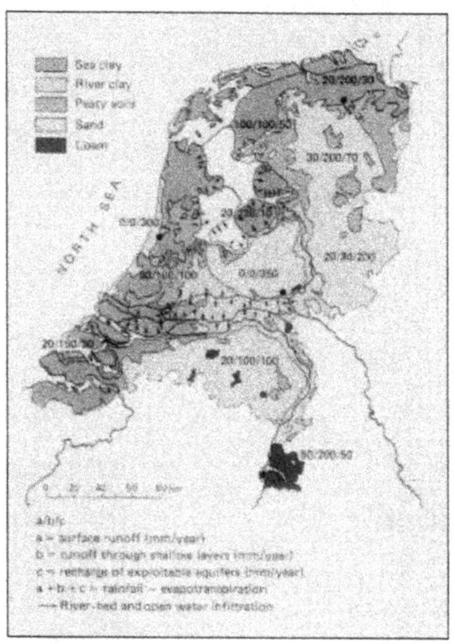

Figure 3. Major soil types of the of the Netherlands, according to Stiboka.

(Figures refer to main annual runoff.)

of the Eastern Scheldt; in stormy weather they are lowered to protect the estuary from high water levels. The method of closure was chosen to conserve the shellfish in the Eastern Scheldt which depend on tidal movement to survive. The Delta Project was completed in 1987. However, recently it was decided to build another special storm surge barrier in the New Waterway, consisting of two revolving circle-sector doors made of steel, which can be closed in times of danger. This added project will be completed in 1997.

THE ZUYDER ZEE PROJECT

In the early 1930s, the Zuyder Zee was closed off from the sea. This entailed the construction of the 30 kilometer Barrier Dam connecting the provinces of Friesland and North Holland. The dam, with sluices, transformed the Zuyder Zee into an inland lake which gradually became a freshwater lake (Yssel Lake).

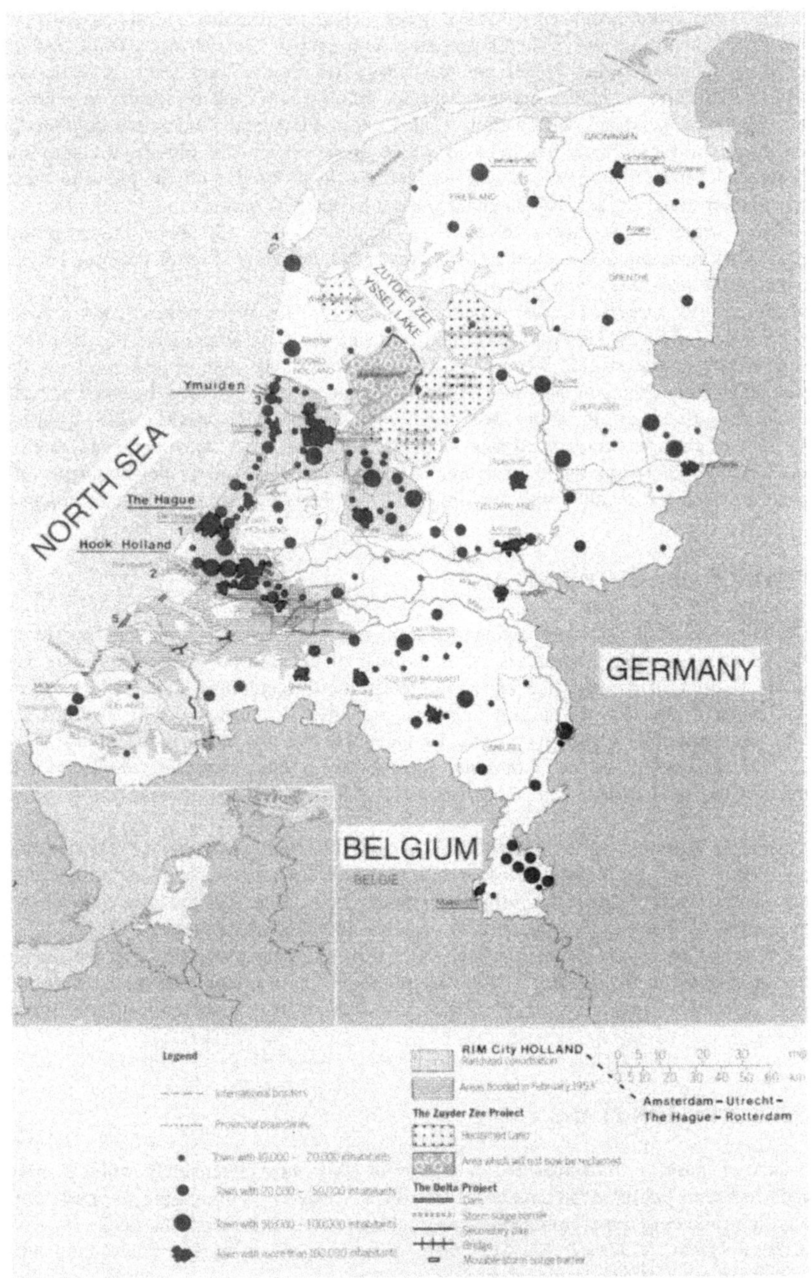

Figure 4. The Zuyder Zee Project and the Delta Project

Once the Yssel Lake had been enclosed, work began on draining enormous polders, four of which have now been completed. They represent a total gain of 165,000 hectares of new land. The two oldest - the Wieringermeer Polder and the North-East Polder - are used for agriculture. The newest, South Flevoland, is being used for housing, employment, and recreation to relieve some of the congestion in the Randstad conurbation. The Eastern Flevoland Polder is a combination of the two. The result of these plans has been to create space within the old Zuyder Zee for several hundred thousand people. In 1986, it was decided not to proceed with the plans to create a fifth polder, the Markerwaard Polder, for financial and environmental reasons.

The Dutch were no strangers to land reclamation before all these major projects were undertaken. They had acquired a great deal of experience and know-how of draining lakes and bogs in the course of the previous seven centuries.

Presently, a third, large-scale plan is being considered. This is the plan to form a new coastline along the Province of South Holland, north and south of the New Waterway, which is the entrance to the largest port in the world, Rotterdam. To the north, the plan would reach as far as The Hague, including Scheveningen and its harbor. The area to be created between the future and present coastline, filled up to above sea level with dunes, will provide new space for many purposes. One of the most interesting aspects of the new plan is its construction method. Another significant factor is its requirement for integrated planning of the many developments in the new coastal zone in which the relation with the hinterland and the bordering sea plays an important role.

THE PROVINCE OF SOUTH HOLLAND

The Province of South Holland is remarkable in many ways. The largest harbor of the world, Rotterdam, is situated within its borders, as is the seat of the Central Government, The Hague. In the Westland District, it contains the largest, uninterrupted greenhouse area of the country (and in the world), while it also has the largest industrial area near Rotterdam. The Province's intensive agricultural and horticultural production and its industrial activity make it the motor of the Dutch economy. In addition, it houses important institutions of education and research, it boasts of fascinating old and new towns, and it has extensive areas of rich scenic beauty for recreation and leisure.

However, the Province of South Holland also has the highest population density, the highest automobile density, and the highest refuse production and energy use, both per capita and in total. There is a serious lack of space for urban expansion, new industrial and office development, and recreation facilities. There is also the need to preserve and expand valuable nature.

The production of waste and its disposal, is causing growing problems, and the unemployment figures are the highest in the country. The plan to enlarge the available coastal areas would provide possibilities to solve many of these problems and create opportunities for specific new developments.

COASTAL EXTENSION PLANS

The idea of pushing outwards the coastline of this most intensively utilized part of The Netherlands has been published in several earlier plans. Briefly summarizing the most important of these plans, the following are mentioned in geographical order, from north to south:

1. The Polyzate 'Village in the Sea' plan, comprising 45 hectares of new land north of the northern mole of Scheveningen harbor.

2. Cadel's and Ten Velden's 'West Dunes' plan, comprising 240 hectares between the southern mole of Scheveningen harbor and Kijkduin.
3. The Volker Stevin 'New Dunes' plan, comprising 1,300 hectares to be reclaimed along the coast between Kijkduin and 's-Gravenzande.
4. In a way, the plan for the fourth harbor extension in Scheveningen, commissioned by the Municipality of The Hague and designed by NEDECO Group-member Haskoning, Royal Dutch Consulting Engineers and Architects, can also be counted among these efforts for it too, comprises reclaimed land for port and industrial development.
5. Extension of the coast to the south of the Hook of Holland figured in the Ballast-Nedam 'Voorne Dam' plan. This plan envisaged the construction of a new peninsula, with an area of 1,250 hectares, connected to the Meuse Plain (Maasvlakte) off the coast of the Island of Voorne. The peninsula would have been available for landing, storage, treatment, and transit of natural gas or the handling of petroleum gas and other hazardous substances.

One common factor of these interesting, former plans, from a constructional point of view, is the application of solid sea walls to protect the new land.

THE PRINCIPLE OF 'BUILDING WITH NATURE'

In 1979, the hydraulic engineer J.N. Svasek, a recognized authority in the field of coastal morphology and harbor construction, launched a new concept for coastal extension. This new method, based on morphological theories, has been named suitably and aptly 'building with nature'. No longer solid stone bulwarks are considered of prime importance, but use is made of the forces acting on the mobile, loose material sand, while creating a new dynamic equilibrium coast. Svasek's plan is based on utilizing the effects of tides, currents, river outflow, waves, wind, and gravity on the sand in the estuarine reaches to restore a coastline which, between the years 1500 and 1600, existed between Scheveningen and Hook of Holland

The Dutch coastline underwent changes during the centuries. This part of the West European continent, including the North Sea floor, is still gradually sinking, whereas the sea level is still rising. In the past, the coastline was breached by action of the sea during periods of a combination of springtide and heavy storms. The position of the various rivers shifted and caused breakthroughs in the coastal defense system. To all this must be added the action of the Dutch engineers in the western part of The Netherlands. All these and other factors combined to make the present coastline.

Now we have the interesting possibility of creating a new, fairly stable coastline via a step-by-step approach, using to the utmost, materials and forces present in nature and utilizing the geographical position. Thus we reach a new phase in creating an optimum coastline position through coastal zone management and development.

THE NEW OVERALL COASTAL EXTENSION PLAN

The present plan, fundamentally based on the concept of Svasek, differs in the alignment of the harbor moles of Scheveningen and Hook of Holland, in the curvature and location of the new coastline inbetween, and in the design and planning of the new land. Furthermore, an analogically designed extension of land south of Hook of Holland has been included in the new plan. This new, overall plan can be divided into a northern (Plan 1) and a southern part (Plan 2). In addition, three newer plans have been developed (Plans 3, 4, and 5), all fitting into the master concept of the new, overall coastal extension plan.

Plan 1 reaches from Scheveningen to Hook of Holland, and is a wedge-shaped addition of new land to the present coastline. It is designed for improved safety from flooding (including the effects of sea level rise) and for increasing the freshwater reservoir under the new dune area. It is multifunctional in design. The functions include: recreation in various forms, a new harbor and harbor-related activities, housing and facilities, a modern office and industrial development, infrastructure, aquaculture, horticulture, and a nature reserve area. A net gain for the environment can be realized, while it strengthens the economy and increases employment.

Plan 2 is situated south of the New Waterway. It is a multifunctional peninsula attached to the greater Europoort/Meuse Plain area of Rotterdam. Plan 2 is primarily designed for the absolute environmental friendly storage and processing of various types of wastes. Additionally, areas are included for industrial and harbor development and for recreation near the new seashore, while provisions are also made for infrastructure and the creation of a nature reserve area.

Plan 3a is a land reclamation south of the southern harbor mole of Ijmuiden, with a yachting harbor, a recreation center, and a nature reserve area with a lake.

Plan 3b is a land reclamation north of the northern harbor mole of Ijmuiden, primarily designed for the storage and processing of various types of wastes and a few other functions.

Plan 4 concerns the alteration and linkage of the island Noorderhaaks to the island of Texel, mainly for coast morphological reasons.

Plan 5 has the shape of a peninsula connected to the Brouwersdam, between Goeree and Schouwen-Duiveland. This new peninsula is a so-called 'Lievense-basin' for pumped storage of energy. In addition, it offers possibilities for recreation.

All these coastal extension plans are multifunctional in character and are based on the principle of building with nature. They all stress the importance of environmental considerations. Most of them can be executed in phases, segment after segment.

TOWARDS AN INTEGRATED COASTAL POLICY FOR SOUTH-HOLLAND

An integrated coastal policy is necessary in order to realize the following:

1. Safety (including coastal protection and protection from salt water intrusion)
2. Building locations (houses/facilities, offices/industries)
3. Public utilities (energy and water supply, sewer system, waste-water purification)
4. Harbors and harbor-related activities
5. Landscape, nature, and environmental values
6. Recreation and tourism
7. Aquaculture
8. Horticulture
9. Infrastructure for public and private transport (roads, canal(s), tram-train-bus, conveyor belt system, pipelines, cables)
10. Water resources management
11. Environmental friendly, controlled storage and processing of harbor silt and waste products
12. Potential possibilities for wind, solar, and water energy
13. Economy and employment

CENTRAL, REGIONAL, AND LOCAL AUTHORITIES

In order to achieve an integrated coastal policy, the following governmental sectors need to be directly involved:

1. Central Government
2. Government of the Province of South-Holland
3. Municipality of The Hague
4. Municipality of Monster
5. Municipality of 's-Gravenzande
6. Municipality of Rotterdam (Hook of Holland)
7. Delfland Water Board
8. Westland Regional Council

PLAN 1

Beginning at Scheveningen-north, the area to be reclaimed between the new and the old coastline is wedge-shaped (Figure 5). In the north it is only a few meters wide and its width gradually increases to 4,500 meters near Hook of Holland. From the top at Scheveningen to the base at Hook of Holland the length is 21 kilometers. The new land has an area of approximately 3,000 hectares and the volume of sand required, amounts to 360 million cubic meters. This amount of sand can either be obtained by widening and deepening the Euro-Meuse Channel and/or by dredging from the seabed of the North Sea parallel to the new coastline, at a distance of 3 to 10 kilometers.

The plan includes the establishment of a primary range of sand dunes parallel to the new coastline with the beach in front, and a secondary range of dunes at an angle to the coast parallel to old dune ridges in the hinterland (which we still can recognize in the orientation of the basic street-pattern of The Hague).

The southern boundary of the wedge-shaped, new land area will be formed by the existing northern harbor mole of Hook of Holland. This harbor mole, with a length of 4.7 kilometers, was originally constructed to protect the entrance of the New Waterway leading to Rotterdam and to provide calm waters for the ships entering and leaving the New Waterway and the Euro-Meuse Channel. The orientation of the new, hollow coastline in relation to existing currents and the movement of sand along the coast could prove to be an improvement on the present situation.

Although it is taken into account that once every five years, a certain amount of beach replenishment will be necessary because of the net loss of sand by coastal drift to the north, coastal protection by groins will most probably not be needed anymore. North of Scheveningen, groins have not proved to be necessary, and in the future, this may also be the case along the new beach.

Another interesting aspect of this plan is that the new land area, being composed mainly of sand, may increase the freshwater reservoir underneath the present dunes. This would improve the protection of the low-lying Westland District with its large greenhouse area and important horticultural activities against saltwater intrusion.

Physical Planning

The essential objective is to reach an integrated coastal policy which can solve many existing and future problems in relation to each other and in relation to the hinterland on one hand and the North Sea on the other. Planning plays an important part in this.

The new area, that will become available behind the new coastline will be needed for many different purposes. Some of the details of the physical planning in various sections going from north to south could be as follows:

Scheveningen-north to Scheveningen Harbor Entrance

1. Most northern part to Von Wied Pavilion
   The plan calls for the widening of the beach in front of the Scheveningen Boulevard, thereby increasing the beach capacity and improving the safety conditions for the beach pavilions during storms and high tide.
2. Von Wied Pavilion to northern harbor mole of Scheveningen
   Primary dunes are to be constructed with the beach in front and behind the dunes, an area for a new model camping site or an educational science park, and a limited number of houses and apartment-buildings. Additionally, there will be facilities for sailing, surfing, and other sports.

Scheveningen Harbor Entrance to Kijkduin

Fourth Scheveningen harbor, with 900 meters quay length and 32 to 55 hectares of land, has been designed for nonpolluting, harbor-related activities. There will also be room for sections of a modern sewage treatment plant based on biotechnological methods, designed to replace the existing plant which serves thirteen municipalities in the province. These developments must be carefully planned so as not to disturb the present Bosjes van Poot (woodland park and its bird sanctuary).

There will also be provisions for public transport and an access road, linked with the horseshoe road system around The Hague, serving the extended harbor and adjacent area. The road will have limited but adequate capacity and can be built at a lower level (for noise protection), with parking areas under the ranges of new dunes.

Additionally, an attractive building site for housing and facilities on The Hague municipal territory is planned. The existing dune area between the already built-up area and the new area (Westduinpark) will be left intact and will be extended to the new beach.

Kijkduin

Planned for this area is a freshwater dune lake of approximately 120 hectares with three lobes and south-oriented beaches. It will have a good microclimate because of the beach orientation and the protection against the prevailing western winds by the primary range of dunes.

This will cause lengthening of the tourist season and provide more recreational variety, because one can always choose between the lake and the new seafront, which is easily accessible. The layout of the lake with its curved, meandering beaches, special depth-profile, bridges, tunnels, water supply, and hygiene, deserves special care. The area between the lake and the new coast can also be used for recreation and tourism. Another building site for an attractive housing area is also possible.

Municipality of Monster

Planned for this area is a golf course and other facilities for not too intensive recreation. These would form a transition to aquaculture and a nature reserve area adjoining Westland Drinking Water Company's terrain.

In this nature reserve, provisions could be made in order to create and preserve living conditions for the various species of flora and fauna specific to the region. Available here are variations of moist-dry, high-low chalk content, high-low nutrition values, varied height of dunes, and differences in microclimate, all of which are essential for creating conditions which will induce a large variety of species.

Additionally, near Ter Heijde and Monster there will be space for housing.

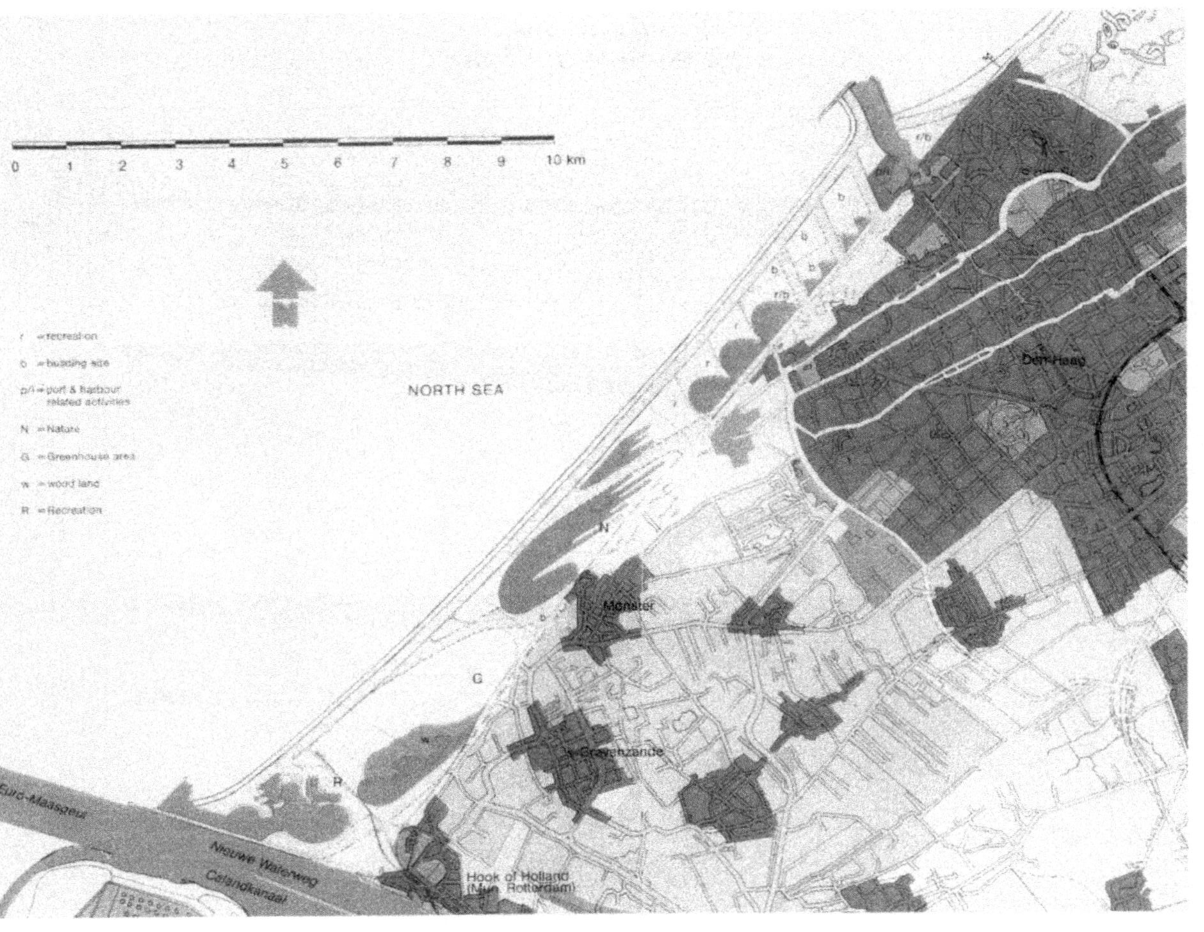

Figure 5. Plan 1: Towards an integrated coastal policy for South Holland

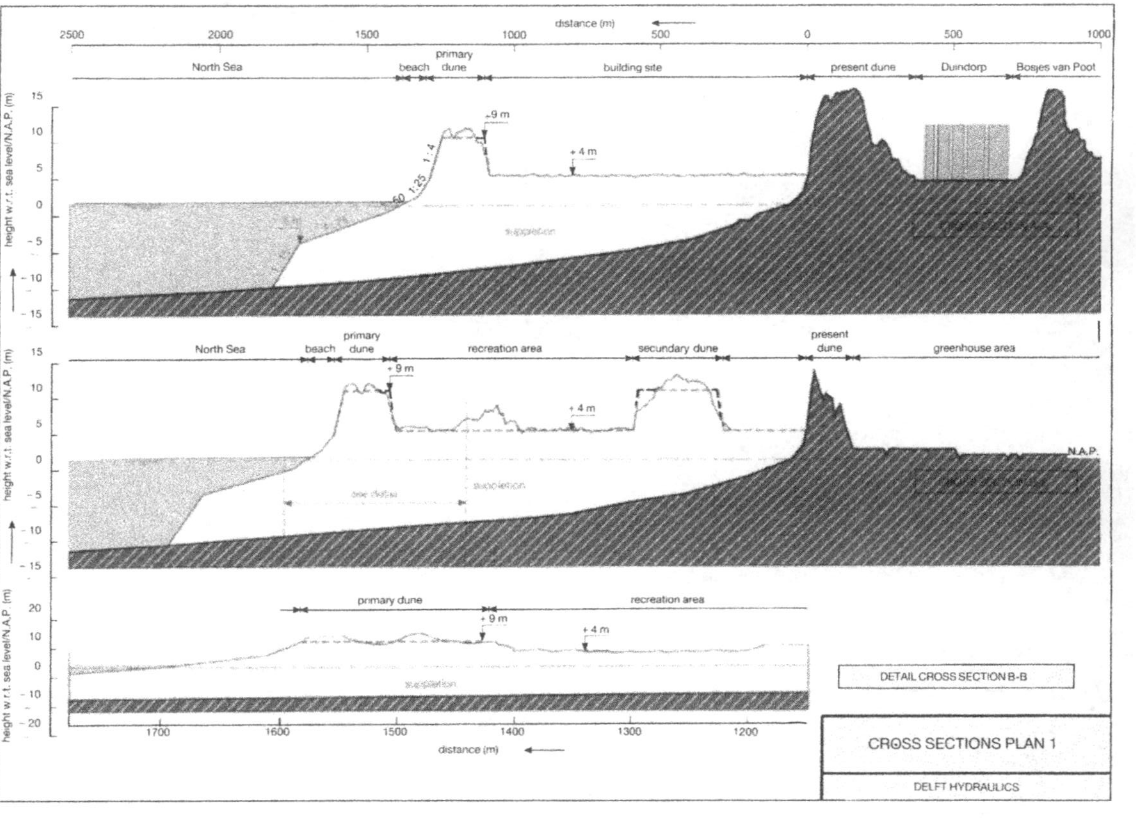

Figure 6. Cross-sections Plan 1
Source: Delft Hydraulics

Municipality of 's-Gravenzande

Urban expansion could be realized near the town's center on the old land, whereas an area between the old and the new coastline could be assigned for horticultural activities (greenhouses) based on modern, energy-productive methods like hydroculture and substratum-culture.

A freshwater dune lake, coupled with a woodland area for tourism and recreation serving Slag Vluchtenburg and the northern part of Hook of Holland is also planned for this section.

Hook of Holland (Municipality of Rotterdam)

The Hook of Holland has a unique position, being situated at the mouth of the rivers Rhine and Meuse, via the New Waterway, and being situated at the North Sea. Because of these features, it deserves a very special development in stages, complete with a saltwater tidal lagoon, a yachting marina, a hotel with rotating top-restaurant, a conference center, housing and facilities, a bungalowpark, offices, and industry as one of the stepping stones in the European market for Great Britain.

To access these new facilities, connections will be realized for public and private traffic, including extending the railroad Rotterdam-Hook of Holland (with a railway station) and extending the A-20 motorway Rotterdam-Hook of Holland.

Physical Construction

The total surface area of the new land (excluding the area occupied by the primary range of dunes) would amount to some 2,500 hectares. This area will be distributed among the four municipalities of The Hague, Monster, 's-Gravenzande, and Rotterdam. The present municipal boundaries will be extended perpendicular to the original coastline. By using sand from the sea the new territory will come into being.

K.P. Vollmer, consultant for urban and rural planning, had an interesting suggestion which could be defined as the three coastlines principle. Instead of creating the new land using sand from the sea and starting from the shore outwards, we first create the new dynamic equilibrium coast consisting of the new primary range of dunes with the new beach in front and on the other side, keeping intact the original beach. Then we plant special Maramm grass (Ammophila Arenaria) and pioneer plants and let nature take hold in the new dunes. After that we either connect the old beach and dune area with the new territory or leave the space in between open, depending on future functions.

Finally, considering the principles of integrating land in sea and water in land, using forces and materials present in nature, it will be necessary that all operations must be carried out in such a way that in the end, nature, landscape, and the environment will benefit. It is essential to use methods that at the same time improve the environment and strengthen the economy.

PLAN 2

Plan 2 is a multifunctional peninsula attached to the greater Europoort/Meuse Plain area of Rotterdam. Its ultimate shape resembles the panhandle peninsula's of Goeree and Schouwen-Duiveland, and to a certain extent of Walcheren, with dominating, southwestern coast orientations typical for the estuarine coast between Hook of Holland and Belgium. Two designs were made, Plan 2a and Plan 2b.

Both Plan 2a and 2b, can be realized in phases, segment after segment. Plan 2 is primarily designed for the absolute, environmental friendly controlled storage and processing of various types of wastes. In addition, areas are included for industrial and harbor development and for recreation near the seashore. Provisions are in place for their infrastructure and a new nature reserve area.

Plan 2a (Figure 7), with the longitudinal axis in SW-NE direction parallel to the coast of Voorne, has the advantage that it can be completely realized by using the method 'building with nature', thereby creating a dynamic equilibrium coast. An important contribution in the design was made by Nethconsult Group-member W.H.A. van Oostrum (ADC).

Plan 2b (Figure 8), with the longitudinal axis in WSW-ENE direction, parallel to the coast of Goeree, is more difficult and expensive to realize. The reason is that in this case, it will be difficult to obtain an overall, dynamic equilibrium coast and the coasts will increase considerably because greater depths have to be met. However, Plan 2b is preferred, taking into account an existing demarcation line (between a recreation/nature zone on one hand and an industrial/harbor zone on the other) and the wish to have more seaspace and an open saltmarsh between Plan 2B and the Voorne coast. In this way, the contact between the Voorne dunes nature reserve and the open sea is maximized (a.o. interaction between wind plus salt-spray and the unique vegetation, which consists of more than 700 species of higher plants).

The Voorne dunes area is unique in Western Europe and therefore the higher costs are fully acceptable. The marsh area and the open sea are very important ecologically, as they also have a very valuable flora and fauna population.

From the viewpoint of construction, the harbor entrance as part of Plan 2b would consist of several, solid sea-wall elements such as breakwaters and harbor moles. The remaining parts, however, could be created by 'building with nature'.

The maximum size of Plan 2b is approximately 3000 hectares. Plan 2a and 2b can be carried out phase after phase, segment after segment. The area they have in common has the contours of the so-called Slufterdam-project, mainly designed by Rotterdam Municipal Works and the Ministry of Transport and Public Works. The volume of sand required can be obtained by dredging in the Euro-Meuse Channel or by excavating the new access channel and harbor to a depth of MSL (Mean Sea Level) minus 15 meters. It can also be obtained by internal excavation, necessary to create storage basins for contaminated harbor silt.

In the latter case, 35 million m^3 of sand can be obtained by creating a storage basin MSL 28 deep, with a surrounding sand dam MSL plus 24 high. This storage basin within the Slufterdam has an internal volume (and storage capacity) of 90 million m^3.

Multidisciplinary planning will be essential to assure that as many interests as possible will be served in a balanced way.

Physical Planning

With regard to physical planning, the zoning of functions is very important.

1. *Southern side of the peninsula*: Along the open salt marsh, a long, fairly narrow zone has been planned as a nature reserve.
2. *An extended Hartel Canal* with loading and unloading quays has been designed.
3. *An area for controlled storage and processing of various types of harbor silt, polluted soils, and industrial waste products has been scheduled.* The storage would take place in excavated basins, the bottom and walls of which are impermeable.

   An internal drainage system could collect all effluent, consolidation, and percolation water. Contaminated water could then be pumped directly to a treatment plant without coming into contact with groundwater. Special basins with semipermeable walls and a controlled drainage system are also a possibility, complete with a monitoring system.

The remaining soil and classified sand and clay types, after being leached and dried, could be used for various purposes. If it is clean clay, it may be used for soil improvement in, for instance agriculture, in clay-lining of dikes, in noise-abating embankments along highways in populated areas, and in visual separation walls.

If it is not sufficiently purified clay, it could still be used for the manufacture of pavings, bricks, clay pebbles, or artificial gravel. The firing process immobilizes the heavy metals and burns off pesticide impurities.

Necessary energy for blending and drying processes can be supplied by various types of energy. In an analogous way, such waste products as fly-ash, slags, gypsum, etc., could be stored and treated, insofar as it is not feasible to treat these products at their site of origin.

Industrial processes should be advanced which manufacture at a higher yield, environment-friendly products using less energy and raw materials, coupled with less polluting emissions to air, water, and soil, and with less solid wastes.

Various types of contaminated silt dredged from harbor areas along the polluted Rhine Basin, other contaminated soils, and toxic wastes that cannot be stored elsewhere, can be brought to this new area for storage and treatment.

A road system, an extended railroad, a conveyor belt system, and subterranean pipelines and cables should be built to facilitate the industry.

4. *Western side of the peninsula*: A harbor (Deltaport) could be planned for landing, storage, and transport of liquid natural gas, petroleum gas, or other hazardous substances far from the population centers, if, in view of safety, the wish and necessity should exist.
5. *Rhine Plain*: This area has been designed to accommodate modern industrial development and harbor-related activities.
6. *Near the seashore*: Recreational activities have been planned for this area.

EXPERIENCE GAINED

Work on both the northern and southern parts of the new plan has already been started and much valuable experience has already been obtained.

As far as Plan 1 is concerned, a wedge of new land has been reclaimed immediately to the north of the harbor mole near Hook of Holland. In 1971-1972, at the instigation of J. van Dixhoorn, the former Director General of the Ministry of Transport and Public Works, some 19 million m^3 of sand were deposited along the stretch from 's-Gravenzande to the Hook of Holland, a length of 3.5 kilometers, with a surface of 100 hectares. The sand had been obtained from dredging the Beer Canal and docks on the Meuse Plain. The principle of 'building with nature' was tried out successfully in this relatively small plan, which included a range of primary dunes parallel to historic dune ridges. A new, dynamically balanced coastline was achieved with a correct orientation. The groins on this part of the old coastline disappeared under the sand, and it has become apparent that they are no longer necessary to the new coastline. This is of importance to recreational sailing and swimming, because the groins with their undertow currents, traditionally constituted a serious source of danger.

Another small-scale experiment took place in Scheveningen after completion of the extension of both harbor moles between 1970 and 1980. The capacity of the beach was improved by beach replenishment.

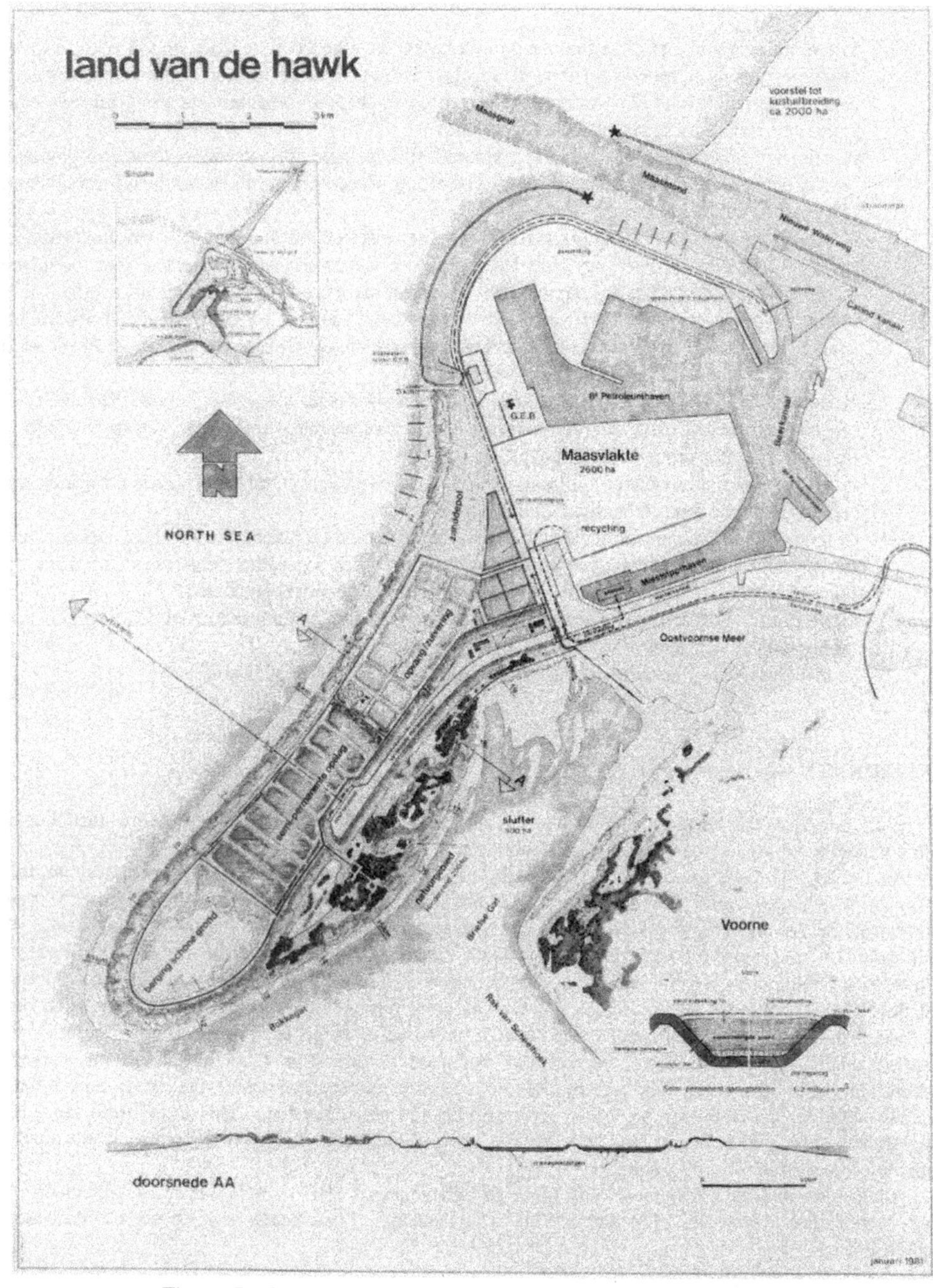

Figure 7. Plan 2A: The Southwestern lobe to The Meuse Plain

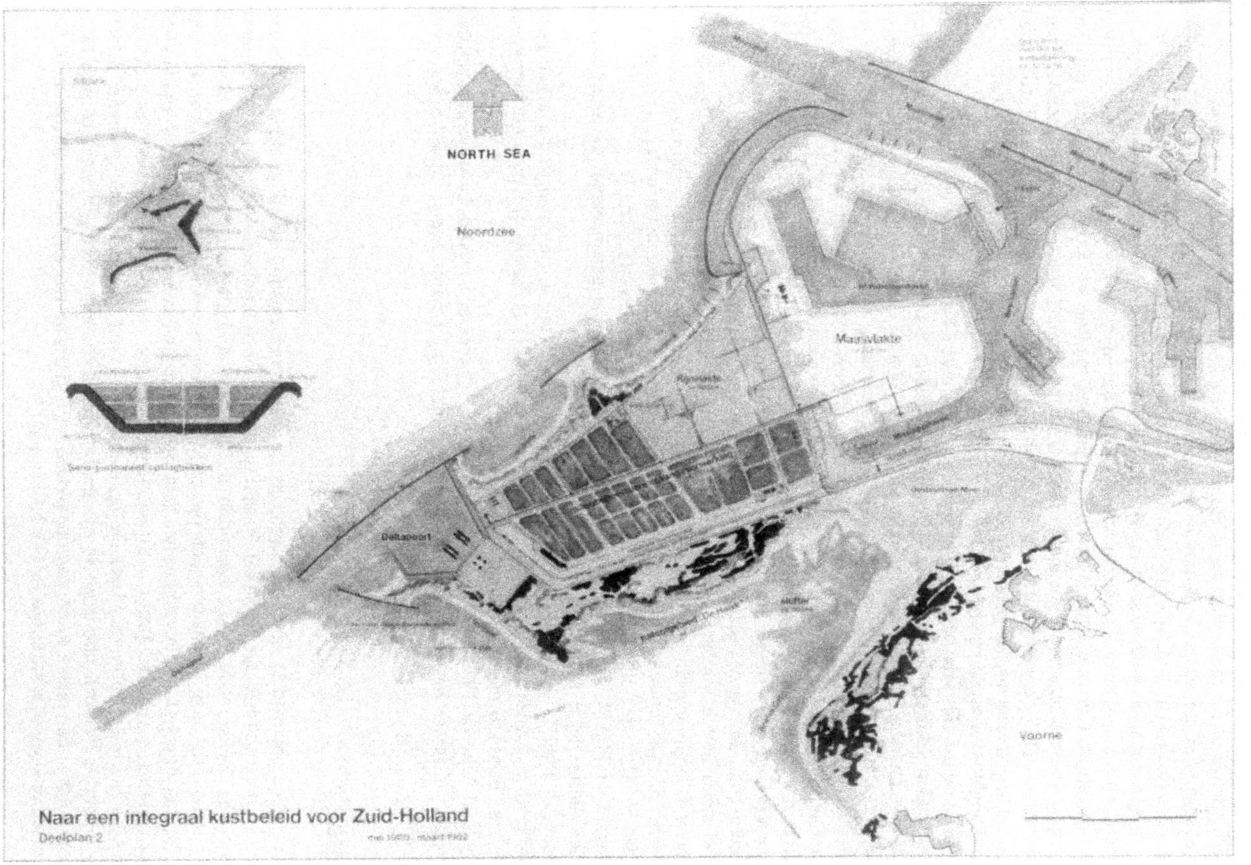

Figure 8. Plan 2B: The Southwestern Lobe to The Meuse Plain

The Municipality of The Hague voted (1989) for an environmental impact assessment concerning a coastal extension for a new Scheveningen-4-harbor, coupled with an improvement and a step-by-step change of functions with regard to the existing harbor and the adjacent area.

As far as Plan 2 is concerned, it should be realized that southwest of the Meuse Plain, new land has already been reclaimed and that this has led to the formation of a dynamically balanced new coastline. This is a strip of land west of the Europe Highway, varying in width from 250 meters to 1.6 kilometers over a length of 4.5 kilometers. Its surface area is 350 hectares. Here, thirteen basins were constructed for the storage of slightly polluted harbor silt from the Caland and Beer Canal silt trap. These basins were not impermeable and the drained water still communicated with the groundwater. The clay soil, after drainage and drying, is already being used for many applications. The experiment has proved to be a successful combination of 'building with nature' and as well thoughtout resources policy.

The realization of this coastal extension plan was followed by another even larger plan. In 1987, the so-called Slufterdam-project was completed, based mainly on a design by Rotterdam Municipal Works and the Ministry of Transport and Public Works. The main purpose of this design was the absolute, environmental friendly storage and processing of contaminated harbor silt and various types of waste products. In addition to that, a new industrial site, with infrastructure, including a road system, railroad extension, extended Hartel Canal, conveyor belt, and pipelines were constructed. Furthermore, recreation along the new beaches and a nature reserve area on the southern side were created. Both executed plans fitted in the larger concept of Subsidiary Plan 2b.

Concerning Plan 3a, we can make the following statement. Nature itself built the larger part of the new land south of the southern harbor mole at Ijmuiden. The new land has a slightly hollow coastline and is triangle-shaped, with a surface area of more than 200 hectares. The north-oriented, net coastal drift of sand caused it to grow. It has been an excellent base upon which to create the complete Plan 3a 'IJmuiden on Sea'. Plan 3a will be completed in 1994 (Figure 9).

With regard to Plan 4, nature itself has a tendency to close the gap between the small island Noorderhaaks and the large island of Texel.

In the case of Plan 5, nature started building a sand deposit (the so-called Middelplaat) attached to the Brouwersdam, which can be used as a base of linking the peninsula for pumped storage of energy (Lievense basin) to the aforementioned Brouwersdam.

In the period between 1980-1993, basic comprehensive studies and reports have been completed concerning many different aspects of the various plans, including environmental impact assessments in several cases.

CONCLUSION

The experience obtained in small-scale projects to change the existing coastline along the North Sea has clearly demonstrated the technical feasibility of the more ambitious plan for a partly new coastline for The Netherlands. The execution of the plan and the creation of new land will undoubtedly reduce the lack of space which presently hampers the development of the area. The possibility, in due time, of net environmental gain concerning the combined terrestrial and marine environment, is also a valuable factor. In addition to general economic expansion, employment opportunities will also be improved, both during construction and also permanently in the future. Challenging opportunities will also arise in the field of research, engineering design, construction, and follow-up work. The plan's successful completion will stimulate the serious consideration of the realization of other similar possibilities in The Netherlands.

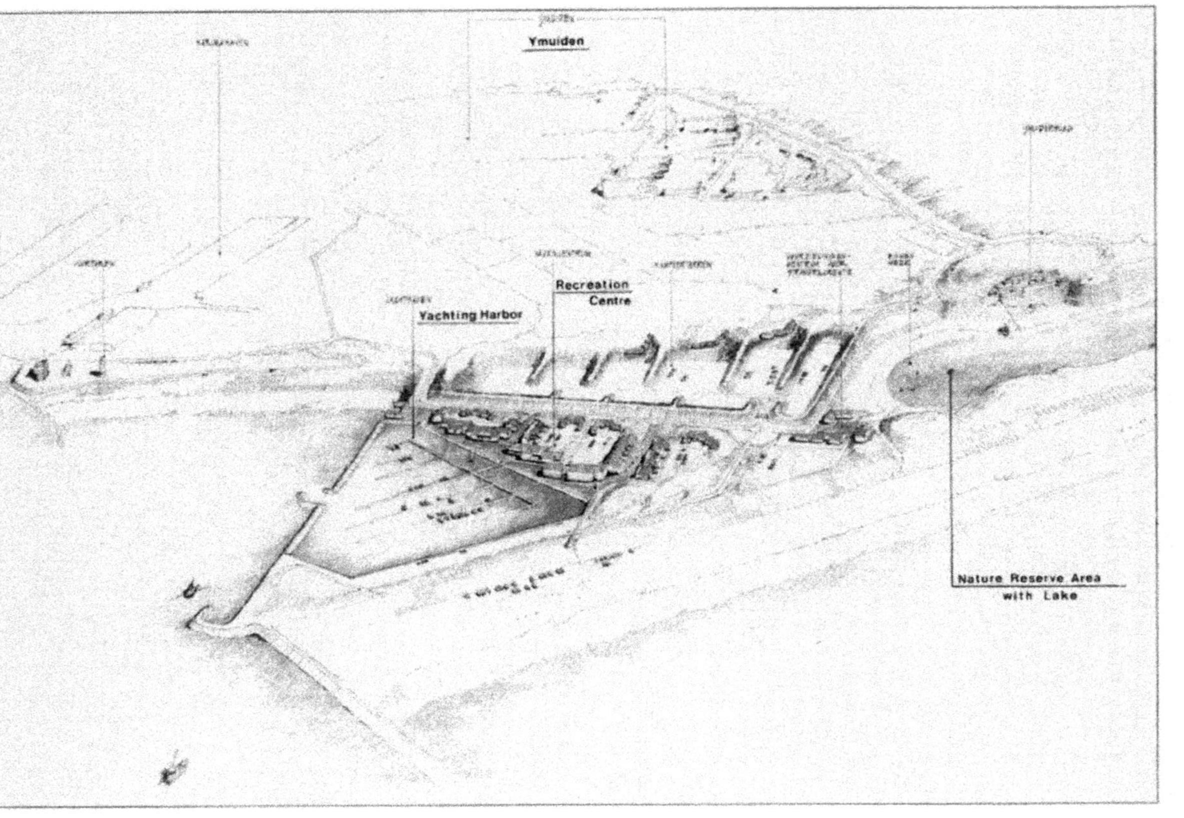

Figure 9. Plan 3a: Ijmuiden on Sea
Source: Municipality of Velsen, November 1985

Finally it should be noted that similar coastal conditions exist elsewhere in the world. The example set in The Netherlands may stimulate planning based on an integrated coastal policy, using the method of building with nature, to be undertaken in other countries.

The export of Dutch know-how to assist in this planning process and its execution can help to solve the existing and future problems in delta's and coastal plains via 'building with nature'.

40

Environmentally Sound Disposal of Wastes in Offshore Island Vaults Developed for Multipurpose Uses

Walter E. Tengelsen
President, Transystems Incorporated
Fort Lee, NJ

ABSTRACT

Urban coastal areas of the industrial nations, especially in the United States, face growing waste disposal problems as increasing populations create more waste. Meanwhile, waste products have become more difficult to handle since the introduction of new chemical materials-including plastics-to the product/waste stream. Waste disposal is further complicated by hazardous and toxic materials that have been and continue to be surreptitiously dumped into landfills, leaching out into the groundwater or into streams and rivers feeding the oceans. Moreover, nuclear waste materials add their own unique difficulties to this disposal problem. One solution to this growing dilemma is to contain all past and future solid waste in waterproof, covered islands–offshore vaults–that can serve other purposes as well. These islands could become the cornerstones of multipurpose offshore activities that meet the needs and alleviate the problems of coastal areas, while producing self-supporting revenues from the "new land" created and services offered. These islands could be designed to incorporate nuclear wastes as well, entombing them (while providing continuous access) within the vault, surrounded by dredged spoil and heavy construction wastes which would act as shielding for the radioactivity. Safe interment of past and future wastes in an offshore vault is technically viable and it would be economically feasible because project costs would be shared by the many productive functions the islands perform. Agreements against ocean dumping (e.g., The London Dumping Convention) prohibit raw or processed materials from being dumped directly into the ocean, no matter the depth. Our concept falls outside of this Convention because of its envisioned series of vaults, standing in the offshore waters, will prevent the waste materials from having any contact with the sea/ocean. These island-vaults would function as a ship, keeping all its "cargo" dry and protected, yet would not pump out bilge or condenser waters.

The principle benefit of this concept is the positive impact it would have on the land, sea, and atmosphere because of its enclosed waste containment. Revenues from the many functions these islands could perform make the concept self-supporting, and therefore so it could be implemented by private capital. This waste disposal concept could be useful in the Baltic, Mediterranean, and North Seas, plus other inland seas across Europe. The recommended steps suggest study and action items that would advance this concept's implementation, including developing alternative financial methodologies, so that the concept does not have to rely solely on government funding.

WASTE DISPOSAL PROBLEMS V. OCEAN DUMPTING BANS IN COASTAL/URBAN AREAS

Our consumer-oriented society has successfully managed to generate more amounts and types of waste than it can easily and safely dispose. Landfill sites are filling up and topping off, while various states (and countries) show a growing reluctance–even active resistance–to accepting any more of the waste from coastal cities. Inland cities are also encountering local opposition to the "taking" of more farms or woodlands for their own landfill sites. Waste products have been accumulating on land for centuries; they were initially disposed of as fill behind bulkheads, as expanding cities created new land along their waterfront (Squires 1983). Some shorefront dump sites were not developed, but continued as waste depositories, growing above grade. They became "sanitary landfills" as new state laws tightened up control over dump sites, but their growing height blocked the view of the water, while their leachate added to the pollution of the coastal waters. The popular wisdom of the past rationalized disposal in landfill sites by assuming it was thus "contained." These wastes were ultimately hidden from view, which seemed to make it acceptable; "out-of-sight, out-of-mind"! But now landfill sites are becoming eyesore mounds, and their continuing pollution of the environment is obvious to even the casual observer. The disposal burden became more troublesome in the post-WWII period when the public's passion for shopping combined with the introduction of new, artificial materials like plastics and other chemicals, to both increase and change the product-waste stream. This greatly complicated the disposal situation as we were no longer dealing with "pure" refuse; many of the chemicals and materials being produced (and thrown out) today are new, man-made compounds, and nature has no natural way of breaking them down or assimilating them. As a further difficulty, lax regulation and supervision led to these hazardous and toxic materials being illegally dumped into landfills. The accumulated waste materials, largely unidentified, and probably containing hazardous and toxic substances thought to have been disposed of in landfill sites are, therefore, continuously polluting the local environment by leachate runoff into rivers and harbors, seepage into the water table, and noxious gas emissions. Unfortunately, heavy metals and toxins in these wastes are easily absorbed into the aquatic food chain, creating a dangerous "feedback" to consumers of seafood. Daily additions to this abundance of stockpiled waste are being generated across the nation. Incineration is not a viable way to rid ourselves of the existing landfill material because of the soil mixed with the refuse and the unknown hazardous/ toxic materials that may have been introduced into the landfill.

Our *past* waste materials, most of which are still being kept in ground-level "accessible" sites (dumps), include the following:

1. Solid waste (from residential, commercial and construction sources)
2. Chemical (hazardous) wastes (from both chemical plants and users)
3. Industrial wastes
4. Toxic wastes (heavy metals/toxins/biologicals/infectious materials/etc., illegally dumped into landfills rather than incinerated)
5. Land reclamation "soils" (usually contaminated)
6. Mining tailings strip mining residue

Other wastes, such as dredging spoil and sewage sludge, were traditionally dumped into the ocean, a process that is now being curtailed because of toxins accumulating in the bottom mud, and a growing body of evidence on the effects of such wastes on the sea's resources (Inter-Governmental Conference 1972; Squires 1983). Dumping in oceans was considered to be acceptable because of dilution, dispersal, and neutralization by the salt water. Additionally, oceans continue to hide the wastes dropped into their depths. But pictures from underwater cameras, plus the retrieval of traumatized, unhealthy fish and crustaceans from the vicinities of ocean dumps, reveals the extent of the damage that has been inflicted on these crucial resources by waste dumping. Arguments have been advanced that the oceans can absorb more of our "selected" waste; after all, it has been absorbing civilization's wastes for millennia. Why, the arguments continue, should we offer "protected status to 71% of the Earth's surface while {the remaining} 29% approaches {a} crisis state? (Champ 1990). This is a valid question and worthy of reasoned debate since it expresses a position that could, by virtue of its expanded perspectives, inspire a more balanced discussion of our waste disposal options. Future needs and problems require that we reevaluate our priorities in the disposal of waste. Can this material only be disposed of on land? Are we fouling our "only nest" while the tree remains untouched? However, before ocean dumping of wastes (directly into the water) is undertaken or renewed -- even for selected wastes—we must have better answers to several questions.

1. What are the limits on how much wastes the ocean can safely absorb and how close are we to those limits now?
2. Will increasing amounts of materials, disposed of in concentrated areas, overwhelm nature's processes locally?
3. What is the final disposition and impact of new, manmade chemical compounds for which nature has no process of breaking down, when released into the oceans? Does this constitute effective "disposal" if we don't know what is happening to it?
4. Can we be certain of not causing small disturbances at critical points in this balanced system, which could be amplified into large changes?
5. Can we contain the spread of dumped materials? (Unlike on land, ocean disposal of some compounds can spread rapidly with disastrous effects; e.g., two supertanker loads of insecticide, spilled into the Pacific Ocean, would disperse and kill all plankton and so eliminate photosynthesis in the Pacific). In November 1972, over sixty nations reached an international consensus to restrict further ocean dumping. The Convention on the Dumping of Wastes at Sea (Inter-Governmental Conference 1972, Article III, Sec. 1) defines the following:
   a) 'Dumping' means:
      (i) any deliberate disposal at sea of wastes or other matter from vessels, ... platforms or other man-made structures at sea;
      (ii) any deliberate disposal at sea of vessels, ... platforms or other man-made structures at sea.
   b) 'Dumping' does not include:
      (i) the disposal at sea of wastes or other material incidental to, or derived from the normal operations of vessels, aircraft, platforms or other man-made structures at sea and their equipment, ...
      (ii) placement of matter for a purpose other than the mere disposal thereof, provided that such placement is not contrary to the aims of this Convention.
   c) The disposal of wastes or other matter directly arising from, or related to the exploration, exploitation and associated offshore processing of seabed mineral resources will not be covered by the provisions of this Convention."

A nonlegal interpretation of these definitions is that while The London Dumping Convention is intended to restrict the direct disposal of waste into international waters where it would freely mix with the waters, it did not really cover the case where a waterproof container of waste is placed in the water. This is expanded upon in Article XXII, Annex I (Inter-Governmental Conference 1972), which lists specific materials and chemicals, including: "6. High-level radio-active wastes or other high-level radio-active matter, ...," that are NOT to be dumped in international waters, but then adds: "8. The preceding paragraphs of this Annex do not apply to substances which are rapidly rendered harmless by physical, chemical, or biological processes in the sea provided they do not: (i) make edible marine organisms unpalatable, or (ii) endanger human health or that of domestic animals." This engineer's interpretation, therefore, is that The London Dumping Convention is principally concerned with materials that are allowed to mix with the seawaters and is not applicable, without amendment or modification, to materials that would be kept within a waterproof shell that stands in the water. Nuclear program waste presents even thornier disposal problems. Low-level radioactive wastes are mostly non-nuclear materials and equipment that have been in contact with radioactivity sources. Their radiation levels are not so high, and their half-lives not as long as actual fissionable material; so, since they are a lesser threat (by comparison), they are simply buried. Unfortunately, burial introduces the possibility of induced radioactivity being carried into the groundwater, by contaminated rust or corrosion particles, and/or the surrounding contaminated soil, being carried along by rainwater seeping through a burial site. When this radioactivity passes into the groundwater and then is absorbed into the food chain, unpredictable genetic damage will result. High-level nuclear waste presents even greater problems because of its higher radiation intensities and radioactivity half-lives of up to 100,000 years for some elements. The hottest materials coming out of the civilian reactor program are the used fuel rods, which contain a variety of transmuted and very radioactive materials. Currently, these used fuel rods are kept in holding pools, usually at the reactor sites, awaiting development of the "ultimate" solution for their disposal.

If it can be harmful to keep hazardous, toxic and radioactive materials in landfill sites, and dumping quantities of them in the oceans is damaging to fish and crustaceans, a crucial food resource, what can we do with all our wastes? One answer is that *all unrecyclable wastes*, accumulated in the past, be collected in the foreseeable future, *and entombed in a leakproof "vault" for safekeeping*. However, to do this simply for containment would probably be unaffordable by most communities, or entail costs that no community would willingly undertake, given today's trade-offs of cost versus risk. One way to make this expense affordable would be to design the vault to serve other purposes as well. On land, vaults would generally serve only one purpose, since other uses are already housed in existing structures now, spread out across the landscape. But in coastal waters, the vault can be part of an integrated, multifunction project that is designed to meet several needs simultaneously.

Past civilizations developed their economies and eased their travels by "bringing water across the land" with canals and aqueducts. Future civilizations, especially those concentrated on coasts, can respond to growing needs by "bringing land across the water" with islands and interlinking pathways. We should not and, indeed, cannot afford to ignore the opportunities and potential benefits of building useful structures out in the offshore waters: it is simply an extension of our past efforts to construct docks and expand our limited, shorefront lands. Structures can be designed to survive the ocean's worst storms; it's only a question of cost versus uses (revenues). By designing modular structures that can be built in large quantities, the per unit costs come down, and by designing them to perform several functions simultaneously, the revenues go up. At some point, the curves will cross and the project becomes economically viable. Industrial nations generated their own problems through overdevelopment and excessive consumption; waste disposal is but one of these, but is the most continuously visible. It is difficult enough to plan for urban areas, especially when so many of the populace choose to settle in cities in hopes of benefiting from urban amenities. This generates a growing demand for public services, which, in turn, requires more land for the myriad functions that accommodate and provide the support required by urban living, thereby creating even greater problems. Problems are further compounded by increasing populations in coastal cities, where expansion is blocked by the sea.

In waste disposal, as in food production and other support functions, we have outgrown nature's abilities to meet our needs. The debate over whether we should undertake large-scale projects was actually decided many years ago, by our population growth. The transition from food gathering to farming, for example, was the introduction of "engineering" to produce enough food for our survival. And as the world's population continues to grow, we will increasingly have to engineer more of our future functions, including the disposal of our wastes.

A SOLUTION: ENTOMB AND UTILIZE WASTE IN AN OFFSHORE, MULTIPURPOSE PROJECT

The most effective and advantageous way of handling the solid waste materials encountered today is to first extract any recyclables, incinerate refuse to recover energy (if the process or site is acceptable for that), and/or neutralize or sanitize any contaminated materials, and finally to compress (possibly solidify) it for permanent entombment in a multipurpose vault. Such processing can be economical if performed on a large scale; by handling large amounts of waste. This material can then be used for construction or as solid fill in a storage vault.

The recommended disposal solution is then to *use these processed waste materials in the construction and filling of offshore islands which become waterproof vaults* for the waste while serving many other purposes. Additionally, the denser waste materials, such as dredging spoil and construction waste, can become an effective shield for radioactive waste. Offshore islands can become the cornerstones of multipurpose activities that produce revenues from the land created and services offered. The recommended location for these vaults, to insure their multiple, revenue-producing uses, is offshore; e.g., the vaults for eastern coastal cities in the United States would be on the continental shelf. As an example of the multiple uses possible, one concept for a coastal project, THE MEGABIGHT BARRIER/MAGLEV WAY (Tengelsen 1991), envisions a string of offshore islands paralleling the East Coast of the United States that offer land and services to meet many of the needs and alleviate the problems of coastal areas. Besides becoming the ultimate disposal sites for waste materials, these islands' many uses include:

1. Waste recycling and incineration sites
2. Storm surge protection (deployable barriers between islands)
3. New sites for commercial and residential development
4. Base for artificial wetlands and mariculture environments
5. Commercial fishing ports
6. New recreation areas and theme parks with ocean access
7. Hazardous material storage (LNG)
8. Power plant and desalination plant sites
9. Oceanography training and research labs
10. Isolation/rehabilitation community sites
11. Environmental research and educational facilities
12. Airport/spaceport sites
13. Deepwater ports/naval base(s)
14. Support for a transportation way (road and/or high speed "ground" transportation [MagLev] in submerged buoyant tunnels)
15. Safe pathway for material, energy and information transfer (pipe or cable; an offshore, coastal infrastructure spine)
16. Leakproof bases for offshore drilling/pumping/refining
17. Wave tripping and/or wave energy recovery

This waste island concept -- a waterproof "vault" standing in offshore waters -- *offers an overall net improvement to the land, ocean, and atmosphere environments by its inclusion and containment of all past and future wastes* while performing other functions. These islands would provide the new land and services needed for other critical activities. The functions and activities would generate revenues to offset the cost of construction, and the cost of disposing of the waste would be "shared" with other functions of the overall project.

BENEFITS AND RISKS OF THIS SOLUTION CONCEPT

The Benefits of Safe Entombment

The benefits of placing all unrecyclable waste in permanent, offshore islands which are actually waterproof "vaults," are substantial.

1. All past waste can be included in these storage islands, which remove and eliminate continuing pollution of the groundwater and air by unidentified toxic and hazardous waste materials that were illegally dumped in the landfill sites.
2. The present landfill sites can be cleaned up, eliminating their pollution of air and groundwater, and put to more productive and valuable uses.
3. All future waste will be centrally processed under rigorous controls, with recyclables removed, and then treated ("sanitized") and possibly solidified before being placed in this regulated environment.
4. The oceans would be protected against surreptitious dumping of wastes; (even past sludge and dredging spoil dumped into the oceans, if deposited in one location and not dispersed, could be dredged up and included in these storage islands; dangerous {i.e., radioactive or toxic} wastes in drums or blocks, that were dropped into the oceans, can also be picked up and properly disposed of this time).
5. The land, ocean, and atmosphere environments would be improved as these waste products and pollutants are isolated and safely stored.
6. The new lands created offshore would provide "waterfront" space and facilities for a wide variety of uses that require access to the ocean or which can benefit from closer proximity to it.
7. Existing onshore lands would receive a "breather," since many uses and functions would be taken over by the islands. This would reduce the pressures and demands on the existing land, while creating some "friendly" competition, inspiring land owners to "clean up their act."

Downside Risks

As a potential downside, artificial islands could impede the tidal flows along the coast and would take water volume and bottom area away from the natural inhabitants. But the islands, individually and in the chain, would be designed to minimize or compensate for any of these possibilities.

Concerning their impact on tidal flows, the islands would be of streamlined design and located offshore to offer minimum resistance to the normal coastal flows. Any flow resistance, creating a turbulent flow as the tide passes the islands, would generate vortices that could be helpful in carrying up bottom nutrients; an additional attraction to fish. Since the islands are expected to be about 20 km offshore, considerably over-the-horizon, their total impact on tidal flows is expected to be minimal.

Concerning their bottom area, the complete island chain would only cover a fraction of 1% of the available area on the continental shelf. The islands, combined with other structures such as artificial reefs and mariculture environments, can more than make up for this by creating new breeding areas for fish on the largely featureless Continental Shelf bottom. United States and international experience in artificial reefs will be incorporated into the project design; there are several artificial reef areas in the New York Bight now, constructed by sportsmen's groups and local governments from tires and rock debris (Squires 1983). With the experience and knowledge now available, designed environments can be tailored to encourage an integrated mix of desired species.

INTERNATIONAL EXPERIENCE IN OFFSHORE WASTE ISLANDS

Japanese Waste Disposal Islands

The Japanese have constructed waste storage islands with interlocking concrete caissons that are sealed to make them essentially leakproof. They are designed with sections, or cells, that are filled sequentially and whose upper surface could be utilized when the cell is compacted and topped with clean fill. Japan's experience and the ocean construction experience in Europe will be sought to guide this concept thinking and to participate in its implementation.

Past U.S. Waste Island Proposals

There have been several proposals for waste storage islands in United States coastal waters, among them the RECAP Island concept put forth by the engineering firm of Pope, Evans & Robbins (Pope 1990). The United States Army Corps of Engineers has built several, large containment islands in coastal waters, principally for dredged spoil. Perhaps the most publicized concept has been the ICONN-Erie plan to build an offshore stormwall of excavated materials (from the redigging of the Erie Canal) that would shelter floating power plants, a deep water port, LNG tanks, etc. (Ehmann 1982). The history of that proposal's progress (or lack thereof) is a classic example of the hurdles and delays awaiting any imposing ideas that require public funding.

ARTIFICIAL ISLAND "VAULT" CONCEPT IN DETAIL

Design and Construction Details

The MEGABIGHT BARRIER/MAGLEV WAY concept envisions a string of islands running along the coast, approximately 20 km offshore (below the horizon), spaced about 8 km apart, so each island is in visual contact with two other islands. The islands must be oblong: "streamlined" to offer minimum resistance to ocean circulation under normal conditions; the longer dimension approximately perpendicular to the coast, the shorter dimension parallel to the coast.

These streamline-shaped islands will be constructed from interlocking, reinforced-concrete cylinders ("caissons") that are sucked into the ocean floor and then grouted for maximum footing support. Rails or tracks on the caissons link one to the other, guiding successive units into place, interlocking them to form a seal which makes the island wall totally leak- and storm-proof. The "pointed bow" of each island will be reinforced (double or triple rows of filled caissons), to ensure its survival under the worst storm conditions. The interior would likely be constructed in cells (probably larger caisson cylinders) to permit controlled filling and waste separation, if desirable, and can also provide structural support to the top.

The completed and sealed island wall is then covered with a permanent, truss-structure top that includes infrastructure connections. The island can then be pumped out, leaving a dry interior. "New land" created on top of the island can be immediately utilized for a variety of purposes while the interior cells are being filled with the waste materials from ashore. The solidified waste, packed inside the island shell, becomes extra mass that adds to the island's strength and permanence. A vent system can be included in the island cover, to recover any residual outgassing, which can probably be used for energy.

If a mixture of the heavier waste materials can be shown to be adequate shielding against radioactivity, then small chambers at the bottom of the islands (an aspirin tablet in a hat box, with separate access tunnels) can be used to store nuclear waste. Such a bottom container could provide continual access, via robotics, to the radioactive waste. Heat exchangers, with isolating loops, could tap off the waste heat from this radioactive material for the benefit of the buildings and activities atop the island.

Each waste islands could be connected to shore by a permanent manifold (multiple corridor) floating or buried tunnel, which allows continuous access to the island's interior. This would permit around-the-clock filling operations and instant access, via robotic vehicles, to any nuclear materials in their own shielded chamber. Having a permanent connection to shore "removes" these islands from T'.e London Dumping Convention jurisdiction.

Within the overall streamlined shape, the islands can be designed to accommodate various uses. Some islands, seen in plan view, can have a cutout (on the lee side) that creates a sheltered harbor. This could be a port for commercial fishermen, a marina for private craft, or a deepwater port for shipping or defense. The topside structures would be designed to support harbor uses: the commercial fishing ports would have processing and storage plants on the islands, the marina would call for a yacht club, and the deepwater ships would require storage and administrative buildings.

Past Resistance to Waste Island Proposals

Past proposals for waste disposal islands in the United States have come up against a variety of objections and bureaucratic hurdles. Their implementation has been frustrated or indefinitely postponed for one or more of the following reasons:

1. The waste storage problem had not reached the present crisis stage.
2. Proposed islands were within sight of shore; "unaesthetic siting."
3. The islands were to be formed by building walls of clean fill, with clay liners, and so would not be "absolutely" leakproof.
4. The islands, or cells, have to be completely filled and contents settled, before any use could be made of the surface; future settling limits the facilities that can be installed on top.
5. The islands were uncapped, except for possible dirt cover after filling, which lets in rain water; liquids leaking out through the bottom or sides could expose the bottom dwelling fish and crustaceans to toxins, etc.
6. The islands would have to be restricted re materials acceptable, because of potential outgassing through the porous cap material and/or the danger of toxic leachate.
7. Proposers expected the government to pay for feasibility studies, for island construction, and operation; they generally did not emphasize the multifunctional "value" of the islands.
8. Political leaders gave up looking ahead beyond their current term in office, and so have little use for long-range solutions.
9. The financial community wanted short-term paybacks; not enthusiastic about such long-term projects (still the case today).
10. The islands were generally intended for only one use and did not attempt to meet other pressing needs facing coastal communities.
11. The island proposers did not anticipate the severe environmental objections, and did not work to build public support for their concepts.
12. No civic leader undertook to champion these concepts.

STORM SURVIVAL AND THE IMPACT OF THIS CONCEPT

Design for Storm Survival

A logical shape for these islands is streamlined, approximately the shape of a ship, with their "bow" pointing offshore into the prevalent storm winds that could be most damaging. Proper design and reinforcement of the "leading edge" of the islands would allow the structure, even when empty, to survive the "100-year-storm." If a storm's winds came perpendicular to their long axis, then the series of islands, like teeth in a comb, will offer protection to each other.

Environmental Impact

The optimal island shape and placement would be chosen to offer minimum resistance to the normal ocean currents and the least interference with estuarine cleansing, combined with "heading into the wind" of major storms. Any bottom surface lost could be compensated for by the addition of artificial reefs, etc.; the reefs, plus the cleaning up of past ocean dumping, will actually make the bottom more productive than it is now. The net impact of the island chain would greatly improve the overall environment, especially the ocean's.

Research Topics Generated by this Concept

Some questions raised by this concept, which identify research areas and/or testing required, include

1. How deep would the caisson tubes have to penetrate the ocean floor, and what geometry, shape, wall thickness, etc., so that the resultant (integral) wall has enough structural integrity to withstand a "100-year" storm?
2. How deep would the caisson tubes have to penetrate the ocean floor to prevent ocean water from seeping up into a pumped-out island?
3. If an anti-seep bottom is required, how to prevent the resultant "tub" from being lifted off the bottom (by buoyancy) before it is filled?
4. What mixtures of wastes are acceptable and which are incompatible for interment in a vault? Which can be combined and which must be kept separated? (This generates the requirements on cell design.)
5. What are the radioactivity shielding properties of the various mixtures of waste that will be available for placement in the islands (e.g., solid waste combined with dredging spoil)? What quantities of radioactive materials, and at what radioactivity levels could be safely stored in an bottom chamber if and when it is completely surrounded by this waste shielding?
6. What is the radiation resistance of reinforced concrete? Can additives improve these characteristics if necessary?
7. What is the present productivity of the ocean floor in the area 20 kilometers offshore? What would be "lost" by constructing an island whose footprint is 400-800 hectares? What artificial reef designs, and areas, would compensate for this loss?

APPROVAL AND FUNDING FOR THIS CONCEPT

Securing Approval Through Public Support

The possibilities of this concept and its multiple functions for alleviating problems in an environmentally acceptable manner, should be presented to the public for their enlightenment and to stimulate a debate on the worth of safely entombing waste in the offshore waters. This can be accomplished by designing museum-quality exhibits at prominent locations in coastal cities and by producing television programs about this approach. By explaining this concept, answering all their questions, and alleviating their doubts, the public will come to see this as a viable, even preferred, alternative. Thus we could secure public support which by itself, can go a long way towards building political support, and counteracting overzealous, unrealistic environmentalists' objections.

Financing Strategies

Even if administrative support of this concept were forthcoming, the financing by government funds would be unlikely because of all the social program demands being made on the public treasuries today. However, this project promises to generate revenues, from land created and services performed, that make it self-supporting. Given this capability, the project could be implemented with private capital; and undertaken as private-initiative. The exhibits and television programs referenced earlier would help by building public support, which can lead to public investment.

SOME PRELIMINARY DESIGN RESULTS FOR THIS CONCEPT

Preliminary Design of this Concept

Consider as an example the New York metropolitan region, which produces about 50 million cubic meters of residential, business, and construction refuse per year (personal contact). Ten years' refuse would be 500 million cubic meters. Assume the internal storage height in an island is 60 meters (the approximate depth of the water at 20 km offshore, on the Continental Shelf). Using these volume and height numbers in the standard volume formula yields a cylindrical island of 1,630 meters radius. So, one 60-meter (internal height) island, with just over a 3 km diameter, could contain ten years of refuse from the NY Metro area. This island's circumference would be 10,233 meters (= 10.2 km), with a top area of 8.33 million square meters (= 833 hectares).

Cost Estimates

Extrapolating from Norway's cost figures on slipforming reinforced concrete tubes (Consultation) we can estimate that caisson tubes, approximately 100 meters long, will cost $300,000 each to fabricate and install in the ocean floor (in 60 meters of water). If the caisson unit is designed in an oval (or racetrack) shape, its longer width being 3 m (tangent to the island's circumference), then the island requires 1,023 caissons. Therefore the total cost of the island wall will be about $307 million. (This cost can probably be reduced if the caissons are extruded and made larger; the great quantities involved lowering the final per-unit costs.) The cost of the top depends on the uses planned. Since the caissons are topped in reinforced concrete, a simple steel truss structure with an outer skin that covers and seals the top, with infrastructure connections built in, could be constructed and installed for approximately $1,000,000 per hectare. The island top would therefore cost $833 million; the total cost of the sealed island would be $1.14 billion.

Revenues and Value-Generated Estimates

The "new land" on the top of the island, if valued at a low to average urban waterfront property value of $1,250,000 per hectare, would be worth $1.1 billion. There are also other revenues that would accrue. The New York region currently pays several hundred dollars per cubic meter to transport waste to regional and out-of-state dumping sites, and these costs are escalating. If this island concept charged only $100 per yard for the transport and entombment of waste, then its annual revenues would be over $5 billion per year. Meanwhile, the cities and states that are disposing of their wastes in the islands would save an equal or greater amount.

Large-Scale Proposal for United States East Coast

THE MEGABIGHT BARRIER/MAGLEV WAY concept envisions a string of smaller islands, of approximately 400 hectares each, spaced 8 km apart, running along the coast. The 8 km spacing allows each island to maintain line-of-sight to its adjacent islands which is of both technical as well as psychological value. The final islands' size, location, orientation, and spacing will be determined in the feasibility study. This means it would take about 420 islands to span the East Coast. These 420 islands, of 400 hectare area and 60 m internal storage height each, would contain a total of 100 billion cubic meters (100 cubic kilometers) of storage volume. If all the solid waste from the East Coast (say 100 million cubic meters per year, an annual installment of one one-thousandth [0.001] of the capacity) were deposited in these islands and the waste sources were charged only $100 per cubic meter for this disposal, the annual income to the island chain would be at least $10 billion, and the communities that are the waste-sources would be saving several times that amount.

The islands are designed to generate early revenues from selling or leasing space on the island and by providing other services, because the sealed top is immediately accessible for use/occupancy. These early numbers on island cost and revenues indicate the self-supporting capability of this concept; a major preliminary design and analysis study is called for, to thoroughly evaluate the concept. The islands, their occupancy structures and internal volumes must be designed and analyzed, including simulation studies and wave-tank tests, etc., to firmly establish the feasibility and cost effectiveness of this approach.

The real value of this concept, beyond its self-supporting revenues, lies in its ability to provide a net improvement in the land, ocean, and atmospheric environments. Additionally, it would be disposing of dangerous waste products which cannot be recycled and doing so for a lower cost than is now being paid. Of greater value is the potential this concept offers the coastal populace by simultaneously alleviating a multitude of problems.

SUMMARY

As the world's population has grown, so our needs have multiplied, long ago exceeding what we could gather from nature's bounty. Obtaining sufficient food for our survival now depends on "farming" and "ranching" to produce more food than would be available naturally. Facing up to this reality, we have to continually engineer our supplies of food, water, housing, etc., to meet our growing needs...but this expanding production produces more waste. As we create more waste than nature can handle, in quantity and types, we must engineer better solutions for its disposal, to prevent drowning in our own over-consumption effluence. Solid waste continues to accumulate at an increasing cost to all municipalities, while becoming a greater health threat because of "new" and/or hazardous/toxic materials that are finding their way into the solid waste stream. Onland storage is now understood to be a continuing source of pollution; existing landfill sites must be cleaned up. Ocean dumping is finally being ruled out because of its threat to our oceanic resources; materials already dumped are still polluting and should be "isolated" from the ocean environment before they do more harm to the aquatic food chain. There are still no totally satisfactory solutions for nuclear waste disposal, especially for the high-level wastes.

We are "running out" of coastal land, for all uses, and current waste disposal methods are polluting the land we have. Solid waste disposal now takes up otherwise-prime land; frequently waterfront property. A sealed vault, on land, would properly isolate waste materials from the environment, but we can't afford the space and costs of constructing them near the coast (where needed), and inland states will not accept them. Additionally, onland vaults would probably have only one use; other functions already have their structures. Therefore they cannot have cost-sharing and would therefore be too expensive to justify.

International agreement calls for a permanent moratorium on dumping in the ocean. This is as it should be; waste material should never be thrown into the life-supporting waters of lakes, rivers, seas or oceans. But the seas can be a repository for a leakproof vault without causing more damage to the oceans. A storm-resistant, sealed vault, imbedded in the ocean floor, on the continental shelf, can be designed to support many other functions and uses which produce the revenues to make this concept self-supporting. *By accepting all past and future waste, they actually improve the land, ocean and atmosphere environments.*

Besides eliminating continuing pollution of the land, groundwater and air, this concept's additional benefits include: allowing present landfill sites to be cleaned up for more productive and valuable uses; future waste being centrally processed under rigorous controls; protecting the ocean against surreptitious dumping of waste; dredging up past sludge and dredging spoil dumped into the oceans, and placing them in these storage islands; picking up radioactive or toxic waste drums dropped into the oceans, for proper disposal; creating new "waterfront" lands offshore that provide space/facilities for a wide variety of uses; and offering a "breather" to existing onshore lands, to aid their cleansing. The ocean and sea bottom lands taken by the islands can be compensated for by introduction of artificial reefs and other mariculture activities, leading to a net improvement in seafood productivity plus a cleaner ocean floor.

RECOMMENDATIONS

The following steps are recommended to advance our knowledge and to help to alleviate this waste disposal problem. Successful solutions will be applicable in coastal areas around the world. The recommended steps, to advance this concept, include:

1. Perform a systems design and analysis of the concept in typical situations, for European sites such as in the Baltic, Mediterranean and North Seas plus other inland waters.
2. Conduct tank testing and simulation studies of candidate designs for the caissons and the islands.
3. Initiate a campaign for public enlightenment, which includes heuristic exhibits in high-visibility locations and television programs, that will build public support.
4. Explore specific waste disposal island-vault applications in selected European sites, compare with standard methods and approaches and identify all revenue-generating services that could be offered.
5. Design pilot projects for selected sites.
6. Develop alternative financing methodologies, especially to develop private funding.

Our waste disposal problems are growing while available financing is shrinking. This self-supporting concept is worth a closer, more detailed look since it could be implemented with private capital. That may be the only way such activities will be funded in these economically distressed times.

REFERENCES

Champ, M.A. 1990. The Ocean and Waste Disposal. *Natural Science* (April): 326-331.

Ehmann, J. 1982. *Chattey's island*. New Haven and New York: Tickner and Fields

Inter-Governmental Conference on the Convention on the Dumping of Wastes at Sea (The London Dumping Convention), October 30-November 13, 1972.

Pope, M. 1990. RECAP Island. Paper presented at Confronting The Infrastructure Crisis Conference, November, at The Cooper Union, NYC.

Squires, D.F. 1983. *The Ocean dumping quandary*. Albany: State University of New York Press.

Tengelsen, W.E. 1991. The Megabight Barrier/Maglev Way. In *IEEE Oceans 91 Proceedings* 896-903.

Consultations with: Engineering staffs of Norwegian companies with slip-formed concrete experience, i.e., MULTICONSULT AS, GRONER OFFSHORE Engineering AS, and Norwegian Contractors AS.

Personal Contact: Solid Waste Specialists from the NYC Department of Sanitation and the Port Authority of New York and New Jersey.

41

Nearshore Resources and Process Studies - Their Application to European Coastal Management

Paolo Ciavola
Coastal Sedimentologist, Coastal Geology Group
British Geological Survey
Keyworth, United Kingdom

ABSTRACT

The main issues in coastal management from the geoscientist's point of view are marine flooding, coastal instability, resource exploitation, pollution and conservation. This paper discusses the United Kingdom's approach to improving knowledge and understanding of coastal zone resources and processes. It also considers the application of this methodology to the Italian North Adriatic coastline and evaluates possible benefits in the formulation of an European Community (EC) Coastal Zone Strategy.

INTRODUCTION

Coastal Zone Management (CZM) is multidisciplinary since it requires inputs from the natural and social sciences. The understanding of physical, chemical, and biological processes is the starting point in the formulation of any strategy that works with nature and not against it.

Marine flooding may threaten life and properties as happened during the floods of 1953 in eastern England and in 1966 in the Venice area of Italy. Because of rising sea levels caused by global warming and subsidence phenomena, areas that previously were not affected by floods are now threatened by tidal surges. Many of these low-lying areas are heavily populated, and effective tidal flood warning systems are not always available.

Coastal instability is normally associated with shoreline erosion and retreat. However, hazards such as coastal landslips and earthquakes can have disastrous consequences for the population living in coastal areas. Earthquakes may produce tsunamis such as the one in 1755 AD in Lisbon or lead to liquifaction of coastal mud deposits.

Coastal conservation is often jeopardized by human activities such as destruction of dunes, reclamation of saltmarshes, and excessive exploitation of seabed resources, such as sand and gravel, for use in the construction industry. Extraction of seabed sediments may create anomalous wave climates on the coastline, causing beach erosion and flooding. It may also seriously affect the marine ecology.

Pollution from land and marine sources threatens both the natural environment and man.

Coastal zone management is of great economical and social importance. Because about 80% of the world's population lives within the coastal zone, there are many conflicts of interest. This paper reviews coastal zone management in the United Kingdom and Italy and considers the role of geoscience in the creation of a CZM plan for the European Community.

DEFINITION OF COASTAL ZONE

The coastal zone has boundaries that change according to the aim of the definition (conservation, planning, scientific research) and to the local setting of the coast. Different countries have differently extended coastal zones. The limits of the coastal zone are marked by an environmental transition. Carter (1988) uses these considerations, defining it as "that space in which terrestrial environments influence marine (or lacustrine) environments and vice versa." The parameters creating this range can be biological, physical, and at times, cultural. In the United Kingdom, the statutory seaward limit of the coastal zone for land use planning is the mean low water mark, despite an acceptance that the limits of the coastal zone are determined by natural processes (Department of the Environment 1992).

For the purposes of this paper the coastal zone is defined using the physical processes acting on a short- and a long-term basis. Figure 1 shows examples from the Netherlands, the United States and the United Kingdom, constructed for different purposes: planning, coastal engineering, and multidisciplinary research.

Figure 1a (Rijkswaterstaat 1990) is based on the geomorphology of the coastal landforms of northern Holland. The area defined as "coastal defense strip" extends from about the -20m depth contour to an inland limit between the dune area and polder (reclaimed land). The seaward boundary is taken as the -20m contour, because natural changes below this depth are not considered relevant to coastal development over a 100-year time scale.

The seaward limit of the coastal area in Figure 1b from the Shore Protection Manual (CERC 1984) is the outer edge of the wave breaker zone. It is not a static line, since the depth at which the waves break varies according to their height.

Figure 1c shows the coastal zone in a wider context. It relates to a multidisciplinary scientific study, the Land Ocean Interaction Study (LOIS), carried out in the United Kingdom by the Natural Environment Research Council (NERC). The definition of the coastal zone has to satisfy all the different disciplines inputting into the project, and therefore takes into account biological, physical, and chemical processes at the land/sea interface. The coastal zone is described as "an indefinite zone of land and sea that straddles the shoreline; it includes all land that is the product of, and/or at risk from (Holocene) marine processes, and extends seaward from the shoreline to a water depth of about 30m." (NERC 1992)

Thus it is not appropriate to produce a rigid definition of the coastal zone. A broad definition that changes according to the local factors, including human influence, should be adopted.

a

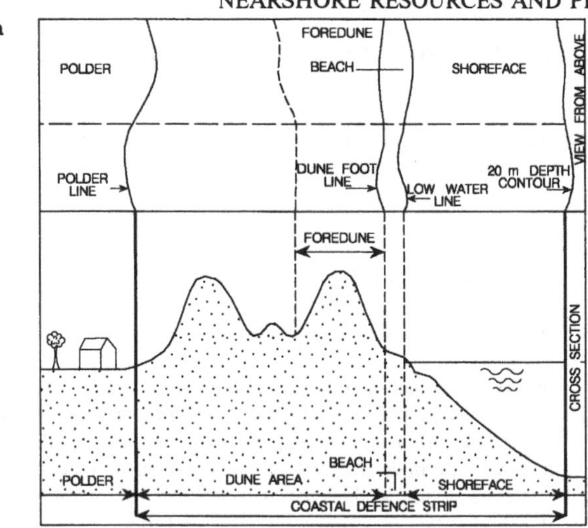

b

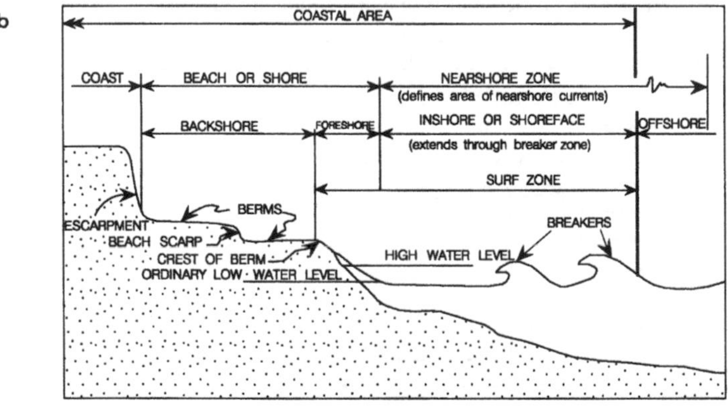

c

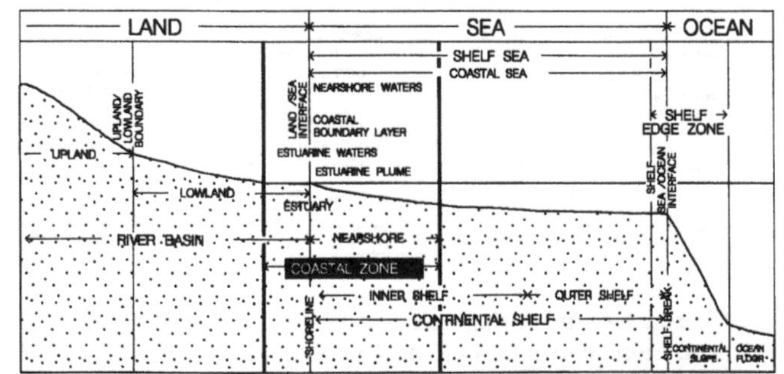

Figure 1. Illustrations of Coastal Zone Nomenclature:
a. The Netherlands (redrawn from Rijkswaterstaat 1990); b. United States (redrawn from CERC 1984); c. United Kingdom (redrawn from NERC 1992).

LEGAL FRAMEWORK IN THE UNITED KINGDOM: REGULATIONS AND RESPONSIBILITIES

Compared to other countries in Europe, the coastal zone of the United Kingdom is heavily regulated. Eight different organizations have statutory responsibilities that overlap spatially.

Coastal Defense

Local authorities are responsible for coastal protection (prevention of coastal erosion). Remedial works on land are subject to environmental assessment and permission from the Crown Estate, if below the mean low water mark. They receive guidance at a national level from the Ministry of Agriculture, Fisheries, and Food (MAFF).

In the past, a piecemeal approach has characterized the coastal policy of the local authorities. However costal authorities are now integrating in Coastal Groups (Figure 2), having members not only from the district councils, ports, and harbor authorities, but also from other authorities having coastal defense responsibilities. This new trend in coastal management reflects an increased awareness among authorities of the physical processes acting in the coastal zone. Because natural forces rarely follow administrative boundaries, it is hoped that planning strategies for future defense works will take into account regional, as well as local, patterns of nearshore sediment transport.

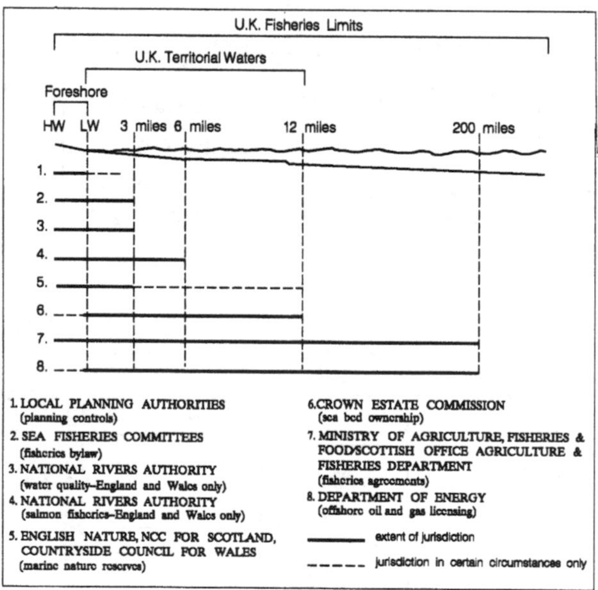

Figure 2. Coastal Groups in England and Wales. National Rivers Authority regions are also shown (modified after National Audit Office 1992)

MAFF determines strategic coastal defense policy in England and Wales and assists the planning authorities on development in areas at risk from flooding. MAFF also funds the national flood and coastal defense research and development program and the Storm Tide Warning Service.

The National Rivers Authority (NRA) has an operational role for sea defenses (to protect the land from flooding by the sea) and is responsible for the tidal barriers in the Thames and in the Humber. It currently spends £26 million per annum on improvement works (House of Commons 1992).

The Scottish and Welsh Offices have wide responsibilities through their departments and cover the same functions as their English counterparts.

Seabed Resources

The seabed stretching between mean low water mark (spring low water mark in Scotland) and the limit of territorial waters (twelve miles offshore) forms part of the Crown Estate (Figure 3). The Crown Estate acts as landlord of the seabed, licencing the extraction of marine sand and gravel and mariculture.

Conservation

English Nature is the statutory adviser to the Government with regard to nature conservation in England and promotes the preservation of wildlife and natural features. Scottish Natural Heritage has a similar role in Scotland, while the Countryside Council for Wales is responsible for the conservation of the natural environment in Wales.

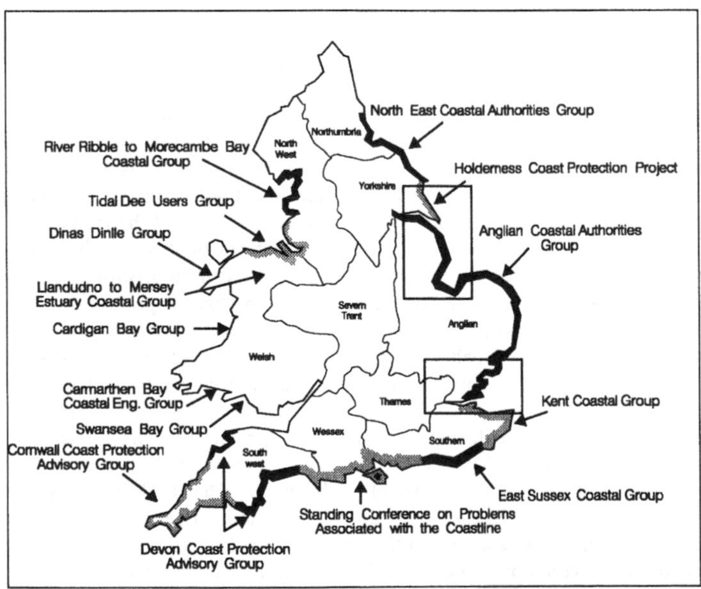

Figure 3. Jurisdiction of statutory organizations in the United Kingdom's coastal zone (redrawn from House of Commons 1992)

UNITED KINGDOM RESEARCH IN NEARSHORE RESOURCES AND PROCESSES. WHAT IS BEING DONE? WHAT HAS BEEN DONE? WHAT WILL BE DONE?

The following paragraphs describe some examples of coastal management from the east coast of England. As the Environment Committee of the House of Commons specified in its recently published report (March 1992), the earth sciences have a significant role to play, and a detailed knowledge of the distribution of nearshore sediments is the background necessary to any management plan.

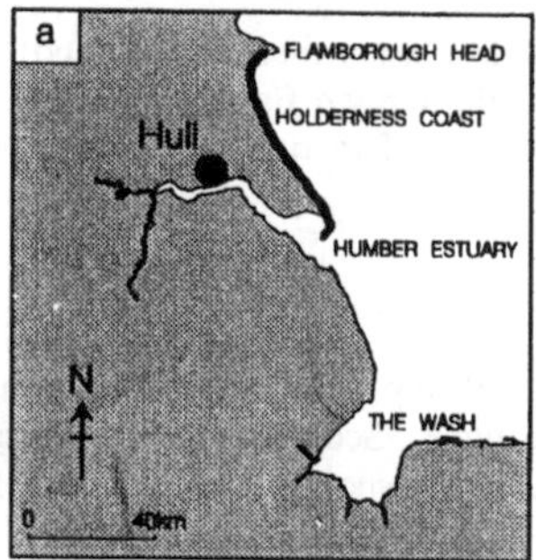

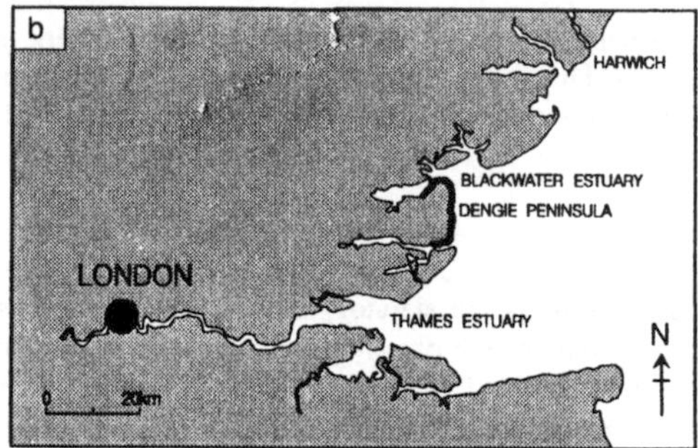

Figure 4. Case studies in coastal management from eastern England: a. Holderness Coast; b. Essex Coast

Erosion of Cliffs

The Holderness coast stretches between Flamborough Head and the mouth of the Humber Estuary (Figure 4a). The area is covered by tills, extremely heterogeneous Quaternary glacial deposits. At the end of the last glaciation (about 10,000 years ago), the rise in sea level initiated the erosion of the tills, a process that is still active. The average erosion rate along the coast is about 1.7 m/year (Valentin 1971), although locally this figure can be much greater. Coastal erosion in this area has been documented since Roman times, and several villages have been lost. Erosion occurs mainly because the cliffs are attacked at their base by waves at high tide, leading to collapse by rotational landslides. Littoral drift is southward, with some sediment deposited at the mouth of the Humber Estuary, creating the sand and shingle spit of Spurn Head. Coastal management of this area is particularly complex because of the existence of cliff-top houses, arable land, and leisure/tourism-related investment.

It is important to bear in mind the consequences of any coastal protection scheme in the area designed to reduce the rate of erosion. Because the whole of Holderness is a single sediment transport cell, the creation of barriers might halt erosion at one site, but will produce an increase downdrift (Figure 5). This could lead to the submission of compensation claims by residents further along the coast (Surveyor 1992).

Thus coastal planners face the dilemma of whether to protect the whole coastline or adopt a do-nothing approach. If the option of protection is chosen, there would be considerable expenditure. Protection would also have serious environmental implications. The Holderness cliffs are an important source of sediments that feed the English coast to the south and further afield. A solution may be to build revetments at the cliff base using scrap tires (Hydraulics Research 1991). This would solve the problem locally, but the overall picture described above regarding the total sediment budget would not change. The most suitable, long-term solution may be to allow the coastal retreat option. Compensation schemes for the local residents could be the most effective management option.

There is a need for a detailed knowledge of the stratigraphy and geotechnical properties of the tills. The identification of sites where the cliff material is particularly soft or where there is a concentration in wave energy, could effectively support coastal planning. Indeed, one of the main aims of NERC's Land Ocean Interaction Study is to study sediment sources, modes of sediment transport, and sediment sinks along the coast.

Figure 5. Coastal erosion at Hornsea on the Holderness Coast (United Kingdom)

Sea Defenses and Conservation

The saltmarshes and mudflats of the Essex coast (Figure 4b) are coastal environments typical of macrotidal conditions and are one of the most important havens for the coastal wildlife of the United Kingdom. They have lost on average almost 25% of their area in the past fifteen years (English Nature 1992). This loss is due to several factors, some natural (relative sea level rise, increase in storminess) and some manmade (reclamation of land with subsequent construction of coastal defenses).

Saltmarshes are presently being studied in detail, not only because of their ecology, but also because of their significance for sea defense. They form natural defenses against relative sea level rise, because they are able to dissipate wave energy and their creeks dissipate tidal energy. If saltmarshes were allowed to migrate inland without constraints from manmade structures, they would respond to sea level rise by adjusting their position along the coast. Because of the great demand for agricultural land, reclamation in Essex started as early as in the 18th century. For example, in the Blackwater Estuary, the construction of earth banks and sea walls was carried out between 1774 and 1799, and by 1840, a great part of saltmarshes were transformed into grazing marsh (English Nature 1992). The reclaimed land is now located below sea level and therefore subject to flooding when the sea walls fail. Saltmarshes in Essex are not free to migrate inland because they are backed by the sea walls that now need upgrading works. Due to increased wave action, the tidal flats have expanded at the expense of saltmarshes. Unless the saltmarsh edge is able to migrate, the intertidal flat will not adjust its profile according to the new energy configuration, worsening the erosion of the saltmarsh (Pethick 1992).

English Nature (1992) considered three scenarios of coastal management for saltmarshes and mudflats. Of prime consideration is the economic value of the land protected by the sea wall, either for agriculture or for housing. Conservation of important wildlife habitats is also a major parameter in the formulation of the plan. These are the options.

Scenario 1. Managed retreat - abandonment of current line of sea defenses protecting land in low-value use

Scenario 2. Partial set-back onto land with existing conservation interest

Scenario 3. Maintenance of existing defences protecting low-lying land in high value use

Managed retreat consists of building a new line of sea defenses further inland. The outer sea wall is then lowered in order to allow flooding at high tide and subsequent deposition of fine-grained sediment. The flooded area is then slowly drained, allowing a run-off rate that permits deposition of silt that will be colonized by vegetation. The effects on coastal geomorphology are yet to be fully assessed. If the managed retreat experiment (in the Blackwater Estuary in Essex), jointly carried out by the NRA and English Nature, proves to be effective, this might become a good way to protect the coastal environment. It is implied that compensation will have to be given to the landowners for the loss.

Partial set-back applies when a freshwater habitat is located behind the sea wall. A low embankment is built seaward in order to decrease wave action and to create a brackish lagoon. This wall is overtopped only at certain high tides. A second sea wall is then built to preserve the freshwater habitat that will have partially retreated landward. The animals and plants of grazing marshes tend to relocate fairly easily (English Nature 1992).

The *maintenance of present coastal defenses* (Scenario 3) does not necessarily mean the building of hard structures. Figure 6 shows how, in the Dengie Peninsula (Essex), some old barges have been recycled as offshore barriers (on the left of the photograph) together with groins made with stakes and geotextile (on the right). English Nature declares that this scheme has produced accretion on the mudflat, even if the effects of these barriers on the adjacent areas are not fully understood.

There is a need to carry out further research, not only with regard to present coastal processes but also to the long-term effects.

Figure 6. Old barges used as offshore barriers in the Dengie Peninsula, (United Kingdom.)

Geotechnical Properties of Intertidal Mud Deposits

The effect of compaction on the muds of the intertidal area is still unknown, and it is not clear how geotechnical properties control the response of saltmarshes to sea level rise. The National Rivers Authority gives an allowance of sea level rise of 6mm/year for the design of coastal defenses in Eastern England (House of Commons 1992). What is going to happen if this rate increases? How many of the coastal habitats are going to be affected? The answer to these questions can only come from a detailed geological and topographical survey of the whole area, since the information presently available is fragmentary and often has limited resolution.

Soft Coastal Engineering and Seabed Resources

In the United Kingdom as elsewhere, soft engineering is gaining favor as a solution to the classical problem of beach erosion. Traditional defenses have high costs and their construction involves significant environmental impact (noise, availability of materials, detrimental aesthetic effects).

One of the major problems in adopting recharge schemes is the availability of suitable recharge material. In order to recharge a beach effectively, it is necessary to use sediment selected according to the critical threshold of cross-shore and longshore currents. In the past, sediment was extracted from land sources, but this involved a high environmental impact. For this reason, where possible, the material should be dredged from the seabed and discharged directly onto the beach. This requires a good knowledge of the distribution of offshore deposits.

The British Geological Survey (BGS) makes available an extensive data base and high quality maps of the seabed sediments on the continental shelf of the United Kingdom. However, these data, with a few exceptions (e.g., the Wash), are not available for shallow waters (above the -10 m contour) because of the size of the vessels used in the data collection. There is a need to extend the surveyed areas further inshore, in order to match the marine geological maps with those produced by surveys on land. The BGS Seabed Sediment maps show that vast areas of the United Kingdom's shelf are covered by sand and gravel deposits. However BGS estimates show that only 10-15% of these deposits can be considered economically workable because of their thickness. BGS has carried out more detailed studies on these deposits on behalf of the Crown Estates and the Department of the Environment, for three areas of eastern and southern England. A great deal of information is also available from the aggregate industry for the areas where prospecting licences have been issued. However, some of the information is not properly fed into an archive widely available (e.g., the BGS Offshore Database). There is a need to record systematically this data and disseminate it, carrying out a full assessment of the United Kingdom seabed resources. An increasing demand for sand and gravel to be used in beach recharge schemes is forecast for the future, and it is necessary to understand the role of possible sources of sediment, such as nearshore banks, in the coastal dynamics.

THE LEGAL FRAMEWORK IN ITALY: REGULATIONS AND RESPONSIBILITIES

Compared to the situation in the United Kingdom, the number of bodies having direct jurisdiction on the Italian coastal zone is much smaller. However the legal structure is complex when looked at in detail.

Coastal Defenses

The law 14-7-1907 n. 542 has distinguished two types of coastal defenses; those that protect built up areas and those to halt erosion of beaches. With regard to the former, the local planning authority (the Comune) is in charge and can obtain financial support from the Ministero dei Lavori Pubblici (Ministry of Public Works), once its operative section (the Genio Civile per le Opere Marittime or G.C.O.M.) has deemed the works to be technically feasible. If coastal erosion does not threaten buildings, both the Comune and the Capitaneria di Porto (Port Authority) can apply for a grant from the Ministero dei Lavori Pubblici in order to build the defenses.

Exploitation of Resources

The Ministero delle Finanze (Ministry of Finance) is the owner of the coast, having a role similar to the Crown Estate in the United Kingdom. It is in charge of any administrative problem related to the coast (called "Demanio Marittimo dello Stato" or governmental marine estate), but does not carry out any monitoring. The Ministero della Marina Mercantile (Ministry of the Navy) is the department in charge of commercial maritime activities and carries out its statutory duties of monitoring and protection of the marine environment through the Capitaneria di Porto. It recommends or halts applications for concession of dredging licences.

Conservation

The Italian coastal zone is, in many ways, very different than that of the United Kingdom. Italy is a country that owes a great part of its wealth to tourism, mainly visitors from Northern Europe. However, the years following the "boom" of the '60s have seen a decrease in the number of foreign visitors, mainly because of the competition from other countries in the Mediterranean area. Nevertheless, the Italians still spend a good part of their summer holidays on the home coast, mainly because many of them own a second house at the seaside. This great demand for housing located in the coastal area has led to high environmental pressures on the coast, often resulting in uncontrolled construction of villas and bungalows that do not always comply with sensible planning criteria.

When the government created its task force in the '60s for the development of southern Italy (Cassa del Mezzogiorno), it produced planning guidelines for the maritime authorities (Stura 1985). Important points from these guidelines include:

1. Natural landforms such as dunes and coastal woodlands should be preserved.
2. Any building should be located at a convenient distance from the shoreline in order to allow free access to the shoreline.
3. Main roads should not be built along the shoreline.
4. Any development should be in harmony with the surrounding landscape.

As Stura (1985) pointed out, these guidelines have been ignored, and all the areas of the coast of southern Italy exploited for housing are in a poor condition. From the geoscientist's viewpoint, this uncontrolled development has undermined the coast, because dune areas that were acting as a natural protection of the hinterland have been destroyed and replaced by houses, roads, and car parks. In 1989, Law *"Galasso"* (1479/89) restricted development to 300 m inland from the high water mark. Stabilization of the dunes in order to protect the hinterland, is now being considered.

The total absence of a statutory body for conservation in Italy has halted the production of a Coastal Zone Management Plan. Although Law 979 defined as marine environments "waters, shores, sea-bed, flora and fauna", the departments which are supposed to manage this heritage, the Ministero dell'Ambiente (Ministry of the Environment) and the Ministero della Marina Mercantile, have failed to act in coordination.

The Unique Legal Framework and Environmental Management of Venice

Venice can justifiably be defined as "the most astonishing monument of Western Europe" (The

Economist 1992), and for this reason, successive Italian governments have spent the last twenty-five years trying to save it from sinking into the heavily polluted water of its lagoon. Following the disastrous flooding of 1966, the Italian government produced in 1973, a special law that declared Venice and the Lagoon to be an area of "national heritage suffering possible destruction." However, the most significant step in the environmental management of this coastal area was the creation of the Consorzio Venezia Nuova in 1984. This is a consortium of twenty-one, state-owned and private companies, designated by the government as managers of the environmental research on the lagoon-barrier island system. The Consorzio has produced a proposal for the creation of tidal barriers at the inlets (Progetto Mosé) to be built by the year 2000 (The Economist 1992).

ITALIAN RESEARCH IN NEARSHORE RESOURCES AND PROCESSES. WHAT IS BEING DONE? WHAT HAS BEEN DONE? WHAT WILL BE DONE?

The different geomorphological domains have been studied in research projects mainly carried out by Consiglio Nazionale delle Ricerche (CNR) institutes and universities, as part of the *Progetto Finalizzato Conservazione del Suolo* (coastal dynamics) in the '70s and the *Progetto Finalizzato Oceanografia e Fondi Marini* (oceanography and seabed sediments) in the '80s. Remarkable products of this research are, for example, the seabed sediment maps of the Adriatic (CNR 1988). One of the main conclusions from these studies was that the majority of the beaches of the northern Adriatic Italian coastline were eroding. In addition, the Venice Lagoon and areas adjacent to the Po Delta (e.g., Ravenna) were suffering from subsidence, mainly due to abstraction of groundwater and hydrocarbons. The Venice Lagoon is currently the target for another research program involving the CNR and universities, *Progetto Venezia*, aiming to refine the conclusions of previous research on the Lagoon. There is also considerable geoscience research carried out by the subsidiaries of the Ente Nazionale Idrocarburi (ENI), the state owned petroleum company.

CASE STUDIES FROM THE ADRIATIC COAST OF NORTHERN AND CENTRAL ITALY. PROBLEMS, PRESENT-DAY SOLUTIONS, AND FUTURE DEVELOPMENTS

The northern Adriatic is a shallow shelf sea, with an average slope from the shoreline to the -30 m contour of 1:800. The coast of northern Italy is under high environmental pressure because of pollution and urbanization. The coastal waters have recently suffered from a marked eutrophication that has caused great concern to the local city councils and the regional authorities, because tourism is one of the main sources of income in the area. The phenomenon was caused by excessive discharge of nutrients into the Po River by factories and farms. Recently the discharge has been more carefully monitored and the presence of mucilage formed by dinoflagellates (unicellular algae) has been reduced.

Pollution is not the only problem along this part of the Adriatic coast. From Venice southwards there is an almost continuous stretch of tourist resorts, with the exception of the Po Delta. Coastal dunes have been destroyed and replaced with housing estates, and coastal erosion has been stopped by constructing breakwaters and rock groins that have altered the littoral sediment budget, producing erosion downdrift.

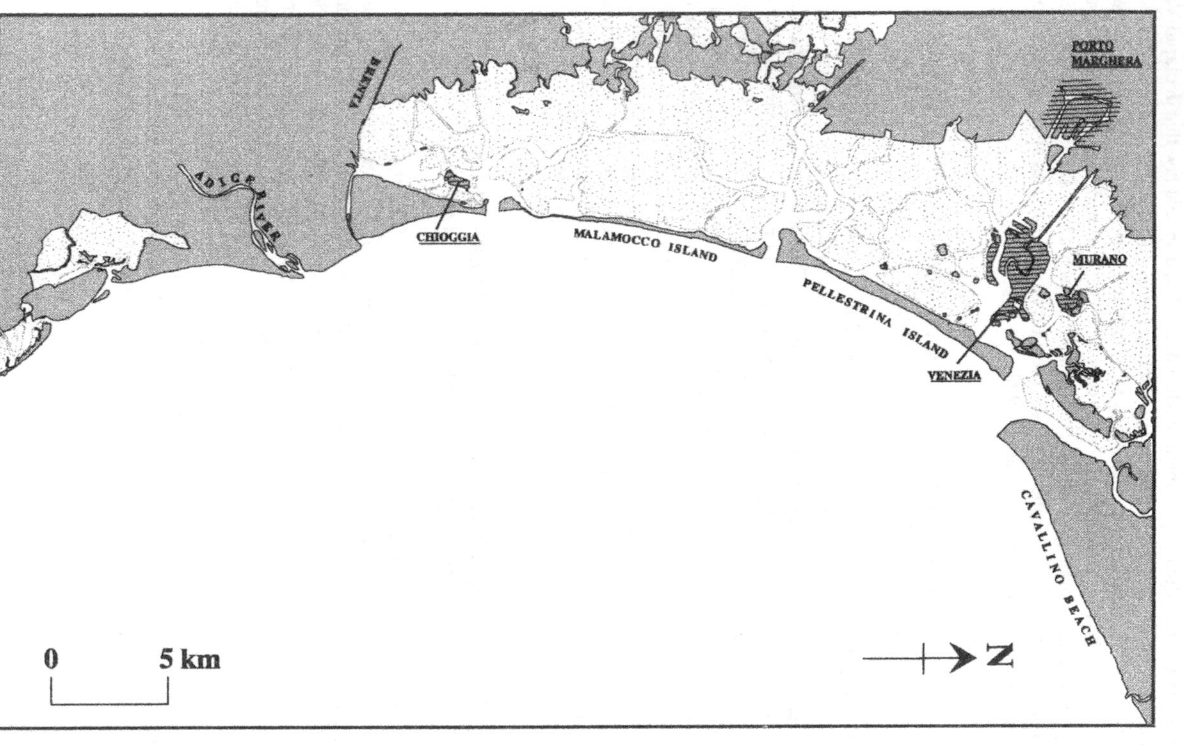

Figure 7. The Venice Lagoon (based on Admiralty Charts No 1483, 1989)

The Venice Lagoon and the "Lidi Veneziani" Barrier Island System

The Venice Lagoon (Figure 7) was formed about 7,000 years ago, when rising sea level flooded the Upper Adriatic paleoplain. The Lagoon is separated fromthe sea by a barrier-island system, the Lidi Veneziani, and has a tidal range of about 0.5 m. Storm surges can increase this range, causing the famous floods.

Because until the 14th century the Lagoon used to have five major rivers discharging a considerable amount of sediment, it rapidly silted up, and the adjacent area suffered from disastrous floods. Between the 14th and 17th century, the Republic of Venice carried out works in order to maintain the Lagoon, diverting the river mouths into the open sea and creating navigable inlets between the islands (Lidi). In 1600, they also diverted the mouth of the Po Delta southwards. Because of the change in freshwater discharge, the ecology of the Lagoon became more saline, favoring the growth of saltmarsh vegetation. The major effects of human-related activities can be seen in the increased subsidence due to abstraction of freshwater (stopped in the early '70s), pollution of the Lagoon, and erosion of the natural creeks of the Lagoon due to the opening of inlets.

The Lidi had to be protected against the force of waves by sea walls (locally called *"murazzi"*). As Marabini (1985) noticed, the sea walls have acted as barriers to the natural migration of the barrier island system, producing a steepening of the submerged beach. This means that storm waves are now able to reach the shore without breaking, removing offshore the beach sediment in the winter months, especially in the central part of the islands.

Here the effect of sea walls is comparable to that observed in Essex in the United Kingdom. If the model proposed by English Nature for the Essex Coast were going to be followed, the coastal management that seems most appropriate is the *Maintenance of existing defenses protecting low-lying land in high value use* (Scenario 3). However, it also seems necessary to carry out beach replenishment schemes in the midsection of the islands, since the construction of long jetties at the inlets has stopped longshore transport. The redistribution of nourished sand offshore during the winter months would allow the formation of offshore bars, acting as natural barriers against storm waves that will break on them. Because a component of longshore transport is still present, as confirmed by the accretion against the jetties at the tidal inlets (Gatto 1984), part of the sand will be accumulated against the jetties. This could be periodically removed and deposited back onto the source area.

The Po Delta Wetlands

The Po Delta covers a surface area of about 73,000 hectares, of which 85% is reclaimed land (Marabini and Veggiani 1991). The landscape is extremely flat and is an alternation of polders and marshes that grow between distributary channels. Several brackish lagoons occupy the coastal zone and are separated from the sea by sandbars. Because of local subsidence produced by the sediment load of the Delta, most of its area lies below sea level (Figure 8b) and is subject to flooding.

The river course has been greatly influenced since the Roman times by engineering works such as artificial embankments and canals. Ciabatti (1966) has recognized two different phases in the evolution of the Delta in modern times. The first phase, up to 1600 AD is characterized by ten, cuspate deltas, followed by a second phase with two, lobate deltas up to the present-day. These morphological changes can be related to the engineering works carried out by the Venetians in the 17th century.

The sandbars gradually grow up to sea level and once above water, they are colonized by plants and extend laterally to join the mainland creating lagoons behind them. The rate of growth of the Delta is controlled by the balance between the sediment load of the river and wave action. Marabini and Veggiani (1991) have considered how changes in rainfall and input of sediment from the catchment area might control the progradation of the Delta. Marabini and Veggiani (1991) also noted that the extraction of sand and gravel from the riverbeds and the construction of dams and artificial basins upstream might have reduced the sediment load, decreasing the potential of the Delta to adjust its morphology in response to increased wave action. The predicted sea level rise scenario has serious implications for the wildlife of the marshes and for the population living in the low-lying areas. Moreover, because of a decreased sediment input, the Delta will not be able to cope with the local subsidence, increasing the net rise in sea level.

It will also be necessary to consider an approach comparable to the one adopted by English Nature for the Essex marshes in the United Kingdom, considering to what extent it is worth mantaining the existing rock embankments that protect the marshy areas which are presently below sea level.

The "Riviera Romagnola" and the Coast of Central Italy

From the Po Delta southward, there are two geomorphological domains. The first stretching from the Po south to the Gabicce headland ("riviera"), comprises a sandy coastline, gently sloping towards the central Adriatic. The second, between Gabicce and Punta Penna, comprises narrow beaches often at the base of high coastal cliffs that are part of the Appennine mountain belt. In this zone the Ancona headland separates two, different, coastal sediment transport cells.

In both cases, coastal erosion has been tackled using hard engineering solutions, such as offshore breakwaters and rock groins. A rather peculiar type of hard coastal defense are star-shaped blocks (Figure 9) used for example, at Porto Recanati. Although these structures might effectively protect the shore (Carbognin and Marabini 1987), they are costly and have a detrimental impact on the recreational use of the beach (low aesthetic value). The construction of closely spaced, offshore breakwaters in the Emilia-Romagna littoral zone has also caused some unpleasant secondary effects, such as a concentration of algae and dirt in the water between the beach and the structure due to decreased water exchange with the open sea in periods of low wave activity (summer). The examples presented could be considered as cases for beach recharge schemes.

A Lack of Information: Nearshore Resources

In contrast to the United Kingdom, the Italian government has not expressed any clear view on coastal management. The implementation of planning strategies is left entirely at a regional (*Regione*) and county level (*Comune*), creating enormous differences in the knowledge of areas often located within the same coastal cell. Presently, the policy pursued is the "emergency measure action", whereby action is taken only when people and structures are endangered. Moreover, the only remedial action taken is often the hard engineering option (because of the gravity of the situation), involving great government expenditure and a decrease in the recreational value of the coastal zone.

In 1985, the government started to support research on the coastal environment, carried out by universities and institutes of CNR. In the near future, the government is likely to publish the draft of a coastal management plan for the whole country (L'Industria Mineraria 1992). However, as the Industria Mineraria reports, it is very unlikely that any great expenditure will be allocated in the near future, because of the present economical climate. The draft has been jointly prepared by the Ministero della Marina Mercantile and the recently created Ministero dell'Ambiente and should clarify the responsibilities for monitoring and acting in the coastal zone. However, it is legitimate to question the way in which the management plan is going to be implemented.

One of the main obstacles to be faced is the lack of geological knowledge of the Italian coastal zone, both onshore and offshore. In Italy, thematic mapping has been left to the regional authorities, who have created a Regional Series of geological maps. It is evident that some regions that are "environmentally aware" have published a significant amount of information, whereas others have virtually no geological information available to the public.

a

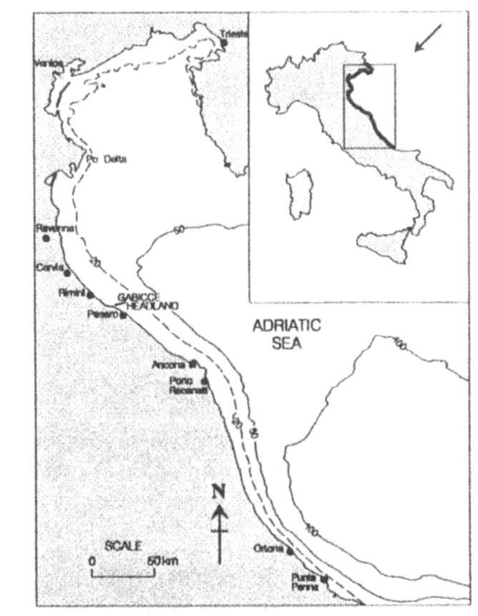

b

Figure 8. a. The Adriatic Coast of northern and central Italy
b. The Po Delta (redrawn from Marabini 1985)

Figure 9. Coastal defenses at Porto Recanati near Ancona (Italy) (Courtesy of F. Marabini and A. Santaniello)

The Servizio Geologico Italiano (Italian Geological Survey) is planning to create 652 map sheets for the whole of the Italian territory at a 1:50.000 scale (Todisco et al. 1992), of which 226 cover marine and coastal areas (CARG Project). The CARG project is still in its early stages, and the geological survey is not going to carry out any surveying directly, but through subcontracting to research institutes and universities. Presently only pilot projects are scheduled and one of these is likely to be in the northern Adriatic.

Beach replenishment, as an effective method of coastal protection, has not achieved the same popularity in Italy as it has in the United Kingdom. Although applied locally in some regions such as Emilia Romagna (Cervia) and Lazio (Lido di Ostia), its design is often handicapped by the lack of knowledge of available sediment sources. Experience in the United Kingdom has shown that a reconnaissance geological study is essential to any management plan. The extraction of sand and gravel from riverbeds should be regulated more carefully, since the negative effects are seen further downstream, in the coastal zone. The practice of using beach sediments for the aggregate industry, as it happens in Calabria in southern Italy (Pratesi and Canu 1993), has to be stopped immediately, and adequate alternative sources have to be sought offshore. It is hoped that the Italian authorities will take this into account when drafting their plan.

CONCLUSIONS

The comparison of coastal management issues and strategies in the United Kingdom and Italy has highlighted many differences and similarities.

Although the northern Adriatic and the western North Sea have different oceanography (exposure to winds, average temperatures, tidal excursion), problems experienced on their coasts are rather similar. In the United Kingdom the government has taken clear action in managing the coastline, while in Italy there is still some confusion.

Coastal defense strategies used in the United Kingdom to protect coastal lowlands, beaches and cliffs, could be applied on the Italian coastline, after adapting them in order to solve local conflicts of interest.

Other countries in the European Community have adopted approaches comparable to both the United Kingdom and Italy. In The Netherlands, priority is given to issues such as dune restoration and effects of sea level rise at a national level (Rijkswaterstaat 1990). In Spain, this strategy is implemented at a regional level, such as in Andalucia, where thematic coastal maps have been produced (Junta de Andalucia 1989) at 1:50.000 scale.

The Commission of the European Communities is presently preparing a Coastal Zone Management Plan for community members. The document is presently in its final stages of review (Comment to the House of Commons by Olivier Bommelaer,DGXI, 1992, 16.1) and should be published shortly.

Such a plan should pay special attention to the following topics:

1. Understanding of coastal processes in order to delimit coastal cells having internal homogeneity with regard to coastal dynamics
2. Knowledge of sediment sources and sinks within the coastal sediment budget
3. Geotechnical assessment of coastal sediments
4. Implementation of soft-engineering solutions having low impact on the environment
5. Full assessment of nearshore seabed sand and gravel resources for use in recharge schemes
6. Full assessment of long and short-term geological hazards, such as sea level rise in coastal lowlands and earthquakes in seismic areas

ACKNOWLEDGMENTS

The author wishes to acknowledge all the institutions and individuals that have given him information on coastal zone management in Europe, presented in this report. Special thanks to Dr. A. Santaniello and Dr. F. Marabini for supplying details of coastal defences in Italy, Miss H. Conefrey and Dr. D. Brew for the help given in editing the manuscript, and Miss A. Hancock for drawing the diagrams. This paper is published with the permission of the Director of the British Geological Survey (NERC).

REFERENCES

Carbognin, L. and F. Marabini. 1987. Environmental impact of some coastal defence works. In *Proceedings of coastal and port engineering in developing countries*, ed. Nanjing Hydraulic Research Institute, 2:1697-1708. China Ocean Press.

Carter, R.G.W. 1988. *Coastal environments*. London, England: Academic Press.

Ciabatti, M. 1966. Ricerche sull'evoluzione del delta padano. In *Giornale di Geologia*, 34: 381-410, Bologna, Italy: Universitá di Bologna.

Coastal Engineering Research Centre (CERC). 1984. *Shore protection manual*. Washington, D.C.: U.S. Government Printing Office.

Consiglio Nazionale delle Ricerche (CNR). 1988. *Carta sedimentologica dell'Adriatico settentrionale*. Ed. A. Brambati and I. Uras. Novara, Italy: Istituto Geografico De Agostini.

Department of the Environment (DoE), Welsh Office. 1992. *Planning policy guidance:coastal planning*. PPG20. London, England: Her Majesty's Stationery Office.

English Nature. 1992. Coastal Zone Conservation: English Nature's rationale, objectives and practical recommendations. In *Campaign for a living coast, English Nature*. Peterborough, England: English Nature.

Gatto, P. 1984. Il cordone litoraneo della laguna di Venezia e le cause del suo degrado. In *Rapporti e Studi*, 9:163-193 ed. Commissione di studio difesa laguna e cittá di Venezia. Venice, Italy: Istituto Veneto Scienze Lettere ed Arti.

House of Commons, Environment Committee. 1992. *Second report coastal zone protection and planning*. Vol.1-2. London, England: Her Majesty's Stationery Office.

Hydraulics Research Ltd (HR). 1991. *A summary guide to the selection of coast protection works for geological sites of special scientific interest*. Wallingford, England: Hydraulics Research Ltd.

Junta de Andalucia. 1989. *Mapa fisiografico del litoral atlantico de Andalucia*. Ed. L. Menanteau, Vanney J.R. and Guillemot E. (Casa de Velasquez). Seville, Spain: Consejería de Obras Públicas y Transportes, Centro de Estudios Territoriales y Urbanos, Agencia de Medio Ambiente, Direccíon General de Planificacíon.

L'Industria Mineraria. 1992. *Migliaia di miliardi per le coste* Serie 3, Anno 12, 6 (December1992):24-5. Roma, Italy: Associazione Mineraria Italiana.

Marabini, F. 1985. Evolutional trend of the Adriatic coast (Italy). In *Proceedings Coastal Zone '85*, pp 13, New York: American Society of Civil Engineers.

Marabini, F., and A. Veggiani. 1991. Evolutional trend of the coastal zone and influence of the climatic fluctuations. In *Proceedings from COSU II*, ed. P.H. Grifman and J.A. Fawcett, pp.17 California: University of Southern California SeaGrant Program.

Natural Environment Research Council (NERC). 1992. *Land-Ocean Interaction Study (LOIS), science plan for a community research project*. Swindon, England: Natural Environment Research Council.

National Audit Office. 1992. *Coastal Defences in England*. London, England: Her Majesty's Stationery Office.

National Rivers Authority (NRA). 1992. *Battling the tide - Flood defences in the Anglian Region*. NRA Information Unit, P69/1/92. Peterborough, England: National Rivers Authority Anglian Region.

Pethick,J.S. 1992. Saltmarsh geomorphology. In *Saltmarshes:morphodynamics, conservation and engineering significance,* ed. Allen J.R.L. and K. Pye, 41-62. Cambridge, England: Cambridge University Press.

Pratesi, I. and A. Canu. 1993. Italy's Coastal Heritage. *Coastline*, 1: 3-6. Leiden, The Netherlands: EUCC.

Rijkswaterstaat. 1990. *A new Coastal Defence Policy for the Netherlands*. JG's-Gravenhage, The Netherlands: Rijkswaterstaat.

Stura, S. 1985. Il ruolo dell'Idraulica marittima nello studio dei litorali. In *La gestione delle aree costiere*, ed. Pranzini E., 46-47. Rome, Italy: Edizioni delle autonomie.

Surveyor. 1992. Clash over coastal works, comment by Jon Reeds. 22 October 1992. Sutton, England: Reed Business Publishing Ltd.

The Economist. 1992. Saving Venice. 8 February 1992: 23-26. London, England: Economist Newspaper Ltd.

Todisco, A., P. Catenacci, P. Lembo, N.A. Pantaleone, and L. Sacchi. 1992. The main aspects of marine geology in the "C.A.R.G. Project" of the National Geological Survey of Italy. In *Proceedings of geological development of the Sicilian-Tunisian Platform*. Congress held in Urbino, Italy, 4-6 November, 1992.

Valentin, H. 1971. Land loss at Holderness. In *Applied Coastal Geomorphology*, ed. Steers J.A., 116-137. Basingstoke, England: MacMillan Publishing Ltd.

42

New Offshore Opportunities in Fish Farming

Sante Scoglio and Tore L. Sveälv
Farmocean International AB, Gothenburg, Sweden

ABSTRACT

A more restrictive attitude to inshore farming, in addition to positive biological findings among offshore farmed fish, have increased the tendency to move fish farms offshore and the demand for suitable offshore fish farming cages/systems. The definition of the term "offshore" in today's fish farming, excludes the open sea where fish farming is negligible.

To develop an offshore cage/system, the requirements of the fish, the fish farmer, the insurance companies, the authorities, and the moorings must be considered. A description and evaluation of the offshore, "Farmocean Offshore Fish Farming Concept," is presented.

INTRODUCTION

Marine fish farming in cages (enclosures, pens, etc.) has increased enormously in popularity during the last two to three decades. During the 1960s and 1970s, the principal cage design consisted of a frame (including floating elements) to which a net bag was connected. The frame also had the function of holding the shape of the net bag. A jump net and a fence, both ~ 1 m high, in combination with a narrow walkway, were also standard. The cage volume was ~ 500 m^3 on average. However, the shape of the cages varied and the most popular ones were octagonal, round, or square. In general, they were manufactured of either wood, plastic, or galvanized steel. With few exceptions, the cages were moored one-by-one in a group or in rows. They could also be moored alongside a gangway that was either connected to land or anchored out at sea. Most of the work carried out at the cages, e.g., feeding the fish, was done manually.

Possible reasons for moving offshore are (1) all suitable inshore sites are already occupied; (2) inshore areas should be protected from industrial activities; (3) concerns about the effects of cage fish farming on the local environment; (4) conflicts between different users of the sea, e.g., fishermen and fish farmers; and (5) better and more stable water quality for the fish. In the beginning, ordinary and/or upgraded cages were moved out to more exposed sites; however, this was a disaster as most of them were destroyed.

The definition of the term "offshore " in this paper includes semisheltered and exposed areas, but totally excludes unsheltered areas and open sea. Today there are no fish farming activities in the open sea; presumbly there will be none in the near future.

The enormous expansion of cage farming the Atlantic salmon, (*Salmo salar*), in Western Europe during the 1980s can illustrate how the technical standards of the business have improved and how the farms have been moved to more and more exposed sites.

The aims of this paper are to: (1) discuss fish; (2) discuss fish farmer cage aspects that must be considered in offshore farming; (3) make a classification on the different offshore fish farms; and (4) present a description and evaluation of "Farmocean Offshore Fish Farming Concept".

THE OFFSHORE ENVIRONMENT

In moving offshore, wave, current, and wind forces increase rapidly. This means that maximum wave heights of 5-10 m, current speed of 2-3 knots, and wind speed of 35 m s^{-1} can occur at the same time and in the same direction. A requirement is, therefore, that the farming unit can withstand conditions like these. However, it is of vital importance that the farm can also satisfy the needs of the fish farmer. Even demands from the insurance company and the authorities must, in some cases, be considered.

Fish Demands

The fish need water of good quality in all aspects, and this is often not a problem in an offshore site. However, a heavy, fouled net can cause problems and for a row of nets, a shielding effect can arise which means that the current passes underneath and alongside the net bags instead of through them (Braaten and Dahle 1988).

The cage size and shape seems to be very important (Berveridge 1987). A large cage with a round shape seems to induce a higher activity and a more natural swimming path among salmonids (Sutterlin et al. 1979), in comparison with a square and smaller cage where a dead space often exists in the corners (Braaten and Hogoy 1982).

A deep cage gives the fish better possibilities to choose or avoid different water layers (Beveridge 1987). However, a very deep cage, especially when it has a small horizontal area, will have a negative influence on growth as it will, for example, be difficult to feed the fish effectively. This, in combination with severe and often unknown stress on the frame and net handling difficulties, means that the net bag depth must be chosen very carefully.

The water movements produced by waves are largely a surface phenomenon and the water particles do not move forward with the wave, but instead rotate in circular orbits (Figure 1).

The amplitude of the water particles decreases rapidly with depth (Beveridge 1987). In deep water, the wave motions at a water depth of more than half the wave length are almost zero. The fish will follow the same motions as the water particles in a constant pressure field as the fish are small compared with the wave length (Newman 1977).

Demands from the Fish Farmer

The cage should be strong enough, it should be easy to run, and the fish should thrive. To fulfill these demands in good weather is not difficult, but problems often occur when the weather worsens.

During work on single cages, the fish farmer needs an adequate work boat and an easy and safe way to board the cage, e.g., a boarding gangway located well away from the fish and always pointing downwind. This means that the cage must be equipped with a superstructure which will allow the farmer to take care of the fish safely for much of the year.

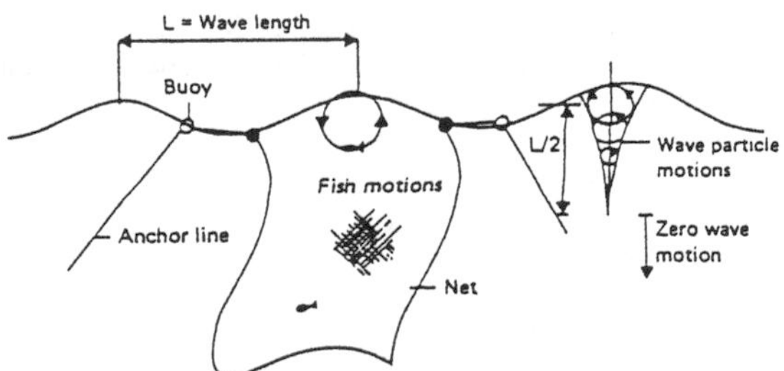

Figure 1. Motions of a cage and of wave particles
Source: Braaten and Dahle 1988

The superstructure should be equipped with a platform for such installations as a feed silo, and monitoring and alarm equipment. Even in good weather there will be days when the fish farmer is unable to visit the cage and feed the fish. This will result in reduced growth and cause stress to the fish. The provision of a feed silo with a dispenser mechanism coupled to a computer, will ensure that the fish are fed even when the farmer cannot be there.

A group of cages could be served from a floating control and feeding station, equipped with all necessary equipment. However, a proper work boat must also be available to make routines like net changing and maintenance easier.

Demands on the Mooring System

It is, of course, of vital importance that the mooring system keeps the cages on location. From the point of view of the fish and the fish farmer, it is also important to take into account the movements in both the horizontal and vertical directions. Therefore, during the planning and construction of a mooring system for a fish farm, the following must be considered so that they satisfy the security and technical aspects; fish needs, and fish farmer demands, such as soft and fexible mooring (especially vertically), simplicity, and as few parts and connections as possible for the ease of inspection of vital parts from the surface.

Other Demands

Concern over the security of stock and staff have increased as offshore fish farming activities have become more common. In Norway, a law has been introduced licensing cages/systems similar to that already governing shipping (Fisjeridirektoratet 1992).

DIFFERENT OFFSHORE FISH FARMING CONCEPTS

Many different concepts for offshore fish farming have been presented during recent years. They may be classified into five main groups (Sveälv 1991).

1. Simple, flexible one-net bag cages
2. Integrated, stiff, semi-submersible one-net bag constructions
3. Simple, stiff poly-net bag cages
4. Integrated, stiff poly-net bag constructions
5. Submersible and submerged cages

Very few of these cages/systems have been fully tested so far. Therefore, experience with respect to the biological and economic aspects cannot be evaluated and compared. Examples of cages/systems representing the different main group are as follows.

1. *Simple Flexible One-Net Bag Cages*
   a) Bridgestone "Hi-Seas Fish Cage"
   b) Fishtechnik Fredelsloh "Offshore Cage"
   c) Polarcirkel "Gigante"
   d) Hydac "Offshore Fish Cage"
   e) Dunlop "Tempest Fish Cage"
2. *Integrated Stiff Semi-Submersible One-Net Bag Constructions*
   a) Farmocean "Offshore Fish Farming System"
3. *Simple Stiff Poly-Net Bag Cages*
   a) Wavemaster "Fisg Farming Cage For Exposed Sites"
   b) Viking "Heavy Duty"
   c) Aqua Service International "Royal Neptun"
   d) UFN "Steel Cage"
4. *Integrated Stiff Poly-Net Bag Constructions*
   a) Aqua Systems International "Aqua System 104"
   b) Seacon Ltd. "Seacon Farm"
   c) Ewos "Giant Cage"
   d) Bomlo Construction Service "Fish Farm"
5. *Submersible and Submerged Cages*
   a) Many different types, but none with a specific trademark.

Description of the Farmocean "Offshore Fish Farming System" as an Example of an Integrated Stiff Semi-Submersible One-Net Bag Construction

In 1982, engineers at the University of Halmstad and at Swedish Maritime Research Institute (SSPA) started to design an offshore fish farm. The aim was to provide the best possible environment for the fish, as well as safe, practical, and profitable operation for the fish farmer. Extensive tests with a fully-instrumented, 1:10 scale model in the large wave tank at SSPA were performed. As the results were only positive, a commercial enterprise called Farmocean AB was established in 1985 to develop, produce, and market the concept. Since 1986, when the first Farmocean "Semi-Submersible Offshore Fish Farming System" was launched, ~ 20 systems have been delivered.

A Farmocean system consists of a tubular, galvanized steel structure mounted on a hexagonal pontoon (Figure 2). The system is available in two sizes, 3500 and 4500 m^3, depending on the size of the sinker tube. In its normal operation position, the farm floats in a semisubmerged position with the pontoon 3 m below the surface. The upper work platform and the feed silo are then located ~ 3 m above the surface.

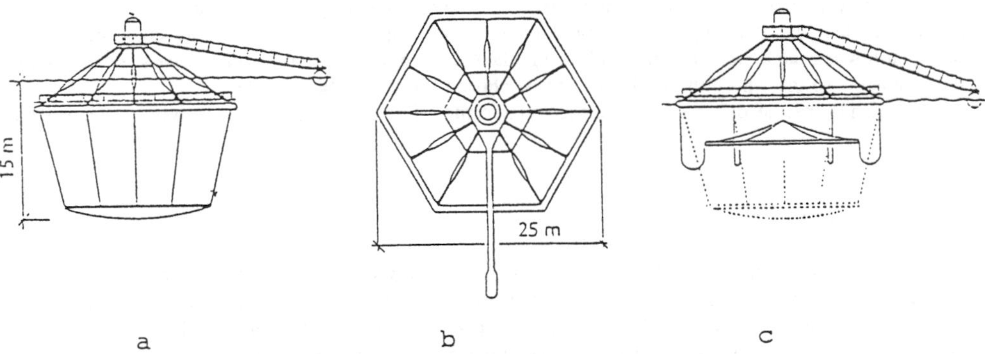

Figure 2. The Farmocean "Offshore Fish Farming System". (a) operating position; (b) seen from above; (c) service position and with the sinker tube hoisted up a little
Source: Farmocean pamphlet

During the development work, the Farmocean design was verified by DnV in a so-called concept verification, valid for 5.5 m waves, 35 m s^{-1} wind, and 2 knots current simultaneously, all acting in the same direction.

The net bag is suspended from the inside of the pontoon and the superstructure. Hanging outside the bottom of the net bag is a heavy sinker tube to which it is attached. This maintains the shape of the net bag and keeps it attached to the pontoon. In addition, twelve predator panels can be fitted to the tubular structures between the pontoon and the upperwork platform.

The Farmocean system is moored with a three-point, soft mooring system (Figure 3). This system corresponds, in principle, to the DnV requirements for safety against breakage, even under extreme circumstances. This mooring arrangement makes it relatively easy to moor a Farmocean system and, if necessary, it is also easy to move the system and its mooring to a new site.

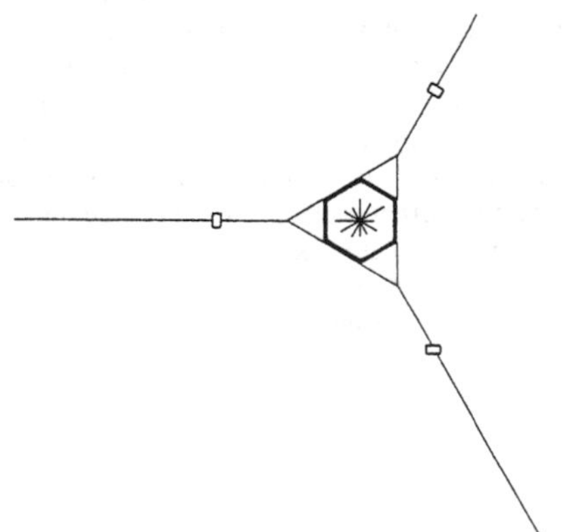

Figure 3. The tree-point mooring system of the farmocean "Offshore Fish Farming system"
Source: Farmocean pamphlet

Assembly of one Farmocean system takes five days, excluding the time for installation of the mooring system and attaching the net bag to the structure. These and other activities could be carried out by following the official manuals. Mooring plans are prepared by Farmocean in close contact with the customer, so that an optimum design can be chosen for any particular site.

The fish farmer needs a moderate service boat to run a Farmocean system for activities such as feed transport.

DISCUSSION

The trend to go offshore with fish farms has been most obvious during recent years. The reasons for this are: fish farms are no longer permitted inshore in many areas; land-based fish farms have become more expensive and exhibit more problems than expected; and the biological results have been very promising in offshore units. That the trend is real is confirmed by the many types of offshore fish farming units which have been developed during a short time period. However, the best evidence is the sales of e.g., the Farmocean systems.

Of the different offshore designs described, only a few cage types and the Farmocean system have been in commercial use for some years. This means that they have been tested in all possible aspects, such as strength, management, and suitability for the fish.

Strength

Frame, Net, and Mooring

The Farmocean system has proved many times that it can withstand periods of very bad weather (Beaufort 12) without any vital damaged. However, nothing is stronger than the weakest link and that is the net.

Farmocean, together with a leading trawl manufacturer in Denmark, has made a specially designed net bag suspended inside the structure. As the Farmocean system is a semisubmersible unit, the movements will be small, even in bad weather, which will minimize the wear and tear on the nets. The sinker tube will also guarantee that the net holds its shape and prevents it from floating up to the surface. The small movements and the low degree of deformation of the net bag will minimize the risk of the fish becoming scaled and/or trapped by the net. The continous movements of the net bags on the simple, flexible one-net bag cages, have often caused wear and tear in the upper parts of the net. To some extent, the brackets on the rubber frame have been the cause. To prevent the net bag bottom from floating up to the surface, in addition to the use of a lead bottom, a weight has been connected to the bottom of each corner. This solution has caused problems during bad weather as a lot of stress, in many directions, was put on the net bottom, often causing tears. Stressed and damaged fish, as a result of net movements, have been reported in some cases (Karlsen 1988).

The shielding effect will, of course, be much more pronounced for the polynet bag concepts, e.g., the "Aqua System 104", compared with the ones with one-net bag such as Farmocean system (Braaten and Dahle 1988).

Management

Fish Transfer

Fish can be moved from a Farmocean system by using either a hand or a fish pump. In addition, it is possible to attach a transfer tunnel to the Farmocean system and the transfer cage so tha the fish can swim into or out of the cage.

Feeding

The Farmocean system is equipped with a feed silo, that holds 3 tons of dry pelletized feed on the upper work platform, 3 m above the surface. As a large amount of feed is needed, a pneumatic transport system has been developed for dry feed. This can, in a gentle way, transfer 3 tons of feed from a service boat up to the silo in forty minutes. Feeding is automatic and computer controlled, with several, built-in compensations for such things as temperature, wave height, and increased biomass for achieving maximum growth and minimum feed consumption. However, without a skillful fish farmer who "knows his fish" and, based on this, can pre-set a suitable program, the result will not be adequate. Despite automatic feeding, it is still very important that the fish farmer checks the fish, e.g., once every day just before dawn. Practical experience shows that the investment in automatic feeding systems will soon be paid back in better growth, reduced feeding, better control, and more time to do other things (Braaten and Schei 1989).

Inspection, Handling, and Slaughter

Observing fish behavior and the net bag bottom are very important for the fish farmer, as they may tell him much about how the fish thrive. By using larger and deeper cages, the "contact" between the fish farmer and the fish is reduced. Diver inspections must therefore be carried out more often that before. The frame and the net bag may be inspected at the same time. While many fish farming companies utilize divers among their own personnel, agreements between professional divers and the fish farming companies will be more common as they need diving service more or less every week. Echo sounders, sonar, and television cameras for monitoring biomass and fish behavior have proved useful, but expensive and are not fully developed for direct use (Karlsen 1988).

Taking out samples of fish, e.g., to treat them against parasites, is often more difficult in a larger cage. This can however be done with different techniques, e.g., fish could be sampled at random using a dip-net. Dead fish could be collected by a diver.

The net bag bottom on the Farmocean system is equipped with a sock in the center, which can be opened from the outside. The dead fish will be collected in the sock and could then be removed from the outside without stressing the rest of the fish. The Farmocean system has a further device for bringing up dead fish without a diver. This is done by means of a hand net that can be slid along the center rope, which is attached to the sock ring in the net bottom and to the winch on the feed silo. However, there are indications that lifting up dead fish inside the net bag can stress the rest of the fish (reduce appetite) and spread diseases. Farmocean therefore recommends that fish farmers remove dead fish from the outside.

Fish could be graded by means of a fish pump and a grader arranged on a boat, in addition to some small cages attached to the boat. This grading arrangement is also well-suited for the Farmocean system.

The fish could be treated for parasites and diseases in the same manner as in smaller cages, with only a slight alteration in method. The treatment procedure must be carried out when the weather is good and the current is weak. A normal salmon lice, (*Lepeophtheirus salmonis*) treatment procedure on a Farmocean system starts with deballasting the farm to the service position. The sinker tube is then hoisted up and thus also the net bag. When a suitable volume is reached, depending on the amount of fish, six tarpaulin "sheets are lowered outside the net bag sides. The "Nuvan" solution is then either poured out evenly over the surface or pumped out through a perforated tube.

Recommended stocking should not be $>$20-30 smolts m^{-3} (Refstie and Kjonnoy 1986). The stocking number is, of course, dependent on the site and type of cage. However, ther are many opinions about what is the optimum fish density. There has been a tendency to increase fish density during the last decade from 10 up to 40-50 kg m^{-3} in some cases (Refstie and Kjonnoy 1986). Too low and too high densities may cause problems, e.g., poor water exchange or reduced growth, respectively (Refstie and Kjonnoy 1986). In general, densities of $\sim$ 20 kg m^{-3} are recommended for different cages/systems. This could, of course, be either lower or higher, depending on site conditions and the time of year. For brief periods, such as just before slaughter, the density may be much higher then recommended. The fish farmer can either choose to slaghter the fish during a short (some weeks) or a long (some months) time period. The result is the same, but the latter is more time consuming and more stressful for the fish. In both cases, the best way is to transfer a certain amount of fish from the offshore cage/systems into a small cage/cages that are towed inshore, where the fish could be starved and then slaughtered. By using a fish pump and lifting the net bottom, fish can be tranferred to a small cage lying adjacent to the Farmocean system. All these activities demand good weather.

Predators of different kinds, such as seals, spiny dogfish, (*Squalus achanthias*), and cormorants, (*Phalacracorax carbo*), can cause problems by killing or damaging fish, damaging nets, stressing fish so that feeding is disrupted, and also, reducing resistance to disease and promoting its spread (Beveridge 1987).

The Farmocean system can be equipped with a predator bag and twelve predator panels to protect the fish from a variety of predators. Predator nets can also protect the net bag against floating objects to some extent.

However, the normal nets are not sufficient protection against theft and sabotage. The shape of Farmocean system can probably reduce the ability of strangers to reach the fish to a certain extent. In addition, there are many possibilities for fitting alarm systems on the structure. Both the Bridgestone cage and the Farmocean are unmanned, as opposed to the stiff poly-net bag constructions where a 24-hour, manning for security reasons is possible, if needed.

Net Change

To ensure that the water exchange in the net bag is sufficient, it is important to change the net before it becomes to fouled. A clean net for the Farmocean system weighs between 500 and 800 kg, and may be placed in the water by hand. A crane is however necessary when the fouled net bag is lifted as it may weight up to ten times more than when in its clean condition. This problem must be considered when a boat for this activity is chosen.

By deballasting the Farmocean system to the service position, fouling down to ~ 3 m can be cleaned off. This possibility reduces the number of times the net bag has to be changed. When a net change is needed, a diver is required to disconnect the net bag bottom corners from the sinker tube. In addition, a service boat with a powerful crane and four men are needed to manage the procedure in ~5 h. The winch attached to the feed silo and the superstructure is very helpful when the fouled net bag is to be pulled and lifted out of the water and on board the boat. To faciliate the cleaning of the large net bags, net washing machines that can manage all sizes of net bags are now available.

Maintenance

As offshore fish farms are always influenced by waves, winds, currents, and marine growth, maintenance and inspections must be performed regularly. Depending on the type of cage/system, the maintenance routines can vary, however, they must be carried out and fouled.

The Farmocean system is divided into a steel structure, net bag, silo, automatic feeder, and mooring system, and there is a maintenance routine for each of these.

Law for Type Approval

Norway provides leadership in many aspects of fish farming, including being the first country to introduce a law for the type approval of floating fish farming units (Fiskeridirektoratet 1992). It was initiated by a request from bank representatives, suppliers of equipment, insurance companies, and fish farmers.

Units which are designed, manufactured, assembled, moored, and equipped in accordance with these rules, qualify for a certificate valid for a certain time period. It is still too early to reach conclusions about the importance of this tentative licensing system and its effect on other fish farming countries.

Economy

Blakstad (1988) made a comparison between an "average inshore fish farm" and an "average offshore fish farm" and found that the latter gave both a better result and a lower cost for producing 1 kg of fish.

Fish Results

Faster growth (15 %), lower mortality (50 %), and a reduced visceral fat content (14 %) have been found among fish in offshore fish farms compared with fish in conventional cages (Karlsen 1988; Sveälv 1988). Experiments with fish forced to swim against a water current have shown that these fish had more protein, less belly fat, and were of a better quality than fish in ordinary cages (Christie 1987). The environmental problems that often occur inshore due to self-pollution from fish farms (Braaten et al. 1983) will be greatly reduced when moving the cages/systems offshore. These results reflect the influence of deeper net bags (less waste), and better water quality and exchange, which results in a lower incidence of disease.

Optimum Volume

By increasing the net bag depth, the volume of a cage may be increased. There is often a temptation to do this, both among cage suppliers and fish farmers, as it in fact lowers the cubic metre cost of the cage and also, supposedly, makes it possible to hold a larger number of fish. However, as a consequence of such a volume increase, severe and often unknown stress on the frame, net handling difficulties, and more complicated husbandry of the fish may negatively affect both the cage/system and the production result.

Objections have been raised against the use of cages/systems enclosing very large volumes, pointing out the risks of holding hundreds of tons of fish in them. Based on the experience of the fish farmers during recent years, a suitable size for a cage/system seems to be $\sim$ 5000 m^3. Looking at a site with some Farmocean systems, each of the units could be considered as more or less an isolated unit. This will strongly reduce the risk of a major disaster that could easily happen in the case of poly-net bag constructions, e.g., during outbreaks of diseases.

As the fish farming business has reached industry status in recent years, it has also become more complex. In today's fish farming, not only the fish farming company, but also their bank and insurance company are highly involved in many important decisions. Therefore, it is essential that the suppliers of cages/systems supply the buyers with relevant and reliable information on their equipment so that they can make correct evaluations and decisions. This will be benificial not only for the individual fish farming company, but also for the fish farming business.

CONCLUSION

In the future, much fish farming will be done offshore. The concepts of farming offshore, being totally different, show very different methods of meeting the demands of the fish, the fish farmer, and the moorings.

Interesting concepts like simple/integrated poly-net bag cages/constructions have also been developed and are now under full-scale testing. Research and projects on submersible and submerged cages are expected to increase in the future. Additionally, rules for the type approval of floating fish farming units will presumably not only be in Norway in the near future. This if it proves to be effective, it will benefit the whole fish farming industry. Faster growth, lowered mortality and reduced self-pollution are examples of benefits from farming offshore. Offshore fish farms in general may also turn out to be economically superior, due to their lower production costs than inshore and land-based farms. The use of single-net bags instead of polynet bag designs will probably be favored because of better health control and lessened risk for insurance of stock and equipment. The Farmocean system can satisfy most of the demands that can be required for an offshore fish farm.

REFERENCES

Beveridge, M., ed. 1987. *Cage aquaculture*. Farnham: Fishing New Books Ltd.

Blakstad, F., 1988. Okonomiske fellesnevere for store merdanlegg i åpent farvann sammenlignet med tradisjonel oppdrett. Föredrad, NITO-konference, 9-11 June, 1988, Stavanger, 24 pp. (Mimeogr.).

Braaten, B., and L.A. Dahle 1988. Largescale offshore farming; Possibilities and limitations. In *Efficiency in aquaculture production:* ed. H. Rosenthal and E. Grimald, 117-43 Bortolazzi Stei, Verona, Italy.

Braaten, B. and Hogoy, I., 1982. Produksjon av matfisk. In; O. Ingebrigtsen, ed. Akvakultur: Oppdrett av laksetisk. NKS-forlaget, Oslo, pp. 146-191.

Braaten, B., and I. Schei. 1989. Culturing technologies: New developments. In *Aquaculture: A biotechnology in progress*, ed. N. De Pauw, E. Jaspers, H. Ackefors, and N. Wilkins, 973- 992. European Aquaculture Society, Vol.2. Bredene, Belgium.

Braaten, B., J. Aure, A. Ervik, and E. Boge, 1983. Pollution problems in Norwegian fish farming. International Council for the Exploration of the Sea, C.M./F:26, 11 pp. Mimeographed.

Christie, M., 1987. Flytende raceway - et nytt og bedre oppdrettsystem. Nor. Fiskeoppdrett, 5:33-35.

Fiskedirektoratet., 1992. Melding fra Fiskeridirektionen, J-154-92. Fiskeridirektoratet P.O.Box 185, 5002 Bergen, 1 pp.

Karlsen, L.I. 1988. Bridgestone oppdrettsmerd. Havmerd-prosjektet. Rapport U42. Fiskeriteknologisk forskningsinstitutt, Foredlingsseksjonen, Tromso, 31 pp.

Newman, J.N. 1977. *Marine hydrodynamics*. Cambridge, MA: The Massachusetts Institute of Technology Press.

Refstie, T., and M. Kjonnoy. 1986. Matfiskanlegg. In *Fiskeoppdrett med framtid* ed. T. Gjedrem, 114-38. Oslo: Landbruksforlaget.

Sutterlin, A.M., K.J. Jokola, and B. Holte. 1979. Swimming behaviour of salmonoid fish in oceans pens. *J. Fish. Res. Board Can.* 36:948-954.

Sveälv, T. 1988. Inshore versus offshore farming. *Aquacult. Eng.* 7:279-287.

Sveälv, T. 1991. Stategies and technologies in offshore farming. *Fisheries Research*. 10:329-349.

Ocean Space Development and Related Technologies

43
Keynote

Coastal Development in Harmony with the Environment

Willard Bascom
Research Associate, Scripps Institution of Oceanography
La Jolla, California

INTRODUCTION

The environment inevitably controls development. From the earliest times, the inhabitants of our world chose to live along the coast because they enjoyed watching the sea and eating seafood as much as we do. Early settlements were often built where rivers reached the sea because such a site had ample water; a good protective barrier on two sides; water transport; and often good fishing, soil, and timber. Originally, this was a harmonious arrangement; now the values of such sites have been degraded by dense population, urban sprawl, and the pollution of air and sea.

It seems appropriate that an international meeting such as this should consider the one environmental matter that will greatly influence the future of not only coastal zones, but of our entire planet. First I will compare the damage done to coasts by the forces of nature with damage caused by man. Then I will discuss the central problem of the future that we are vaguely aware of, but choose mostly to ignore while concentrating on trivia.

First, let us agree on definitions. Development means growth or increasing changes. Harmony means a combination of things that go well together. Thus, the question before us is whether coastal growth in the next thirty years will go well with our vision of what the environment should be. Environment means not only green grass, blue sea, clean sky, and unendangered animals, but the quality of human life. The last of these is in great danger unless we concentrate on the essential problem.

HISTORY AND GEOMORPHOLOGY

We live in a diverse world that contains many bodies of water; each different in size, depth, waves, currents, temperature, and sea life. Their coastal areas vary in structure and appearance from low sandy foreshores to steep rocky mountains rising from the sea; from coral rims to frozen beaches. On some coasts, rivers carry much sediment and build the coast seaward; on other coasts, the shoreline is retreating. At every site the problems are different, but there are useful lessons in geomorphology.

Consider the principal effect of nature on shorelines. After the last ice age, worldwide sea level began rising; it stabilized about 4,000 years ago and many coasts are still adjusting to the new level. The rising water drowned many valleys and caused waves to attack newly exposed headlands. The result is that the submerged valleys have been steadily filling in and headlands have been retreating. Longshore currents use the rocky debris to construct bay-mouth bars and enclose lagoons. As a coast straightens, the lagoons become steadily shallower, changing to wetlands and eventually to dry land. These effects are observable in a few decades and the use of new coastal lands formed in this way is under consideration in Spain, Turkey, and the United States. Coastal plains that at one time graded smoothly into the sea, now end against abrupt cliffs cut by waves. A number of places around the British Isles have reported sea cliffs of unconsolidated material retreating 10 to 20 meters in a single storm. In four years the city of Encinitas, California, atop a cliff 20 meters high, lost six city blocks to wave erosion. Much of America's sandy eastern shoreline is retreating with the resultant loss of valuable beachfront land.

Large rivers create deltas, expanding the land at their mouth. Where population pressures are great, people move onto the new land when it is barely above sea level. One result of this behavior can be great loss of life, as in Bangladesh at the mouth of the Ganges whenever storm tides sweep in from the Indian Ocean.

Erosion of inland mountains produces great quantities of sand and gravel that steadily move downward and extend the coast seaward. Much of the west coast of Italy has widened greatly in historic times. Pisa, once a coastal city, is now some 10 kilometers from the sea. Piombino and Monte Argentino, once islands, are now tied to the mainland by huge tombolos. Ostia, once the port of Rome, was overwhelmed by sediment a thousand years ago and is now inland. Ephesus and Cyzycus, in Turkey, met similar fates.

Earthquakes in the Aegean region toppled the buildings of many ancient coastal cities in Greece and Turkey. Associated tectonic changes have raised or lowered the land as much as 6 meters in some areas, with the result that harbors and parts of cities became unusable.

A great tsunami from the explosion of Santorini destroyed cities around the Aegean about 1470 BC, and water from a great wave went over the walls of Constantinople in 1509. Cities in Chile, Japan, Indonesia, Newfoundland, and Hawaii also have been wiped out by great waves. Port Royal, Jamaica, at the tip of a sandpit, suddenly subsided into the Caribbean during an earthquake in 1672, much as Helice, in the Gulf of Corinth had done in about 400 BC. Volcanic ash or lava has filled in harbors in Iceland and the Philippines in the last few years and, in Roman times, destroyed Pompeii and Herculaneum on the Bay of Naples.

Somehow those many violent acts of nature do not seem to have decreased man's desire to live by the sea. Humans have short memories and a fatalistic, tolerant attitude about living under a volcano, or in dry stream beds, or building houses at the edge of cliffs being undermined by the sea. The chance of natural destruction seems small. But one would expect more concern about man-made disasters where the outcome is certain.

THE EFFECTS OF MAN

For comparative purposes, let me cite some problems caused when man's developments defied nature's laws. For example: Tillamook, Oregon, in order to maintain a channel into its bay, built a single jetty that dammed sand moving south along the coast. This cut off the flow of sand to Kinchaloe Spit and thus destroyed what had been prime land containing a village and luxury hotel. Many jetties, intended to keep channels open for navigation, have been built on the coasts of Florida and California with similar damming, erosion, and losses.

In order to maintain sand on the famous Waikiki Beach in Hawaii, sand was removed from several other beaches which were relatively unpopulated at the time. Venice, Italy has taken enough household water from an aquifer beneath the city to cause it to subside so that storm-tide flooding results. At Long Beach, California, oil was removed from beneath its harbor, causing shipyards and an oil field to subside as much as 7 meters.

The damming of the Nile and the increased use of its water has reduced the supply of sand to the delta which is retreating rapidly; the loss of nutrients from the Nile has reduced the fish population of the eastern Mediterranean. The shrinking of the Aral Sea that resulted when its water supply was taken for cotton farming has made much of its former coastal zone unusable.

Pollution is a word that seems to mean something different to everyone. I define pollution as a "damaging excess." The concept of damage is important, because with modern analysis techniques, extremely small amounts of toxic substances that are not harmful can be found almost everywhere.

To some persons, pollution means trash or garbage; to others it is an oil spill, or loud noise, or bad odors, or pathogenic bacteria, or chemical contaminants. All these are associated with cities and, of course, they are our own fault. This kind of pollution is usually short-lived and local. It is well to remember that only the excess causes problems and it is unwise to spend a great effort trying to achieve excessively difficult standards of cleanliness. So much for man's minor environmental problems.

Now we come to a really serious problem. Development means growth and growth is caused by additional people. This "damaging excess" of people is the most difficult kind of pollution to deal with and much of the world still tries to ignore it. Virtually all human caused problems are traceable to the increase in population that is using up the world's space and resources. Finally, man can rival nature in doing serious environmental damage.

In 1930, when I was in high school, the population of the earth was about two billion. Since then it has tripled. In the four days of this conference, world population has increased by a million people. It now increases by 10,000 an hour, 260,000 a day, 95 million a year and a billion people in a decade. And the next decade, and the next. This fantastic growth (development, if you prefer) is, by far, the world's most serious environmental problem.

The various measures now being taken in hope of maintaining the world's forests, and animals, and fish, and farm lands, and water supplies, and fuels, cannot succeed in keeping up with the onrush of new people for more than a few decades. The world's coastal zones will be overrun by this onrushing mass of people as surely as the ancient cities mentioned earlier were overrun by tsunamis, or sediments, or volcanoes, or rising sea level.

The reduction in the quality of living caused by crowding and excessive urban development in much of the world is apparent. Large areas of major cities have become slums that cannot support themselves. Utilities and infrastructure are decaying.

In rural areas, farms are being divided into impossibly small fields, forests are being destroyed for firewood, and water supplies diverted to the extent that rivers are going dry. It is quite clear that the growth of population is the most serious threat to the concept of living in harmony with the environment.

ACTION REQUIRED

There is no point in our going about our ordinary business of coastal studies with eyes diverted, pretending that population increases are not our problem. The civilization we have known will be swept away by the deluge of people. In my lifetime I have seen a huge degradation of the environment caused by the addition of three times as many people; those of you who survive will see a very much worse effect in the next thirty years.

By then there will probably be nine billion people. By then the world's tropical forests will be mostly gone; there will be really serious energy shortages; some major rivers will no longer flow into the sea (the Nile and the Colorado are already in that condition), in the equatorial region, where most of the new people will be born, there will be far too little arable land. The new people will not be able to sustain themselves.

What can be done to slow this increase in people? Experts on this subject, which I am not, believe that education, especially of women, is the most promising possibility. This raises the question of how that can best be done.

Population Communications International (PCI) believes that by mobilizing the broadcast systems of the developing world (where 90% of world population growth takes place) audiences can be motivated to choose small families. They point to a dramatic success in Mexico where population growth dropped 40% in ten years. The motivating influence was "exciting, long running, family planning dramas broadcast in prime time on radio and television."

The experience gained in this experiment is now being put to use. PCI is taking this soap opera approach to leading broadcasters in developing countries who reach half the people in the world. Family planning soap operas are now in production or on the air in Brazil, India, Kenya, and Nigeria, as well as Mexico. I wish them success.

There are many other techniques for slowing population growth and all should be utilized. This is not the place for a discussion of details. The important thing that we can all do is to keep reminding others that population is the world's highest priority environmental problem and it must be attacked much more intensively, at once.

If we coastal specialists do not join with others who wish to protect the world environment, we may find too late, that we have spent our efforts on environmental trivia and missed the really important problem. Harmony will be impossible.

44

A Very Large Platform for Floating Offshore Facilities

Hajime Inoue and Reisaku Inoue
Ship Research Institute, Ministry of Transport
Tokyo, Japan

ABSTRACT

Many kinds of conceptual plans for huge floating platforms have been presented as an effective utilization of the ocean space.

Though a floating type of structure has a lot of merits in comparison with a reclamation type and is technically possible enough to construct, a huge floating platform has not yet become a reality. However, conceptual plans for such platforms have been recently presented by the Floating Structures Association of Japan.

The Ship Research Institute focused on designing the huge floating platform in 1977. In its first research activities, it has been investigating for about two years whether the construction of the Kansai International Airport as a floating type is technically possible (Ando et al. 1985).

After this feasibility study on the floating airport, fundamental investigations on the technical problems surrounding a huge floating platform were continued from 1982 to 1986 in order to ensure its successful construction (Ando et al. 1985; Takaishi et al. 1985).

The technology of the basic design elements of a huge floating platform were established by these fundamental and applied research projects in various fields. After that, in order to verify the element technology and confirm the safety and reliability of a huge floating platform, the at-sea experiment was carried out at 3 km offshore of Yura port, Yamagata prefecture in the Japan Sea from 1986 to 1990 (Hara et al. 1988).

FEASIBILITY STUDY ON FLOATING KANSAI INTERNATIONAL AIRPORT

Outline of Study

The technical feasibility of the floating Kansai airport model as shown in Figure 1 that had been proposed by Japan Ship Building Industrial Association, was investigated in both 1977 and 1978. This floating platform model was 5,000 m in length and 840 m in breadth, and supported by a lot of floating "columns." In this project, the following technical items concerned with the construction of the floating airport were investigated:

1. Determination of the environmental condition in the area
2. Estimation of the exciting forces acting on the floating bodies
3. Estimation of the mooring forces and behavior of floating bodies
4. Investigation of the structural strength of floating airport

The unsolved problems were clarified through the works of model testing and computer simulation. As the result of the these studies, it was concluded that the construction of the floating airport would be technically possible with the techniques available at that time. In addition to the above-mentioned technical feasibility studies, synthetic investigations on the floating airport were also performed.

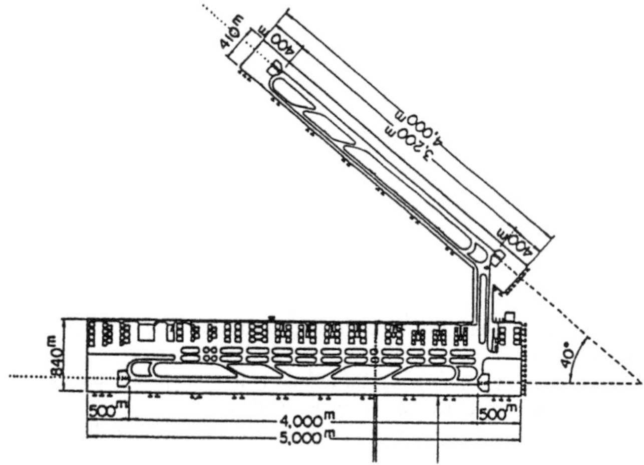

Figure 1. Proposed floating airport

Experimental Studies on Exciting Force

The Wind tunnel and basin tests were carried out in order to estimate accurately wind, wave, and current forces acting on the floating airport as follows:

1. The partial models on a scale ratio of 1/30 and the general model on a scale ratio of 1/1000 were used in the wind tunnel tests to investigate the wind forces acting on the upper structure or the columns between the upper deck and sea surface.
2. The two-dimensional basin tests were performed using a lot of column models on a scale ratio of 1/100 in order to investigate the mutual interaction of waves and the decrease in wave height when it passes among the columns of the floating body.
3. The basin tests, using the large structure model on a scale ratio of 1/30, were carried out to improve the accuracy of theoretical estimation of wave forces and wave drifting forces acting on the floating airport.
4. The two-dimensional basin tests were carried out using column models on a scale ratio of 1/17 in order to estimate the current forces acting on the columns of the floating body.

Experimental Studies on Motion and Mooring Forces

The basin tests were carried out to investigate the motion of the floating airport in waves using the large scale model as shown in Figure 2. the scale ratio of the model was 1/30 and the elasticity of the structure of the model was similar to that of the proposed floating airport.

It was concluded that the Dolphin type mooring would be the most suitable for the floating airport after investigating the three kinds of mooring: the Catenary type, the Tension-Leg type, and the Dolphin type.

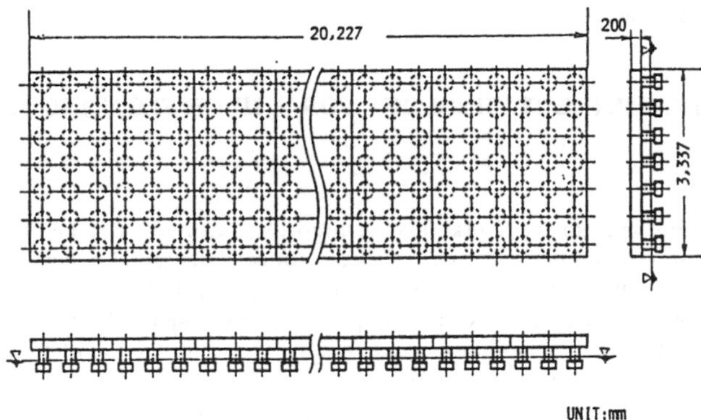

Figure 2. Floating structure model

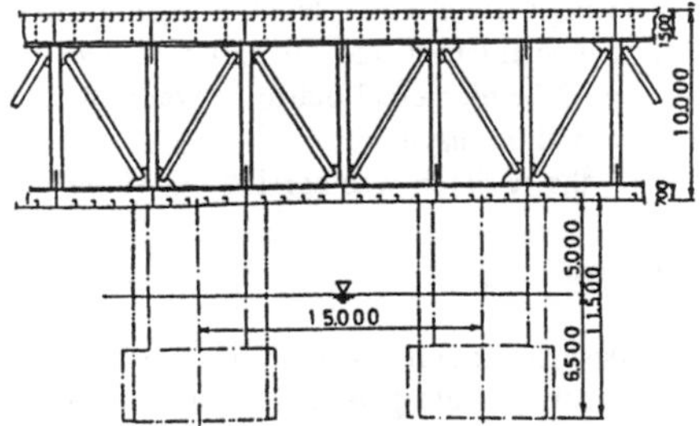

Figure 3. Structure of floating body

Structure of Floating Airport

The dimensions of main and subsidiary runway structures for the proposed floating airport were 5,000 m and 4,000 m in length and 840 m and 410 m in breadth respectively.

The two kinds of floating columnar, both with and without footing, would be adopted. Their number would be 18,000 under the main runway structure and 7,500 under the subsidiary one respectively.

The material to construct the upper structure would be steel and the floating bodies would be made of steel and concrete as shown in Figure 3.

The investigations on the strength of structure were as follows:

1. The estimation of the deformation and stress of the structure due to wave force
2. The estimation of the deformation and stress of the structure due to solar radiation
3. The estimation of the deformation and stress of the structure due to moving, taking-off, and landing of aircraft
4. The estimation of the maximum deformation and stress of structure due to the combined forces of wind, wave, and current

STUDY ON FUNDAMENTAL TECHNOLOGY FOR A HUGE FLOATING PLATFORM

Outline of Study

After the feasibility study on the Kansai floating airport was carried out, a lot of technical problems still remained to be solved in order to ensure the successful construction of a huge floating platform in severe sea conditions. This fundamental study was summarized as follows:

1. Accurate understanding and exact description of environmental conditions, including the prediction of the severest storm expected to be encounter in the life of the structure
2. Accurate estimation of external forces excited by wind, current, and waves, or of hydrodynamic forces associated with the motions of the structure
3. Accurate calculation or simulation of motions excited by the external forces, taking into account the elastic responses of the structure which cannot be ignored for such a large platform
4. Optimum design of mooring systems to achieve the equilibrium of tension forces acting on the mooring lines of the multi-anchoring system within the allowable mooring capacity
5. Development of an anchoring system which has the capacity to resist tension forces
6. Development of several construction and installation techniques, such as the towing, assembling, or connecting of the elemental platform which would be built at docks and towed to the sea area where it would be installed

At the ship Research Institute, the research was carried out along the following lines:

Preliminary Design of the Platform

The floating platforms were designed assuming several purposes. The floating bodies supporting the platform were of various types, corresponding to the loading capacities of the columns with buoyant footings, to the columns with a lower hull, or to the barges chosen as the fundamental shapes as depicted in Figure 4.

Investigation of External Forces Excited by Waves

In the first stage of the study, the wave exciting forces acting on the individual or the assembled floating bodies were measured, using the three, medium-sized models shown in Figure 5.

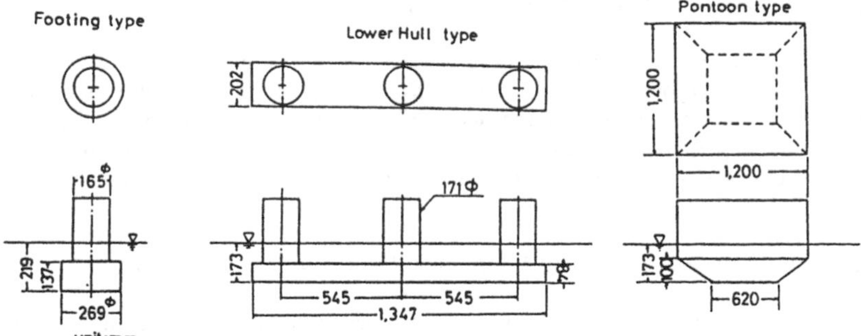

Figure 4. Floating bodies supporting platform

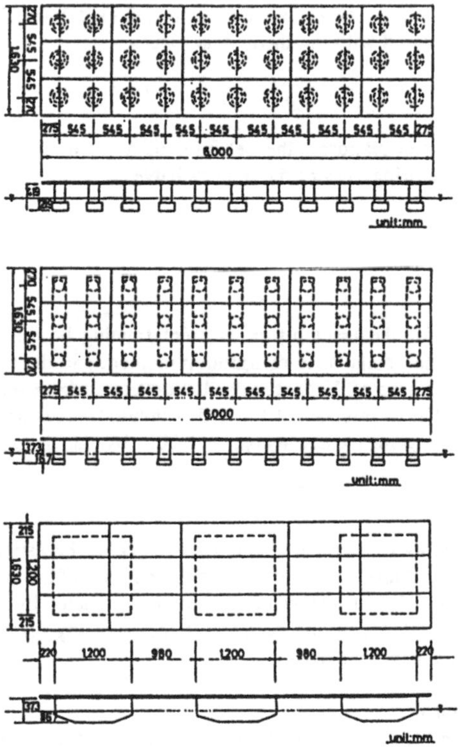

Figure 5. Medium-sized models

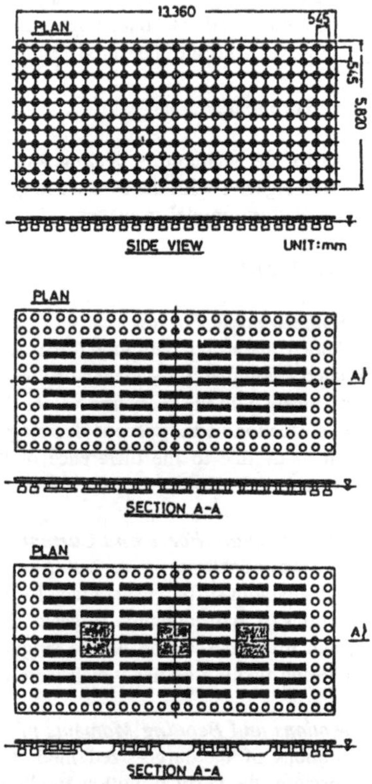

Figure 6. Large-sized models

Investigation of Responses of the Platform in Waves

The motion of the platform in waves were measured by use of elastic platform models as shown in Figure 6. Elastic responses, i.e., the deflections of the platform, were also measured by using accelerometers or photoelectric position sensors distributed on the platform models. The influence of the rigidity of the upper structure on the dynamic or quasi-static deflection or bending stress induced by the external forces was investigated through model experimentation, as well as theoretical analyses.

Mooring Forces in the Sea Conditions Combining Wind, Current, and Wave

The platform models were moored in the basin and the tension forces acting on the multi-mooring lines were obtained experimentally to grasp the non-uniformity of the induced tensions. The direction of the models against external forces was chosen randomly.

Connecting Forces in Assembling the Element of the Platform

The connecting forces and moments acting on the jointed parts between individual platforms which had been assembled into one, huge-scale floating platform at sea, were investigated experimentally in regular waves, and the measured results were compared with the theoretical values.

Access Technology

As for the technology dealing with access to the platform, the behavior of moored ships were measured together with the forces acting on their mooring lines or the fenders between the model ships and the platform model in waves.

Results of Study

Hydrodynamic Forces

Added mass and damping force coefficients for the assembly of element floating bodies were estimated theoretically by means of three-dimensional, singularity distribution method, and the results agreed nicely with the measured values derived by means of the forced oscillation tests, taking account of the correction of the viscous damping forces. The differences of the hydrodynamic forces on each element body due to the difference of its position were clearly recognized both by the theory and experiment.

Wave Exciting Force and Current Force

The wave-exciting forces acting on the assembly of element floating bodies could be estimated theoretically with sufficient accuracy for their practical use, in spite of an initial ignorance of the mutual interactions of hydrodynamic forces. The drag force by current velocity could be evaluated by using drag and lift forces acting on the single element body by taking account of the shielding effects of neighboring floating bodies which were determined experimentally.

Motions and Bending Moments of the Platform

Motions of the large-sized models in waves could be calculated theoretically with accuracy. Furthermore, the vertical motion amplitude and the bending moment distribution of the platform were measured and compared with calculated values which took into account the elasticity of upper structures. Both calculated and measured values agreed well with each other, both qualitatively and in general.

Mooring Forces Acting on the Multi-Anchoring Systems

Steady and oscillating forces acting on the mooring lines of the large-sized models were measured experimentally, both in regular and in irregular waves, together with the additional drifting forces which corresponded to the current or wind forces that were applied mechanically on the platform, and to the distribution of mooring forces along the four sides of the platform. They were accurately estimated. The so-called slow-drift motion of a moored structure played a significant role in the estimation of probability distribution of maxima of mooring forces which, still interesting enough, remains unsolved because of the non-linearity of the phenomena, especially for such complex and huge structures.

AT-SEA EXPERIMENT OF FLOATING STRUCTURE

Outline of At-Sea Experiment

Both the fundamental and applied research on the huge floating platforms have been carried out using basin tests and theoretical calculations. In the course of the research, an at-sea experiment using a prototype floating surface was planned. The main purpose of this field of measurement was to validate the element technologies which had been developed so far, as well as to evaluate the reliability of the floating platform in the real ocean environment.

The test area, 3 km offshore from Yura port at the Japan sea on northeastern coast of Honshu, was chosen by considering the meteorological data of the test area in the past. The sea floor of the test area was almost flat, and thus suitable for mooring the structure. The water depth was about 41 m.

The test structure was named POSEIDON representing the initials of "Platform for Ocean Space Exploitation"–its original purpose. Figure 7 shows the structure, itself, while Figure 8 shows the schematic view and Table 1 shows the principal dimensions of POSEIDON. The upper structure of POSEIDON was supported by columns with footings which were arranged in three rows of four, each row equidistant from the others. An instrument house was arranged in the central part on the upper deck for machinery and data acquisition equipment. The electric power used for measurement was supplied through the underwater cable from the shore.

Figure 7. Floating structure POSEIDON

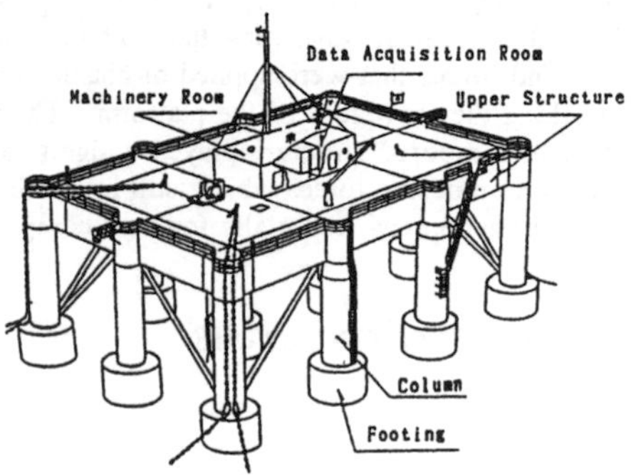

Figure 8. Floating structure POSEIDON

Table 1

Dimensions of POSEIDON

ITEM	DIMENSIONS
Length overall	34.0 m
Breadth overall	24.0 m
Height of main structure	13.5 m
Draft	5.5 m
Distance between columns	10.0 m
Column diameter	2.0 m (partially 2.5 m)
Column height	8.5 m
Footing diameter	4.0 m
Footing height	2.5 m
Displaement (Δ)	530.805 ton

POSEIDON was slackly moored by the four lines to its front and two lines to its back in a WNW direction, corresponding to the prevailing winter wind direction. The mooring lines consisted of grade-3 chains of ∅55 mm from POSEIDON, down to the anchoring position at about 225 m.

Measuring Items and Devices

The measuring items and devices were classified into the following groups:

1. Environmental conditions
2. Motions of the structure
3. Mooring line tension
4. Strength of the structure

5. Anticorrosion of the structure

The measuring items and devices were chosen as shown in Table 2. The three, ultrasonic, wave height meters were installed as a linear array on the sea floor at about 180 m offshore of POSEIDON. The system, using ultrasonic waves to measure the slow drift motion of the structure, consisted of two transmitters installed at the footing of POSEIDON and three receivers installed on the sea floor. The current meter was hung from upper deck at the position of 10 m below the sea surface.

Table 2

Measuring Items and Devices

ITEMS	NO.	DEVICES & REMARKS
Wind	2	ultrasonic, 3 axes, 19.5 m above W.L ultrasonic, 1 axis, 10.0 m. above W.L.
Wave	3	ultrasonic, sea floor of 180 m. offshore of POSEIDON line array
Current	1	impeller type, under the POSEIDON
Temperature air	1	resistance temperature device, on the top of house
water surface	1	semiconductor type, at the footing
bottom	1	semiconductor type, on the sea floor
Humidity	1	thin film capacitive polymer type, on the top of house
Solar radiational	1	thermopile type, on the top of house
Temperture deck plate	1	resistance temperature device
house wall	1	resistance temperature device
Relative water level	1	ultrasonic, at the center column of offshore side
Impact pressure	3	strain type, 1 m, 3 m and 6 m above W.L. on the center column of offshore side
Wind pressure	2	semiconductor sensor, difference of wind pressures on fore and aft side of upperstructure and starbord and port side of upperstructure
Acceleration	5	servo type, surge, sway, heave (center, fore, and aft)
Roll	1	vertical gyroscope
Pitch	1	vertical gyroscope
Yaw	1	directional gyroscope
Slow drift motion	6	ultrasonic, 2 transmitters on footings and 3 receivers on sea floor
Mooring line tension	8	load cell type, strain gauge type
Structural strain	12	strain-guage installed indirectly

Data Acquisition System

Forty-eight measurements, including wave, wind, motion of structure, mooring line tension, structural strains, wind pressure, water pressure, and relative wave height, were automatically recorded by a personal computer at regular intervals of 6 hours, 34 minutes, and 8 seconds. The sampling time interval was 0.5 seconds; thus 4,096 pieces of data were recorded for each measurement. This continuous measurement could also be started by a command sent by an operator from the measuring house onshore using the telemeter. The sampling time interval was 1 second in this case. The recorded data were stored onboard on hard disc of 40 MB. Some of the data sent by the telemeter could be monitored at any time by an observer at the measuring house onshore. The data of waves and wind were especially analyzed by a personal computer every hour at the measuring house onshore. The data collected as mentioned above were brought to Ship Research Institute in Tokyo and analyzed by either the supercomputer or personal computers, depending on the measurement items.

Results of Experiment

The field measurements were carried out from September 1986 to July 1990. The analyses of vast data are still going on. The following is a summary of the results to date.

Wind Spectrum

(Kato et al. 1991). The characteristics of wind fluctuation were investigated, using data measured at 19.5 m above water level. Then, a new formula of representing wind spectrum on the sea was proposed.

$$fS_u(f)\ /\sigma_u = 0.4751\ X\ (1 + X)^{-5/6}$$

where $X = f/\alpha$ $\quad \alpha = 0.0623\ (u^*/\sigma_u)^3\ a^{3/2}\ U_{19.5}$

$$= 0.0623\ (\sqrt{C_D/I})^3\ a^{3/2}\ U_{19.5}$$

u^* : friction velocity
C : friction coefficient
I : turbulence intensity ($=\sigma_u/U_{19.5}$)
a : Kolmogorov const.

Figure 9 shows an example of a measured spectrum compared to other formulations.

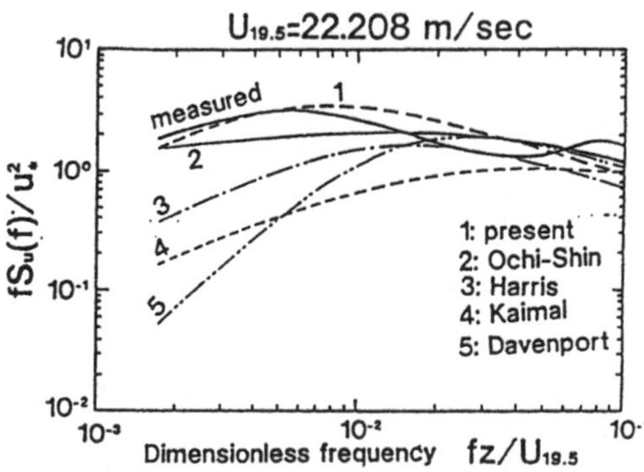

Figure 9. Measured and proposed wind spectra

Directional Wave Spectrum

(Yoshimoto et al. 1992). Three, ultrasonic wave height meters were installed as a line array at the sea bottom 180 m ahead of POSEIDON. The line array was adopted to accurately detect waves coming from a WNW direction.

The characteristics of frequency spectrum and directional functions were investigated using the maximum likelihood method (MLM).

1. The shape of the frequency spectrum in the high-frequency range is not proportional to f^{-5} but to f^{-4} as shown in Figure 10, and the JONSWAP spectrum was modified to satisfy the f^{-4} rule.
2. The measured directional functions can be approximated by the Mitsuyasu-Type function. The spreading parameters show the tendency of frequency dependency and their peak values are at the peak frequency, as suggested by Mitsuyasu. The change rate of measured spreading parameters with respect to frequency is not as rapid as the formula given by Mitsuyasu (Figure 11).

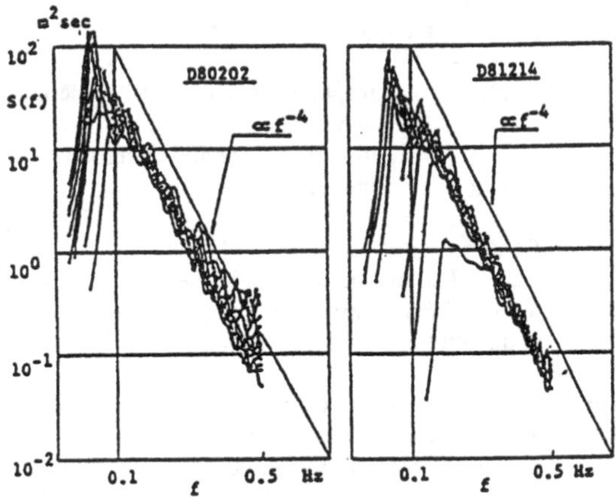

Figure 10. Wave frequency spectra in every two hours

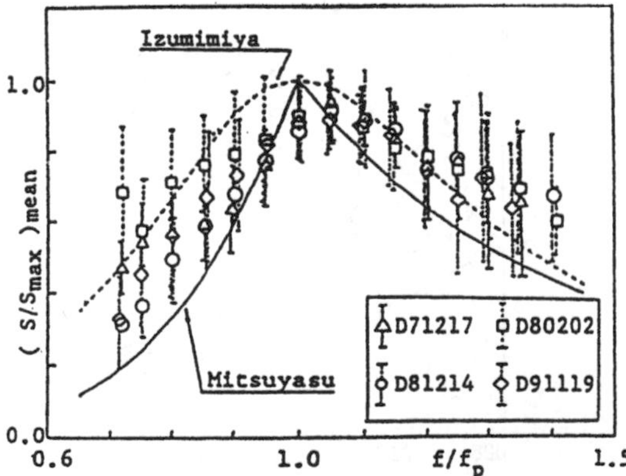

Figure 11. Spreading parameter

Motion Response in Directional Waves

(Ohmatsu et al. 1992). The directional wave spectra were analyzed using measured data. The 6° motions of POSEIDON were measured by accelerometers and vertical/directional gyroscopes. Then we tried to estimate the directional frequency response functions in the wave frequency range as shown in Figure 12. The results estimated by the presented method were compared with the theoretical response amplitude operator, and it was found that many experimental rungs of various wave directions would be needed to obtain a fairly good response amplitude operator.

The-low frequency motion with the period from 60 to 90 seconds was due to waves occurring regularly. The extreme values in N peaks of surge motion is shown in Figure 13. The measured data are scattered between the lines of the result calculated by the linear theory of Longuet-Higgins (Cartwright and Longuet-Higgins 1956) and the nonlinear theory of Kato (Kato et al. 1992). The latter gives the upper limit to the measured data. Therefore, the maximum expected values of low-frequency motion due to waves can be estimated using this method.

$$S_i(\omega) = S(\omega)\int_{-\chi}^{\chi} |H_i(\omega,\chi)|^2 D(\omega,\chi)\, d\chi$$

where $S_i(\omega)$; response spectrum of i-mode motion
$S(\omega)$; wave spectrum
$H_i(\omega,\chi)$; directional frequency response function
$D(\omega,\chi)$; directional function

Strength of Structure and Fatigue Analysis

(Yago et al. 1991). The structural strains induced by waves were measured at twelve locations on main structural members over a total of four years. The fatigue damage of the structure was estimated from the measured stress range distributions applying Miner's law. The results were also compared with the existing design values.

Figure 14 shows an example of measured and calculated response function of strain. The measured values were obtained in the same manner as the motion responses, considering wave directionality. Figure 15 shows the fatigue ratio per month and the progress of accumulated fatigue ratio, estimated by measured strain distribution, for four years. In this estimation, the Rainflow method and Amplitude method were used for counting the stress range distribution, but almost the same results were obtained.

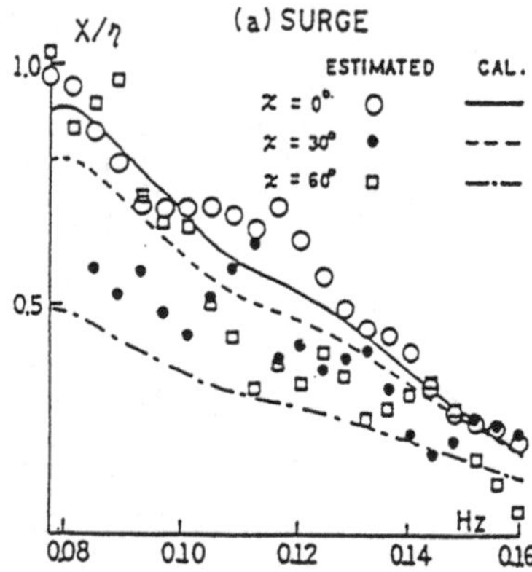

Figure 12. Directional frequency response functions

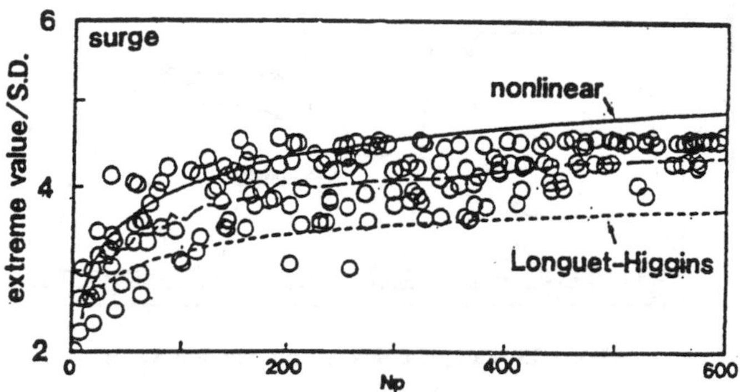

Figure 13. Extreme value/S.D. in N peaks

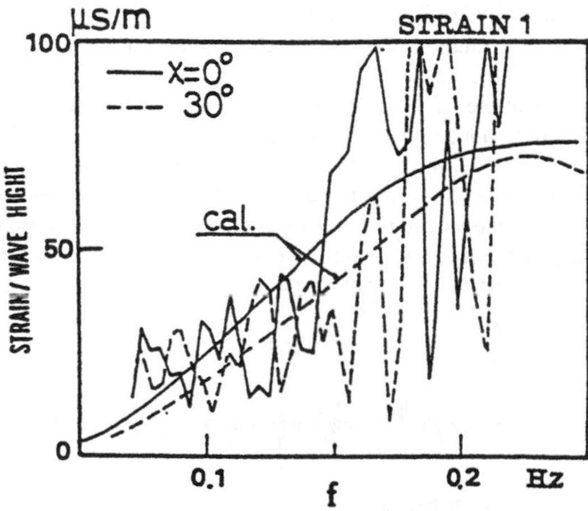

Figure 14. Response function of strain

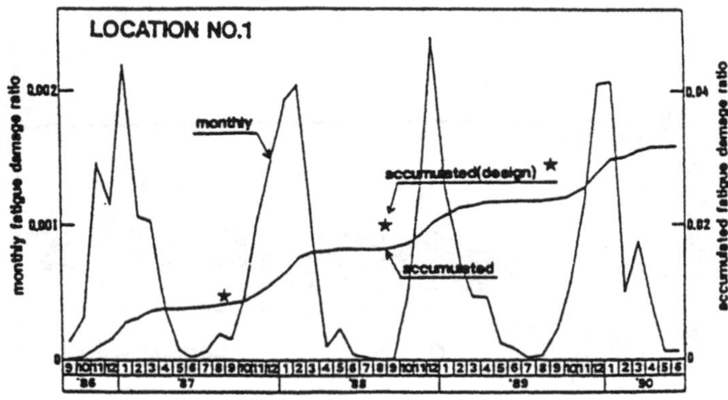

Figure 15. Fatigue damage ratio for four years

Tension of Mooring Line

(Ohmatsu et al. 1992). POSEIDON was moored by six chain lines, and the tensions of two of these lines were measured by underwater force transducers. The low-frequency motions of the structure were measured by an underwater ultrasonic device, while high-frequency motions were measured by accelerometer sand vertical/directional gyroscopes.

The numerical simulation method and statistical characteristics of mooring line tensions were investigated using the field measured data.

Figure 16 shows an example of a measured time history of tension and a simulated one. The mooring line tension has strong nonlinear and asymmetric properties when the amplitude becomes large. The statistical estimation method for such phenomena is not established yet. Then we applied the statistical analysis to measured tension in various cases and tried to fit some existing statistical model to analyze the results. As it happened, it was found that the Hermite moment model can be applied quite well. It was developed recently and is said to have the most flexibility for non linear phenomena. Figures 17 and 18 shown examples of instantaneous distribution and amplitude distribution f tension.

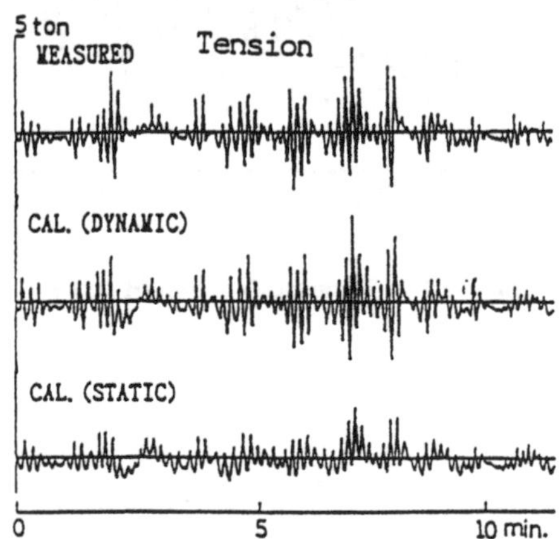

Figure 16. Measured and simulated tensions

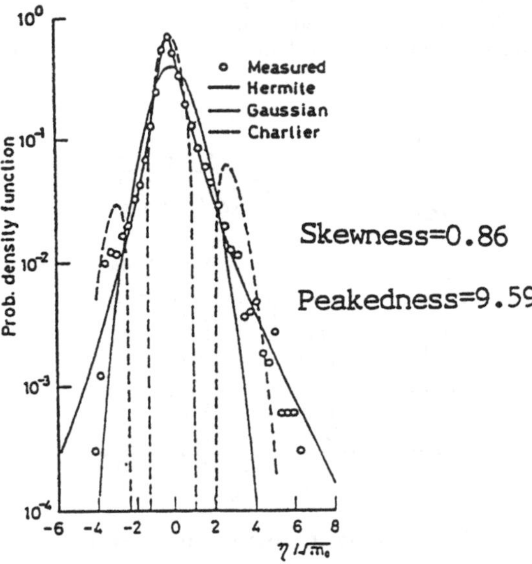

Figure 17. Instantaneous distribution of tensions

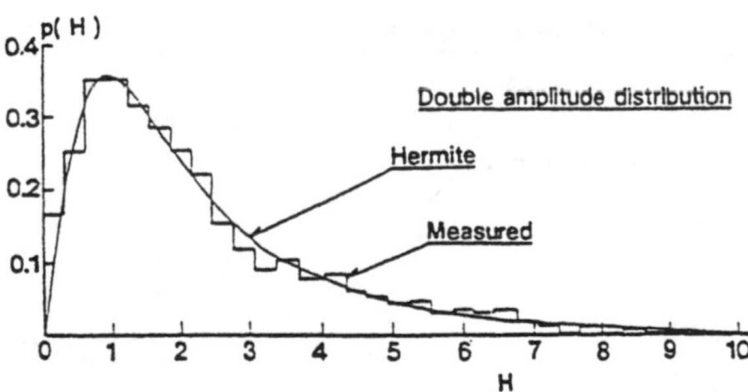

Figure 18. Amplitude distribution of tensions

Deformation and Stress Due to Solar Radiation

(Hoshino et al. 1992) In order to effectuate the construction of a huge platform, investigations of structural deformation and stress distribution caused by nonuniform thermal distribution from solar radiation are indispensable.

The long, continuous data of solar radiation, temperature, and strains of steel at various points were obtained in this experiment. Using those field data, the estimation methods of solar radiation, thermal distribution, and stress distribution were investigated, and thus validity of those methods was confirmed.

Figure 19 shows the measured and estimated solar radiation on the upper deck on some typical, fine days. For this estimation, the theory and experimental formulae which have been well established in the field of meteorology and architecture were applied. Figure 20 shows the measured strain of structure and estimated values. For this estimation, the finite element method of thermal stress analysis was applied. These figures all indicated the validity of these estimation methods.

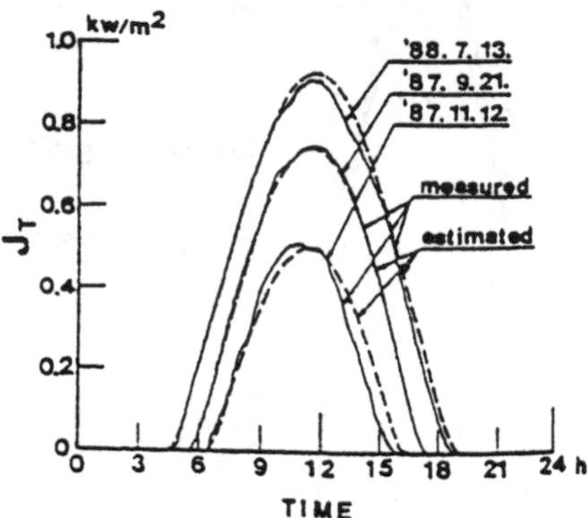

Figure 19. Solar radiation on upper deck

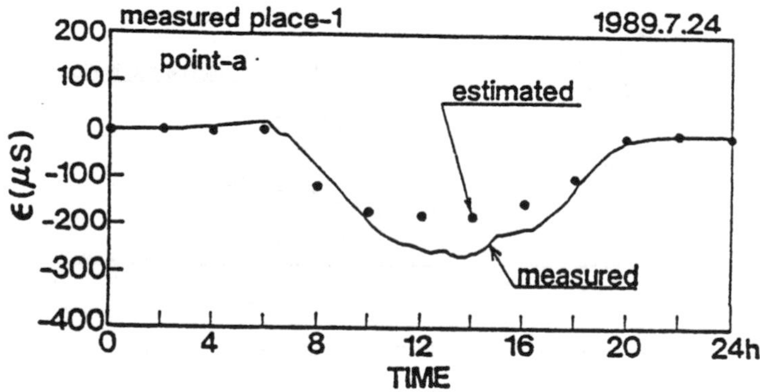

Figure 20. Strain due to thermal stress

FUTURE RESEARCH ACTIVITIES

It will be technically possible to construct a huge floating platform with present techniques. However, we are challenging our research component to develop the technologies needed for more advanced floating platforms.

Development of Advanced Floating Platform

The research thrusts are as follows:

1. More accurate estimation of the external forces acting on floating bodies
2. Active control of the motions of floating platform by oil pressure devices
3. Passive control of the motions of floating platform by changing its shape or by altering the arrangements of its floating bodies

Construction of Pulsation Wind Tunnel with Water Channel

A water channel has been installed under the test section of the wind tunnel so that waves and currents can be generated in it. In this manner we can perform the experiments on a floating platform under the combined external forces of wind, wave, and current. This wind tunnel will be completed in August 1993. A schematic view and principal dimensions are shown in Figure 21 and Table 3 respectively.

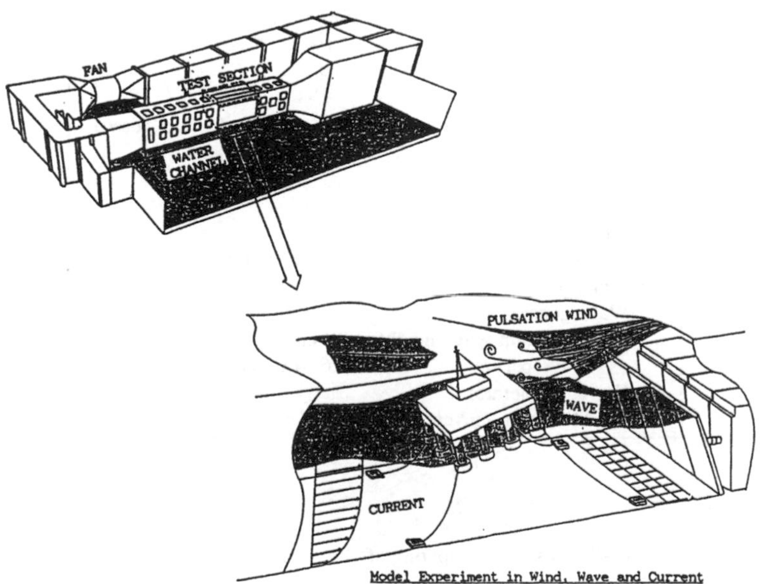

Figure 21. Schematic view of pulsation wind tunnel with water channel

Table 3

Principal Dimensions

Principal Dimensions	
1. Wind tunnel	
Type	
Götingen circulating flow type	
Test Section	
Closed section	
breadth x depth x length	3m x 2 m x 15 m
One part (3m length) can be used as open section.	
Performance	
Steady Wind	
Maximum wind speed	30m/s
Velocity uniformity	± 1%
Turbulence intensity	0.5%
Pulsation Wind	
Sinusoidal pulsation	13m/s±50% (period 50sec)
Irregular pulsation	by external signal
2. Water Channel	
Test Section	
breadth x depth x length	3m x 1.5m x 10m
Performance	
Wave generator	
Maximum wave height	20 cm
Wave period	0.6 4.0 Sec
Current generator	
Maximum current speed	0.3m/s

REFERENCES

Ando, S. et al. 1983. Feasibility study of floating offshore airport. Report of Ship Reseaerch Institute, Supplement No. 4.

Ando, S. et al. 1985. Abstract of the fundamental technology concerned with construction of a huge-scale floating platform. Report of SRI, Supplement No. 6.

Cartwright, D.E. and M.S. Longuet-Higgins 1956. The statistical distribution of the maxima of a random function, *P.R.S.* 239.

Hara, S. et al. 1988. At-sea experiment of a floating platform. Report of 15th UJNR/MFP Meeting.

Kato, S. et al. 1991. At-sea experiment of a floating offshore structure–wind characteristics at the test field. *Naval Architecture and Ocean Engineering 29*.

Kato, S. et al. 1992. At-sea experiment of a floating offshore structure–wind characteristics at the test field. *Naval Architecture and Ocean Engineering* 29.

Ohmatsu, S. et al. 1991. On the tensions of mooring ilnes of floating platform "POSEIDON." *Proceedings of the 11th OMAE symposium.*

Takaishi, Y. et al. 1985. Fundamental study of the huge-scale floating platform for use of sea space. Report of 13th UJNR/MFP Meeting.

Yago, K. et al. 1991. At-sea experiment of a floating offshore structure. Part 5: Measurement of stress range distribution for fatigue analysis. *J.SNAJ* 170.

Yoshimoto, H. et al. 1992. Characteristics of directional wave spectrum measured at Japan Sea. In *Proceedings of the 11th OMAE symposium.*

45

Technology for Coastal Development Activities in Japan

Hajime Tsuchida
President, Coastal Development Institute of Technology
Tokyo, Japan

ABSTRACT

One of the major approaches to efficient utilization of coastal ocean space is the construction of artificial islands. Such islands can be used for seaports, airports, industrial production, residential and leisure areas, and the like. In this paper, major artificial islands in Japan are described.

Many of these artificial islands are used as ports and because of that, the technologies which are necessary for ocean space utilization with an artificial island are almost identical to port technology. The long-term policy for development of the port technology established by the Ministry of Transport is described, together with background data. This paper also describes some of the major organizations working for the development of port technology.

INTRODUCTION

Coastal ocean space is utilized in many different ways. One of those is to create an artificial island and use it as land space for recreational facilities, residential development, transportation facilities, and industries. In Japan, many artificial islands have been constructed and effectively utilized. In addition, there are many projects under study to construct artificial islands and utilize them for multiple purposes.

In the first part of this paper, those artificial islands will be introduced. Technologies to construct and utilize the artificial islands are quite similar to port and harbor technologies. Therefore, the creation of technologies for port development holds significant importance to coastal ocean space utilization. The Ports and Harbors Bureau of the Ministry of Transport has established a long-term policy for the development of port technology. The second part of this paper will outline and describe this long-term policy.

HISTORY OF ARTIFICIAL ISLANDS IN JAPAN

The construction of artificial islands in Japan began with an artificial island for the purpose of defense at Tokyo Bay in the early 19th century. In the 1950s, artificial islands were constructed for mining marine coal. During the period of strong economic growth in the 1960s, reclaimed land for industrial use was created in various parts of Japan. Recent land reclamation has, however, shifted to the creation of artificial coastal islands for environmental reasons. Further, their use is gradually being applied to deal with problematic sea areas with unfavorable natural conditions, such as deep water, rough waves, and/or a seabed of soft soil.

USE OF ARTIFICIAL ISLANDS

Artificial island projects can spur economic activity as part of local development programs through their use as sites for industrial development, fishery promotion, tourism development, and uses related to energy and natural resources. Artificial islands also can be utilized for new construction or improvement of urban and transportation facilities.

Particularly notable are applications based on social improvements, i.e., enhancing habitability and convenience by making better use of this land and by providing civic amenities. Equally important are applications based on national policies for the supply of energy.

In its case studies, the following objectives and concepts were investigated for the planning of artificial islands:

1. Distribution bases of various kinds
2. Islands for sightseeing and leisure activities
3. Educational and scientific research bases
4. New cities (residential areas and hotels)
5. Land for industrial development and for waste dumping
6. Offshore mining bases
7. Energy bases
8. Farming and aqua-farming areas (marine farms, ranches, and processing plants)
9. Refuge during natural disasters
10. Base for building deep-sea structures
11. Others

Clearly, multipurpose use of artificial islands is more advantageous than single-purpose use. An important approach in planning multipurpose use is to view the artificial island and the calmed sea area around it as a single unit of newly usable space.

DEVELOPMENT PROJECTS OF OFFSHORE ARTIFICIAL ISLANDS

The Ministry of Transport and local governments have been studying thirteen projects to develop artificial islands. One of those projects is already under construction: 'Wakayama Marina City'. This one and some of the other projects will be described briefly.

Wakayama Marina City

In Wakayama Prefecture, an international marine recreation base was planned and studied, the major facility of which was a marina. In addition to the marina, facilities for various sports, human relations, transportation and leisure, and residential accommodation have been planned. Construction of the island started in 1989. It is expected that the facilities will be partly opened for public use in 1994.

Offshore Shimonoseki Artificial Island

In the Kitaura Sea area of Shimonoseki City, an artificial island is being planned and designed to serve as a general cultural exchange center for East Asia. For this purpose, the following are planned: a transportation and distribution complex with an international ferry base; a container terminal; an international exchange and information center; a marine industry complex with an East Asia international marine ranch center and fishery products processing factory; and a technological and cultural exchange complex with an international convention hall, exhibition facilities, and marine leisure land.

NEW CONCEPT OF PORT IN JAPAN

The Ports and Harbors Bureau of the Ministry of Transport established a long-term policy for port and harbor development, "Ports and Harbors Toward the Twenty-first Century" in 1985, in which the creation of an affluent waterfront was planned. This was supplemented by another long-term policy, "Coastal Area Toward Twenty-first Century" in 1990. The new, long-term policy was subtitled "Creation of affluent and environment friendly waterfront."

Those long-term policies consider a port and harbor area as a space for comprehensive uses and intend to improve the quality of such space where an ideal balance is maintained among "transportation and distribution," "industrial production," and "human life." Therefore, the meaning of the Japanese term "port and harbor" was expanded from being merely a place for transportation and distribution to a space where the three components described above could be accommodated. Because of such circumstances, in Japan, promotion of the coastal ocean space utilization significantly overlaps with development of the ports and harbors.

LONG-TERM POLICY FOR DEVELOPMENT OF PORT TECHNOLOGY

Background

As it has been described, the Ministry of Transport has established the long-term policy for future development of coastal areas, including ports. To support this type of development, the expansion of related technologies is encouraged. If organizations and people who are professionals in the realm of port development share a clear understanding of the technologies required to achieve future coastal developments, those technologies would be developed efficiently and effectively. From this point of view, the Ministry of Transport has established a long-term policy for development of port technology entitled, "Port Technology Which is Friendly to Human Beings and the Globe."

Process

The policy was drafted by staffs of the Bureau for Ports and Harbors and the Port and Harbor Research Institute, both of which belong to the Ministry of Transport. During the course of the study and the drafting, the staffs were advised by the technical committee which consisted of university professors and specialists of various fields related to port development and operation. The policy was approved by the Ports and Harbors Bureau in June 1992 and published in September 1992 by the Government Printing Office.

Contents

The long-term policy reviews historical background and the present state of Japanese ports and related technologies. It also describes the long-term policy for port development which has been presented as "Ports and Harbors Toward the Twenty-First Century" and "Coastal Area Toward Twenty-First Century." Then it presents two, fundamental principles to be taken into consideration when developing port technology and meaningful use subjects. There were ten, identified use subjects which were given high priority from the standpoints of importance and urgency. The policy also presents a guideline for sharing such subjects among the governmental, academic, and private sectors, as well as outlining the measures which the government can take to help and promote the development of technology.

Fundamental Principles

Two principles are recognized as a base when future development of coastal ocean space is considered.

1. Creating waterfronts which are friendly to human beings and the Globe.
2. Helping and promoting international communication through efficient flow of people and information and by contributing to the world.

SUBJECTS ON WHICH FURTHER TECHNOLOGY DEVELOPMENTS ARE REQUIRED

The numbered items below represent the social needs to be addressed, while the lettered entries are the technological subjects that respond to those needs.

1. Reorganization of coastal space utilization
    a) Technology to improve quality of port space
    b) Technology for intensive utilization of port space
    c) Technology for creating human friendly facilities
    d) Technology for maintenance and management of port space
    e) Technology to create new land and calm water areas for coastal space utilization
2. Efficient transportation
    a) Technology to develop high-efficiency transportation terminals
    b) Technology to increase the number of workable days in construction and operation of ports
    c) Technology for traffic facilities in port space
3. Increasing time for recreation

a) Technology for waterfront facilities allowing human access to water
b) Technology for planning parks and maintaining green fields in coastal area

4. Increase of aged people
   a) Technology creating safe coastal space for aged and handicapped people
   b) Technology to increase reliability of facilities in coastal area
5. Minimizing manpower and mechanization
   a) Technology to minimize manpower in investigation and execution
   b) Technology for high-efficiency execution
   c) Technology for high-grade execution management
6. Informational society
   a) Technology to process and utilize information
   b) Technology for information network
   c) Technology for communication system
7. Conservation and improvement of environment
   a) Technology for reconnaissance, prediction, and assessment
   b) Technology for countermeasures to global warming
   c) Technology for countermeasures to wastes
   d) Technology for improving qualities of seawater and soil of seabed
   e) Technology for protecting ocean from pollution
   f) Technology to create better environment
   g) Technology to minimize effects of construction execution to environment
8. Resources and energy
   a) Technology for planning and constructing ports for handling energy-related materials such as oil, natural gas, coal, etc.
   b) Technology for minimizing energy consumption in ports
   c) Technology to utilize renewable energy
   d) Technology for recycling wastes from construction
9. Facilities of multipurposes-purposes
   a) Technology for multipurpose facilities such as breakwater with wave power generation
10. Maintenance and renovation
    a) Technology to increase durability of facilities
    b) Technology to maintain and reinforce existing facilities
    c) Technology for renovating facilities
    d) Technology for preventing siltation at navigation channel and efficient dredging
    e) Technology for countermeasures against ice in sea
11. Level-up of fundamental technologies
    a) Technology for port planning
    b) Technology for reconnaissance and prediction of change on natural environment
    c) Technology to design facilities in port and coastal area
    d) Technology for disaster prevention and preparedness
12. International communication and contribution
    a) Technology to promote international technological cooperation
    b) Technology required in cooperation with developing countries

The technological subjects above are not entirely new ones, so some work has been done on them. Therefore, it should be understood that improvements of existing technologies or development of new ones are required to meet the social needs.

In the long-term policy, high priority is given to the following ten subjects:

1. Technology to improve quality of port space
2. Technology to create new land area and calm water area for coastal space utilization
3. Technology to develop a high-efficiency transportation terminal
4. Technology to minimize manpower in investigation and execution
5. Technology for countermeasures to global warming
6. Technology for countermeasures to wastes
7. Technology to create better environment
8. Technology to increase durability of facilities
9. Technology for disaster prevention and preparedness
10. Technology required in cooperation with developing countries

ROLES OF TECHNOLOGY DEVELOPMENT ENTITIES

The Ministry of Transport (central government), port management bodies (local governments or port authorities), third sectors, private enterprises, universities, and public service corporations, should respectively develop the technologies which they consider to be necessary and appropriate for their services and technological development faculties. At the same time, these technology-related entities should, as a rule, exert their autonomous efforts for securing the monetary and human resources necessary for such technological development.

In the Ministry of Transport, the Port and Harbor Research Institute (PHRI), which is a governmental research institution, the Investigation and Design Offices, the Machinery Offices, and the Port Construction Offices of the District Port Construction Bureaus are responsible for the port construction works in each district. They cooperate with one another to develop comprehensive port and harbor technologies, thus making the most use of their special characteristics.

In particular, PHRI, as a comprehensive research organ covering the entire spectrum of technologies concerning ports and harbors, has been conducting extensive, technical research and development and technical exchanges, ranging from fundamental fields to applied ones. PHRI has a role of systematic accumulation of the data of port and harbor technologies in Japan. The development results and accumulated data have been offered to the national organs and port management bodies, as well as to the private sector and developing countries (in the appropriate form), as part of the international technological cooperation.

However, in reference to technological developments in a new field, advanced or systematic technology development may be required, which means computer, new materials, biotechnology, and others, applying the particular technologies of ports and harbors. Thus, for technological development to blossom where comprehensive, technical faculties are required, close cooperation among participating entities (which have a common objective) may bring satisfactory results efficiently, economically, and in a relatively short period of time. In such a case, the Ministry of Transport would utilize the public service corporations as required and call on extensively, at home and abroad, not only the private, academic, and governmental circles related to ports and harbors, but other fields as well, in order to promote joint research and technological development.

Table 1
Artificial Islands in Japan

Name	Use	Water Depth (m)	Wave Height (m)	*Bulkhead Length (km)	Soil Type	Construction Period	Area (1,000m^2)
Ogishima	Industrial land	0-15	3.4	0.4	Clay	1971-1975	5,150
Higashi Ogishima	Harbor and transporation facility	1-10	-	0.7	Silt	1972-1984	4,340
Yokohama Daikoku	Harbor facility, coastal park	12	5.5-6.0	0.5	Silt	1963-1985	3,210
Nagoya Port Island	**	6-7.5	2.0	1.2	Clay, Silt	1975-1987	1,140
Nagoya Kinjo	Harbor, foreign trade facility, exhibition hall, park	0-5	1.0	1.4	Soft soil	1963-1985	1,910
Yokkaichi Kasumigaura	Harbor facility, industrial land	4.5-12	4.0	0.1	Soft soil	1967-1988	3,870
Gobo Thermal Power Plant	Power plant site	5-18	17.5	0.2	sandstone, sandy soil	1980-1983	350
Osaka South Port	Harbor and commercial facility, urban dev. site, coastal park	10	3.3	0	Clay	1958-1984	9,370
Osaka North Port	Harbor and waste disposal facility, industrial land	10	3.3	0.45	Clay	1972-1988	6,150
Kobe Port Island	Harbor and international exchange facility, urban dev. site	10-13	-	0.4	Clay	1966-1981	4,360
Rokko Island	Harbor facility, urban dev. site	10-14	-	0.2	Clay	1971-1985	5,830
Kanda Earth Dump	Dredged sand dump, park	7.5	2.5	3.5	Soft soil	1977-1986	1,530
Nagasaki Airport	Airport	10-18	2.0	1.5	Clay, Basalt	1971-1974	1,630
Nagasaki Mitsui-Miikie Island (#3)	***	10	3.3	6.0	Soft soil	1969-1970	6
Kansai International Airport	Airport	20	-	5.0	Soft soil	1985-	1,200

Notes. * Distance from the coast
** Dumping space for soil excavated in urban area
*** Vertical ventilation shaft for undersea, coal mine

46

Advances in Coastal Ocean Space Utilization: Artificial Islands and Floating Cities

Takeo Kondo
Nihon University, Tokyo, Japan

Nobuo Hirai and Akihide Tada
Nishimatsu Co., Tokyo, Japan

ABSTRACT

World population is steadily increasing and continues to favor living, working, and recreating in the coastal region. Development of marine resources is also very prominent within coastal ocean space. The infrastructure supporting these increased coastal and offshore activities and the future needs for further expansion are clearly noted.

Artificial islands and floating offshore facilities provide a means for fulfilling many of these needs. Over the last three decades, many new coastal facilities have been constructed or planned.

Of course, not all proposals have been undertaken and none are pending for the future. This paper will cover fifty-four such facilities to illustrate the trends and plans for the future. The time involved is covered in three phases: 1960s to Mid-1970s, Mid-1970s to Mid-1980s, and Mid-1980s-onward. The later period is emphasized in this paper. Many of the infrastructure improvements in the 21st century will provide for the expansion of the coastal region and the coastline with artificial islands and the use of floating structures in the coastal ocean space beyond the shore.

The new coastal infrastructure will tend toward providing multifunctional needs in an integrated manner for maximum utility, efficiency, and enjoyment with special regard to conservation and protection of the coastal environment.

CHANGES IN UTILIZATION OF COASTAL AND OCEAN SPACE

From the early 1960s, when John F. Kennedy was President of the United States, the movement to probe and develop various resources existing in the ocean became active. In his inaugural speech, President Kennedy proposed three frontiers to be conquered - outer space, ocean, and cancer. Since then, the series of activities for probing and developing the resources existing in the ocean produced remarkable results. For example, manganese nodule and polymetallic sulfide were discovered on the deep ocean floor, indicating that the Earth is producing various mineral resources.

Progress in space development provided a broader perspective to confront new problems as a result of being able to take an objective view of the Earth. Global warming, the ozone hole, acid rain, and marine pollution were some of the problems. Progress in oceanic and space development further clarified the seriousness of these problems. The global population has been rapidly growing, and it is clear that at this rate, it will surpass 10 billion by the mid-21st century. As a result, the destruction, on a global scale, of arable land and the food crisis triggered by such a process, have become realistic scenarios.

The oil crisis that began in 1973 highlighted oil as a new, international, strategic commodity, and environmental issues are now being positioned as the new, strategic concept for developing countries. This is attributable to the fact that these developing countries, as typified by China, have rapidly started treading the path towards heavy industrialization, a choice of development that consumes a large volume of energy. This tends, following the example of China, to transform the country from an agricultural society to an industrial one. The problem here is the fact that during this transition, a large volume of carbon dioxide and hydrogen sulfide is emitted into the atmosphere as a result of using the so-called carbonic energy, such as coal and oil.

With the oil crisis as the turning point, oceanic development favored offshore oil development, including great depths and the Arctic Ocean. New ocean structures and engineering systems were developed for use in those sea areas to overcome harsh natural conditions, while efforts were made at the same time to develop alternative sources of energy. As a result, ocean energy, including Ocean Thermal Energy Conversion (OTEC) and wave energy conversion, were studied and various systems were developed for utilization of the enormous energy resources possessed by ocean space.

In the meantime, industrial society had been creating new sites by reclaiming the space along the coast of urban areas in order to secure more space for heavy industry. At the same time, harbors and airports have also been built in the coastal zone based on the consideration of the efficient transportation of raw materials and products. In addition, now that the advanced countries have become information-oriented and concerned about the environment, examining the utilization of ocean space appears to be an important task. Therefore, this paper will extract futuristic projects generated from 1985 onward, that have been proposed by private enterprises in Japan involving ocean development, and analyze their trends in an effort to identify a major approach to coastal and ocean development in the 21st century.

In groping for a coastal and ocean space utilization plan for the future, a history of space utilization has been summarized.

Primary Phase of Coastal and Ocean Development: 1960s to Mid-1970s

This was the period of exploration and production development for ocean floor oil (energy source), fishery products (food resources), and manganese nodules (mineral resources) that existed in the sea. Offshore oil development was the major focus, while other resources were in the research stage. During this period, the majority of coastal and ocean space utilization was on a comparatively small scale and was of a single purpose.

Secondary Phase of Coastal and Ocean Development: Mid-1970s to Mid-1980s

With the oil crisis as the turning point, the development of alternative sources of energy to oil was stressed, and many systems utilizing energy in the ocean, including OTEC and wave power generation, were proposed. Redevelopment of coastal cities attracted attention from all over the world and urban waterfront redevelopment became intensified. Moreover, resort development became a worldwide trend and many marine resort developments were carried out in coastal countries with abundance of natural vistas.

Tertiary Phase of Coastal and Ocean Development: Mid-1980s-Onward

The global warming issue started to attract attention, giving rise to the question of how coastal and ocean space shall be utilized in harmony with the environment. This is no small problem to tackle, since it means developing a plan for the sustainable conservation of the ocean environment which accounts for 70% of the earth surface. In addition, the ideal state of infrastructure corresponding to the new age has become a question along with the globalization of the economy as the social system shifts from an industrial society to an information one.

SELECTION AND ANALYSIS OF NEW OCEAN PROJECTS

It appears that the tertiary coastal and ocean development bandwagon started on an international scale around 1985. Its typical examples included new developments in the United States of the 2020 Project in Los Angeles, OTEC Project in Hawaii, and various government-sponsored ocean control projects that have since been proposed in Japan. Many new and original ocean space utilization projects are currently being proposed by private companies and Institutes in response to this trend. These latest of coastal and ocean related projects had, and still have, as their backdrop, the following social and technical scenarios.

Special System-Related Impact Factors

1. Transition from industrial society to information society
2. Globalization of economy and formation of economic blocks, e.g., EC and NAFTA
3. Collapse of the socialist system and formation of a new global order
4. Concentration of population, economy, and information in expanding coastal cities and depopulation of inland cities
5. Massive generation of waste from cities and the problems of disposal and treatment
6. Availability of real-time information through development of satellite communication network
7. Polarization of population increase in the world and the aging problem in the population structures of developed countries
8. Regional strife and refugee problems worldwide

Global Environment-Related Impact Factors

1. Manifestation of global warming problem due to enormous generation of carbon dioxide
2. Melting of continental glacier caused by global warming and the problem of sea level rise
3. Destruction of the ozone layer caused by the use of freon gas and the threat of ultraviolet rays
4. Destruction of the tropical rain forest due to acid rain and the threat of forest destruction on a global scale
5. Reduction of arable land and the threat of reduced food production
6. Enormous growth of tropical cyclones and the threat of natural disasters
7. Rapid increase of world population and its threat on stable food supply
8. Global environmental issues become the strategic concept in the global economy

Scientific and Technological Growth Impact Factors

1. Production and control of various organisms through development of biotechnology and development of applied technology
2. Changes in form of communication through development of computer hardware and software
3. Changes in international distribution network through development in means of high-speed, large volume transportation
4. Development of new transport system through advancement in linear motor
5. Development of real-time communication system through advancement in satellite communication network
6. Emergence of enforceable crises due to development of new chemical substances
7. Development of new energy through plutonium fusion power generation and its problems
8. Increase of energy demand on a global scale and development of new energy systems that are environmentally sound

Numerous new ideas on utilization of coastal and ocean space are being presented in response to these emerging demands. As a matter of fact, Japan has led the world in proposed, coastal and ocean space utilization projects during the last five years. These projects have been generated by private companies and institutions and all involve original concepts. The fifty-four projects that were included in the study are shown in Table 1. Many of the selected projects take the aforementioned background into full consideration and propose coastal infrastructures and cities that are compatible with the new age.

Moreover, there are new proposals that take global environmental issues into consideration such as the utilization of ocean space in harmony with the environment and treatment of urban waste. A pattern classification method called Quantification Theory Class III, was used for analyzing these projects. In other words, it is an attempt to collect similar reaction patterns and typify them according to how the selected projects responded to several categories. Further, it is a method that performs interpretation of the meaning and classification of properties by plotting characteristic vectors that correspond to the characteristic values in minimum dimension space. In performing the analysis, attention was given to scale, function, form of space, and location to clarify the characteristic of each project using the evaluation factors as shown in Table 2.

TREND OF TERTIARY (1985 AND BEYOND) COASTAL AND OCEAN SPACE UTILIZATION PROJECTS

The properties of the X and Y axes on the fifty-four projects that have been proposed will be illustrated by the category quantity scattering diagram. A category quantity scattering diagram that groups characteristic vectors on two-dimensional coordinates of the X and Y axes from the solutions obtained through the Quantification Theory Class III, is shown in Figure 1. Incidentally, characteristic value was 0.42 for the first root and 0.38 for the second root.

Examining the category quantity scattering diagram, those in the positive direction on the X-axis can be positioned as projects that are located in the coastal region and are large-scale and integrated urban functions. Those in the negative direction can be positioned as social base-related projects that are being developed offshore. Thus, the X-axis can be understood as an axis that signifies the scale and location characteristics of the proposed project. Moving on to the examination of the category characteristics of the Y axis, the positive direction represents the construction and design characteristics of marine structures such as these used for reclamation and reconstruction. On the other hand, the negative direction shows the method of floating and signifies projects at greater depths. As a result, the Y-axis can be understood as an axis that signifies the functional characteristics and structural form characteristics of the project.

Table 1

Projects Selected for Analyses

	Co. Name	Project Name	Year	Type of Facility	Location (km Offshore)	Size (Hectares)	Construction Method
1	OBAYASHI GUMI	Millennium Tower	1991	Business & Residential	Coastal Zone	20.0 ha	Reclamation Super High-Rise Building 800m
2		Airo Polis 2001	1989	Complex City	Urayasu 10 km	1100.0 ha	Reclamation Super High-Rise Building 2001m
3		Under Grado	1989	Infrastructure	Continental Shelf		Underground
4		Japan-Korea Linkaged Tunnel	1989	Infrastructure	Tsushima Strait		Submerged Tunnel
5		Pacific Airport 21	1989	Offshore Airport	Hura 9 km depth: 100m	572.0 ha	Reclamation & Floating
6		Eco-Land		Complex City	Osaka Bay Area	1000.0 ha	Reclamation
7	OKUMURA GUMI	Marine Community 21	1990	Infrastructure	Osaka Bay 5 km	50.0 ha	Reclamation
8	KAJIMA	DIB-200	1990	Business & Residential	Waterfront	150.0 ha	
9		Marine Polis	1989	Infrastructure & Complex City	Tokyo Bay Area depth: 10-15m	4.0 ha	Caisson Foundation
10		Kuzyukuri 7-Island	1990	Complex City	Kuzyukuri 10 km depth: 20-25m	6000.0 ha	Reclamation
11		Fruits	1989	Resort	Offshore 10 km depth: 50-100 m	1000.0 ha	Floating
12		Marine Colosseum	1990	Recreation	Coastal Zone depth: 100-200 m	4.0 ha	Reclamation & Soft Ground*1 & Caisson Foundation
13		Waste Disposal on the Sea	1989	Waste Disposal	Tokyo Bay Area depth: 10-15 m	1.1 ha	Soft Ground
14	KUMAGAI GUMI	Genesis Teleport	1991	International Information City	Continental Shelf	3.1 ha	Reclamation & Floating
15		Odysseia 21	1990	Underground City	Tokyo	3.1 ha	Underground
16		Flying Triton 21	1991	Offshore Airport	5 km		Soft Ground

*1: Soft Ground = Structure wide footing on soft ground

Table 1 (continued)

Projects Selected for Analyses

	Co. Name	Project Name	Year	Type of Facility	Location (km Offshore)	Size (Hectares)	Construction Method
17	KONOIKE GUMI	Bay Area Canal	1990	Infrastructure	Waterfront	5000.0 ha	Reclamation
18	JIMSTEF	Marine Corridor	1991	Infrastructure	Osaka Bay 1-6 km depth: 30-50		Tunnel
19	GOYO	Penta H-SST	1990	Recreation	Offshore		Soft Ground
20	SATO KOGYO	Aqua Neo Polis	1991	Complex City		12000.0 ha	Floating
21	CITY COAD	Marine Free City	1992	Complex City	Japan Sea	4500.0 ha	Floating
22	SHIMIZU	TRY 2004	1991	Complex City	Waterfront	8800.0 ha	Reclamation Super High-Rise Building 2004 m
23		Big Pan Island	1991	Recreation	Coastal Zone depth: 30-40 m	56.0 ha	Floating
24		Marination	1988	R & D of Marine Resources	Pacific Ocean depth: 100 m	70650.0 ha	Reclamation by Drainage
25		ARC Airport 21	1989	Offshore Airport	Yokohama depth: 20 m	910.0 ha	Reclamation & Caisson Foundation
26		Aqua Amusement Land	1991	Waste Disposal	Coastal Zone	6.0 ha	Floating
27		Genesis Project		Recreation	Tranquil Sea	15.0 ha	Floating & Soft Ground
28		Step Over Tower	1991	Business & Residential	On the Ground	132.0 ha	Super High-Rise Building 800 m
29	TAISEI	X-SEED 4000	1990	Complex City	Coastal Zone	5000.0 - 7000.0 ha	Reclamation Super High-Building 4000 m
30		Linear Motor Catapult	1988	Space Port	Pacific Ocean		
31		Johnason	1990	Power Plant Marine Ranching	Bay - Offshore depth: 50-150 m	17.1 ha	Mooring
32		Never Never Land	1990	Leisure	depth: 15-30 m	2.4 ha	Soft Ground

Table 1 (continued)

Projects Selected for Analyses

	Co. Name	Project Name	Year	Type of Facility	Location (km Offshore)	Size (Hectares)	Construction Method
33	TAISEI	Dyna City	1989	Sewerage Diposal & Energy Plant	Tokyo Osaka depth: 15-30 m	1.0 ha	Caisson Foundation
34		Tokyo Bay City Island		Complex City	Tokyo (River) depth: 10 m	23000.0 ha	Reclamation
35		Marine Plantation Island		Leisure	Coastal Zone		Reclamation
36		Blue Bay Plan	1989	Recreation	Tokyo Bay		Reclamation
37	TAKENAKA	Sky City 1000	1989	Business & Residential	On the Ground	800.0 ha	Super High-Rise Building 1000 m
38		Floating Mariner	1989	Recreation	Offshore 1-2 km		Floating
39	TOA	Harmony-21	1991	Complex City	500 m	100.0 ha	Soft Ground
40	TOYO	Marilin	1989	Recreation & Sport	1-2 km	200.0 ha	Reclamation
41		Osaka-Bay Green Belt	1990	Infrastructure	Osaka Bay Area	1700.0 ha	Reclamation
42	TODA	Tokio 21 Project Bass	1990	Offshore Airport	Kuzyukuri 10 km depth: 20-25 m	2400.0 ha	Reclamation
43	TOBISHIMA	Top of the Japan Sea Rim	1991	International Ocean City	Japan Sea	314.0 ha	Floating & Soft Ground
44	NIKKEN	Soft Landing Island	1990	Leisure & City	Tokyo Bay 1-2 km	600.0 ha	Floating & Soft Ground
45	NISHIMATU	Marine Uranus	1990	Complex City	Offshore 10 km	45.1 ha	Floating & Soft Ground
46		Honeycomb Island	1990	Complex City	Tokyo Bay 1 km	94.7 ha	Reclamation & Floating & Caisson Foundation
47	JAL	Romantic City 2020	1990	Offshore Airport & Leisure	Coastal Zone 15 km	200.0 ha	Reclamation
48	KOKUDO-KAIHATSU	The Road on Sagami Bay	1989	Insfrastructure	Sagami 1-3 km depth: 30-40 m	7500.0 ha	Reclamation & Pile & Floating
49	HAZAMA GUMI	Marinas	1988	Complex City	Tokyo Bay Area 100 m	370.0 ha	Floating & Pile & Soft Ground
50		Super Roof	1990	Marine Ranching & Leisure		5026.5 ha	Reclamation

Table 1 (continued)

Projects Selected for Analyses

	Co. Name	Project Name	Year	Type of Facility	Location (km Offshore)	Size (Hectares)	Construction Method
51	FUDO	Deep Ocean Frontier 21	1990	Deep See Probing	Deep Ocean depth: 3000-5000 m		Soft Ground
52	MAEDA	Costa	1992	Leisure	Bay - Offshore	90.0 ha	Floating
53	MARINE ODYSEEY 21	Tokyo Neo Atlantis	1991	Office & City	Tokyo Bay Area	314.0 ha	Reclamation
54	MITSUI	Recycle Long Island	1992	Waste Disposal	Coastal Zone	240.0 ha	Reclamation

Table 2

Major Factors Considered

Category / Item	1	2	3	4	5
Surface Area	Over 500 ha	Over 50 ha	Unspecified		
Population Size	Over 100,000	Over 10,000	Unspecified		
Main Use	Complex City	Infrastructure	Leisure use	Others	
Utilization Space	On the sea and over the sea	On the sea	Under the sea	Complex space	Underground/others
Location	Waterfront	Bay area	Inland sea	Open sea	On the ground/others
Construction Method	Land reclamation type	Floating type	Soft landing type	Complex type	
Supplementary Facilities	Waste disposal/water quality	Energy	Complex	Unspecified	

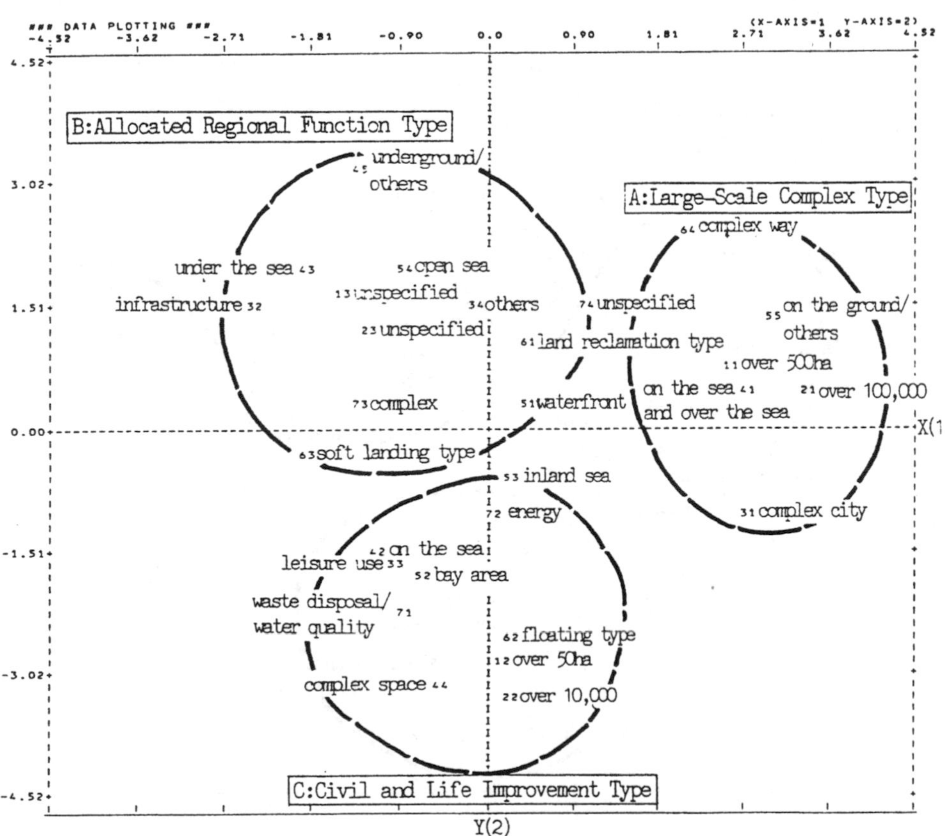

Figure 1. Category Quantity Scattering Diagram

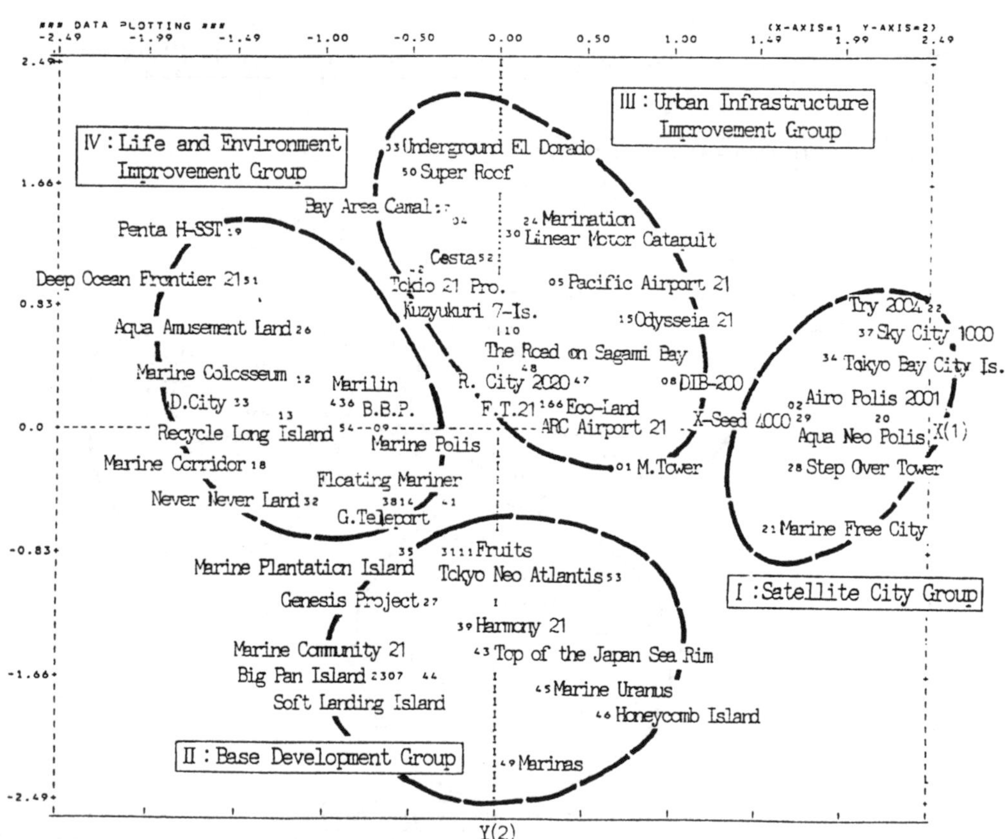

Figure 2. Category Quantity Scattering Diagram

Therefore, as shown in the diagram, the characteristic vector of the categories can be roughly divided into: (1) a group of large-scale, integrated projects; (2) a group of base-specified projects; and (3) a group of society and livelihood improvement projects.

In Figure 2, sample value was obtained from the category weight of each project and arranged in a two-dimensional space. They can be categorized into the following four patterns by taking into consideration the meaning and interpretation of each axis and the arrangement of category score.

Satellite City-Type Projects: Large Scale Integrated Cities Located in the Coastal Region

This group of projects focuses on solving the problems of big cities with regard to concentration of population, environment, and energy. As typified by SKY CITY 1000, these projects propose super, high-rise cities exhibiting a high-level of comfort and functionality and located adjacent to a big city in the coastal region.

Marine Base-Type Projects: Base Development and Building Oriented City Located Offshore

This group of projects aims to build an economic network base in view of the international situation for cities and nations that are adjacent to a bay or inland sea. As represented by MARINE URANUS and PAN SEA of JAPAN TOP CONCEPT, these projects propose floating marine cities that are located in the East China Sea and Sea of Japan and contribute to economic exchange among the surrounding countries.

Social Base Improvement-Type Projects: Transportation and Information Infrastructure Located on the Sea

This group of projects attempt to build new global, physical, distribution and information bases by reflecting the transition towards the information society. As represented by the linear motor catapult, these projects propose an innovative, marine rocket launching system for broadcast-type satellites.

Livelihood and Environment Improvement-Type Projects: Based on Recreation and Environmental Conservation

This is a group of projects with a goal to create a new environment for recreation by utilizing the abundance of the ocean in conjunction with the proposal for a system that could help ameliorate global environmental problems. As represented by NEVER NEVER LAND and the MARINE CORRIDOR CONCEPT, these projects propose active utilization of the environmental potential of the ocean, promotion of living resources, and purification of the sea, while providing integrated use patterns of the area such as recreation and cultural preservation.

Thus, the analysis of Tertiary Coastal and Ocean Space Utilization Projects reveals many proposals that take into consideration the present social situation, as well as the state-of-the-art science and technology. However, the problem that is common among many of these projects is the fact they are not taking into full consideration the biological, energy, and chemical/physical characteristics that are intrinsic to coastal ocean space. Many projects are located in ocean space as substitute for land, to utilize all the open space available in the ocean because of limited space for land development, or that the price of space on land is higher. This trend is particularly strong among Class I and Class II projects. A trend of this sort may be especially peculiar for Japan because of high-density land space and history of the land capital system. However, it can be seen from Class II and Class IV projects that proposals are being made through numerous advanced and motivated projects to take the natural characteristics and environmental issues of the sea area as well as the geopolitical circumstances surrounding the ocean space into full consideration.

OUTLOOK OF COASTAL AND OCEAN SPACE UTILIZATION IN THE 21ST CENTURY

The important item that must be first considered when thinking about the new coastal and ocean space utilization projects for the 21st century is the proposal of a concept which is created through state-of-the-art technology (hardware) and state-of-the-art logic (software). This is an important guideline for engineers who are ambitiously attempting to engage in emerging technological development. The problem with ordinary engineers up to now has been their excessively strong tendency to try and solve the current social problems. In other words, an engineering approach that responded to social needs was the mainstream.

However, it has become important for the new types of future-oriented engineers to learn the technique for setting his or her own goals, creating a hypothesis, and preparing a scenario. The story that incorporates these phenomena and steps becomes a concept. A concept is not a mere flash of idea or thinking. Rather, it is a root philosophy or thought, a strategic base to which engineers should return when perplexed or confronted with a problem. Viewed from a different perspective, it can be considered dogmatic and prejudiced. However, the dogmatism and prejudice referred to here are basically the development of foresighted thinking and proposals for distinguished new forms that are substantiated by objectivity and variety of progressive and state-of-the-art information. The second most important item is how to obtain a variety of progressive and state-of-the-art information which is necessary for contemplating the new concept. As mentioned earlier, one must have access to a human network for sensitively and promptly considering the evolving background and social needs. Moreover, one must know how to analyze the information obtained, and the level at which to position it. In other words, it is important to determine which level of the formation of a project concept the information is on, a strategic level, tactics level, or armor level.

The third most important item is systematic analyses of the range of factors under the influence of the concept and the information network that can be accessed. For this purpose, it is important to create an ideological space concept with due consideration for the levels involved.

1. *Global level*: Searches for social needs in globalization
2. *International level*: Positions the social urgency in the international community
3. *National level*: Proposes future visions that improves living standards, quality of life, and national pride
4. *Regional level*: Builds various networks that contribute to activation and development of the region
5. *Community level*: Creates a safe, fulfilling, and comfortable system for the living environment
6. *Human level*: Understands basic desires of people for comfort security, and intellectual growth

The case analysis of the Tertiary Coastal and Ocean Space Utilization Projects discussed in this paper indicated that, with some exceptions, there were few projects that addressed, integrated, and complemented the six levels. Many projects fell into the trap of focusing on a singular objective. Said another way, there were many projects that boasted a proposed, unique technology and artificial projects that tried to attract special attention. Then, what is the project concept that is being anticipated as we approach the-21st century? The key words are divided in three levels.

Key Words on Strategy Level - Arrival of an Information-Oriented Society and Preparation of New Base

1. Global warming, ozone, and acid rain issues and development of new sources of energy

2. Population increase problems and associated food crisis
3. Regional economic differences and formation of economic blocks

Key Words on Tactics Level

1. Develop national bases for social, economic, and environmental reasons for locating on the sea
2. Utilization of marine resources for food, energy, and minerals
3. Creation of a new order network using ocean space as a starting base

Key Words on Armor Level

1. Construction of intercontinental tunnels and bridges
2. Construction of intercontinental canals
3. Development of Arctic Ocean routes and development of new means of transportation
4. Construction of marine rocket launching stations
5. Marine research production base that creates satellite software programs and intellectual property
6. Development of carbon dioxide sequestering system utilizing marine organisms
7. Development of power generation system utilizing ocean energy resources
8. Promotion of resource management-oriented fisheries on a global level
9. Building a futures market for fisheries products and building a system for mutual aid

CONCLUSION

The ideal approach for utilization of coastal and ocean space which is compatible with the information society and social system of the 21st century has been examined through the analysis of state-of-the-art projects in Japan. This can provide some direction for various projects that will be emerging. The proposals for the armor level project concept are only a small example of ideas that can be conceived at present. At the same time, the reason for describing the project concept proposals using key words was that this concept has yet to fully ripen, although some have already been proposed by the authors and others.

The conclusion of this paper, as can be inferred from the key words of the proposals on project concept, is that many of the infrastructure improvement projects in the 21st century will naturally be located offshore and will be based on floating structures. In addition, the space utilization functions sought will be multipurpose and integrated, unlike previous simple purpose ocean projects. There will be a great tendency to form cities because of the need for people to work and live in that space. There is much opposition to project concepts for artificial islands created by reclaiming the coastal region because of environmental contamination during construction and destruction of habitats for marine organisms. However, the increase in artificial islands through new reclaiming will extend the coastline and provide an aeration effect corresponding to the coastline that has been extended, which, in turn, can be designed to have a positive impact on water purification and organism production volume.

Furthermore, it will contribute to the sequestering of carbon dioxide by marine organisms and may be of some help in controlling global warming. For this purpose, it is necessary to select a design using natural sand and rocks in the construction of an artificial island for compatibility with the natural environment, instead of building a vertical concrete wharf along the coast. All future developments must always emphasize harmony with the experiment.

ACKNOWLEDGMENT

The authors wish to acknowledge Joseph R. Vadus of the National Oceanic and Atmospheric Administration, Washington D.C. for many valuable suggestions and editorial remarks which were incorporated in this paper.

47

Changing Nature Of Ports In Coastal Management

Gerhardt Muller
Adjunct Professor, City University of New York-Baruch College and Nassau Community College, New York State

ABSTRACT

Waterfronts in major ports, especially in the United States and in at least one instance in Canada, continue to struggle with the equally strong forces of those who wish to maintain traditional, waterfront maritime activities and those that favor redevelopment proposals that embrace mixed uses, including recreational, residential, and commercial activities. Although it is not the intent of this paper to evaluate the economic impact of cargo-related activities on a port and its environs, at stake are jobs, economic enhancement of valuable real estate, and sometimes conflict with existing coastal zone directives that might favor one activity over another. Some ports have been successful in resolving these differences, while others are still looking for a reasonable compromise.

INTRODUCTION

For thousands of years, most of the world's great ports were considered major economic and/or military assets. The inner part of the port was usually lined with piers and had a constant movement of ships and related commercial activity. Manufacturing firms located close to these activities because waterborne transportation was usually the main form of transportation at the time. Located there were also waterfront activities that catered strictly to the residents and visitors. Further out from the center of the port were less important, and perhaps, fewer glamorous port activities, such as bulk handling facilities, including raw materials and in some cases waste.

Note: A large part of this paper was based on research developed by the author and his colleagues at The Port Authority of New York and New Jersey who collaborated on several studies dealing with economic revitalization of the region during the late 1970s and early 1980s. It is important to note, however, that the updating of the information, analysis, interpretations and other observations expressed in this paper are not those of the Port Authority or any of its employees, but strictly those of the author.

Advances in transportation technologies, changing trading patterns, and economic and political deregulation altered the economic focus for some of these ports. These simultaneous forces, especially in the past decade, aided in part by economic recessions and environmental concerns, not only placed severe constraints on traditional waterfront activities, but also prolonged their development and maintenance process.

Simultaneously, there was and continues to be the need for ports to identify potential, new waterfront projects and activities that conform with existing and emerging coastal management regulations and practices that represent appropriate and desirable public and private participation in port and regional revitalization. Logically, this cooperative effort should utilize assets of both the public and private sector as tools to improve the social and economic welfare of the region. If successful, this will be accomplished by reusing, if not rebuilding, deteriorated and underutilized waterfronts for new commercial, recreational, and residential uses that support profitable and socially beneficial activities. That challenge appears to have mixed results.

CHANGES ALONG PORT WATERFRONTS

Today's port maritime waterfronts and their related activities, especially in the inner harbor where those activities represented at one time or another the core of the region's economic and social makeup, involve economic, physical, political, institutional, and environmental factors.

Economic

Since the mid to late 1960s, most ports around the globe embraced and promoted the development of maritime related innovations. Even under continuing competitive pressures from neighboring ports, most of these ports will continue with this goal for the foreseeable future.

However, containerization and other forms of advanced transportation management systems such as intermodalism as it is linked to logistics management, caused the gradual shift of general cargo activity away from the inner harbor. Traditional, break-bulk pier facilities were no longer in great demand, except along certain portions of the inner harbor were alternative activities were limited. Often, such noncompetitive, traditional port facilities and activities were abandoned or misused. Longshoremen and other cargo-operation-related jobs began to vanish or relocate to the newer and more modern facilities. The result was a rising tide of unemployment which placed an additional strain on municipal social and welfare programs and funds.

Those areas of the inner harbor, and sometimes whole ports, that were not torn down to make room for other productive uses, were left to the ravages of time and vandalism. Some remain as scars on the waterfront, while at the same time, the resultant debris serves as a major threat to navigation for large and small vessels alike. In addition, this visual blight and decay often contributed to the decline of adjacent communities that at one time served the needs of those employed on the adjacent piers.

Rapid growth of containerization and its ancillary services also contributed to redundancy of facilities. In many ports, wasteful and ruinous competition between port-related activities was the logical development of stiff price, service competition, and duplication of operations at the expense of alternative waterfront activities. This was certainly true in some ports where there was a lack of control by at least one public entity such as a port authority. This is not to say that container facilities could have been built and located according to textbook theories, but at least some coordination would have minimized the negative aspects of the present situation.

Equally affected by technological changes were the manufacturing plants and facilities that were originally located on the waterfront. Activities established there depended on waterborne transportation for supplies and finished products. Many such facilities moved elsewhere once this dependence no longer existed, and it became more economical to operate at inland sites. Such businesses often left behind empty shells that eventually fell prey to the same consequences of decay experienced by outdated maritime structures and piers. Very often these abandoned buildings were located in the same neighborhoods as the piers, and together, all contributed to the general deterioration of the economic and social fabric of what were once prosperous and exciting places where people had lived and worked.

The existence of this deterioration and decay, both in terms of physical and social considerations and as visual pollution, further detracted from the waterfront's ability to attract new forms of industrial or commercial activity. Companies were not willing to risk their financial investments as well as personal security of their labor forces in an environment that appeared to be hostile. Thus, large sectors of the inner harbor waterfront and its neighborhoods experienced the classic circular effects of self-reinforcing disinvestment with no hope of revitalization in view.

Another economic constraint is the costly, complex, and time-consuming construction techniques associated with the waterfront as compared to similar facility construction in upland areas. Shoreline retention and protection structures and supports, if not already in place, must be constructed, often at a relatively high cost per square foot, and must be maintained thereafter at equally high rates. Consequently, greater capital resources are generally required for waterfront construction. As a result, projects lacking a special need for a waterfront location or that do not derive a special benefit from such a location often cannot economically justify being located there.

Physical

The economic inactivity and perhaps political default regarding a ports' inner harbor waterfront often leads to obvious physical deterioration of a formerly developed and vibrant waterfront. There are many examples, especially in some larger American ports like New York, San Francisco, and Seattle. Sometimes, and despite some small signs of progress towards physical redevelopment, the extent of existing physical deterioration contributes to the self-defeating belief that the problem is too big in scope and is, therefore, hopeless.

For example, on the west side of Manhattan, from the area just north of the Battery Park City Authority site, up to Riverside Park, there are many piers and shore structures that have been abandoned. Some reached such a state of deterioration that they had to be removed for the protection of the public. Consequently, local neighborhoods, the Borough of Manhattan, the City and the region are all left without potential, vibrant uses. There are plans for eventual development along this waterfront, most of which is tightly controlled by a newly created waterfront development authority. Meanwhile, and despite the fact that some usable piers have been, through redevelopment, converted into Recreational areas, no large, permanent investment by either the public or private sector will move ahead until all plans are completed and survive the inevitable onslaught of court cases.

Over time, the waterfront has been also subject to physical abuses that can destroy natural and man-made waterfront assets. Examples include waste disposal and water pollution. Additionally, some sectors of the waterfront were devoted to salvage yards and unsightly storage activities that contribute to air, water, and noise pollution. Highways and rail lines were often built along shorelines of some of the harbor's more important waterfront reaches, thus creating physical and psychological barriers between the local community and the waterfront.

Political and Institutional

Some ports are comprised of many counties and municipalities. For example, the Port District of New York and New Jersey has seventeen counties and 211 municipalities. Even the inner harbor waterfront of some 75 linear miles spans seven counties and nine municipalities. These numbers still do not reflect the many "waterfront communities" within each municipality that in recent years have become particularly influential in determining what should take place on their waterfront properties. At the same time, some particularly ambitious neighborhoods have succeeded in decentralizing themselves from their larger municipality, thus reinforcing the community/neighborhood sense of importance and strength. This is something that will most likely remain, if not increase in the foreseeable future, especially in the age of deregulation of commercial activities, especially transportation.

In addition, unions, despite the difficulties and decline in membership (mostly due to automation), are still a power to be reckoned with along the inner harbor. The result is that besides the physical divisions of the inner harbor that are created by the existence of many waterways and other natural barriers, the division of the urban waterfront into so many political parts slows or even prevents a region from viewing the waterfront as a whole. This situation has also affected the coordination of planning and development or redevelopment of inner harbor reaches.

But this idea of planning has other challenges to it that often impact actual development. Additional layers of regulation, permit applications, and extensive planning studies not backed by concrete investment proposals have resulted in little or no activity.

Closely associated to the political factors and constraints affecting waterfront redevelopment are institutional factors and constraints. Perhaps at no other geographic location than on the waterfront do so many public institutions have a possible regulatory, administrative, advisory, and/or development role. It has been estimated that in the United States there are as many as thirty-five agencies on the federal level alone that have been identified as having some degree of responsibility for this area.

Federal, state, and local governmental differences of opinion, together with the self-interests of private industry, labor, as well as the community's sense of self-preservation all contribute to constraining waterfront redevelopment. This is particularly noticeable where each level of government, through many overlapping agencies, tries to put its own identity on policies and programs. Layers of bureaucracy and vested interests with no incentive to make "trade-offs" stop ideas and concepts that could result in meaningful programs for the benefit of the local community, the inner harbor, and the region as a whole.

The complex institutional patterns have caused the private sector to reach the opinion that any returns on a waterfront development investment can be and often are jeopardized by all the institutional "red tape" restrictions and deregulations as to make the risk of waterfront redevelopment too great. Lack of quick response and lack of flexibility by government agencies deters private investment in waterfront projects. It has often been suggested that the idea of public/private partnerships could insulate the private developers from as many bureaucratic, administrative, and regulatory hassles as possible.

Environment

The decades leading up to and including the 90s are increasingly recognized as the decades of the environment. In the United States, as in other countries around the world, especially after the collapse of communism, the full force of several important, national environment protection policy acts took hold. These policy actions often result in other national, state and, local legislation being forced on environmental issues affecting water quality and waterfront development. Usually these actions have been and continue to be effective. In many instances, especially in the United States, the result is an improvement in water quality. Very strong measures require that any waterfront development, including the deepening and maintenance of navigable waterways, be conducted under conditions that would minimize harm or danger to the environment. Other signs of progress were shown in controlling the discharge of effluent material into port waterways, including assuring that some form of treatment to the affluent was made before it reached the waterway.

The impact of these measures and sense of commitment have had a visible proof of success. Fish are returning to waterways where not so long ago there were none. Recreational swimming is now becoming a reality that once belonged only in the memory of those who swam in those same waters many decades earlier. Alternatively, certain marine creatures such as the marine borers, a small waterborne animal that feeds on underwater wooden structures such as piers and other pilings, is understood to be returning to inner harbor areas because of improved water quality.

The downside of these achievements is often considerable delay and sometimes refusal to approve projects that have the potential of helping in the economic redevelopment of regional waterfronts. Furthermore, through the process of seemingly endless public hearings, more studies, and sometimes through a lengthy and often very expensive court procedures, significant delays occur in the development process and force changes in the developer's plans. Often the projects are forced to be delayed or abandoned.

It should be recognized, however, that in recent years there has been a noticeable shift in making a real effort to achieve balanced reviews of development among the leadership of environmental groups. Costs and benefits of certain environmental goals are being scrutinized with a view toward a recognition of the need for some trade-offs that will recognize the value of development at some relatively small environmental cost. If developmental proposals incorporate early recognitions of the need for environmental sensitivity, projects in the future will reach faster and less costly conclusions than those being considered today.

It also should be recognized that there remains the continued importance of environmental concerns and constraints for so-called natural or undeveloped areas. These areas are often considered off-limits to any form of development, a stance that will most likely continue, given growing awareness of the need to protect the environment. This assessment is reinforced by current trends in coastal zone management.

Other

There are other issues that are equally important which relate to land costs and ownership, the ability to change waterfront property uses, and the right to develop now underutilized property.

The purchasing price of land is often dependent on the tax policies affecting the land, including developed and underdeveloped. Such policies affect the use of shoreline property and the unique vistas from the waterfront on that property.

Often, waterfront property, some of it underutilized, is under the ownership or control of the estates of financially weak companies. This adds complexity to the process of acquiring such property for new uses at reasonable costs. Determining land ownership on the waterfront is a process that varies from region to region, and certainly from country to country. Some major waterfront municipalities like New York have computerized waterfront properties. Computerization will make the identification, planning, and monitoring of such properties easier and much faster.

The issue of ownership is further confused when one considers the right to use riparian land (land over which tidewater flowed at sometime) is subject to laws and grants given at another time and place. This is particularly true in cases where charters granted by royal degree in a claimed territory is subject to the laws and interpretations of those laws as understood in present times.

Another, but not necessarily the last issue, is the voluntary transfer of property in designated park lands in and near port areas. Often there is the issue of what limits are placed on these properties before inner harbor, underutilized waterfront property could be converted into more economically viable assets.

EXPERIENCES OF OTHER PORTS

Fortunately, there are several examples of ports around the globe that show the ability of dealing effectively when economic, technical, and regulatory changes seem to combine at the same time. In doing so, it was the coordination and implementation of several public and private redevelopment programs and projects that were brought together for one purpose: to make a change along the waterfront that continues the economic and social benefits that previous activities at the same location once had. Usually, these projects responded to existing and emerging coastal management issues.

In identifying these ports, it should be understood that what might work successfully in one port, might not be as successful in another. Nevertheless, elements of each example port should be examined closely and applied later to meet the characteristics of the port wishing to make a change.

New York

One of the leading ports in the world in terms of total cargo tonnage, the Port of New York and New Jersey metropolitan region (an area approximately 25 miles in radius from the Statue of Liberty) continues to struggle with redevelopment of its unused portions of its waterfront. Except for small waterfront areas in Brooklyn and Staten Island, most general cargo activities shifted to the New Jersey side of the river once the era of containerization was firmly underway. In the wake of that shift, many of the port's once thriving piers and other waterfront activities dwindled to almost nothing. What was left, except for general cargo activities in Brooklyn, Staten Island, and passengership operations on the west side of Manhattan, fell into disuse and eventually became a visual and physical blight on the waterfront (Muller 1990)[1].

To remedy this situation, New York City recently completed a comprehensive waterfront plan (City of New York 1992). The plan builds on the experience of past historical economic and political events and lessons learned. At the same time, it addresses today's conditions and works within a myriad of legal and regulatory parameters affecting the use and development of the city's waterfront. Although still in the development stage, the plan focuses on the concept of "public trust," which establishes that certain waterfront benefits are held in trust for all the city's residents and visitors. The plan was initiated as partial compliance with the federal Coastal Zone Management Act of 1972. Because of the studies that lead up to the final plan, four principal functions of the waterfront were identified.

1. The Natural Waterfront, comprising beaches, wetlands, wildlife habitats, sensitive ecosystems, and the water itself
2. The Public Waterfront, including parks, esplanades, piers, street ends, vistas, and waterways that offer public open spaces and waterfront views.
3. The Working Waterfront, where water-dependent, maritime,and industrial uses cluster or where various transportation and municipal facilities are dispersed
4. The Redeveloping Waterfront, where land uses have recently changed or where vacant and underutilized properties suggest potential for beneficial change

The plan recommends over 100 sites for new or improved waterside public spaces; nearly fifty public parks and existing parks where new attractions could be created at the water's edge; twenty-five public street ends that, with modest improvements, could provide points of access for nearby residents and workers; and another forty sites where public access would be a mandatory component of new residential or commercial development.

In response to the decline in manufacturing and the continued deterioration of many of its

[1] The Port of New York and New Jersey was actually the site of the birth of waterborne containerization when SeaLand Services initiated their service between New York and Galveston in 1956.

waterfront industrial properties, the plan recommends that some 500 acres of manufacturing-zoned land be rezoned for residential, commercial, and recreational use. Even with these bold initiatives, the plan assures that sufficient land will be available to meet the needs of industry and the working waterfront. Thirty percent of the city's 578-mile shoreline is presently zoned for industrial use.

As an essential counterpart to land use guidelines, the plan proposes comprehensive zoning reforms that address the unique qualities of waterfront property. These changes would streamline the waterfront regulatory process, increase public access, facilitate water-dependent uses, and encourage appropriately scaled waterfront development with a compatible and lively mix of uses.

Boston

In many respects the City of Boston is a smaller version of the New York-New Jersey Port in terms of both geography and climate. Break-bulk and container cargo almost disappeared a decade ago, but recent cargo handling developments through containerization and taking advantage of improved rail services to the New England area, have raised the expectation that cargo will not disappear completely. Nevertheless, most of its waterfront was left to deteriorate as the volume of cargo declined.

Then in the mid-70s, particularly under the banner of the "Spirit of '76," Boston's waterfront serves today as a symbol of what can be done if the will, determination, and considerable funding are available. Under the direction of the Boston Redevelopment Authority, redevelopment of the waterfront began with the badly deteriorated, downtown area near City Hall. Overall, over $420 million in urban renewal funds was leveraged into $15 billion in private investment, with Faneuil Hall Marketplace as a major focal point of the program.

Until the mid-1970s the complex of buildings collectively known as Faneuil Hall were either vacant or underutilized and deteriorating. Under the guidance of the Boston Redevelopment Authority and the specific development and management of a private developer, Faneuil Hall Marketplace, which focuses on retail sales of a wide array of products, especially food, is considered one of the largest tourist attractions along the East Coast of the United States. The success of that project inspired other redevelopment projects to take place along Boston's waterfront including Charleston Naval Shipyard and the Fish Pier.

Despite these signs of progressive movement in waterfront rehabilitation, Boston is still subject to objections from different groups who fear that continued development will negatively affect the environment. Recently, connections to one of Boston's sewer systems that was part of a larger development project on the waterfront, was halted. Affected was the ability to rent empty space in the newest commercial and residential buildings in areas where waterfront development was planned. It was claimed that the sewer system, which eventually empties into Boston Harbor, is a major cause of pollution of the harbor. The resolution of that dilemma is still undecided at the time this report is being written (Disenhouse 1991).

San Diego

The Port of San Diego and its neighboring communities have come the closest to achieving a total waterfront development and redevelopment program. The program realistically reflects the Port's relatively limited, commercial marine cargo activities, especially since the start of the container era. Several decades ago, when the program was developed and implemented, comparative advantages of development of major projects for recreation, commercial, and fish handling were emphasized.

During the 1950s, the Port of San Diego and its waterfront began to lose the large volume of cargo activity that it enjoyed during World War II and the Korean conflict. In response to this trend, a comprehensive, waterfront development master plan was created with the cooperation of the five major municipalities that border San Diego. This agreement eventually led to the creation of the San Diego

Unified Port District in 1962. In essence, this plan granted all waterfront property up to the high water line to the District in trust. Although political and social differences existed during the formation and implementation phase, it was recognized by all concerned that without such a plan and an agency to implement it, the waterfront as it then existed would continue to degenerate through underutilization, abandonment, and incompatibility of adjacent uses.

The result of this process is evident today. The waterfront has been divided into several separate sectors, each of which either reflects the predominant prior uses of its waterfront or has been designated to accommodate other uses that include recreational and light commercial activities. These planning considerations were based on the District's authority to undertake project in four major areas: commerce, navigation, fisheries, and recreation. However, each of these areas of authority to operate has been broadly interpreted, thus permitting the construction and operation of public parks, commercial fishing piers and processing facilities, along with the leasing of land and facilities for use by hotels, restaurants, and marina operators. It should be noted, however, that severe economic pressures on the U.S. flag have reduced the commercial tuna industry fleet and its activities as originally planned.

Although the district has many successful commercial operations within its jurisdiction, including Lindberg Field, San Diego's regional airport, certain projects such as parks and promenades obviously are engaging in revenue generating activities. Yet, the ability to balance the profit-making operations such as hotels, restaurants, commercial fishing facilities, and marina against public-use projects has given the entire waterfront a deeply embedded aura of "success." Moreover, San Diego's waterfront has developed into a major tourist attraction and simultaneously has improved the quality of life for the area's residents.

Baltimore

Historically, the inner harbor area of Baltimore had been the main center of maritime activity for that large East Coast city. Then, about the mid-70s, a new container terminal outside of the inner harbor was constructed at Dundalk. Shortly after this terminal was opened, the inner harbor area suffered the abandonment of activity. The decay of Baltimore's inner harbor echoed the symptoms of the general downtown Baltimore area which had suffered similar problems after business began to relocate to more modern industrial facilities and office complexes skirting the city.

With renewed interest in the downtown area, a private firm began to plan a major redevelopment of the inner harbor area that eventually would reflect both commercial and recreational activities. Today, Harborside is a 240-acre project that surrounds the harbor basin. This is the same location in which the City of Baltimore was originally founded and where it flourished in later years.

The first phase of the plan comprised 95 acres of land along three sides of the harbor basin which would be used for an aquarium, promenades, small boat rentals, a river steamer converted into a restaurant, a marina, and playing fields. The cost of this phase was about $260 million of which almost three-quarters was from private investments. Public funding constituted the balance, most of which came from federal grants and City bond issues.

Toronto

Unlike many northern cities, Toronto, during the past two decades, has experienced population growth and expansion beyond its traditional boundaries. Particularly noteworthy has been the City's attention to lakefront redevelopment, emphasizing "adaptive re-use" of waterfront facilities.

As part of an overall, updated master plan, most of the redevelopment of Toronto's waterfront during the past two decades has largely been focused on the creation of both active and passive recreational parks and commercial uses. However, the Toronto Harbor Commissioners, the region's port authority, was concerned with the growth and protection of their marine cargo facilities that have made the port one of the major port cities on the Great Lakes. The parks by themselves are not self-supporting, but add to the much broader concept of total waterfront development which included other more profitable projects.

The Toronto Harbor Commissioners, in cooperation with the Province of Ontario and other environmental agencies, created a number of parks that stretch for several miles along the shoreline where previously, decay and misuse were evident. These efforts in turn stimulated other projects that included the removal of old and marginally used piers. Eventually, a hotel/apartment/office complex known as Harbor Castle was built. It is located almost within the shadow of the most conspicuous structure in Toronto, namely the 1,826 foot high CN Tower, one of the tallest, free-standing structures in the world. Pier 4, a renovated finger pier located slightly to the east of Harbor Castle, now boasts a number of reasonably priced restaurants, shops, and marine accessory suppliers. On both sides of this pier, that was dilapidated until a few years ago, is a small marina that, although reportedly not profitable by itself, provides the kind of atmosphere that stimulates repeated visits by Toronto citizens.

Toronto also can boast about another landmark type of development: Ontario Place. This 96-acre recreational center which was constructed by the Province of Ontario and supervised by the Toronto Harbor Commissioners, was opened next to the site of the annual Canadian Exposition. Open from May to September, Ontario Place features almost every possible waterfront activity except housing and office space and is supplemented by elegant restaurants, big screen movies, an amphitheater, and even a specially designed children's park where no one over four feet tall is admitted. There is a small charge for admission to the park. It is one of the most heavily visited recreational activity centers in the Toronto area. It should be noted that although the park it serves operates with a slight deficit, it has managed to encourage investment in other year-round, waterfront activities that generate revenue for all the various governments in the region.

Seattle

This large and important northwest port city and its residents have agreed on how to shape the harbor's future. The port of Seattle is proceeding with a nearly $600 million development program for the colorful Elliott Bay waterfront (King 1991). Although a substantial share of these funds has been set aside for the continued development of marine container facilities, the program is also moving into commercial real estate development. The port was set up as a public body by Seattle and King County residents in 1911 with a five-member commission, independent of the city and county. Members of the commission, however, are elected directly by the citizens.

The port's piers and terminals ring Elliott Bay in the shadow of the Space Needle and stretch inland along the Duwamish river. Northward on Salmon Bay, it operates Fisherman's Terminal, where a 700-boat fishing fleet is wedged between the prestigious Queen Anne and Magnolia neighborhoods. Nearby, its Shilshole Bay Marina moors 1,500 pleasure boats.

Historically, the downtown harbor was crowded with cargo ships. Piers accommodated huge warehouses, mostly for general cargo and fish-processing, much of it tied to the Alaska and Far East trade. However, about two and a half decades ago, the entire waterfront transportation industry changed in Seattle. The larger- sized ships, especially containerships calling at the port, required deeper waters and larger terminals; facilities that had to be constructed further away from the main hub of activity near the inner harbor area.

The empty piers near downtown Seattle began to age quickly because of little or no use or maintenance. Some of the aging piers were demolished, others fell into the water, or have been boarded up for years. Private owners in the city converted many others into brightly painted restaurants, import shops, parks and an aquarium that cater to both residents and tourists. The port operates cruise ship docks and the state oversees the fleet of passenger and car ferries.

In a controversial step, the port is moving forward in a new direction with its first, major venture into commercial real estate development. Some residents opposed any commercial development by the port, while others contended that the port's land should be used only for marine-related development. The port has planned private marine businesses and public access areas.

The planning of these projects recognized that perhaps in contrast with other ports, the development in Seattle has been moving toward the waterfront. The expectation is that formerly blighted streets next to port properties will become desirable as places to work and live. For example, one biotechnology company this summer bought a parking lot two blocks south of the proposed Pier 66 upland complex. It is considering building 300,000 square feet of laboratories and offices there.

Another goal is to tie in the waterfront with the downtown just a few blocks away that is now cut off by railway tracks, a major highway, and a steep hillside.

CONCLUSION

Setting aside an in-depth analysis of the economic value of maritime activities on a region affected by those activities, it becomes important to recognize that in those ports where a shift of cargo handling operations and technologies from one location to another has taken place, in part or in total, there has and will continue to be long-term implications on the future of the port and its environs. Economic, physical, institutional, and environmental pressures from all directions are forcing ports to deal with change that is sometimes slow in coming. But change, especially when viewed from the lessons learned by some of the largest ports in the United States and Canada, is possible. Decision makers of today and those who will follow, need to recognize all of the possibilities available to them. Compromises need to made, and if successful, the economic and social welfare of the port and its environs will be enhanced in the generations to come.

ACKNOWLEDGEMENTS

The author wants to acknowledge his sincere appreciation for having the opportunity of collaborating with some of his colleagues over the years at the Port Authority of New York and New Jersey on several studies dealing with economic revitalization of the New York-New Jersey region. These studies, of which this paper is a direct result, were made as part of a team effort and subsequent assignments at the Port Authority between the late 1970s and early 1980s. In particular most of that appreciation goes to Rosemary Scanlon, Mike Krieger, and Ted Kleiner. Also, special recognition must go to the author's friend and colleague, Carl Sobremisana, someone who always recognized the changes taking place in port maritime/mixed used developments.

REFERENCES

City of New York. 1991. *New York comprehensive waterfront plan; Reclaiming the city's edge.* Department of City Planning. Summer.

Diesenhouse, Susan. 1991. Boston Economy Stymied by Harbor Effort. *The New York Times*, 29 April, sec. D:1.

Dower, Rick. 1992. Making San Diego shipshape for major cruise trade. *San Diego Business Journal* (March 30): 9.

Kelly, Brian and Roger K. Lewis. 1992. What's right (and wrong) about the inner harbor. *Planning* (April): 28-32.

King, Harriet. 1992. For Seattle, A wave of harbor projects. *The New York Times*, 1 September, Real Estate Section:3.

King, Elliot. 1992. World class ports: The year 2000 and beyond. *Global Trade* (January): 10-13.

Muller, E.J. 1992. Ports: Development is stuck in the mud. *Distribution* (July): 48-51.

Muller, Gerhardt. 1990. *Intermodal Freight Transportation, 2nd Edition*. Westport, Connecticut. Eno Foundation for Transportation, Westport, Connecticut.

Tabor, Mary. Despite Opposition, Agency Plans to Sell Brooklyn Piers to Private Developer. *The New York Times*, 3 April, 1992: B3(L).

Tessier, J.M. 1991. Port-city relations -- In harmony. *Portus* (June):16-20.

Volk, David. 1992. Seattle port project stirs pot: Proposed waterfront development fails to win over all the port's critics. *Puget Sound Business Journal*. (11 December):22.

48

Development of Artificial Reclaimed Lands and Their Integrated Planning in Taiwan, R.O.C.

Ho-Shong Hou, Ph.D., P.E.
Director, Transportation Engineering Department
Institute of Transportation, Ministry of
Transportation and Communications, Taipei, Taiwan, R.O.C.

ABSTRACT

During the past three decades, industrial and population development and economic prosperity have crowded land areas and made their acquisition much more difficult. Therefore people are searching for ocean and coastal area and expanding reclaimed land in order to build artificial islands. This is the modern tendency for the 21st century. Because of the reduced land area and growing population density in Japan, Hong Kong, and Taiwan, R.O.C. etc., the need for such undertakings has increased. Japan has created more than twenty Artificial Islands, and Hong Kong is going to build the Chek-Lap-Kok Artificial Island as an international airport.

Paralleling the waterfront industrial and recreational development is offshore, heavy industrial town and Deep Water Port Development Planning, etc. Their plans deal with artificial island development. Therefore, they are under the governing auspices of the Ministry of Economic Affairs, Council of Agriculture, and Ministry of Transportation and Communications.

Coastal resources are limited in the Taiwan district; therefore, systematic, integrated planning is necessary to solve environmental impact problems and to promote the quality of life. A number of artificial islands are planned for various functions along the west coast of Taiwan.

INTRODUCTION

In the past thirty years, because of prosperous economic activities and booming business opportunities, living space is becoming crowded and land acquisition has become more difficult. People, therefore, turn to seashore areas to develop more living space. Thus, reclamation of the seashore to make more land and development of reclaimed land become the tendency everywhere, especially in overpopulated countries such as Japan, Taiwan, and Hong Kong.

These peoples are desperately concentrating on this. Up to now, Japan has finished twenty more artificial islands, Hong Kong is building an international airport on Chek-Lap-Kok Artificial Island (Figure 1) and Taiwan is also planning to develop an artificial island concomitant with the development of shore industrial recreational zones, as well as the harbor. The authorities include the Ministries of Economics, Agriculture, and Transportation and Communications. Taiwan has only limited seashore treasures, therefore it needs systematic planning to solve the environment impact as well as to raise the quality of living. This article emphasizes the problems of seashore environmental impact and hopefully, will be submitted to all the authorities for reference and guidance.

PREFACE

The following facts highlight the serious problems due to insufficient land:

1. Polluted city environment due to lower living qualities
2. Huge costs of purchasing land for public construction purposes
3. Lack of parking lots
4. Insufficient parks, and recreation areas

There are two methods to solve the problem. One is to increase the utilization of land to underground and toward the sky. The other is to consider reclamation of the seashore to make more land, the most obvious example of which is Taichung, an artificial harbor. This involves building breakwaters and jetties, deepening and widening navigational channels (augmented by more than 4,000 hectares of new harbor land), and construction of a separate artificial island for number of purposes, similar to Japan's Kansayi.

An new international airport, under current planning, is aimed at solving many problems of noise, exhaust fumes, air pollution, and dangers of plane collisions. This can be defined as a large, successful reclamation project. At present, 511 hectare have been reclaimed. This can also be a model for planning an international airport on the seashore area of our country (Figure 2). In the past thirty years, we devoted our best efforts to developing our economic activities while ignoring environmental protection. The results is that our living environment have been seriously polluted, however, in the past ten years, our people have acquired a sense of environmental protection and are requesting better living quality. All kinds of manufacturing or industrial facilities that may create pollution are rejected. Such facilities may include, inter alia, airports, nuclear power plants, fire electricity plants, refinery plants, and petro-chemical industries: also they may be as small as incendiary waste treatment plants, cement plants, chemical plants, used non-ferrous metals disposal plants, etc.

The actions of people are simple and effective; they merely surround the suspected enterprise protesting and claiming compensation. Eventually, such action causes all factories to lose a lot because they either have to stop their operation; pay tremendous compensation; or even move or close their business. The alternative is to move the factory or facilities to some place far from residential communities because nobody likes to be the neighbor of those plants.

The Artificial Island has the advantage of distance from human communities; it minimizes pollution to the human community; and it lowers opposition to the choice of location of the plants. Besides, the Artificial Island can also be a "city function site", and "industrial zone", a "port land", a "deep water harbor", an "ocean recreation place", a "storage place for petroleum, coal and ore". Compared to certain areas which have the problems of land storage and serious pollution, the Artificial Island costs less, and the advantages become more obvious.

Up until now, Japan and United States have been establishing airports and electricity plants on artificial islands. Therefore, with meager land and overpopulation in Taiwan, we certainly need a comprehensive program of reclamation and development of artificial islands to solve problems of insufficient land and environment pollution before they worsen.

Figure 1. Chek-Lap-Kok new airport in Hong Kong

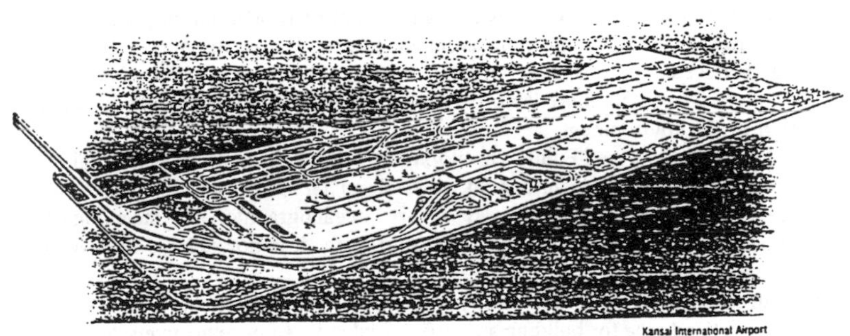

Figure 2. Kansai new airport in Japan

THE FORMATION OF ARTIFICIAL ISLAND

To form an artificial island, we need to look for natures geographical data, such as "offshore barriers" (which are accumulated by the waves from the flat coasts, such as "sand dunes"), "lagoons", natural tiny islands, etc. Then we need to remodel such places to form an artificial island.

The main characteristic of "flat coasts" is a low sea bottom slope. W. Johnson (1919) believes that the "flat coasts" might be formed by coastal emergence, but it only applies to flat sea bottom areas. The evidence shows that most of the flat sea bottom areas are spread around some low latitude and mean latitude costs; the best way to reduce the impact of wave energy on lower,rocky coasts is to build an artificial barrier.

Usually, barrier islands are more easily formed at "low tide difference", but sometimes it also can be formed on "high tide range" when other conditions are admitted. Examples are Taichung Harbor and the Chang-hua shore area. Waves are the main influence in forming barrier islands, especially at the sites of surge waves. Therefore, the main conditions for forming sandbars are: (a) use of sufficient sand fill (b) and identification of an adequate beach which has a low degree slope. A wide open seashore area with a more constructive swell characteristic would be the best area to create a barrier island.

Reviewing the above conclusions with respect to Taiwan, we can then understand that only west coast seashores belong to a so-called "continental shelf" style, flat shore category, which has accumulated enough material for use in planning an artificial island. As for the east coast, while its sea bottom slope is $i=10\%$ in most places, only 4 to 5 km from the seashore the depth reaches to 1,000 meters. It is therefore not the right place to do reclamation work for artificial islands.

Artificial islands are being developed in Europe, the United States and Australia. As for Japan, because of its limited land with high population density, artificial islands are used for airports, power plants, wharfs, industrial zones, or activities of commercial or residential purposes.

Taiwan experiences the same condition as Japan, but the European countries are more concerned with oil drilling, piers etc., purposes for artificial islands. In any case, the advantages of artificial islands are: (a) low cost to obtain more land, (b) distance from heavily populated communities, and (c) reduced pollution of the environment.

Taiwan's West Coast ("Arca-Future Development") is an example of successful artificial island construction. Surrounding the Taiwan district, this island faces both the Pacific Ocean and Taiwan Strait. The east coast's steep slope causes severe oceanographic and meteological conditions. Hurricanes and heavy swells often attack this area; therefore, it is not suitable to develop an artificial island along the east coast.

The Taiwan west coast is a flat, mostly deposited coast, therefore, it is more suited to the reclamation of the wetland area for building an artificial island. At present, many projects of reclaimed land planned by central and provincial governments, are listed here and considered for integrated planning and development in order to obtain coastal balance, i.e., reasonable and systematic development.

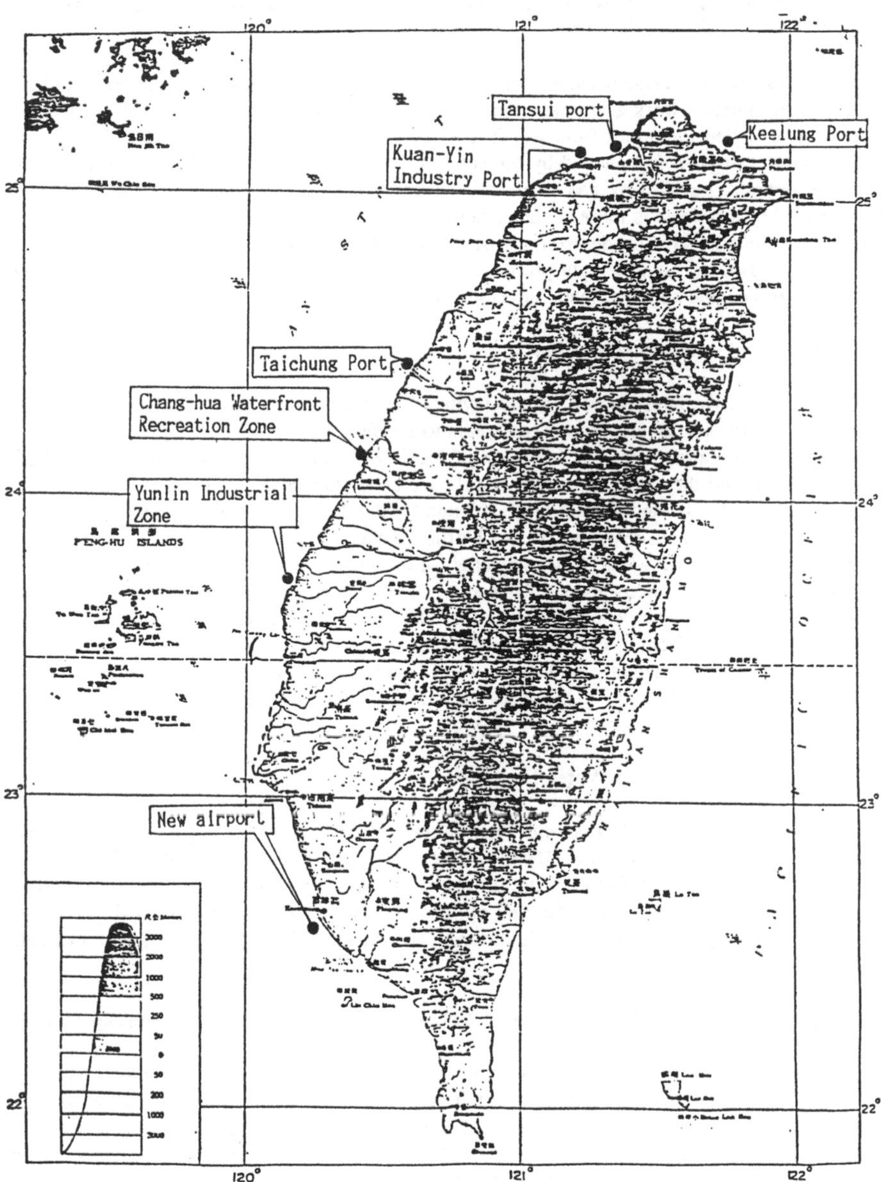

Figure 3. Location of artificial reclaimed land in Taiwan

THE FUTURE DEVELOPMENT OF RECLAMATION AND ARTIFICIAL ISLANDS AT TAIWAN WEST COAST AREA

In the last twenty years, whenever the Japanese developed or enlarged their ports and harbors, they did the reclamation work at the same time. They built about 20,000 hectares of artificial islands, and these are excellent examples from which to learn.

Although Taiwan is surrounded by the sea, its east coast is characterized not only by cliffs and steep slope terrains, but also by bad weather conditions, typhoons, swells, etc. For artificial island development the west coast is more suitable, because of its flat sea bottoms and continental shelf-type seashores. Hereunder are the descriptions of a few intensive plans presented by central and local governments. Also, we submitted our proposals on a few suitable locations only after considering their balance with reasonable and systematic developments.

Keelung New Port - Reclamation Plan

In July 1984, Keelung Harbor Bureau invited China's Engineering Consultant to start planning Keelung's new port operations. It took two years to finish the entire job. The preliminary choices of Keelung's new port range from the west side of Keelung Harbor, all the way to Yehliu (Figure 3), Wanli, Dawulun, and the Waimushan fish wharf, are all within this coastline area. East of Marshou Creek, the Wanli Beach, and Feitsei Gulf belongs to the Yehliu Peninsula. The above-mentioned coastlines constitute a vital recreation area of North Taiwan.

The new port ranges about 1,300 hectares. Following reclamation, the artificial island shape wharf (or piers) will encompass about 400 hectare. The reclaimed mountain dirt can be used to build a container wharf in the new developing city, just like the port island of Kobe, Japan.

Because the new Keelung Port faces deep water on its west, middle, and east sides (the deepest breakwater reaches to 60 meters), the techniques of building contour facilities should be studied more deeply and carefully. We can take Pusan, Japan, and Portugal Sines harbors as good, imitative examples.

The Planning of Tamsui Deep-Water Harbor

This location is selected from our "preliminary planning of building a deep-water harbor", including the total zone area from the Tamsui River mouth northward. It is about 7.5 km in length (apart from the seashore) and about 2.5 km in width. The harbor zone and the coasts are separated by 150 meters of green belt. As such, the total artificial island pier zones are about 1,500 hectares. As it were, the priority yields only to the Kaohsiung deep-water harbor. Yet we still are evaluating its necessity (Figure 4).

Tamsui Special Harbor for Sand and Stone Delivery-Tamsui Domestic Commercial Harbor

North of Taiwan there is great demand for sand and stone for construction purpose, and the rivers north of Taiwan cannot create enough of this material. Furthermore, the difficulty of transferring sand and stone from inland and the plethora of sand and stone in the rivers of East Taiwan, create a demand to transfer these products from East of Taiwan to the north. The alternative of building a special pier for sand and stone at the mouth of Tamsui becomes a special issue. In response to the request from local government to enlarge the size of land and scale as a commercial harbor, the artificial island zone of about 250 hectares will become a pier for sand and stone, cement, storage sites, a connection road, etc. (Figure 5).

Kuan-Yin Industrial Harbor for Petroleum Importing

To solve the demand from relevant petro-industries, construction of a special harbor to import materials, both dry commodities (coal, salt, and sulfate) and liquid commodities (crude oil and petroleum), becomes a necessity. A 1,000 hectare space is estimated to be necessary for all facilities to set up and operate. In addition, a 400 hectare space of artificial island is needed for handling oil tanks and vessels under 150,000 tons (Figure 6).

Taichung Harbor - A Successful Artificial Island Man-Made By Our Own Country People

Taichung Harbor has a total space about 4,300 hectares all are "contour facilities", such as breakwaters, bulkheads, groins, seawalls, and other forms of protection. Those walls are surrounding the sea and land areas, thus successfully utilizing the sand from dredging channels to backfill and reclaim the piers as well as new born land.

This is a good example of a convenient, economic method to accomplish the reclamation and build an artificial island. This is a very high rate of return on investment, especially when this year's (1992) total throughout (17,000,000 tons) is over the 12,000,000 tons prospective target.

The Development of Chang-hua Waterfront Industrial and Recreation Zones

In the Chang-hua Waterfront Industrial Zone the artificial space of Sane-hsi, Lun-Wei, and Ru-Kang etc., total around 3,643 hectares. Of this 2,300 ha are for industrial purposes, 1,248 ha are for public facilities (including wind break); and 65 ha are for liaison roads. Since 1979, this space has been under development by the Industrial Development Bureau (IDB) of the Ministry of Economics.

In the Chang-hua Waterfront Recreation Zone the total space (including Sungkang, Fuhsing, and Hangbin) is about 2,844 ha. At present, the plans are in programming process by the government of the Taiwan province. They also set up a committee of development to put through this program, the goals of which are sight-seeing area, swim racing pool, recreation zone, sporting park, horse-racing ground, and refuse disposal place. The above mentioned zones total 6,500 hectare. All the relevant problems caused by the impact of the sea should be handled by a coordinating authority. (Figure 7).

The Developing of Yuenlin Chaiyi Tainan Coasts Area and the Planning of an Artificial Island for a Base of Industries

Building a base of industries in Yuenlin and the developing of a parasol-shaped bar outside Yuenlin and Chaiyi should be a combined planning effort.

To offer a suitable environment for basic industries and sub-industries to insure their efficient operations, and to combine them into a multifunction base (such as industrial, residential, and environmental), the Industrial Development Bureau of Ministry of Economic Affairs chose to develop a 7,000 ha site at Yuenlin and Chaiyi. The parasol-shaped bar area is intended for a petroleum refinery, steel refinery, and all relevant high-tech industries. They also intend to develop a 4,000 ha site for residence, commerce, education, recreation, wind break, green belt, environment protection center, industrial harbor, storage pool, and other public facilities. Therefore, its total space needed for reclamation, i.e., the artificial island, would be about 10,000 to 15,000 ha. (Figures 8 and 9).

Under present planning by the Industrial Bureau of Ministry Economics for Yuenlin, an off-island basic industrial zone will be developed from North of Mailiao southward, step by step.

According to the theory of "shore deformation," the sand drift derived from shore deposit sand from the Chushui River will flow south to form a parasol-shaped bar, in a few years. If the planning starts from Mailiao, the flow of deposited sand from the Chushui River will be stopped, and the parasol-shaped bar will eventually disappear.

To summarize the above viewpoint, the author believes that land created by reclamation for a basic industrial town should develop from south to north. Not only can we thus obtain new land from reclamation, but we can also more effectively confront the natural geo-environment. Furthermore, we can fully accomplish the integrated planning of the 15,000 ha sand bar at low tide on the parasol-shaped bar area, the 1,000 ha man-made land at Paimen and a 2,800 ha man-made land site at Chiku on the Tainan coast. This last site can be used as a petrochemical, industrial zone and was chosen as a candidate for an industrial town. According to the calculations of the Industrial Bureau however, the total development space for a petrochemical zone should reach to 10,000 ha.

Because Yuen-Chia near the coast includes about 1,000 ha belonging to the Taiwan Sugar Co., the technique of obtaining land is much easier. Further, judging from the already formed, man-made land area, we can conclude that the most suitable place to develop a petrochemical zone is Yuen-Chia. However, reclamation and other engineering techniques are needed to be introduced from Japan and Holland.

The Integrated Planning of Southern International Airport and Kaoshung Harbor

In order to comply with our future demands for sea and air transportation, and recognizing the trend to increasingly large, ocean bulk liners, the study of deepwater harbor feasibility becomes increasingly important. Among all of the harbors along the west coasts of Taiwan, Kaohsiung is the most suitable deepwater harbor. The location starts from the south breakwater of the No. 2 harbor entrance to the Kao-Ping estuary; this means extending the present coastline out to -28 meter depth, its contour, about 6 km from shore, and the length of shoreline about 13 km. Thus, the harbor zone needs 6,500 ha of reclamation works. This planning must actually take place in two parts, i.e., a deep harbor zone and a reclamation effort. As indicated on the chart, this includes the breakwater wall, sea wall, loading and unloading facilities, navigation aid facilities, shore protection, devices navigation channels, and a turning basin, etc.

Reclaimed land will be used for a borrow pit, temporally isolated wall, green belt, isolated belt, and anti-pollution facilities. Other public facilities will include a water discharge system, power system, and transportation system. A man-made land zone still need to be well planned for purposes of an international airport, ranging from 1,000 ha to 1,500 ha (Figure 10).

Southern International Airport

The following environmental protection problems must be faced when building a new international airport (although they can nearly all be solved by an island international airport):

1. *Jet Noise*: The severest environmental impact is the jet noise when airports are near harbors. The methods to reduce noise include:
   a) Regulations of air transportation
   b) Improvement of aircraft engineering
   c) Movement of harbors and air routes far away form cities.

   Criteria of Environment and Noise Prediction Methods, entail predicting the sources and bases of noise by scientific methods; surveying the directions and numbers of routes; and integrating wind direction, wind velocity etc., into the planning for the location of the airport.
2. *Air Pollution from Aircraft*: To minimize pollution from aircraft, it is advisable not only to test the density of gas per se, but also to use air currents to dilute this density. Generally speaking, if the location of new airports and harbors are chosen incorporating environmental principles, choosing wider spaces will reduce pollution.

3. *Noise and Air Pollution:* Incoming and outgoing traffic of airports and harbors will create considerable noise and air pollution. Shifting the interior traffic of airports and harbors to railways will generally reduce this problem. Another alternative is to move the road traffic facilities to seabottom tunnels and use an exhaust tower to dispel all the waste air.
4. *Variation in Tidal Current and Water Pollution:* Currents and wave action are the main items for study in siting sea shore facilities. Pollution will be partly dispersed by currents. Furthermore, it would be advisable to consider the effects of embankments upon reclamation in relation to disposing of hazardous materials.
5. *Reclamation Works*: The reclamation, ecology, landscaping, and environmental conditions should be investigated to maximize the ratio of benefit to harm. An overall detailed and judicious plan for coping with natural disasters is most important.
6. *An International Airport Island*
   a) Knowledge of oceanographics, geology, and meteorology is important for international airports constructed on artificial land. The foundation may have to be as deep as 40 m below the sea bottom in consideration of settling.
   b) The airport island and the mainland should be linked by a bridge or sunken tunnel.
   c) Special maintenance requirements of the airport facilities must be taken into account.

INTEGRATED PLANNING OF ARTIFICIAL ISLAND

Construction of nearshore and offshore artificial islands causes obvious offshore environmental impacts such as water quality, fishery activities, bird-resting, current changes, pollution, offshore coastal morphological changes, etc. Therefore, it is necessary to conduct integrated planning.

Taiwan's west coast is a continuous, sandy coast and represents high wave energy area. The prevailing littoral transport is from north to south. Artificial breakwater construction for the artificial island will affect the surrounding coastal deposition and erosion. In a normal situation, the existing coastal structure will cause sand deposition in the upstream direction and coastal erosion in the downstream direction. Formation of an artificial island will have obvious effects on deposition and erosion. The serious problem will be the scouring phenomenon in the downstream side.

Based on the above analyses, the reclamation plan for coastal artificial islands need to reflect effective compromise among central government (each ministry) and provincial government. It must also illustrate planning for multiple objectives.

For the grand project of coastal development, it is necessary to plan in both vertical and horizontal dimensions and to integrate all possible national benefits. All of this simply refers to the need for a systems approach.

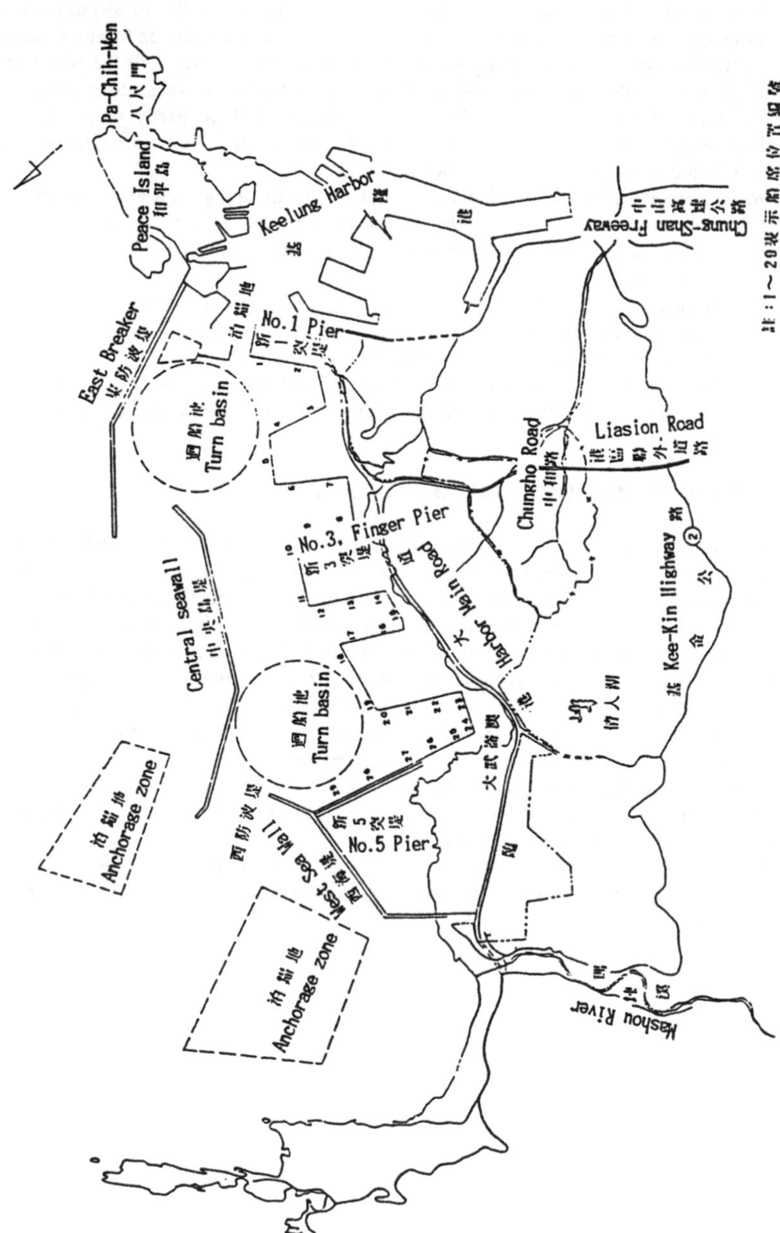

Figure 4. The planning of Keelung Harbor for Reclamation

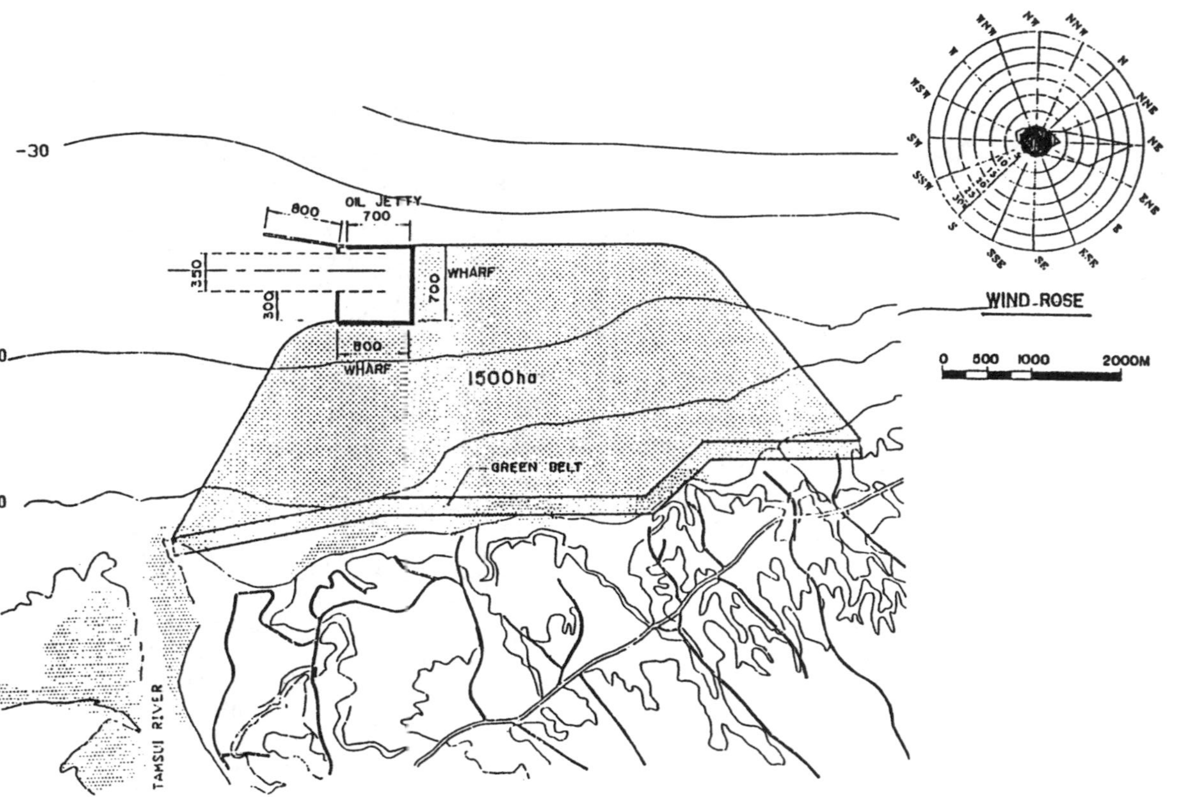

Figure 5. The planning of Deep Harbor, Tamsui

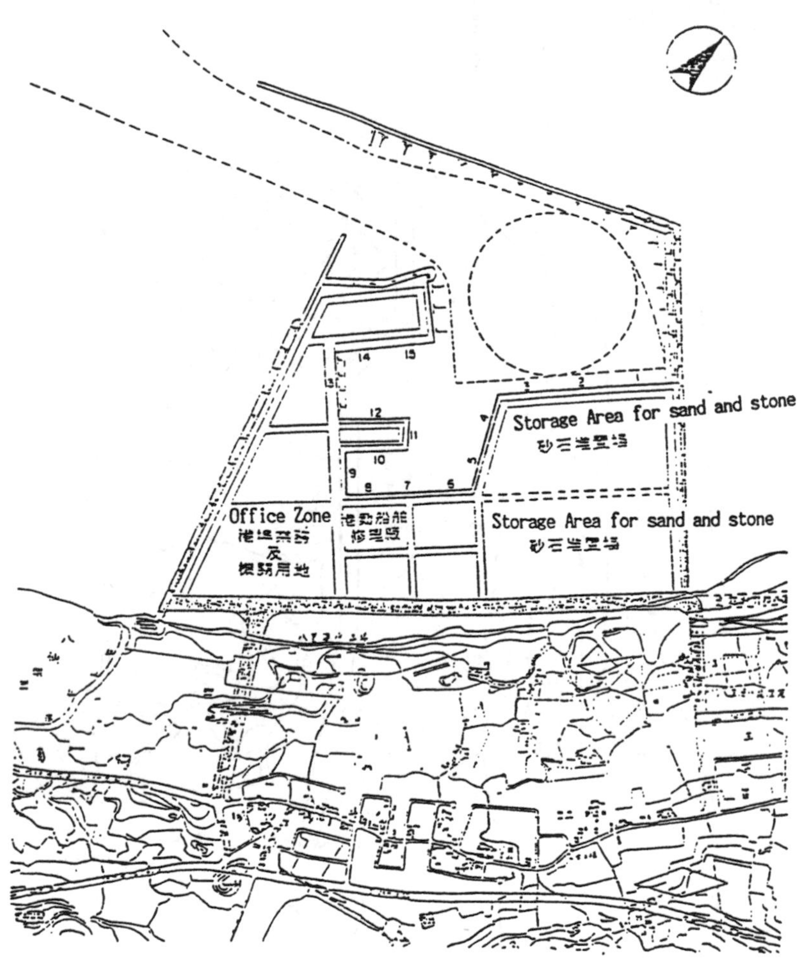

Figure 6. The planning of Tamsui Harbor for gravel and stone

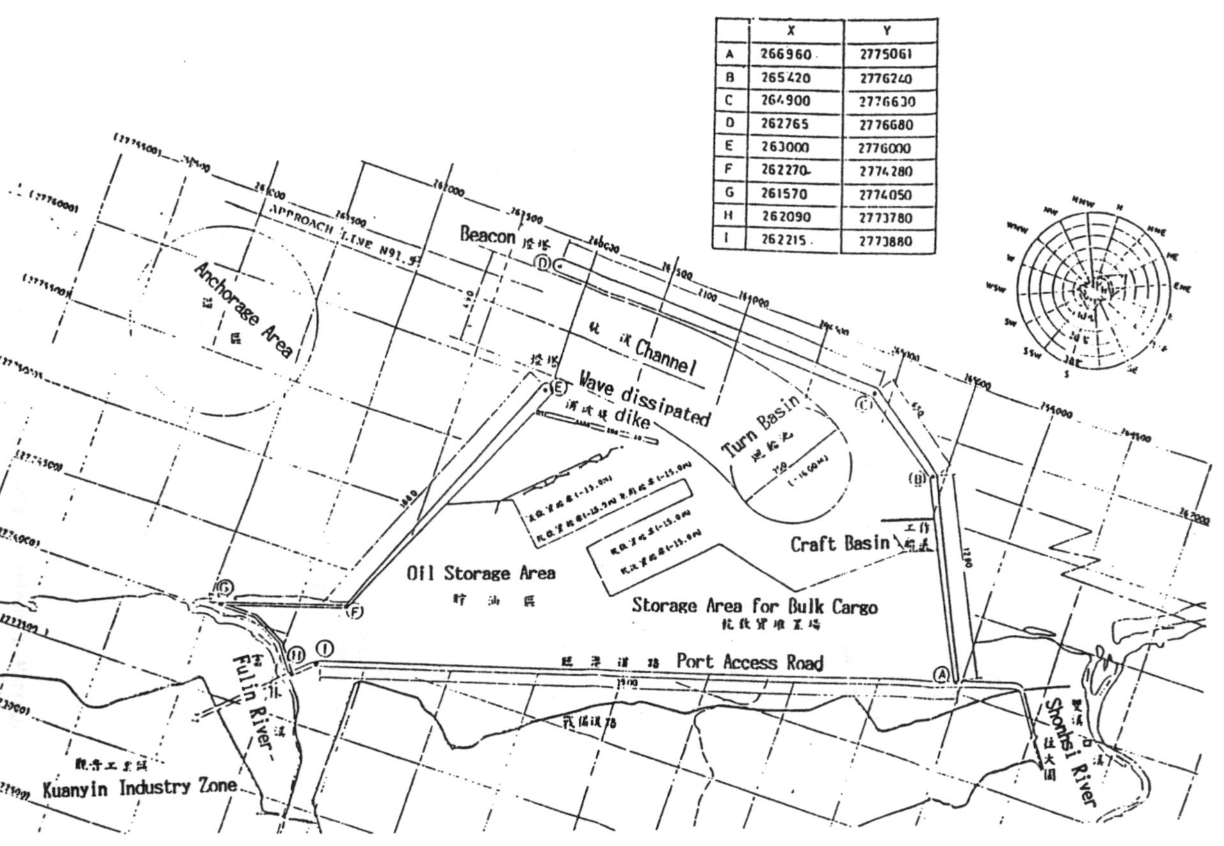

	X	Y
A	266960	2775061
B	265420	2776240
C	264900	2776630
D	262765	2776680
E	263000	2776000
F	262270	2774280
G	261570	2774050
H	262090	2773780
I	262215	2773880

Figure 7. Yuanyin Harbor, Petroleum

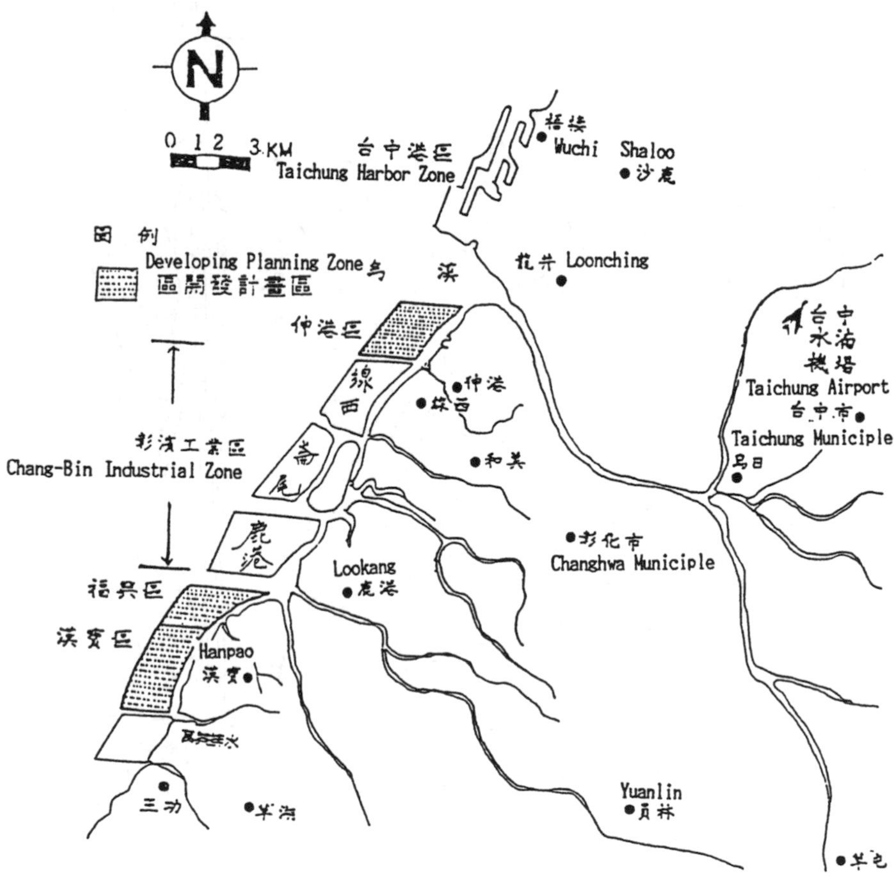

Figure 8. Planning for Chung-Bin Industrial & Recreation Zone

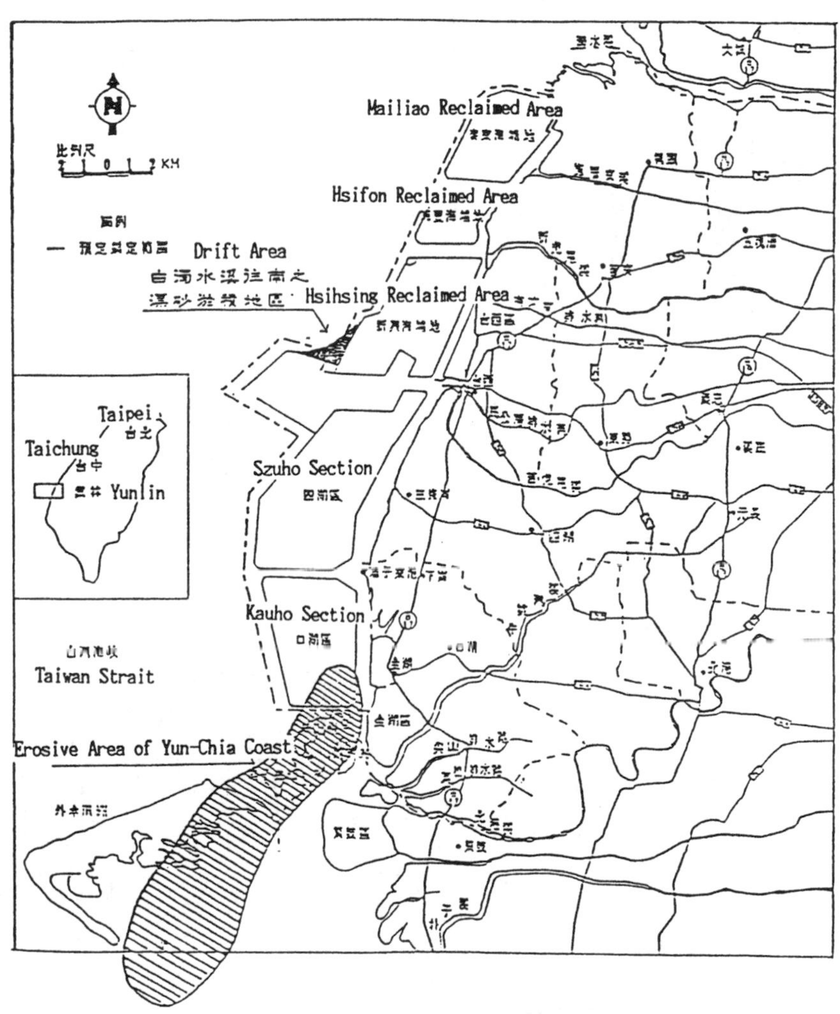

Figure 9. Possible Erosive Area of Yunlin Industrial Harbor

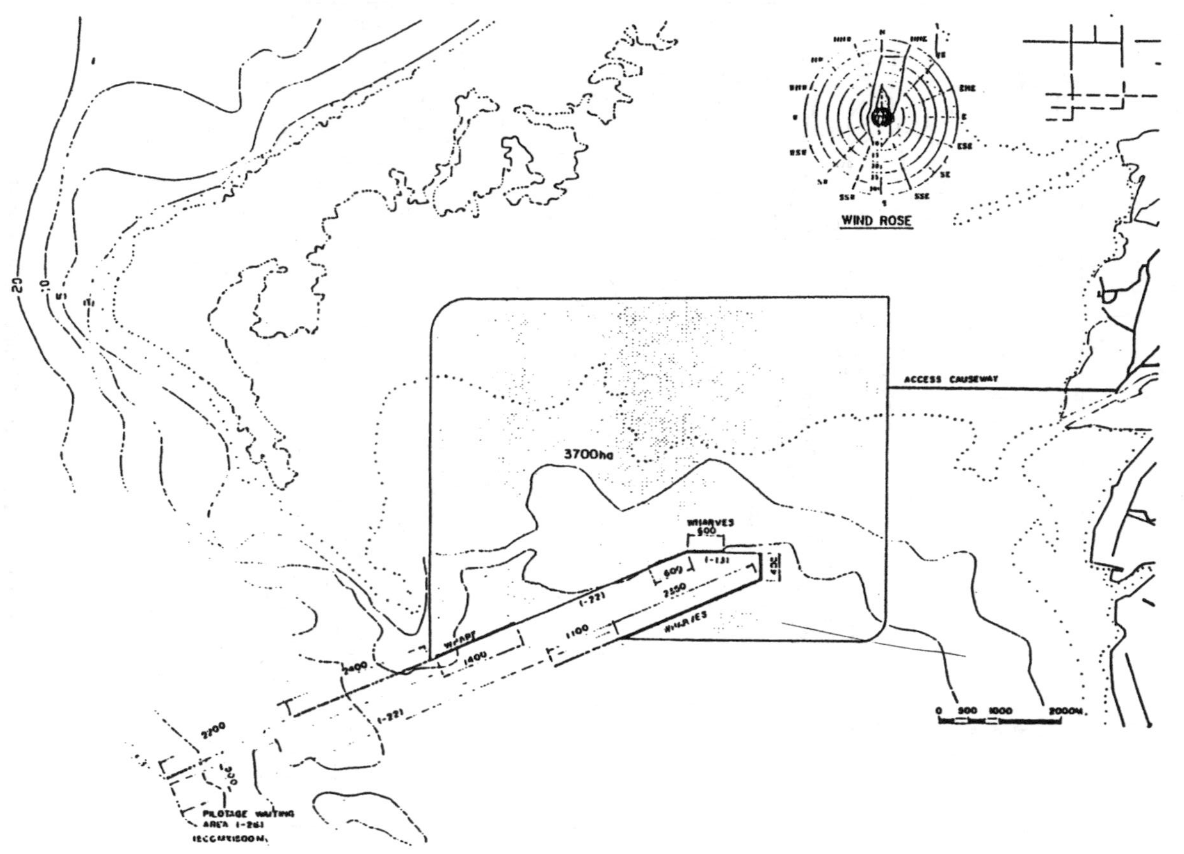

Figure 10. Yun-Chia Integrated Development Plan

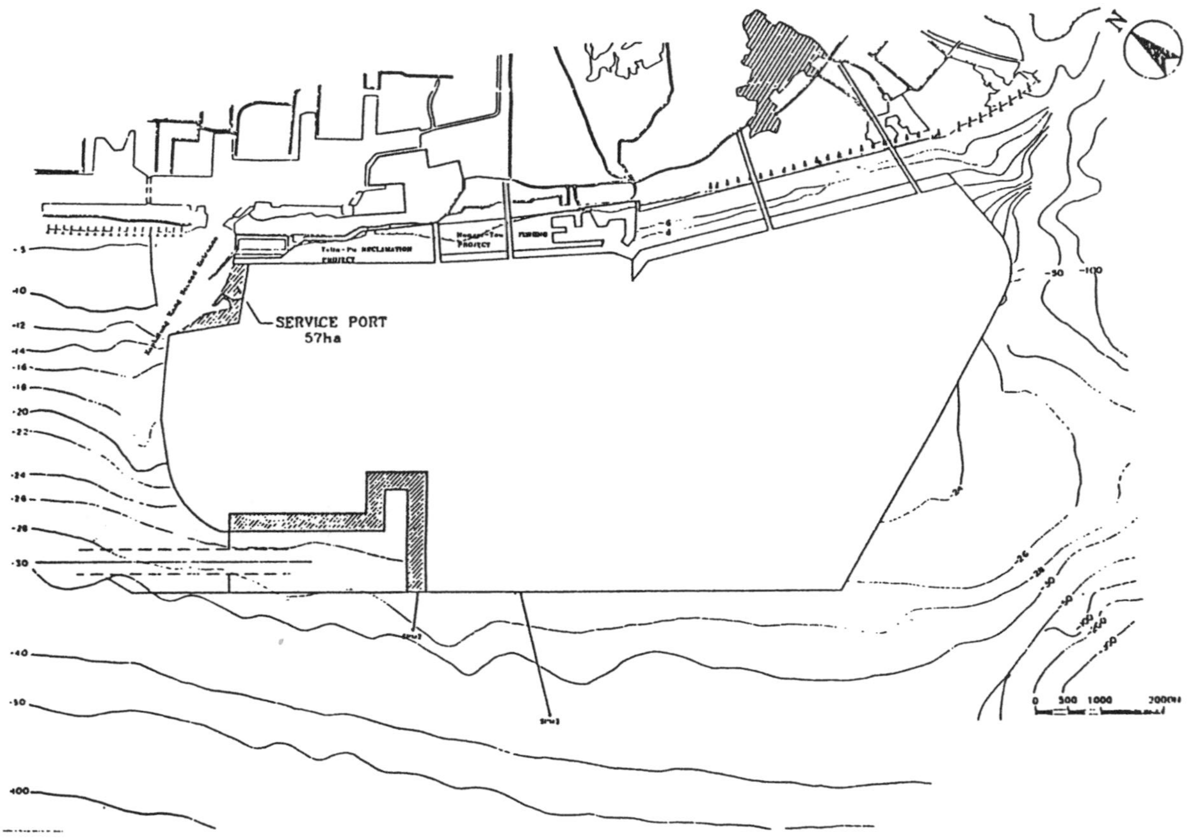

Figure 11. Integrated Planning of Southern International Airport and Kaohsiung Habor

49

Underwater Road Link in the Gulf of Naples: A Case Study for E.I.A. in Coastal Zone

Marco Berta
Manager R&D Environment, Tecnomare SpA, Venezia, Italia

Renzo Piantoni
Senior Technologist R&D Environment, Tecnomare SpA, Venezia, Italia

ABSTRACT

The enactment of laws and regulations issued by Italy and the European Community on Environmental Impact Assessment (EIA) designed to avoid unacceptable environmental consequences of proposed projects, as well as growing public involvement in environmental protection, spurred Tecnomare to develop a methodology for carrying out EIA studies related to its own projects in shallow and deep water. In this paper the research project, "Underwater Road Link in the Gulf of Naples," has been selected as "a case study" in order to (1) describe for a specific application, the overall methodology; (2) discuss how the impact study process worked in conjunction with the project's development; and (3) highlight the importance of this decision-making support tool as an attempt to reconcile the attainment of both developmental and environmental protection objectives.

INTRODUCTION

The protection of coastal areas is a key issue within the framework of marine environmental defense. Along the coastal strip, the natural balance is complex and precarious, and the balance of the sea and the land is at its most delicate. Yet it is in this zone that the shore and the sea are mainly exploited.

Regulations introduced with the establishment of the Ministry of the Environment in 1986, and with the approval in 1988 of the two Prime Minister's Decrees governing categories of works to undergo E.I.A. and technical norms for the compiling of the relative studies, could offer effective guidelines for operating properly in this delicate environment. These regulations have formed a network intended to cope with the whole environmental question but, as regards the marine environment, they have left much undone (Wood and Lee 1989; Jorissen and Coenen 1991).

Tecnomare has recently developed a research project concerning subsea civil infrastructures. One of the related activities deals with the definition of a EIA methodology and its application to the case of Naples' underwater road link.

The project proposes the construction of a road tunnel, buried under the seabed in the Gulf of Naples, to connect the western part of the city with the eastern one, offering the city traffic a center bypass. The system is complemented by four underwater car parks, installed along the coast and connected with the main road, which permit pedestrian access to the city just on the seaside promenade.

Hereafter an hypothesis of application of EIA's methodology to this case study is presented. The EIA provides a qualitative assessment of the impact of the underwater road and car parks project on the biological, physical, and human environments. The purpose is to identify the positive and negative effects the development could have. It identifies the interactions between the project and the environment and then assesses the potential impacts of these interactions.

METHODOLOGY FOR ENVIRONMENTAL IMPACT ASSESSMENT

The methodology has been tailored to Italian environmental legislation and makes it possible to formulate environmental and socioeconomic compatibility studies for a single project or to compare several projects with the aim of selecting the options which minimize the impacts. It utilizes the so-called "interaction matrix" to identify the interactions between the project activities and the biological, physical, and human environment and then assesses the potential impacts, either positive or negative, of these interactions.

The presentation of the results allows a direct comparison among the environmental components, indicating the most endangered ones. In this way it will be possible to define an acceptable "environmental compatibility" and, operating alongside other design planning activities, to allow design or operations to be modified at an early stage as unacceptable environmental effects become evident. The whole process is thus interactive.

The term "Environmental Impact Assessment" (EIA) usually refers to a procedure consisting of a rigorous analysis of the potential compatibility of a certain project with a certain environment. This is done on the basis of information, including the identification of potential polluting elements and their synergism and antagonisms and the knowledge, on the other hand, of the ability of the environment to assimilate the potential effects due to the interactions with the project activities. Therefore, the EIA must take care not only of the descriptive and estimate phase (phase of "analysis"), but it must reach the formulation of a judgment on the environmental performance of the project as a whole, in order to aid the decision-making process.

On the other hand, an Environmental Impact Assessment is a complex of activities supporting the decision-making process which accompanies the design of a work. The impact study has to be started simultaneously with the first phases of the design activities and implemented side-by-side with them. It is only in this way that all the relations between the characteristics of the project and the natural and human environments can be evaluated from the first design steps. Thus EIA is a sequential process: there are several stages, many overlap and there is necessary feedback from one stage to another. This evaluation makes it possible to direct technological choices and design criteria towards an acceptable environmental compatibility.

In other words, EIA is a tool which enables a better quality of design activities, with the ultimate objective of increasing environmental protection.

The general architecture of the methodology proposed includes the followings items (Figure 1):

1. Identification of interactions between the project and the environment
2. Analysis and estimate of the impacts
3. Evaluation of the impacts (assessment)

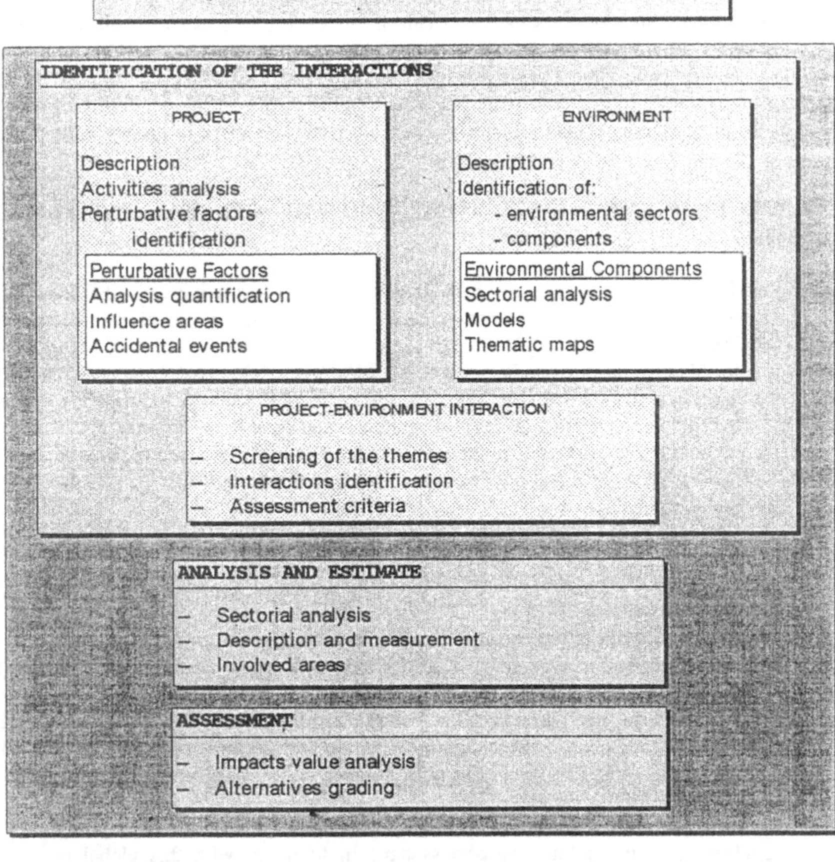

Figure 1. General methodology for impact analysis and assessment

This kind of organization is efficient for the evaluation model itself, which being based on multicriteria techniques, manages to identify the best among the different alternatives. The screening of the most important environmental problems produced by the kind of construction and the site in question enables the environmental sectors affected to be identified.

This kind of organization is functional to the assessment model, which tries to measure the efficiency of each of the possible alternatives. The general objective is "the reduction of environmental impact". The global objective has been divided into sub-objectives. The importance given to each of these subdivisions is variable, depending either on the preference expressed by the public or on programming policy (Goicochea et al. 1982; Colorni et al. 1991).

A relevant aspect which deserves particular care in the decisional process, is related to the need to consider the socioeconomic aspects involved in the construction of the work. This is equivalent to considering the term "environment" in a wider sense (Integrated evaluation model) (Pellizzoni 1991).

Another methodological aspect to point out is its flexibility, due to the possibility of following all the steps of the methodological path separately, from the particular work in project to its design level.

THE SUBJECT OF THE IMPACT STUDY AND THE GENERAL ENVIRONMENTAL PROBLEMS

During the last few years, many cities have witnessed such a traffic increase, that local authorities have taken some drastic steps, such as roads closed to vehicular traffic in central streets, "alternate rating" circulation, etc.

According to this general picture, Italian coastal cities, above all, those compressed between hills and sea, are in a particularly critical state regarding infrastructural renewal and mobility organization. These cities, often having developed and grown around a maritime port, are scarcely livable because of city traffic and heavy transportation. In these cities, the possibility of using the coastal zone and the seaside to build underwater roads and parking systems should not be ignored. This choice can produce the following advantages:

1. Traffic decrease in the central areas
2. Improvement of park supply in nodal sites
3. Possibility of decreasing atmospheric pollution by means of implementing monitoring systems and air purification plants
4. Lower risks during the design phases, in costs and construction time assessment.

The city of Naples is an excellent example of this kind of stressed environment. The coastal road link between the eastern and the western part of the city is one of the unresolved questions in the city's history. Bearing in mind the city's structure, one can say that the coastal traffic axis represents one of the essential elements which characterize urban mobility; in fact the urban mobility naturally finds its main route on this coastal axis. The level of vehicular traffic on this road is near capacity over its total length during the entire day.

To solve this important problem, town planning instruments suggest a new coastal road, coupled with a reorganization of the public transport system. In harmony with this global plan for urban mobility, SAIPEM ITALIA and TECNOMARE have proposed an underwater road link to connect the planned third tunnel under Posillipo (west side) with the eastern highways network. The road will have an overall length of about 4000 m. In addition, the proposed solution includes four underwater car parks, capable of holding 2,500 cars. Some sketches of the road plot and the parking plan are presented in Figure 2.

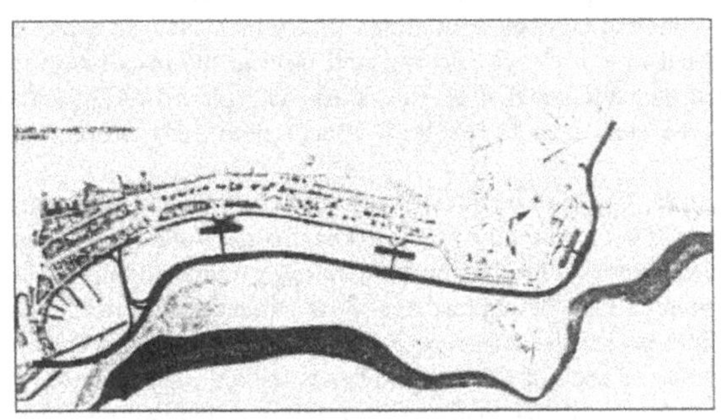

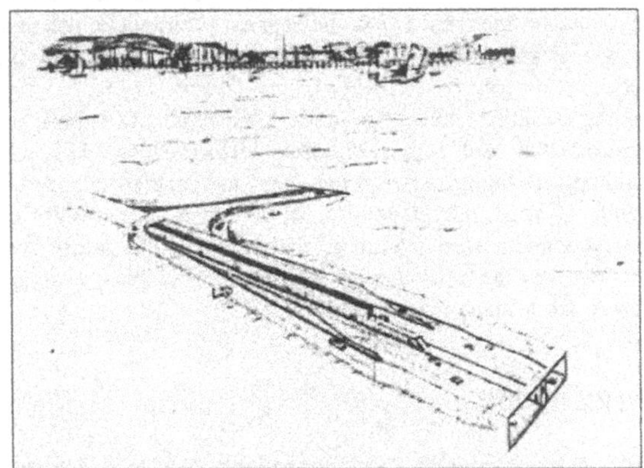

Figure 2. Underwater tunnel and car parks: plan and sections

For the underwater road, the solution of a completely buried tunnel is proposed. A design alternative might be a tunnel partially emerging from the seabed. The tunnel will be built of prefabricated steel and concrete composite modules that will be laid in a sea bottom trench. Module construction will be started in a dry dock and finished next to the installation site in sheltered waters. Then they will be towed over the installation site, sunk, and placed on the bottom of the trench. The ends of the tunnel will be connected to the local road network by proper spurs, built with similar technologies.

Regarding the car parks, they will be placed on the sea bottom too, but nearer to the coastline than the tunnel. Pedestrian junctions to the city public transport network are proposed in proximity to the underwater car parks, sited between the proposed underwater tunnel and the coastline. The sites selected for the car parks correspond to the existing terraces. After their construction, over each submerged park top, a sheltered little slip between the terrace and the protective breakwater will be created.

The detailed knowledge of each of the design phases makes it possible to recognize the sources of potential impact. These sources will be considered in relation to the various environmental components.

The project's implementation, if able to solve the urban mobility question, would reach another important objective: the recovery of the seaside promenade, rescued from traffic jams and again available for recreational and tourist uses. This would make it possible to bring out the peerless amenity of the Gulf area.

The development of the feasibility study has made it necessary to submit the project to a specific analysis about its environmental and socioeconomic performances. This need derives from the dimensions and the complexity of the proposed works and the particular context in which they have to be realized. The building of road infrastructures in a coastal metropolis could have significant consequences on the social/economic environment which is, by its nature, particularly sensitive to change. Furthermore, it involves the coastal zone surrounding Naples; a geographic area constantly under considerable pressure from man and his activities.

MAIN RESULTS OF THE STUDY

The "feasibility study" nature of this project did not lend itself to a detailed analysis of the design activities. The sometimes incomplete definitions of the technical solutions proposed, allowed only qualitative estimates of the impacts and their spatial and temporal evolution. In particular, the involvement of the so-called sector's experts was only simulated. In spite of this, the methodological application has led to the identification of a synoptic outline in which one can find both positive and negative elements related to the interactions between the works and the environment.

The Prime Minister's Decree of the 27th December 1988, makes it compulsory to compile impact studies with regard to three reference frameworks: the Programmed, the Planning, and the Environmental.

The Programmed Reference Framework

The consistency of the project with town planning instruments was checked. The project of the coastal, underwater road link between Posillipo-Mergellina and the Port area is consistent with the objectives pursued by town planning decisions. In addition, the underwater car parks are intended to be realized separately from the implementation of the tunnel, accessible from the ordinary road system until the tunnel is completed. In this way the proposed design makes it possible to respect directives included in the Urban Planning Plan (PUP), whose main objective is the reduction of vehicular traffic and the consequent pollution.

The Planning Reference Framework

The aspects concerning the destination of mud coming from sea bottom dredging were not considered in the feasibility study, but in an impact study they must be considered for their potential ability to affect the receiving site. The judgment of environmental compatibility will be based upon the following aspects:

1. Identification of disposal sites
2. Physical, chemical, and biological description of mud
3. Technical and operative choices
4. Effects on environmental components and mitigation measures

Another open question is represented by risk analysis. The characteristics of the proposed work (underwater road tunnel) and of the natural and human environment (highly urbanized area; presence of geodynamic and seismic activity) strengthen the need to consider risk analyses.

The Environmental Reference Framework

The heart of the evaluation process is represented by the analysis and estimation of the interactions within the environmental reference framework. The assessment of the global impact of the project on all the environmental factors needs to consider different criteria, related to each of the following environmental components:

1. Natural environment
   a) Atmosphere
   b) Marine environment
   c) Land environment
   d) Public health
   e) Noise and vibrations
   f) Landscape
2. Socioeconomic environment
   a) Demography
   b) Housing
   c) Economic structure
   d) Tourism
   e) Transports
   f) Social consent
   g) Services

The effects of each of the impact sources upon all the environmental components were assessed. These estimations are summed up in matrix form. The study has enabled four matrices to be written, two for the tunnel (natural environment and socioeconomic environment) and two for the underwater car parks (natural environment and socioeconomic environment). The two matrices concerning the tunnel case are presented in Tables 1 and 2. In them, on the horizontal axis are the sources of potential impact and on the vertical axis, the involved environmental components. After compiling the matrices and identifying the potential impacts, we gave a value to each matrix. This value expresses a judgment based on three criteria:

1. *Sign*. An interaction is assumed to be "positive" when it gives rise to an impact on a single environmental component, assumed useful for that component.
2. *Strength*. A qualitative scale is used which considers three levels: minor, moderate, and major.
3. *Reversibility*. Intended as the possibility to recover a potentially endangered situation due to project implementation.

Table 1

Interaction Matrix: Tunnel Project Versus Biophysical Environment

PHASE		CONSTRUCTION				SEABOTTOM DREDGING						INSTALL.			OPERATIONAL				
ENVIRONMENTAL COMPONENTS	INDICATORS / IMPACT SOURCES	NOISE	URBAN ROADS NETWORK USE	LABOUR SUPPLY	PRESENCE OF SUPPLY VESSEL	TRENCH	SEDIMENT UPSETTING	NOISE	VIBRATIONS	PRESENCE OF SUPPLY VESSEL	MARINE AREA INTERDICTION	SEDIMENT UPSETTING	NOISE	PRESENCE OF SUPPLY VESSEL	PRESENCE OF STRUCTURES	EXHAUST GAS EMISSIONS	NOISE	SERVICES SUPPLY	LABOUR SUPPLY
ATMOSPHERE	AIR QUALITY		▨1													▨1		○3	
MARINE ENVIRONMENT	CORRENT PATTERNS														■1				
	WATER QUALITY						▨3					▨1							
	SEABED STABILITY					▨1													
	PELAGIC COMMUNITY						▨2	▨1					▨1						
	BENTHIC COMMUNITY					▨2	▨3												
LAND ENVIRONMENT	HYDROGEOLOG. CHARACT.														■1				
	COASTAL MORPHOLOGY														■1				
PUBLIC HEALTH	DIRECT CONSEQUENCES																	○2	
	INDIRECT CONSEQUENCES																	○2	
NOISE	NOISE LEVEL	▨1						▨1	▨1				▨1				▨1	○3	
LANDSCAPE	VISUAL ASPECTS				▨1		▨2			▨1				▨1				○3	

IMPACT RATING

1 = MINOR

2 = MODERATE

3 = MAJOR

	POSITIVE IMPACT 1	2	3	NEGATIVE IMPACT 1	2	3
REVERSIBLE	◌1	◌2	◌3	▨1	▨2	▨3
IRREVERSIBLE	○1	○2	○3	■1	■2	■3

Table 2

Interaction Matrix: Tunnel Project Versus Socioeconomic Environment

SOCIO ECONOMIC COMPONENTS	PHASE / IMPACT SOURCES / INDICATORS	CONSTRUCTION				SEABOTTOM DREDGING						INSTALL.			OPERATIONAL				
		NOISE	URBAN ROADS NETWORK USE	LABOUR SUPPLY	PRESENCE OF SUPPLY VESSEL	TRENCH	SEDIMENT UPSETTING	NOISE	VIBRATIONS	PRESENCE OF SUPPLY VESSEL	MARINE AREA INTERDICTION	SEDIMENT UPSETTING	NOISE	PRESENCE OF SUPPLY VESSEL	PRESENCE OF STRUCTURES	EXHAUST GAS EMISSIONS	NOISE	SERVICES SUPPLY	LABOUR SUPPLY
DEMOGRAPHY	POPULATION STRUCTURE																	+	
HOUSING	DIMENSIONS																	+	
ECONOMIC	LABOUR FORCE			◍ 2														○ 2	○ 1
	TERTIARY DEVELOPMENT																	+	
TOURISM	SIGHT SEER						▨ 1	▨ 1										○ 2	
	PLEASURE BOAT						▨ 1			▨ 2	▨ 1			▨ 2					
TRANSPORTS	URBAN TRAFFIC		▨ 1															○ 3	
	SHIPPING									▨ 2				▨ 2					
SOCIAL CONSENT	PROJECT ACCEPTABILITY			◍ 2		▨ 1	▨ 1	▨ 1	▨ 1		▨ 2							○ 3	
SERVICES	AREA ACCESSIBILITY										▨ 1							○ 3	

IMPACT RATING

1 = MINOR

2 = MODERATE

3 = MAJOR

	1	2	3
REVERSIBLE	◍	◍	◍
IRREVERSIBLE	○	○	○

POSITIVE IMPACT

	1	2	3
REVERSIBLE	▨	▨	▨
IRREVERSIBLE	■	■	■

NEGATIVE IMPACT

By these judgments, the matrixes are able to sum up impact estimations.

The utilization of a final matrix to summarize all the data pertinent to these interactions has made it possible to effectively organize a large amount of information. Above all, it offers a global view that makes it possible to evaluate the project in a less sectorial way. This result enables the identification of the most stressed environmental components, the most critical perturbing factors, the impact associated to each implementation phase, and the sectors towards which the project expresses its greatest uses.

The main consequence is the identification of the critical items, showing the need to examine carefully those projected activities which could mean relevant impacts.

Major Results

Natural Environment

Positive interactions are all located in the operational phase. They are correlated to the services offered by the tunnel and the car parks.

The operational phase of the tunnel will generate some positive impacts, qualified as “major” on the sectors “air quality”, “sound level”, and “landscape”, with indirect consequences of the same sign on “public health”. Similar benefits, to a lesser extent, are connected to the operation of the submarine car parks. It may be important to emphasize that the utility associated with these benefits is strengthened by the present state of the components involved, which is particularly critical.

With regard to the tunnel, negative interactions can be found both during the construction phase and the operational one. As regards the car parks, negative impacts concentrate themselves in the operational phase. The environmental sector which is principally penalized is the “marine environment”. The values related to the potential impacts are highly different, depending on what the factors are and the environmental component that are considered. Main negative interactions are due to sediment displacement caused by dredging and trench filling on the seabed. Even if these potential impacts are all qualified as “reversible”, they represent a potentially critical element. Therefore, in an Environmental Impact Study aimed at obtaining the authority’s approval for the project, they need closer investigation.

In the operational phase, the “presence of structures” may cause changes to coastal streams and thus negatively impact the marine environment. Their potential intensity should be low, but they have an “irreversible” character. In addition, they could have some negative consequences on other sectors, in particular, water quality and some socioeconomic components (tourism and social consent, above all).

Socioeconomic Environment

As for the case of the natural environment, impacts due to tunnel construction are generally stronger than those generated by underwater car parks.

Impacts on socioeconomic components are concentrated during the operational phase. These are principally due to the services offered by planned works on urban mobility and area accessibility. Socioeconomic sectors which will potentially receive major disruptions, both positive and negative, from the implementation of the whole project are “tourism”, “traffic” (both on land and sea) and “area accessibility”. It must be said that the use of the most recent underwater technologies makes it possible to work without disturbing ordinary city life (yard openings, road closures, etc.), thus limiting negative impacts during the construction phase.

REFERENCES

Colorni, A., and E. Laniado. 1991. SILVIA: A decision support system for environmental impact assessment. In *Environmental impact assessment*, Vol. 1, ed. A. G. Colombo. Netherlands: Kluwer Academic Publishers.

Goicoechea, A., D. R. Hansen, and L. Duckstein. 1982. *Multiojective decision analysis with engineering and business applications*. New York: John Wiley.

Jorissen, J., and R. Coenen. 1991. The EEC directive on EIA and its implementation in the EC Member States. In *Environmental impact assessment*, Vol. 1, ed. A. G. Colombo. Netherlands: Kluwer Academic Publishers.

Pellizzoni, L. 1991. Sociological aspects of environmental impact assessment. In *Environmental impact assessment*, Vol. 1, ed. A. G. Colombo. Netherlands: Kluwer Academic Publishers.

Wood, C., and N. Lee. 1989. EIA in EEC countries. Manchester University, EIA Centre, Dept. of Planning and Landscape, United Kingdom. *Newsletter n. 2 and 3*.

50

Seagrass Restoration in Venice Lagoon

Giovanni Cecconi
Study Department, Venezia Nuova Consortium,
Venezia, Italy

Andrea Rismondo, Francesco Scarton
Wetland Biologists, Progetto Adriatico Association,
Venezia, Italy

ABSTRACT

The ecological value of seagrasses, i.e., flowering plants living in brackish and marine waters, is well known. Over the last twenty years there have been plans in several areas around the world, aimed at restoring the seagrass beds damaged by human activity, such as pollution, dredging, and the building of roads, ports, and industrial facilities. Most of these interventions of restoration include careful analysis of the proposed sites; monitoring of environmental parameters, e.g., current velocity and water transparency; and transplanting of the selected species. In the Lagoon of Venice, a project of seagrass restoration began in 1990 with the goal of stopping the ecosystem's deterioration.

After a survey of seagrass distribution, an analysis of their ecological requirements, and the selection of suitable sites, the next step will entail transplanting *Zostera marina* L. and *Cymodocea nodosa* (Ucria) Asch. in several areas.

INTRODUCTION

Seagrasses are marine plants occurring in most coastal and marine regions. They are angiosperms, i.e., flowering plants which have evolved to a marine environment from a terrestrial existence. There are some sixty species of seagrasses in the world, living in environments ranging from shallow waters (as in the temperate latitudes) to a depth of approximately 60 m, in tropical areas.

Seagrass beds are highly productive habitats; they serve parallel functions as nursery grounds and critical habitat for fish. They are also important because they retard or slow currents, enhancing sediment stability and increasing the accumulation of organic and inorganic matter. Moreover their roots, by binding sediments, reduce erosion and stabilize bottom sediments (Fonseca 1989; Thoraugh 1990). In spite of their well-known importance and value, seagrass beds have been heavily affected by human activities such as dredging of canals, ports and marinas, and filling of wetland for industrial purposes. Urban wastes coming from coastal cities and discharged near seagrass beds include oil, fertilizers, and chemicals of all nature. Agricultural effects include soil erosion, which increases turbidity, thereby killing seagrasses that no longer receive enough light in deeper waters. Fertilizers, pesticides, and other substances can also adversely affect estuarine submerged vegetation (Thoraugh 1990). In order to recreate the once existing beds and to enhance the extension of those already present, many plans of seagrass restoration have been executed in different countries. The most important goals were the following (Fonseca 1989):

1. Restore areas previously impacted by poor water quality that once had seagrasses, after the improvement of water quality.
2. Convert filled or dredged areas that were once seagrass meadows back to their original condition.
3. Recreation of acreage of vegetative cover equivalent to that lost.
4. Develop faunal population structure and abundance in the new bed equivalent to natural beds.

A review of seagrass restoration projects showed that twenty-one groups of workers have made over 165 attempts of restoration in sites from Europe and the Americas, to Australia and the Philippines. Techniques adopted included the use of plugs, seeds, turf/sods, and shoots (Thoraugh 1990).

Most of these attempts were successful. Failures were usually due to lack of knowledge about the general physiology of seagrasses and especially the physical environmental factors that affected the restoration plans (Fonseca 1989; Thoraugh 1990). A careful environmental analysis of the proposed sites for the intervention is considered the most important factor in every restoration plan.

The object of this note is to present the ongoing activities devoted to recreating the seagrass beds in the Lagoon of Venice.

PLANNING THE RESTORATION OF SEAGRASSES IN THE LAGOON OF VENICE

The Lagoon of Venice, with an extension of 555 km^2, is the largest coastal lagoon in Italy. There are three entrance channels which allow water exchange with the sea. The Lagoon is characterized by large areas of very shallow waters (less than 1 m) which are flooded and drained through an intricate network of channels, mud flats, and marshes that are covered with alophytic vegetation.

In the framework of activities financed by the Italian Government with the aim of stopping the morphological degradation of the Lagoon of Venice, a general intervention of seagrass restoration has been planned. This intervention has been scheduled in the following way:

1. Mapping of seagrasses living in the Lagoon
2. Studying the possible sites selected for transplanting seagrasses
3. Characterization of environmental conditions
4. Starting the field work

Until a few years ago, very little was known about seagrass distribution and ecology in the Lagoon of Venice. According to the general knowledge, two or three species of seagrasses were present, mostly near the entrance channels. Even if there were no maps showing the past distribution, seagrass beds were thought to be disappearing year after year. No data were available about sediment texture preference, growth pattern, or preferred water depth.

As a first step for the restoration plan, Venezia Nuova Consortium (a pool of Italian companies that operates as a state concessionaire) launched an intensive mapping project performed from March to July of 1990, and the main results (Caniglia et al. 1992; Consorzio Venezia Nuova 1991) may be summarized as follows:

1. Three seagrass species, *Zostera marina*, *Zostera noltii* Hornem and *Cymodocea nodosa* live in the Lagoon of Venice.

2. The most widespread species is *Z. noltii*, which covers 4,235 ha, followed by *Z. marina* (3,635 ha) and *C. nodosa* (1,560 ha). The first species can be observed in several different environments, from the inner, silty-clay mudflats to the entrance channels. *Z. marina* is also widely distributed, whereas *C. nodosa* is restricted to the sandy bottoms close to the barrier islands of Lido and Pellestrina.
3. The southern part of the Lagoon seems the most favorable area for seagrasses, with its large beds often plurispecific. In the northern part, seagrasses have a scattered distribution and in the central part they are almost completely absent. This is probably due, as it appears from ongoing research, to the extreme water turbidity in this area, the sediment transportation, excessive loads of nutrients, and the overgrowth of macroalgae (notably *Ulva laetevirens* and *Gracilaria confervoides*).

At the same time, data on the ecology of each species were gathered in several areas. Here the seasonal biomass trends showed clear differences among the three species studied. *C. nodosa* presented a steady increase with only one maximum in September, while the highest values for *Z. marina* and *Z. noltii* were found respectively in June and August-September. July was characterized by low values particularly for *Z. marina*. In several stations, grain size analyses of sediments inhabited by seagrasses were also performed. Sediments with *Z. noltii* showed the highest percentages of clay (up to 15%), whereas sand was usually between 10% and 50% and silt between 11% and 77%. *C. nodosa* was observed in much more sandy muds, where sand ranged between 61% and 97% and clay was always less than 2%. *Z. marina* grows in a broader range of sediment texture, from almost pure sand (90% sand) to silty mud (63% silt). In Table 1 the preferred depth and sediment textures observed in the Lagoon of Venice are presented.

Table 1

Preferred Water Depth and Sediment Texture Observed for Seagrasses in the Lagoon of Venice.

Species	Depth (m)	Sediment Texture
Z. noltii	0-1.5	silty
Z. marina	0.5-2.5	sandy-silty
C. nodosa	1-1.5	sandy

Along with other activities developed to restore the Lagoon's environment, such as the recreation of small canals once existing to enhance water circulation, collecting macroalgae biomass to reduce nutrients recycling, and the recreation of salt marshes using dredged material, an important role is also played by the restoration of seagrass beds.

In accordance with the results of the mapping, several possible sites have been selected for starting the program. It is clear that the extension of this intervention in the inner areas of the Lagoon must be accompanied by a significant improvement of water quality. Two categories of possible sites have been indicated.

1. *Bare bottoms close to existing, well-established seagrass beds*. It is likely these areas were, until recently, covered with seagrasses, but human activities, e.g., dredging of canals with sediment resuspension and deposition in the nearby areas; seagrass beds scraping due to vessels passage; and illegal fishing, such as trawling, destroyed them. These areas can be found in several zones of the Lagoon.
2. *Large bottoms near the entrance channels and along the barrier islands*. These areas were inhabited by seagrasses, but now unsuitable environmental conditions prevent these bottoms

from being colonized by marine phanerogams. With the expected, improved water quality due to the ameliorating activities previously mentioned, it is likely the seagrasses could reoccupy these sites in the near future. Human intervention, in this case, is geared to speeding up the natural processes of recolonization.

The suitability of the sites will be evaluated with an accurate, environmental monitoring project that will take into account temperature; salinity; current velocity and direction; sediments; and probably the most important parameter, light penetration to the bottom, in terms of photosynthetic active radiation (PAR). Indeed, the accurate measure of photosynthetically active radiation reaching the bottom in both time and space is of outstanding importance in evaluating site suitability.

The selected species for transplanting are *Z. marina*, largely used in similar restoration plans and *C. nodosa*, which is particularly effective in stabilizing sediments in Venice Lagoon (Rismondo and Scarton 1990).

At the same time, a pilot project on a small area has been financed by the local county administration, in order to evaluate transplanting methods and production of transplanted plants (Curiel and Rismondo, 1992).

CONCLUSIONS

The results of the research that has been recently carried out and those that are presently ongoing in the Lagoon of Venice form the basis for all future interventions of seagrass transplanting and environmental restoration.

It is important to add that the transplanting method (turf/sods, shoots or plugs) has yet to be selected, since a period of experimentation is needed for this purpose.

Considering the complexity of the Venice Lagoon ecosystem, the large number of possible limiting factors, and the number of interventions planned for the near future, it seems reasonable to start transplanting experiments on a small scale before developing larger interventions.

REFERENCES

Caniglia, G., S. Borella, D. Curiel, P. Nascimbeni, A.F., Paloschi, A. Rismondo, F. Scarton, D. Tagliapietra, and L. Zanella. 1992. Distribuzione delle fanerogame marine (*Zostera marina L., Zostera noltii Hornem, Cymodocea nodosa* (Ucria) Asch. in Laguna di Venezia. *Lavori Societa' Veneziana Scienze Naturali 17:137-150.*

Consorzio Venezia Nuova. 1991. Studio A.3.16-I fase. *Distribuzione delle comunita' biologiche in laguna di Venezia. Venice*: internal report.

Curiel, D., and A. Rismondo. 1992. Reimpianto di fanerogame marine in laguna di Venezia. Comune di Venezia, Assessorato all'Ecologia: internal report.

Fonseca, M.S. 1989. Regional analysis of the creation and restoration of seagrass systems. In *Wetland creation and restoration: the status of the science*, ed. J.A. Kusler and M.E. Kentula, 175-198. San Francisco:Island Press.

Rismondo, A., and F. Scarton, 1992. Praterie a fanerogame marine in laguna di Venezia. *Ambiente Risorse e Salute* 4: 22-26.

Thoraugh, A. 1990. Restoration of mangrove and seagrasses: economic benefits for fisheries and mariculture. In *Environmental restoration*, ed. J.Berger, 265-281. San Francisco: Island Press.

51

Seabed Physiography and Morphology and Their Impact on the Installation and Protection of Man-Made Structures

Giulio E. Melegari
Saipem S.p.A., San Donato Milanese (MI)

INTRODUCTION

In the context of modern marine engineering, underwater structures include

1. Pipelines
2. Cables
3. Fixed platforms and other seabed completions (i.e., templates, well-heads, storage units, Pipeline End Manifold [PLEMS], etc.)

The installation of subsea structures, their stability, and protection depend greatly on seabed morphology and physiography. Quite often the mere presence of subsea structures has an effect on the existing tides and currents and, therefore, the seabed. The solid profile of the structure may increase the local current velocity and cause the seabed to go into shear. This is commonly referred to as scour. The effects of scour may, in time, interact with local geological conditions and have a detrimental effect on the stability of structures, particularly of pipelines which can be brought to unacceptable stresses.

A considerable amount of damage to pipelines and other artificial structures have occurred because of unpredicted and sudden changes in the original morphology of the sea floor on which they rest. The evolving shape of the sea floor is dependent not only on the local geological setting and nature of the substrate (as generally thought), but it is strongly influenced by the dynamics of the overlying water mass that makes each marine basin a complex, dynamic system. Hydrodynamic conditions of submarine environments are very important factors responsible for the endless modification of sea floor morphology, since submarine currents produce erosion, transport, and deposition of sediment.

PHYSICAL PROCESSES RESPONSIBLE FOR HIGH HYDRODYNAMISM IN UNDERWATER ENVIRONMENTS

High-energy conditions of the water masses are common in shallow seas. They lie between those parts of the sea dominated by nearshore processes and those dominated by oceanic processes, at depths between 10 m and 200 m. Shallow seas include continental shelves and enclosed basin such as the

North Sea, Yellow Sea, and Bering Sea. High-energy conditions in shallow seas arise from the interaction of waves and currents. Four main types of currents exist: (1) oceanic circulation currents, (2) tidal currents, (3) meteorological currents, and (4) density currents.

1. *Oceanic currents* (Gulf Stream, Agulhas Current, Florida Current) are the result of differences in temperature between high and low latitudes. Although the major currents lie oceanward of the shelf edge, they induce active interchange between oceanic and shelf waters. Velocities of currents that encroach the shelf can reach 250 cm/s, and they can cause erosion, e.g., Blake Plateau (Swift 1969b) or migration of sand waves, as on the Agulhas and outer Saharan shelves.
2. *Tidal currents* are the result of gravitational attraction by the moon on the surface waters of the Earth. They are bidirectional, rotary or rectilinear, and can reach velocities of 100-200 cm/s (Knight 1980). The most powerful currents can cause erosion of the seafloor and migration of bedforms. When tidal current velocities exceed 150 cm/s, they scour erosional features which range in size from hollows 150 km long by 5 km wide and about 150m below the surrounding sea floor, e.g., Hurd Deep, English Channel (Hamilton and Smith 1972), to small longitudinal furrows parallel to the peak tidal current that can be 8 km long, 30 km wide, and 1 m deep. In addition to these erosional features, sand waves and sand ridges are constructed.
3. *Meteorological currents* include wind-driven currents, storm surge, and nearshore wave-induce currents. The latter form a system of cell circulation, comprising rip-currents and longshore currents (Shepard and Inman 1950). Rip-currents are narrow, high velocity, seaward-directed currents, that are potentially hazardous for pipelines because of their high velocities (200 cm/s) and because they cause erosion of shallow channels perpendicular to the coastline and migration of bedforms.

   Wind and wave-driven currents can be very effective in modifying the morphology of shallow seas (e.g., Bering Sea, Washington-Oregon shelf, Barents Sea). Powerful, unidirectional currents generated during storms with velocities exceeding 100 cm/s, cause the migration of large sandwaves.
4. *Density currents* are generated by variations in temperature, salinity, and concentration of suspended sediment. When bottom currents develop, they can produce erosion of the seafloor and bedform migration.

Until 1945 it was a common notion that high-energy conditions were restricted to shallow waters and that deep seas beyond the shelf edge were stagnant, moving only at the surface in response to wind stresses. The systematic study of deep-sea sediments by the Deep Sea Drilling Project and by the Ocean Drilling Program, shows that the deep seas are dynamic systems of great complexity, in which surface currents coexist with deep currents of some strength that are able to substantially reshape the morphology of the sea floor. High hydrodynamism in deep waters is provided by the occurence of normal bottom currents and by the resedimentation processes triggered by slope failure, such as turbidity currents, debris flows, slumps, etc. All of these processes are capable of eroding, transporting, and depositing sediment, giving rise to the development in deep waters of a series of morphologic features such as channels, scours, flutes, bedforms, slump scars, etc., that represent potentially dangerous features for pipeline stability.

Normal bottom currents include all those deep currents that are not driven by sediment suspension, and may therefore flow along the slope, downslope, and upslope.

1. *Internal waves and tides*. These are large-scale oscillations at density discontinuities between water layers in the upper few hundreds of meters, notably at the thermocline (Lafond 1962). Internal waves and tides can cause erosion and migration of bedforms at the shelf break, on top of seamounts, or in relatively shallow slope and shelf basins.

2. *Contour currents, deep ocean contour currents*. These are formed by the cooling and sinking of surface waters at high latitudes and the deep slow thermocline circulation of these polar masses throughout the world's oceans (Neuman 1968). Whereas much of the deep sea floor is swept by very slow currents (2cm/s), some currents can be greater than 100 cm/s where the flow is restricted throughout narrow passages, such as straits. The effects of contour currents in the morphology of deep sea floors include the erosion of channels, moats and furrows, resuspension and transport of fine-grained sediment, and the migration of bedforms.

SEDIMENTARY PROCESSES AND FEATURES POTENTIALLY HAZARDOUS

The main sedimentary processes responsible of morphological changes on the sea floor that are potentially dangerous for pipeline stability include: slope failure, bedform aggradation and migration, and scouring.

Slope Failure

Slope failure is the primary process that alters the morphology of slopes. It includes mass movement of material that occurs below the shoreline of a body of water for which gravity provides the driving force (Dott 1963). Mass flows are more commmon on slopes located in areas characterized by high sedimentation rates, such as in deltaic regions, e.g., Missisipi Delta (Prior and Coleman 1980) and deltas in Norwegian fjords (Terzaghi 1956), where the sedimentation rate can reach 3000 cm/y. Mass movements are the most common depositional process in Gilbert-type deltas; that is deltas characterized by very steep, subaqueous, sandy and gravelly delta fronts, that can reach angles up to 35°. These are very common in restricted marine basins, such as Norwegian fjords, the Corinto Basin and the Messina Strait (Colella 1988).

By the early 1960s, the offshore oil and gas industry had experienced a high number of pipeline damages, especially in the upper delta slopes in water depths of less than 30 m. Survey and diver's reports indicate that pipelines were both locally displaced in a downslope direction and appeared to have sunk within the sediment. Moreover, mass movements can also occur frequently in deep water settings and on very low-angle slopes, less than 1°, such as on isolated rises (e.g., Madeira Rise).

The main types of slope features and processes include: (1) growth faults, (2) pseudo-diapirs, (3) rockfalls, (4) slides and slumps, (5) collapse depressions, and (6) sediment gravity flows. Sediment gravity flows can be further subdivided on the basis of their internal mechanical behavior and dominant, sediment support mechanism.

1. *Growth faults* are faults in sediments along which the movement is contemporaneous to sedimentation. In most instances these faults, which are common in deltaic regions, tend to cut the recent sediment surface, forming abrupt scarps on the sea floor. These scarps can provide localized areas for additional, downslope mass movement of material by slumping.
2. *Pseudo-diapirs*. Pseudo-diapiric intrusions of mud into overlying sands, giving rise to a sudden development of islands in the sedimentary surface, occur frequently near the distributor mouths of some deltas, e.g., Missisipi Delta (Morgan 1961; Coleman 1976). These pseudo-diapirs, or mudlumps, are formed because of the rapid deposition of dense fluvial sands on top of less dense prodelta and marine shelf muds. The differentially weighted muds are squeezed out and forced upwards, producing vertical displacements up to 200 m. Formation of new mudlumps or rejuvenation of old mudlumps follow major river discharges when large amounts of sand are rapidly deposited at the river mouth. It is not uncommon that mudlumps are pushed vertically to 6-9 m during a single river flood.

3. *Rockfalls* are rapid accumulations of clasts by freefall. In marine environments, they occur at the base of steep slopes such as carbonate reefs, fault scarps, and canyon walls. Displaced clasts may be very large (10m) and bounce or roll downslope up to several hundreds of meters before coming to rest (Abbate, Bortolotti, and Passerini 1970).
4. *Slumps and slides* involve downslope displacement of a semi-consolidated sediment mass along a basal shear plane while retaining some internal coherence. Slides have a planar shear plane and little internal disturbance. Slumps or rotational slides exhibit concave-upward shear planes, a backward rotation of the slumped body and high internal disturbance. Slumped and slided deposits are very widespread and can range in volume from less than 1 m to over 100 m. A large slump generally shows three morphological areas:
   a) A source area, or head, characterized by tensional structures such as faults and slump scars
   b) A central transport zone represented by a channel
   c) A distal depositional area, or toe, displaying compressional features and overlapping lobes of remolded debris.

   A man-induced reason for increased concern in areas of potentially unstable sediments on submarine slopes is the post-trenching of existing pipelines after laying/. Trenching operations generally cause local stress concentrations within the sediments and induce excess pore pressure within the sediments. The results of these stress concentrations and increased pore pressure is often the spreading of submarine slumps. An effect very similar to the consequences generated by pipeline trenching is often produced by anchors and anchor wires during operations offshore carried out by conventional work barges and lay-barges in areas of submarine slopes.
5. *Collapse depressions.* These features, which occur mainly in shallow, deltaic areas (Prior and Coleman 1980), are bowl-shaped depressions of the sea floor, ranging in size from 50 m to 150 m. Typically, the depressions are bounded by curved scarps up to 3 m in height, within which the bottom is depressed and filled with irregular blocks of sediment. These features are the result of liquefaction and decrease in the volume of the sediment-gas-water system. They are triggered by sedimentary loading from river deposition, cyclic loading by passage of storm waves, and production of methane gases within the sediment by biochemical degradation of incorporated organic debris.
6. *Sediment gravity flows* occur commonly in deep seas and are associated with slides and slumps. They are divided into: debris flows, grain flows, liquefied and fluidized flows, and turbidity currents.
   a) Debris flows are highly concentrated, highly viscous, sediment dispersions that possess yield strength and plastic flow behavior (Hampton 1972). They are slurry-like or glacier-like laminar flows, advance downslope continuously or intermittently, and the front shows a scarp up to 30 m or more in height. They can be triggered by earthquakes, but they also can be generated by rapid sedimentation, as it occurs in deltaic regions.
   b) Grain flows are visco-elastic flows of sand and gravel characterized by grain-to-grain collision. They require slopes steeper than 18° and represent a very localized process in the subaqueous environment. They are frequent on the steep, sandy, and gravely deltaic slopes of Gilbert-type deltas (Colella et al. 1987).
   c) Liquefied and fluidized flows include the collapse of loosely packed silts and sands by upwards moving pore-fluid. Fluidized sand behaves like a fluid of high viscosity and can flow rapidly down slopes in excess of 2° - 3°.
   d) Turbidity currents are catastrophic flows that can reach velocities of up to 100 km/h, where the sediment is kept in suspension by fluid turbulence. They are the most common sediment gravity flows occurring in the deep sea and can be extremely dangerous for

stability of pipelines, since they produce sudden erosion and deposition of sediment. Turbidity currents have been responsible for severe damages to submarine cables. The typical example is the Grand Banks earthquake of 1929 that triggered an enormous slump and a turbidity current traveling downslope from hundreds of kilometers to the Sohm Abyssal Plain (Heezen and Ewing 1952). The maximum velocity attained by this current was some 70 km/h. Other examples have occurred off the coast of Algeria from the canyon systems off the mouths of the Congo and Magdalena Rivers, in the western New Britain Trench (Heezen and Hollister 1971), and in the Messina Strait after the earthquake of 1908 (Ryan and Heezen 1965). Such currents can be up to several kilometers wide and several hundreds of meters thick (Komar 1969) and can travel as fast as 4000-5000 km. The frequency with which turbidity currents are generated depends on the nature of the area from which the currents originate, the proximity of the source area, and the seismicity of the source area. Turbidity currents generated in deltaic areas after periods of high river discharge may occur as often as once every two years (Heezen and Hollister 1971). Areas located at the base of slopes, that is, in proximity of the source areas of turbidity currents, can be affected by such flows once every ten years (Gorsline and Emery 1959). In the Messina Strait, a frequency of one current every seventy-two years has been calculated (Selli et al. 1978-79). More distal settings, such as basin plains, are affected by turbidity currents once every 1,000-3,000 years (Rupke and Stanley 1974).

Bedform Aggradation and Migration

Bedforms include undulations of the sandy and gravelly sedimentary surface, produced by tractive currents. They can be symmetric when produced by wave motion and asymmetric when produced by unidirectional and tidal currents. Bedforms include, among others: (1) tidal sand ridges, (2) sediment waves, and (3) sand waves.

1. *Tidal sand ridges* are a very prominent feature of the southern North Sea. They are large-scale, linear ridges parallel to the direction of tidal currents. They are up to 40 m high, 2 km wide, 60 km long and have spacings between 5 km and 12 km. The ridges are asymmetric, with the steep face inclined at a maximum of about 6°.
2. *Sediment waves* are asymmetric undulations of a muddy, sedimentary surface with amplitudes of dozens of meters and heights of several meters. They can occur on abyssal plains and continental rises and slopes and have frequently been attributed to the action of contour currents origin for sediment waves in channel levees of deep-sea fans (Damuth 1979; Normark et al. 1980).
3. *Sand waves and dunes* represent two big groups of large-scale, asymmetric bedforms. These are common in the continental shelf at shallow water depths, where they are produced by powerful tidal currents, strong unidirectional flows, surface waves and meteorologically-driven currents (Swift and Ludwick 1976; Flemming 1981). However, they can occur also in deep seas (e.g., Messina Strait), where high-energy conditions occur. Because of the confusion existing in the literature about the criteria that distinguish these two types of bedforms, for brevity we will group them in the class of sand waves.

   Sand waves are more or less regularly spaced undulations of a sedimentary surface, and represent the product of the movement of grains under the effects of unidirectional and tidal currents. When these currents are active, sand waves migrate in the direction of the currents. Sand wave wavelength varies generally from 0.6 m to many hundreds of meters and the height commonly ranges from 0.05 m to 15 m or more. Sand waves can have straight or sinuous, continuous or discontinuous crestlines and occur in groups that form fields of varying shape and size.

Migrating sand waves can be responsible for a rapid evolution of the morphology of the sea floor and can be very critical for pipeline stability. Potential risks for pipelines crossing fields of active sand waves are

a) Development of free spans, the length of which is dependent on the bedform wavelength.
b) Vibrations of pipelines due to occurrence of macroturbulence on the lee side of sand waves.
c) Collapse of sand waves. This process has been documented by a sedimentologic study of ancient sand wave deposit outcropping along the margins of the Messina Strait (Montenat and Barrier 1980), which are similar to the sand waves that are presently migrating on the sea floor of the Strait. The collapse of sand waves can be triggered by earthquakes and can produce mass flows, such as liquefied flows and/or turbidity currents. This process has been also proved by the breaking of cables in the Messina Strait after the earthquake of 1908. Breaking, in fact, developed where the cables were crossing sand wave fields and not where cables where lying on bedrock.

Scouring

Erosion results in truncation of the sea floor and development of depressions of varying shapes and sizes that can be potentially risky for pipelines because of the development of free spans. Erosional features of water origin occur over a wide range of scales, up to hundred of meters deep and kilometers wide, and over a wide range of water depths.

There are two main processes that produce erosion.

1. Erosion by mass movement downslope creates a feature commonly arcuate along the slope, i.e., a slump scar. This type of structure has been already described in the slope failure section.
2. Erosion by water scour creates features elongated in the direction of fluid movement, e.g., channels, chutes, flutes, etc.

A third type of process which is not so widespread as the previous ones, but which can create hazardous conditions for pipelines located at high latitudes, is represented by ice scouring. The fact that large, deep draft icebergs may contact the seabed raises a number of concerns in the design and the protection of seabed structures such as pipelines and related installations (wellheads, etc.). The interaction between the iceberg keel and the seabed slope can be modelled as a ploughing phenomenon in which the sediment in front of the iceberg keel is heaved up in a series of passive failure surfaces and displaced to the sides as the iceberg keeps moving forward under the action of current and wind. In general, the resulting scour resembles a long linear or irregular trench or furrow of gradually increasing depth with berms on both sides.

Critical conditions for pipeline stability arise mainly by the sudden and unpredicted development of erosional features on an originally flat surface. Scouring can be triggered by small irregularities of the sea floor of biogenic (e.g., fish scouring) and sedimentary origin (e.g., sand volcanoes). Small bumps or depressions on the sedimentary surface will cause acceleration of the flow which gives rise to flow separation. The associated, higher shear stresses lead to erosion which, in turn, emphasizes the relief near the irregularity, and so on.

Fish Scouring

Local scouring can be triggered and/or caused by the presence of fishes permanently living in proximity of the seabed and in close proximity of natural and artificial structures. Fish scouring can lead to the formation of hollows generated by the movement of the fins (active fish scouring) in search of food near a pipeline or for the excavation of a more protected shelter. In some environmental conditions, the mere presence of the fish in close proximity of the seabed is sufficient to induce scouring effects depending upon the current speed (passive fish scouring).

Sand Volcanoes

Sand volcanoes are conical features with a crater-like depression on their top, ranging in diameter from 10 cm to several meters and up to 50 cm high. They are the surface aspect of clastic dikes and result from liquified sand being extruded through a local vent at the sediment surface. They reflect release of pressure from a liquified sand, generally following a shock. Sand volcanoes can be common in seismically active areas and in deltaic areas.

PIPELINE SURVEY AND HAZARD IDENTIFICATION

To identify the hazard which may exist along a proposed pipeline route, data must be gathered regarding waves and surface and subsurface currents, etc. Environmental and natural hazards can be classified into two main categories: hazards which pre-exist and can be encountered during the construction (i.e., pipeline installation) on the seabed and hazards which can appear or occur during its planned operational life. The specific hazards and their severity depend on the pipeline-site location, while the protection works to be performed and the corrective actions to be taken depend mainly on the water depth and on the type of hazard.

In gulf and delta areas, the pipeline may be exposed to mud slides and turbidity currents, as well as to potential severe storm consequences induced underwater, and other major bottom instabilities. In the nearshore areas, the pipeline is normally exposed to strong hydrodynamic forces and actions if it is just installed on the seabed without any protective trenching. In other areas, depending upon local situations and conditions, pipelines may have to be designed and installed considering earthquakes, as well as active faulting, which may occur in the area. When installed across straits, channels, and narrow seas communicating with larger basins, pipelines may be subject to strong bottom currents, migration and collapse of dunes, and sand waves, slumps, and other slope instabilities. When laid across wide areas of shallow or relatively shallow seas, the pipeline is very likely to rest across a sequence of aggradation bedforms and sedimentary, rhythmic, mounded obstructions and parallel depressions which may cause unsupported free spans. Sand waves, dunes, and other large, current generated bedforms with a ripple shape may become a major problem in the stability and protection of pipelines crossing wide seabed areas where these structures are present and subject to current induced migration. One of the most critical and still uncertain questions to geotechnical and construction engineers is the stability of these larger bedforms in relationship with the pipelines and the related offshore installations they support. Although the actual size of these sand bedforms may not form an important direct threat to piling of vertical offshore structures, the turbulence and detachment of flow can induce vibrations on tubular components. If the resonance in different pilings varies, damaging stresses can occur in pipelines and in the case of fixed structures, in horizontal beam connections at the lower levels.

Scouring normally is not considered important as a structural threat and as a potential source of structural damage to vertical installations if pilings are set deep enough into underlaying formations. Scour affects large structures in two ways. Global scour occurs around the whole structure, while local scour, which occurs around the legs, spool pieces, and bracings, forms smaller, deeper, scour pits. The resulting erosion of material around the jacket base increases, undermining the structural integrity of the foundation in the case of gravity structures and of anchoring pilings too shallow.

Based on the current concepts for identifying the various hazards along a proposed route, the basic criteria in selecting pipeline routes, particularly on unstable seabottom, include special environmental considerations and surveys. Whether developing a deep water prospect or laying pipelines across deep water zones, detailed geophysical surveys are undertaken to establish if any geological or geotechnical problems exist which could affect operations. Basic criteria for the selection of the most appropriate and safest pipeline route across unstable areas also include basic guidelines. First minimize pipe length in unstable seafloors and route the pipeline into a more stable area. Secondly, in mud-flow areas, minimize any soil movement and slump risks of damage to the pipeline by routing the pipe in such a way that it runs in the same direction as the ascertained or most probable mud flow. This can normally be accomplished by having the pipeline routed in a direction perpendicular to the bottom depth contours in the area.

A major aspect of this work is to map the seabed morphology to identify potential hazardous areas such as rock outcrops, sand wave fields, and slump zones. These surveys should be carried out using deep tow geophysical sensors to obtain the optimum seabed information. The sensors, typically side scan sonars and sub bottom profilers (SSS and SBP), can either be installed in a passive towfish, or on a remotely operated vehicle (ROV) or manned submersible. In all cases, one of the critical requirements is accurate position monitoring of the geophysical sensor.

Similar relative accuracy is required to control pipe laying operations through critical areas defined by the geophysical survey. This can also be achieved using the long base line (LBL) technique and the pipeline position monitored continuously to ensure it is laid on target. A further role is also played by acoustics in monitoring the "as-laid" position of pipelines in unstable, seabed areas relative to fixed reference points.

The objective of a pre-route, underwater survey consists of several aspects inclusive of the following purposes and aims:

1. Gather data on wind, wave, and current conditions existing along the planned pipeline route.
2. Determine water depth, bathymetry, and longitudinal profiles along the planned route.
3. Identify obstructions and depressions which may be present along the proposed pipeline route.
4. Ascertain sub bottom features and structures and stratigraphy along the route and in the area adjacent to both sides of it.
5. Assess the stability and the possible evolution of the area morphology, including erosion, deposit, and transport of sediments.
6. Establish a general understanding of the geotechnical conditions and situations in the area.

Main investigation activities in a pre-route survey include:

1. Echo sounder/bathymetry data gathering (seafloor profiling)
2. Side scan sonar (seafloor mapping)
3. Sub bottom profiler (seafloor buried features)
4. Soil investigations

Echo sounders are used to measure the water depth and to obtain a longitudinal profile of the planned pipeline route. Water depth can be measured with an accuracy of $\pm$ 01.% to 1 % of water depth, depending on the precision of the specific instrument used. The side scan sonar (SSS) sends a wide band of ultrasonic pulses from objects or from degradation/aggradation structures on the seabed which are received by the same transducer and recorded continuously on a chart aboard the towing vessels. By studying the intensity of the reflected signals and acoustic images on the recorded chart, it is possible to evaluate the sonar reflections in a geological/geomorphic manner and to obtain information on depressions, obstructions, rock outcrops, sand waves, and mud flow features. The sub bottom profiler relies on a continuous-reflection profiling technique which provides data on the geological structure and composition beneath the seabed. Whenever the acoustic signal emitted by the towed transducer strikes an acoustic interface, a portion of the signal is reflected and detected by hydrophone. The various interfaces normally mean a change in the geological properties of the sediments, indicated by a change in the acoustic velocity. The type of sediments and the geotechnical properties of the seabed between the interface boundaries can then be evaluated in correlation with soil coring samples obtained in the field.

The purpose of obtaining soil samples is to ascertain sediment properties along the planned pipeline route and to determine soil stability in the area, load bearing capability of the upper layers, soil strength deterioration due to cyclic wave loadings, and resistance of the seabed to pipeline movement and loads. Various devices and methods normally used for soil investigation on the seabed include: sediment samplers (grab sampler, mud snapper, etc.), corers (gravity corer, piston gravity corer, drilled core sampler, vibrocorer, etc.), and penetrometers. Soil properties normally needed for the pipeline protection studies include: general soil classification and grain size distribution, consistency (Atterberg limits), permeability, specific gravity, moisture content, and undisturbed (clavey soils) and disturbed (remolded) shear strength (sensitivity). Underwater system and direct visual inspection by geomorphologists (manned submersible/one atmosphere diving systems) are of paramount importance in the examination and study of seabed sediments and features, particularly for the identification of degradation and aggradation features and of outcrops and boulders.

STABILITY AND PROTECTION OF PIPELINES

To minimize potential risks of damage to the pipeline, the environmental hazards must first be identified in the specific site, then measures taken to protect the pipeline from these hazards. The protection methods include trenching the pipeline below the seabed, anchoring of the pipeline, increased concrete coating, installation of supports, installation of load/protection mattresses, gravel dumping, and strengthening the pipeline.

In the choice and the application of the most adequate and effective protection method, water depth plays a relevant role as a determining factor. There are methods suitable for installation and actuation by a manipulator arm operated from inside a submarine or from the surface on a remotely controlled vehicle, while there are other methods which require a hyperbaric diver and can be actuated by the direct intervention of the human hand only. Some protection methods, on the other hand, can be put into action in more than one way, depending upon the water depth and whether the human hand or a remotely controlled work system is used. Of course, the different installation system and procedure has an economical impact basically dependent on whether man-in-the-sea techniques are employed or not.

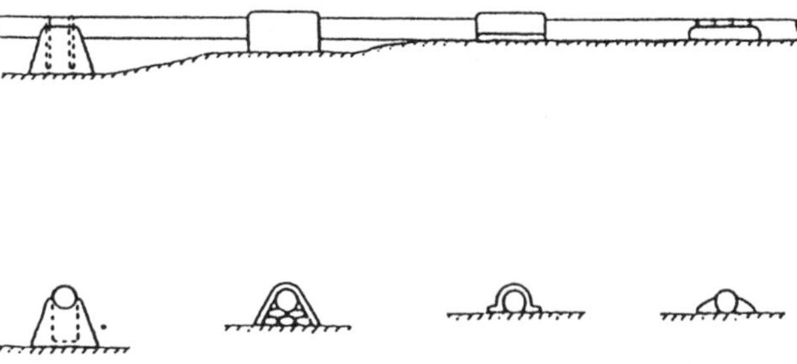

Figure 1. Pipeline stabilization

For stability and protection against mechanical damages, most offshore pipelines rely on the application of concrete coating of variable thickness. The design thickness of the concrete coating depends on factors such as seabed currents, pipeline size, buoyancy, concrete density, required corrosion protection, and resistance to mechanical damage and spanning. Natural irregularities in the seabed topography, geomorphic discontinuities, or scour under the pipeline during service normally result in substantial free lengths of unsupported pipeline which cannot be tolerated. The concrete coating becomes overstressed and spans, leading to loss of protection and instability, and eventually, to permanent damage to the pipeline structure. In common practice, freespan reduction and correction is achieved by providing intermediate supports, augmented where necessary by additional weight and perimeter protection.

In this regard, purpose-made and custom-tailored engineering solutions have been devised and introduced which consist of the following systems, specifically designed to minimize installation time and simplify the role of the diver.

Pipeline Support	Pipeline Protection
- sand bags	- grout mattresses
- grout bags	- bitumen mattresses
- gravel dumping	- concrete mattresses
- jack-ups	- concrete saddles
(mechanical support)	- anchoring systems
	- artificial seaweed mats

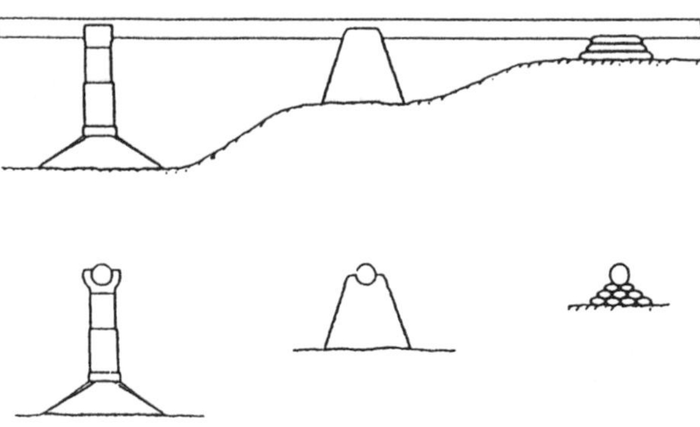

Figure 2. Freespan correction

Sandbags

Sandbagging is carried out to support free spans of cover-exposed pipelines. It is very often the simplest and most cost-effective method for depths down to 50 m. Experience and tests in the North Sea have shown that the conventional sandbags give a rather low level of protection against mechanical damage, while Gulf Coast experience using sand/cement mixtures has been effective.

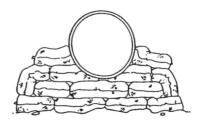

Figure 3. Sand bagging (freespan correction)

Grout Bags

This engineering solution, to be installed by hyperbaric divers under normal environmental and technical conditions at diveable depths, is based on a well-proven, flexible fabric which is constructed to from bags or mattresses when filled with cement grout. The formwork is made from a purpose-woven, polypropylene fabric; the support/underpin form being based on several interconnected compartments with vent pipes to ensure correct filling and maintenance of contact with the underside of the pipeline for a standard distance/clearance. The system is tailored to inflate with grout into a pyramid shape beneath the pipeline and the bags can be adapted to accommodate varying heights of undercut beneath the pipeline. The filling grout is pumped from the surface support vessels through a grout umbilical consisting of two pressure hoses which provide a return line to the surface in order that the grout can be circulated when necessary.

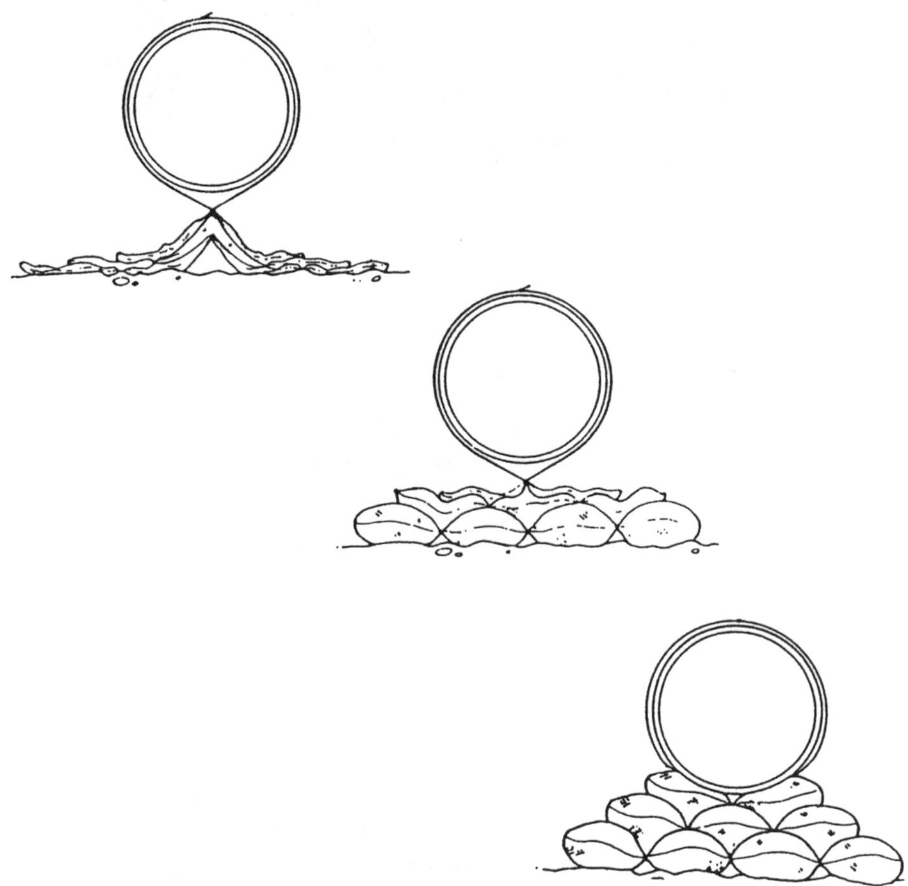

Figure 4. Grout bags (freespan correction)
Source: COLOS, SAIPEM/SNAM Files

Jack-Ups (mechanical supports)

In several cases, specially designated, active supports have been used to reduce the length of free spans with variable bottom clearance. These supports can, by means of releasable legs, adapt themselves perfectly to unevenness of the seabed. They are equipped with hydraulic jacks and compressed air cylinders that allow the pipe to be jacked-up. In this case they become "active supports". These types of support are today suitable for installation at depths exceeding 500 m with the assistance of a dynamic position vessel and a manned submarine or a remotely operated vehicle.

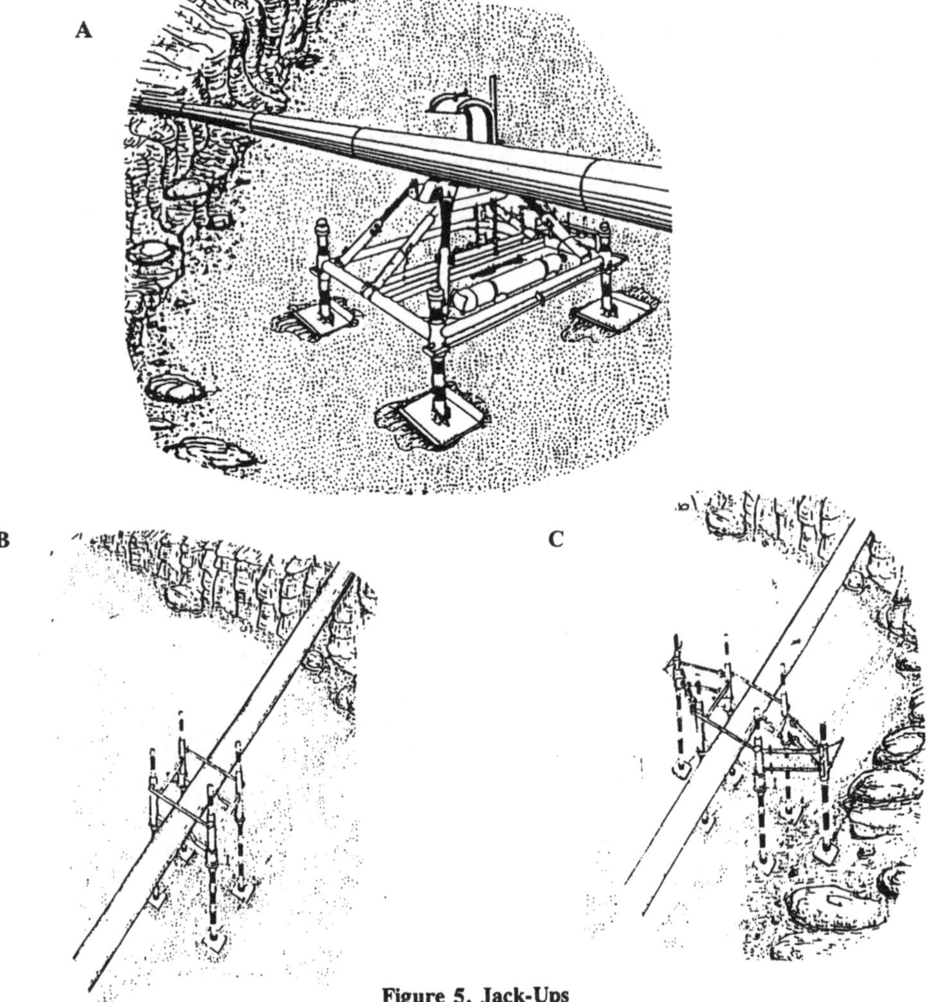

Figure 5. Jack-Ups
A. active type, B. mechanical support (4 legs), C. mechanical support (6 legs)
Source: SAIPEM/SNAM Files

Gravel Dumping

This method is widely used for the correction of free spans. In this way, a free span can be covered along its total length or only at certain points by heaps of gravel. Gravel backfilling can also be used in combination with pipeline steel supports. The practice is to cover the supports after installation so as to provide more stability and protection from scouring.

One important factor for the engineered backfilling is to determine the correct mixture and grain size of the backfill material. The backfill material must remain in place during the different environmental conditions that may occur. Also, it must not prevent fishing with bottom-towed fishing gears in the area.

For the water depths to 50 m, backfilling from side-or split-dumping barges can be used, depending on the current in the actual area. However, owing to the spread of the dumped gravel, placement can be inaccurate and consequently unacceptable. On these vessels, the gravel is carried in compartments at deck level from where hydraulically operated shelves push the dumping material progressively over the side of the vessel, resulting in a continuous flow. During dumping, the vessel can keep station or move along on a predetermined track.

Figure 6. Gravel dumping

In deeper waters, the most efficient way of dumping gravel is through a guide pipe. The lower end of fall-pipe is kept about 10 m above the pipeline, and by the use of conventional, navigation positioning control equipment, the vessel maintains its position over the pipeline. The vessels are perfectly dynamically positioned and equipped with acoustic control devices and navigation systems.

Grout-Mattresses

The fabric formwork used for the installation of grouted supports can also be tailored in the form of a saddlebag to provide additional weight coating or protection over the pipeline. In such circumstances the same fabric material is tailored and adapted to suit the pipeline size, height from the seabed, and specific weight.

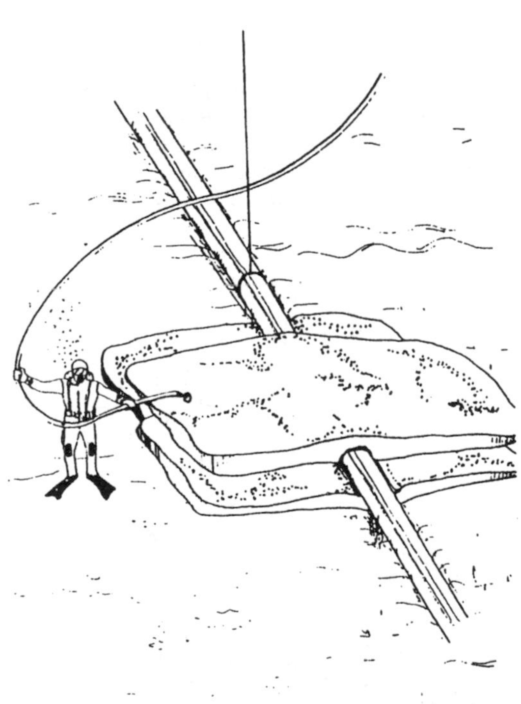

Figure 7. Grout bags/Mattress filling procedure
Source: COLOS, SAIPEM/SNAM Files

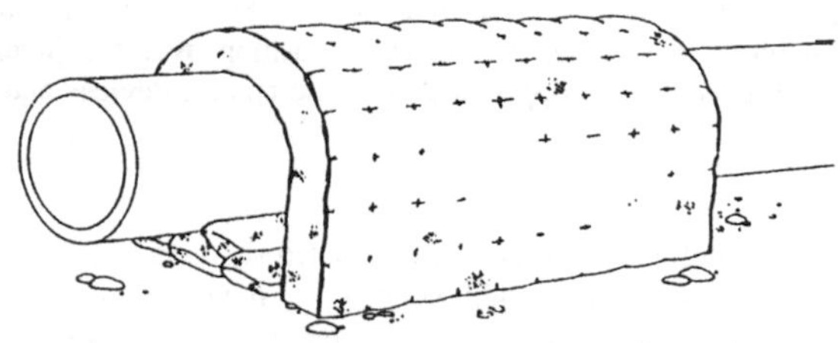

Figure 8. Grout bags/Mattresses (pipeline stability and protection)
Source: COLOS, SAIPEM/SNAM Files

Bitumen Mattresses

Bitumen mattresses are considered more suitable for the protection/stabilization of exposed sections of pipelines in deep water. Whenever an unsupported span is of considerable length and its clearance from the seabed is very little (centimetres), its correction can be achieved by increasing the pipe's negative buoyancy (i.e., by pushing the pipeline onto the seabed), provided that it is maintained within the allowable stress limits.

These mattresses can also be used in combination with other methods whenever negative reactions in the pipeline are required (i.e., support stability, etc.). Bitumen mattresses, to be fully effective, need to be properly sized in relation to pipeline diameter. They are extremely heavy for their size and with their flexibility, they can provide a very effective cure to difficult stabilization problems. The bituminous filler, combined with dense aggregates, is used to provide weight, flexibility, and long-lasting protection to the pipeline.

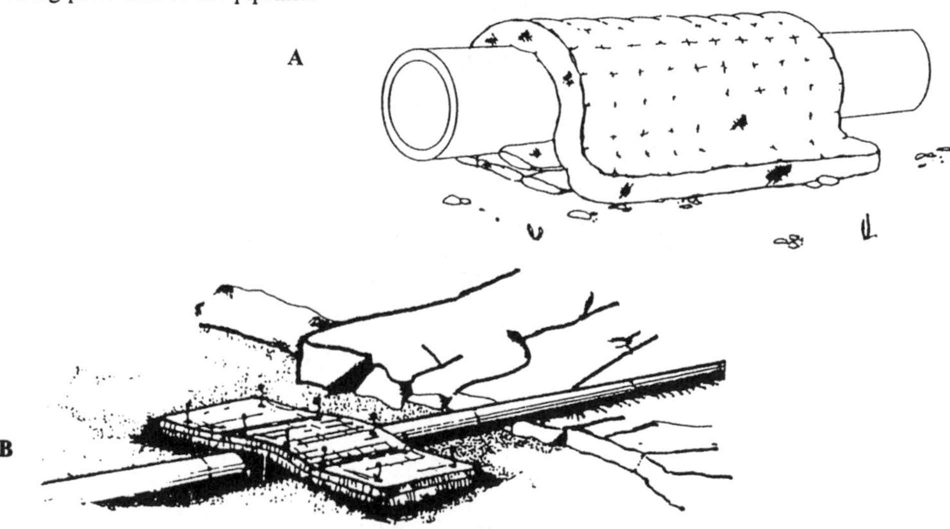

Figure 9: Bitumen mattresses
A. pipeline protection; B. pipeline stabilization

Concrete Mattresses

These mattresses are constructed from reinforced concrete bars and interconnected by steel or polyprolylene ropes which provide flexibility and capability to cope with an uneven seabed profile. This type of protection is suitable for multiple applications in remedial works on pipelines (repair of weight coating, mechanical protection, pipeline stabilization, etc.) and in the prevention of scour.

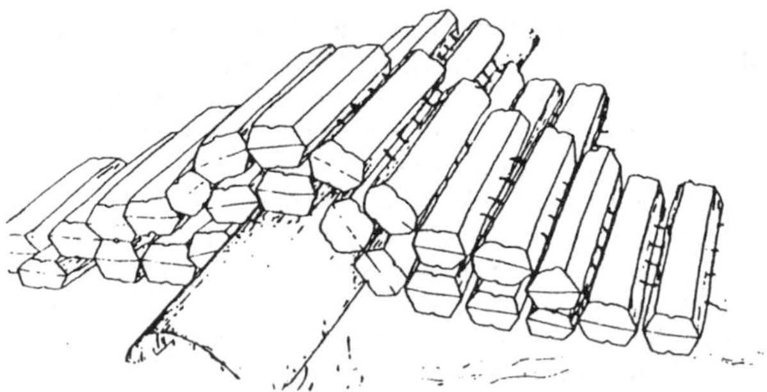

Figure 10. Concrete mattress

Concrete Saddles

Concrete saddles can be used instead of sub mattresses to ensure additional weight coating and protection on the pipeline and to protect it from local mechanical damage. These saddles are suitable for repair of weight coating and local scour prevention. They obviously provide good "mechanical" protection to a pipeline and can be easily installed by divers or remotely controlled vehicles.

Anchoring System

In critical areas of a pipeline, such as a shore approach or near platforms, it may be necessary to anchor the pipeline by physically fixing it to the seabed to eliminate longitudinal or lateral movements. Two piles are driven into the soil, one on each side of the pipeline, and a clamp is fitted around the top of the pipe and to the piles. This system is independent of the seabed soil because the anchors can be piled, drilled, or screwed to the depth required to provide adequate restraint.

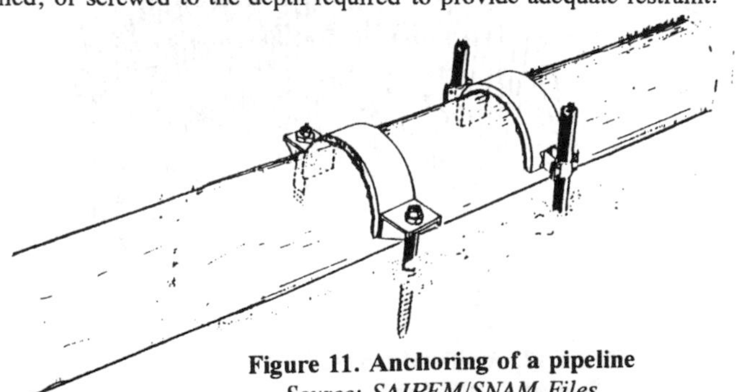

Figure 11. Anchoring of a pipeline
Source: SAIPEM/SNAM Files

Artificial Seaweed Mats

This method has been devised in order to overcome the drawbacks of the two most commonly used scour protection techniques: gravel dumping and sub mats. Gravel dumping, under certain local conditions, requires further maintenance and control due to settlement of the stones. With sub mats, as with any rigid profile object, edge scour still occurs, resulting in slow settlement and requiring further intervention.

The fundamental idea of this method is based on a technique which has been used for centuries, whereby dunes and quicksand areas were stabilized by means of grass. Following this principle, artificial seaweed sewn to weighted mats is utilized in areas with a moveable seabed to reduce local flow velocity and turbulences in the vicinity of the pipeline, thus not only preventing erosion, but also causing the building-up of sand between the synthetic seawed fronds.

This system is based on building stable, mass fiber reinforced banks. These banks are created by mats of polypropylene fronds that have a dual action. They apply viscous drag which reduces the current velocity so that particles of sands are deposited into the mats, thereby building up a cohesive underwater sand bank around the structure to be protected. Experience from various locations and environmental conditions shows that this equipmnent is not always fully effective and the reasons will not be fully known until more research into seabed scouring has been carried out.

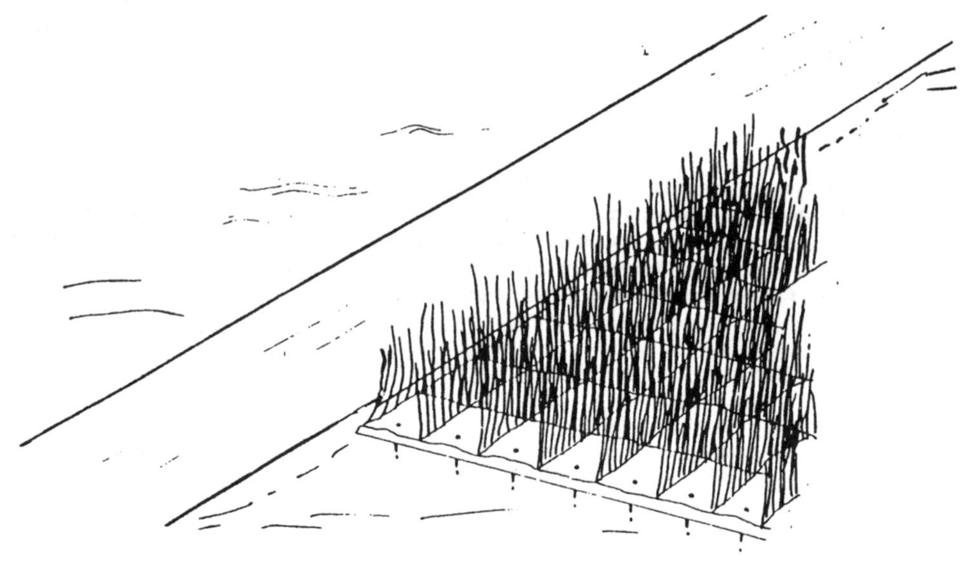

Figure 12. Artificial seaweed mat
Source: SAIPEM /SNAM Files

Contributing Authors

Abel, Robert A.
Albertelli, Giancarlo C.
Ambrogi, Romano
Balchen, Jens G.
Balostro, Roberto
Bandarin, Francesco
Bascom, Willard
Bavestrello, Giorgio
Ben-Amotz, Ami
Berta, Marco
Berti, Dario
Biondi, Esteban L.
Bookman, Charles A.
Broadus, James M.
Buttino, Isabella
Camano, Andres
Caplan, Norman
Catteneo-Vietti, Riccardo
Cecconi, Giovanni
Chiappori, Andrea
Ciavola, Paolo
Cid, Gonzalo A.
Cironi, Romeo
Colwell, Rita R.
Conversi, Alessandra
Della Croce, Norberto
Ehler, Charles N.
Fanara, Alfredo
Fierro, Domenia
Gagliano, S. Thomas
Gallardo, Victor A.
Gallino, Giovanni
Garbuglia, Emanuele
Greco, Nicola
Gulnick, Jeanne
Hempel, Gotthilf
Hill, Russell T.
Hirai, Nobuo
Hong, Seoung-Yong
Hou, Ho-Shong
Huh, Hyung-Tack
Inoue, Hajime
Inoue, Reisaku
Ioannilli, Edmondo
Jarry, Jean
Kondo, Takeo
Lara, Albina L.
Lara, Ascensio C.
Laughlin, Thomas L.
Lonardi, Alberto G.
McKinley, Kelton
Melegari, Giulio E.
Merola, Daniele
Muller, Gerhardt
Olivari, Enrico
Oyediran, O.A.
Park, Douglas L.
Patel, Dipen
Petrillo, Mario
Piantoni, Renzo
Piroi, Marco
Pontremoli, Sandro
Psuty, Norbert P.
Queirazza, Guilio
Rismondo, Andrea
Ruol, Piero
Sansone, Giovanni
Santillo, David
Scarton, F.
Schubel, Jerry R.
Scoglio, Sante
Sertorio, Tecla Zunini
Sherman, Kenneth
Silva, Eduardo
Svealv, Tore L.
Tada, Akihide
Takahashi, Patrick
Tengelsen, Walter E.
Thake, Brenda
Thornton, Daniel
Tirelli, Lino
Tomasino, Mario
Tondello, Massimo
Tsuchida, Hajime
Vadus, Joseph R.
Vallega, Adalberto
Van der Weide, J.
Vartanov, Raphael V.
Veness, Terry
Vitali, Roberto
Wahbeh, Mohammed I.
Walsh, Don
Waterman, Ronald E.
Yuen, Paul